Soybean Seed Composition

Moulay Abdelmajid Kassem
Editor

Soybean Seed Composition

Protein, Oil, Fatty Acids, Amino Acids, Sugars, Mineral Nutrients, Tocopherols, and Isoflavones

Springer

Editor
Moulay Abdelmajid Kassem
Department of Biological and Forensic Sciences
Fayetteville State University
Fayetteville, NC, USA

ISBN 978-3-030-82905-6 ISBN 978-3-030-82906-3 (eBook)
https://doi.org/10.1007/978-3-030-82906-3

This Springer imprint is published by the registered company Springer Nature Switzerland AG
The registered company address is: Gewerbestrasse 11, 6330 Cham, Switzerland

This book is dedicated to the souls of my lovely mother Ait Elhaj Ijja (1941–2018) and father My Abderrahmane Kassem (1937–1993), to my lovely wife Sanaa Rajaallah, and our children Zakariah, Safiah, and Sakinah.
Thank you. Without your support and patience, I would have never achieved my dream.

Foreword

The book *Soybean Seed Composition: Protein, Oil, Fatty Acids, Amino Acids, Sugars, Mineral Nutrients, Tocopherols, and Isoflavone* is a timely update for those interested in seed composition and the healthful effects of phytochemicals. Protein, oil, and Isoflavone can improve human and animal health and also behavior.

Chapter 1 covers seed protein, oil, fatty acids, and amino acids and the effects of genetic and environmental factors on them. Chapter 2 covers QTL that control seed protein, oil, and fatty acids contents, and Chap. 3 covers seed amino acids, macronutrients, micronutrients, sugars, and other compounds that are key to selection for crop improvement. Chapter 4 covers two decades of QTL mapping of mineral deficiencies in soybean, which sheds light on the importance of a balanced mineral nutrition in soybean and other crops. Chapter 5 covers 16 years of salt stress tolerance QTL mapping, which is another, challenge that faces soybean and other crop production worldwide. Chapters 6, 7, 8, 9, and 10 cover in great detail the important soybean seed Isoflavone from their biosynthesis and quantification methods, locations, and variations in seeds, roots, leaves to their QTL mapping for over two decades, and Chap. 11 covers lunasin, a bioactive anticancer peptide in soybean seeds, which will help farmers and breeders to develop soybean cultivars with improved seed Isoflavone and lunasin contents.

The book makes a great primer for those new to the field, including undergraduate and graduate students, and also serves as a great assistance to the alleged experts. It should help direct policy and funding agencies alike.

Carbondale, IL, USA　　　　Khalid Meksem

Reviewers

I wholeheartedly thank the following colleagues who took time from their busy schedules to review one or more chapters of this book. All chapters are peer-reviewed by at least two of these experts:

Khalid Meksem, Department of Plant, Soil, and Agricultural Systems, Southern Illinois University at Carbondale, Carbondale, IL, USA
Istvan Rajcan, Plant Agriculture, University of Guelph, Guelph, ON, Canada
Cevdet Akbay, Department of Chemistry, Physics, and Materials Sciences, Fayetteville State University, Fayetteville, NC, USA
Shubo Han, Department of Chemistry, Physics, and Materials Sciences, Fayetteville State University, Fayetteville, NC, USA
Mohammad Atik Rahman, Center for Viticulture and Small Fruit Research, Biotechnology Unit, Florida A&M University, Tallahassee, FL, USA
Naoufal Lakhssassi, Department of Plant, Soil, and Agricultural Systems, College of Agriculture, Southern Illinois University, Carbondale, IL, USA
Tri Vuong, Molecular Genetics & Soybean Genomics Laboratory, Division of Plant Sciences, College of Agriculture, Food and Natural Resources, University of Missouri, Columbia, MO, USA
Patil Gunvant, Department of Plant and Soil Science, Institute of Genomics for Crop Abiotic Stress Tolerance, Texas Tech University, Lubbock, TX, USA
Prakash Niraula, Department of Biological Sciences, Mississippi State University, Mississippi State, MS, USA
Xiaofang Xie, Fujia Agriculture and Forestry University, Fuzhou, China
Guanshui Chen, Fujia Agriculture and Forestry University, Fuzhou, China

Preface

Soybean [*Glycine max* (L.) Merr.] seeds are a great source of Isoflavone (mainly daidzein, genistein, and glycitein), protein, oil, fatty and amino acids, nutrients, and many other beneficial compounds for human and animal consumption.

The idea of writing this book started when I wrote what was intended to be a review paper entitled "Isoflavone Quantitative Trait Loci (QTL) Mapping" back in fall 2014. In this review paper, I decided to cover soybean seed Isoflavone QTL mapping for over one and a half decade and the paper was getting too large; therefore, I changed my mind to add a few chapters on soybean seed Isoflavone such as "Isoflavone Biosynthetic Pathways and Methods of Quantification," "Isoflavone Locations and Variations in Seeds, Roots, Leaves, and Other Plant Parts," "Isoflavone Positive and Negative Effects on Humans, Animals, and Plants," and "Environmental Factors Affecting Isoflavone Contents" and make it a book about seed Isoflavone only. However, with the passage of time, I decided to expand the book and change its title to "Soybean Seed Composition: Protein, Oil, Fatty Acids, Amino Acids, Sugars, Mineral Nutrients, Tocopherols, and Isoflavone," added the chapters "Seed Protein, Oil, Fatty Acids, and Amino Acids: Effect of Genetic and Environmental Factors," "QTL That Control Seed Protein, Oil, and Fatty Acids Contents," "Seed Amino Acids, Macronutrients, Micronutrients, Sugars, and Other Compounds," "Two Decades of QTL Mapping of Mineral Deficiencies in Soybean," "Sixteen Years (2004–2020) of Salt Stress Tolerance QTL Mapping in Soybean," and "Bioactive Anticancer Peptides in Soybean Seeds," and decided to cover literature up to December 2020 and end the book by this date.

I hope that this book will add to the knowledge of soybean seed composition, especially Isoflavone, their biosynthesis, effects on humans and animal health and food, and their genetic mapping but also protein, oil, fatty acids, amino acids, mineral nutrients and other compounds and that it will benefit undergraduate and graduate students, faculty, scientists, and other individuals interested in these subjects.

Fayetteville, NC, USA Moulay Abdelmajid Kassem

Acknowledgments

I would like to thank Prof. Dr. Khalid Meksem of Southern Illinois University at Carbondale (SIUC) for reviewing the book and writing its foreword, Dr. Ali Siamaki of Fayetteville State University (FSU) for drawing several figures, and all reviewers who reviewed one or more chapters of this book.

Chapter 10 was written while I served as the Dean of the School of Arts and Sciences (SAS) at the American University of Ras Al Khaimah (AURAK) in 2014–2015; therefore, I thank the President Prof. Hassan Al Alkim and AURAK. I also thank Fayetteville State University (FSU) for providing a great academic atmosphere to compete the book by December 2020.

Contents

About the Editor

Moulay Abdelmajid Kassem is currently working as a Professor of Botany and Chair of the Department of Biological and Forensic Sciences at Fayetteville State University, Fayetteville, NC, USA. He earned a Bachelor of Science in Plant Biology from Faculty of Sciences, University Mohamed V, Morocco, in 1992; a Master of Science in Enzymatic Engineering, Bioconversion, and Microbiology from University of Picardie Jules Verne, Amiens, France, in 1995; and a PhD in Plant Biology from Southern Illinois University, Carbondale, IL, USA, in 2003.

Dr. Kassem worked as a high school teacher with Chicago Public Schools, Chicago, IL, from 2001 to 2004 and as an Assistant Professor of Botany in the Department of Biology at Kean University, Union, NJ. He then moved to Fayetteville State University in fall 2006 as an Associate Professor of Botany. He was promoted to Full Professor in 2009 and to Department Chair in 2010.

Dr. Kassem has 28+ years of experience in plant genetics, genomics, and biotechnology. His main areas of research include genetic and quantitative trait loci (QTL) mapping of important agronomic traits in soybean with a focus on yield, yield components, and seed composition traits. Dr. Kassem published 60+ research articles in high-quality peer-reviewed international journals, 80+ abstracts in national and international conferences, and 4 book chapters and received over $1.4 million in grants, including federal grants from the Department of Defense and the Department of Education.

Dr. Kassem is the founder and CEO of Atlas Publishing, LLC (www.atlas-publishing.org). He serves as the Editor-in-Chief of *Atlas Journal of Biology*, a member of the Editorial Board of the *Journal of Biotech Research*, and a reviewer for several international journals and granting agencies. He also co-organizes the American Moroccan Agricultural Sciences Conference (AMAS Conference; www.amas-conference.org) and organizes a workshop entitled "Teaching Genetics, Genomics, Biotechnology, and Bioinformatics" at the prestigious International Plant and Animal Genome Conference (PAG Conference, www.intlpag.org).

Contributors

Nacer Bellaloui USDA, Agriculture Research Service, Crop Genetics Research Unit, Stoneville, MS, USA

Meriam Bousselham Plant Genomics and Biotechnology Lab, Department of Biological and Forensic Sciences, Fayetteville State University, Fayetteville, NC, USA

Moulay Abdelmajid Kassem Plant Genomics and Biotechnology Lab, Department of Biological and Forensic Sciences, Fayetteville State University, Fayetteville, NC, USA

Jiazheng Yuan Plant Genomics and Biotechnology Lab, Department of Biological and Forensic Sciences, Fayetteville State University, Fayetteville, NC, USA

Abbreviations

4CL	4-coumarate-CoA-ligase
AAP	American Academy of Pediatrics
Ab	Antibody
ACNA	Acyl-CoA n-acyltransferase
ACP	Acyl carrier protein
AFLP	Amplified fragment length polymorphism
Agq	Antigen
Al	Aluminum
ALA	Alanine
ANOVA	Analysis of variance
ARD	Average root diameter
ARG	Arginine
ARV	Average root volume
ASN	Asparagine
ASP	Aspartate
B	Boron
BBI	Bowman-Birk protease inhibitor
BC	Backcross
BMI	Body mass index
C	Carbon
C4H	Cinnamate-4-hydroxylase
Ca	Calcium
CHI	Chalcone isomerase
CHR	Chalcone reductase
CHS	Chalcone synthase
CIM	Composite interval mapping
Cl	Chlorine
CO_2	Carbon dioxide
CSPS	Conventional soybean planting system
Cu	Copper
Cys	Cysteine

DAG	Diacylglycerol
DMPBQ	2,3-dimethyl-6-phytyl-plastoquinol
ELISA	Enzyme-linked immunosorbent assay
ER	Endoplasmic reticulum
ESPS	Early soybean planting system
F6H	Flavonone-6-hydroxylase
FA	Fatty acid
FAD	Fatty acid desaturase
FatB	Fatty acyl-ACP thioesterase B
FDA	Food and Drug Administration
Fe	Iron
G3P	Glycerol-3-phosphate
GC	Gas chromatography
GC-MS	Gas chromatography mass-spectrometry
GLN	Glutamine
GLU	Glutamate
GLY	Glycine
GRAS	Generally Recognized as Safe
GT	Glucosyltransferase
GWAS	Genome-wide association study
H	Hydrogen
H_2O	Water
HDL	High-density lipoproteins
His	Histidine
HPLC	High-performance liquid chromatography
HPT	Phytyl transferase
IDC	Iron deficiency chlorosis
IFS	2-hydroxyisoflavanone synthase
ILE	Isoleucine
IM	Interval mapping
IMT	Isoflavone methyl-transferase
K	Potassium
lacZ	Operon lactose Z gene
LASH	Leaf ash
LCC	Leaf chloride content
LC-MS	Liquid chromatography mass-spectrometry
LCO	Lipo-chitooligosaccharides
LDL	Low-density lipoproteins
Leu	Leucine
LN	Leaf number
LPA	Lysophosphatidic acid
LSS	Leaf scorch score
Lys	Lysine
MEP	2-C-methyl-d-erythritol-4-phosphate (MEP) pathway
Met	Methionine

Mg	Magnesium
Mn	Manganese
Mo	Molybdenum
MPBQ	Methyl-6-phytyl-1,4-benzoquinone
MPBQ-MT	MPBQ-methyltransferase
MPK	Mitogen-activated protein kinase
MRL	Maximum root length
MT	Malonyl-transferase
N	Nitrogen
Na	Sodium
Ni	Nickel
NIL	Near isogenic line
NIRS	Near-infrared reflectance spectroscopy
NPC	Nonpolar compounds (urinary)
O	Oxygen
OVX	Ovariectomized
OZR	Obese Zucker (rats)
P	Phosphorus
PAL	Phenylalanine ammonia-lyase
PBS	Phosphate-buffered saline
PC	Phosphatidylcholine
PDHK	Pyruvate dehydrogenase kinase
PDW	Plant dry weight
PE	Phosphorus efficiency
PH	Plant height
PHE	Phenylalanine
PHOs	Partially hydrogenated oils
PLC	Phospholipase C
PLD	Phospholipase D
PLE	Pressurized liquid extraction
PPAR	Peroxisome-proliferator activated receptor
PPS	Percentage of plant survival
PRO	Proline
PSE	Pressurized-solvent extraction
PYR	Pyruvate
QTL	Quantitative trait loci
QTN	Quantitative trait network
RDW	Root dry weight
RFLP	Restriction fragment length polymorphism
RFW	Root fresh weight
RIL	Recombinant inbred line
RL	Root length
RNA-Seq	Ribonucleic acid sequencing
ROD1	Reduced oleate desaturation 1
RRE	Relative root elongation

RRG	Root relative growth
RSA	Root surface area
RTE	Root tap extension
RTI	Root tolerance index
RT-PCR	Reverse-transcription polymerase chain reaction
RT-qPCR	Real-time quantitative reverse transcription polymerase chain reaction
RV	Root volume
S	Sulfur
SAD	Stearoyl-acyl-carrier-protein desaturase
SDW	Shoot dry weight
SER	Serine
SFE	Supercritical carbon dioxide fluid extraction
SFW	Shoot fresh weight
Si	Silicon
SNP	Single nucleotide polymorphism
SPAD	Leaf chlorophyll (SPAD value)
SSR	Simple sequence repeats
STR	Salt tolerance rating
TAG	Triacylglycerol
TC or TOC	Tocopherol (α–, β–, γ–, and δ–tocopherol)
THR	Threonine
TPO	Thyroid peroxidase
Trp	Tryptophan
Tyr	Tyrosine
Val	Valine
VIGS	Virus-induced gene silencing
WD	Water deficit
WDT	Water deficit tolerance
WUE	Water use efficiency
Zn	Zinc

Chapter 1
Seed Protein, Oil, Fatty Acids, and Amino Acids: Effects of Genetic and Environmental Factors

Nacer Bellaloui and Moulay Abdelmajid Kassem

1.1 Introduction

Soybean is a major crop in the world and a source of high-quality protein, oil, and other nutrients. Soybean seed nutrients (seed composition constituents) include protein (40–45%), oil (18–24%) (Medic et al. 2014), and fatty acids such as palmitic (C16:0, 8–12%), stearic (C18:0, 3–5%), oleic (C18:1, 18–24%), linoleic (C18:2, 48–58%), and linolenic (C18:3, 5–10%) acids. Soybean seed fatty acids biosynthetic pathway is shown in Fig. 1.1 (Fang et al. 2017). Soybean seed contains high-quality protein for human nutrition and livestock meal. Also, it contains amino acids, isoflavones, and minerals. Health benefit to human has been previously reported by various studies (Hu et al. 1997; Maestri et al. 1998; Federal Register 2003; Fehr 2007; Business Sphere 2007; Clemente and Cahoon 2009). Breeders goal for desirable seed composition constituents include high oleic and low linolenic acids, low phytic acid, high sucrose, and low raffinose and stachyose levels. It has been shown that soybeans with higher levels of mono-unsaturated fatty acids such as oleic acid or lower levels of polyunsaturated fatty acids such as linoleic or linolenic are more desirable for human consumption than saturated fatty acids such as palmitic and stearic acids (Haun et al. 2014). However, higher levels of monounsaturated fatty acids such as oleic acid and low levels of polyunsaturated fatty acids such as linoleic and linolenic acid are desirable by the industry as they contribute to oil oxidative stability, short shelf life, and less rancidity. This trait is desirable because it can minimize hydrogenation of the oil. Hydrogenation has been reported

N. Bellaloui
USDA, Agriculture Research Service, Crop Genetics Research Unit, Stoneville, MS, USA

M. A. Kassem (✉)
Plant Genomics and Biotechnology Lab, Department of Biological and Forensic Sciences, Fayetteville State University, Fayetteville, NC, USA
e-mail: mkassem@uncfsu.edu

M. A. Kassem (ed.), *Soybean Seed Composition*,
https://doi.org/10.1007/978-3-030-82906-3_1

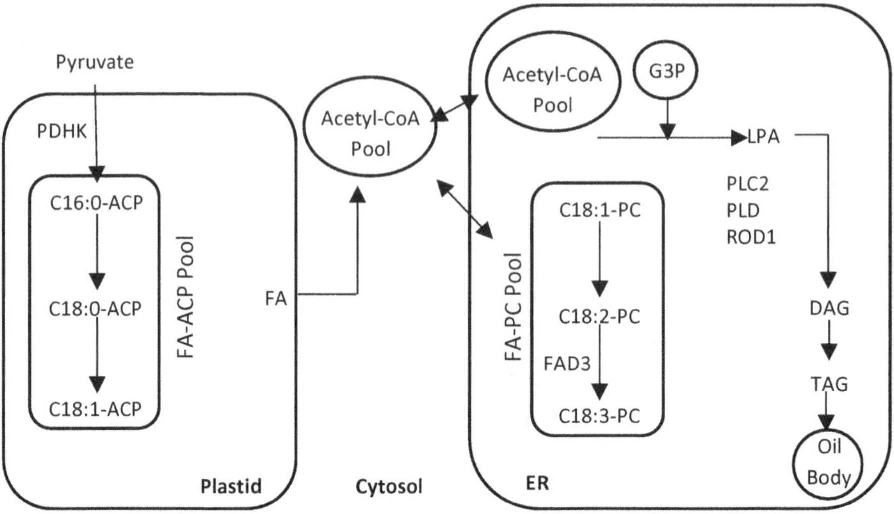

Fig. 1.1 Fatty acids biosynthetic pathways. ACP acyl carrier protein, DAG diacylglycerol, G3P glycerol-3-phosphate, FA fatty acid, LPA lysophosphatidic acid, PC phosphatidylcholine, PYR pyruvate, TAG triacylglycerol, ACNA acyl-CoA n-acyltransferase, FAD fatty acid desaturase, FatB fatty acyl-ACP thioesterase B, PDHK pyruvate dehydrogenase kinase, PLC phospholipase C, PLD phospholipase D, ROD1 reduced oleate desaturation 1, SAD stearoyl-acyl-carrier-protein desaturase, ER endoplasmic reticulum. (Adopted from Fang et al. 2017)

to have undesirable health effects by increasing the risk of coronary heart disease due to higher LDL-cholesterol and lower HDL-cholesterol (Federal Register 2003; Business Sphere 2007; Clemente and Cahoon 2009). The partial hydrogenation of polyunsaturated fatty acids such as linolenic acid results in the conversion of linolenic acid to oleic and stearic acids, thereby reducing polyunsaturated fatty acids to about 18% and linolenic acid to below 2% (Gerde et al. 2007; Clemente and Cahoon 2009; Haun et al. 2014).

Therefore, the main goal of breeders is to breed soybean with low linolenic acid and high oleic acid so as to reduce the hydrogenation process, thereby minimizing the level of trans-fatty acids in foods. In 2015, the Food and Drug Administration determined that partially hydrogenated oils (PHOs) are not generally recognized as safe (GRAS). This determination was based on research and input from stakeholders during the public comment. The PHOs are considered the primary dietary source of artificial transfat in processed foods. Removing PHOs from processed foods could prevent thousands of heart attacks and deaths each year (US Food and Drug Administration 2018: https://www.fda.gov/food/food-additives-petitions/final-determination-regarding-partially-hydrogenated-oils-removing-trans-fat) "GRAS: is an acronym for the phrase Generally Recognized As Safe. Under sections 201(s) and 409 of the Federal Food, Drug, and Cosmetic Act (the Act), any substance that is intentionally added to food is a food additive, that is subject to premarket review and approval by FDA, unless the substance is generally recognized, among qualified experts, as having been adequately shown to be safe under the conditions of its intended use, or unless the use of the substance is otherwise excepted from the definition of a food additive" (US Food and Drug Administration 2019: https://www.fda.gov/food/food-ingredients-packaging/generally-recognized-safe-gras)

Also, higher raffinose and stachyose are not desirable as these sugar fractions are indigestible and cause flatulence or diarrhea in human and monogastric animals such as swine and chicken (Liu 1997; Obendorf et al. 1998). High phytic acid is another antinutritional component in soybean. Phytic acid is the major storage form of phosphorous in soybean, known as a food inhibitor. Phytic acid chelates micronutrient such as Fe and Zn and prevents phosphorus from absorption and bioavailability for monogastric animals, including humans. This occurs due to lack of enzyme phytase in their digestive tract. Therefore, genetic improvement of soybean with low phytic acid is still among the goals of soybean breeders. Phytic acid pretreatment method such as enzymatic treatment of soybean seed with phytase enzyme is a regular method used in industries (Gupta et al. 2015). This chapter will focus on reviewing soybean seed protein, oil, and fatty acids and highlight the main research conducted on improving soybean seed nutritional traits from the perspectives of genetic, environmental, and agricultural practices.

1.2 Interactions of Seed Composition Constituents with Genetics and Environmental Factors

Protein, oil, and fatty acids (oleic, linoleic, linolenic, stearic, and palmitic acids) are among the top compounds available in soybean seeds, consumed by humans, and highly desirable traits to improve in breeding programs (Yazdi-Samadi et al. 1977; Burton 1987; Blackman et al. 1992). Their contents vary depending on the genotype, abiotic and biotic stresses, and environmental conditions (Table 1.1) (Badami et al. 1984; Hartwig and Kilen 1991; Bellaloui et al. 2014a; Rincker et al. 2014; Gulluoglu et al. 2018; Wijewardana et al. 2019).

Several studies have reported that crop rotation affects seed yield and seed composition including protein, oil, fatty acids, and nutrient contents (Temperly and Borges 2006; Bellaloui et al. 2010b). It was shown that soybean seed protein content decreased from 357 g kg^{-1} to 351 g kg^{-1} from the 1st to 5th year of soybean consecutive growth after five years of corn consecutive growth (Temperly and Borges 2006).

It is well known that elevated temperatures and CO_2 levels affect plant growth and development including seed composition (Thomas et al. 2003; Long et al. 2004; Prasad et al. 2005; Hay and Porter 2006; Taub et al. 2008; Bellaloui et al. 2016). In another study, authors investigated the effects of increased temperatures (28 °C, 32 °C, 36 °C, 40 °C, 44 °C) and CO_2 levels (350, 700 μmol mol^{-1}) on seed composition. They found that seed oil, linolenic acid, and carbohydrates contents decreased

Table 1.1 Soybean seed protein, oil, and fatty acids (oleic, linoleic, linolenic, stearic, and palmitic acids) contents

Trait	Content (seed dry weight basis or %)	Reference
Protein	36.3–41.9% 41% 380–420 g kg^{-1} 35.16–37.35%	Badami et al. (1984) Hartwig and Kilen (1991) Bellaloui et al. (2014a) Bolon et al. (2014)
Oil	15.6–25.8% 21% 190–230 g kg^{-1} 15.29–17.43%	Badami et al. (1984) Hartwig and Kilen (1991) Bellaloui et al. (2014a) Bolon et al. (2014)
Saturated fatty acids	*8.6–16.7%*	Badami et al. (1984)
Stearic acid	2.2–7.2% - 4%	Badami et al. (1984) Bellaloui et al. (2014a) Hartwig and Kilen (1991)
Palmitic acid	120–130 g kg^{-1} 10%	Bellaloui et al. (2014a) Hartwig and Kilen (1991)
Unsaturated fatty acids	*83.3–91.4%*	Badami et al. (1984)
Oleic acid	22% 30–40 g kg^{-1}	Hartwig and Kilen (1991) Bellaloui et al. (2014a)
Linoleic acid	54% 480–580 g kg^{-1}	Hartwig and Kilen (1991) Bellaloui et al. (2014a)
Linolenic acid	10% 50–80 g kg^{-1}	Hartwig and Kilen (1991) Bellaloui et al. (2014a)

when temperatures increased, and seed oleic acid, N, and P contents increased when temperatures increased. However, CO_2 increase had a minimal effect on seed composition (Thomas et al. 2003). On the other hand, other researchers have predicted that the changes in seed yield and composition were due to global climate changes including high CO_2, affecting stomatal conductance and photosynthetic rates (Long et al. 2004; Prasad et al. 2005; Hay and Porter 2006; Taub et al. 2008). A recent research on the effect of high temperatures [26 °C and 45 °C] and CO_2 [360 and 700 μmol mol^{-1}] on seed composition showed that seed protein, linolenic acid, and mineral nutrient (P, K, N, Fe, Zn, and B) contents decreased and seed oil and oleic acid contents increased with the increase of temperatures and CO_2 levels. They also found that seed fructose, glucose, and sucrose contents increased with increased temperature and CO_2 levels; however, seed raffinose and stachyose contents did not change (Bellaloui et al. 2016).

The effects of soybean–corn crop rotation on seed composition in Stoneville, MS, USA (Bellaloui et al. 2010c) were also studied. They showed that three-year rotation increased the level of fatty acids content from 61% to 68%, P content from 60% to 75%, Fe content from 70% to 71%, B content from 34% to 69%, and oleic acid content from 22.63% to 30.22%; however, a decrease in linoleic acid content was noticed (Bellaloui et al. 2010c).

Several studies showed the effect of row spacing and seeding rate (SDR) on soybean seed yield and composition (Al-Tawaha and Seguin 2006; Ragin et al. 2014;

Bellaloui et al. 2014a; Bellaloui et al. 2020). The effect of row spacing (RS) and seeding rate (SDR) on seed composition was investigated in four soybean cultivars (P 93M90, AG 3906, P 94B73, and V 52N3) over two years (2006–2007) in a field in Stoneville, MS (Bellaloui et al. 2014a). The results showed that SDR, cultivar, and year significantly affected seed Fe, P, B, sucrose, protein, and oil contents, but no stachyose and linolenic acid contents were observed. Similarly, RS significantly affected Fe, B, raffinose, and sucrose contents, but no protein and oil contents were recorded (Bellaloui et al. 2014a).

In an experiment, Gulluoglu et al. (2018) investigated the effects of two cropping systems (main and double cropping systems) on seed composition traits (fatty acids and oil contents) in several soybean varieties adapted to Europe (Turkey). The temperatures ranged from 19.5 °C to 28.6 °C and from 19.5 °C to 28.6 °C during the growing seasons. They found that soybean plants grown in the main cropping system showed an increase in seed oil contents from 18.45% to 19.99%, while those grown in the double cropping system from 17.11% to 19.37%. They also found that a positive correlation between high temperatures and high seed oil content was observed, in agreement with those reported by others (Belalloui et al. 2015a, b). Fatty acids such as oleic, linoleic, and linolenic acids were also affected in the following ways. The content of seed palmitic acid in double cropping system ranged from 10.76% to 12.23% – higher than its range of 10.59% to 12.04% in the main cropping system. The content of seed stearic acid in double cropping system ranged from 3.94% to 4.87% – higher than its range of 3.11% to 4.52% in the main cropping system. Similarly, for seed oleic acid, the content ranged from 22.69% to 29.51% – lower than its range of 27.02% to 34.09% in the main cropping system; For seed linoleic acid, the content ranged from 48.40% to 54.14% of total oil in double cropping system – higher than its range of 44.5% to 51.80% in the main cropping system. Similar pattern was also observed for seed linolenic acid content that ranged from 5.41% to 6.62% in double cropping system – higher than its range of 4.44% to 5.61% in the main cropping system (Gulluoglu et al. 2018).

The accumulation levels of protein, oil, and fatty acids in seeds are genetically controlled, although environmental conditions and agricultural practices can significantly alter the levels of these seed constituents in seeds. For example, Dardanelli et al. (2006) investigated the effects of maturity group (MG) and environment (E) on protein and oil in a 3-year experiment using six maturity groups (IIIII, IV, V, VI, VII, and VIII-IX) and in 14–24 environments each year. They found that the environment was the most important source of variation for protein and oil content, and the main effect of MG was higher than that of MG × E interaction for oil content and oil + protein content. They also found that oil content was higher in seeds from MGs II-III and IV. Protein content was higher in MG VI in some environments, whereas it was higher in MG II-III in some other environments. The high temperatures during seed-fill period could cause consistent higher oil across years and environments in early MGs (Dardanelli et al. 2006). Heat effects on seed protein, oil, and fatty acids were also evaluated by using a growth chamber where one soybean was grown at normal temperature (25/20 °C), higher temperature (36/30 °C), and at heat stress (40/36 °C). Light intensity was about 1000 μmol $m^{-2}\cdot s^{-1}$, which was supplied with a combination of 10 high pressure sodium and metal halide lights, each of 400 W. Results showed that soybean grown under high temperature showed a lower content of protein and linolenic acid, but higher content of oil and oleic acid (Figs. 1.2a, b, 1.3, and 1.4a, b). Although high temperature may promote oil production, this pattern cannot be generalized as moderate high temperature can increase protein, but severe high temperature can decrease protein. The effect of temperature and maturity on seed protein and oil was also investigated by previous researchers. For example, in an experiment conducted by Piper and Boote (1999), they evaluated the effects of field experiments across 60 environments in 20 cultivars from 10 MGs. Temperatures across these environments ranged from 14.6 to 28.7 °C. Results obtained from these field experiments showed a quadratic relationship between protein and mean daily temperature during seed fill with higher concentrations of protein with temperatures below 20 °C. Other research conducted to investigate the effects of mean temperature during the developmental period of soybean on seed composition showed a negative correlation between oil content and mean temperature during seed maturation (Maestri et al. 1998), but no effect of temperature on protein or fatty acids was recorded. It was concluded that the variability of seed composition constituents was due to MG, genotype within MG, environment, and their interactions (Bellaloui et al. 2009c).

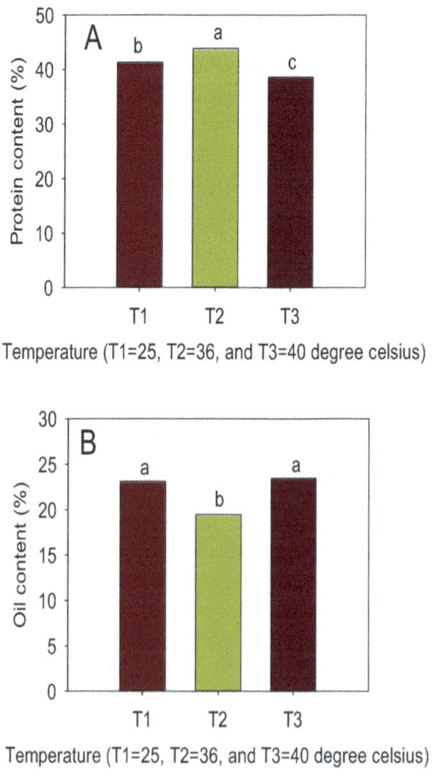

Fig. 1.2 Effect of temperature (normal, 25/20 °C; higher temperature, 36/30 °C; and heat stress, 40/36 °C) on seed protein (**a**) and oil (**b**) in Hutcheson cultivar. Bars are mean of three replicates. Different letters on the bars indicate significant differences at $P \leq 0.05$

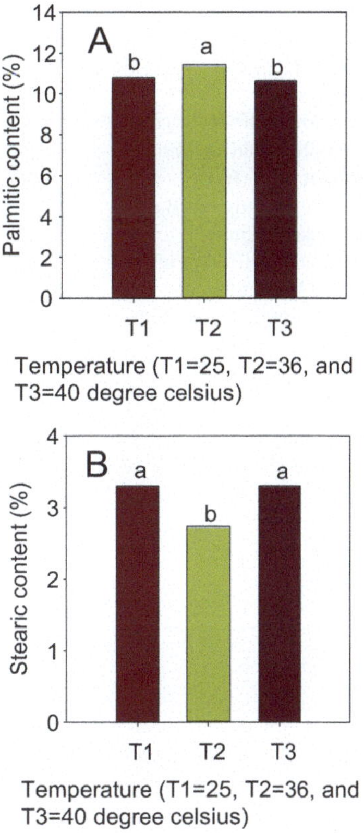

Fig. 1.3 Effect of temperature (normal, 25/20 °C; higher temperature, 36/30 °C; and heat stress, 40/36 °C) on seed palmitic acid (**a**) and stearic acid (**b**) in Hutcheson cultivar. Bars are mean of three replicates. Different letters on the bars indicate significant differences at $P \leq 0.05$

Rincker et al. (2014) studied MG II, MG III, and MG IV soybean cultivars over a period of 80 years and reported a decrease of 0.16–0.22 g kg^{-1} $year^{-1}$ in seed protein contents and a decrease of 0.05–0.14 g kg^{-1} $year^{-1}$ in seed oil contents due to the fact that protein and oil contents are negatively correlated. Other studies reported similar results of decrease in seed protein and oil contents over times depending on the genetic backgrounds (Voldeng et al. 1997; Wilcox 2001; Wilson 2004).

A recent study showed that the use of harvest aids such as paraquat, carfentrazone-ethyl (AIM), glyphosate, and sodium chlorate affect seed composition, especially seed protein, fructose, oleic acid, oil, and fructose contents; however, they had small effects on seed amino acids contents during development stages R6 and R7 depending on the year of growth (Bellaloui et al. 2020). For example, seed palmitic acid, stearic acid, and linolenic acid have not been affected by the addition of harvest aids at the R7 growth stage; however, the addition of paraquat plus AIM or paraquat caused a decrease in seed oleic acid content and a high increase in seed oil content. The addition of glyphosate increased the seed protein content, while the addition of $NaClO_3$ decreased it. The addition of both AIM and glyphosate decreased the seed fructose content; however, a small non-significant effect of the addition of these harvest aids was observed for seed amino acids contents (Bellaloui et al. 2020).

Fig. 1.4 Effect of temperature (normal, 25/20 °C; higher temperature, 36/30 °C; and heat stress, 40/36 °C) on seed oleic acid (**a**) and linoleic acid (**b**) in Hutcheson cultivar. Bars are mean of three replicates. Different letters on the bars indicate significant differences at $P \leq 0.05$

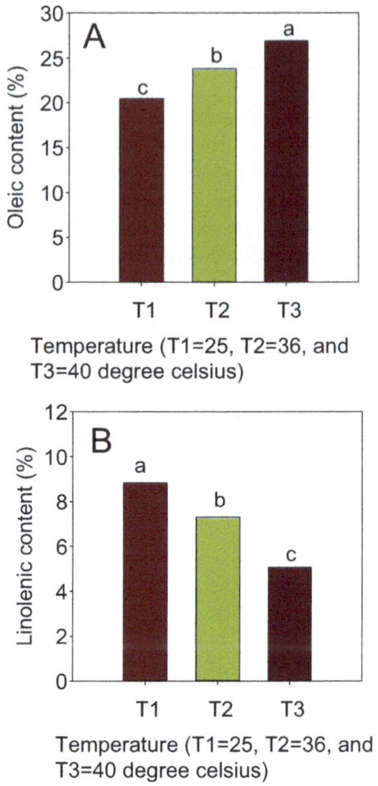

To separate the effects of genotype from environment and maturity effects, a 2-year experiment was conducted to study the effect of maturity on seed composition without the bias of the confounding factors of genotype and maturity (Bellaloui et al. 2009a, b, c). In this research, two sets of near-isogenic lines that developed with different maturities within a common genotypic background were used. One set with nine isolines was derived from "Clark" (Johnson 1958) and the other isoline set with seven lines was derived from "Harosoy" (Weiss and Stevenson 1955). The maturity in each set was different due to the combination of maturity genes (E1, E2, E3, E5) (the maturity of each line within a set varied, but all had a common genotypic background). This experiment allowed investigating the effects of maturity among and between the Clark and Harosoy isoline sets on seed composition in the Early Soybean Production System of the midsouth (Bellaloui et al. 2009a, b, c). They found that there were positive linear relationships between protein content and maturity among isolines of the Clark set in 2004 ($R^2 = 0.75$; $P \leq 0.001$) and 2005 ($R^2 = 0.63$; $P \leq 0.001$). However, in Harosoy isolines, there was no relationship between protein and maturity. On the other hand, there were negative linear relationship between oil content and maturity for Clark (in 2004, $R^2 = 0.82$, $P \leq 0.001$; in 2005, $R^2 = 0.91$, $P \leq 0.0001$) and Harosoy (in 2004, $R^2 = 0.19$, $P \leq 0.05$; in 2005, $R^2 = 0.36$, $P \leq 0.01$). Also, they found that maturity had a bigger influence on seed composition than maximum temperature. They concluded that the relationship between seed composition and maturity was different between the Clark and Harosoy sets of isolines, depending on the range of temperature, as the range of maximum temperature during the last 20 days before maturity for the Harosoy isoline set (early isolines) was from 31.6 to 33.6 °C in 2004 and from 33.5 to 35.5 °C in 2005. However, the maximum temperature for the Clark isoline set was from 31.8 to 33.5 °C in 2004 and from 33.2 to 36 °C in 2005. The lowest protein concentrations were found when temperature ranges from 20 °C to 25 °C and the highest protein contents were found at temperatures lower than 20 °C or greater than 25 °C (Piper and Boote 1999; Dardanelli et al. 2006). Moreover, temperature during seed-fill was the main reason for maximal protein in MG II-III and MG VI in some environments where MG II-III generally matured under higher temperatures than MG VI. The total oil content increased as temperature increased to a certain point, then decreased as temperature increased (Gibson and Mullen 1996; Dornbos and Mullen 1992).

Effects of drought on seed protein, oil, and fatty acids were investigated in a greenhouse experiment. The experiment was conducted to evaluate the responses of slow-wilting trait to heat and drought using NC-Roy (fast wilting: Control1), Boggs (intermediate in wilting: Control2), and NTCPR94–5157 (slow-wilting: SW1) and N04–9646 (slow-wilting: SW2) genotypes. Plants were either well watered or drought stressed. Soil of watered plants was kept between −15 and −20 kPa and this was considered the field capacity and used as control. Drought-stressed plants were kept between −90 and −100 kPa (Bellaloui et al. 2010a) and were considered fully matured when they reached R8 (full maturity) according to Fehr and Caviness (1977). At full maturity, 95% of pods reached full maturity. Three replicates were used in each treatment. Greenhouse temperature conditions were kept at 34 °C ± 9 °C during the day and 28 °C ± 7 °C at night. Photosynthetic photon flux density (PPFD) during the day of about 800–2300 $\mu mol \cdot m^{-2} \cdot s^{-1}$ was measured by a quantum meter (Spectrum Technologies, Inc., Aurora. Illinois, USA). The wide range of light intensity reflects a bright, sunny, or cloudy day. The experiment was conducted during the normal growing season (from April to September) to simulate the growing season photoperiod of soybean production in the midsouth USA. The fully expanded leaves at seed-fill stages (R5-R6) were analyzed for water potential and mineral nutrition. Mature seeds at R8 were harvested for seed protein, oil, and fatty acids. The results from this experiment showed that protein and oleic acid were higher in slow-wilting soybeans than the controls (Figs. 1.5a,b, 1.6, 1.7, 1.8, 1.9, 1.10, and 1.11a, b) because of the inverse relationships between protein and oil and between oleic acid and linolenic acid. This was explained by the fact that slow-wilting soybean has the ability to maintain its cell water turgor and higher leaf water potential compared to controls.

The genetic modification of the fatty acid composition of soybean oil has been previously reported by Fehr (2007). For example, modified soybean oils have been sold commercially, including oils in which linolenic acid (18:3) content has been reduced from 8 to 1% and oleic acid (18:1) has been increased from 25 to >80%. This oil composition reduces or eliminates the need for hydrogenation to achieve

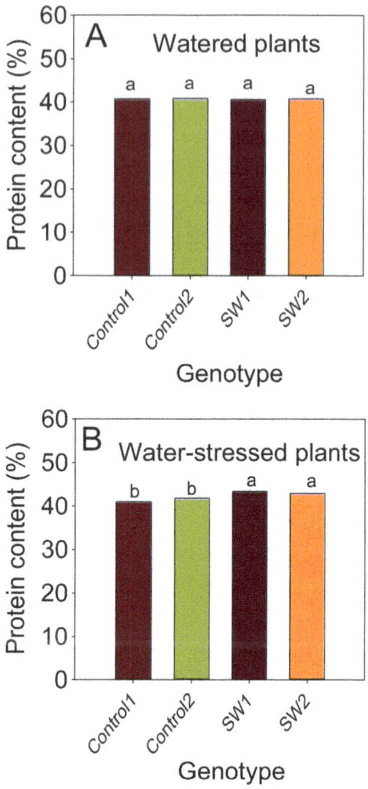

Fig. 1.5 Effect of water stress (drought) on seed protein in NC-Roy (fast wilting: Control1), Boggs (intermediate wilting: Control2), NTCPR94–5157 (slow wilting: SW1), and N04–9646 (slow wilting: SW2) genotypes. Plants were either well watered (**a**) or drought stressed (**b**). Soil of watered plants was kept between −15 and −20 kPa and this was considered the field capacity and used as control. Drought-stressed plants were kept between −90 and −100 kPa. Bars are mean of three replicates. Different letters on the bars indicate significant differences at $P \leq 0.05$

stability, minimize or eliminate transfats production, and prolong shelf life. Further, oil with palmitic acid (16:0) levels reduced from 11 to <4% has been achieved, resulting in oil low in saturated fatty acids content, a desirable oil trait for cardiovascular health. Genetic modification of oils has been achieved by genetic engineering and conventional breeding. It was reported that mutagenesis was the conventional breeding method used to develop the major genes contributing to reduced palmitic and linolenic acids, while genetic engineering was used to increase oleic acid to >80%. Leamy et al. (2017) reported that soybean (*Glycine max*) is a major crop in the world (Medic et al. 2014; Leamy et al. (2017). It was reported that decreased levels of saturated palmitic acid and increased levels of unsaturated oleic acid in soybean oil are beneficial for human cardiovascular health. Therefore, these oil traits became one of the major goals of soybean breeders.

In a recent study, Del Conte et al. (2020) found strong positive correlations between seed numbers, pod numbers, and plant node numbers and seed oil content, and these agronomic traits can be used for indirect selection for increased seed oil content.

Soybean [*Glycine max* (L.) Merr.] is grown worldwide due to the high protein and oil contents of its seed (Medic et al. 2014), and the characterization of soybean

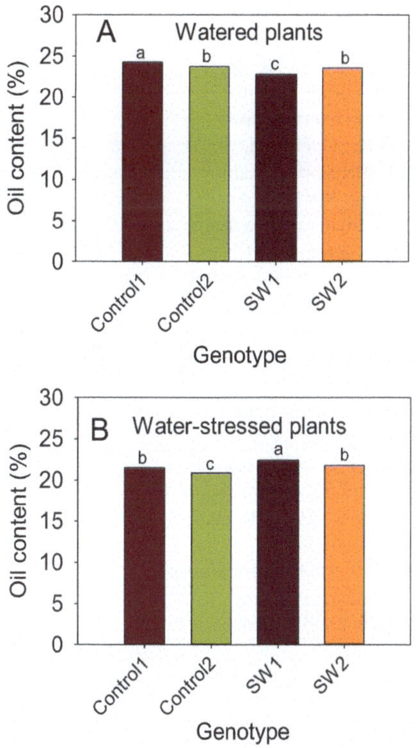

Fig. 1.6 Effect of water stress (drought) on seed oil in NC-Roy (fast wilting: Control1), Boggs (intermediate wilting: Control2), NTCPR94–5157 (slow wilting: SW1), and N04–9646 (slow wilting: SW2) genotypes. Plants were either well watered (**a**) or drought stressed (**b**). Soil of watered plants was kept between −15 and −20 kPa and this was considered the field capacity and used as control. Drought-stressed plants were kept between −90 and −100 kPa. Bars are mean of three replicates. Different letters on the bars indicate significant differences at $P \leq 0.05$

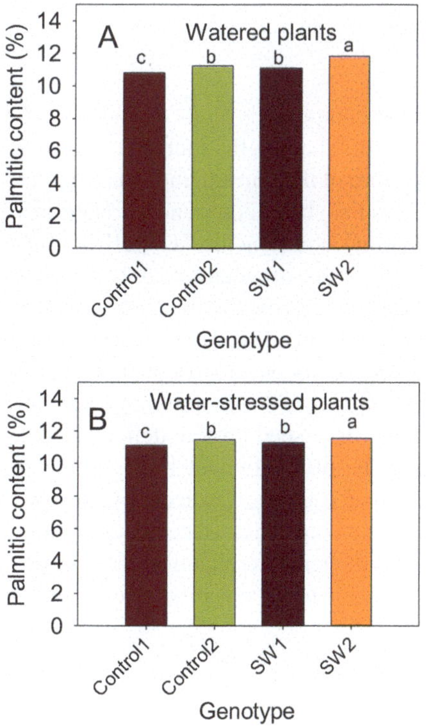

Fig. 1.7 Effect of water stress (drought) on seed palmitic acid in NC-Roy (fast wilting: Control1), Boggs (intermediate wilting: Control2), and NTCPR94–5157 (slow wilting: SW1), and N04–9646 (slow wilting: SW2) genotypes. Plants were either well watered (**a**) or drought stressed (**b**). Soil of watered plants was kept between −15 and −20 kPa and this was considered the field capacity and used as control. Drought-stressed plants were kept between −90 and −100 kPa. Bars are mean of three replicates. Different letters on the bars indicate significant differences at $P \leq 0.05$

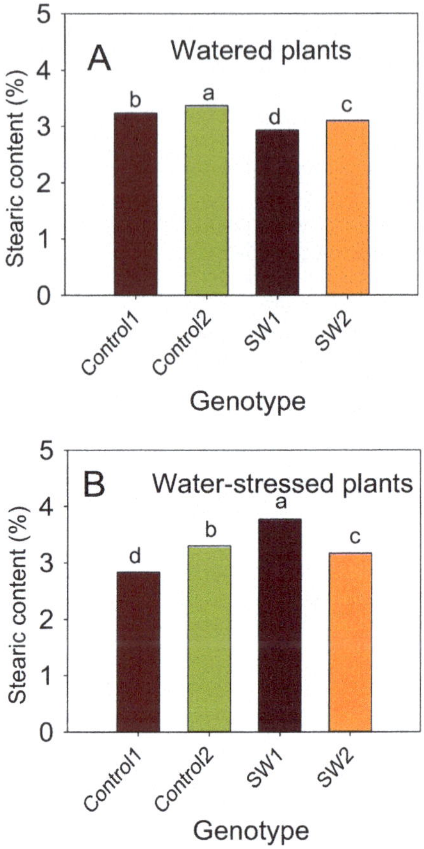

Fig. 1.8 Effect of water stress (drought) on seed stearic acid in NC-Roy (fast wilting: Control1), Boggs (intermediate wilting: Control2), NTCPR94–5157 (slow wilting: SW1), and N04–9646 (slow wilting: SW2) genotypes. Plants were either well watered (**a**) or drought stressed (**b**). Soil of watered plants was kept between −15 and −20 kPa and this was considered the field capacity and used as control. Drought-stressed plants were kept between −90 and −100 kPa. Bars are mean of three replicates. Different letters on the bars indicate significant differences at $P \leq 0.05$

seed components leads to understanding the needs of a growing world population. For this chapter, the literature was reviewed and condensed to create a well-rounded picture of the current understanding of structural, functional, and nutritional properties of soybean components. Natural variation in soybean protein, lipid, and carbohydrate components, as well as the minor constituents such as phytic acid and isoflavones, are mentioned. Environment- or genetic-induced shifts in natural variation are described with respect to nutrition and functional improvements in soybean. Orlowski et al. (2017) reported that high protein meal, used for animal feed, is achieved by seed protein concentrations above 380 g kg^{-1} (dry weight basis) (Hurburgh 1994; Brumm and Hurburgh 2006). International soybean markets currently face seed protein deficits (Dardanelli et al. 2006; Medic et al. 2014), impacting the production of high protein meal required for profitable marketing. They evaluated these 97 selected genotypes during the 2010 to 2011 growing season at two sowing dates (December 6 and December 27). Planting dates were used as environmental replication for testing genotypic differences under conditions that are known to severely affect seed protein concentration (Medic et al. 2014).

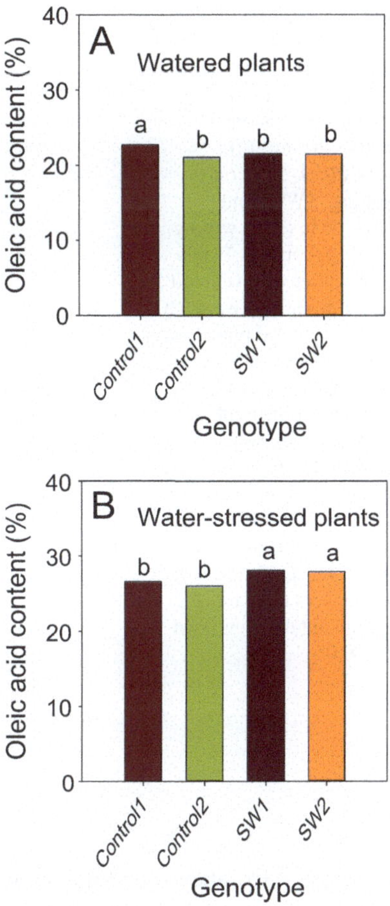

Fig. 1.9 Effect of water stress (drought) on seed oleic acid in NC-Roy (fast wilting: Control1), Boggs (intermediate wilting: Control2), NTCPR94–5157 (slow wilting: SW1), and N04–9646 (slow wilting: SW2) genotypes. Plants were either well-watered (**a**) or drought stressed (**b**). Soil of watered plants was kept between −15 and −20 kPa and this was considered the field capacity and used as control. Drought-stressed plants were kept between −90 and −100 kPa. Bars are mean of three replicates. Different letters on the bars indicate significant differences at $P \leq 0.05$

Reduction of protein content in seed will lead to low soymeal quality in the international market, impacting meat quality for optimum human health. Medic et al. (2014) reported that factors associated with reduced seed protein included environmental conditions and management practices (Medic et al. 2014). For example, early planting dates or abundant water availability during seed filling is usually associated with reduced protein concentration (Bellaloui et al. 2011; Gillen and Bellaloui 2018). Also, it was reported that seed composition is also influenced by the genotype x environment interaction (Dardanelli et al. 2006; Medic et al. 2014). An attempt was made to develop early- and late-maturing lines with consistent > 50% oleic acid content in Mississippi (Gillen and Bellaloui 2018). They selected early and late segregants resulted from three genetically different breeding populations segregating for mid-oleic acid derived from crosses to germplasm N98-4445A and non-transgenic freely available line with > 50% oleic acid. They found that no late-maturing lines (MG V) met the targeted mid-oleic acid content. However, MG III and early MG IV lines with more than 50% oleic acid lines were

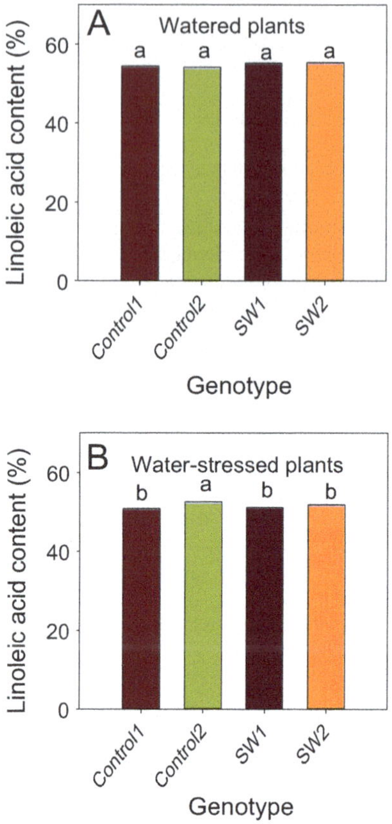

Fig. 1.10 Effect of water stress (drought) on seed linoleic acid in NC-Roy (fast wilting: Control1), Boggs (intermediate wilting: Control2), NTCPR94–5157 (slow wilting: SW1), and N04–9646 (slow wilting: SW2) genotypes. Plants were either well watered (**a**) or drought stressed (**b**). Soil of watered plants was kept between −15 and −20 kPa and this was considered the field capacity and used as control. Drought-stressed plants were kept between −90 and −100 kPa. Bars are mean of three replicates. Different letters on the bars indicate significant differences at $P \leq 0.05$

achieved. Also, they found that there was no effect of maturity alone on oleic acid content. They concluded that breeders are able to develop early soybeans with oleic acid levels greater than 50%. Bachleda et al. (2017) reported that high oleic acid content in soybean seed contributes to oxidative stability and increases oil functionality and shelf life. Bachleda et al. (2017) used near-isogenic lines (NILs: have various combinations of *FAD2-1A* and *FAD2-1B* alleles that were derived from the same backcrossing populations) of G00–3213 for the high oleic acid content and found that G00–3213 NILs with both homozygous mutant *FAD2-1A* and *FAD2-1B* alleles produced an average of 78.8% oleic acid content. They found that G00–3213 NILs with both homozygous mutant *FAD2-1A* and *FAD2-1B* alleles produced an average of 78.8% oleic acid content. They also found that possessing these mutant alleles did not result in yield reduction and modified seed composition for oleic acid did not affect seed germination when seed germination tests across 12 temperatures (12.8–32.0 °C) were conducted, although a reduction in seed germination vigor was observed when high oleic seeds were planted in cold soil, and no effects on either seed or plant development was observed in the resulted mutant *FAD2-1A* and *FAD2-1B* alleles. De Vries et al. (2011) used chemical mutagenesis and developed

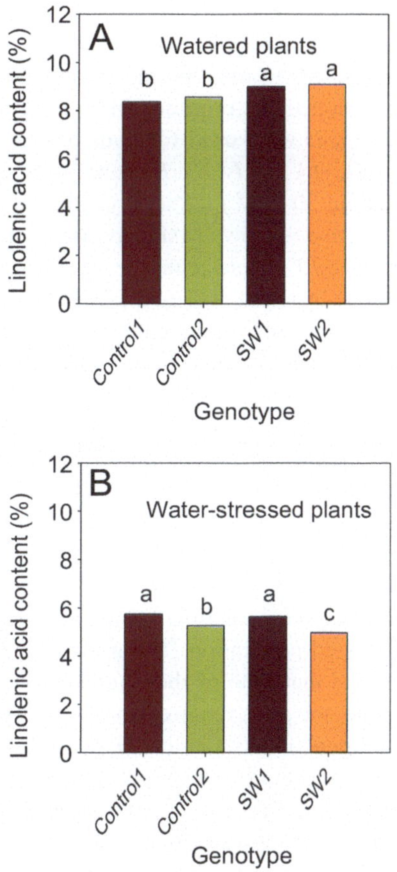

Fig. 1.11 Effect of water stress (drought) on seed linolenic acid in NC-Roy (fast wilting: Control1), Boggs (intermediate in wilting: Control2), NTCPR94–5157 (slow wilting: SW1), and N04–9646 (slow wilting: SW2) genotypes. Plants were either well watered (**a**) or drought stressed (**b**). Soil of watered plants was kept between −15 and −20 kPa and this was considered the field capacity and used as control. Drought-stressed plants were kept between −90 and −100 kPa. Bars are mean of three replicates. Different letters on the bars indicate significant differences at $P \leq 0.05$

mutants with elevated palmitate contents and designated them as fap2(A21), fap4(A24), fap5(A27), fap6(A25), and fap7(A30). In their research, they attempted to determine whether the elevated palmitate phenotypes were associated with mutations in the two known 3-ketoacyl-ACP synthase II (KAS II) genes of soybean, GmKAS IIA and GmKAS IIB. Using deoxyribonucleic acid sequence analysis, they were able to detect single nucleotide polymorphisms (SNPs) that differentiated mutant and wild-type alleles in one of the GmKAS genes in lines A21, A25, A27, and A30 but not in A24. They also found that the fap2 (A21) allele was associated with two consecutive SNPs within the *GmKAS IIA* gene. However, the allele designated fap5 (A27) also had an SNP within the GmKAS IIA gene. They concluded that the fap6 (A25) and fap7 (A30) alleles contained a single SNP at different locations within the GmKAS IIB gene and proposed that the designation of the mutant allele in A25 remains *fap6(A25)*, but the allele in A30 needs to be changed from fap7(A30) to fap6(A30). Thapa et al. (2016) reported that high oleic soybean contributes to oxidative stability so that they are used in food, fuel, and other applications. They were able to evaluate high oleic acid soybean lines derived from new gene combinations for seed yield and seed composition traits. Soybean lines with about 75% to 80% oleic acid were developed by combining mutant allele S117N *FAD2-1A* from 17D and mutant allele P137R *FAD2-1B* from PI 283327. Lines developed by crossing M23 with a different FAD2-1A mutation × (Jake × PI 283327) were used for comparison. Forty $F_{4:7}$ high-oleic lines with these mutated *FAD2-1A* and *FAD2-1B* genes were compared with forty $F_{4:7}$ normal oleic acid lines (about 20–25%) for seed yield, five fatty acids, total oil, and protein contents from six crosses grown in six environments. Genotypes with high oleic acid contents averaged >79% oleic acidoleic acid; however, palmitic and linolenic acids contents were lower by about 30% in these high-oleic acid contents lines than in their normal oleic acid contents line in each environment. Protein in high oleic acid was higher than that of normal oleic acid lines in all six populations, and oil was higher in the high-oleic lines than normal oleic lines from the 17D populations but was lower in the high-oleic lines derived from M23. They concluded that high-oleic soybeans derived by combining mutant allele S117N FAD2-1A allele with mutant P137R FAD2-1B allele were comparable in yield, but showed higher oil and higher protein meal than soybeans with normal oleic acid content.

Thapa et al. (2016) showed that the source used for genetic variation for high oleic acid was soybean [*Glycine max* (L.) Merr.] lines containing a higher fraction of oleic acid identified through a forward genetic screening of a chemically mutagenized population. The mutant lines used contained 30 to 40% oleic acid. It was found that nine of the lines identified had novel point mutations in the FAD2-1A gene, the gene required for the conversion of oleic acid to linoleic acid. In addition, the genetic association of the novel polymorphisms with the increased oleic acid trait was confirmed by mutation-specific markers, developed to detect the mutant alleles in segregating populations. Thapa et al. (2016) concluded that the identified lines with elevated oleic acid can be used in breeding programs by combining these lines traits with other genes to develop new soybean germplasm with high levels of oleic acid for human nutrition and edible oil market.

Leamy et al. (2017) used a genome-wide association (GWA) approach with nearly 30,000 single nucleotide polymorphisms (SNPs). They investigated the genetic basis of protein, oil, and fatty acids content in soybean seeds in 570 wild soybeans (*Glycine soja*), the progenitor of domesticated soybean, to identify quantitative trait loci (QTLs) affecting these seed composition traits. They were able to identify 29 SNPs located on ten different chromosomes that are significantly associated with protein, oil, and fatty acids. Among these, eight SNPs were associated with QTLs previously identified in linkage or association mapping studied with cultivated soybean samples, in addition to other novel locations. They also identified 24 of the SNPs associated with fatty acid variation on chromosomes 14 (6 SNPs) and 7 (8 SNPs). They found two SNPs that were common for two or more fatty acids, indicating loci with pleiotropic effects, and some candidate genes that were involved in fatty acid metabolism and regulation. They concluded that this research was the first GWA study conducted on seed composition traits in wild soybean and suggested that the QTLs found in wild soybean can be useful to breeders who select for modified protein, oil, and fatty acids content.

Wijewardana et al. (2019) investigated the effect of soil moisture on several seed composition traits such as protein, oil, fatty acids, and amino acids as well as sugars (sucrose, raffinose, and stachyose) and several mineral nutrients (N, P, and K) in two maturity group V cultivars, Asgrow AG5332 and Progeny 5333RY, using five irrigation regimes of 20% ET, 40% ET, 60% ET, 80% ET, and 100% ET (percent of evapotranspiration – ET). Their results showed that seed yield and protein contents increased with an increase in soil moisture for both cultivars (Asgrow AG5332 and Progeny 5333RY), while oil contents decreased with an increase in soil moisture for both cultivars. Another multiple-year study reported that higher irrigation regimes increased seed protein contents (Specht et al. 2001). This increase of seed oil content due to water deficit was also reported earlier (Bellaloui et al. 2008, 2012a). It is well known that seed protein and oil contents are negatively correlated (Bellaloui et al. 2008, 2009a, b, c, 2012a, b). Seed palmitic acid contents were found to be high with higher soil moisture for both cultivars, but seed oleic content was found to be higher with lower soil moisture for both cultivars. Similar pattern was found for seed linoleic acid contents that increased with increased soil moisture for both cultivars. Seed linolenic acid contents were shown to be high with higher soil moisture for both cultivars. It is well documented that oleic and linoleic acids biosynthesis is stimulated by water stress deficit (Martínez-Force et al. 1998; Mertz-Henning et al. 1998); under water stress, soil boron was found to stimulate nodule growth and enhance nitrogen fixation efficiency (Bellaloui et al. 2014b, 2017).

Seed sugars of both cultivars showed higher levels of sucrose followed by stachyose and then raffinose; seed sucrose contents increased with increased soil moisture for both cultivars. Seed raffinose contents were found to be high with increased soil moisture for both cultivars. On the other hand, seed stachyose contents showed the opposite trend in that it decreased with increased soil moisture for both cultivars. Previous research showed that water deficit (WD) negatively correlated with seed stachyose content, and sucrose seed content positively correlated with raffinose seed content (Bellaloui et al. 2012a, b). It was explained that under WD, sucrose synthase activity decreases drastically, thereby decreasing sucrose content and influencing the activities of other enzymes involved in stachyose and raffinose biosyntheses (Gonzalez et al. 1995). Sucrose biosynthesis and accumulation are highly sensitive to WD than the other sugars since sucrose translocation from sources (leaves) to sinks (developing seeds and fruits) is much affected by WD (Taiz and Zeiger 1998; Bellaloui et al. 2012a, b). Although stachyose and raffinose play a major role during seed germination, they are not desirable for animal and human consumption due to the production of phytate complexes, tannins, and trypsin inhibitors, leading to reduced performance of gastrointestine (Arendt and Zannini 2013; Avilles-Gaxiola et al. 2017).

With regard to seed mineral nutrients, seed N contents and seed P contents increased with increased soil moisture for both cultivars, but the increase was higher for Progeny P5333RY compared to Asgrow AG5332. Seed K content increased with increased soil moisture for Asgrow AG5332; however, seed K content for Progeny P5333RY increased, reached a plateau, and then decreased with increased soil moisture. Seed Ca content increased with increased soil moisture for both cultivars;

however, the increase was higher for Asgrow AG5332 compared to Progeny P5333RY. Seed Fe content increased with increased soil moisture for both cultivars; however, the decrease was sharp for Asgrow AG5332 compared to Progeny P5333RY. Seed Mn content increased with increased soil moisture for both cultivars; however, the decrease was higher for Asgrow AG5332 compared to Progeny P5333RY. Seed Mg content increased, reached a plateau, and then decreased with increased soil moisture for both cultivars. Seed Zn content decreased sharply with increased soil moisture for both cultivars. Seed Cu content decreased sharply with increased soil moisture for both cultivars. Content of B in seeds of both cultivars decreased with increased soil moisture (Wijewardana et al. 2019).

In an experiment conducted by YunHo et al. (2020), growing three cultivars ["Pungsannamulkong," "Deawonkong," and "Deapungkong"] in four temperature regimes in growth chambers, they found that flowering time, 100-seed weight, pod numbers, and protein and oil contents are all affected by increasing temperatures. Also, a rapid decrease in protein and oil contents was observed during the grain filling period (YunHo et al. 2020).

1.3 Modified Altered Soybean Oils

Breeding selection or gene transformation became essential to achieve specific desirable oil, fatty acids, or protein, although manipulation of these seed composition components has not been completely mastered. For example, modified soybean with desirable oils (high oleic acid, low linolenic acid, or oil with high performing oil lubricant and oil engine with low impact on environment) was achieved by either breeding selection or gene transformation. Vistive gold soybean is a genetically modified soybean (68–74% oleic acid, 3% palmitic, 3% stearic, 15% linoleic acid, and <3% linolenic acid) developed by Monsanto. The oil produced is a high-performing biosynthetic oil lubricant, is used as oil engine that has low impact on the environment, is renewable, is non-toxic, and equals or exceeds high-quality petroleum-based oils. This oil has high viscosity and is better used at high temperature. In 2014, Biosynthetic Technologies with support of USB has received American Petroleum Institute (API) certification on two motor oil. Pioneer® brand Plenish® is also a genetically modified soybean with high oleic acid and low linolenic acid, less saturated fat, and high monounsaturated fat compared to commodity soybean oil developed by Pioneer which showed yield and stable fatty acid profile across a range of environments even with significant weather stress in both full-season and double-crop systems. Plenish® high oleic soybeans were made by suppressing the enzyme omega 6 desaturase, which converts oleic acid to linoleic acid. Therefore, suppressing this enzyme by insertion of a soy-based gene would lead to the accumulation of oleic acid in the oil. Plenish also contains a transgene conferring glyphosate resistance. Currently, plant breeding and gene modification strategies are being adopted to develop soybean seed with desirable individual fatty acids in seed oil for human health or industrial processing purposes. The Plenish™ has zero transfat with high stability for extended high temperature use with nutritional benefits. Vistive® is a low linolenic acid content soybean line with low levels of linolenic acid and low transfat in processed oil which increase shelf life and flavor stability in processed foods. A non-GMO high oleic soybean (SOYLEIC™), established in 2018, was developed and owned by USDA-Agricultural Research Service and Curators of the University of Missouri. The patents were exclusively licensed to the Missouri Soybean Merchandising Council (Missouri Soybean 2021; https://mosoy.org/soyleic/). The trait was supported by soybean farmers through Checkoff program and expanded to maturity groups I to VII through partnerships with the University of Arkansas, University of Georgia, University of Illinois, Michigan State University, University of Minnesota, University of Tennessee, and others. SOYLEIC contains 80–85 percent oleic acid, exceeding the industry's 70 percent standard for high oleic. As the genetic modification technologies and breeding program goals focus on desirable healthy fatty acids continue, soybeans with designed fatty acid profiles in seed oil may dominate the market in the future long term.

Acknowledgments Technical support provided by S. Mosley is appreciated. Mention of trade names or commercial products in this publication is solely for the purpose of providing specific information and does not imply recommendation or endorsement by the US Department of Agriculture (USDA). USDA is an equal opportunity provider and employer.

Funding This research was partially funded by the US Department of Agriculture, Agricultural Research Service Project 6066–21220-014-00D.

References

Al-Tawaha, A.M., and P. Seguin. 2006. Seeding date, row spacing, and weed effects on soybean isoflavone concentrations and other seed characteristics. *Canadian Journal of Plant Science* 86: 1079–1087.

Arendt, E.K., and E. Zannini. 2013. Cereal grains for the food and beverage industries. Food Science Technology and Nutrition: 409–438.

Aviles-Gaxiola, S., C. Chuck-Hernandez, and S.O. Saldivar. 2017. Inactivation methods of trypsin inhibitor in legumes: A review. *Journal of Food Science* 83: 17–29.

Bachleda, N., T. Grey, and Z. Li. 2017. Effects of high oleic acid soybean on seed yield, protein and oil contents, and seed germination revealed by near-isogeneic lines. *Plant Breeding* 136: 539–549. https://doi.org/10.1111/pbr.12497.

Badami, R.C., J.K. Thakkar, and N. Bhat. 1984. Characterization of twelve genotype varieties of soybean for oil content, protein content and fatty acid composition. *European Journal of Lipid Science and Technology* 86 (6): 246–247.

Bellaloui, N., R.M. Zablotowicz, K.N. Reddy, and C.A. Abel. 2008. Nitrogen metabolism and seed composition as influenced by glyphosate application in glyphosate-resistant soybean. *Journal of Agricultural and Food Chemistry* 56: 2765–2772. https://doi.org/10.1021/jf703615m.

Bellaloui, N., H.K. Abbas, A.M. Gillen, and C.A. Abel. 2009a. Effect of glyphosate-boron application on seed composition and nitrogen metabolism in glyphosate-resistant soybean. *Journal of Agricultural and Food Chemistry* 57: 9050–9056. https://doi.org/10.1021/jf901801z.

Bellaloui, N., J.E. Hanks, D.K. Fisher, and A. Mengistu .2009b. Soybean seed composition is influenced by within-field variability in soil nutrients. *Crop Management*. https://doi.org/10.1094/CM-2009-1203-01-RS.

Bellaloui, N., J.R. Smith, J.D. Ray, and A.M. Gillen. 2009c. Effect of maturity on seed composition in the early soybean production system as measured on near-isogenic soybean lines. *Crop Science* 49: 608–620. https://doi.org/10.2135/cropsci2008.04.0192.

Bellaloui, N., K.N. Reddy, A.M. Gillen, and C.A. Abel. 2010a. Nitrogen metabolism and seed composition as influenced by foliar boran application in soybean. *Plant and Soil* 336: 143–155. https://doi.org/10.1007/s11104-010-0455-6.

Bellaloui, N., H.A. Bruns, A.M. Gillen, H.K. Abbas, R.M. Zablotowicz, A. Mengistu, and R.L. Paris. 2010b. Soybean seed protein, oil, fatty acids, and mineral composition as influenced by soybean-corn rotation. *Agricultural Sciences* 1 (3): 102–109.

Bellaloui, N., J.R. Smith, A.M. Gillen, and J.D. Ray. 2010c. Effect of maturity on seed sugars as measured on near-isogenic soybean (*Glycine max*). Lines. *Crop Science* 50: 1978–1987. https://doi.org/10.2135/cropsci2009.10.0596.

———. 2011. Effects of maturity, genotypic background, and temperature on seed mineral composition in near- isogenic soybean lines in the early soybean production system. *Crop Science* 51: 1161–1171. https://doi.org/10.2135/cropsci2010.04.0187.

Bellaloui, N., A. Mengistu, D.K. Fisher, and C.A. Abel. 2012a. Soybean seed composition constituents as affected by drought and Phomopsis in phomopsis susceptible and resistant genotypes. *Journal of Crop Improvement* 26: 428–453. https://doi.org/10.1080/15427528.2011.651774.

Bellaloui, N., J.R. Smith, A.M. Gillen, D.K. Fisher, and A. Mengistu. 2012b. Effect of shade on seed protein, oil, fatty acids, and minerals in soybean lines varying in seed germinability in the early soybean production system. *American Journal of Plant Sciences* 3: 84–95. https://doi.org/10.4236/ajps.2012.31008.

Bellaloui, N., A. Mengistu, E.R. Walker, and L.D. Young. 2014a. Soybean seed composition as affected by seeding rates and row spacing. *Crop Science* 54: 1782–1795.

Bellaloui, N., A. Mengistu, M.A. Kassem, C.A. Abel, and L.H.S. Zobiole. 2014b. Role of boron nutrient in nodule growth and nitrogen fixation in soybean genotypes under water stress conditions. In *Advances in biology and ecology of nitrogen fixation*, ed. T. Ohyama. IntechOpen. https://doi.org/10.5772/56994.

Bellaloui, N., L. Khandaker, M. Akond, S.K. Kantartzi, K. Meksem, A. Mengistu, and M.A. Kassem. 2015a. Identification of QTLs underlying seed macronutrients accumulation in 'MD96-5722′ by 'Spencer' recombinant inbred lines of soybean. *Atlas Journal of Biology* 3 (2): 224–235. https://doi.org/10.5147/ajb.v3i2.34.

Bellaloui, N., L. Khandaker, M. Akond, S.K. Kantartzi, K. Meksem, A. Mengistu, D.A. Lightfoot, and M.A. Kassem. 2015b. Identification of QTL underlying seed micronutrients accumulation in 'MD 96-5722' by 'Spencer' recombinant inbred lines of Soybean. *Atlas Journal of Plant Biology* 1 (3): 39–49. https://doi.org/10.5147/ajpb.v1i3.112.

Bellaloui, N., Y. Hu, A. Mengistu, H.K. Abbas, M.A. Kassem, and M. Tigabu. 2016. Elevated atmospheric carbon dioxide and temperature affect seed composition, mineral nutrition, and 15N and 13C dynamics in Soybean genotypes under controlled environments. *Atlas Journal of Plant Biology* 2016: 56–65. https://doi.org/10.5147/ajpb.2016.0157.

Bellaloui, N., A. Mengistu, H.K. Abbas, and M.A. Kassem. 2017. Effects of drought and elevated atmospheric carbon dioxide on seed nutrition and 15N and 13C natural abundance isotopes in soybean under controlled environments. In *Soybean – The basis of yield, biomass and productivity*, ed. M. Kasai. Croatia: IntechOpen. https://doi.org/10.5772/67511.

Bellaloui, N., H.A. Bruns, H.K. Abbas, D.K. Fisher, and A. Mengistu. 2020. Effects of harvest-aids on seed nutrition in soybean under midsouth USA conditions. *Plants* 2020 (9): 1007. https://doi.org/10.3390/plants9081007.

Blackman, S.A., R.L. Obendorf, and A.C. Leopold. 1992. Maturation proteins and sugars in desiccation tolerance of developing soybean seeds. *Plant Physiology* 100: 225–230.

Bolon, Y.T., D.L. Hyten, J.H. Orf, C.P. Vance, and G.J. Muehlbauer. 2014. eQTL networks reveal complex genetic architecture in the immature soybean seed. *Plant Genome* 7: 1–14. https://doi.org/10.3835/plantgenome2013.08.0027.

Brumm, T.J., and C.R. Hurburgh. 2006. Changes in Long-term Soybean compositional patterns. *Journal of the American Oil Chemists' Society* 83 (11): 981–983.

Burton, J.W. 1987. Quantitative genetics: Results relevant to soybean breeding. In *Soybeans: Improvement, production, and uses*, ed. J.R. Wilcox, 2nd ed. Madison: Agron. Monogr. 16. ASA, CSSA, and SSSA.

Business Sphere. 2007. www.notrans.iastate.edu. 19(2). Verified on September 5, 2019.

Clemente, T.E., and E.B. Cahoon. 2009. Soybean oil: Genetic approaches for modification of functionality and total content. *Plant Physiology* 151: 1030–1040. https://doi.org/10.1104/pp.109.146282.

Dardanelli, J.L., M. Balzarini, M.J. Martínez, M. Cuniberti, S. Resnik, S.F. Ramunda, R. Herrero, and H. Baigorri. 2006. Soybean maturity groups, environments, and their interaction define mega-environments for seed composition in Argentina. *Crop Science* 46: 1939–1947. https://doi.org/10.2135/cropsci2005.12-0480.

De Vries, B.D., W.R. Fehr, G.E. Welke, and R.E. Dewey. 2011. Molecular analysis of mutant alleles for elevated palmitate concentration in soybean. *Crop Science* 51: 2554–2560. https://doi.org/10.2135/cropsci2011.03.0156.

Del Conte, M.V., P.C.S. Carneiro, M.D.V. de Resende, F.L. da Silva, and L.A. Peternelli. 2020. Overcoming collinearity in path analysis of soybean [*Glycine max* (L.) Merr.] grain oil content. *PLoS One* 15 (5): e0233290. https://doi.org/10.1371/journal.pone.0233290.

Dornbos, D.L., and R.E. Mullen. 1992. Soybean seed protein and oil contents and fatty-acid composition adjustments by drought and temperature. *Journal of the American Oil Chemists' Society* 69: 228–231.

Fang, C., Y. Ma, S. Wu, L.C. Fang, Y. Ma, S. Wu, Z. Liu, Z. Wang, R. Yang, G. Hu, Z. Zhou, H. Yu, M. Zhang, Y. Pan, G. Zhou, H. Ren, W. Du, H. Yan, Y. Wang, D. Han, Y. Shen, S. Liu, T. Liu, J. Zhang, H. Qin, J. Yuan, X. Yuan, F. Kong, B. Liu, J. Li, Z. Zhang, G. Wang, B. Zhu, and Z. Tian. 2017. Genome-wide association studies dissect the genetic networks underlying agronomical traits in soybean. *Genome Biology* 18: 161. https://doi.org/10.1186/s13059-017-1289-9.

Federal Register. 2003. Food labeling; trans fatty acids in nutrition labeling, consumer research to consider nutrient content and health claims and possible footnote or disclosure statements; final rule and proposed rule. - 68 FR 41433 July 11.

Fehr, W.R. 2007. Breeding for modified fatty acid composition in soybean. *Crop Science* 47: 72–87.

Fehr, W.R., and C.E. Caviness. 1977. Stages of soybean development. Special Report. 87. Iowa State University. https://lib.dr.iastate.edu/specialreports/87.

Gerde, J., C. Hardy, W.R. Fehr, and P. White. 2007. Frying performance of no-trans, low-linolenic acid soybean oils. *Journal of the American Oil Chemists' Society* 84: 557–563.

Gibson, L.R., and R.E. Mullen. 1996. Soybean seed composition under high day and night growth temperatures. *Journal of the American Oil Chemists Society* 73: 733–737.

Gillen, A.M., and N. Bellaloui. 2018. Effect of maturity and temperature on breeding for mid-oleic soybean lines under the high heat conditions of the Mississippi Delta, USA. *Euphytica* 214:115: 1–21. https://doi.org/10.1007/s10681-018-2181-8.

Gonzalez, E.M., A.J. Gordon, C.L. James, and C. Arrese-lgor. 1995. The role of sucrose synthase in the response of soybean nodulesto drought. *Journal of Experimental Botany* 46: 1515–1523.

GRAS. 2019. Generally recognized as safe (GRAS). Verified on 3/15/2021. https://www.fda.gov/food/food-ingredients-packaging/generally-recognized-safe-gras.

Gulluoglu, L., H. Bakal, and H. Arioglu. 2018. Oil content and composition of soybean genotypes grown in different growing seasons under Mediterranean conditions. *The Journal of Experimental Biology* 39: 211–215.

Gupta, R.K., S.S. Gangoliya, and N.K. Singh. 2015. Reduction of phytic acid and enhancement of bioavailable micronutrients in food grains. *Journal of Food Science and Technology* 52: 676–684. https://doi.org/10.1007/s13197-013-0978-y.

Hartwig, E.E., and T.C. Kilen. 1991. Yield and composition of soybean seed from parents with different protein, similar yield. *Crop Science* 31 (2): 290–292.

Haun, W., A. Coffman, B.M. Clasen, Z.L. Demorest, A. Lowy, E. Ray, A. Retterath, T. Stoddard, A. Juillerat, F. Cedrone, L. Mathis, D.F. Voytas, and F. Zhang. 2014. Improved soybean oil quality by targeted mutagenesis of the fatty acid desaturase 2 gene family. *Plant Biotechnology Journal* 12: 934–940.

Hay, R., and J. Porter. 2006. *The physiology of crop yield*. 2nd ed. Oxford: Blackwell.

Hu, F.B., M.J. Stampfer, J.E. Manson, E. Rimm, G.A. Colditz, B.A. Rosner, C.H. Hennekens, and W.C. Willett. 1997. Dietary fat intake and the risk of coronary heart disease in women. *NEJM* 337: 1491–1499. https://doi.org/10.1056/NEJM199711203372102.

Hurburgh, C.R. 1994. Identification and segregation of high-value soybeans at a country elevator. *Journal of the American Oil Chemists Society* 71: 1073–1078.

Johnson, H.W. 1958. Registration of soybean varieties: VI. *Agronomy Journal* 50: 690–691.

Leamy, L.J., H. Zhang, C. Li, C.Y. Chen, and B.H. Song. 2017. A genome-wide association study of seed composition traits in wild soybean (Glycine soja). *BMC Genomics* 18: 18. https://doi.org/10.1186/s12864-016-3397-4.

Liu, K. 1997. *Soybeans chemistry, technology, and utilization*. New York: Chapman & Hall.

Long, S.P., E.A. Ainsworth, A. Rogers, and D.R. Ort. 2004. Rising atmospheric carbon dioxide: Plants FACE the future. *Annual Review of Plant Biology* 55: 591–628.

Maestri, D.M., D.O. Labuckas, J.M. Meriles, A.L. Lamarque, J.A. Zygadlo, and C.A. Guzman. 1998. Seed composition of soybean cultivars evaluated in different environmental regions. *Journal of the Science of Food and Agriculture* 77: 494–498.

Martínez-Force, E., R. Álvarez-Ortega, S. Cantisán, and R. Garcés. 1998. Fatty Acid composition in developing high saturated sunflower (Helianthus annuus) Seeds: Maturation changes and temperature effect. *Journal of Agricultural and Food Chemistry 46 (9): 3577–3582*. https://doi.org/10.1021/jf980276e.

Medic, J., C. Atkinson, and C.R. Hurburgh. 2014. Current knowledge in soybean composition. *Journal of the American Oil Chemists' Society* 91 (3): 363–384.

Mertz-Henning, L.M., L.C. Ferreira, F.A. Henning, J.M.G. Mandarino, E.D. Santos, M.C.N.D. Oliveira, A.L. Nepomuceno, J.R.B. Farias, and N. Neumaier. 1998. Effect of water deficit-induced at vegetative and reproductive stages on protein and oil content in soybean grains. *Agronomy 2018, 8(1): 3*. https://doi.org/10.3390/agronomy8010003.

Missouri Soybean. 2021. Verified on 3/18/2021 https://mosoy.org/soyleic/.

Obendorf, R.L., M. Horbowicz, A.M. Dickerman, P. Brenac, and M.E. Smith. 1998. Soluble oligosaccharides and galactosyl cyclitols in maturing soybean seeds. In planta and in vitro. *Crop Physiology & Metabolism* 38 (1): 78–84.

Orlowski, J.M., W.R. Serson, M. AL-Amery, G.L. Gregg, and C.D. Lee. 2017. Early-season stress can have small effect on soybean seed protein and oil content. *Crop Forage, and Turfgrass Management* 3 (1): 2016-12-0082. https://doi.org/10.2134/cftm2016.12.0082.

Piper, E.L., and K.J. Boote. 1999. Temperature and cultivar effects on soybean seed oil and protein concentrations. *Journal of the American Oil Chemists' Society* 76: 1233–1242.

Prasad, P., V. Vara, L.H. Allen, and K.J. Boote. 2005. Crop responses to elevated carbon dioxide and interaction with temperature: Grain legumes. *Journal of Crop Improvement* 13: 113–155.

Ragin, B., M. Akond, S.K. Kantartzi, K. Meksem, H. Herrera, C. Akbay, D.A. Lightfoot, and M.A. Kassem. 2014. Effect of row spacing on seed Isoflavone contents in Soybean [*Glycine max* (L.) Merr.]. *American Journal of Plant Sciences* 5: 4003–4010. https://doi.org/10.4236/ajps.2014.526418.

Rincker, K., R. Nelson, J. Specht, D. Sleper, T. Cary, S.R. Cianzio, S. Casteel, S. Conley, P. Chen, V. Davis, C. Fox, G. Graef, C. Godsey, D. Holshouser, G.L. Jiang, S.K. Kantartzi, W. Kenworthy, C. Lee, R. Mian, L. McHale, S. Naeve, J. Orf, V. Poysa, W. Schapaugh,

G. Shannon, R. Uniatowski, D. Wang, and B.W. Diers. 2014. Genetic improvement of U.S. Soybean in Maturity Groups II, III, and IV. *Crop Science* 54: 1419–1432.

Specht, J.E., K. Chase, M. Macrander, G.L. Graef, J. Chung, J.P. Markwell, M. Germann, J.H. Orf, and K.G. Lark. 2001. Soybean Response to water: A QTL analysis of drought tolerance. *Crop Science 41 (2): 493–509.*

Taiz, L., and E. Zeiger. 1998. Mineral nutrition. In *Plant physiology*. Sunderland: Sinauer Associates Inc.

Taub, D., B. Miller, and H. Allen. 2008. Effects of elevated CO_2 on the protein concentration of food crops: A meta-analysis. *Global Change Biology* 14: 565–575.

Temperly, R.T., and R. Borges. 2006. Tillage and crop rotation impact on soybean grain yield and composition. *Agronomy Journal* 98: 999–1004.

Thapa, R., M. Carrero-Colon, M. Crowe, E. Gaskin, and K. Hudson. 2016. Novel FAD2–1A alleles confer an elevated oleic acid phenotype in soybean seeds. *Crop Science* 56: 226–231. https://doi.org/10.2135/cropsci2015.06.0339.

Thomas, J.M.G., K.J. Boote, L.H. Allen, M. Gallo-Meagher, and J.M. Davis. 2003. Elevated temperature and carbon dioxide effects on soybean seed composition and transcript abundance. *Crop Science* 43: 1548–1557.

U.S. Food and Drug Administration. Final determination regarding partially hydrogenated oils (removing trans-fat) (2018) final determination regarding partially hydrogenated oils (removing trans-fat). Verified on 3/15/2021. https://www.fda.gov/food/food-additives-petitions/final-determination-regarding-partially-hydrogenated-oils-removing-trans-fat.

Voldeng, H.D., E.R. Cober, D.J. Hume, C. Gillard, and M.J. Morrison. 1997. Fifty-eight years of genetic improvement of shortseason soybean cultivars in Canada. *Crop Science* 37: 428–431. https://doi.org/10.2135/cropsci1997.0011183X003700020020x.

Weiss, M.G., and T.M. Stevenson. 1955. Registration of soybean varieties: V. *Agronomy Journal* 47: 541–543.

Wijewardana, C., K.R. Reddy, and N. Bellaloui. 2019. Soybean seed physiology, quality, and chemical composition under soil moisture stress. *Food Chemistry* 278: 92–100. https://doi.org/10.1016/j.foodchem.2018.11.0.35.

Wilcox, J.R. 2001. Sixty years of improvement in publicly developed elite soybean lines. *Crop Science* 41: 1711–1716. https://doi.org/10.2135/cropsci2001.1711.

Wilson, R.F. 2004. Seed composition. In *Soybeans: Improvement, production, and uses*, ed. H.R. Boerma and J.E. Specht, 3rd ed., 621–677. Madison: Agronomy Monographs 16. ASA, CSSA, and SSSA.

Yazdi-Samadi, B., R.W. Rinne, and R.D. Seif. 1977. Components of developing soybean seeds: Oil, protein, sugars, starch, organic acids, and amino acids. *Agronomy Journal* 69: 481–486.

YunHo, L., S. WanGyu, C. JungIl, and S. MyungChul. 2020. Effects of high temperature on soybean physiology, protein and oil content, and yield. *Korean Journal of Crop Science* 64 (4): 395–405.

Chapter 2
QTL That Control Seed Protein, Oil, and Fatty Acids Contents

Moulay Abdelmajid Kassem

2.1 Introduction

To develop soybean cultivars and lines with high seed protein and oil contents, high oleic acid contents, and low palmitic, stearic, and linolenic acid contents is highly desirable for plant breeders and farmers (Graef et al. 1988; Reinprecht et al. 2006; Stevenson et al. 2007; Fehr 2007; Kim et al. 2015). However, these traits are complex in nature, and several are negatively correlated with each other (Graef et al. 1988; Reinprecht et al. 2006; Stevenson et al. 2007; Fehr 2007; Kim et al. 2015).

Different types of molecular markers and mapping populations have been used to genetically map quantitative trait loci (QTL) that control seed protein, oil, and fatty acids in soybean (SoyBase: http://www.soybase.org, Grant et al. 2010; Brown et al. 2020). The main approaches are the use of biparental recombinant inbred line populations and the use of genome-wide association studies (GWAS). There are approximately 1348+ QTL that control seed protein, oil, and fatty acids that have been identified and mapped on all soybean 20 chromosomes (Table 2.1) and reported in SoyBase (Grant et al. 2010; Brown et al. 2020); however, several others have not been reported in SoyBase (Tajuddin et al. 2003; Akond et al. 2014). In this chapter, we will discuss and summarize all studies that reported QTL that control seed protein, oil, and fatty acids from 1992 to 2020, that is, almost two decades of studies on this subject.

M. A. Kassem (✉)
Plant Genomics and Biotechnology Lab, Department of Biological and Forensic Sciences, Fayetteville State University, Fayetteville, NC, USA
e-mail: mkassem@uncfsu.edu

M. A. Kassem (ed.), *Soybean Seed Composition*,
https://doi.org/10.1007/978-3-030-82906-3_2

Table 2.1 Numbers of protein, oil, fatty acids, and amino acids QTL reported in SoyBase (Grant et al. 2010; Brown et al. 2020; http://www.soybase.org)

	Chromosome#/linkage group (LG)										
	1	**2**	**3**	**4**	**5**	**6**	**7**	**8**	**9**	**10**	
Trait	**D1a**	**D1b**	**N**	**C1**	**A1**	**C2**	**M**	**A2**	**K**	**O**	
Protein	9	8	14	16	12	21	13	10	16	11	
Amino acids (combined)	26	12	4	7	6	19	9	8	32	20	
Oil	14	13	18	12	24	19	20	9	14	9	
Fatty acids (combined)	6	28	22	13	14	13	22	6	18	14	
Total	55	61	58	48	56	72	64	33	80	54	
	Chromosome#/linkage group (LG)										
	11	**12**	**13**	**14**	**15**	**16**	**17**	**18**	**19**	**20**	
Trait	**B1**	**H**	**F**	**B2**	**E**	**J**	**D2**	**G**	**L**	**I**	**Total**
Protein	12	10	17	13	25	6	13	18	18	40	130
Amino acids (combined)	2	0	29	14	29	7	24	13	21	12	143
Oil	5	11	11	20	31	10	21	24	26	48	152
Fatty acids (combined)	14	4	23	63	30	18	18	31	20	16	156
Total	33	25	80	110	115	41	76	86	85	116	1348

2.2 QTL that Control Seed Protein, Oil, and Fatty Acids' Contents

In SoyBase database (Grant et al. 2010; Brown et al. 2020), there are approximately 1348 QTL for protein, oil, and fatty acids (mainly palmitic acid, stearic acid, oleic acid, linoleic acid, and linolenic acid) as of December 2020 (Table 2.2) (Diers and Shoemaker 1992; Diers et al. 1992; Lark et al. 1994; Mansur et al. 1993, 1996; Lee et al. 1996; Brummer et al. 1997; Qiu et al. 1999; Orf et al. 1999; Sebolt et al. 2000; Specht et al. 2001; Csanadi et al. 2001; Vollmann et al. 2002; Chung et al. 2003; Chapman et al. 2003; Tajuddin et al. 2003; Stombaugh et al. 2004; Kabelka et al. 2004; Hyten et al., 2004a,b; Fasoula et al. 2004; Spencer et al. 2004; Panthee et al. 2005, 2006; Nichols et al. 2006; Reinprecht et al. 2006; Gai et al. 2007; Chen et al. 2007; Jun et al. 2008; Shibata et al. 2008; Palomeque et al. 2009a, b; Bachlava et al. 2009; Liang et al. 2010; Kim et al. 2010; Bolon et al. 2010, 2014; Moongkanna et al. 2011; Pandurangan et al. 2012; Xie et al. 2012; Eskandari et al. 2013; Lu et al. 2013; Rossi et al. 2013; Mao et al. 2013; Yesudas et al. 2013; Pathan et al. 2013; Ha et al. 2014; Wang et al. 2012, 2014a, 2014b, 2020; Akond et al. 2014; Hwang et al. 2014; Haun et al. 2014; Qi et al. 2011, 2014; Zhou et al. 2015; Priolli et al. 2015; Bandillo et al. 2015; Sonah et al. 2015; Zhou et al. 2015; Han et al. 2015; Fan et al. 2015; Leite et al. 2016; Leamy et al. 2017; Fang et al. 2017; Teng et al. 2017; Zhang et al. 2004, 2017, 2018; Yang et al. 2017; Hacisalihoglu et al. 2018; Patil et al. 2018; La et al. 2019; Li et al., 2002, 2010, 2011, 2015, 2018a,b, 2019a,b,c, 2020; Karikari et al. 2019).

Here, I summarized most studies that used the biparental RIL or other populations. For example, Diers et al. (1992) genotyped 60 F_2 lines derived from a cross between

Table 2.2 Details of QTL that control seed protein, oil, and fatty acids (palmitic, stearic, oleic, linoleic, and linolenic) content in soybean from 1992 to 2020

Trait	QTL	LG/Chr	Position (cM) / bp	Marker/interval	LOD/*P*	R^2 (%)	Environment	Reference
Protein	*Seed protein 1-1*	I/20	37.10–39.10	K011_1	$p = 0.0001$	42.00	Ames, IA	Diers et al. (1992)
	Seed protein 1-2	I/20	38.4–40.40	A407_1	$p = 0.0001$	39.00	Ames, IA	
	Seed protein 1-3	I/20	31.40–33.40	A144_1	$p = 0.0007$	24.00	Ames, IA	
	Seed protein 1-4	I/20	31.40–33.40	A688_1	$p = 0.001$	25.00	Ames, IA	
	Seed protein 1-5	E/15	5.30–7.30	SAC7_1	$p = 0.003$	24.00	Ames, IA	
	Seed protein 1-6	B2/14	32.13–34.13	A242_1	$p = 0.004$	19.00	Ames, IA	
	Seed protein 1-7	L/19	35.70–37.70	A023_1	$p = 0.01$	16.00	Ames, IA	
	Seed protein 1-8	G/18	88.97–90.97	A245_1	$p = 0.01$	12.00	Ames, IA	
Oil	*Seed oil 1-1*	I/20	37.10–39.10	K011_1	$p = 0.0002$	27.00	Ames, IA	
	Seed oil 1-2	I/20	38.40–40.40	A407_1	$p = 0.0005$	28.00	Ames, IA	
	Seed oil 1-3	E/15	12.60–14.60	Pb	$p = 0.0001$	27.00	Ames, IA	
	Seed oil 1-4	E/15	29.90–31.90	A454_1	$p = 0.0008$	23.00	Ames, IA	
	Seed oil 1-5	E/15	5.30–7.30	SAC7_1	$p = 0.0001$	43.00	Ames, IA	
	Seed oil 1-6	B2/14	32.13–34.13	A242_1	$p = 0.0001$	39.00	Ames, IA	
	Seed oil 1-7	L/19	35.70–37.70	A023_1	$p = 0.0001$	32.00	Ames, IA	
	Seed oil 1-8	E/15	27.27–29.27	K229_1	$p = 0.001$	22.00	Ames, IA	
	Seed oil 1-9	E/15	33.62–35.62	A203_1	$p = 0.006$	18.00	Ames, IA	

(continued)

Table 2.2 (continued)

Palmitic	*Seed palmitic 1-1*	B2/14	35.99–37.99	A343_1	$p = 0.0010$	0.24	Ames, IA	Diers and Shoemaker (1992)
	Seed palmitic 1-2	B2/14	52.54–54.54	A018_1	$p = 0.0050$	0.19	Ames, IA	
	Seed palmitic 1-3	J/16	67.30–69.30	K375_1	$p = 0.0050$	0.18	Ames, IA	
Stearic	*Seed stearic 1-1*	J/16	82.15–84.15	A233_1	$p = 0.0050$	0.19	Ames, IA	
Oleic	*Seed oleic 1-1*	A1/5	101.30–103.30	A082_1	$p = 0.0001$	0.28	Ames, IA	
	Seed oleic 1-2	A1/5	91.30–93.30	A104_1	$p = 0.0003$	0.26	Ames, IA	
	Seed oleic 1-3	A1/5	91.30–93.31	A170_1	$p = 0.0010$	0.23	Ames, IA	
	Seed oleic 1-4	E/15	10.02–12.02	A242_2	$p = 0.0020$	0.20	Ames, IA	
	Seed oleic 1-5	B2/14	64.00–66.00	A617_1	$p = 0.0040$	0.19	Ames, IA	
	Seed oleic 1-6	E/15	12.60–14.60	Pb	$p = 0.0010$	0.21	Ames, IA	
Linoleic	*Seed linoleic 1-1*	A1/5	101.30–103.30	A082_1	$p = 0.0001$	0.38	Ames, IA	
	Seed linoleic 1-2	A1/5	91.30–93.30	A104_1	$p = 0.0001$	0.33	Ames, IA	
	Seed linoleic 1-3	A1/5	91.56–93.56	A170_1	$p = 0.0001$	0.30	Ames, IA	
	Seed linoleic 1-4	E/15	10.02–12.02	A242_2	$p = 0.0020$	0.21	Ames, IA	
	Seed linoleic 1-5	B1/11	57.91–59.91	A118_1	$p = 0.0060$	0.20	Ames, IA	
	Seed linoleic 1-6	E/15	12.60–14.60	Pb	$p = 0.0060$	0.20	Ames, IA	
Linolenic	*Seed linolenic 1-1*	E/15	5.30–7.30	SAC7_1	$p = 0.0050$	0.31	Ames, IA	

	Seed linolenic 1-2	L/19	35.70–37.70	A023_1	$p = 0.0050$	0.26	Ames, IA	
	Seed linolenic 1-3	E/15	33.62–35.62	A203_1	$p = 0.0020$	0.22	Ames, IA	
	Seed linolenic 1-4	E/15	10.02–12.02	A242_2	$p = 0.0080$	0.23	Ames, IA	
	Seed linolenic 1-5	E/15	29.90–31.90	A454_1	$p = 0.0010$	0.22	Ames, IA	
	Seed linolenic 1-6	K/9	52.90–54.90	A065_3	$p = 0.0030$	0.20	Ames, IA	
	Seed linolenic 1-7	E/15	27.27–29.27	K229_1	$p = 0.0030$	0.20	Ames, IA	
	Seed oil 1-1	A2/8	43.90–45.90	A111_1	5.5	36.00	Ames, IA	Mansur et al. (1993)
	Seed oil 1-2	K/9	2.40–28.72	A315_1	2.9	24.00	Ames, IA	
Oil/prot	*Seed oil/prot 1-1*	A1/5	41.41–43.41	R183_1	$p = 0.0018$	14.40	MN, Chile	Lark et al. (1994)
Protein	*Seed protein 2-1*	A1/5	92.60–94.60	T155_1	$p < 0.001$	9.10	Minnesota	Mansur et al. (1996)
	Seed protein 2-2	L/19	91.00–93.00	Satt006	$p < 0.001$	7.80	Minnesota	
	Seed protein 2-3	A1/5	29.30–31.30	A329_2	$p < 0.001$	5.20	Minnesota	
Oil	*Seed oil 3-1*	L/19	91.00–93.00	Satt006	$p < 0.001$	8.50	Minnesota	
	Seed oil 3-2	A1/5	92.60–94.60	T155_1	$p < 0.001$	7.30	Minnesota	
	Seed oil 3-3	A1/5	29.30–31.30	A329_2	$p < 0.001$	5.20	Minnesota	
Protein	*Seed protein 4-1*	C1/4	120.00–122.00	gc197_1	$p < 0.005$	12.50	GA, NC, SC	Lee et al. (1996)

(continued)

Table 2.2 (continued)

	Seed protein 4-2	C1/4	125.50–127.50	EV3_1	$p < 0.005$	11.10	GA, NC, SC	
	Seed protein 4-3	C1/4	96.00–98.00	A338_1	$p < 0.005$	10.10	GA, NC, SC	
	Seed protein 4-4	C1/4	20.04–55.04	A463_1	$p < 0.005$	6.80	GA, NC, SC	
	Seed protein 4-5	E/15	25.02–27.02	A517_1	$p < 0.005$	7.40	GA, NC, SC	
	Seed protein 4-6	E/15	26.30–28.30	cr167_1	$p < 0.005$	6.50	GA, NC, SC	
	Seed protein 4-7	J/16	26.63–28.63	B166_1	$p < 0.005$	7.60	GA, NC, SC	
	Seed protein 4-8	N/3	29.35–31.35	A071_2	$p < 0.005$	11.20	GA, NC, SC	
	Seed protein 4-9	N/3	33.40–35.40	gc34_2	$p < 0.005$	7.60	GA, NC, SC	
	Seed protein 4-10	B2/14	42.60–45.60	B142_1	$p < 0.005$	10.20	GA, NC, SC	
	Seed protein 4-11	B2/14	28.19–30.19	A352_1	$p < 0.005$	10.00	GA, NC, SC	
	Seed protein 4-12	–	–	A199_3	$p < 0.005$	13.80	GA, NC, SC	
	Seed protein 4-13	E/15	44.00–46.00	cr274_1	$p < 0.005$	8.00	GA, NC, SC	
Oil	*Seed oil 5-1*	E/15	22.70–24.70	A069_1	$p < 0.005$	6.70	GA, NC, SC	
	Seed oil 5-2	J/16	56.20–58.20	B122_1	$p < 0.005$	7.00	GA, NC, SC	
	Seed oil 5-3	L/19	35.70–37.70	A023_1	$p < 0.005$	7.40	GA, NC, SC	
	Seed oil 5-4	D2/17	78.40–80.40	cr142_1	$p < 0.005$	12.90	GA, NC, SC	

	Seed oil 5-5	D2/17	72.50–74.50	K258_2	$p < 0.005$	8.80	GA, NC, SC	
	Seed oil 5-6	D2/17	78.60–80.60	cr326_1	$p < 0.005$	8.80	GA, NC, SC	
Protein	*Seed protein 3-1*	A2/8	131.31–133.31	A505_1	$p < 0.005$	11.20	Minnesota	Brummer et al. (1997)
	Seed protein 3-2	B1/11	28.17–30.17	A109_1	$p < 0.005$	39.00	Minnesota	
	Seed protein 3-3	C1/4	89.70–91.70	A063_1	$p < 0.005$	12.00	Minnesota	
	Seed protein 3-4	D1a/1	5.43–7.43	A398_1	$p < 0.005$	16.90	Minnesota	
	Seed protein 3-5	D1a/1	39.80–41.80	A691_1	$p < 0.005$	10.30	Minnesota	
	Seed protein 3-6	E/15	29.88–31.88	B174_1	$p < 0.005$	11.10	Minnesota	
	Seed protein 3-7	F/13	46.63–48.63	K002_1	$p < 0.005$	8.70	Minnesota	
	Seed protein 3-8	G/18	66.53–68.53	A816_1	$p < 0.005$	12.00	Minnesota	
	Seed protein 3-9	G/18	66.70–68.70	A890_1	$p < 0.005$	15.60	Minnesota	
	Seed protein 3-10	G/18	96.24–98.24	A235_1	$p < 0.005$	8.30	Minnesota	
	Seed protein 3-11	H/12	32.17–34.17	A069_1	$p < 0.005$	7.20	Minnesota	
	Seed protein 3-12	I/20	31.42–33.42	A144_1	$p < 0.005$	27.50	Minnesota	
Oil	*Seed oil 4-1*	A1/5	52.36–54.36	K400_1	$p < 0.005$	14.40	Minnesota	
	Seed oil 4-2	A1/5	74.41–76.41	A975_1	$p < 0.005$	10.90	Minnesota	

(continued)

Table 2.2 (continued)

	Seed oil 4-3	A1/5	91.30–93.30	A104_1	$p < 0.005$	18.90	Minnesota	
	Seed oil 4-4	A2/8	131.31–133.31	A505_1	$p < 0.005$	31.30	Minnesota	
	Seed oil 4-5	B1/11	28.20–30.20	A109_1	$p < 0.005$	8.50	Minnesota	
	Seed oil 4-6	C2/6	96.20–98.20	L148_1	$p < 0.005$	11.40	Minnesota	
	Seed oil 4-7	G/18	64.60–66.60	A584_1	$p < 0.005$	18.50	Minnesota	
	Seed oil 4-8	G/18	66.53–68.53	A816_1	$p < 0.005$	11.30	Minnesota	
	Seed oil 4-9	G/18	66.70–68.70	A890_1	$p < 0.005$	14.90	Minnesota	
	Seed oil 4-10	H/12	32.20–34.20	A069_1	$p < 0.005$	18.00	Minnesota	
	Seed oil 4-11	K/9	97.90–99.9	K387_1	$p < 0.005$	15.90	Minnesota	
Protein	*Seed protein 6-1*	H/12	123.10–125.10	B072_1	$p = 0.0018$	32.00	Missouri	Qui et al. (1999)
	Seed protein 6-2	F/13	104.84–106.84	B148_1	$p = 0.0001$	17.00	Missouri	
Oil	*Seed oil 7-1*	H/12	123.10–125.10	B072_1	$p = 0.0020$	21.00	Missouri	
Protein	*Seed protein 7-1*	M/7	38.00–40.00	R079_1	3.2	6.00	MN, Chile	Orf et al. (1999)
	Seed protein 7-2	C1/4	64.10–66.10	Satt578	6.4	12.00	MN, Chile	
	Seed protein 8-1	L/19	65.50–67.50	Satt166	3.3	11.00	MN, Chile	
	Seed protein 9-1	A1/5	92.60–94.60	T155_1	4.2	15.00	MN, Chile	
	Seed protein 9-2	C1/4	9.34–11.34	SOYGPATR	3.7	12.00	MN, Chile	
Oil	*Seed oil 8-1*	A1/5	87.60–89.60	Satt174	4.9	10.00	MN, Chile	
	Seed oil 8-2	C1/4	9.34–11.34	SOYGPATR	3.3	11.00	MN, Chile	
	Seed oil 9-1	C1/4	9.34–11.34	SOYGPATR	3.4	7.00	MN, Chile	
	Seed oil 9-2	C2/6	37.10–39.10	Satt432	3.3	11.00	MN, Chile	

	Seed oil 9-3	L/19	94.40–96.40	A489_1	6.1	19.00	MN, Chile	
	Seed oil 10-1	A1/5	92.60–94.60	T155_1	4.2	15.00	MN, Chile	
Protein	*Seed protein 10-1*	I/20	34.35–36.35	Satt127	$p = 0.0001$	65.00	Michigan	Sebolt et al. (2000)
	Seed protein 11-1	I/20	31.42–33.42	A144_1	$p = 0.0001$	65.00	Michigan	
Oil	*Seed oil 11-1*	I/20	31.42–33.42	A144_1	$p = 0.0001$	39.00	Michigan	
	Seed oil 12-1	I/20	31.42–33.42	A144_1	$p = 0.0001$	15.00	Michigan	
Protein	*Seed protein 12-1*	A1/5	93.92–95.92	B170_1	4.98	4.60	Nebraska	Specht et al. (2001)
	Seed protein 12-2	C1/4	32.30–34.30	K001_1	4.05	7.10	Nebraska	
	Seed protein 12-3	K/9	39.90–41.90	Satt178	3.35	3.30	Nebraska	
	Seed protein 12-4	M/7	32.50–34.50	Satt567	12.78	27.10	Nebraska	
	Seed protein 12-5	O/10	70.10–72.10	Satt478	2.99	6.30	Nebraska	
Oil	*Seed oil 13.1*	A1/5	93.92–95.92	B170_1	4.62	4.10	Nebraska	
	Seed oil 13.2	D1a/1	68.90–70.90	Satt468	4.66	9.10	Nebraska	
	Seed oil 13.3	F/13	70.41–72.41	Satt510	3.3	6.30	Nebraska	
	Seed oil 13.4	I/20	34.22–36.22	BLT002_1	5.12	4.50	Nebraska	
	Seed oil 13.5	I/20	118.10–120.10	L026_1	2.58	4.50	Nebraska	
Protein	*Seed protein 13-1*	D1a/1	76.50–78.50	Satt077	$p = 0.0009$	4.70	Austria	Csanadi et al. (2001)
	Seed protein 13-2	C2/6	121.02–123.02	Sct_028	2.7	6.60	Austria	

(continued)

Table 2.2 (continued)

	Seed protein 13-3	M/7	32.50–34.50	Satt567	$p = 0.0020$	7.10	Austria	
	Seed protein 13-4	K/9	103.80–105.80	Satt196	$p = 0.0009$	4.70	Austria	
Oil	*Seed oil 14-1*	B2/14	71.10–73.10	Satt020	$p = 0.0013$	3.10	Austria	
	Seed oil 14-2	K/9	103.80–105.80	Satt196	$p = 0.0004$	6.50	Austria	
	Seed oil 14-3	I/20	21.84–23.84	Satt562	$p = 0.00140$	5.70	Austria	
Protein	*Seed protein*	A2/8	149.00–151.00	Ti	$p < 0.05$	3.00	Austria	Vollmann et al. (2002)
Palmitic	*Seed palmitic 2-1*	A1/5	2.54–4.54	Satt684	$p = 0.0010$	33.00	Greenhouse	Li et al. (2002)
	Seed palmitic 2-2	M/7	65.99–67.99	Satt175	$p = 0.0100$	9.00	Greenhouse	
Protein	*Seed protein 15-1*	I/20	36.40–36.94	OPAJ13a-OPAW13a	16.92	45.30	Nebraska	Chung et al. (2003)
	Seed protein 15-1	I/20	36.40–36.94	OPAJ13a-OPAW13a	12.18	27.70	Nebraska	
Oil	*Seed oil 15-1*	I/20	36.40–36.94	OPAJ13a-OPAW13a	10.46	27.80	Nebraska	
	Seed oil 15-2	I/20	36.40–36.94	OPAJ13a-OPAW13a	7.94	24.20	Nebraska	
Protein	*Seed protein 16-1*	B1/11	35.50–37.50	Satt251	$p < 0.05$	3.00	Knoxville	Chapman et al. (2003)
	Seed protein 16-2	L/19	106.24–108.24	Satt373	$p < 0.05$	4.00	Knoxville	
	Seed oil 16-1	B1/11	35.50–37.50	Satt251	$p < 0.05$	4.00	Knoxville	
	Seed oil 16-2	D2/17	27.40–29.40	Satt014	*p*0.01	5.00	Knoxville	
Protein	*Seed protein 30-1*	I/20	24.70–26.70	Satt239	9.27	20.90	Matsudo, Japan	Tajuddin et al. (2003)
			28.61–30.61	Satt239	15.97	29.70	Kashiwa, Japan	

	Seed protein 30-2	D2/17	89.30–91.30	Satt310	3.24	7.40	Matsudo, Japan	
			78.23–80.23	Satt301	3.08	5.50	Kashiwa, Japan	
	Seed protein 30-3	E/15	18.62–20.62	Satt384	3.56	6.80	Matsudo, Japan	
			18.62–20.62	Satt384	4.56	8.20	Kashiwa, Japan	
	Seed protein 30-4	A2/8	0.00–2.00	*I*	2.59	4.90	Matsudo, Japan	
	Seed protein 30-5	C2/6	37.90–39.90	Satt281	3.71	6.20	Kashiwa, Japan	
	Seed protein 30-6	A2/8	2.50–4.50	A104	3.31	7.50	Matsudo, Japan	
	Seed protein 30-7	L/19	47.90–49.90	Satt156	2.19	5.60	Matsudo, Japan	
	Seed protein 30-8	K/9	8.00–10.00	GM195	2.94	8.40	Kashiwa, Japan	
	Seed protein 30-9	N/3	5.00–7.00	Q026	2.11	4.20	Kashiwa, Japan	
	Seed protein 30-10	G/18	1.50–3.50	A378	2.12	3.40	Kashiwa, Japan	
Lipid	*Seed oil 34-1*	A2/8	54.90–56.90	A111	2.90	9.30	Matsudo, Japan	
	Seed oil 34-2	B2/14	60.10–66.10	A519	2.18	9.00	Matsudo, Japan	
	Seed oil 34-3	I/20	27.53–35.16	Satt127	3.36	8.60	Matsudo, Japan	

(continued)

Table 2.2 (continued)

	Seed oil 34-4	J/16	25.12–31.12	K384	3.36	8.60	Matsudo, Japan	
	Seed oil 34-5	M/7	130.80–140.20	Satt308	3.26	7.10	Kashiwa, Japan	
	Seed oil 34-6	O/10	35.40–49.40	GM214b	2.28	7.00	Kashiwa, Japan	
	Seed oil 34-7	M/7	94.70–100.80	Satt250	2.20	6.30	Matsudo, Japan	
Protein	*Seed protein 19-1*	C1/4	47.10–50.10	Satt578	3.90	15.40	Minnesota	Stombaugh et al. (2004)
Protein	*Seed protein 21-1*	A2/8	125.10–127.10	Satt409	3.60	6.00	IL, IN, IA	Kabelka et al. (2004)
	Seed protein 21-2	C1/4	100.22–102.22	Satt338	$p < 0.0001$	16.00	IL, IN, IA	
	Seed protein 21-3	C2/6	88.71–90.71	Satt363	$p < 0.0001$	5.00	IL, IN, IA	
	Seed protein 21-4	D1b/2	45.23–47.23	Satt157	4.30	14.00	IL, IN, IA	
	Seed protein 21-5	O/10	4.42–6.42	Satt358	3.00	9.00	IL, IN, IA	
	Seed protein 21-6	F/13		Satt510	$p < 0.0001$	16.00	IL, IN, IA	
	Seed protein 21-7	K/9	37.62–39.62	Satt544	3.70	3.00	IL, IN, IA	
	Seed protein 21-8	B2/14	45.87–47.87	Satt168	2.70	1.00	IL, IN, IA	
	Seed protein 21-9	N/3	59.17–61.17	Satt339	4.40	13.00	IL, IN, IA	
	Seed protein 21-10	H/12	82.80–84.80	Satt142	2.80	3.00	IL, IN, IA	

	Seed protein 21-11	M/7		Satt308	$p < 0.0001$	10.00	IL, IN, IA	
Oil	*Seed oil 20-1*	C1/4	122.80–124.80	Satt338	$p < 0.0001$	5.00	IL, IN, IA	
	Seed oil 20-2	D1b/2	36.10–38.10	Satt157	5.10	10.00	IL, IN, IA	
	Seed oil 20-3	J/16	24.50–26.51	Satt285	$p < 0.0001$	16.00	IL, IN, IA	
Protein	*Seed protein 24-1*	C2/6	117.80–121.80	Satt277	9.80	27.60	Jackson, TN	Hyten et al. (2004b)
	Seed protein 24-2	F/13	79.70–99.70	Satt335	4.40	18.10	Jackson, TN	
	Seed protein 24-3	K/9	5.80–25.80	Satt539	4.30	24.40	Jackson, TN	
	Seed protein 24-4	M/7	35.85–50.10	Satt463/Satt540	3.00	13.30	Jackson, TN	
Oil	*Seed oil 23-1*	C2/6	107.60–117.8	Satt277/Satt460	3.70	8.80	Jackson, TN	
	Seed oil 23-2	D1a/1	17.52–20.52	Satt179/Satt184	3.30	8.10	Jackson, TN	
	Seed oil 23-3	D2/17	24.52–30.52	Satt154/Satt458	3.00	7.80	Jackson, TN	
	Seed oil 23-4	L/19	66.51–70.51	Dt1/Satt166	3.40	8.00	Jackson, TN	
	Seed oil 23-5	L/19	91.10–101.90	Satt229	3.30	8.30	Jackson, TN	
	Seed oil 23-6	M/7	35.85–49.90	Satt540	3.60	11.00	Jackson, TN	
Palmitic	*Seed palmitic 3-1*	D2/17	24.50–46.50	–	3.00	9.60	Jackson, TN	Hyten et al. (2004a)
	Seed palmitic 3-2	K/9	66.60–78.60	–	3.20	8.30	Jackson, TN	
	Seed palmitic 3-3	L/19	70.50–93.10	–	5.40	13.10	Jackson, TN	
Stearic	*Seed stearic 2-1*	C2/6	117.80–125.30	–	4.70	13.10	Jackson, TN	

(continued)

Table 2.2 (continued)

	Seed stearic 2-2	L/19	54.60–64.20	–	5.50	16.80	Jackson, TN	
Oleic	*Seed oleic 2-1*	D1b/2	97.20–116.90	–	3.10	7.20	Jackson, TN	
	Seed oleic 2-2	L/19	68.50–89.10	–	15.10	35.30	Jackson, TN	
Linoleic	*Seed linoleic 2-1*	F/13	83.70–106.10	–	4.20	10.00	Jackson, TN	
	Seed linoleic 2-2	L/19	66.50–88.50	–	20.20	50.50	Jackson, TN	
Linolenic	*Seed linolenic 2-1*	F/13	7.90–17.40	–	4.80	8.10	Jackson, TN	
	Seed linolenic 2-2	L/19	42.60–62.20	–	4.60	13.60	Jackson, TN	
	Seed linolenic 2-3	L/19	72.50–91.10	–	8.20	24.80	Jackson, TN	
Protein	*Seed protein 32-1*	B2/14	96.91–98.91	A953_1-Satt560	3.50	12.40	China	Zhang et al. (2004)
Oil	*Seed oil 26-1*	D1b/2	7.63–9.63	A725_1	2.50	7.40	China	
Protein	*cqPRO-001*	E/15	28.9–30.9	A454_1	$p = 0.0010$	12.30	Athens, GA	Fasoula et al. (2004)
Oil	*cqOil-001*	C1/4	89.72–91.72	A063_1	$p = 0.0011$	8.00	Athens, GA	
	cqOil-002	H/12	85.80–87.80	A566_2	$p = 0.0008$	8.30	Athens, GA	
	cqOil-003	L/19	30.58–34.54	Satt398	$p = 0.0025$	–	Athens, GA	
Linolenic	*Seed linolenic 10-1*	B2/14	86.26–88.58	Satt534	$p = 0.0001$	0.59		Spencer et al. (2004)
	Seed linolenic 10-2	B2/14	96.91–98.91	Satt560	$p = 0.0001$	0.59		
Protein	*Seed protein 20-1*	G/18	11.74–13.74	Satt570	3.50	20.20	Knoxville, TN	Panthee et al. (2005)
Oil	*Seed oil 19-1*	D1b/2	115.35–117.35	Satt274	3.00	11.80	Knoxville, TN	

	Seed oil 19-2	H/12	88.52–90.52	Satt317	2.60	9.40	Knoxville, TN	
	Seed oil 19-3	O/10	49.71–54.20	Satt420	3.50	15.00	Knoxville, TN	
	Seed oil 19-4	O/10	53.20–55.20	Satt479	3.10	12.00	Knoxville, TN	
Palmitic	*Seed palmitic 5-1*	A2/8	124.37–126.37	Satt133	2.50	12.20	Knoxville, TN	Panthee et al. (2006)
	Seed palmitic 5-2	D1b/2	74.66–76.66	Satt537	3.20	19.70	Knoxville, TN	
Stearic	*Seed stearic 4-1*	B2/14	55.20–75.34	Satt168	4.20	18.10	Knoxville, TN	
	Seed stearic 4-2	J/16	10.74–12.74	Satt249	2.80	11.80	Knoxville, TN	
Oleic	*Seed oleic 4-1*	E/15	44.40–46.40	Satt263	2.60	10.00	Knoxville, TN	
Linoleic	*Seed linoleic 4-1*	E/15	43.75–45.75	Satt185	3.40	13.80	Knoxville, TN	
Linolenic	*Seed linolenic 4-1*	E/15	44.40–46.40	Satt263	3.30	12.30	Knoxville, TN	
	Seed linolenic 4-2	G/18	20.88–22.88	Satt235	4.60	22.50	Knoxville, TN	
Protein	*cqPRO-003*	I/20	35.40–37.40	Satt496	$p < 0.001$	15.40	Urbana, IL	Nichols et al. (2006)
Oil	*cqOIL-004*	I/20	35.40–37.40	Satt496	$p < 0.001$	11.10	Urbana, IL	
Protein	*Seed protein 26-1*	A2/8	53.91–55.91	Satt187	2.60	5.00	Canada	Reinprecht et al. (2006)
	Seed protein 26-2	D2/17	78.23–80.23	Satt389	3.60	8.00	Canada	
	Seed protein 26-3	I/20	17.50–19.50	Satt571	4.10	9.00	Canada	
	Seed protein 26-4	I/20	20.90–22.90	Satt419	3.50	7.00	Canada	

(continued)

Table 2.2 (continued)

	Seed protein 26-5	I/20	49.11–51.11	Satt270	3.10	9.00	Canada	
	Seed protein 26-6	B1/11	99.87–101.87	Sat_123	2.90	11.00	Canada	
	Seed protein 26-7	C2/6	106.58–108.58	Satt277	2.90	8.00	Canada	
	Seed protein 26-8	G/18	42.40–44.40	Satt394	3.60	10.00	Canada	
	Seed protein 26-9	I/20	85.73–87.73	Satt162	2.10	7.00	Canada	
	Seed protein 26-10	B2/14	92.50–94.50	Satt063	3.40	13.00	Canada	
	Seed protein 26-11	F/13	129.63–131.63	Sat_090	2.30	6.00	Canada	
	Seed protein 26-12	G/18	3.53–5.53	Satt309	2.00	6.00	Canada	
	Seed protein 26-13	F/13	2.35–4.35	Satt569	2.20	6.00	Canada	
	Seed protein 26-14	G/18	42.40–44.40	Satt394	2.20	8.00	Canada	
Oil	*Seed oil 27-1*	C2/6	106.60–108.60	Satt277	3.50	12.00	Canada	
	Seed oil 27-2	E/15	31.30–33.30	Satt212	2.20	10.00	Canada	
	Seed oil 27-3	I/20	17.50–19.50	Satt571	3.20	8.00	Canada	
	Seed oil 27-4	I/20	49.11–51.11	Satt270	2.50	7.00	Canada	
	Seed oil 27-5	L/19	13.03–15.03	Satt182	2.50	7.00	Canada	
	Seed oil 27-6	C2/6	112.40–114.40	Satt489	2.30	8.00	Canada	
	Seed oil 27-7	I/20	27.00–29.00	Satt367	2.50	10.00	Canada	
	Seed oil 27-8	I/20	57.82–59.82	Satt049	2.40	5.00	Canada	

	Seed oil 27-9	I/20	85.73–87.73	Satt162	2.50	8.00	Canada	
	Seed oil 27-10	G/18	47.80–49.80	Satt394	2.10	8.00	Canada	
	Seed oil 27-11	–	–	Lox3	2.50	6.00	Canada	
	Seed oil 27-12	–	–	UBC300-800	2.20	5.00	Canada	
Palmitic	*Seed palmitic 4-1*	C2/6	112.38–114.38	Satt489	2.20	0.06	Canada	
	Seed palmitic 4-2	D1b/2	36.07–38.07	Satt157	3.00	0.07	Canada	
	Seed palmitic 4-3	D2/17	56.50–58.50	Sat_092	2.60	0.11	Canada	
	Seed palmitic 4-4	G/18	75.76–77.76	Satt288	2.30	0.07	Canada	
	Seed palmitic 4-5	O/10	7.75–9.75	Sat_132	2.20	0.06	Canada	
	Seed palmitic 4-6	B2/14	71.80–73.80	Satt070	3.80	0.10	Canada	
	Seed palmitic 4-7	–	–	UBC444-2300	4.10	0.13	Canada	
Stearic	*Seed stearic 3-1*	F/13	129.63–131.63	Sat_090	3.60	0.10	Canada	
	Seed stearic 3-2	G/18	75.76–77.76	Satt288	3.90	0.11	Canada	
	Seed stearic 3-3	M/7	6.84–8.84	Satt590	2.20	0.04	Canada	
	Seed stearic 3-4	G/18	42.38–44.38	Satt394	2.50	0.07	Canada	
Oleic	*Seed oleic 3-1*	C2/6	112.38–114.38	Satt489	2.60	0.07	Canada	

(continued)

Table 2.2 (continued)

	Seed oleic 3-2	G/18	0.00–2.00	Satt163	4.50	0.10	Canada	
	Seed oleic 3-3	G/18	75.76–77.76	Satt288	3.00	0.10	Canada	
	Seed oleic 3-4	–	–	UBC440-700	2.20	0.08	Canada	
Linoleic	*Seed linoleic 3-1*	G/18	0.84–2.84	Satt038	2.60	0.03	Canada	
	Seed linoleic 3-2	K/9	41.38–43.38	Satt349	2.30	0.08	Canada	
	Seed linoleic 3-3	B2/14	98.24–100.24	Satt063	32.80	0.70	Canada	
	Seed linoleic 3-4	–	–	UBC440-700	2.90	0.03	Canada	
	Seed linoleic 3-5	–	–	UBC300-800	2.60	0.04	Canada	
Linolenic	*Seed linolenic 3-1*	E/15	43.75–45.75	Satt185	2.10	0,02	Canada	
	Seed linolenic 3-2	K/9	41.38–43.38	Satt349	2.10	0.06	Canada	
	Seed linolenic 3-3	B2/14	92.48–94.48	Satt063	41.40	0.76	Canada	
Protein	*Seed protein 25-1*	B1/11	46.40–58.91	Satt197-A118_1	2.43	10.80	China	Gai et al. (2007)
	Seed protein 25-2	B1/11	58.91–59.10	A118_1-A520_1	2.18	5.80	China	
	Seed protein 25-3	D1b/2	7.63–9.63	A725_1	2.21	6.70	China	
Oil	*Seed oil 25-1*	B2/14	2.00–24.80	A953_1H-B221T	2.20	9.10	China	
	Seed oil 25-2	M/7	0.80–16.00	A60V-AACCAA08	2.38	13.30	China	
	Seed oil 25-3	M/7	2.00–6.80	AACCAA08-AACCAA09	2.43	13.90	China	
	Seed oil 25-4	M/7	4.0–12.60	AACCAA09-AAGCAT11	2.27	13.90	China	

Protein	*Seed protein 27-1*	D1b/2	70.65–75.65	Satt537	2.53	5.16	China	Chen et al. (2007)
	Seed protein 27-2	E/15	44.93–45.40	Satt151	5.87	14.14	China	
	Seed protein 27-3	E/15	44.93–45.40	Satt151	5.12	14.29	China	
	Seed protein 27-4	N/3	91.55–114.55	Satt257	4.05	13.46	China	
	Seed protein 27-5	O/10	58.40–64.40	Satt581	2.81	12.09	China	
Oil	*Seed oil 28-1*	B2/14	72.20–75.40	Satt556	5.04	21.98	China	
	Seed oil 28-2	N/3	91.55–114.55	Satt257	4.20	13.17	China	
	Seed oil 28-3	N/3	101.05–115.05	Satt022	4.08	12.73	China	
	Seed oil 28-4	D1a/1	–	Sat_001-Sat_114	8.21	18.10	China	
Oil+prot	*Seed oil + prot 1-1*	A2/8	108.83–110.83	Satt327	2.60	5.56	China	
	Seed oil + prot 1-2	C2/6	112.83–116.83	Satt202-Satt134	5.98	15.44	China	
	Seed oil + prot 1-3	C2/6	107.60–112.30	Satt289	7.05	16.80	China	
	Seed oil + prot 1-4	E/15	44.93–46.93	Satt263	3.22	6.62	China	
	Seed oil + prot 1-5	E/15	44.93–45.77	Satt117	3.18	6.58	China	
Protein	*Seed protein 41-1*	A1/5	93.10–95.10	Satt385	$p = 0.00007$	15.00	Korea	Jun et al. (2008)
	Seed protein 41-2	E/15	30.40–32.40	Satt268	$p = 0.00006$	11.10	Korea	

(continued)

Table 2.2 (continued)

	Seed protein 41-3	G/18	67.20–69.20	Satt564	$p = 0.00006$	15.60	Korea	
	Seed protein 41-4	I/20	34.80–36.80	Satt571	$p = 0.00000$	65.00	Korea	
	Seed protein 41-5	J/16	27.20–29.20	Satt405	$p = 0.00003$	7.60	Korea	
	Seed protein 41-6	J/16	82.03–84.03	Satt431	$p = 0.00006$	–	Korea	
	Seed protein 41-7	K/9	31.30–33.30	Satt242	$p = 0.00002$	10.20	Korea	
	Seed protein 41-8	L/19	36.20–38.20	Satt723	$p = 0.00009$	16.00	Korea	
	Seed protein 41-9	M/7	89.50–91.50	Satt551	$p = 0.00001$	–	Korea	
	Seed protein 41-10	N/3	29.80–31.80	Satt159	$p = 0.00002$	11.20	Korea	
	Seed protein 41-11	O/10	70.60–72.60	Satt653	$p = 0.00002$	6.30	Korea	
Oil	*Seed oil 32-1*	E/15	15.71–22.86	Satt384	4.66	22.7	Japan	Shibata et al. (2008)
	Seed oil 32-2	H/12	46.00–48.00	Satt442	3.17	14.4	Japan	
	Seed oil 32-3	I/20	30.80–42.00	Satt496	6.23	27.6	Japan	
Linolenic	*Seed linolenic 9-1*	C2/6	–	AGG/GGC380	5.23	22.40	Japan	
	Seed linolenic 9-2	E/15	18.30–22.60	Satt384	8.21	36.90	Japan	
	Seed linolenic 9-3	H/12	27.80–29.80	Sat_127	2.55	15.70	Japan	
	Seed linolenic 9-4	O/10	106.60–108.60	Sat_274	3.51	20.80	Japan	

Protein	*Seed protein 29-1*	C2/6	112.95–114.95	Satt100	$p = 0.0009$	15.00	Tavistock-04	Palomeque et al. (2009a, b)
	Seed protein 29-1	C2/6	112.95–114.95	Satt100	$p = 0.0062$	11.00	Woodstock-05	
Oil	*Seed oil 33-1*	C2/6	112.95–114.95	Satt100	$p = 0.0004$	16.00	Tavistock-04	
	Seed oil 33-2	C2/6	106.60–108.60	Satt277	$p = 0.0055$	9.00	Tavistock-05	
	Seed oil 33-3	I/20	85.73–87.72	Satt162	$p = 0.0055$	12.00	Woodstock-05	
Oleic	*Seed oleic 6-1*	N/3	77.09–87.09	Satt022	LOD > 2.5	11.20	North Carolina	Bachlava et al. (2009)
	Seed oleic 6-2	A2/8	119.58–131.97	Satt333	LOD > 2.5	11.60	North Carolina	
	Seed oleic 6-3	F/13	26.98–41.47	Sat_309	LOD > 2.5	16.50	North Carolina	
	Seed oleic 6-4	F/13	84.48–99.83	Sat_299	LOD > 2.5	7.40	North Carolina	
	Seed oleic 6-5	A1/5	46.45–69.20	Satt050	LOD > 2.5	3.70	North Carolina	
	Seed oleic 6-6	E/15	32.09–47.50	Sat_273	LOD > 2.5	4.50	North Carolina	
	Seed oleic 6-7	M/7	56.29–65.79	Sat_226	LOD > 2.5	11.70	North Carolina	
	Seed oleic 6-8	G/18	76.76–94.83	Satt472	LOD > 2.5	5.00	North Carolina	
	Seed oleic 6-9	D2/17	87.55–93.70	Sat_300	LOD > 2.5	2.70	North Carolina	
	Seed oleic 6-10	O/10	118.30–129.30	Sat_108	LOD > 2.5	2.75	North Carolina	
	Seed oleic 6-11	I/20	43.68–48.58	Satt354	LOD > 2.5	9.80	North Carolina	
	Seed oleic 6-12	F/13	41.47–50.77	Sat_133	LOD > 2.5	18.30	North Carolina	
Linoleic	*Seed linoleic 10-1*	N/3	77.09–87.09	Satt022	LOD > 2.5	9.40	North Carolina	
	Seed linoleic 10-2	A2/8	100.08–119.58	Satt333	LOD > 2.5	20.60	North Carolina	
	Seed linoleic 10-3	F/13	26.98–41.47	Sat_309	LOD > 2.5	16.70	North Carolina	

(continued)

Table 2.2 (continued)

	Seed linoleic 10-4	I/20	84.48–99.83	Sat_299	LOD > 2.5	10.10	North Carolina	
	Seed linoleic 10-5	A1/5	95.16–101.57	Sat_127	LOD > 2.5	3.30	North Carolina	
	Seed linoleic 10-6	D2/17	47.73–67.70	Satt002	LOD > 2.5	5.90	North Carolina	
	Seed linoleic 10-7	O/10	117.29–131.31	Sat_108	LOD > 2.5	3.05	North Carolina	
	Seed linoleic 10-8	I/20	43.77–48.68	Satt354	LOD > 2.5	7.10	North Carolina	
	Seed linoleic 10-9	F/13	41.47–50.77	Sat_133	LOD > 2.5	20.90	North Carolina	
	Seed linoleic 10-10	G/18	76.76–94.83	Satt472	LOD > 2.5	4.00	North Carolina	
Linolenic	*Seed linolenic 7-1*	L/19	0.00–11.12	Satt495	LOD > 2.5	6.10	North Carolina	
	Seed linolenic 7-2	A2/8	131.97–145.57	Sat_294	LOD > 2.5	5.60	North Carolina	
	Seed linolenic 7-3	D1b/2	46.61–67.62	Sat_423	LOD > 2.5	10.60	North Carolina	
	Seed linolenic 7-4	B2/14	104.96–122.26	Satt687	LOD > 2.5	35.90	North Carolina	
	Seed linolenic 7-5	A1/5	3.54–14.37	Sat_368	LOD > 2.5	5.10	North Carolina	
	Seed linolenic 7-6	B2/14	66.23–78.83	Satt066	LOD > 2.5	23.20	North Carolina	
	Seed linolenic 7-7	J/16	65.04–89.48	Sat_395	LOD > 2.5	16.30	North Carolina	

Protein	*Seed protein 42-1*	O/10	55.10–59.43	Satt345	7.34	8.70	China	Li et al. (2010)
Oil	*Seed oil 41-1*	D1b/2	76.10–77.34	Satt282-Satt428	16.34	5.24	China	
Protein	*Seed protein 28-1*	C2/6	97.40–99.40	Satt557-Satt100	2.16	5.89	China	Liang et al. (2010)
	Seed protein 28-2	G/18	119.40–121.40	Satt199-Satt012	2.14	6.73	China	
	Seed protein 28-3	H/12	30.80–32.80	Satt568-Satt442	2.56	8.05	China	
	Seed protein 28-4	B2/14	77.40–80.40	Satt063	2.25	9.58	China	
	Seed protein 28-5	G/18	1.84–10.64	Satt275	2.24	9.58	China	
Oil	*Seed oil 30-1*	A1/5	85.60–96.60	Satt599	5.14	11.72	China	
	Seed oil 30-2	A2/8	50.40–53.40	Sat_162	12.61	32.76	China	
	Seed oil 30-3	A2/8	54.91–78.21	Sat_129	11.81	55.43	China	
	Seed oil 30-4	B2/14	72.80–89.90	Satt063	3.70	7.88	China	
	Seed oil 30-5	C2/6	100.80–107.60	Satt277	2.20	5.41	China	
	Seed oil 30-6	D2/17	105.44–113.74	Satt186	2.01	4.32	China	
Oil	*Seed oil 31-1*	D1b/2	86.20–88.20	Satt546	$p = 0.0070$	6.67	Korea	Kim et al. (2010)
	Seed oil 31-2	C2/6	112.95–114.95	Satt100	$p = 0.0160$	5.59	Korea	
	Seed oil 31-3	L/19	29.93–31.93	Satt418	$p = 0.0093$	6.27	Korea	
Palmitic plus	*Seed palm + steal-1*	D2/17	45.80–47.80	Satt154	$p = 0.0005$	10.54	Korea	
Stearic	*Seed palm + steal-2*	L/19	24.60–26.60	Satt523	$p = 0.0003$	12.27	Korea	
Oleic	*Seed oleic 7-1*	M/7	6.75–8.75	Satt590	$p = 0.0312$	4.27	Korea	
	Seed oleic 7-2	B1/11	48.10–50.10	Satt197	$p = 0.0024$	8.26	Korea	

(continued)

Table 2.2 (continued)

	Seed oleic 7-3	B2/14	62.30–64.30	Satt556	$p = 0.0158$	5.27	Korea	
	Seed oleic 7-4	J/16	47.10–49.10	Satt183	$p = 0.0160$	5.21	Korea	
	Seed oleic 7-5	L/19	–	Sct_010	$p = 0.0032$	7.70	Korea	
Linoleic	*Seed linoleic 7-1*	D1b/2	81.40–83.40	Satt537	$p = 0.0426$	3.79	Korea	
	Seed linoleic 7-2	B1/11	48.10–50.10	Satt197	$p = 0.0019$	8.66	Korea	
	Seed linoleic 7-3	B2/14	62.30–64.30	Satt556	$p = 0.0041$	7.39	Korea	
	Seed linoleic 7-4	J/16	47.10–49.10	Satt183	$p = 0.0343$	4.04	Korea	
	Seed linoleic 7-5	L/19	–	Sct_010	$p = 0.0037$	7.48	Korea	
Linolenic	*Seed linolenic 8-1*	A2/8	30.70–32.70	Satt177	$p = 0.0250$	4.69	Korea	
	Seed linolenic 8-2	O/10	52.40–54.40	Satt173	$p = 0.0473$	3.73	Korea	
	Seed linolenic 8-3	L/19	17.10–19.10	Satt238	$p = 0.0032$	7.71	Korea	
Oil	*Seed oil 29-1*	C1/4	63.0–66.2	GNE176b-GNB004	$f = 11.4$	5.5	China	Li et al. (2011)
	Seed oil 29-2	E/15	90.1–94.7	Satt685-LI37 V	$f = 12$	4.3	China	
	Seed oil 29-3	O/10	125.2–132.6	Sat_231-Sat_109	$f = 10.6$	5.9	China	
Palmitic	*Seed palmitic 6-1*	A2/8	26.0–27.8	Satt525-A117_2H	$f = 12.9$	3.3	China	
	Seed palmitic 6-2	B1/11	62.3–68.1	A118T-A520T	$f = 15.3$	6.7	China	
	Seed palmitic 6-3	E/15	0.0–2.0	Satt651-L34_1I	$f = 8.3$	4	China	

	Seed palmitic 6-4	G/18	16.5–19.7	GNB012-Sat_308	$f = 9.2$	3.5	China	
	Seed palmitic 6-5	J/16	19.7–24.1	Sat_228-GMKF073	$f = 8.7$	3.2	China	
	Seed palmitic 6-6	K/9	57.4–59.6	Sat_043-GNE054	$f = 12.8$	6.2	China	
Stearic	*Seed stearic 5-1*	B2/14	20.6–24.5	Sat_189-Satt272	$f = 44.9$	13.2	China	
	Seed stearic 5-2	C2/6	30.7–32.7	B131 V-Satt277	$f = 14.3$	3.7	China	
	Seed stearic 5-3	D1b/2	106.2–111.2	Satt558-GNE174	$f = 16.9$	6.3	China	
	Seed stearic 5-4	G/18	37.80–38.80	GMKF012	$f = 12.1$	3.4	China	
	Seed stearic 5-5	K/9	57.40–68.60	GNE054-Satt260	$f = 12.0$	3.1	China	
Oleic	*Seed oleic 5-1*	B1/11	87.2–89.8	GNE307-Sat_331	$f = 10.5$	3.9	China	
	Seed oleic 5-2	G/18	51.5–55.5	GNB022b-Satt472	$f = 9.9$	2.7	China	
	Seed oleic 5-3	I/20	58.8–62.0	Sat_419-B39 V	$f = 7.8$	3.8	China	
Linoleic	*Seed linoleic 5-1*	G/18	51.5–54.5	GNB022b-Satt472	$f = 12.4$	4.3	China	
	Seed linoleic 5-2	N/3	19.6–23.5	Satt675-OPF16	$f = 10.2$	4.3	China	
	Seed linoleic 5-3	O/10	116.1–119.1	LE23T-GNE035	$f = 14.1$	3	China	
Linolenic	*Seed linolenic 6-1*	B1/11	87.2–89.8	GNE307-Sat_331	$f = 32.1$	8.3	China	

(continued)

Table 2.2 (continued)

	Seed linolenic 6-2	C1/4	63.0–66.2	GNB004-GNE002	*f* = 13.8	5.6	China	
	Seed linolenic 6-3	D1b/2	17.7–21.6	Satt537-Satt350	*f* = 13.6	3.4	China	
	Seed linolenic 6-4	J/16	57.5–59.7	GNE332-GNE263	*f* = 10.7	4	China	
	Seed linolenic 6-5	K/9	46.3–52.6	GNE097b-A199H	*f* = 11.3	3.1	China	
	Seed linolenic 6-6	O/10	69.6–71.3	Sat_291-Satt633	*f* = 9.2	1.6	China	
	Seed linolenic 6-7	O/10	117.1–123.2	GNE035-Sat_231	*f* = 10.2	2.5	China	
Oil	*Seed oil 35-1*	A1/5	92.88–101.60	Satt217	3.50	27.16	Thailand	Moongkanna et al. (2011)
	Seed oil 35-2	A1/5	95.54–101.60	Satt258	3.50	45.18	Thailand	
Oil	*mqSeed oil 001*	A1/5	92.56–94.20	A170_1-Satt511	–	10.83	Various locations	Qi et al. (2011)
	mqSeed oil 002	A1/5	28.08–30.28	Satt454-A329_2	–	29.18	Various locations	
	mqSeed oil 003	A1/5	85.58–91.24	Satt599-Bng077_1	–	38.41	Various locations	
	mqSeed oil 004	A2/8	67.33–77.70	A111_1-Satt341	–	39.15	Various locations	
	mqSeed oil 005	B2/14	68.98–78.66	Sat_083-Sat_009	–	9.56	Various locations	
	mqSeed oil 006	C1/4	5.36–15.70	Satt690-A351_2	–	8.67	Various locations	
	mqSeed oil 007	C1/4	122.63–123.79	AI794821-Satt338	–	13.60	Various locations	

	mqSeed oil 008	C2/6	96.12–99.18	R092_3-Sat_076	–	5.50	Various locations	
	mqSeed oil 009	D1a/1	56.40–59.00	Satt254-Satt203	–	8.56	Various locations	
	mqSeed oil 010	D1a/1	68.62–72.30	Satt198-Satt439	–	9.10	Various locations	
	mqSeed oil 011	D2/17	75.29–80.20	Sat_292-Satt461	–	10.17	Various locations	
	mqSeed oil 012	E/15	22.84–28.3	OP_M12b-K221_1	–	14.35	Various locations	
	mqSeed oil 013	E/15	28.27–35.80	K229_1-Satt573	–	20.50	Various locations	
	mqSeed oil 014	F/13	1.09–3.04	M8E6mr1-Satt343	–	12.17	Various locations	
	mqSeed oil 015	F/13	16.08–20.56	Satt252-Satt423	–	13.59	Various locations	
	mqSeed oil 016	G/18	68.56–73.40	A073_1-Sat_143	–	12.53	Various locations	
	mqSeed oil 017	G/18	94.40–96.60	Sct_199-Satt191	–	13.27	Various locations	
	mqSeed oil 018	H/12	86.49–90.30	Satt142-A748_2	–	9.63	Various locations	
	mqSeed oil 019	I/20	30.56–34.70	B214_2-A352_2	–	27.00	Various locations	
	mqSeed oil 020	I/20	36.03–49.30	Sat_219-Sat_105	–	18.22	Various locations	
	mqSeed oil 021	K/9	97.13–107.00	R-M7E8mr3	–	11.20	Various locations	

(continued)

Table 2.2 (continued)

	mqSeed oil 022	L/19	33.71–36.10	Satt497-Satt613	–	19.50	Various locations	
	mqSeed oil 023	L/19	90.34–95.40	DUBC015-A489_1	–	11.83	Various locations	
	mqSeed oil 024	N/3	68.76–77.10	A808_1-Sat_305	–	6.57	Various locations	
	mqSeed oil 025	N/3	92.56–102.06	Satt257-Satt022	–	37.31	Various locations	
Protein	*Seed protein 31-1*	I/20	35.40–37.40	Satt496	$p < 0.05$	56.00	Illinois	Pandurangan et al. (2012)
	Seed protein 31-2	E/15	2.30–4.30	Satt575	$p < 0.05$	18.00	Illinois	
	Seed protein 31-3	D1a/1	76.50–78.50	Satt077	$p < 0.05$	15.00	Illinois	
	Seed protein 31-4	C2/6	41.40–43.40	Satt520	$p < 0.05$	13.00	Illinois	
Palmitic	*Seed palmitic 8-1*	D2/17	78.23–80.23	Satt389	$p = 0.023$	3.67	China	Xie et al. (2012)
	Seed palmitic 8-2	J/16	39.04–41.04	Sat_144	$p < 0.0001$	7.78	China	
	Seed palmitic 8-3	G/18	67.66–69.66	Sat_164	$p = 0.0008$	2.72	China	
	Seed palmitic 8-4	K/9	45.79–47.79	Satt727	$p = 0.0032$	1.77	China	
Stearic	*Seed stearic 7-1*	D1b/2	39.04–41.04	Satt701	$p = 0.0032$	13.48	China	
Oleic	*Seed oleic 8-1*	J/16	39.04–41.04	Sat_144	$p < 0.0001$	10.29	China	
	Seed oleic 8-2	G/18	67.66–69.66	Sat_164	$p < 0.0001$	13.91	China	

	Seed oleic 8-3	K/9	42.34–44.34	Satt544	$p = 0.0013$	8.95	China	
	Seed oleic 8-4	D2/17	78.23–80.23	Satt389	$p = 0.0008$	7.80	China	
Linolenic	*Seed linolenic 8-1*	G/18	67.66–69.66	Sat_164	$p = 0.002$	17..68	China	
	Seed linolenic 8-2	D2/17	78.23–80.23	Satt389	$p = 0.002$	5.34	China	
	Seed linolenic 8-3	J/16	39.04–41.04	Sat_144	$p = 0.0032$	6.29	China	
	Seed linolenic 8-4	B2/14	99.55–101.55	Satt726	$p = 0.0001$	7.76	China	
Linoleic	*Seed linoleic 11-1*	B2/14	99.55–101.55	Satt726	$p = 0.0001$	5.80	China	
	Seed linoleic 11-2	B2/14	92.48–94.48	Satt063	$p = 0.0001$	37.30	China	
	Seed linoleic 11-3	–	–	FAD3b-1	$p = 0.0032$	11.20	China	
	Seed linoleic 11-4	–	–	FAD3c-2	$p = 0.0001$	7.20	China	
	Seed linoleic 11-5	D1b/2	39.04–41.04	Satt701	$p = 0.0007$	6.67	China	
	Seed linoleic 11-6	G/18	67.66–69.66	Sat_164	$p = 0.0001$	15.42	China	
Oil	*Seed oil 36-1*	A1/5	0.00–4.95	Sat_137	2.93	7.68	South Dakota	Wang et al. (2012)
	Seed oil 36-2	C2/6	4.22–8.34	AW734043	2.50	6.28	South Dakota	
	Seed oil 36-3	E/15	45.10–53.34	Satt452	2.34	11.90	South Dakota	
	Seed oil 36-4	F/13	25.60–32.32	Sat_240	2.30	13.60	South Dakota	
	Seed oil 36-5	F/13	59.60–65.20	Sat_297	2.06	6.15	South Dakota	

(continued)

Table 2.2 (continued)

	Seed oil 36-6	G/18	94.83–99.78	Satt472	3.39	9.08	South Dakota	
	Seed oil 36-7	K/9	42.40–45.70	Satt544	3.26	8.06	South Dakota	
	Seed oil 36-8	K/9	93.63–99.53	Sat_293	2.92	7.01	South Dakota	
	Seed oil 36-9	N/3	97.92–102.05	Satt022	3.04	8.21	South Dakota	
Palmitic	*Seed palmitic 7-1*	A1/5	0.00–4.95	Sat_137	8.28	23.01	South Dakota	
	Seed palmitic 7-2	C2/6	4.22–8.34	Sat_240	2.24	5.79	South Dakota	
	Seed palmitic 7-3	E/15	38.40–46.65	Satt452	2.32	13.38	South Dakota	
	Seed palmitic 7-4	F/13	25.60–32.32	Sat_240	2.32	13.81	South Dakota	
	Seed palmitic 7-5	F/13	59.60–68.67	Sat_297	2.11	4.79	South Dakota	
	Seed palmitic 7-6	G/18	94.83–99.78	Satt472	3.34	7.55	South Dakota	
	Seed palmitic 7-7	K/9	42.38–47.33	Satt544	3.41	7.55	South Dakota	
	Seed palmitic 7-8	K/9	99.09–101.09	Sat_293	2.54	12.44	South Dakota	
Stearic	*Seed stearic 6-1*	A1/5	0.00–5.36	Sat_137	3.79	19.41	South Dakota	
	Seed stearic 6-2	A1/5	95.68–103.93	Sat_374	2.55	13.66	South Dakota	
	Seed stearic 6-3	F/13	1.92–16.08	Satt252	2.36	14.16	South Dakota	
	Seed stearic 6-4	G/18	94.40–108.69	Sat_064	3.18	10.57	South Dakota	

	Seed stearic 6-5	N/3	92.15–102.15	Satt022	2.23	6.78	South Dakota	
Protein	*Seed protein 33-1*	H/12	81.04–89.51	Satt302	3.00	8.00	Canada	Eskandari et al. (2013)
	Seed protein 33-2	F/13	71.41–77.70	Satt510	3.00	5.00	Canada	
	Seed protein 33-3	K/9	71.41–77.70	Satt001	3.40	7.00	Canada	
	Seed protein 33-4	L/19	14.03–27.92	Satt001	3.00	5.00	Canada	
	Seed protein 33-5	M/7	50.10–60.10	Satt323	2.80	5.00	Canada	
Oil	*Seed oil 37-1*	D2/17	29.80–31.80	Sat_284	$p = 0.0006$	5.00	Canada	
	Seed oil 37-2	D1a/1	108.66–110.66	Satt129	$p = 0.0006$	5.00	Canada	
	Seed oil 37-3	K/9	102.60–104.60	Sat_020	$p = 0.0037$	6.00	Canada	
	Seed oil 37-4	B2/14	77.83–79.83	Satt066	$p0.01$	4.00	Canada	
	Seed oil 37-5	H/12	81.04–89.51	Satt302	5.70	9.00	Canada	
	Seed oil 37-6	L/19	13.98–27.92	Satt523	3.60	7.00	Canada	
	Seed oil 37-7	F/13	50.70–77.70	Satt510	3.10	6.00	Canada	
	Seed oil 37-8	F/13	81.97–97.97	Satt490	3.00	5.00	Canada	
	Seed oil 37-9	J/16	41.63–46.53	Satt215	2.60	6.00	Canada	
	Seed oil 37-10	K/9	50.56–56.61	Satt273	2.70	5.00	Canada	
	Seed oil 37-11	M/7	50.10–60.10	Satt323	3.90	6.00	Canada	
Protein	*Seed protein 34-1*	A1/5	92.45–95.95	Satt211	4.97	9.70	China	Lu et al. (2013)
	Seed protein 34-2	C2/6	69.70–90.93	Sat_213	2.79	9.10	China	

(continued)

Table 2.2 (continued)

	Seed protein 34-3	M/7	130.75–136.75	Satt336	4.63	8.30	China	
	Seed protein 34-4	A2/8	52.60–62.60	GNENOD2B	4.06	12.10	China	
	Seed protein 34-5	A2/8	52.60–54.60	Sat_215	4.09	8.10	China	
	Seed protein 34-6	K/9	67.90–80.68	Satt475	2.53	8.80	China	
	Seed protein 34-7	B1/11	24.66–35.00	Sat_156	2.64	5.70	China	
	Seed protein 34-8	H/12	53.34–58.91	Satt469	5.57	4.50	China	
	Seed protein 34-9	G/18	56.20–61.41	Satt564	4.61	7.20	China	
	Seed protein 34-10	L/19	29.20–33.20	Satt143	4.41	5.90	China	
	Seed protein 34-11	I/20	35.16–50.11	AB002807	2.97	12.60	China	
Protein	*Seed protein 35-1*	C2/6	106.60–108.60	Satt277	$p = 0.01$	8.70	US/Canada/China	Rossi et al. (2013)
	Seed protein 35-2	C2/6	112.95–114.95	Satt100	$p = 0.01$	10.60	The USA/Canada/China	
	Seed protein 35-3	I/20	83.73–85.73	Satt162	$p = 0.01$	7.50	The USA/Canada/China	
	Seed protein 35-4	K/9	55.61–57.61	Satt273	$p = 0.01$	15.50	The USA/Canada/China	
Oil	*Seed oil 38-1*	A1/5	28.00–30.00	Satt364	$p = 0.01$	10.60	The USA/Canada/China	

	Seed oil 38-2	C2/6	112.95–114.95	Satt100	$p = 0.01$	15.80	The USA/ Canada/China	
	Seed oil 38-3	D1b/2	74.29–76.29	Satt005	$p = 0.01$	7.80	The USA/ Canada/China	
	Seed oil 38-4	F/13	74.97–76.97	Sat_120	$p = 0.01$	7.00	The USA/ Canada/China	
Protein	*Seed protein 36-1*	A1/5	27.78–43.23	Satt449	3.60	23.00	China	Mao et al. (2013)
	Seed protein 36-2	B1/11	26.80–28.80	Satt197	2.37	7.00	China	
	Seed protein 36-3	B2/14	72.91–72.80	Sat_189	2.66	7.00	China	
	Seed protein 36-4	C1/4	72.80–75.23	Satt070	3.30	7.00	China	
	Seed protein 36-5	C1/4	5.74–10.74	Satt565	2.79	31.00	China	
	Seed protein 36-6	C1/4	0.00–10.84	Sat_404	3.14	7.00	China	
	Seed protein 36-7	C2/6	101.75–126.23	Satt202	3.34	20.00	China	
	Seed protein 36-8	C2/6	101.75–126.23	Satt316	2.10	6.00	China	
	Seed protein 36-9	D1a/1	85.50–109.66	Satt129	2.90	11.00	China	
	Seed protein 36-10	D1a/1	69.91–85.48	Sat_414	2.90	9.00	China	
	Seed protein 36-11	D1b/2	102.60–116.34	Satt274	2.75	6.00	China	

(continued)

Table 2.2 (continued)

	Seed protein 36-12	D1b/2	37.07–52.61	Satt296	3.10	7.00	China	
	Seed protein 36-13	D1b/2	100.88–116.34	Satt296	3.59	18.00	China	
	Seed protein 36-14	D2/17	85.15–93.70	Satt301	2.22	7.00	China	
	Seed protein 36-15	D2/17	84.62–85.15	Satt226	7.98	17.00	China	
	Seed protein 36-16	D2/17	88.01–92.01	Satt615	2.10	7.00	China	
	Seed protein 36-17	D2/17	67.70–87.38	Sat_365	5.85	17.00	China	
	Seed protein 36-18	F/13	67.70–87.38	Sct_033	2.28	8.00	China	
	Seed protein 36-19	F/13	3.04–9.69	Satt343	2.19	8.00	China	
	Seed protein 36-20	F/13	3.42–13.42	Satt193	2.80	17.00	China	
	Seed protein 36-21	F/13	18.12–22.35	Sat_074	2.20	8.00	China	
	Seed protein 36-22	F/13	3.04–9.69	Sat_262	2.80	10.00	China	
	Seed protein 36-23	F/13	67.70–87.38	Sct_033	2.36	12.00	China	
	Seed protein 36-24	F/13	74.12–78.05	Satt334	3.30	15.00	China	
	Seed protein 36-25	G/18	4.53–10.75	Sat_372	4.59	10.00	China	
	Seed protein 36-26	I/20	28.00–32.00	Satt367	2.38	11.00	China	

	Seed protein 36-27	K/9	4.85–14.35	Satt242	3.30	10.00	China	
	Seed protein 36-28	K/9	30.30–35.30	Satt102	7.02	17.00	China	
	Seed protein 36-29	K/9	14.35–20.35	Sat_293	4.08	9.00	China	
	Seed protein 36-30	K/9	46.25–56.61	Satt273	2.75	13.00	China	
	Seed protein 36-31	L/19	87.41–92.66	Sat_286	3.31	27.00	China	
	Seed protein 36-32	L/19	87.41–92.66	Sat_286	2.54	7.00	China	
	Seed protein 36-33	M/7	13.60–18.60	Satt150	2.10	4.00	China	
	Seed protein 36-34	N/3	40.81–50.81	Satt022	2.43	7.00	China	
	Seed protein 36-35	N/3	95.00–102.05	Sat_295	4.80	17.00	China	
	Seed protein 36-36	N/3	32.84–34.52	Satt683	2.19	17.00	China	
	Seed protein 36-37	N/3	70.60–79.51	Sat_091	2.34	8.00	China	
	Seed protein 36-38	O/10	67.93–72.75	Satt241	3.82	21.00	China	
	Seed protein 36-39	O/10	58.40–62.81	Satt173	2.84	6.00	China	
	Seed protein 36-40	O/10	39.82–53.81	Sat_282	2.57	20.00	China	

(continued)

Table 2.2 (continued)

Oil	*Seed oil 43-1*	A2/8	39.82–53.81	Satt207	3.41	17.00	China	
	Seed oil 43-2	B2/14	72.91–93.48	Sat_189	3.50	8.00	China	
	Seed oil 43-3	B2/14	72.80–75.80	Satt070	3.32	15.00	China	
	Seed oil 43-4	B2/14	71.68–72.12	Satt272	2.50	4.00	China	
	Seed oil 43-5	B2/14	87.58–100.09	Sat_424	2.94	0.15	China	
	Seed oil 43-6	B2/14	73.20–80.15	Satt202	2.87	12.00	China	
	Seed oil 43-7	C2/6	145.47–151.91	Satt357	3.70	9.00	China	
	Seed oil 43-8	D1a/1	85.48–109.66	Satt129	2.89	21.00	China	
	Seed oil 43-9	D1b/2	102.60–116.34	Satt274	3.25	10.00	China	
	Seed oil 43-10	D1b/2	37.07–52.61	Satt296	4.51	12.00	China	
	Seed oil 43-11	D1b/2	75.41–75.94	Satt579	2.11	8.00	China	
	Seed oil 43-12	D2/17	57.07–67.70	Satt669	3.20	9.00	China	
	Seed oil 43-13	D2/17	87.87–105.44	Satt662	4.10	22.00	China	
	Seed oil 43-14	D2/17	87.87–105.45	Satt226	2.11	5.00	China	
	Seed oil 43-15	E/15	84.62–85.15	Satt706	3.09	7.00	China	
	Seed oil 43-16	G/18	47.54–49.54	Satt340	2.10	7.00	China	
	Seed oil 43-17	I/20	27.98–34.98	Satt367	2.62	11.00	China	
	Seed oil 43-18	I/20	55.09–60.09	Sat_268	2.03	13.00	China	
	Seed oil 43-19	J/16	53.68–58.40	Satt621	2.68	6.00	China	
	Seed oil 43-20	J/16	32.09–39.63	Satt596	2.69	6.00	China	
	Seed oil 43-21	K/9	4.85–10.85	Satt212	3.27	22.00	China	
	Seed oil 43-22	K/9	30.28–40.28	Satt102	3.17	7.00	China	
	Seed oil 43-23	L/19	28.27–31.21	Satt278	2.21	5.00	China	
	Seed oil 43-24	L/19	11.12–27.92	Satt523	2.00	3.00	China	
	Seed oil 43-25	L/19	30.93–32.36	Sat_320	2.70	6.00	China	
	Seed oil 43-26	L/19	54.57–87.41	Sat_286	3.34	22.00	China	
	Seed oil 43-27	L/19	0.00–10.84	Satt495	2.33	8.00	China	

	Seed oil 43-28	L/19	70.36–87.41	Sat_286	2.10	5.00	China	
	Seed oil 43-29	M/7	13.56–20.80	Satt201	2.09	3.00	China	
	Seed oil 43-30	N/3	40.81–50.81	Satt022	2.09	16.00	China	
	Seed oil 43-31	N/3	22.67–27.12	Satt159	2.88	12.00	China	
	Seed oil 43-32	N/3	32.84–34.52	Satt683	2.74	7.00	China	
	Seed oil 43-33	O/10	59.43–63.81	Satt345	2.40	8.00	China	
	Seed oil 43-34	O/10	27.78–40.23	Satt173	3.12	7.00	China	
	Seed oil 43-35	O/10	50.11–55.09	Satt420	3.21	10.00	China	
Protein	*Seed protein*	C2/6	127.93–133.35	Satt371-Satt357	$p = 0.0015$	18.00	Illinois	Yesuda et al. (2013)
	Seed protein	D1b/2	–	SaagB13L17	$p = 0.0076$	7.00	Illinois	
	Seed protein	E/15	85.2–96.42	Satt369-Satt231	$p = 0.0035$	14.00	Illinois	
	Seed protein	H/12	77.95–79.95	Satt302	$p = 0.0015$	18.00	Illinois	
	Seed protein	I/20	–	Sat_B17E19	$p = 0.0064$	7.00	Illinois	
Oil	*Seed oil*	A2/8	30.52–34.0	Satt589-BLT65	$p = 0.0071$	6.00	Illinois	
	Seed oil	I/20	–	Sat_B17E19	$p = 0.0009$	11.00	Illinois	
Protein	*Seed protein*	N/3	21.00–23.60	BARc-016199-02307	3.30	5.60	Missouri	Pathan et al. (2013)
	Seed protein	C2/6	33.30–35.50	BARc-059985-16,274	3.10	5.30	Missouri	
	Seed protein	M/7	42.60–46.00	BARc-014705-01623	3.70	6.80	Missouri	
	Seed protein	C2/6	38.20–39.60	Satt281	2.70	4.80	Missouri	
	Seed protein	E/15	9.90–13.90	BARc-018923-03037	4.00	7.50	Missouri	
	Seed protein	M/7	43.30–45.30	BARc-014705-01623	3.00	7.00	Missouri	
	Seed protein	N/3	19.70–22.50	BARc-028645-05979	3.40	6.70	Missouri	
	Seed protein	A1/5	79.60–82.00	Satt236	3.20	6.40	Missouri	
	Seed protein	C2/6	113.6–116.2	Satt433	3.30	7.60	Missouri	
	Seed protein	A2/8	41.60–51.00	Satt632	6.70	14.20	Missouri	
	Seed protein	A1/5	76.80–80.80	BARc-041257-07953	4.00	7.70	Missouri	

(continued)

Table 2.2 (continued)

	Seed protein	C2/6	31.70–37.10	BARc-059985-16,274_Satt291	4.70	9.00	Missouri	
	Seed protein	A2/8	44.7–47.9	Satt632_BARc-010097-00518	3.60	6.80	Missouri	
Oil	*cqSeed oil 005*	N/3	26.50–28.70	BARc-055637-13,558	3.10	5.60	Missouri	
	cqSeed oil 006	C2/6	32.20–36.60	BARc-059985-16,274	4.20	7.40	Missouri	
	cqSeed oil 007	E/15	17.70–19.70	BARc-039687-07541	3.00	5.30	Missouri	
	cqSeed oil 008	A1/5	76.60–81.00	BARc-041257-07953	4.20	7.10	Missouri	
	cqSeed oil 009	C2/6	30.80–38.00	BARc-059985-16,274	5.60	9.30	Missouri	
	cqSeed oil 010	E/15	15.10–22.30	BARc-039687-07541	5.60	9.70	Missouri	
	cqSeed oil 011	D1b/2	124.00–128.40	BARc-044747-08795	4.20	10.50	Missouri	
	cqSeed oil 012	A1/5	77.40–80.20	BARc-041257-07953	3.40	8.40	Missouri	
	cqSeed oil 013	E/15	46.00–48.40	BARc-018901-03270	3.20	9.20	Missouri	
	cqSeed oil 014	D1b/2	116.00–117.00	Sat_198	2.50	6.10	Missouri	
	cqSeed oil 015	A1/5	78.30–79.30	BARc-041257-07953	2.50	5.70	Missouri	
	cqSeed oil 016	C2/6	32.80–38.00	Sat_336	4.60	12.30	Missouri	
Protein	*Seed protein 37-1*	B1/11	29.14–36.48	BARC-032333-08951	2.90	15.70	South Dakota	Wang et al. (2014a)
	Seed protein 37-2	C1/4	123.80–129.10	BARCSOYSSR_04	3.00	14.60	South Dakota	
	Seed protein 37-3	C1/4	47.11–65.60	BARC-064861-18,829	3.30	17.10	South Dakota	
	Seed protein 37-4	D1b/2	116.34–131.91	Sat_289	3.40	16.40	South Dakota	
	Seed protein 37-5	D2/17	82.47–87.55	Sat_300	3.40	16.30	South Dakota	
	Seed protein 37-6	D2/17	60.44–61.37	BARCSOYSSR_17_0621	2.70	26.00	South Dakota	
	Seed protein 37-7	I/20	14.80–22.84	Satt562	7.90	37.00	South Dakota	

	Seed protein 37-8	I/20	31.49–46.22	Satt354	9.50	33.80	South Dakota	
	Seed protein 37-9	I/20	20.23–22.28	BARC-057049-14,550	6.50	31.90	South Dakota	
	Seed protein 37-10	K/9	54.14–55.63	BARCSOYSSR_09_1171	3.00	14.60	South Dakota	
	Seed protein 37-11	N/3	36.86–37.98	Satt584	3.80	17.90	South Dakota	
Oil	*Seed oil 39-1*	A1/5	95.70–97.60	Sat_374	7.60	23.00	South Dakota	
	Seed oil 39-2	B1/11	43.30–53.42	BARCSOYSSR_11_0545	2.80	16.20	South Dakota	
	Seed oil 39-3	C1/4	41.43–53.67	BARCSOYSSR_04_0366	2.40	15.00	South Dakota	
	Seed oil 39-4	C1/4	48.62–49.10	BARC-040777-07848	2.10	10.50	South Dakota	
	Seed oil 39-5	D1a/1	60.40–62.40	BARC-054393-12,560	3.20	15.80	South Dakota	
	Seed oil 39-6	D1b/2	116.34–126.44	Staga002	2.70	13.10	South Dakota	
	Seed oil 39-7	D2/17	64.60–76.70	BARCSOYSSR_17_0690	5.10	11.90	South Dakota	
	Seed oil 39-8	E/15	45.10–46.70	Satt045	6.40	26.50	South Dakota	
	Seed oil 39-9	G/18	93.86–94.83	BARCSOYSSR_18_1699	5.90	13.50	South Dakota	
	Seed oil 39-10	I/20	22.84–31.94	Satt614	3.90	20.10	South Dakota	
	Seed oil 39-11	I/20	20.34–22.03	BARC-057049-14,550	4.00	20.00	South Dakota	
	Seed oil 39-12	J/16	55.73–78.57	Sat_350	3.60	17.70	South Dakota	
	Seed oil 39-13	K/9	53.90–54.12	BARCSOYSSR_09_1170	4.80	11.60	South Dakota	
	Seed oil 39-14	N/3	28.52–29.28	Satt641	3.30	15.90	South Dakota	
	Seed oil 39-15	N/3	37.98–40.63	Satt584	3.90	18.90	South Dakota	
	Seed oil 39-16	O/10	106.04–108.84	BARC-050013-09288	5.30	22.90	South Dakota	
Oleic	*qOLE-E-1*	E/15	69.5–71.5	Satt606	3.10	14.90	South Dakota	Wang et al. (2014b)
	qOLE-E-2	E/15	75.9–77.9	BARC-SOYSSR_15_1073-Satt452	6.00	19.00	South Dakota	

(continued)

Table 2.2 (continued)

	qOLE-G	G/18	85–87	Satt472-BARC-017867-02359	3.70	19.70	South Dakota	
	qOLE-I	I/20	39.75–41.75	Satt354	2.30	13.20	South Dakota	
	qOLE-J	J/16	59.72–61.72	Sat_350	2.80	14.70	South Dakota	
	qOLE0O	O/10	106.3–108.3	Satt243	6.30	30.30	South Dakota	
Linoleic	*qLLE-A1*	A1/5	40.6–42.6	Sat_356-Sat_050	7.50	17.76	South Dakota	
	qLLE-G-1	G/18	24.96–36.97	Satt235-Sat_403	9.60	22.60	South Dakota	
	qLLE-G-2	G/18	100.8–102.8	BARCSOYSSR_18_1820-Sat_064	10.70	29.10	South Dakota	
	qLLE-O	O/10	95.6–107.3	Satt243-Satt581	10.40	38.60	South Dakota	
Linolenic	*qLLN-C2*	C2/6	51.5–53.5	Satt457	3.50	20.40	South Dakota	
	qLLN-D1a	D1a/1	48.4–50.4	BARC-052429-11,439	4.10	22.70	South Dakota	
	qLLN-D1b	D1b/2	83.5–85.5	Sat_135-GmFAD3B	19.80	43.40	South Dakota	
	qLLN-F	F/13	10.5–12.5	BARCSOYSSR_13_0119-Sat_039	4.10	19.30	South Dakota	
	qLLN-G	G/18	24.96–36.97	Satt235-Sat_403	9.30	18.70	South Dakota	
Linolenic	*Seed linolenic 12-1*	A1/5	52.01–54.01	BARC-026129-052	5.30	8.00	Missouri	Ha et al. (2014)
	Seed linolenic 12-2	B1/11	41.48–44.82	BARC-050091-09377	3.80	5.90	Missouri	
	Seed linolenic 12-3	B2/14	91.29–96.64	BARC-013273-00464	8.20	13.80	Missouri	
	Seed linolenic 12-4	E/15	18.69–19.37	BARC-028907-06042	9.60	15.90	Missouri	
	Seed linolenic 12-5	D2/17	19.27–26.05	BARC-056107-14,093	4.30	7.00	Missouri	
	Seed linolenic 12-6	C2/6	84.13–86.43	SSR06-682	5.40	10.00	Missouri	

	Seed linolenic 12-7	E/15	14.27–17.96	BARC-054257-12,408	4.00	7.30	Missouri	
	Seed linolenic 12-8	B1/11	29.13–29.17	BARC-030863-06956	4.40	7.60	Missouri	
	Seed linolenic 12-9	H/12	81.00–81.03	BARC-050003-09283	5.80	10.50	Missouri	
	Seed linolenic 12-10	N/3	91.94–102.20	BARC-065687-19,660	3.80	6.90	Missouri	
	Seed linolenic 12-11	A1/5	46.12–52.17	BARC-049091-10,809	4.10	8.10	Missouri	
	Seed linolenic 12-12	H/12	66.01–70.52	BARC-018895-03034	5.40	10.50	Missouri	
Protein	*qPRO001*	B2/14	0.00–1.40	ss248293401-ss248275088	4.13	7.00	North Carolina	Akond et al. (2014)
Oil	*qOIL001 (45-1)*	N/3	22.04–25.56	ss244929361-ss244932961	2.66	47.00	North Carolina	
	qOIL002 (45-2)	N/3	32.04–39.91	ss244980861-ss245114421	2.97	97.00	North Carolina	
	qOIL003 (45-3)	A1/5	96.47–97.51	ss245763524-ss245763185	2.75	10.00	North Carolina	
	qOIL004 (45-4)	K/9	45.39–46.11	ss246875694-ss246860974	2.69	9.90	North Carolina	
	qOIL005 (45-5)	F/13	3.09–3.13	ss247937603-ss247940172	3.34	12.00	North Carolina	
	qOIL006 (45-6)	B2/14	19.87–52.03	ss248293401-ss248275088	2.72	11.20	North Carolina	
	qOIL007 (45-7)	B2/14	37.04–64.98	ss248278947-ss248330608	2.51	11.60	North Carolina	
	qOIL008 (45-8)	B2/14	57.88–57.92	ss248308943-ss248309108	3.40	11.60	North Carolina	

(continued)

Table 2.2 (continued)

	qOIL009 (45-9)	J/16	22.96–23.03	ss248981433-ss248982430	2.69	51.00	North Carolina	
	qOIL010 (45-10)	J/16	31.46–33.46	ss248982430-ss248982430	3.81	74.00	North Carolina	
	qOIL011 (45-11)	J/16	32.46–40.97	ss248982430-ss249037210	4.67	72.00	North Carolina	
Palmitic	*qPAL001 (9-1)*	N/3	25.81–28.04	ss244942610-ss244947608	3.21	54.00	North Carolina	
	qPAL002 (9-2)	G/18	45.51–46.07	ss249629157-ss249621644	4.80	62.00	North Carolina	
Stearic	*qSTEL001 (9-1)*	J/16	31.57–31.96	ss248977568-ss248979552	3.67	23.00	North Carolina	
Oleic	*qOLE001 (10-1)*	F/13	3.04–3.57	ss247950750-ss247942156	3.56	8.00	North Carolina	
	qOLE002 (10-2)	J/16	32.30–34.67	ss248981433-ss248993887	3.60	17.00	North Carolina	
	qOLE003 (10-3)	J/16	40.97–43.07	ss249037210-ss249039670	3.39	32.00	North Carolina	
Linoleic	*qLINL001 (9-1)*	N/3	25.81–28.04	ss24494261-ss244947608	2.57	62.00	North Carolina	
	qLINL002 (9-2)	J/16	31.57–31.96	ss248977568-ss248979552	3.01	61.00	North Carolina	
	qLINL003 (9-3)	J/16	32.30–34.67	ss248981433-ss248993887	3.51	61.00	North Carolina	
Linolenic	*qLINN001 (13-1)*	F/13	1.95–2.64	ss247957446-ss247965852	3.81	10.00	North Carolina	
	qLINN002 (13-2)	F/13	2.09–3.57	ss247970215-ss247950750	3.81	11.00	North Carolina	
	qLINN003 (13-3)	B2/14	57.74–57.85	ss248308274-ss248308822	29.70	7.00	North Carolina	

	qLINN004 (13-4)	E/15	40.11–42.76	ss248604753-ss248616287	2.58	5.00	North Carolina	
	qLINN005 (13-5)	J/16	32.19–32.37	ss248980834-ss248981847	2.74	23.00	North Carolina	
	qLINN006 (13-6)	J/16	38.10–43.07	ss249010538-ss249039670	3.02	23.00	North Carolina	
Protein	*Seed protein*	A2/8	–	BARC_1.01_9683120_C_T	(–log*P* = 3.22)	–	MD, NE	Hwang et al. (2014)
	Seed protein	A2/8	–	BARC_1.01_9741542_T_C	(–log*P* = 3.28)	–	MD, NE	
	Seed protein	K/9	–	BARC_1.01_4921120_C_T	(–log*P* = 3.06)	–	MD, NE	
	Seed protein	I/20	–	BARC_1.01_29395999_T_C	(–log*P* = 5.52)	–	MD, NE	
	Seed protein	I/20	–	BARC_1.01_295112680_A_G	(–log*P* = 4.96)	–	MD, NE	
	Seed protein	I/20	–	BARC_1.01_29594697_A_G	(–log*P* = 5.52)	–	MD, NE	
	Seed protein	I/20	–	BARC_1.01_29983050_A_G	(–log*P* = 4.82)	–	MD, NE	
Oil	*Seed oil*	A2/8	–	BARC_1.01_9683120_C_T	(–log*P* = 3.25)	–	MD, NE	
	Seed oil	A2/8	–	BARC_1.01_9741542_T_C	(–log*P* = 3.55)	–	MD, NE	
	Seed oil	K/9	–	BARC_1.01_4921120_C_T	(–log*P* = 4.71)	–	MD, NE	
	Seed oil	I/20	–	BARC_1.01_29395999_T_C	(–log*P* = 3.93)	–	MD, NE	
	Seed oil	I/20	–	BARC_1.01_295112680_A_G	(–log*P* = 3.69)	–	MD, NE	
	Seed oil	I/20	–	BARC_1.01_29594697_A_G	(–log*P* = 3.93)	–	MD, NE	
	Seed oil	I/20	–	BARC_1.01_29983050_A_G	(–log*P* = 3.03)	–	MD, NE	
Protein	*qPRO6-2*	C2/6	205.91	Sct_033-sat_097	2.30	6.55	2007JMS, China	Qi et al. (2014)
	qPRO10-1	O/10	7.11	Satt151-satt117	2.08	5.79	2007JMS	HRB=Harbin
	qPRO3-1	N/3	61.5	Satt251-satt229	5.53	10.10	2007JMS	HXL = Hongsinglong
	qPRO7-1	M/7	138.6	Satt220-sat_112	2.43	5.30	2007JMS	JMS = Jiamusi
	qPRO10-1	O/10	6.9	Satt151-satt117	5.08	6.40	2007JMS	

(continued)

Table 2.2 (continued)

	qPRO6-1	C2/6	155.91	Satt243-satt341	2.74	7.24	2007HXL	
	qPRO8-1	A2/8	51.11	Sat_069-sat_135	2.58	10.92	2007HXL	
	qPRO11-1	B1/11	2.01	Satt252-satt269	2.30	4.17	2007HXL	
	qPRO16-1	J/16	0.01	Satt349-satt555	2.76	4.71	2007HXL	
	qPRO19-1	L/19	38.91	Satt257-satt551	2.05	9.09	2007HXL	
	qPRO1-1	D1a/1	157.1	Sat_119-satt242	2.97	4.20	2007HXL	
	qPRO4-1	C1/4	90.1	Sct_094-sat_083	2.19	4.30	2007HXL	
	qPRO8-1	A2/8	49.1	Sat_069-sat_135	2.32	11.80	2007HXL	
	qPRO11-1	B1/11	6	Satt252-satt269	3.85	12.40	2007HXL	
	qPRO16-1	J/16	0.1	Satt349-satt555	2.99	3.90	2007HXL	
	qPRO3-2	N/3	2.01	Satt233-satt426	2.61	8.53	2007HRB	
	qPRO7-2	M/7	61.21	Satt528-satt175	4.18	14.94	2007HRB	
	qPRO7-3	M/7	203.11	Satt482-sat_106	2.97	7.54	2007HRB	
	qPRO12-1	H/12	6.41	Satt505-sat_088	3.04	10.51	2007HRB	
	qPRO14-1	B2/14	10.91	Satt530-satt440	2.17	9.32	2007HRB	
	qPRO3-2	N/3	6	Satt233-satt426	3.21	9.30	2007HRB	
	qPRO7-2	M/7	61.2	Satt528-satt175	4.80	11.00	2007HRB	
	qPRO7-3	M/7	206.1	Satt482-sat_106	4.10	8.60	2007HRB	
	qPRO12-1	H/12	7.4	Satt505-sat_088	3.85	11.10	2007HRB	
	qPRO14-1	B2/14	10.9	Satt530-satt440	2.67	4.70	2007HRB	
	qPRO7-6	M/7	240.41	Satt203-satt198	2.01	7.68	2008JMS	
	qPRO8-2	A2/8	105.71	Satt141-satt041	2.31	4.75	2008JMS	
	qPRO12-2	H/12	11.11	Sat_088-satt138	2.53	5.03	2008JMS	
	qPRO7-5	M/7	42	Satt358-sat_001	6.61	14.50	2008JMS	
	qPRO7-6	M/7	241.4	Satt203-satt198	6.36	15.10	2008JMS	
	qPRO7-7	M/7	283.1	Satt273-satt502	2.55	14.60	2008JMS	
	qPRO8-2	A2/8	106.7	Satt141-satt041	3.49	4.50	2008JMS	

	qPRO12-2	H/12	11.2	Sat_088-satt138	2.37	6.30	2008JMS	
	qPRO1-2	D1a/1	199.01	Satt270-satt390	2.89	5.88	2008HXL	
	qPRO6-3	C2/6	142.11	Satt289-satt277	2.18	3.31	2008HXL	
	qPRO15-1	E/15	76.71	Sat_093-sat_076	2.38	7.48	2008HXL	
	qPRO1-3	D1a/1	89	Satt164-satt042	3.72	5.30	2008HXL	
	qPRO1-2	D1a/1	199.1	Satt270-satt390	6.50	9.50	2008HXL	
	qPRO5-1	A1/5	0.1	Satt195-sat_042	2.23	2.50	2008HXL	
	qPRO6-3	C2/6	142.2	Satt289-satt277	2.20	0.80	2008HXL	
	qPRO7-4	M/7	43.1	Sat_001-sat_114	6.01	8.60	2008HXL	
	qPRO7-2	M/7	55.2	Satt528-satt175	6.40	10.90	2008HXL	
	qPRO15-1	E/15	80.7	Sat_093-sat_076	2.04	0.40	2008HXL	
	qPRO5-1	A1/5	0.01	Satt195-sat_042	2.46	4.27	2008HRB	
	qPRO7-4	M/7	45.01	Sat_001-sat_114	2.40	4.57	2008HRB	
	qPRO4-2	C1/4	60.4	Satt272-satt168	2.17	10.50	2008HRB	
	qPRO5-1	A1/5	0.1	Satt195-sat_042	2.96	4.70	2008HRB	
	qPRO7-4	M/7	47	Sat_001-sat_114	2.12	4.50	2008HRB	
	qPRO7-8	M/7	202.11	Sat_062-satt482	2.23	6.52	2009JMS	
	qPRO15-3	E/15	27.41	Satt244satt547-satt431	2.02	3.91	2009JMS	
	qPRO17-1	D2/17	3.01	Satt398-sat_134	2.11	5.93	2009JMS	
	qPRO20-1	I/20	8.01	Satt331-satt173	2.15	9.95	2009JMS	
	qPRO1-4	D1a/1	46.4	Satt545-sat_087	2.31	3.50	2009JMS	
	qPRO1-3	D1a/1	97.9	Satt164-satt042	5.84	16.30	2009JMS	
	qPRO1-5	D1a/1	167.4	Satt242-sat_105	6.13	8.80	2009JMS	
	qPRO15-3	E/15	27	Satt244satt547-satt431	3.72	3.80	2009JMS	
	qPRO17-1	D2/17	3	Satt398-sat_134	3.91	8.70	2009JMS	
	qPRO3-2	N/3	0.01	Satt233-satt426	2.51	4.51	2009HXL	

(continued)

Table 2.2 (continued)

	qPRO14-1	B2/14	2.91	Satt530-satt440	2.60	4.40	2009HXL	
	qPRO15-2	E/15	0.01	Satt457-satt244satt547	2.52	4.98	2009HXL	
	qPRO20-1	I/20	2.01	Satt331-satt173	2.72	9.78	2009HXL	
	qPRO3-2	N/3	0.1	Satt233-satt426	3.16	0.80	2009HXL	
	qPRO8-3	A2/8	124.1	Satt428-satt266	3.76	1.40	2009HXL	
	qPRO11-2	B1/11	10.9	Satt269-GMRUBP	3.22	8.80	2009HXL	
	qPRO14-1	B2/14	3	Satt530-satt440	2.75	3.40	2009HXL	
	qPRO15-2	E/15	0.1	Satt457-satt244satt547	2.08	1.40	2009HXL	
	qPRO20-1	I/20	2	Satt331-satt173	2.74	4.40	2009HXL	
	qPRO6-4	C2/6	141.31	Satt134-satt289	2.44	5.96	2009HRB	
	qPRO6-1	C2/6	157.91	Satt243-satt341	4.64	15.02	2009HRB	
	qPRO8-3	A2/8	124.01	Satt428-satt266	2.91	8.33	2009HRB	
	qPRO10-2	O/10	0.01	Satt263-satt151	2.25	6.45	2009HRB	
	qPRO10-3	O/10	7.51	Satt117-satt452satt355	4.53	11.41	2009HRB	
	qPRO6-4	C2/6	141.4	Satt134-satt289	2.48	2.90	2009HRB	
	qPRO6-1	C2/6	158.9	Satt243-satt341	2.10	9.40	2009HRB	
	qPRO8-3	A2/8	124.1	Satt428-satt266	2.43	4.50	2009HRB	
	qPRO10-3	O/10	7.6	Satt117-satt452satt355	2.60	7.20	2009HRB	
	qPRO13-1	F/13	28.9	Satt191-satt293	2.53	5.90	2009HRB	
	qPRO20-1	I/20	4	Satt331-satt173	2.10	5.60	2009HRB	
	qPRO1-1	D1a/1	162.8	Sat_119-Satt242	3.10	9.00	2010JMS	
	qPRO7-3	M/7	204.2	Satt482-Sat_106	4.20	14.00	2010JMS	
	qPRO7-5	M/7	0.1	Satt358-Sat_001	3.38	7.20	2010JMS	
	qPRO7-3	M/7	203.2	Satt482-Sat_106	2.78	5.30	2010JMS	
	qPRO1-5	D1a/1	168.3	Satt242-Sat_105	3.57	8.90	2010JMS	
	qPRO12-1	H/12	3.2	Satt505-Sat_088	3.10	8.00	2010HXL	
	qPRO12-2	H/12	12.5	Sat_088-Satt138	3.00	7.00	2010HXL	

	qPRO13-1	F/13	26.20-43.00	Satt191-Satt293	3.40	20.00	2010HXL	
	qPRO12-3	H/12	14.3	Satt138-Sat_094	1.81	4.60	2010HXL	
	qPRO13-1	F/13	33.9	Satt191-Satt293	2.10	10.70	2010HXL	
	qPRO20-1	I/20	0.1	Satt331-Satt173	3.36	9.30	2010HXL	
	qPRO11-3	B1/11	52.9	Satt146-Satt538	1.94	6.00	2010HRB	
	qPRO7-5	M/7	9	Satt358-Sat_001	1.56	8.00	2010HRB	
	qPRO7-7	M/7	270.1	Satt273-Satt502	2.42	18.90	2010HRB	
Protein	*Seed protein*	A1/5	bp = 31,114,364	–	$p = 9.72 \times 10^{-7}$		Canada	Sonah et al. (2015)
	Seed protein	A2/8	bp = 45,563,060	–	$p = 6.53 \times 10^{-7}$		Canada	
	Seed protein	O/10	bp = 1,335,189	–	$p = 1.00 \times 10^{-7}$		Canada	
	Seed protein	B2/14	bp = 120,437	–	$p = 3.36 \times 10^{-8}$		Canada	
	Seed protein	J/16	bp = 4,183,401	–	$p = 1.30 \times 10^{-6}$		Canada	
	Seed protein	L/19	bp = 50,418,986	–	$p = 2.48 \times 10^{-7}$		Canada	
	Seed protein	I/20	bp = 10,052,759	–	$p = 1.02 \times 10^{-5}$		Canada	
	Seed protein	I/20	bp = 30,619,328	–	$p = 3.55 \times 10^{-5}$		Canada	
Oil	*Seed oil*	A1/5	bp = 31,114,364	–	$p = 1.61 \times 10^{-7}$		Canada	
	Seed oil	A2/8	bp = 45,563,060	–	$p = 6.09 \times 10^{-6}$		Canada	
	Seed oil	O/10	bp = 1,811,225	–	$p = 3.86 \times 10^{-7}$		Canada	
	Seed oil	B2/14	bp = 58,684	–	$p = 1.95 \times 10^{-6}$		Canada	
	Seed oil	J/16	bp = 4,183,401	–	$p = 1.62 \times 10^{-6}$		Canada	
	Seed oil	L/19	bp = 50,418,986	–	$p = 1.09 \times 10^{-6}$		Canada	
	Seed oil	I/20	bp = 10,052,759	–	$p = 1.02 \times 10^{-6}$		Canada	
	Seed oil	I/20	bp = 30,728,503	–	$p = 1.18 \times 10^{-5}$		Canada	
Oil	*Seed oil*	F/13	bp = 33,982,053	–	$(-\log P = 10.88)$		China	Zhou et al. (2015)
	Seed oil	N/3	bp = 43,382,349	–	$(-\log P = 10.37)$		China	
	Seed oil	D2/17	bp = 37,805,372	–	$(-\log P = 9.93)$		China	

(continued)

Table 2.2 (continued)

	Seed oil	H/12	bp = 36,685,538	–	(−log*P* = 9.34)		China	
	Seed oil	B1/11	bp = 30,407,369	–	(log*P* = 9.11)		China	
	Seed oil	E/15	bp = 11,120,747	–	(−log*P* = 8.93)		China	
FA	Oil 23-2	D1a/1	bp = 25,343,314–25,345,506	–	–		China	
	–	D1b/2	bp = 34,074,074–34,081,893	–	–		China	
	Oil 31-1	D1b/2	bp = 43,743,770–43,747,502	–	–		China	
	–	C2/6	bp = 8,469,468–8,483,404	–	–		China	
	–	M/7	bp = 4,175,785–4,179,336	–	–		China	
	–	M/7	bp = 17,502,600–17,506,321	–	–		China	
	Oil 34-5	M/7	bp = 43,970,385–43,974,648	–	–		China	
	–	A2/8	bp = 687,470–689,239	–	–		China	
	–	A2/8	bp = 45,223,982–45,229,809	–	–		China	
	–	O/10	bp = 49,416,645–49,418,380	–	–		China	
	–	F/13	bp = 33,848,800–33,850,851	–	–		China	
	Oil 24-17, Oil 30-4	B2/14	bp = 34,322,879–34,325,575	–	–		China	

		E/15	bp = 276,266–278,043	–	–		China	
	Oil 32-1, Oil 2-3	E/15	bp = 4,116,287–4,120,919	–	–		China	
	Oil32-1, Oil2-3, Oil2-9	E/15	bp = 5,561,199–5,563,177	–	–		China	
	Oil 2-3, Oil 2-9, Oil2-8	E/15	bp = 8,229,206–8,233,359	–	–		China	
	Oil 2-9, Oil 24-24	E/15	bp = 11,724,822–11,730,419	–	–		China	
	Oil 23-3	D2/17	bp = 9,856,959–9,861,288	–	–		China	
	Oil 32-3, Oil 11-1	I/20	bp = 17,523,090-17,528,873	–	–		China	
Palmitic	*Seed palmitic*	A1/5	–	Map-0751	3.10E-04	4.00	China	Li et al. (2015)
	Seed palmitic	A1/5	–	Map-6064	2.40E-08	7.70	China	
	Seed palmitic	M/7	–	Map-1323	6.80E-03	2.50	China	
	Seed palmitic	K/9	–	Map-6356	7.70E-03	2.40	China	
	Seed palmitic	H/12	–	Map-6395	5.80E-03	2.50	China	
	Seed palmitic	E/15	–	Map-6540	4.20E-03	2.00	China	
	Seed palmitic	E/15	–	Map-6590	1.40E-03	3.20	China	
Stearic	*Seed stearic*	B2/14	–	BARC-013927-01275	1.40E-07	7.80	China	
	Seed stearic	B2/14	–	Map-6506	4.00E-08	8.50	China	
	Seed stearic	E/15	–	Map-6568	6.50E-03	1.80	China	
	Seed stearic	G/18	–	Map-3670	4.90E-03	1.90	China	
	Seed stearic	G/18	–	Map-6776	4.40E-03	2.60	China	
Oleic	*Seed oleic*	N/3	–	Map-6056	4.70E-03	2.00	China	

(continued)

Table 2.2 (continued)

	Seed oleic	C1/4	–	BARC-042719-08393	7.60E-03	1.70	China	
	Seed oleic	C2/6	–	Map-6135	7.70E-03	1.70	China	
	Seed oleic	A2/8	–	Map-6325	1.30E-03	2.50	China	
	Seed oleic	G/18	–	Map-3670	6.50E-03	1.80	China	
Linoleic	*Seed linoleic*	N/3	–	BARC-028709-05992	3.90E-03	2.00	China	
	Seed linoleic	C2/6	–	Map-6135	7.60E-03	1.70	China	
	Seed linoleic	C2/6	–	Map-6146	4.10E-03	2.70	China	
	Seed linoleic	A2/8	–	Map-6283	6.20E-03	1.80	China	
	Seed linoleic	A2/8	–	Map-6325	7.40E-03	1.70	China	
	Seed linoleic	A2/8	–	Map-6326	1.00E-02	2.20	China	
	Seed linoleic	O/10	–	Map-6383	1.70E-03	2.40	China	
	Seed linoleic	E/15	–	Map-6520	7.50E-03	2.40	China	
Linolenic	*Seed linolenic*	D1a/1	–	Map-0008	2.70E-03	2.20	China	
	Seed linolenic	D1a/1	–	BARC-017909-02439	2.80E-03	2.20	China	
	Seed linolenic	D1a/1	–	Map-0076	3.60E-03	2.80	China	
	Seed linolenic	D1b/2	–	Map-6017	2.90E-04	3.20	China	
	Seed linolenic	A1/5	–	Map-6077	2.60E-03	2.20	China	
	Seed linolenic	A2/8	–	Map-6270	2.10E-03	3.00	China	
	Seed linolenic	E/15	–	Map-6520	7.60E-03	2.40	China	
	Seed linolenic	G/18	–	Map-6738	9.40E-03	1.60	China	
	Seed linolenic	G/18	–	Map-3569	3.00E-03	2.20	China	
	Seed linolenic	G/18	–	Map-3580	2.80E-03	2.90	China	
	Seed linolenic	G/18	–	Map-3665	9.10E-03	2.30	China	
	Seed linolenic	G/18	–	Map-6782	7.50E-03	1.70	China	
Oil	*Seed oil*	C1/4	–	Satt294	1.80E-02	–	Brazil	Priolli et al. (2015)
	Seed oil	C2/6	–	AW734137	1.63E-02	–	Brazil	
	Seed oil	C2/6	–	Satt227	2.75E-03	–	Brazil	

	Seed oil	C2/6	–	Satt281	7.03E-04	–	Brazil	
	Seed oil	D1a/1	–	Satt129	2.31E-03	–	Brazil	
	Seed oil	D2/17	–	Satt186	2.82E-03	–	Brazil	
	Seed oil	G/18	–	Satt130	1.16E-02	–	Brazil	
	Seed oil	I/20	–	Satt239	8.73E-03	–	Brazil	
	Seed oil	I/20	–	Satt571	2.51E-04	–	Brazil	
	Seed oil	L/19	–	Satt182	1.24E-06	–	Brazil	
	Seed oil	L/19	–	Satt373	9.31E-03	–	Brazil	
	Seed oil	M/7	–	Satt308	3.23E-04	–	Brazil	
	Seed oil	M/7	–	Satt463	1.68E-03	–	Brazil	
	Seed oil	N/3	–	Sat_266	8.88E-03	–	Brazil	
Palmitic	*Seed palmitic*	A2/8	–	Satt429	4.58E-03	–	Brazil	
	Seed palmitic	B2/14	–	Satt168	1.35E-04	–	Brazil	
	Seed palmitic	C1/4	–	Satt294	3.05E-03	–	Brazil	
	Seed palmitic	D1a/1	–	Satt129	2.53E-08	–	Brazil	
	Seed palmitic	D2/17	–	Satt386	1.02E-03	–	Brazil	
	Seed palmitic	E/15	–	Satt045	9.62E-04	–	Brazil	
	Seed palmitic	F/13	–	Satt334	2.85E-04	–	Brazil	
	Seed palmitic	F/13	–	Satt586	1.30E-05	–	Brazil	
	Seed palmitic	G/18	–	Satt303	9.26E-03	–	Brazil	
	Seed palmitic	G/18	–	Satt324	7.94E-03	–	Brazil	
	Seed palmitic	I/20	–	Satt270	1.88E-02	–	Brazil	
	Seed palmitic	I/20	–	Satt562	1.34E-07	–	Brazil	
	Seed palmitic	I/20	–	Satt571	2.50E-10	–	Brazil	
	Seed palmitic	J/16	–	Sat_144	2.65E-06	–	Brazil	
	Seed palmitic	J/16	–	Sat_224	1.16E-03	–	Brazil	

(continued)

Table 2.2 (continued)

	Seed palmitic	K/9	–	Satt001	4.16E-04	–	Brazil	
	Seed palmitic	K/9	–	Satt196	2.65E-03	–	Brazil	
	Seed palmitic	L/19	–	Satt182	1.00E-04	–	Brazil	
	Seed palmitic	L/19	–	Satt229	1.83E-02	–	Brazil	
	Seed palmitic	L/19	–	Satt448	6.93E-03	–	Brazil	
	Seed palmitic	M/7	–	Satt540	2.77E-03	–	Brazil	
	Seed palmitic	N/3	–	Satt530	2.67E-04	–	Brazil	
	Seed palmitic	O/10	–	Satt345	1.53E-03	–	Brazil	
	Seed palmitic	O/10	–	Satt487	1.09E-14	–	Brazil	
Oleic	*Seed oleic*	B1/11	–	Satt665	9.13E-04	–	Brazil	
	Seed oleic	B2/14	–	Satt070	1.42E-04	–	Brazil	
	Seed oleic	B2/14	–	Satt168	5.94E-04	–	Brazil	
	Seed oleic	B2/14	–	Satt534	1.29E-02	–	Brazil	
	Seed oleic	C1/4	–	Satt294	5.56E-08	–	Brazil	
	Seed oleic	D1a/1	–	Satt129	2.96E-08	–	Brazil	
	Seed oleic	D1b/2	–	Satt216	9.23E-04	–	Brazil	
	Seed oleic	D1b/2	–	Satt274	6.07E-03	–	Brazil	
	Seed oleic	D1b/2	–	Satt600	3.56E-03	–	Brazil	
	Seed oleic	D2/17	–	Satt002	6.33E-04	–	Brazil	
	Seed oleic	D2/17	–	Satt386	3.94E-12	–	Brazil	
	Seed oleic	D2/17	–	Satt389	1.54E-02	–	Brazil	
	Seed oleic	E/15	–	Satt263	1.97E-03	–	Brazil	
	Seed oleic	F/13	–	Satt335	3.24E-04	–	Brazil	
	Seed oleic	F/13	–	Satt586	2.21E-09	–	Brazil	
	Seed oleic	F/13	–	Sct_188	7.98E-03	–	Brazil	
	Seed oleic	G/18	–	Satt038	3.90E-03	–	Brazil	
	Seed oleic	G/18	–	Satt303	6.93E-05	–	Brazil	

	Seed oleic	I/20	–	Satt270	1.10E-04	–	Brazil	
	Seed oleic	I/20	–	Satt292	4.77E-03	–	Brazil	
	Seed oleic	I/20	–	Satt562	7.71E-04	–	Brazil	
	Seed oleic	I/20	–	Satt571	2.80E-14	–	Brazil	
	Seed oleic	J/16	–	Sat_144	1.28E-07	–	Brazil	
	Seed oleic	J/16	–	Sat_224	6.12E-05	–	Brazil	
	Seed oleic	J/16	–	Satt596	9.64E-04	–	Brazil	
	Seed oleic	K/9	–	Satt001	1.66E-03	–	Brazil	
	Seed oleic	K/9	–	Satt196	1.92E-03	–	Brazil	
	Seed oleic	L/19	–	Satt182	4.73E-04	–	Brazil	
	Seed oleic	L/19	–	Satt448	1.51E-03	–	Brazil	
	Seed oleic	M/7	–	Satt540	4.26E-03	–	Brazil	
	Seed oleic	N/3	–	Satt080	1.20E-03	–	Brazil	
	Seed oleic	N/3	–	Satt387	2.05E-03	–	Brazil	
	Seed oleic	O/10	–	Satt487	1.78E-21	–	Brazil	
Linoleic	*Seed linoleic*	B1/11	–	Satt665	3.76E-03	–	Brazil	
	Seed linoleic	B2/14	–	Satt070	1.97E-04	–	Brazil	
	Seed linoleic	B2/14	–	Satt534	2.62E-03	–	Brazil	
	Seed linoleic	C1/4	–	Satt294	1.36E-06	–	Brazil	
	Seed linoleic	D1a/1	–	Satt129	4.34E-06	–	Brazil	
	Seed linoleic	D1b/2	–	Satt216	1.25E-03	–	Brazil	
	Seed linoleic	D1b/2	–	Satt600	1.50E-02	–	Brazil	
	Seed linoleic	D2/17	–	Satt002	1.67E-03	–	Brazil	
	Seed linoleic	D2/17	–	Satt386	2.78E-12	–	Brazil	
	Seed linoleic	E/15	–	Satt263	1.53E-02	–	Brazil	
	Seed linoleic	F/13	–	Satt335	1.77E-04	–	Brazil	

(continued)

Table 2.2 (continued)

	Seed linoleic	F/13	–	Satt586	2.78E-07	–	Brazil	
	Seed linoleic	F/13	–	Sct_188	1.21E-02	–	Brazil	
	Seed linoleic	G/18	–	Satt038	1.41E-02	–	Brazil	
	Seed linoleic	G/18	–	Satt303	1.04E-03	–	Brazil	
	Seed linoleic	I/20	–	Satt270	1.34E-03	–	Brazil	
	Seed linoleic	I/20	–	Satt292	1.33E-02	–	Brazil	
	Seed linoleic	I/20	–	Satt571	1.34E-12	–	Brazil	
	Seed linoleic	J/16	–	Sat_144	1.71E-05	–	Brazil	
	Seed linoleic	J/16	–	Sat_224	6.67E-04	–	Brazil	
	Seed linoleic	J/16	–	Satt596	2.11E-03	–	Brazil	
	Seed linoleic	K/9	–	Satt196	8.70E-03	–	Brazil	
	Seed linoleic	L/19	–	Satt448	1.72E-03	–	Brazil	
	Seed linoleic	N/3	–	Satt080	6.13E-03	–	Brazil	
	Seed linoleic	N/3	–	Satt387	8.79E-03	–	Brazil	
	Seed linoleic	O/10	–	Satt487	8.39E-13	–	Brazil	
Linolenic	*Seed linolenic*	B1/11	–	Satt665	7.05E-03	–	Brazil	
	Seed linolenic	B2/14	–	Satt168	6.19E-03	–	Brazil	
	Seed linolenic	C1/4	–	Satt294	1.44E-02	–	Brazil	
	Seed linolenic	C1/4	–	Satt578	2.15E-04	–	Brazil	
	Seed linolenic	D2/17	–	Satt002	1.03E-02	–	Brazil	
	Seed linolenic	D2/17	–	Satt386	3.10E-05	–	Brazil	
	Seed linolenic	F/13	–	Satt586	4.80E-10	–	Brazil	
	Seed linolenic	I/20	–	Satt270	7.93E-03	–	Brazil	
	Seed linolenic	I/20	–	Satt571	3.75E-03	–	Brazil	
	Seed linolenic	J/16	–	Sat_144	9.13E-03	–	Brazil	
	Seed linolenic	J/16	–	Sat_224	7.09E-03	–	Brazil	
	Seed linolenic	J/16	–	Satt596	4.17E-03	–	Brazil	

	Seed linolenic	O/10	–	Satt487	1.07E-10	–	Brazil	
Oil	*Seed oil 42-1*	A1/5	2.54–4.54	Satt684	–	–	China	Han et al. (2015)
	Seed oil 42-2	A1/5	26.98–28.98	Satt526	–	–	China	
	Seed oil 42-3	A1/5	27.08–29.08	Satt454	–	–	China	
	Seed oil 42-4	A1/5	29.93–31.93	Satt300	–	–	China	
	Seed oil 42-5	A1/5	93.19–95.19	Satt511	–	–	China	
	Seed oil 42-6	M/7	49.09–51.09	Satt463	–	–	China	
	Seed oil 42-7	M/7	55.29–57.29	Satt220	–	–	China	
	Seed oil 42-8	M/7	75.41–77.41	Satt655	–	–	China	
	Seed oil 42-9	M/7	76.19–78.19	Satt680	–	–	China	
	Seed oil 42-10	B2/14	16.77–18.77	Satt467	–	–	China	
	Seed oil 42-11	B2/14	71.12–73.12	Satt467	–	–	China	
	Seed oil 42-12	D2/17	65.26–67.26	Satt447	–	–	China	
	Seed oil 42-13	D2/17	86.66–88.66	Satt574	–	–	China	
	Seed oil 42-14	D2/17	86.66–88.67	Satt464	–	–	China	
	Seed oil 42-15	D2/17	111.84–113.84	Sat_326	–	–	China	
	Seed oil 42-16	I/20	19.34–21.34	Satt451	–	–	China	
	Seed oil 42-17	I/20	20.90–22.90	Satt419	–	0.330	China	
	Seed oil 42-18	I/20	21.84–23.84	Satt562	–	0.330	China	
	Seed oil 42-19	I/20	30.94–32.94	Satt614	–	0.390	China	
	Seed oil 42-20	D1a/1	39.86–41.86	Satt531	–	0.044	China	
	Seed oil 42-21	D1a/1	69.69–71.69	Satt436	–	0.045	China	
	Seed oil 42-22	M/7	34.84–36.84	Satt540	–	0.050	China	
	Seed oil 42-23	M/7	52.54–54.54	Satt245	–	–	China	
	Seed oil 42-24	M/7	55.29–57.29	Satt220	–	–	China	
	Seed oil 42-25	M/7	55.29–57.30	Satt323	–	0.037	China	

(continued)

Table 2.2 (continued)

	Seed oil 42-26	K/9	29.28–31.28	Satt102	–	0.050	China	
	Seed oil 42-27	B2/14	72.20–74.20	Satt556	–	–	China	
	Seed oil 42-28	B2/14	26.62–28.62	Satt126	–	–	China	
	Seed oil 42-29	E/15	33.20–35.20	Satt598	–	–	China	
	Seed oil 42-30	D2/17	83.17–85.17	GMHSP179	–	–	China	
	Seed oil 42-31	G/18	4.53–6.53	Satt309	–	–	China	
	Seed oil 42-32	G/18	50.68–52.68	Satt427	–	0.043	China	
	Seed oil 42-33	G/18	51.93–53.93	Satt594	–	0.035	China	
	Seed oil 42-34	G/18	52.41–54.41	Satt303	–	–	China	
	Seed oil 42-35	G/18	75.76–77.76	Satt288	–	0.037	China	
	Seed oil 42-36	I/20	17.50–19.50	Satt571	–	0.370	China	
	Seed oil 42-37	I/20	19.34–21.34	Satt451	–	0.380	China	
	Seed oil 42-38	I/20	21.84–23.84	Satt562	–	0.390	China	
	Seed oil 42-39	I/20	73.26–75.26	Sat_418	–	0.034	China	
Palmitic	*Seed palmitic 10-1*	G/18	10.92–43.38	Satt610-Satt394	3.60	9.90	China	Fan et al. (2015)
	Seed palmitic 10-2	I/20	18.50–50.11	Satt571-Satt270	3.10	9.10	China	
Stearic	*Seed stearic 8-1*	C2/6	42.36–45.68	Satt520-Satt365	3.70	13.50	China	
	Seed stearic 8-2	A2/8	125.37–130.37	Satt133-Satt429	16.00	40.00	China	
	Seed stearic 8-3	F/13	3.95–26.70	Satt030-Satt659	5.80	11.90	China	
	Seed stearic 8-4	I/20	63.33–75.00	Satt650-Sat_170	9.30	23.60	China	
	Seed stearic 8-5	N/3	28.52–37.98	Satt009-Satt584	11.40	28.30	China	

	Seed stearic 8-6	M/7	13.56–21.77	Satt201-Satt680	10.90	20.80	China	
Oleic	*Seed oleic 9-1*	N/3	32.84–34.65	Satt530-Satt675	3.30	9.50	China	
	Seed oleic 9-2	O/10	55.09–60.09	Satt550-Satt331	7.80	26.50	China	
	Seed oleic 9-3	O/10	8.75–12.68	Sat_132-Satt585	4.80	25.30	China	
Linoleic	*Seed linoleic 10-1*	N/3	32.84–34.65	Satt530-Satt675	3.60	11.50	China	
	Seed linoleic 10-2	K/9	46.20–49.52	Satt441-Satt326	3.90	9.40	China	
	Seed linoleic 10-3	O/10	55.09–60.09	Satt550-Satt331	8.80	32.30	China	
	Seed linoleic 10-4	O/10	8.75–12.68	Sat_132-Satt585	6.00	33.90	China	
	Seed linoleic 10-5	H/12	53.34–60.48	Satt541-Satt142	7.70	18.00	China	
Linolenic	*Seed linolenic 14-1*	N/3	53.25–60.48	Satt387-Satt255	6.60	13.70	China	
	Seed linolenic 14-2	C2/6	26.65–31.87	Satt227-Satt079	8.40	22.80	China	
	Seed linolenic 14-3	M/7	13.56–21.19	Satt201-Satt680	3.10	6.10	China	
	Seed linolenic 14-4	M/7	60.04–62.31	Sat_003-Satt323	4.70	11.30	China	
	Seed linolenic 14-5	A2/8	125.37–130.37	Satt133-Satt429	4.70	17.60	China	
	Seed linolenic 14-6	N/3	28.52–37.98	Satt009-Satt584	3.50	8.70	China	

(continued)

Table 2.2 (continued)

	Seed linolenic 14-7	M/7	62.31–65.31	Sat_003-Satt336	7.30	18.60	China	
Oil	*Seed oil 44-1*	K/9	58.00–71.01	Sat_044-Satt499		7.80	Brazil	Leite et al. (2016)
	Seed oil 44-2	H/12	8.48–12.48	Satt353-Satt629		46.75	Brazil	
Protein	*Seed protein*	B2/14	bp = 24,367,698	ss715617976	3.10E-05	3.41	Mississippi	Leamy et al. (2017)
	Seed protein	G/18	bp = 45,438,677	ss715631057	3.42E-05	3.09	Mississippi	
Oil	*Seed oil*	N/3	bp = 34,839,398	ss715585518	9.21E-06	3.53	Mississippi	
	Seed oil	I/20	bp = 36,705,945	ss715637730	1.31E-05	3.31	Mississippi	
	Seed oil	I/20	bp = 36,710,220	ss715637732	3.24E-05	2.99	Mississippi	
Palmitic	*Seed palmitic*	C2/6	bp = 5,303,094	ss715595323	1.91E-05	3.87	Mississippi	
	Seed palmitic	M/7	bp = 37,911,072	ss715597684	5.07E-06	3.4	Mississippi	
	Seed palmitic	B2/14	bp = 22,537,916	ss715617910	4.15E-05	3.13	Mississippi	
Stearic	*Seed stearic*	B2/14	bp = 17,499,955	ss715618427	5.17E-05	2.95	Mississippi	
	Seed stearic	B2/14	bp = 17,561,477	ss715618430	3.37E-05	3.15	Mississippi	
	Seed stearic	J/16	bp = 6,804,502	ss715625341	5.31E-06	3.76	Mississippi	
Oleic	*Seed oleic*	M/7	bp = 11,603,310	ss715596070	1.25E-05	3.39	Mississippi	
	Seed oleic	B2/14	bp = 22,537,916	ss715617910	2.06E-05	3.13	Mississippi	
	Seed oleic	J/16	bp = 1,258,943	ss715623399	2.23E-06	2.93	Mississippi	
	Seed oleic	J/16	bp = 1,260,003	ss715623400	6.66E-06	2.76	Mississippi	
	Seed oleic	L/19	bp = 12,874,336	ss715633271	5.24E-05	2.49	Mississippi	
Linoleic	*Seed linoleic*	D1b/2	bp = 39,934,224	ss715582510	8.71E-06	3.6	Mississippi	
	Seed linoleic	D1b/2	bp = 39,940,256	ss715582512	1.57E-05	3.41	Mississippi	
	Seed linoleic	M/7	bp = 11,584,261	ss715596075	2.00E-05	2.72	Mississippi	
	Seed linoleic	M/7	bp = 11,585,878	ss715596074	1.39E-05	3.35	Mississippi	
	Seed linoleic	M/7	bp = 11,592,971	ss715596072	9.15E-06	3.52	Mississippi	
	Seed linoleic	M/7	bp = 11,602,516	ss715596071	1.88E-05	3.64	Mississippi	
	Seed linoleic	M/7	bp = 11,603,310	ss715596070	2.10E-06	3.38	Mississippi	

	Seed linoleic	M/7	bp = 11,756,838	ss715596058	3.62E-06	4.17	Mississippi	
	Seed linoleic	K/9	bp = 44,197,778	ss715604488	3.20E-05	4.04	Mississippi	
Linolenic	*Seed linolenic*	D1b/2	bp = 573,749	ss715583655	5.04E-05	3.72	Mississippi	
	Seed linolenic	D1b/2	bp = 578,327	ss715583662	1.46E-05	3.34	Mississippi	
	Seed linolenic	B2/14	bp = 22,514,991	ss715617909	3.99E-05	3.01	Mississippi	
	Seed linolenic	B2/14	bp = 2,537,916	ss715617910	6.71E-06	2.92	Mississippi	
SBA=	*SbaHZA1-1*	A1/5	10.56–12.56	Satt300–Satt545	6.13	26.89	China	Yang et al. (2017)
Soybean	*SbaHDA1-1*		15.82–17.82	Satt300–Satt073	4.23	7.99	China	
Agglutinin	*SbaHTA1-1*		20.28–22.28	Satt545–Satt300	5.95	22.64	China	
	SbaHZC1-1	C1/4	2.65–4.65	Satt139–Sat_042	3.14	19.09	China	
	SbaHDC1-1		13.96–15.96	Satt139–Sat_042	5.09	7.32	China	
	SbaHTC1-1		2.67–4.67	Satt139–Satt578	3.53	12.04	China	
	SbaHZD1b-1	D1b/2	13.78–14.78	Satt189–Satt157	2.89	28.57	China	
	SbaHDD1b-1		8.04–10.04	Satt189–Satt537	4.87	9.37	China	
	SbaHTD1b-1		13.17–15.17	Satt537–Satt189	8.89	19.28	China	
	SbaHZD1a-1	D1a/1	10.37–12.37	Satt402–Satt221	6.56	5.79	China	
	SbaHDD1a-1		7.69–9.69	Satt402–Satt198	3.95	6.71	China	
	SbaHZA1-2	A1/5	3.12–5.12	Satt382–Satt591	2.65	11.32	China	
	SbaHZK-1	K/9	13.82–15.82	Satt375–Satt264	4.76	6.35	China	
	SbaHTH-1	H/12	3.65–5.65	Satt302–Satt541	3.13	16.24	China	
	SbaHDA1-2	A1/5	1.78–3.78	Satt382–Satt155	2.63	21.29	China	
	SbaHDC2-1	C2/6	12.27–14.27	Satt520–Satt134	5.65	7.98	China	
Palmitic	*Seed palmitic*	A1/5	bp = 925,502–1,476,699	Chr05_1050109	(−log$P$ = 13.63)	–	China	Fang et al. (2017)
Stearic	*Seed stearic*	D1b/2	bp = 19,886,706–19,985,370	Chr02_19955907	(−log$P$ = 7.14)	–	China	

(continued)

Table 2.2 (continued)

	Seed stearic	N/3	bp = 12,501,628–12,954,110	Chr03_12704607	(−log$P$ = 9.02)	–	China	
	Seed stearic	B2/14	bp = 15,199,586–19,630,147	Chr14_17456087	(−log$P$ = 10.26)	–	China	
	Seed stearic	B2/14	bp = 34,771,615–40,356,659	Chr14_39684297	(−log$P$ = 9.94)	–	China	
Linoleic	*Seed linoleic*	C1/4	bp = 5,365,622–5,369,762	Chr04_5369762	(−log$P$ = 6.84)	–	China	
	Seed linoleic	B2/14	bp = 45,909,173–46,019,811	Chr14_45985528	(−log$P$ = 10.79)	–	China	
Palmitic ratio	*Seed palmitic ratio*	A1/5	bp = 925,502–1,476,699	Chr05_1140198	(−log$P$ = 15.07)	–	China	
	Seed palmitic ratio	A1/5	bp = 1,692,088–1,716,595	Chr05_1701656	(−log$P$ = 7.40)	–	China	
	Seed palmitic ratio	F/13	bp = 14,070,515–14,114,485	Chr13_14109681	(−log$P$ = 8.22)	–	China	
Stearic ratio	*Seed stearic ratio*	D1b/2	bp = 19,851,000–19,997,509	Chr02_19886706	(−log$P$ = 8.94)	–	China	
	Seed stearic ratio	N/3	bp = 12,500,458–12,989,558	Chr03_12702388	(−log$P$ = 12.66)	–	China	
	Seed stearic ratio	B2/14	bp = 14,985,465–19,660,594	Chr14_17456087	(−log$P$ = 14.09)	–	China	
	Seed stearic ratio	B2/14	bp = 34,208,617–40,536,866	Chr14_39684297	(−log$P$ = 13.37)	–	China	
Linoleic ratio	*Seed linoleic ratio*	B2/14	bp = 45,927,234–45,985,528	Chr14_45985528	(−log$P$ = 9.69)	–	China	
Ratio st/pal	*Ratio stearic/palm*	D1b/2	bp = 19,851,000–19,961,433	Chr02_19951148	(−log$P$ = 7.59)	–	China	

	Ratio stearic/palm	N/3	bp = 12,525,686–12,958,942	Chr03_12917268	(−logP = 10.20)	–	China	
	Ratio stearic/palm	A1/5	bp = 1,116,077–1,120,475	Chr05_1116077	(−logP = 6.85)	–	China	
	Ratio stearic/palm	B2/14	bp = 15,191,684–19,630,147	Chr14_17083726	(−logP = 10.80)	–	China	
	Ratio stearic/palm	B2/14	bp = 34,381,036–40,409,746	Chr14_39684297	(−logP = 10.62)	–	China	
Ratio lin/linn	*Ratio linolenic/linol*	O/10	bp = 45,310,798–45,310,798	Chr10_45310798	(−logP = 6.94)	–	China	
and linoleic	*Ratio linolenic/linol*	B2/14	bp = 45,907,223–46,067,489	Chr14_45938204	(−logP = 17.11)	–	China	
SFA	*SFA content*	A1/5	bp = 925,502–1,476,699	Chr05_1050109	(−logP = 11.96)	–	China	
SFA/USFA	*SFA/USFA*	A1/5	bp = 925,502–1,476,699	Chr05_1095346	(−logP = 13.12)	–	China	
	SFA/USFA	A1/5	bp = 1,701,656–1,715,940	Chr05_1715940	(−logP = 6.95)	–	China	
Protein cont.	*GqPC3*	N/3	bp = 37,773,722	AX-93995056	–	22.90	China	Zhang et al. (2017)
PC	*GqPC4-1*	C1/4	bp = 34,743,951	AX-94274580	–	29.20	China	
	GqPC4-2	C1/4	bp = 46,200,673	AX-93920058	–	22.70	China	
	GqPC5	A1/5	bp = 5,637,601	AX-94012201	–	17.40	China	
	GqPC6	C2/6	bp = 26,098,086	AX-93731783	–	26.90	China	
	GqPC9	K/9	bp = 11,138,977	AX-93764734	–	25.80	China	
	GqPC10	O/10	bp = 40,504,375	AX-93933852	–	25.40	China	
	GqPC11-1	B1/11	bp = 24,782,059	AX-94089116	–	25.80	China	

(continued)

Table 2.2 (continued)

	GqPC11-2	B1/11	bp = 37,461,551	AX-93795201	–	25.90	China	
	GqPC12	H/12	bp = 34,037,114	AX-93804391	–	26.00	China	
	GqPC13	F/13	bp = 23,091,289	AX-94109235	–	22.00	China	
	GqPC14	B2/14	bp = 7,715,347	AX-93822697	–	3.20	China	
	GqPC15-1	E/15	bp = 7,681,400	AX-93649790	–	29.20	China	
	GqPC15-2	E/15	bp = 11,235,816	AX-93837099	–	22.60	China	
	GqPC15-3	E/15	bp = 16,417,543	AX-94136114	–	25.80	China	
	GqPC19	L/19	bp = 47,615,818	AX-94196006	–	26.70	China	
Water-soluble	*GqWSPC1*	D1a/1	bp = 9,078,593	AX-93961814	–	9.50	China	
Protein cont.	*GqWSPC7*	M/7	bp = 3,484,506	AX-94283226	–	4.90	China	
WSPC	*GqWSPC8*	A2/8	bp = 8,643,359	AX-94048210	–	19.30	China	
	GqWSPC10	O/10	bp = 40,504,375	AX-93933852	–	13.00	China	
	GqWSPC11-2	B1/11	bp = 37,461,551	AX-93795201	–	8.90	China	
	GqWSPC12-1	H/12	bp = 2,583,421	AX-94268787	–	3.80	China	
	GqWSPC12-2	H/12	bp = 36,952,663	AX-93805526	–	12.30	China	
	GqWSPC13	F/13	bp = 30,988,071	AX-94112235	–	11.00	China	
	GqWSPC15	E/15	bp = 48,823,741	AX-93649393	–	5.80	China	
	GqWSPC16	J/16	bp = 28,496,401	AX-94150758	–	3.00	China	
	GqWSPC17	D2/17	bp = 7,962,112	AX-94156641	–	12.30	China	
	GqWSPC18-1	G/18	bp = 6,745,767	AX-93870261	–	8.00	China	
	GqWSPC18-2	G/18	bp = 17,076,547	AX-94170844	–	8.30	China	
	GqWSPC18-3	G/18	bp = 56,190,147	AX-94182257	–	1.10	China	
	GqWSPC19	L/19	bp = 47,615,818	AX-94196006	–	7.90	China	
	GqWSPC20	I/20	bp = 4,049,700	AX-94198333	–	2.20	China	

Protein	*Seed protein 47-1*	M/7	70.70–72.70	Satt494	$p < 0.0001$	5.43	China	Teng et al. (2017)
	Seed protein 47-2	K/9	50.00–52.00	Satt710	$p < 0.0001$	2.35	China	
	Seed protein 47-3	H/12	60.33–62.33	Sat_122	$p = 0.0005$	9.54	China	
	Seed protein 47-4	B2/14	5.05–7.05	Satt577	$p = 0.0005$	6.82	China	
	Seed protein 47-5	D2/17	87.01–89.01	Satt543	$p = 0.0004$	6.64	China	
	Seed protein 47-6	G/18	17.25–19.25	Satt217	$p < 0.0001$	2.89	China	
	Seed protein 47-7	G/18	0.00–2.00	Satt163	$p = 0.0007$	3.08	China	
	Seed protein 47-8	I/20	30.94–32.94	Satt614	$p < 0.0001$	15.53	China	
Protein	*qPRO001*	G/18	0.00–3.90	ss249909538-ss249919445	2.50	10.00	North Carolina	Akond et al. (2018)
Oil	*qOIL001*	C2/6	7.00–11.70	ss246100375-ss245879277	39.47	76.00	North Carolina	
Palmitic	*qPAL001*	C2/6	7.70–12.90	ss246100375-ss245879277	2.71	21.00	North Carolina	
	qPAL002	C2/6	26.50–27.30	ss245914593-ss245908292	2.98	21.00	North Carolina	
Oleic	*qOLE001*	N/3	59.10–67.80	ss245246964-ss245259860	2.65	9.00	North Carolina	
	qOLE002	C2/6	26.30–27.30	ss245914593-ss245908292	2.83	15.00	North Carolina	
	qOLE003	G/18	0.40–5.40	ss249909538-ss249506152	5.00	12.00	North Carolina	
Linolenic	*qLIN001*	C2/6	36.30–40.00	ss245914593-ss245790648	2.66	13.00	North Carolina	
Glu	*qGLU001*	6/C2	26.30–27.30	ss245914593-ss245908292	2.51	14.00	North Carolina	
Ala	*qALA001*	6/C2	7.00–9.40	ss246100375-ss245879277	50.24	74.00	North Carolina	
Cys	*qCYS001*	6/C2	25.90–27.30	ss245898080-ss245908292	2.53	11.00	North Carolina	

(continued)

Table 2.2 (continued)

Val	*qVAL001*	18/G	0.00–4.00	ss249909538-ss249919445	2.62	11.00	North Carolina	
His	*qHIS001*	6/C2	26.00–27.30	ss245898080-ss245908292	2.70	12.00	North Carolina	
Protein	*Seed protein 48-1*	I/20	61.20–63.20	BARC-025987-05207	5.09	4.06	MN & Turkey	Hacisalihoglu et al. (2018)
	SeedwtQTL4.1	C1/4	36.00–38.00	044331-08673	4.50	7.03	MN & Turkey	
	SeedwtQTL6.1	C2/6	101.00–103.00	031099-06997	6.00	17.70	MN & Turkey	
	PodwtQTL4.2	C1/4	51.00–53.00	061273-17,145	7.66	21.00	MN & Turkey	
	SeedwtQTL4.3	C1/4	51.00–53.00	061273-17,145	5.06	19.10	MN & Turkey	
	VolQTL6.2	C2/6	101.00–103.00	031099-06997	6.56	17.30	MN & Turkey	
Protein	*qPro_06*	C2/6	0–0.31	bin-6-21,984,752	3.90	9.70	BREC-12	Patil et al. (2018)
	qPro_19	L/19	2.9–4.9	bin-19-1,024,102	3.40	8.40	BREC-12	
	qPro_20	I/20	*	bin-20-2,558,940	2.90	6.90	BREC-12	
	qPro_20	I/20	0–0.32	bin-20-33,975,596	2.90	5.40	BREC-12	
	qPro_19	L/19	2.9–4.9	bin-19-1,024,102	5.20	10.70	BREC-13	
	qPro_20	I/20	0–0.32	bin-20-33,975,596	6.60	14.00	BREC-13	
	qPro_08	A2/8	0.50–2.50	bin-8-7,951,579-bin-8-8,602,440	3.10	4.60	BREC-13	
	qPro_13	F/13	3.6–5.6	bin-13-33,162,248	4.50	7.40	BREC-13	
	qPro_19	L/19	0–0.3	bin-19-1,024,102	3.90	6.40	BREC-13	
	qPro_20	I/20	0.032	bin-20-33,975,596	8.30	14.30	BREC-13	
	qPro_19	L/19	2.9–4.9	bin-19-1,147,261	2.10	4.60	CR13	
	qPro_20	I/20	0–0.32	bin-20-33,975,596	7.80	19.60	CR13	
Oil	*qOil_02*	D1b/2	0–0.61	bin-2-45,571,767	3.10	7.10	BREC12	
	qOil_17	D2/17	0–0.21	bin-17-2,091,529	3.00	7.60	BREC12	
	qOil_20	I/20	0–0.32	bin-20-33,975,596	3.50	9.00	BREC12	
	qOil_02	D1b/2	0–0.31	bin-2-45,571,767	5.10	10.10	BREC13	

	qOil_08	A2/8	0–0.30	bin-8-7,542,280-bin-8-7,564,536	3.40	6.60	BREC13	
	qOil_15	E/15	0–0.30	bin-15-5,286,513	3.20	6.20	BREC13	
	qOil_19	L/19	0–0.91	bin-19-1,021,102	4.00	7.70	BREC13	
	qOil_20	I/20	0–0.63	bin-20-33,975,596	3.00	5.30	BREC13	
	qOil_02	D1b/2	0–0.31	bin-2-45,571,767	7.30	13.50	BREC14	
	qOil_08	A2/8	0–0.30	bin-8-7,951,579-bin-8-8,095,904	7.20	13.30	BREC14	
	qOil_09	K/9	0–0.31	bin-9-5,819,186-bin-9-5,896,549	4.60	8.10	BREC14	
	qOil_20	I/20	0–0.32	bin-20-33,975,596	3.70	6.50	BREC14	
Protein	*Seed protein*	E/15	bp = 4,685,574	ss715622170	1.86E-08	4.00	South Dakota	Zhang et al. (2018)
	Seed protein	E/15	bp = 10,102,255	ss715623324	2.19E-07	2.30	South Dakota	
	Seed protein	J/16	bp = 35,721,993	ss715624781	6.41E-07	2.50	South Dakota	
Oil	*Seed oil*	N/3	bp = 42,377,595	ss715586293	2.07E-07	2.40	South Dakota	
	Seed oil	A1/5	bp = 29,109,248	ss715590407	1.19E-06	0.80	South Dakota	
	Seed oil	A1/5	bp = 41,524,771	ss715591687	2.96E-08	2.80	South Dakota	
	Seed oil	A1/5	bp = 41,883,826	ss715591638	2.65E-08	2.80	South Dakota	
	Seed oil	A1/5	bp = 4,192,5052	ss715591630	8.32E-07	2.20	South Dakota	
	Seed oil	A1/5	bp = 42,119,049	ss715591621	1.20E-06	2.10	South Dakota	
	Seed oil	C2/6	bp = 16,753,938	ss715593620	6.90E-07	0.80	South Dakota	
	Seed oil	M/7	bp = 3,363,285	ss715597212	1.39E-06	2.10	South Dakota	
	Seed oil	M/7	bp = 3,713,267	ss715597475	1.20E-07	2.50	South Dakota	
	Seed oil	M/7	bp = 3,879,026	ss715597771	1.37E-06	2.10	South Dakota	
	Seed oil	A2/8	bp = 1,831,933	ss715600039	2.17E-11	1.60	South Dakota	
	Seed oil	A2/8	bp = 7,792,993	ss715602707	8.26E-09	3.00	South Dakota	

(continued)

Table 2.2 (continued)

	Seed oil	A2/8	bp = 9,188,633	ss715602839	4.74E-12	4.00	South Dakota	
	Seed oil	A2/8	bp = 18,036,698	ss715600011	1.07E-06	0.80	South Dakota	
	Seed oil	F/13	bp = 43,777,586	ss715616488	8.06E-07	2.20	South Dakota	
	Seed oil	B2/14	bp = 44,417,018	ss715619006	6.55E-07	2.20	South Dakota	
	Seed oil	E/15	bp = 3,720,749	ss715621742	5.46E-08	2.70	South Dakota	
	Seed oil	E/15	bp = 3,846,538	ss715621777	2.24E-08	2.80	South Dakota	
	Seed oil	E/15	bp = 6,853,445	ss715622952	4.72E-09	3.10	South Dakota	
	Seed oil	J/16	bp = 7,101,528	ss715625387	5.71E-11	3.90	South Dakota	
	Seed oil	J/16	bp = 35,760,955	ss715624783	7.94E-07	0.90	South Dakota	
	Seed oil	D2/17	bp = 8,842,317	ss715628327	5.65E-11	1.50	South Dakota	
	Seed oil	G/18	bp = 5,543,903	ss715631635	1.43E-06	2.10	South Dakota	
	Seed oil	G/18	bp = 18,452,921	ss715629475	7.50E-08	2.60	South Dakota	
	Seed oil	G/18	bp = 48,627,264	ss715631395	3.92E-07	2.30	South Dakota	
Palmitic	*Seed palmitic*	A1/5	bp = 1,049,939	*ss715592503*	5.38E-11	9.90	South Dakota	
	Seed palmitic	C2/6	bp = 45,759,557	*ss715594516*	3.03E-07	3.50	South Dakota	
	Seed palmitic	A2/8	bp = 7,815,158	*ss715602709*	1.43E-06	5.20	South Dakota	
	Seed palmitic	B2/14	bp = 17,499,955	*ss715618427*†	1.69E-06	5.10	South Dakota	
	Seed palmitic	B2/14	bp = 35,662,926	*ss715618299*	1.45E-10	6.20	South Dakota	
	Seed palmitic	B2/14	bp = 46,215,024	*ss715619208*	4.39E-09	4.60	South Dakota	
	Seed palmitic	E/15	bp = 3,032,666	*ss715621537*	1.89E-10	5.90	South Dakota	
Stearic	*Seed stearic*	A1/5	bp = 1,180,877	*ss715592495*	7.07E-08	4.20	South Dakota	
	Seed stearic	B2/14	bp = 17,499,955	*ss715618427*‡	1.38E-05	2.70	South Dakota	
	Seed stearic	B2/14	bp = 40,524,189	*ss715618720*	1.63E-08	3.90	South Dakota	
Oleic	*Seed oleic*	A1/5	bp = 37,237,915	*ss715591404*	7.32E-08	4.40	South Dakota	
	Seed oleic	A2/8	bp = 8,276,381	*ss715602786*	5.57E-08	7.00	South Dakota	
	Seed oleic	B2/14	bp = 47,398,552	*ss715619390*	1.52E-10	5.20	South Dakota	
	Seed oleic	L/19	bp = 36,429,269	*ss715634472*	7.69E-08	6.10	South Dakota	

Linoleic	*Seed linoleic*	D2/17	bp = 36,411,792	*ss715627021*	2.92E-07	3.80	South Dakota	
Linolenic	*Seed linolenic*	B2/14	bp = 45,971,865	*ss715619179*	4.09E-07	4.60	South Dakota	
	Seed linolenic	E/15	bp = 2,831,678	*ss715621481*	1.13E-06	5.00	South Dakota	
	Seed linolenic	E/15	bp = 3,846,538	*ss715621777*	1.98E-08	5.60	South Dakota	
Palmitic	*Palmitic*	I/20	bp = 35,574,654	ss715637594	$p = 6.25 \times 10^{-7}$	–	Missouri	La et al. (2019)
Oleic	*Oleic*	N/3	bp = 36,701,848	ss715585752	$p = 7.16 \times 10^{-7}$	–	Missouri	
	Oleic	M/7	bp = 40,940,646	ss715598001	$p = 4.10 \times 10^{-7}$	–	Missouri	
	Oleic	A2/8	bp = 7,864,681	ss715602718	$p = 2.99 \times 10^{-10}$	–	Missouri	
	Oleic	B2/14	bp = 43,083,528	ss715618857	$p = 9.92 \times 10^{-7}$	–	Missouri	
Linoleic	*Linoleic*	D1b/2	bp = 9308,478	ss715584162	$p = 1.05 \times 10^{-6}$	–	Missouri	
	Linoleic	C1/4	bp = 46,118,547	ss715588279	$p = 1.01 \times 10^{-9}$	–	Missouri	
	Linoleic	B1/11	bp = 27,298,217	ss715609933	$p = 1.55 \times 10^{-6}$	–	Missouri	
	Linoleic	B1/11	bp = 32,304,429	ss715610306	$p = 3.32 \times 10^{-7}$	–	Missouri	
	Linoleic	F/13	bp = 17,001,366	ss715616705	$p = 1.29 \times 10^{-6}$	–	Missouri	
Aspartic	*Aspartic*	F/13	bp = 38,026,151	ss715616006	$p = 5.84 \times 10^{-7}$	–	Missouri	
Glutamic	*Glutamic*	F/13	bp = 38,026,151	ss715616006	$p = 9.26 \times 10^{-7}$	–	Missouri	
Protein	*qPRO-J*	J/16	4.50–6.50	ss715624939-ss715624938	15.68	23.2	China	Li et al. (2019a)
		J/16	4.50–6.50	ss715624939-ss715624938	13.45	14.1	China	
	qPRO-K	K/9	7.40–9.40	ss715603464-ss715604608	6.31	13.7	China	
		K/9	17.70–19.70	ss715604608-ss715605140	8.3	11.1	China	
	qPRO-F	F/13	22.00–24.00	ss715614511-ss715614527	2.76	2.6	China	
	qPRO-O	O/10	21.50–23.50	ss715606777-ss715606801	2.93	4.6	China	
		O/10	20.20–22.20	ss715607012-ss715606777	3.58	3.5	China	
	qPRO-A2	A2/8	86.30–88.30	ss715602883-ss715602895	5.25	5.1	China	
		A2/8	86.30–88.30	ss715602883-ss715602895	3.46	4	China	
	qPRO-B1	B1/11	110.50–112.50	ss715609500-ss715609581	3.69	3.6	China	

(continued)

Table 2.2 (continued)

		B1/11	110.50–112.50	ss715609581-ss715610190	4.51	7.3	China	
	qPRO-C1	C1/4	15.30–17.30	ss715587753-ss715587789	3.58	3.4	China	
	qPRO-N	N/3	21.10–23.10	ss715586273-ss715586259	3.31	4.5	China	
Oil	*qOIL-A1*	A1/5	92.20–99.20	ss715591822-ss715591799	18.07	22.1	China	
		A1/5	92.20–99.20	ss715591799-ss715591790	13.23	19.3	China	
		A1/5	92.20–99.20	ss715591799-ss715591790	6.11	10.9	China	
		A1/5	92.20–99.20	ss715591790-ss715591780	3.57	4.8	China	
	qOIL-J	J/16	4.50–6.50	ss715624939-ss715624938	5.26	7.1	China	
	qOIL-C1	C1/4	17.50–19.50	ss715587789-ss715587843	3.67	4	China	
	qOIL-N	N/3	22.80–24.80	ss715586252-ss715586200	3.39	4.5	China	
P + O	*qP + O-A1*	A1/5	92.20–99.20	ss715591790-ss715591780	5.95	6.6	China	
		A1/5	92.20–99.20	ss715591799-ss715591790	13.23	16.4	China	
		A1/5	92.20–99.20	ss715591799-ss715591790	6.8	8.1	China	
		A1/5	92.20–99.20	ss715591954-ss715591831	5.76	4.8	China	
	qP + O-J	J/16	4.50–6.50	ss715624939-ss715624938	24.87	23.3	China	
		J/16	4.50–6.50	ss715624939-ss715624938	5.48	6.4	China	
	qP + O-K	K/9	9.80–11.80	ss715603464-ss715604608	5.21	7.1	China	
		K/9	18.10–24.10	ss715604608-ss715605140	2.84	2.7	China	
		K/9	18.10–24.10	ss715604608-ss715605140	6.83	9	China	
		K/9	34.90–36.90	ss715605140-ss715603353	3.21	5.7	China	
	qP + O-I	I/20	48.50–56.60	ss715638252-ss715638616	9.07	13.2	China	
		I/20	48.50–56.60	ss715638209-ss715638252	4.28	3.3	China	
		I/20	48.50–56.60	ss715638190-ss715638209	2.76	3.1	China	
	qP + O-F	F/13	23.60–25.60	ss715614578-ss715614642	3.59	4.1	China	
		F/13	23.60–25.60	ss715614567-ss715614578	4.34	5.1	China	
	qP + O-O	O/10	17.20–19.20	ss715607012-ss715606777	4.18	5.3	China	
Protein	*Seed protein*	D1a/1	bp = 53,140,888	rs53140888	($-\log P = 3.68$)	16.03	China	Li et al. (2019b)

	Seed protein	D1a/1	bp = 53,141,323	rs53141323	(−log*P* = 3.62)	15.88	China	
	Seed protein	D1a/1	bp = 53,141,571	rs53141571	(−log*P* = 3.29)	15.12	China	
	Seed protein	D1b/2	bp = 41,860,441	rs41860441	(−log*P* = 3.01)	14.46	China	
	Seed protein	C1/4	bp = 4,187,467	rs4187467	(−log*P* = 3.15)	14.79	China	
	Seed protein	C1/4	bp = 4,553,067	rs4553067	(−log*P* = 3.4)	15.38	China	
	Seed protein	F/13	bp = 18,150,116	rs18150116	(−log*P* = 3.56)	15.75	China	
	Seed protein	F/13	bp = 19,485,676	rs19485676	(−log*P* = 3.45)	15.49	China	
	Seed protein	I/20	bp = 24,787,338	rs24787338	(−log*P* = 3.74)	16.17	China	
	Seed protein	I/20	bp = 29,457,452	rs29457452	(−log*P* = 3.18)	14.85	China	
	Seed protein	I/20	bp = 29,979,450	rs29979450	(−log*P* = 3.27)	15.07	China	
	Seed protein	D1a/1	bp = 53,140,888	rs53140888	(−log*P* = 3.01)	13.58	China	
	Seed protein	D1b/2	bp = 39,587,099	rs39587099	(−log*P* = 3.55)	14.85	China	
	Seed protein	D1b/2	bp = 41,989,469	rs41989469	(−log*P* = 3.57)	14.90	China	
	Seed protein	H/12	bp = 6,121,092	rs6121092	(−log*P* = 3.19)	14.01	China	
	Seed protein	F/13	bp = 19,485,676	rs19485676	(−log*P* = 3.41)	14.53	China	
	Seed protein	I/20	bp = 24,787,338	rs24787338	(−log*P* = 3.5)	14.72	China	
Oil	*Seed oil*	C1/4	bp = 204,699	rs204699	(−log*P* = 3.12)	29.28	China	
	Seed oil	C1/4	bp = 32,697,665	rs32697665	(−log*P* = 3.21)	29.45	China	
	Seed oil	A1/5	bp = 35,315,668	rs35315668	(−log*P* = 3.11)	29.25	China	
	Seed oil	M/7	bp = 15,774,585	rs15774585	(−log*P* = 4.48)	31.96	China	
	Seed oil	M/7	bp = 15,783,346	rs15783346	(−log*P* = 4.27)	31.54	China	
	Seed oil	C1/4	bp = 14,297,133	rs14297133	(−log*P* = 3.14)	33.03	China	
	Seed oil	A1/5	bp = 12,023,370	rs12023370	(−log*P* = 3.1)	32.95	China	
	Seed oil	M/7	bp = 15,774,585	rs15774585	(−log*P* = 4.26)	35.11	China	
	Seed oil	M/7	bp = 15,783,346	rs15783346	(−log*P* = 3.44)	33.58	China	

(continued)

Table 2.2 (continued)

	Seed oil	A2/8	bp = 40,944,487	rs40944487	(−log$P$ = 3.15)	33.04	China	
	Seed oil	H/12	bp = 35,333,667	rs35333667	(−log$P$ = 3.27)	33.26	China	
	Seed oil	G/18	bp = 1,452,418	rs1452418	(−log$P$ = 3.04)	32.85	China	
	Seed oil	G/18	bp = 55,473,106	rs55473106	(−log$P$ = 3.12)	32.99	China	
	Seed oil	L/19	bp = 49,097,495	rs49097495	(−log$P$ = 3.21)	33.15	China	
Protein	*qProt-1-1*	D1a/1	*	Gm01_50257226	(−log$P$ = 11.7)	2.80	China	Li et al. (2019c)
	qProt-2-1	D1b/2	*	Gm02_ BLOCK_34241156_34302885	(−log$P$ = 3.8)	1.00	China	
	qProt-3-1	N/3	*	Gm03_46437265	(−log$P$ = 2.9)	0.60	China	
	qProt-4-1	C1/4	*	Gm04_ BLOCK_2157397_2345500	(−log$P$ = 8.2)	2.40	China	
	qProt-4-2	C1/4	*	Gm04_8725710	(−log$P$ = 2.1)	0.40	China	
	qProt-4-3	C1/4	*	Gm04_ BLOCK_39987192_40167695	(−log$P$ = 2.8)	1.20	China	
	qProt-6-2	C2/6	*	Gm06_ BLOCK_46724398_46724570	(−log$P$ = 5.3)	1.40	China	
	qProt-7-1	M/7	*	Gm07_ BLOCK_8657174_8845556	(−log$P$ = 15.2)	4.70	China	
	qProt-8-1	A2/8	*	Gm08_ BLOCK_16237107_16278243	(−log$P$ = 24.1)	8.00	China	
	qProt-8-2	A2/8	*	Gm08_ BLOCK_42026048_42200505	(−log$P$ = 3.2)	1.20	China	
	qProt-10-1	O/10	*	Gm10_ BLOCK_856393_1044958	(−log$P$ = 3.0)	1.00	China	
	qProt-10-2	O/10	*	Gm10_ BLOCK_37907908_38079904	(−log$P$ = 13.2)	3.60	China	
	qProt-10-3	O/10	*	Gm10_ BLOCK_38679572_38679818	(−log$P$ = 5.1)	1.50	China	

	qProt-13-1	F/13	*	Gm13_ BLOCK_2883529_3082036	(−log$P$ = 5.3)	1.50	China	
	qProt-15-1	E/15	*	Gm15_3633885	(−log$P$ = 3.0)	0.60	China	
	qProt-15-2	E/15	*	Gm15_31011761	(−log$P$ = 5.9)	1.30	China	
	qProt-17-1	D2/17	*	Gm17_ BLOCK_6574500_6577175	(−log$P$ = 7.8)	2.00	China	
	qProt-18-1	G/18	*	Gm18_ BLOCK_35961794_36027902	(−log$P$ = 3.0)	0.80	China	
	qProt-19-1	L/19	*	Gm19_37500961	(−log$P$ = 3.0)	0.60	China	
	qProt-19-2	L/19	*	Gm19_ BLOCK_38886258_38944167	(−log$P$ = 3.1)	0.90	China	
	qProt-20-1	I/20	*	Gm20_5531497	(−log$P$ = 2.5)	0.50	China	
	qProt-20-2	I/20	*	Gm20_ BLOCK_27111387_27111623	(−log$P$ = 3.6)	0.90	China	
	qProt-20-3	I/20	*	Gm20_ BLOCK_30995685_31177423	(−log$P$ = 56.1)	16.00	China	
	qProt-20-4	I/20	*	Gm20_ BLOCK_43288485_43465351	(−log$P$ = 5.7)	1.60	China	
Oil	*qOil-1-1*	D1a/1	*	Gm01_ BLOCK_41522087_41713586	(−log$P$ = 5.9)	1.90	China	
	qOil-1-2	D1a/1	*	Gm01_50257226	(−log$P$ = 9.5)	2.10	China	
	qOil-3-1	N/3	*	Gm03_ BLOCK_4553710_4553720	(−log$P$ = 5.0)	1.20	China	
	qOil-3-2	N/3	*	Gm03_ BLOCK_11906061_11923443	(−log$P$ = 7.1)	1.70	China	
	qOil-4-1	C1/4	*	Gm04_21748147	(−log$P$ = 5.3)	1.10	China	
	qOil-4-2	C1/4	*	Gm04_41026444	(−log$P$ = 3.2)	0.60	China	

(continued)

Table 2.2 (continued)

	qOil-4-3	C1/4	*	Gm04_ BLOCK_47957394_47957714	(−log$P$ = 2.3)	0.40	China	
	qOil-6-1	C2/6	*	Gm06_44035193	(−log$P$ = 3.3)	0.60	China	
	qOil-7-1	M/7	*	Gm07_38954920	(−log$P$ = 3.0)	0.60	China	
	qOil-8-1	A2/8	*	Gm08_ BLOCK_14242705_14306849	(−log$P$ = 10.7)	3.20	China	
	qOil-8-2	A2/8	*	Gm08_ BLOCK_16237107_16278243	(−log$P$ = 15.4)	5.20	China	
	qOil-8-3	A2/8	*	Gm08_ BLOCK_18015046_18031943	(−log$P$ = 5.7)	1.60	China	
	qOil-10-1	O/10	*	Gm10_ BLOCK_5509737_5559675	(−log$P$ = 4.1)	1.00	China	
	qOil-10-2	O/10	*	Gm10_22446436	(−log$P$ = 2.9)	0.60	China	
	qOil-10-3	O/10	*	Gm10_ BLOCK_38679572_38679818	(−log$P$ = 9.9)	2.60	China	
	qOil-10-4	O/10	*	Gm10_ BLOCK_46662161_46730774	(−log$P$ = 18.4)	4.60	China	
	qOil-13-1	F/13	*	Gm13_ BLOCK_30751302_30790418	(−log$P$ = 3.8)	1.30	China	
	qOil-15-1	E/15	*	Gm15_ BLOCK_10984687_11112792	(−log$P$ = 2.6)	0.60	China	
	qOil-16-1	J/16	*	Gm16_31945745	(−log$P$ = 11.7)	2.70	China	
	qOil-17-1	D2/17	*	Gm17_ BLOCK_5346273_5356960	(−log$P$ = 8.0)	1.80	China	
	qOil-17-2	D2/17	*	Gm17_9078832	(−log$P$ = 5.1)	1.10	China	
	qOil-18-1	G/18	*	Gm18_ BLOCK_11985000_12125810	(−log$P$ = 4.9)	1.40	China	

	qOil-20-1	I/20	*	Gm20_ BLOCK_30995685_31177423	(−log$P$ = 55.0)	15.10	China	
Protein	*qPro-1-I*	D1a/1	37.9–44.4	bin29-bin39	2.74	6.32	China	Karikari et al. (2019)
	qPro-4-1	C1/4	27.1–41.4	bin365-bin371	2.55	5.80	China	
	qPro-6-1	C2/6	56.1–65.4	bin516-bin519	6.17	15.09	China	
	qPro-6-2	C2/6	65.4–74.8	bin19-bin521	5.08	13.23	China	
	qPro-6-3	C2/6	163.5–172.7	bin516-bin521	3.47	11.16	China	
	qPro-7-1	M/7	34.2–44.7	bin769-bin776	5.62	13.59	China	
	qPro-7-2	M/7	48.8–51.9	bin778-bin780	5.30	15.01	China	
	qPro-8-1	A2/8	12.4–25.9	bin877-881	2.60	8.21	China	
	qPro-9-1	K	61.6–69.5	bin1048-bin1051	3.51	7.70	China	
	qPro-9-2	K	70.3–81.8	bin1051-bin1053	3.54	11.94	China	
	qPro-9-3	K	91.3–105.4	bin1055-bin-1059	2.78	6.40	China	
	qPro-10-1	O/10	22.3–27.5	bin1152-bin1147	8.93	21.52	China	
	qPro-10-2	O/10	32.9–33.9	bin1147-bin1151	5.46	13.62	China	
	qPro-10-3	O/10	58.0–61.3	bib1141-bin1139	3.66	8.10	China	
	qPro-10-4	O/10	62.7–70.8	bin1139-bin1138	3.53	7.84	China	
	qPro-13-1	F/13	0.0–06.6	bin1438-bin1436	3.24	10.27	China	
	qPro-13-2	F/13	75.6–81.1	bin1449-bin1450	3.05	10.78	China	
	qPro-14-1	B2/14	62.9–67.0	bin1609-bin1611	2.70	8.46	China	
	qPro-14-2	B2/14	104.4–105.5	bin1642-bin1643	2.66	5.74	China	
	qPro-16-1	J/16	89.2–97.2	bin1810-bin1814	3.20	6.96	China	
	qPro-17-1	D2/17	34.7–39.3	bin1865-bin1867	3.99	9.32	China	
	qPro-18-1	G/18	56.4–61.6	bin2000-bin2004	4.52	10.09	China	
	qPro-18-2	G/18	62.2–69.2	bin2004-bin2014	3.11	7.15	China	
	qPro-19-1	L/19	10.7–17.2	bin2069-bin2072	3.52	8.05	China	

(continued)

Table 2.2 (continued)

	qPro-20-1	I/20	0.0–2.9	bin2177-bin2178	3.34	10.71	China	
Oil	*qOil-1-1*	D1a/1	37.4–39.5	bin28-bin31	4.88	10.58	China	
	qOil-2-1	D1b/2	138.0–149.4	bin193-bin200	4.08	13.31	China	
	qOil-2-2	D1b/2	110.8–121.9	bin213-bin218	3.10	7.73	China	
	qOil-3-1	N/3	1.09 0.9–15.4	bib228-bin238	2.83	8.96	China	
	qOil-6-1	C2/6	55.0–75.6	bin516-bin521	2.74	5.44	China	
	qOil-8-1	A2/8	9.7–17.6	bin878-bin886	2.77	6.03	China	
	qOil-8-2	A2/8	36.3–37.0	bin891-bin892	2.93	7.98	China	
	qOil-8-3	A2/8	40.7–43.8	bin895-bin901	4.08	10.49	China	
	qOil-8-4	A2/8	50.1–55.2	bin900-bin904	4.70	12.69	China	
	qOil-10-1	O/10	16.1–19.3	bin1150-bin1149	3.54	9.64	China	
	qOil-10-2	O/10	19.0–26.1	bin1148-bin1147	3.70	10.94	China	
	qOil-10-3	O/10	30.4–30.8	bin1147-bin1151	3.70	9.90	China	
	qOil-10-4	O/10	32.2–33.6	bin1147-bin1151	5.66	14.50	China	
	qOil-11-1	B1/11	46.0–55.5	bin1268-bin1274	4.85	12.61	China	
	qOil-13-1	F/13	32.3–43.1	bin1430-bin1427	3.35	10.01	China	
	qOil-16-1	J/16	87.7–97.8	bin1809-bin1814	3.92	8.69	China	
	qOil-20-1	I/20	0.0–13.8	bin2177-bin2223	3.09	9.86	China	
	qOil-20-2	I/20	66.3–81.8	bin2236-bin2239	3.15	9.27	China	
	qOil-20-3	I/20	92.7–102.2	bin2242-bin2244	3.92	8.28	China	
Protein	*qPro-A1-1*	A1/5	27.7–51.6	Sat_171-Sat_369	3.94	6.37	China	Li et al. (2020)
	qPro-A1-2	A1/5	26.8–28.8	Satt042-Sat_410	4.02	7.44	China	
	qPro-A1-3	A1/5	5.8–15.5	Satt572-Satt684	2.53	6.43	China	
	qPro-A1-4	A1/5	3.6–5.8	Satt684-Sat_137	3.26	8.85	China	
	qPro-A2-1	A2/8	29.5–31.5	AW132402-Satt589	4.05	8.12	China	
	qPro-A2-2	A2/8	66.8–69.9	Sat_115-Satt341	2.67	7.26	China	

	qPro-B1-1	B1/11	107.4–109.4	Satt453-BARCSOYSSR_11_0442	2.91	11.58	China	
	qPro-B2-1	B2/14	14.5–16.5	Sat_342-Sat_264	3.06	8.39	China	
	qPro-B2-2	B2/14	56.4–63.5	Sat_355-Sat_189	4.63	8.78	China	
	qPro-B2-3	B2/14	4.9–6.9	Sat_189-Sat_177	2.56	5.65	China	
	qPro-C1-1	C1/4	63.4–65.4	Sat_207-Satt338	2.51	4.27	China	
	qPro-C2-1	C2/6	107–115.6	Satt460-Satt316	2.57	3.62	China	
	qPro-D1b-1	D1b/2	112.2–114.2	Sat_183-Sat_096	3.17	7.35	China	
	qPro-D1b-2	D1b/2	–	Sat_373-Satt701	3.01	5.05	China	
	qPro-D1b-3	D1b/2	47.4–53	Sat_173-Satt558	2.52	5.39	China	
	qPro-D2-1	D2/17	96.1–100.9	Sat_022-Satt413	3.29	4.27	China	
	qPro-D2-2	D2/17	79.6–100	Sat_354-Satt256	3.00	9.09	China	
	qPro-D2-3	D2/17	47.2–63.9	Sat_222-Satt582	2.58	7.89	China	
	qPro-F-1	F/13	7.00–9.00	Sat_298-AW186493	2.66	5.92	China	
	qPro-F-2	F/13	22.3–24.3	Satt149-BE806387	2.80	4.49	China	
	qPro-G-1	G/18	10.1–14	Sat_141-Satt309	3.58	8.51	China	
	qPro-G-2	G/18	71.6–85.6	Satt288-Sct_199	2.60	5.82	China	
	qPro-H-1	H/12	55.1–63.3	Satt253-Satt469	2.92	7.20	China	
	qPro-H-2	H/12	83–100.7	Satt293-Sat_180	2.71	10.14	China	
	qPro-J-1	J/16	81.5–83.5	Sat_228-Satt431	2.51	4.05	China	
	qPro-K-1	K/9	38.2–65.1	Sat_167-Satt349	2.64	5.40	China	
	qPro-K-2	K/9	14.7–18.5	Satt242-Sat_119	2.61	2.85	China	
	qPro-K-3	K/9	39.2–40.1	Satt247-Satt727	5.25	10.09	China	
	qPro-L-1	L/19	55.7–60.6	Sat_245-Satt076	4.36	13.83	China	
	qPro-M-1	M/7	124.5–126.5	Satt308-Satt336	2.57	7.73	China	

(continued)

Table 2.2 (continued)

	qPro-N-1	N/3	23.2–24.3	Sat_186-Satt641	3.11	5.62	China	
	qPro-N-2	N/3	28.6–34.5	Satt125-Satt624	2.87	6.54	China	
	qPro-N-3	N/3	23.2–39.7	Satt641-Sat_266	3.16	12.61	China	
	qPro-N-4	N/3	57.2–61.6	Satt549-Satt255	2.87	6.12	China	
Oil	*qOil-A1-1*	A1/5	14.5–16.5	BARCSOYSSR_05_0513-Satt572	2.66	5.65	China	
	qOil-A1-2	A1/5	80.1–85.8	Sat_217-Satt200	3.72	6.81	China	
	qOil-A1-3	A1/5	78.7–78.8	Sat_374-Satt211	3.09	4.93	China	
	qOil-A1-4	A1/5	26.9–28.9	Satt042-Sat_410	4.43	9.11	China	
	qOil-A1-5	A1/5	46.2–48.2	Satt200-Satt717	2.74	2.07	China	
	qOil-A1-6	A1/5	3.6–5.8	Satt684-Sat_137	3.03	8.05	China	
	qOil-A2-1	A2/8	69.9–91.6	Sat_392-Sat_115	3.44	8.05	China	
	qOil-B1-1	B1/11	82.3–96.1	Satt583-Sat_123	2.64	5.75	China	
	qOil-B2-1	B2/14	14.5–16.5	Sat_342-Sat_264	3.02	3.09	China	
	qOil-C1-1	C1/4	46–49.8	Satt139-Satt646	3.76	3.78	China	
	qOil-C1-2	C1/4	59.5–61.5	Satt161-Satt195	3.11	6.47	China	
	qOil-C2-1	C2/6	72.2–74.2	BARCSOYSSR_06_0148-Satt322	3.19	7.86	China	
	qOil-C2-2	C2/6	101.3–103.3	BARCSOYSSR_06_1462-Satt557	2.92	9.58	China	
	qOil-C2-3	C2/6	132.4–134.4	Satt357-GMAC7L	3.00	4.39	China	
	qOil-D1b-1	D1b/2	52–54	Satt558-BARCSOYSSR_02_0607	2.68	4.39	China	
	qOil-D1b-2	D1b/2	47.4–49.7	Satt701-Sat_173	2.67	3.97	China	
	qOil-D2-1	D2/17	98.5–114.3	Sat_326-Sat_220	2.65	2.65	China	

Glycine max and *Glycine soja* with 252 mainly restriction fragment length polymorphism (RFLP) markers. They have constructed one of the first genetic linkage maps of soybean based mainly on RFLP markers, covered 2147 cM with 31 linkage groups, and an average marker distance of 8.5 cM between two adjacent markers. For seed protein and oil QTL analysis, they used single-marker analysis method. Eight QTL that control seed protein content have been identified and mapped on chromosomes 14, 15, 18, 19, and 20. Nine QTL that control seed oil content have been identified and mapped on chromosomes 14, 15, 19, and 20 (Fig. 2.2; Table 2.2) (Diers et al. 1992).

Diers and Shoemaker (1992) grew an F_2 population derived from a cross between "A81-356022" and "PI 468916" ($n = 60$) in Ames, IA, genotyped it with 252 markers (RFLP, isozymes, morphological, and storage protein markers), and used single-marker (ANOVA) analysis for QTL data analysis. The "A81-356022" by "PI 468916"-based genetic linkage map covered 2147 cM. Three QTL that control seed palmitic acid content have been identified and mapped on chromosomes 14 and 16; 1 QTL that controls seed stearic acid content and 6 QTL that control seed oleic content have been identified and mapped on chromosome 16; 6 QTL that control seed linoleic content have been identified and mapped on chromosomes 5, 11, and 15; and 7 QTL that control seed linolenic content have been identified and mapped on chromosomes 9, 15, and 19 (Fig. 2.2; Table 2.2) (Diers and Shoemaker 1992).

Mansur et al. 1993 grew a RIL population derived from a cross between "Minsoy" and "Noir 1" in Rosemont, MN, genotyped it with 156 markers, and identified QTL for seed protein and oil contents among QTL for several other traits. Two QTL for seed oil content have been identified and mapped on chromosomes 8 and 9 (Fig. 2.2; Table 2.2) (Mansur et al. (1993). In a subsequent study, Mansur et al. (1996) used the same RIL population and genotyped it with 132 markers. In this study, 3 QTL for seed protein content and 3 QTL for seed oil content have been identified and mapped on chromosomes 5 and 19 (Mansur et al. 1996).

Lark et al. (1994) grew the same RIL population derived from "Minsoy" and "Noir 1" in fields in Minnesota and Chile, genotyped it with 156 markers (Mansur et al. 1993), and used single-marker analysis (ANOVA) for QTL data analysis. QTL that control seed oil to protein ratio (*oil/prot 1–1*) was identified and mapped on chromosome 5 (Fig. 2.2; Table 2.2) (Lark et al. 1994).

Similarly, Lee et al. (1996) used two populations and RFLP markers in their QTL analysis study. Both populations were grown in several fields in North Carolina, South Carolina, and Georgia, USA. The first population was composed of 120 F_4 lines ($n = 120$) derived from a cross between cultivar "Young" and PI 416937. This population was genotyped with 155 RFLP markers. The second population was composed of 111 F_2 lines ($n = 111$) derived from a cross between PI 97100 and cultivar "Coker 237." This population was genotyped with 153 RFLP markers. Thirteen (13) QTL for seed protein content have been identified and mapped on chromosomes 3, 4, 14, 15, and 16, and 6 QTL for seed oil content have been identified and mapped on chromosomes 15, 16, 17, and 19 across multiple locations using single-marker analysis method (Fig. 2.2; Table 2.2) (Lee et al. 1996).

Brummer et al. (1997) used eight RIL populations that were grown in the Midwestern states of the USA, Indiana, Minnesota, and Nebraska to identify QTL for seed protein and oil contents. The first RIL population derived from a cross between "A87-296011" and "CX1039-99" (*n* = 100) was genotyped with 38 RFLP markers; the second RIL population derived from a cross between "C1763" and "CX1159-49-1" (*n* = 89) was genotyped with 42 RFLP markers; and the third RIL population derived from a cross between "C1763" and "CX1039-99" (*n* = 83) was genotyped with 32 RFLP markers. These RIL populations 1–3 were grown in Indiana. The fourth RIL population derived from a cross between "M82-806" and "HHP" (*n* = 71) was genotyped with 85 RFLP markers; the fifth RIL population derived from a cross between "M84-492" and "Sturdy" (*n* = 92) was genotyped with 44 RFLP markers; and the sixth RIL population derived from a cross between "McCall" and "PI445.815" (*n* = 92) was genotyped with 65 RFLP markers. These RIL populations 4–6 were grown in Minnesota. The seventh RIL population derived from a cross between "M81-382" and "PI423.949" (*n* = 81) was genotyped with 46 RFLP markers, and the eighth RIL population derived from a cross between "LN83-2356" and "PI360.843" (*n* = 69) was genotyped with 21 RFLP markers. These RIL populations 7–8 were grown in Nebraska. Fourteen QTL that control seed protein content have been identified and mapped on chromosomes 1, 4, 8, 12, 13, 14, 15, 18, and 20. Twelve QTL that control seed oil content have been identified and mapped on chromosomes 5, 6, 8, 9, 11, 12, and 18 (Fig. 2.2; Table 2.2) (Brummer et al. 1997).

Qiu et al. (1999) grew a RIL population derived from a cross between "Peking" and "Essex" (n = 200) in Columbia, MO, and genotyped it with 216 RFLP markers. Two QTL that control seed protein content have been identified and mapped on chromosomes 12 and 13, and one QTL that control seed oil content has been identified and mapped on chromosome 12 at the same location as the QTL for seed protein content (Fig. 2.2; Table 2.2) (Qiu et al. 1999).

Orf et al. (1999) used three RIL populations that were grown in four environments in Rosemont, MN, USA, and Chile to identify QTL for seed protein and oil contents among other important agronomic traits. The first RIL population was derived from a cross between "Minsoy" and "Archer" (*n* = 233); the second RIL population was derived from a cross between "Minsoy" and "Noir 1" (*n* = 240); and the third RIL population was derived from a cross between "Noir 1" and "Archer" (*n* = 240). These three RIL populations were genotyped with 400+ RFLP and SSR markers, and a combined composite map was constructed. Five QTL for seed protein content have been identified and mapped on chromosomes 4, 5, 7, and 19, and five QTL for seed oil content have been identified and mapped on chromosomes 4, 5, 6, and 19 (Fig. 2.2; Table 2.2) (Orf et al. 1999).

Sebolt et al. (2000) used three backcrosses (BC), BC1F1, BC2F1, and BC3F1. They first crossed *G. max* "A81-356022" and *G. soja* "PI 468916" and then used the line "A81-356022" for backcrossing for three generations to develop BC1F1, BC2F1, and BC3F1 populations. One QTL that controls seed protein content has been identified and mapped on chromosome 20, and one QTL for seed oil content has been identified and mapped on the same location on chromosome 20 (Fig. 2.2; Table 2.2) (Sebolt et al. 2000).

Specht et al. (2001) grew a RIL population derived from a cross between “Minsoy” and “Noir 1” (n = 236); genotyped it with 665 RFLP, SSR, and classical markers; and investigated water use efficiency among other important agronomic traits. Five QTL that control seed protein content have been identified and mapped on chromosomes 4, 5, 7, 9, and 10, and five QTL that control seed oil content have been identified and mapped on chromosomes 1, 5, 13, and 20 (Fig. 2.2; Table 2.2) (Specht et al. 2001).

In another study, Csanadi et al. 2001 grew the “Ma.Belle” by “Proto” F_2 population ($n = 82$) in 5 environments in Austria; genotyped it by 113 SSR, 6 RAPD, and 1 RFLP markers; and identified 4 QTL for protein content, 3 QTL for oil content, and 8 QTL for seed weight using one-way ANOVA and single-marker analysis for QTL data analysis. The QTL for protein content were p4.7 (Satt077) on chromosome 1, p6.6 (Sct_028) on chromosome 6, and p4.7 (Satt196) on chromosome 9. The QTL for oil content were o5.7 (Satt562) on chromosome 20 and o6.5 (Satt196) on chromosome 9. The QTL on chromosome 9 controls both protein and oil contents (Csanadi et al. (2001).

Vollmann et al. (2002) grew a RIL population derived from a cross between “M91-212006” and “SZG9652” lines ($n = 128$) in Raasdorf, Austria, and one QTL for seed protein content was identified and mapped on chromosome 8 (Fig. 2.2; Table 2.2) (Vollmann et al. 2002).

Li et al. (2002) grew an F_2 population derived from a cross between “Cook” (normal palmitic acid) and “N87-2122-4” (reduced palmitic acid) lines ($n = 121$) and four near-isogenic lines (NILs) derived from [“Benning” (normal palmitic acid) by [Cook by N87-2122-4]] in the greenhouse, genotyped it with SSR markers, and used single-marker (ANOVA) analysis for QTL data analysis. Two QTL that control seed palmitic acid content have been identified and mapped on chromosomes 5 and 7 (Fig. 2.2; Table 2.2) (Li et al. 2002).

Chung et al. (2003) grew a RIL population derived from the cross of “A3733” and “PI437088A” ($n = 76$) in multiple environments in Mead, NE. The study identified and mapped a major QTL that controls both seed protein and oil contents on chromosome 20 (Fig. 2.2; Table 2.2) (Chung et al. 2003).

Chapman et al. (2003) grew a RIL population derived from the cross between “Essex” and “Williams” lines ($n = 177$) in multiple environments in Knoxville, TN. Two QTL that control seed protein content have been identified and mapped on chromosomes 11 and 19, and two QTL that control seed oil content have been identified and mapped on chromosomes 11 and 17 (Fig. 2.2; Table 2.2) (Chapman et al. 2003).

In another study, Tajuddin et al. (2003) grew a RIL population derived from the cross of “Misuzudaizu” and “Moshidou Gong 503” ($n = 156$) in two environments in Japan, used a previous genetic linkage map composed of RFLP and SSR markers (Watanabe et al. 2002), and used single-marker analysis (one-way ANOVA), interval mapping (IM), and composite interval mapping (CIM) for QTL data analysis. Ten QTL that control seed protein content have been identified and mapped on chromosomes 3, 6, 8, 9, 15, 17, 18, 19, and 20, and seven QTL that control seed lipid contents have been identified and mapped on chromosomes 7, 10, 14, 16, and 20 (Fig. 2.2, Table 2.2) (Tajuddin et al. 2003).

Stombaugh et al. (2004) used a RIL population derived from a cross between "Minsoy" and "Archer" (n = 108). Three QTL controlling seed cell wall polysaccharide (CWP) have been identified and mapped on chromosomes 5, 6, and 9. One QTL that controls seed fructose content has been identified and mapped on chromosome 9. Fructose, arabinose, and galactose sugars are incorporated into CWP. One QTL that controls seed protein content has been identified and mapped on chromosome 4 (Fig. 2.2; Table 2.2. and Table 8.3) (Stombaugh et al. 2004).

Kabelka et al. (2004) grew the RIL population derived from a cross between "BSR 101" and "LG82-8379" lines (n = 167) in 12 environments in Illinois, Indiana, and Iowa, USA, genotyped it with 145 SSR markers, and used QTL Cartographer for QTL data analysis. Eleven QTL for seed protein concentration have been identified and mapped on chromosomes 2, 3, 4, 6, 7, 8, 9, 10, 12, 13, and 14 explaining up to 53% of the phenotypic variation in protein concentration. Three QTL for seed oil content have been identified and mapped on chromosomes 2, 4, and 16 explaining up to 10–16% of the phenotypic variation in oil content. The study also identified 15 QTL for seed yield, and 12 QTL for plant height, 4 QTL for days to maturity, and 1 QTL for lodging (Fig. 2.2; Table 2.2) (Kabelka et al. 2004).

Hyten et al. (2004b) grew a RIL population derived from a cross between "Essex" and "Williams" lines (n = 131) in three environments in Jackson, TN, genotyped it with 568 SSR markers, and used CIM for QTL data analysis. For QTL that control seed protein content have been identified and mapped on chromosomes 6, 7, 9, and 13, and six QTL that control seed oil content have been identified and mapped on chromosomes 1, 6, 7, 17, and 19 (Fig. 2.2; Table 2.2) (Hyten et al. 2004b). In another study, Hyten et al. (2004a) grew a RIL population derived from a cross between "Essex" and "Williams" lines (n = 131) in two environments in Jackson, TN, genotyped them with 100 SSR markers, and used single-marker analysis (ANOVA) and CIM of QTL Cartographer for QTL data analysis. Three QTL that control seed palmitic acid content have been identified and mapped on chromosomes 9, 17, and 19; two QTL that control seed stearic acid content have been identified and mapped on chromosomes 6 and 19; two QTL that control seed oleic acid content have been identified and mapped on chromosomes 2 and 19; two QTL that control seed linoleic acid content have been identified and mapped on chromosomes 13 and 19; and three QTL that control seed linolenic acid content have been identified and mapped on chromosomes 13 and 19 (Fig. 2.2; Table 2.2) (Hyten et al. 2004a).

Zhang et al. (2004) grew a RIL population derived from a cross between "Kefeng No.1" and "Nannong 1138-2" lines (n = 184) in several environments in China, genotyped it with 452 markers, and constructed a genetic linkage map covering 3595.9 cM. For QTL data analysis, they used the method of CIM of WinQTL Cartographer. One QTL that controls seed protein content has been identified and mapped on chromosome 14, and one QTL that controls seed oil content has been identified and mapped on chromosomes 2 and 19 (Fig. 2.2; Table 2.2) (Zhang et al. 2004).

Fasoula et al., (2004) grew two RIL populations in two to three environments in Athens, GA. The first RIL population was derived from a cross between "PI97100" and "Coker 237" lines (n = 176) and the second RIL population was derived from "Young" and "PI416937" lines (n = 176). Both populations were genotyped with SSR and RFLP markers, and authors used single-marker analysis (ANOVA) and CIM for QTL data analysis. One QTL that controls seed protein content has been identified and mapped on chromosome 15, and three QTL that control seed oil content have been identified and mapped on chromosomes 4, 12, and 19 (Fig. 2.2; Table 2.2) (Fasoula et al. 2004).

Spencer et al. (2004) grew a RIL population derived from a cross between "N97-3708-13" (<5.4% linolenic acid content) and "Anand" (9.7% linolenic acid content) lines (n = 90), in the greenhouse, genotyped it with 780 SSR markers, and used CIM for QTL data analysis. Two QTL for seed linolenic acid content have been identified and mapped on chromosome 14 (Fig. 2.2; Table 2.2) (Spencer et al. 2004).

Panthee et al. (2005) grew a RIL population derived from a cross between "N87-984-16" and "TN93-99" (n = 101) in X, genotyped it with 94 SSR markers, and used single-marker analysis (ANOVA) and composite interval mapping (CIM) for QTL data analysis. One QTL that control seed protein content has been identified and mapped on chromosome 18, and four QTL that control seed oil content have been identified and mapped on chromosomes 2, 10, and 12 (Fig. 2.2; Table 2.2) (Panthee et al. 2005). In subsequent study, using the same RIL population ("N87-984-16" by "TN93-99," n = 101), two QTL that control seed palmitic acid content have been identified and mapped on chromosomes 2 and 8; two QTL that control seed stearic acid content have been identified and mapped on chromosomes 14 and 16; one QTL that controls seed oleic acid content has been identified and mapped on chromosome 15; one QTL that controls seed linoleic acid content has been identified and mapped on chromosome 15; and two QTL that control seed linolenic acid content have been identified and mapped on chromosomes 15 and 18 (Fig. 2.2; Table 2.2) (Panthee et al. 2006).

Nichols et al. (2006) used two backcross populations to fine map the QTL that controls both seed protein (*cqPRO-003*) and oil (*cqOIL-004*) contents identified and mapped on chromosome 20 (Diers et al. 1992; Sebolt et al. 2000; Chung et al. 2003). They introduced the high-protein allele from *Glycine soja* into the line A81-356022. The study narrowed the QTL region from 10 cM (Sebolt et al. 2000) to a 3 cM region (Nichols et al. 2006).

Reinprecht et al. 2006 grew the RIL population derived from a cross between "RG10" and "OX948" lines (n = 169) in three environments in Canada, genotyped it with 270 SSR and 18 STS markers (the actual genetic linkage map contained only 118 markers and covered 1247.5 cM only), and used single-marker analysis (one-way ANOVA) and composite interval mapping (CIM) for QTL data analysis. Fourteen QTL for seed protein content have been identified and mapped on chromosomes 6, 8, 11, 13, 14, 17, 18, and 20; 12 QTL for seed oil content have been identified and mapped on chromosomes 6, 15, 18, 19, and 20; 7 QTL that control seed palmitic acid have been identified and mapped on chromosomes 2, 6, 10, 14,

17, and 18; 4 QTL that control seed stearic acid have been identified and mapped on chromosomes 7, 13, and 18; 4 QTL that control seed oleic acid have been identified and mapped on chromosomes 6 and 18; 5 QTL that control seed linoleic acid have been identified and mapped on chromosomes 9, 14, and 18; and 3 QTL that control seed linolenic acid have been identified and mapped on chromosomes 9, 14, and 15 (Fig. 2.2; Table 2.2) (Reinprecht et al. (2006).

Gai et al. (2007) grew the RIL population derived from a cross between "Kefeng No. 1" and "1138-2" lines (*n* = 184) in China, genotyped it with 307 markers (2 morphological, 60 SSR, 79 AFLP, and 166 RFLP markers), used CIM of QTL Cartographer, and identified 3 QTL for protein content (*pt1*, *pt2*, and *pt3*) and 4 QTL for oil content (*ol1*, *ol2*, *ol3*, and *ol4*) (Fig. 2.2; Table 2.2) (Gai et al. 2007). Interestingly, the three QTL for oil (*ol2*, *ol3*, and *ol4*) on chromosome 7 overlap with each other and overlap with a QTL for stearic acid (Reinprecht et al. 2007) and a QTL for lipids (LIP5, Tajuddin et al. 2003) which proves QTL and gene clustering on the soybean genome (Mansur et al. 1996; Meksem et al. 1999; Reinprecht et al. 2006).

Chen et al. (2007) grew the RIL population derived from a cross between "Charleston" and "Dongnong 594" (high-protein) lines (*n* = 154) in China, genotyped it with 164 SSR markers, and used CIM of QTL Cartographer for QTL data analysis. Five QTL that control seed protein content [*Qspro 1*, *Qspro 2*, *Qspro 3*, *Qspro 4*, and *Qspro 5*] have been identified and mapped on chromosomes 2, 3, 10, and 15; four QTL that control seed oil content [*Qsoil 1*, *Qsoil 2*, *Qsoil 3*, and *Qsoil 4*] have been identified and mapped on chromosomes 1, 3, and 14; and five QTL that control combined seed protein and oil contents have been identified and mapped on chromosomes 6, 8, and 15 (Fig. 2.2; Table 2.2) (Chen et al. 2007).

Shibata et al. (2008) grew the RIL population derived from a cross between "TK 780" and "Hidaka 4" (wild soybean) lines (*n* = 96) in Japan, genotyped it with 282 markers (Liu et al. 2007), and used IM of QTL Cartographer for QTL data analysis. Three QTL for seed oil content have been identified and mapped on chromosomes 12, 15, and 20, and four QTL for linolenic acid concentration have been identified and mapped on chromosomes 6, 10, 12, and 15 (Fig. 2.2; Table 2.2) (Shibata et al. 2008).

Palomeque et al. (2009b) grew the RIL population derived from a cross between "OAC Millennium" and "Heinong 38" lines (*n* = 98) in China and Canada, genotyped it with 105 SSR markers (Palomeque et al. 2009a), used ANOVA and single-marker analysis, and identified a QTL that controls both protein and oil contents on chromosome 6 around the SSR marker Satt100 in the Canadian environment; however, the same QTL controls only oil content in the Chinese environment. A second QTL was identified on the same chromosome 6 around the marker Satt277, and it controls both protein and oil contents. A third QTL for seed oil content was identified and mapped on chromosome 20 around the SSR marker Satt162 (Fig. 2.2; Table 2.2) (Palomeque et al. 2009b).

Bachlava et al. (2009) grew two RIL populations in three environments (Clayton, Kinston, and Clinton) in North Carolina over 2 years (2006–2008). The first RIL population (FAF population) was derived from a cross between "N97-3363-3" and

"PI 423893" lines (n = 118), and the second RIL population (FAS population) was derived from a cross between '' and "PI 423893" lines (n = 231). FAF RIL population was genotyped with 164 SSR markers, and FAS RIL population was genotyped with 88 SSR markers. Multiple-interval mapping (MIM) method of QTL Cartographer was used for QTL data analysis in FAF population, while FAS population was used to confirm identified QTL. Twelve QTL that control seed oleic acid content have been identified and mapped on chromosomes 3, 5, 7, 8, 10, 13, 15, 17, 18, and 20; QTL that control seed linoleic acid content have been identified and mapped on chromosomes 3, 5, 8, 10, 13, 17, 18, and 20; and QTL that control seed linolenic acid content have been identified and mapped on chromosomes 2, 5, 8, 14, 16, and 19 (Fig. 2.2; Table 2.2) (Bachlava et al. 2009).

Li et al. (2010) grew a RIL population derived from a cross between "Hefeng 25" and "Maple Arrow" lines (n = 149), genotyped it with 109 SSR markers, and used CIM of QTL Cartographer for QTL data analysis. One QTL that controls seed protein content has been identified and mapped on chromosome 10, and one QTL that controls seed oil content has been identified and mapped on chromosome 2. The study also identified and mapped two QTL for plant height on chromosomes 2 and 13; two QTL for node numbers on chromosomes 6 and 20; two QTL for pod numbers on chromosomes 17 and 20; two QTL for seed numbers on chromosomes 2 and 4; four QTL for seed weight on chromosomes 2, 4, 7, and 17; one QTL for maturity on chromosome 2, and one QTL for flowering time on chromosome 2 (Fig. 2.2; Table 2.2) (Li et al. 2010).

Liang et al. (2010) used a RIL population derived from a cross between "Jindou 23" and "Huibuzhi" lines (n = 474) and genotyped it with SSR markers. Five QTL that control seed protein content have been identified and mapped on chromosomes 6, 12, 14, and 18, and six QTL that control seed oil content have been identified and mapped on chromosomes 5, 6, 8, 14, and 17 (Fig. 2.2; Table 2.2) (Liang et al. (2010).

Kim et al. 2010 grew the RIL population derived from a cross between "Keunolkong" and "Iksan10" lines (n = 115) in Korea, genotyped it with SSR and morphological markers, and used single-marker analysis (ANOVA) and MapManager QT to analyze QTL data. Three QTL for seed oil content have been identified and mapped on chromosomes 2, 6, and 19. For fatty acids, the study identified 15 QTL. One QTL that controls seed palmitic acid content has been identified and mapped on chromosome 17, and one QTL that controls seed stearic acid content has been identified and mapped on chromosome 19. Five QTL that control seed oleic content have been identified and mapped on chromosomes 7, 11, 14, 16, and 19; five QTL that control seed linoleic acid content have been identified and mapped on chromosomes 2, 11, 14, 16, and 19; and three QTL that control seed linolenic acid content have been identified and mapped on chromosomes 8, 10, and 19 (Fig. 2.2; Table 2.2) (Kim et al. (2010).

Bolon et al. 2010 used a near-isogenic line (NIL) to fine map the region on chromosome 20 that contains the major QTL controlling both seed protein and oil contents (*cqPRO-003* and *cqOIL-004*) (Diers et al. 1992; Sebolt et al. 2000; Chung et al. 2003; Nichols et al. 2006). They narrowed the QTL region into a region of only 8.4 Mbp [from 24.54 Mb (Sat_174) to 32.92 Mb (ssrpqtl_38)]. Moreover, using transcriptome analysis, they identified 12 genes that are differentially expressed in the NILs in this same region (Bolon et al. (2010).

Li et al. (2011) grew the RIL population derived from a cross between "KF1" and "NN1138-2" lines (*n* = 184) in China, genotyped it with 834 markers, and used QTLNetwork 2.0 for QTL data analysis. Three QTL for seed oil content have been identified and mapped on chromosomes 4, 10, and 15; five QTL that control seed palmitic acid content have been identified and mapped on chromosomes 8, 9, 11, 15, 16, and 18; six QTL that control seed stearic acid content have been identified and mapped on chromosomes 2, 6, 9, 14, and 18; three QTL that control seed oleic content have been identified and mapped on chromosomes 11, 18, and 20; three QTL that control seed linoleic acid content have been identified and mapped on chromosomes 3, 10, and 18; and seven QTL that control seed linolenic acid content have been identified and mapped on chromosomes 2, 4, 9, 10, 11, and 16 (Fig. 2.2; Table 2.2) (Li et al. 2011).

Moongkanna et al. 2011 grew a RIL population derived from a cross between "Pak Chong 2" and "Laos 7122" lines (*n* = 186), genotyped it with 138 SSR markers, and used single-marker analysis (ANOVA) and CIM for QTL data analysis. Two QTL that control seed oil content have been identified on chromosome 5 at the same location. Other QTL for other agronomic traits are as follows: three QTL for seed width, two QTL for seed length, two QTL for 100-SW, and one QTL for node number (Fig. 2.2; Table 2.2) (Moongkanna et al. (2011).

Qi et al. (2011) performed a meta-analysis on 130 QTL that control seed oil contents, published from 1992 to 2011 and reported in SoyBase (Grant et al. 2010; Brown et al. 2020; http://www.soybase.org) using the software BioMercator 2.1 (Arcade et al. 2004). One hundred and eight (108) QTL have been projected to the consensus map but not 28 QTL. The results showed 25 consensus QTL with map distance range of 1.49–12.35 cM and R^2 value range of 5.5–39.15 (Table 2.2). They also identified 1876 genes within these QTL regions. Gene ontology of the 1876 genes using the software InterProScan identified 6 genes involved in oil and fatty acids biosynthetic pathways. These genes encode for microsomal omega-6 fatty acid desaturase, beta-ketoacyl-acyl carrier protein synthase III, stearoyl-ACP desaturase, beta-ketoacyl-ACP synthetase I, and two long-chain fatty acid-CoA ligases (Table 2.3) (Qi et al. 2011).

Pandurangan et al. (2012) used two RIL populations to identify QTL for protein and oil contents. The first RIL population adapted to Illinois, USA, was derived from a cross between "Williams 82" and "LG00-13365" lines (*n* = 73) and was grown in eight environments in Illinois and identified four QTL for protein content on chromosomes 1, 15, 18, and 20 which explained 68% of the variation of protein content in this RIL population. The second RIL population adapted to Ontario, Canada, was derived from a cross between "X3145" (high-protein) and "AC Brant" lines (*n* = 96); was grown in Ontario, Canada; was genotyped with 130 SSR markers; and identified four QTL for protein content on chromosomes 1 (Satt077), 6 (Satt520), 15 (Satt575), and 20 (Satt496) which explained 68% of the variation of protein content in this RIL population (Fig. 2.2; Table 2.2) (Pandurangan et al. 2012). The study also investigated correlations between the amino acid asparagine and protein and oil contents and other studied compounds. In Williams 82 X LG00-13365, protein and oil contents showed a strong negative correlation, and protein and asparagine contents showed a strong positive correlation; however, oil and asparagine contents showed a negative correlation (Pandurangan et al. 2012).

Xie et al. (2012) grew a RIL population derived from a cross between "Hefeng 25" and "Dongnong L-5" lines (*n* = 125) in 6 environments in China, genotyped it with 600 SSR and 8 FAD3 SNP markers, and used single-factor analysis for QTL data analysis. The genetic linkage map was constructed with 112 SSR and 3 SNP markers and covered 2718 cM. Four QTL that control seed palmitic acid content have been identified and mapped on chromosomes 9, 16, 17, and 18; one QTL that controls seed stearic acid content has been identified and mapped on chromosome 2; four QTL that control seed oleic acid content have been identified and mapped on chromosomes 9, 16, 17, and 18; four QTL that control seed linolenic acid content have been identified and mapped on chromosomes 14, 16, 17, and 18; and six QTL that control seed linolenic acid content have been identified and mapped on chromosomes 2, 14, and 18 (Fig. 2.2; Table 2.2) (Xie et al. 2012).

Wang et al. (2012) grew the RIL population derived from a cross between "SD02-4-59" and "A02-381100" lines (*n* = 87) in Aurora, SD, genotyped it with 294 SNP and 238 SSR markers, and used QTL Cartographer for QTL data analysis. Nine QTL that control total seed saturated fatty acid (palmitic acid + stearic acid) contents have been identified and mapped on chromosomes 3, 5, 6, 9, 13, 15, and 18; eight QTL that control seed palmitic acid content have been identified and mapped on chromosomes 5, 6, 9, 13, 15, and 18; and five QTL that control seed stearic acid content have been identified and mapped on chromosomes 3, 5, 13, and 18; (Fig. 2.2; Table 2.2) (Wang et al. 2012).

Eskandari et al. (2013) grew a RIL population derived from a cross between "AOC Wallace" and "AOC Glencoe" lines (*n* = 203), genotyped it with 90 SSR markers and 3 gene-specific markers (GmDGAT1B, GmDGAT2B, and GmDGAT2C), and used single-marker analysis (ANOVA) and multiple-interval mapping (MIM) methods for QTL data analysis. Five QTL that control seed protein content have been identified and mapped on chromosomes 7, 9, 12, and 13, and 11 QTL that control seed oil content have been identified and mapped on chromosomes 1, 2, 7, 9, 12, 13, 14, 16, 17, and 19. Nineteen QTL that control other important agronomic traits have been identified by the study such as 4 QTL for seed yield, 2 QTL for 100-SW, and 1 QTL for each of plant height and days to maturity (Fig. 2.2; Table 2.2) (Eskandari et al. (2013)
).

Lu et al. (2013) grew a RIL population derived from a cross between "ZDD09454" and "Yudou 12" lines (n = 212) in 6 environments in China, genotyped it with 301 markers, and used CIM for QTL data analysis. Eleven QTL that control seed protein content have been identified and mapped on chromosomes 5, 6, 7, 8, 9, 11, 12, 18, 19, and 20 (Fig. 2.2; Table 2.2) (Lu et al. 2013).

Rossi et al. (2013) used two RIL populations. The first RIL population was derived from a cross between "OAC Millennium" and "Heinong 38" lines (*n* = 92) (Palomeque et al. 2009a), and the second RIL population was derived from a cross between "Pioneer 9071" and "#8902" lines (n = 131). Both populations have been grown in different environments in the USA (MN), Canada, and China and used single-marker analysis (ANOVA) and IM of MapQTL5 (Van Ooijen 2004) for QTL data analysis. Four QTL that control seed protein content have been identified and mapped on chromosomes 6, 9, and 20, and four QTL that control seed oil content have been identified and mapped on chromosomes 2, 5, 6, and 13 (Rossi et al. 2013).

Mao et al. (2013) identified QTL for protein and oil contents in three RIL populations. The first RIL population was derived from a cross between "Hefeng 47" and "Heinong 37" lines (n = 148); the second RIL population was derived from a cross between "Hefeng 47" and "Heinong 44" lines (n = 109); and the third RIL population was derived from a cross between "Dongnong 46" and "KenJian 23" lines (n = 153). All the three RIL populations and their parents have been grown in eight environments in China. Nine and four universal QTL (QTL_U) that control protein and oil contents, respectively, have been identified in one or two RIL populations. Likewise, 40 specific QTL (QTL_{SP}) that control seed protein content have been identified and mapped on chromosomes 1, 2, 3, 4, 5, 6, 7, 9, 10, 11, 13, 14, 17, 18, 19, and 20; and 35 QTL_{SP} that control oil content have been identified and mapped on chromosomes 1, 2, 3, 6, 7, 8, 9, 10, 14, 15, 16, 17, 18, 19, and 20 (Fig. 2.2; Table 2.2) (Mao et al. 2013).

Yesudas et al. 2013 grew the RIL population derived from a cross between "Essex" and "Forrest" lines (n = 94) in Carbondale, IL, genotyped it with 413 markers, and used ANOVA and QTL Cartographer for QTL data analysis. Five QTL that control seed protein content have been identified and mapped on chromosome 2, 6, 8, 12, 15, and 20, and two QTL that control seed oil content have been identified and mapped on chromosomes 8 and 20. The study showed also a strong negative correlation between protein and oil contents (Fig. 2.2; Table 2.2) (Yesudas et al. (2013).

Pathan et al. (2013) grew two RIL populations in two environments in Missouri. The first RIL population was derived from a cross between "Magellan" and "PI 438489B" lines (MPB, n = 216), and the second RIL population was derived from a cross between "Magellan" and 'PI 567516C' lines (MPC, n = 156). The MPB- and MPC-based genetic linkage maps used in this study have been described earlier (Vuong et al. 2010, 2011). Thirteen QTL that control seed protein content have been identified and mapped on chromosomes 3, 5, 6, 7, 8, and 15. All these QTL are confirmed in other studies. Twelve QTL that control seed oil content have been identified and mapped on chromosomes 2, 3, 5, 6, and 15 (Fig. 2.2; Table 2.2) (Pathan et al. 2013).

Wang et al. (2014a) used two RIL populations to identify QTL for oil and protein contents as well as seed yield. The first RIL population was derived from a cross between SD02-4-59 and A02-381100 lines (n = 87); was grown in Aurora, SD; and genotyped with 294 SNP and 238 SSR markers (Wang et al. 2012). The validation RIL population was derived from a cross between SD02-911 and SD00-1501 lines (n = 196). Both populations have been grown in three environments (2009, 2010, and 2011) in Aurora, SD, USA. The genetic linkage map used was composed of 399 SSR markers, 311 SNP markers, and 3 *GmFAD3* genes and covered approximately

2099.9 cM (Wang et al. 2014a). The method of CIM and/or IM of QTL cartographer was used for QTL data analysis. Eleven QTL that control seed protein content have been identified and mapped on chromosomes 2, 3, 4, 9, 11, 17, and 20 as well as 16 QTL that control seed oil content on chromosomes 1, 2, 3, 4, 5, 9, 10, 11, 15, 16, 17, 18, and 20 (Fig. 2.2; Table 2.2) (Wang et al. 2014a). In another study, Wang et al. (2014b) grew the same RIL population (SD02-4-59 by A02-381100, $n = 87$) in two environments in South Dakota; genotyped it with 516 SNP markers, 477 SSR markers, and 3 genes; and used IM and CIM methods for QT: data analysis. Six QTL that control seed oleic acid content have been identified and mapped on chromosomes 10, 15, 16, 18, and 20; four QTL that control seed linoleic acid have been identified and mapped on chromosomes 5, 10, and 18; and five QTL that control seed linolenic acid have been identified and mapped on chromosomes 1, 2, 6, 13, and 18 (Fig. 2.2; Table 2.2) (Wang et al. 2014b).

Ha et al. (2014) grew the RIL population derived from a cross between "PI483463" and "Hutcheson" lines ($n = 188$) in 4 environments (2008, 2009, 2010, and 2012) in Portageville, MO, USA; genotyped it with 546 SNP and 16 SSR markers, used CIM of QTL Cartographer for QTL analysis, and identified 12 QTL that control alpha-linolenic (ALA) acid content on chromosomes 3, 5, 6, 11, 12, 14, 15, and 17 (Fig. 2.2; Table 2.2) (Ha et al. 2014).

In another study, we grew the RIL population derived from "MD 96-5722" and "Spencer" ($n = 54$) in a field in Fayetteville, NC, in 2012 and genotyped it with 550 SNPs (Akond et al. 2013c). The method of CIM of QTL Cartographer was used for QTL data analysis. One QTL that controls seed protein content has been identified and mapped on chromosome 14; 11 QTL that control seed oil content have been identified and mapped on chromosomes 3, 5, 9, 13, 14, and 16; 2 QTL that control seed palmitic acid content have been identified and mapped on chromosomes 3 and 18; 1 QTL that controls seed stearic acid content has been identified and mapped on chromosome 16; 3 QTL that control seed oleic acid content have been identified and mapped on chromosomes 13 and 16; 3 QTL that control seed linoleic acid content have been identified and mapped on chromosomes 3 and 16; and 6 QTL that control seed linolenic acid content have been identified and mapped on chromosomes 13, 14, 15, and 16 (Fig. 2.2; Table 2.2) (Akond et al. 2014).

Han et al. (2015) grew a RIL population derived from a cross between "Zhongdou 27" (42.84% protein, 18.73% oil, and 3791 $\mu g.g^{-1}$ total isoflavones) and "Jiunong 20" (34.05% protein, 21.42% oil, and 2061 $\mu g.g^{-1}$ total isoflavones contents) lines ($n = 130$) in three environments (Harbin, Hulan, and Suihua 2005–2007) in China in 2005–2007 and genotyped it with 606 SSR markers; however, the genetic linkage map used was based on 194 polymorphic SSR markers and covered 2020 cM. They performed conditional and unconditional analyses for QTL data using the MINQUE method (Zhu et al. 1992) to identify QTL for seed protein, oil, daidzein, genistein, glycitein, and total isoflavone contents. Five QTL for seed daidzein (DZ) content have been identified and mapped on chromosomes 3, 6, 9, 10, 13, 17, 18, and 20; three QTL for seed genistein (GC) contents have been identified and mapped on chromosomes 7, 9, 10, 15, and 20; six QTL for seed glycitein (GT) contents have been identified and mapped on chromosomes 3, 7, 8, 10, 13, 17, and 19; and seven

QTL for seed total isoflavone (TI) contents have been identified and mapped on chromosomes 2, 4, 6, 7, 8, 10, 12, 13, 15, and 19. Thirty nine (39) QTL that control seed oil content have been identified and mapped on chromosomes 1, 5, 7, 9, 14, 15, 17, 18, and 20. However, when the QTL for isoflavones contents are conditioned with QTL for protein and oil contents, they identified 37 QTL: 9 QTL controlling daidzein content, 6 QTL controlling glycitein content, 9 QTL controlling genistein content, and 13 QTL controlling total isoflavones content. Seven among these QTL control also see protein and oil contents (Han et al. 2015). These QTL were identified and mapped on all the 20 chromosomes of soybean along with QTL for protein and oil contents, and clusters of these QTL have been observed on most chromosomes (Han et al. 2015).

Fan et al. (2015) grew a RIL population derived from a cross between "Luheidou2" and "Nanhuizao" lines (n = 100) in three environments in China over 3 years (2009–2011), genotyped it with 161 SSR markers, and used inclusive composite interval mapping (ICIM) of IciMapping software v3.3 for QTL data analysis. Two QTL that control seed palmitic acid content have been identified and mapped on chromosomes 18 and 20; six QTL that control seed stearic acid content have been identified and mapped on chromosomes 3, 6, 7, 8, 13, and 20; three QTL that control seed oleic acid content have been identified and mapped on chromosomes 3 and 10; five QTL that control seed linoleic acid content have been identified and mapped on chromosomes 3, 9, 10, and 12; and seven QTL that control seed linolenic acid content have been identified and mapped on chromosomes 3, 6, 7, and 8 (Fig. 2.2; Table 2.2) (Fan et al. 2015).

Leite et al. (2016) grew a RIL population derived from a cross between "Lineage 69" and "Tucunare" lines (n = 244) in three environments in Brazil and genotyped it with 358 markers; however, the genetic map was constructed with 56 markers only. They used CIM of QTL Cartographer for QTL data analysis. Two QTL that underlie seed oil content have been identified and mapped on chromosomes 9 and 12 (Fig. 2.2; Table 2.2) (Leite et al. 2016).

Teng et al. (2017) grew a RIL population derived from a cross between "Dongnong 46" and "L-100" lines (n = 129) in 10 environments in China, genotyped it with 213 markers, and used single-marker analysis (ANOVA) for QTL data analysis. Eight QTL that control seed protein content have been identified and mapped on chromosomes 7, 9, 12, 14, 17, 18, and 20 (Fig. 2.2; Table 2.2) (Teng et al. 2017).

In another study, we grew the RIL population derived from "Hamilton" and "Spencer" (n = 93) in a field in Fayetteville, NC, in 2012, genotyped it with 1510 SNP markers, and used the method of CIM of QTL Cartographer for QTL data analysis. One QTL that controls seed protein content (*qPRO001*) has been identified and mapped on chromosome 18; one QTL that controls seed oil content (*qOIL001*) has been identified and mapped on chromosome 6; two QTL that control seed palmitic acid content (*qPAL001* and *qPAL002*) have been identified and mapped on chromosome 6; three QTL that control seed oleic acid (*qOLE001*) content have been identified and mapped on chromosome 3, 6, and 18; and one QTL that controls seed linolenic acid content (*qLIN001*) has been identified and mapped on chromosome 6. In addition, several QTL that control several amino acids' contents have also been identified. QTL that control seed glutamic acid (Glu), alanine (Ala), cysteine (Cys), valine (Val), and histidine (His) have been identified and mapped on chromosomes 6 [*qGLU001*, *qALA001*, *qCYS001*, and *qHIS001*] and chromosome 18 [*qVAL001*] (Fig. 2.2; Table 2.2) (Akond et al. 2018).

Hacisalihoglu et al. (2018) grew a RIL population derived from a cross between "Fiskeby III" and "Mandarin" (Ottawa) lines (*n* = 244) in Minnesota, USA, (*n* = 244) and in a greenhouse in Turkey (*n* = 103), genotyped it with 501 SNP markers, and used single-marker analysis (ANOVA) and CIM for QTL data analysis. The subset RIL population (*n* = 103) grown in the greenhouse in Turkey was grown in low-phosphorus (low-P) and high-phosphorus (high-P) soils to identify and map QTL that control seed weight and seed composition under these phosphorus levels. One QTL that controls seed protein content has been identified and mapped on chromosome 20 (Fig. 2.2; Table 2.2). Two QTL for seed weight (SW) under low-P have been identified and mapped on chromosomes 4 and 6; one QTL for SW under optimal-P has been identified and mapped on chromosome 4; one QTL for pod weight (PW) under low-P has been identified and mapped on chromosome 4; one QTL for seed volume (SV) under low-P has been identified and mapped on chromosome 6; and one QTL for seed protein content under low-P has been identified and mapped on chromosome 20 (Fig. 2.2; Table 2.2) (Hacisalihoglu et al. 2018).

Patil et al. (2018) grew an interspecific RIL population derived from a cross between "Williams 82" (*Glycine max*) and "PI 483460B" (*Glycine soja*) (*n* = 188) in Portageville, MO, in 2012–2014, genotyped it with 6000+ SNP markers, used a genetic linkage map of 2925 cM, and used CIM of WinQTL Cartographer for QTL data analysis of seed protein, oil, and sucrose contents. Twelve QTL that control seed protein content have been identified and mapped on chromosomes 6, 8, 13, 19, and 20; 16 QTL that control seed oil content have been identified and mapped on chromosomes 2, 8, 9, 15, 17, and 20; and 8 QTL that control seed sucrose content have been identified and mapped on chromosomes 6, 8, and 20 (Fig. 2.2; Table 2.2) (Patil et al. 2018).

Li et al. 2019a grew two RIL populations in four environments in China (E1–E4). The first RIL population was derived from a cross between "Zheng92116" (high-protein and low oil content line) and "Qihuang30 (ZQ)" (low-protein and high oil content line) (*n* = 289), and the second RIL population was derived from a cross between "Zheng92116" and "Liaodou14 (ZL)" (*n* = 345). ZL RIL population and a natural population (*n* = 320) that consists of different landraces and varieties have been used to verify QTL identified in ZQ RIL population. The RILs were genotyped with 6000 SNPs (SoySNP 6 K Illumina Infinium Bead Chip), but ZL population was genotyped with an additional 1015 SSR markers (Li et al. 2018b). The study identified a total of ten QTL in ZQ RIL population in various environments among which four QTL have been also identified (verified) in the ZL RIL population. Eight QTL that control seed protein content have been identified and mapped on chromosomes 3, 4, 8, 9, 10, 11, 13, and 16. For QTL that control seed oil content have been identified and mapped on chromosomes 3, 4, 5, and 16. Eight QTL that control both seed protein and oil (P + O) contents have been identified and mapped on chromosomes 5, 9, 10, 13, 16, and 20. Among these QTL are two major QTL, one QTL controlling oil content (*qOIL-A1*) identified in all four environments (E1–E4) and explained up to 22.1% of the phenotypic variation in seed oil content and the other QTL controlling both seed protein and oil contents (*qP + O-A1*) also identified in all four environments (E1–E4) and explained up to 16.4% of the phenotypic variation in seed protein plus oil contents. Both QTL have been mapped on the same location of chromosome 5. Another major QTL that control seed protein content mapped to chromosome 16 explained up to 23.2% of the phenotypic variation in seed protein content (Fig. 2.2; Table 2.2) (Li et al. (2019a).

On the other hand, we will summarize here the most important genome-wide association studies (GWAS) that aimed at identifying QTL or markers associated with seed protein, oil, and fatty acids contents in soybean (Jun et al. 2008; Hwang et al. 2014; Bandillo et al. 2015; Sonah et al. 2015; Zhou et al. 2015; Li et al. 2015, 2018a,b, 2019b; Priolli et al. 2015; Leamy et al. 2017; Fang et al. 2017; Zhang et al. 2017, 2018; La et al. 2019).

Jun et al. (2008) grew 96 USDA accessions originally from Japan, Korea, and China. These accessions were divided into two groups: 48 accessions with high-protein content (HP, 50–57.4%) and 48 accessions with low-protein content (LP, 31.7–38.7%). They genotyped the accessions with 150 SSR markers, used GWAS mapping method, and identified 11 QTL for seed protein content on chromosomes 3, 5, 7, 9, 10, 15, 16, 18, 19, and 20. Nine of these QTL have been previously reported, and two were novel QTL on chromosome 7 (Satt551) and chromosome 16 (Satt431)) (Fig. 2.2; Table 2.2) (Jun et al. 2008).

Interestingly, QTL (*seed protein 41-4*) on chromosome 20 overlaps with QTL for protein and oil contents (Reinprecht et al. 2006), QTL for lipids and protein contents (*Seed protein 30-1/Seed oil 34-3*, Tajuddin et al. 2003), and QTL for protein content and seed yield (Fig. 2.2; Table 2.2) (Csanadi et al. 2001) which again proves QTL and gene clustering on the soybean genome. Hwang et al. 2014 used 298 soybean accessions, genotyped them with 31,954 SNP markers, and performed a GWAS to investigate protein and oil contents. The study and identified 40 SNP markers highly associated with protein content and 25 SNP markers highly associated with oil content. The GWAS identified most QTL for protein and oil contents previously reported, narrowed the genomic regions of these reported QTL, and identified several candidate genes within these QTL regions (Hwang et al. 2014). Moreover, the major genomic region of 8.4 Mbp reported on chromosome 20 containing and about 12 candidate genes controlling both seed protein and oil contents (Bolon et al. 2010) is narrowed further to a region of 2.4 Mbp containing 6 candidate genes of the 12 reported earlier. These candidate genes are *Glyma.20 g19620*, *Glyma.20 g19630* [putative ammonium transporter AMT1], *Glyma.20 g19680* [Hsp22.5], *Glyma.20 g21030*, *Glyma.20 g21040*, and *Glyma.20 g21080* [ATP synthase D chain] (Fig. 2.2; Table 2.2) (Hwang et al. (2014).

Qi et al. (2014) grew a RIL population derived from a cross between "Charleston" and "Dongnong 594" ($n = 147$) in three environments (Harbin (HRB), Hongxinglong (HXL), and Jiamusi (JMS)) in China. The genetic linkage map used in this study was constructed using 164 SSR markers and covered 1913.5 cM (Chen et al. 2005). They used CIM and MIM of WinQTL Cartographer for QTL data analysis. Forty three QTL that control seed protein content have been identified and mapped on 17 out of 20 chromosomes of the soybean genome (Fig. 2.2; Table 2.2) (Qi et al. 2014).

Bandillo et al. (2015) used 12,000 soybean accessions, genotyped them with 50,000 SNP markers, and used GWAS for QTL data analysis. Nineteen SNPs (QTL) associated with seed protein content have been identified and mapped on chromosomes 6, 13, 15, and 20, and 18 SNPs (QTL) associated with seed oil content have been identified and mapped on chromosome 5, 15, and 20. Within two narrow regions of chromosome 15 (3.82–3.96 Mb) and chromosome 20 (29.59–31.97 Mb), high numbers of associated SNP markers have been identified (Bandillo et al. 2015).

Sonah et al. (2015) grew a population of early maturity lines (n = 304) in three environments in Canada and used GWAS to identify QTL for seed protein and oil contents among other important agronomic traits (plant height, seed weigh, and maturity). A subpopulation (n = 139) has been genotyped with more than 17,172 SNP markers, and data analysis identified 8 QTL regions (21 SNPs) associated with seed protein contents on chromosomes 5, 8, 10, 14, 16, 19, and 20, and most of them overlap with previously reported QTL controlling seed protein content using biparental populations. Likewise, eight QTL regions associated with seed oil content have been identified on chromosomes 5, 8, 10, 14, 16, 19, and 20, and among them six have been mapped to the same locations of seed protein QTL (Fig. 2.2; Table 2.2) (Sonah et al. 2015).

Zhou et al. (2015) re-sequenced a population of wild and cultivated soybean accessions (n = 302), used GWAS for QTL data analysis, and identified a total of 230 selected regions among which 96 fall within QTL controlling seed oil content and 21 fall within QTL controlling fatty acid contents. Six SNPs associated with six QTL that control seed oil content have been identified and mapped on chromosomes 3, 11, 12, 13, 15, and 17 (Table 7.1). Twenty-one candidate genes involved in fatty acid biosynthetic pathways have been identified by this study, and 10 of them fall within previously identified QTL controlling oil content. The 19 QTL that control seed fatty acids contents and candidate genes have been identified and mapped on chromosomes 1, 2, 6, 7, 8, 10, 13, 14, 15, 17, and 20 (Tables 7.1 and 7.3). However, no QTL for oil content (*qOC10*) was identified using a biparental population (Fig. 2.2; Table 2.2) (Zhou et al. 2015).

Li et al. (2015) used a population of accessions (n = 421), genotyped them with 1536 SNP markers, and used GWAS to identify SNPs associated with the 5 major fatty acid (FAs = palmitic, stearic, oleic, linoleic, and linolenic acids) contents. The study identified 37 QTL (33 SNPs) associated with fatty acids contents, and a third of them mapped to previously reported QTL controlling seed fatty acids contents. Seven QTL controlling seed palmitic acid content have been identified and mapped on chromosomes 5, 7, 9, 12, and 15; 5 QTL controlling seed stearic acid content have been identified and mapped on chromosomes 14, 15, and 18; 5 QTL controlling seed oleic acid content have been identified and mapped on chromosomes 3, 4, 6, 8, and 18; 12 QTL controlling seed linolenic acid content have been identified and mapped on chromosomes 1, 2, 5, 8, 15, and 18; and 8 QTL controlling seed linoleic acid content have been identified and mapped on chromosomes 3, 6, 8, 10, and 15 (Table 2.2). Candidate genes within these QTL regions are shown in Table 2.3 (Li et al. 2015).

Priolli et al. 2015 grew a population of 94 accessions (*n* = 94) from Brazil, the USA, Japan, South Korea, China, and India in 210 and 2011 in a field in Sao Paulo, Brazil, genotyped them with 142 SSR markers, used GWAS, and identified 110 SSRs associated with seed oil and fatty acid contents. Fourteen QTL controlling seed oil content have been identified and mapped on chromosomes 1, 3, 4, 6, 7, 17, 18, 19, and 20; 24 QTL controlling seed palmitic acid content have been identified and mapped on chromosomes 1, 3, 4, 7, 8, 9, 10, 13, 14, 15, 16, 17, 18, 19, and 20; 33 QTL controlling seed oleic acid content have been identified and mapped on chromosomes 1, 2, 3, 4, 7, 9, 10, 11, 13, 14, 15, 16, 17, 18, 19, and 20; 26 QTL controlling seed linoleic acid content have been identified and mapped on chromosomes 1, 2, 3, 4, 9, 10, 11, 13, 14, 15, 16, 17, 18, 19, and 20; and 13 QTL controlling seed linolenic acid content have been identified and mapped on chromosomes 4, 10, 11, 13, 14, 16, 17, and 20 (Table 2.2) (Priolli et al. (2015).

Leamy et al. (2017) used a population of wild soybean accessions (*Glycine soja*) (*n* = 570) and studied seed protein, oil, fatty acid, and amino acid contents using GWAS. The population was genotyped with more than 29,969 SNP markers, and the study identified 29 SNPs associated with protein, oil, or fatty acid contents. Two QTL controlling seed protein content have been identified and mapped on chromosomes 14 and 18; three QTL controlling seed oil content have been identified and mapped on chromosomes 3 and 20; four QTL controlling seed palmitic acid content have been identified and mapped on chromosomes 6, 7, and 14; three QTL controlling seed stearic acid content have been identified and mapped on chromosomes 14 and 16; five QTL controlling seed oleic acid content have been identified and mapped on chromosomes 7, 14, 16, and 19; nine QTL controlling seed linoleic acid content have been identified and mapped on chromosomes 2, 7, and 9; and four QTL controlling seed linolenic acid content have been identified and mapped on chromosomes 2 and 14 (Table 2.2) (Leamy et al. 2017). The study also identified several candidate genes within these QTL regions among which are genes encoding for a phospholipase D alpha 1 and plant stearoyl-acyl-carrier-protein desaturase (Leamy et al. 2017).

Soybean agglutinin (SBA) has a negative effect on human and animal nutrition, and developing soybean cultivars with reduced seed SBA content is a highly desirable trait by scientists, breeders, and farmers alike (Liener 1996; Lalles and Tukur 1996). To genetically map QTL for SBA, Yang et al. (2017) grew three RIL populations in three environments in China. The first RIL population was derived from a cross between "Hefeng 45" and "Zhongdon 27" lines (*n* = 167), the second RIL population was derived from a cross between "Hefeng 45" and "Dongnong 48" lines (*n* = 189), and the third RIL population was derived from a cross between "Hefeng 45" and "Taipingchuan Blacksoy" lines (*n* = 201). In Hefeng 45 by Zhongdon 27 RIL population, six QTL that control SBA have been identified and mapped on chromosomes 1, 2, 4, 5, and 9. In Hefeng 45 by Dongnong 48 RIL population, 5 QTL that control SBA have been identified and mapped on chromosomes 1, 2, 4, 5, and 6. In "Hefeng 45" and "Taipingchuan Blacksoy" RIL population, six QTL that control SBA have been identified and mapped on chromosomes 2, 4, 5, and 12 (Fig. 2.2; Table 2.2) (Yang et al. 2017).

Fang et al. (2017) used a genome-wide association study (GWAS) on 809 soybean accessions collected worldwide, grown in 2 environments China in 2013–2014 and 2014–2015, and studied 84 traits among them seed lipids and fatty acids contents. One QTL that controls seed palmitic acid content has been identified and mapped on chromosome 5; four QTL that control seed stearic acid content have been identified and mapped on chromosomes 2, 3, and 14; two QTL that control seed linoleic acid content have been identified and mapped on chromosomes 4 and 14; three QTL that control seed palmitic acid ratio have been identified and mapped on chromosomes 5 and 13; four QTL that control seed stearic acid ratio have been identified and mapped on chromosomes 2, 3, and 14; one QTL that control seed linoleic acid ratio has been identified and mapped on chromosome 14; five QTL that control seed stearic acid to palmitic acid ratio (stearic/palmitic) have been identified and mapped on chromosomes 2, 3, 5, and 14; two QTL that control seed linolenic acid to linoleic acid ratio (linolenic/linoleic) have been identified and mapped on chromosomes 10 and 14; one QTL that control seed saturated fatty acids (SFA) have been identified and mapped on chromosome 5; and two QTL that control seed saturated fatty acids (SFA) to unsaturated fatty acids (USFA) ratio (SFA/USFA) have been identified and mapped on chromosome 5 (Table 7.1). In addition to the three main genes involved in fatty acids biosynthesis: FatB gene encoding fatty acyl-ACP thioesterase, SAD gene encoding plant stearoyl-acyl-carrier protein desaturase, and FAD3 gene encoding fatty acid desaturase; the study identified five additional genes involved in fatty acid biosynthetic pathways and six genes involved in lipid biosynthetic pathways. Among these genes are genes encoding for phospholipase C2, phospholipase D alpha 1, reduced oleate desaturation 1, 3-ketoacyl-CoA synthase 11, diacylglycerol kinase family protein, fatty acid desaturase 5, O-acyltransferase (WSD1-like) family protein, allene oxide cyclase 3, glycolipid transfer protein 2, and several others (Tables 7.2 and 7.3) (Fang et al. 2017).

Zhang et al. (2017) grew a population of accessions (n = 219) and a RIL population derived from a cross between the low-protein content "Nannong 94–156" and the high-protein content "Bogao" lines (n = 152) in six environments (E1–E6) in China and used GWAS method for data analysis to identify QTL for water-soluble protein content (WSPC) and seed protein content (PC). The two populations were genotyped with 201,994 SNP markers and the "Nannong 94–156" by "Bogao" genetic linkage map contained 6159 SNPs and covered 3020.59 cM (Zhang et al. 2017). Sixteen QTL that control PC have been identified and mapped on chromosomes 3, 4, 5, 6, 9, 10, 11, 12, 13, 14, 15, and 19, and 16 QTL that control WSPC been identified and mapped on chromosomes 1, 7, 8, 10, 11, 12, 13, 15, 16, 17, 18, 19, and 20 (Tables 7.2 and 7.3) (Zhang et al. 2017).

Zhang et al. (2018) used a genome-wide association study (GWAS) on 313 soybean accessions and studied seed protein, oil, fatty acid, and amino acid contents. Three QTL that control seed protein content have been identified and mapped on chromosomes 15 and 16; 25 QTL that control seed oil content have been identified and mapped on chromosomes 3, 5, 6, 7, 8, 13, 14, 15, 16, 17, and 18; 18 QTL that control seed palmitic acid content have been identified and mapped on chromosomes 5, 6, 8, 14, and 15; QTL that control seed stearic acid content have been identified and mapped on chromosomes 5 and 14; 4 QTL that control seed oleic acid content have been identified and mapped on chromosomes 5, 8, 14, and 19; 1 QTL that controls seed linoleic acid content has been identified and mapped on chromosome 17; 3 QTL that control seed linolenic acid content have been identified and mapped on chromosomes 14 and 15 (Table 2.2); and 92 QTL that control seed amino acids contents have been identified and mapped on all 20 soybean chromosomes (Table 8.1) (Zhang et al. 2018).

Li et al. (2018a) used GWAS on a population of diverse soybean cultivars and landraces ($n = 421$) from China, genotyped it with 1536 SNP panel, and investigated seed protein and oil contents. The study identified 12 SNPs associated with QTL that control seed protein and oil contents, and these QTL (QTNs; quantitative trait nucleotides) have been identified and mapped on chromosomes 2, 3, 5, 8, 12, 15, and 18. The following candidate genes have been identified on chromosome 2 (*Glyma02g03870*), on chromosome 5 (*Glyma05g33790*, *Glyma05g05680*, *Glyma05g37520*, and *Glyma05g34030*), on chromosome 8 (*Glyma08g13060*), on chromosome 12 (*Glyma12g01380*), on chromosome 15 (*Glyma15g17090*), and on chromosome 18 (*Glyma18g02690* and *Glyma18g02710*) (Li et al. 2018a).

Several other candidate genes have been identified on the same regions or very close to these identified QTL. Among these candidate genes are *GmSAT1* encoding the *Glycine max* symbiotic ammonium transporter 1 involved in N_2 fixation identified on chromosome 15: *Glyma.15 g049900*, *Glyma.15 g049700*, and *Glyma.15 g049800* encoding the nodulin MtN21 protein family (Table 2.3) (Zhang et al. 2018). La et al. (2019) grew 79 *Glycine soja* plant introduction (PI) accessions from the USDA germplasm collection in Clayton, NC, and Columbia, MO, in 2013. The 79 accessions have been genotyped with 35,285 SNP markers, and GWAS analysis identified 10 SNP markers highly associated with fatty acid contents but no association with seed protein or oil contents. One QTL controlling seed palmitic acid (C16:0) content has been identified and mapped on chromosome 20; four QTL controlling seed oleic acid (C18:1) content have been identified and mapped on chromosomes 3, 7, 8, and 14; and four QTL controlling seed linoleic acid content have been identified and mapped on chromosomes 2, 4, 11, and 13. One QTL that controls both seed aspartic acid (Asp) and glutamic acid (Glu) contents has been identified and mapped on chromosome 13 (Fig. 2.2, Table 2.2) (La et al. 2019).

Li et al. (2019b) grew a population of accessions ($n = 185$) in two environments in Harbin City in 2015–2016, China, used specific length amplified fragment sequencing (SLAF–sequencing), and genotyped the population with 12,072 SNP markers. They used GWAS method for data QTL data analysis to identify QTL for seed protein and oil contents. Seventeen QTL that control seed protein content have been identified and mapped on chromosomes 1, 2, 4, 12, 13, and 20; and 14 QTL that control seed oil content have been identified and mapped on chromosomes 4, 5, 7, 8, 12, 18, and 19 (Table 2.2) (Li et al. 2019b).

Li et al. (2019c) used restricted two-stage multi-locus genome-wide association analysis (RTM-GWAS) on a population of soybean accessions (n = 279) from China, genotyped it with 8148 SNPLDBs (SNP linkage disequilibrium blocks), and investigated seed protein and oil contents. Twenty six QTL that control seed protein content have been identified and mapped on chromosomes 1, 2, 3, 4, 6, 7, 8, 9, 10, 13, 15, 17, 18, 19, and 20, and 23 QTL that control seed oil content have been identified and mapped on chromosomes 1, 3, 4, 6, 7, 8, 10, 13, 15, 16, 17, 18, and 20. Ten of these are novel QTL. Likewise, 55 candidate genes have been identified within the QTL regions controlling seed protein contents, and 51 candidate genes have been identified within the QTL regions that control seed oil contents (Table 2.2) (Li et al. 2019c).

Li et al. (2020) grew a four-way RIL population derived from a cross between (Kenfeng14 by Kenfeng15) and (Heinong 48 by Kenfeng19) lines (n = 160) in eight environments in China and genotyped it with 275 SSR markers. The genetic linkage map based on this FW-RIL population covered 3636.26 cM, and they used inclusive complete interval mapping (ICIM) and interval mapping (IM) for QTL data analysis. The study reported 52 conditional QTL and 59 unconditional QTL that control seed protein and oil contents that have been identified and mapped on all 20 chromosomes of soybean. Among these QTL, two QTL that control seed protein content and five QTL that control seed oil contents were novel QTL (Table 2.2) (Li et al. 2020).

Karikari et al. (2019) grew a RIL population derived from a cross between "Linhefenqingdou" (has high seed protein and low oil contents) and "Meng 8206" (has low seed protein and high oil contents) lines (n = 104) (LM6 population) in 6 environments in China and genotyped it with 2267 bin markers. They used the LM6-based genetic link map that covered 2453.79 cM and CIM of WinQTL Cartographer for QTL data analysis. Twenty five QTL that control seed protein content have been identified and mapped on chromosomes 1, 4, 6, 7, 8, 9, 10, 13, 14, 16, 17, 18, 19, and 20; and 19 QTL that control seed oil content have been identified and mapped on chromosomes 1, 2, 3, 6, 8, 10, 11, 13, 16, and 20 (Fig. 2.2; Table 2.2). Within the QTL (*qPro-7-1*), authors identified the genes *Glyma07g09260* involved in regulation of cyclin-dependent protein kinase activity and stomatal lineage progression, *Glyma07g09270* involved in transmembrane transport, and *Glyma07g09630* involved in systemic acquired resistance and phenylpropanoid biosynthetic process. Within the QTL (*qOil-8-3*), authors identified the genes *Glyma08g14920* that encodes for an ABC1 family, *Glyma08g15750* that encodes for a zinc finger protein, *Glyma08g15921* that encodes for a protein kinase domain, and *Glyma08g16072* that encodes for a protein serine/threonine kinase. Within the QTL (*qOIl-10-2*), authors identified the genes *Glyma10g06800* that encodes for a phosphatidylinositol-4-phosphate 5-Kinase, *Glyma10g06810* that encodes for amidohydrolase, and *Glyma10g06820* that encodes for a glutamine synthetase. Within the QTL (*qOil-10-4*), authors identified the genes *Glyma10g25630* that encodes for a TCP-1/cpn60 chaperonin family, *Glyma10g26510* that encodes for a lipase (class 3), *Glyma10g26770* that encodes for a tetratricopeptide repeat, *Glyma10g26870* that encodes for a WD domain, *Glyma10g26920* that encodes for an RNA recognition motif, and many other genes (Table 2.3) (Karikari et al. 2019).

Recently, Wang et al. (2020) used a genome-wide association study (GWAS) on a population of soybean natural accessions (*n* = 260) in Northeast China over 2 years (2018–2019), genotyped it with 2,311,337 SNM markers, and studied seed oleic acid content's genetic architecture. Twenty-one and 8 candidate genes that control seed oleic content have been identified in 2018 and 2019, respectively. The functions of these genes include cell wall, metal handing, transport, amino acid metabolism, lipid metabolism, hormone metabolism, redox, protein, development, signaling, cell signaling, N-metabolism, tetrapyrrole synthesis, and amino acid metabolism (Table 2.3) (Wang et al. 2020) (Fig. 2.1).

2.3 Candidate Genes Identified Within the QTL Regions

Several recent studies attempted to fine map the genomic regions containing the QTL that control seed protein, oil, and fatty acids contents and identify candidate genes involved in the biosynthetic pathways of these beneficial compounds (Bolon et al. 2010; Hwang et al. 2014; Zhou et al. 2015; Li et al. 2015, 2018a, 2019b, 2019c; Fang et al. 2017; Zhang et al. 2018). Major candidate genes identified by these studies within the QTL regions that control seed protein, oil, fatty acid, and amino acid contents are shown in Table 2.3.

Bolon et al. 2010 used a near-isogenic line (NIL) to fine map the region on chromosome 20 that contains the major QTL controlling both seed protein and oil contents (*cqPRO-003* and *cqOIL-004*) (Diers et al. 1992; Sebolt et al. 2000; Chung et al. 2003; Nichols et al. 2006). They narrowed the QTL region into a region of only 8.4 Mbp [from 24.54 Mb (Sat_174) to 32.92 Mb (ssrpqtl_38)]. Moreover, using transcriptome analysis, they identified 12 genes that are differentially expressed in the NILs in this same region (Bolon et al. (2010). In a subsequent study, Bolon et al. (2014) performed a genome-wide analysis study (GWAS) and gene expression profiling of a RIL population derived from a cross between "Minsoy" and "Noir" (*n* = 93). Early-developing to mid-mature green seeds of the parents and RILs have been studied. The RIL was genotyped with 557 SNP markers and the constructed MxN genetic linkage map covered about 2500 cM. Two QTL that control seed oil content has been identified and mapped on chromosome 8, and one QTL that controls seed protein content has been identified and mapped on chromosomes 4 and 6 (Bolon et al. 2014). Interestingly, the QTL for protein content on chromosome 4 was mapped at the same location of a seed protein QTL previously reported (Stombaugh et al. 2004). The QTL for seed protein and oil contents on chromosome 6 was also mapped at the same location of other QTL for protein and oil contents previously reported (Kabelka et al. 2004; Hyten et al. 2004).

QTL clustering is widely reported in soybean and other crops. Hwang et al. (2014) used 298 soybean accessions, genotyped them with 31,954 SNP markers, and performed a GWAS to investigate protein and oil contents. The study identified 40 SNP markers highly associated with protein content and 25 SNP markers highly associated with oil content. The GWAS identified most QTL for protein and oil

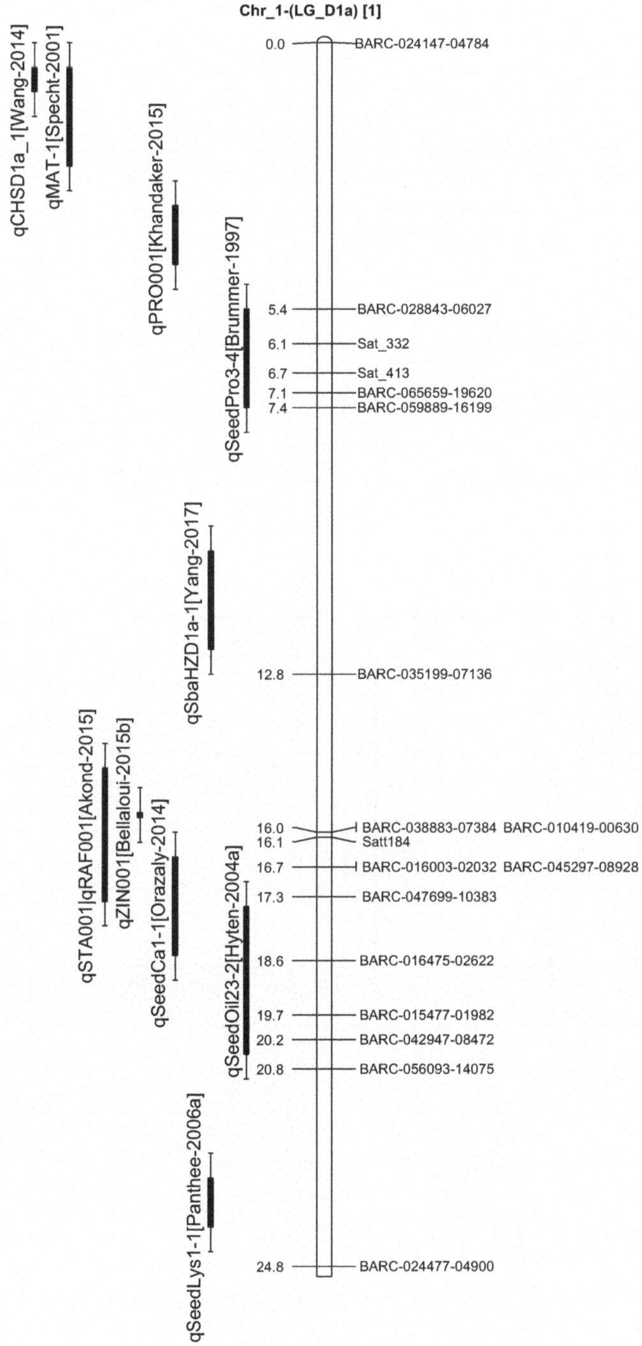

Fig. 2.1 Soybean consensus map (SoyBase: www.soybase.org; Grant et al. 2010; Brown et al. 2020) and location of QTL that control seed protein, oil, fatty acids (palmitic, stearic, oleic, linoleic, and linolenic), amino acids, macronutrients, micronutrients, sugars, isoflavones (daidzein, genistein, glycitein, total isoflavones, daidzin, genistin, glycitin, acetyldaidzin, acetylglycin, acetylgenistin,

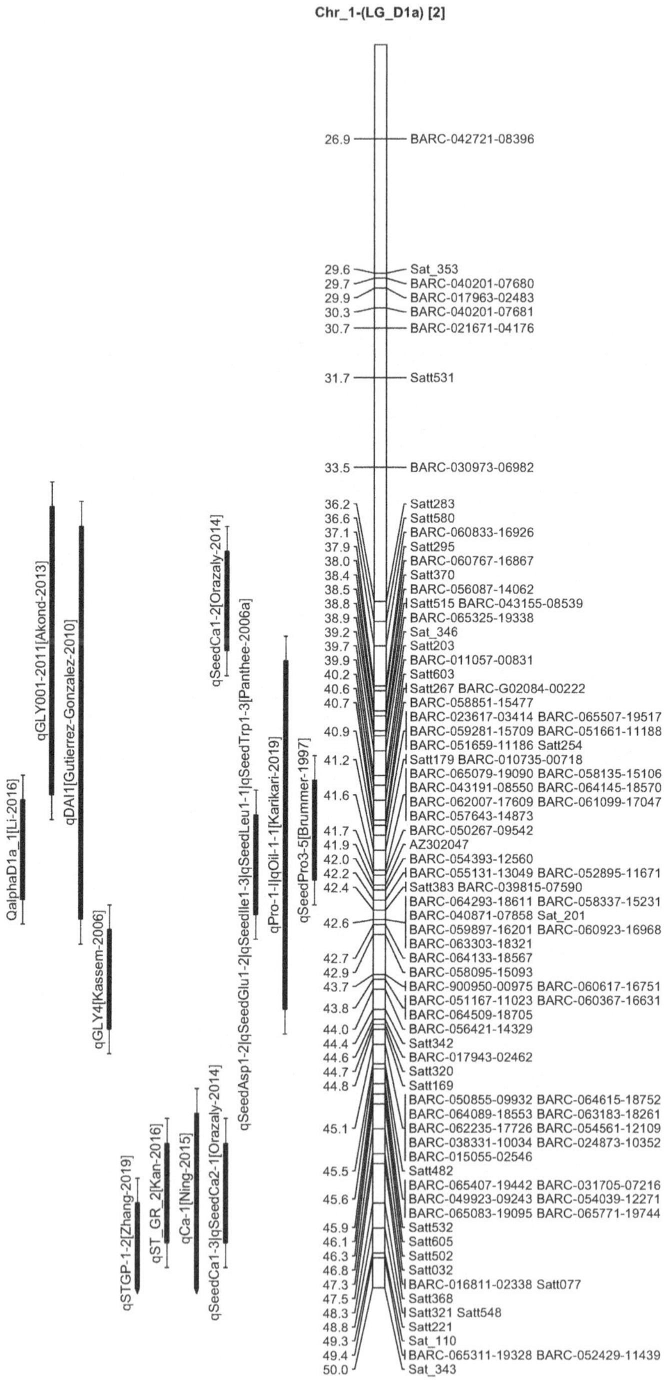

Fig 2.1 (continued) malonyldaidzein, malonylgenistein, and malonylglycitein), α-tocopherol (α-Toc), β-tocopherol (β-Toc), γ-tocopherol (γ-Toc), δ-tocopherol (δ-Toc), total tocopherol (T-Toc), and other compounds' contents from 1992 to 2020. Also QTL that control mineral deficiencies and salt tolerance are included. Chromosomes were drawn using MapChart (Voorrips 2002)

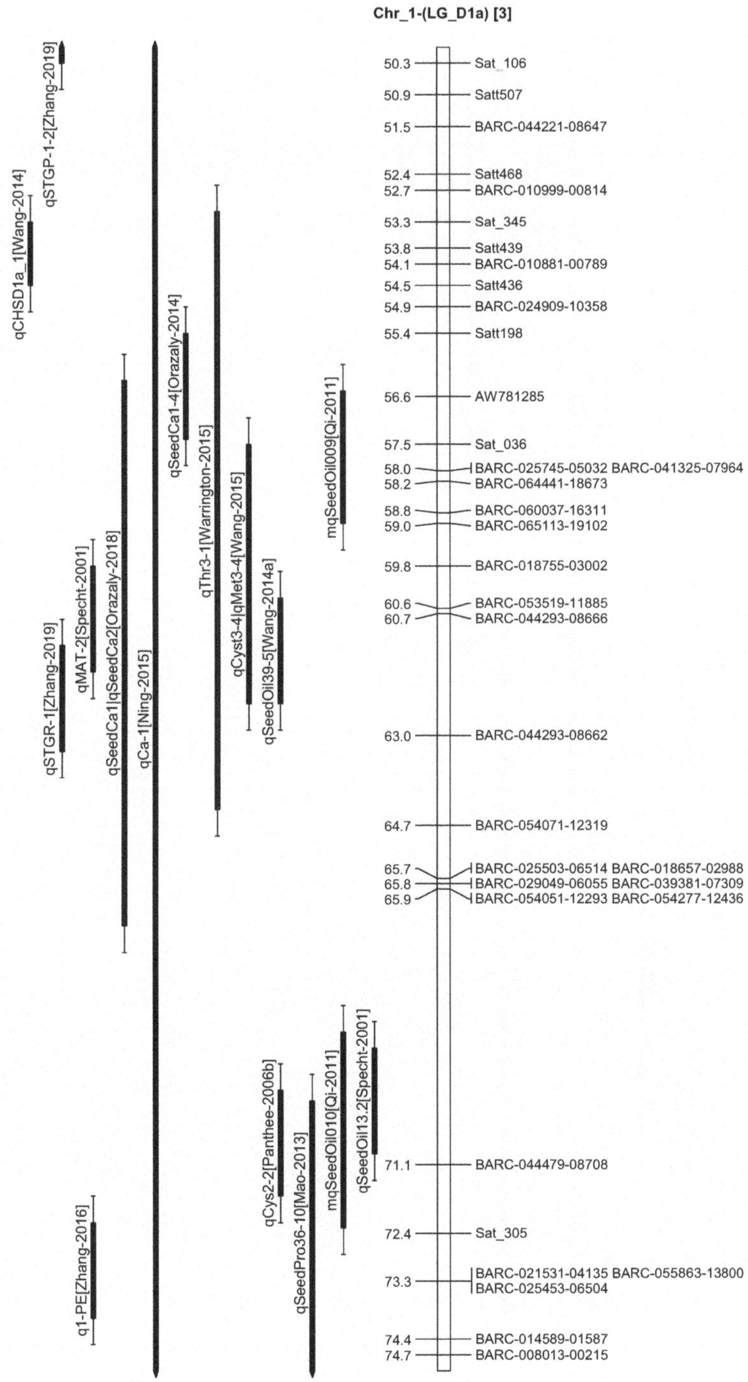

Fig 2.1 (continued)

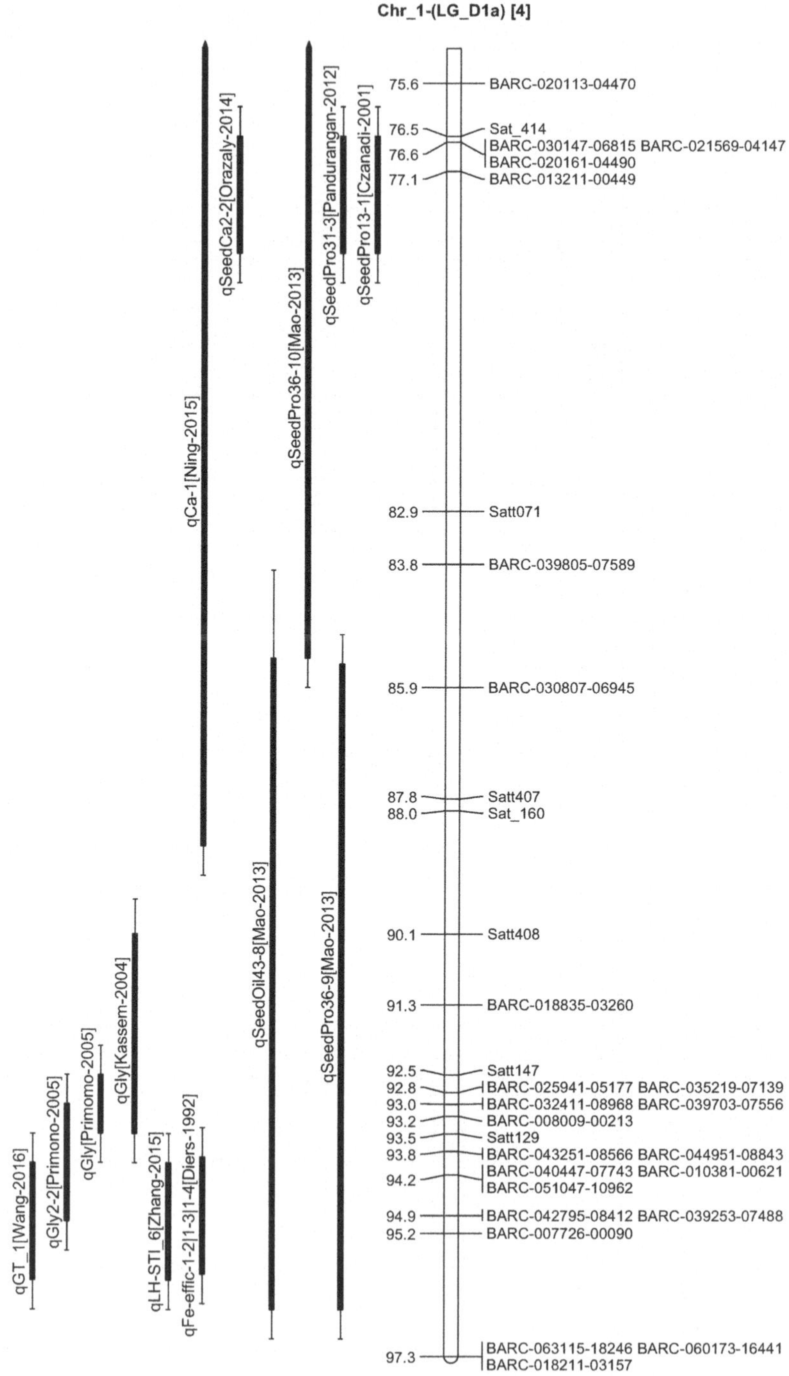

Fig 2.1 (continued)

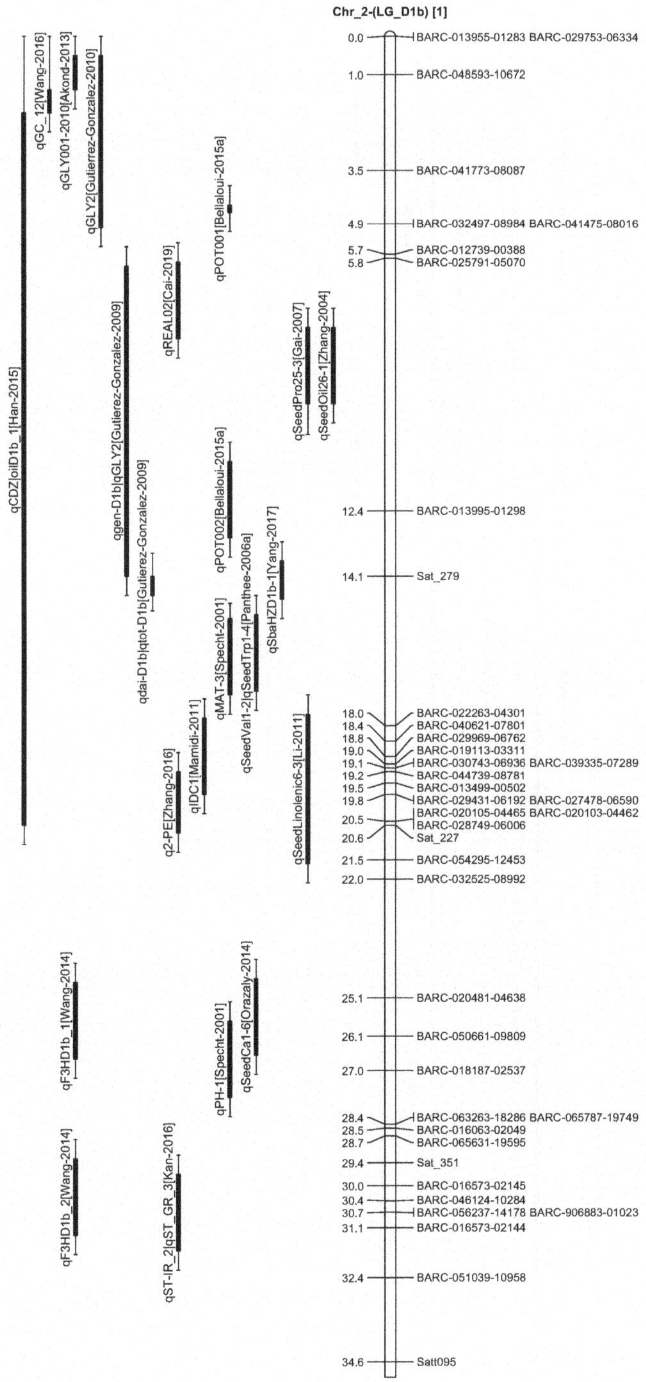

Fig 2.1 (continued)

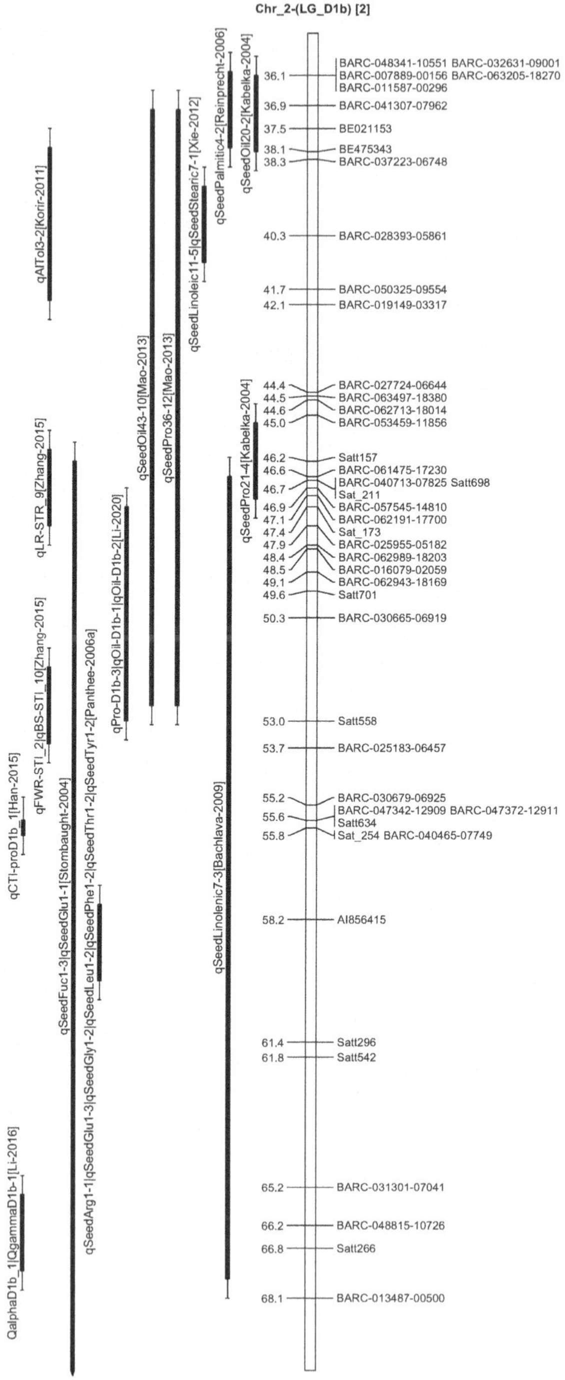

Fig 2.1 (continued)

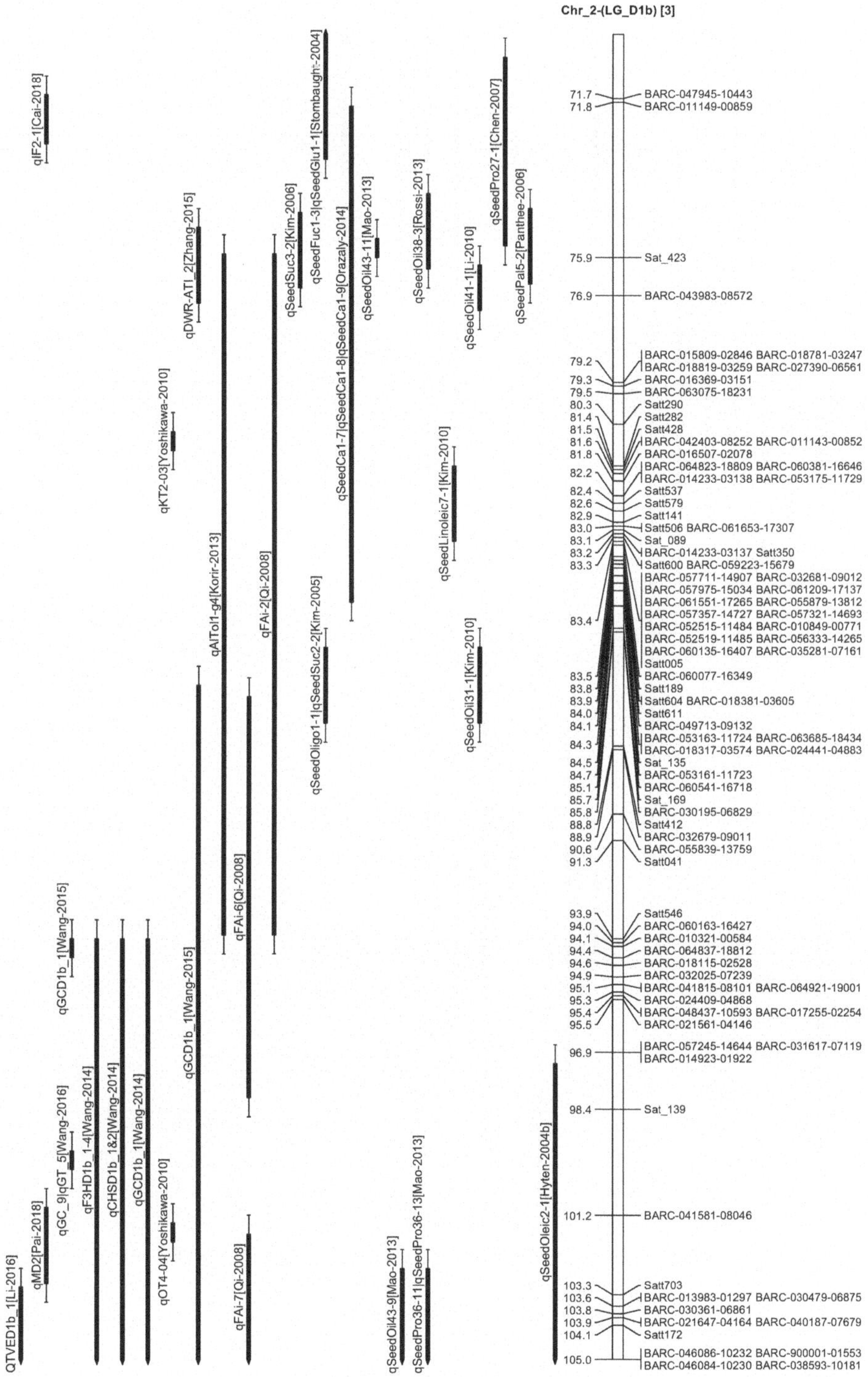

Fig 2.1 (continued)

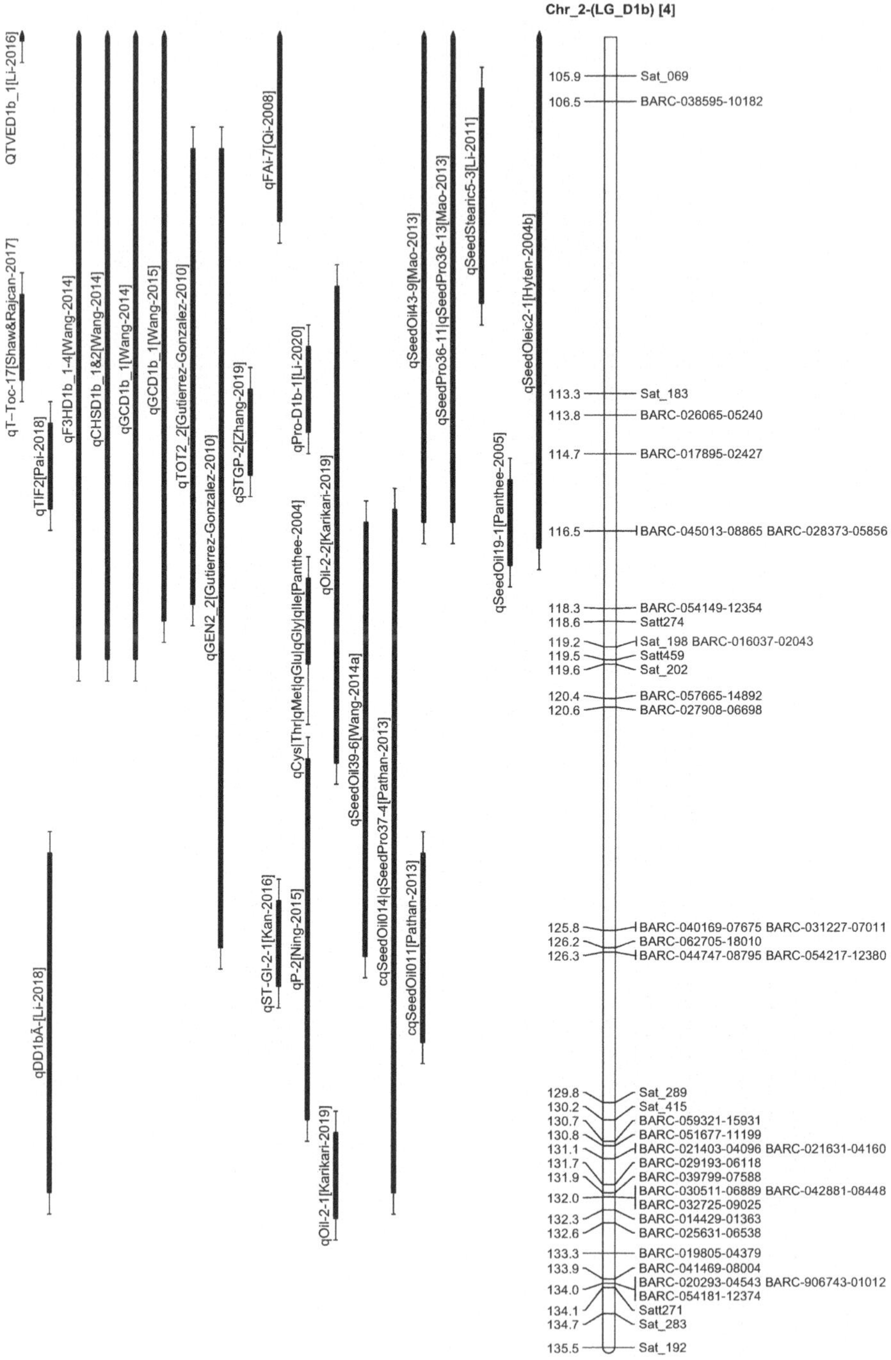

Fig 2.1 (continued)

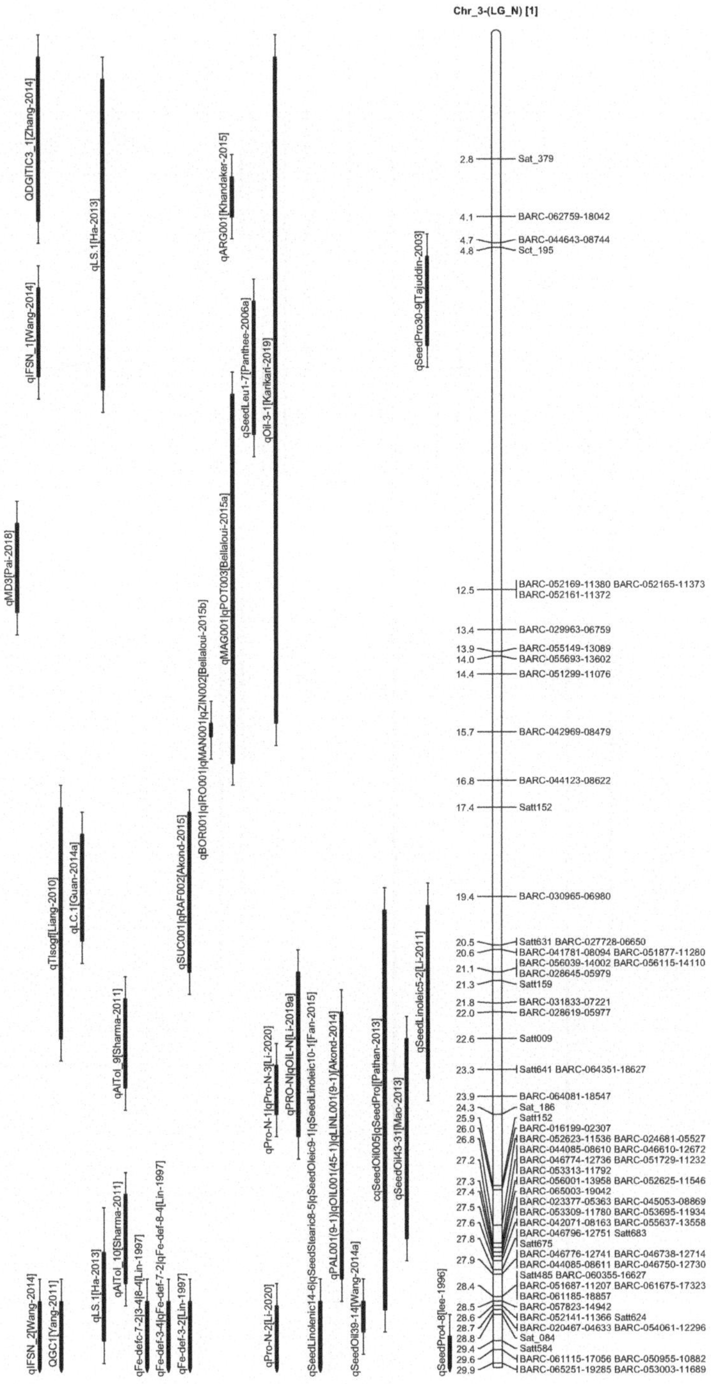

Fig 2.1 (continued)

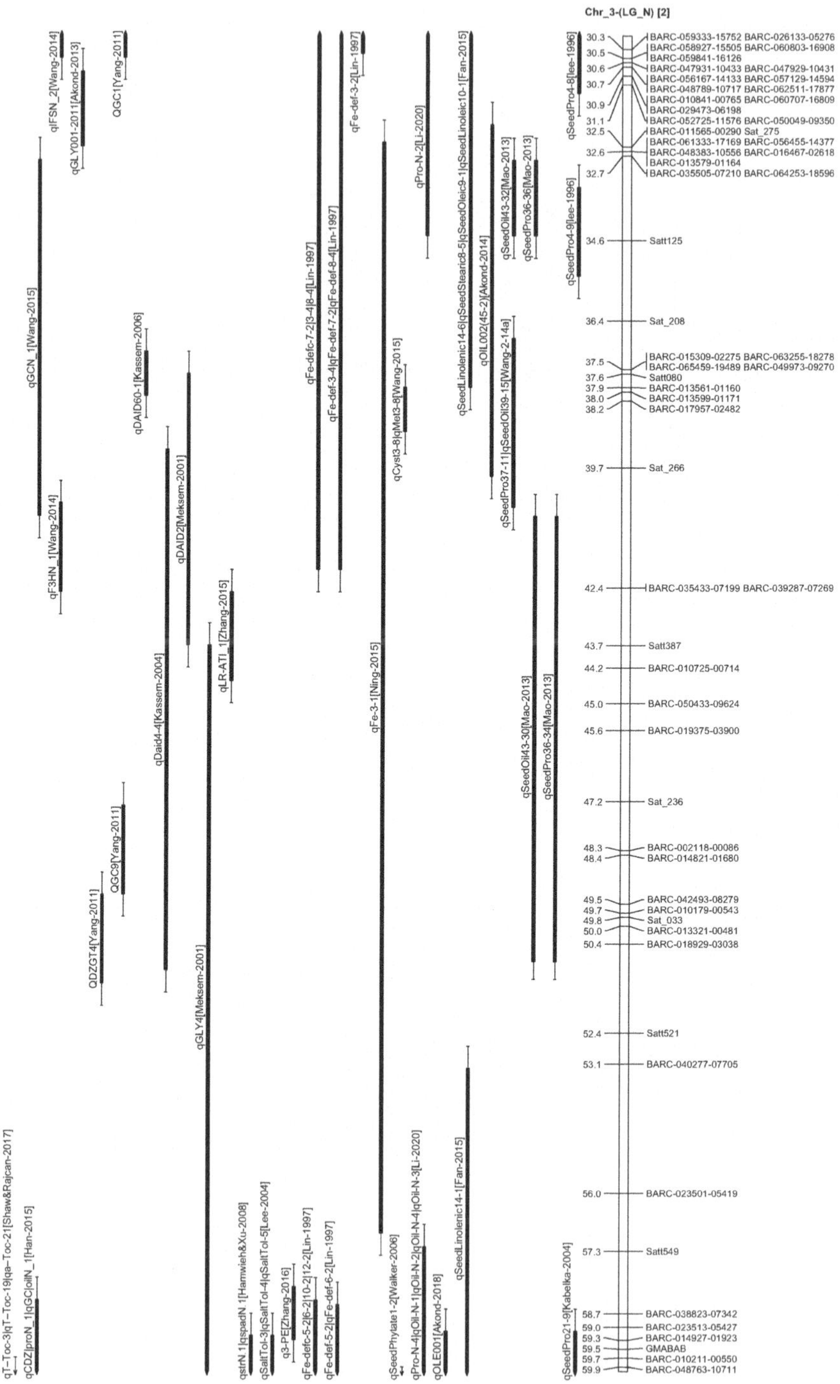

Fig 2.1 (continued)

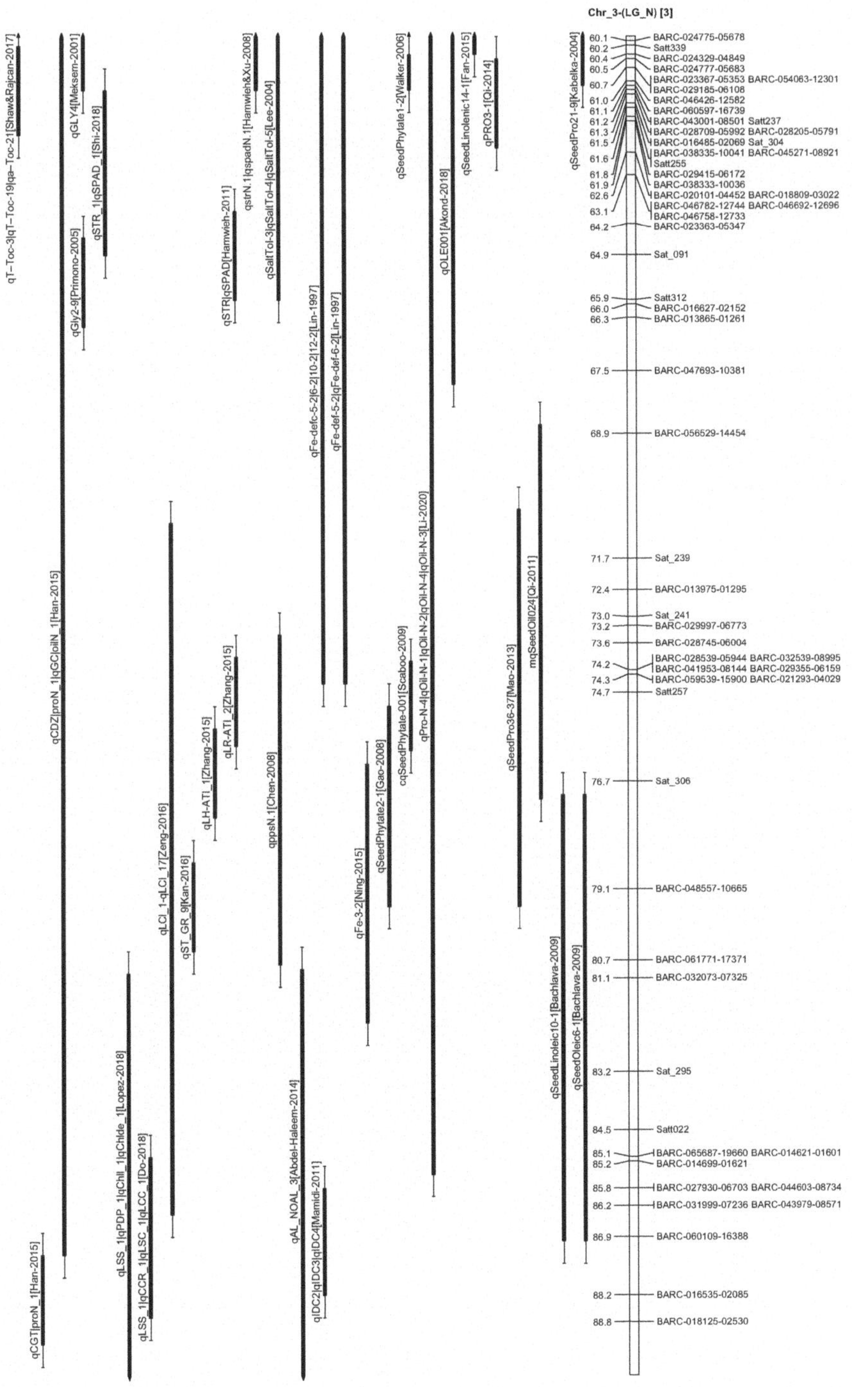

Fig 2.1 (continued)

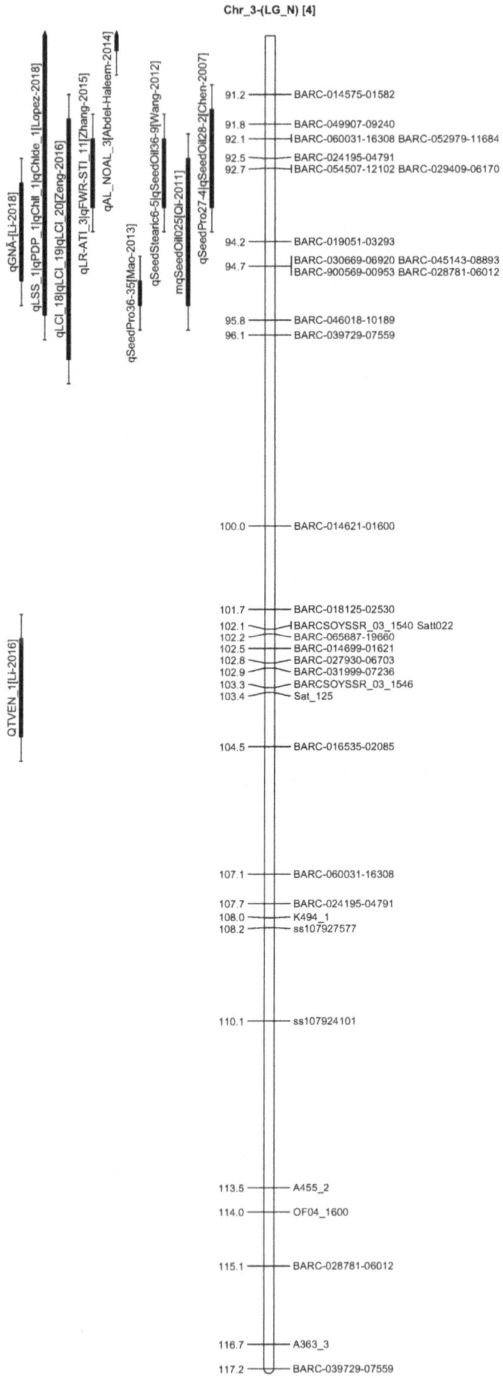

Fig 2.1 (continued)

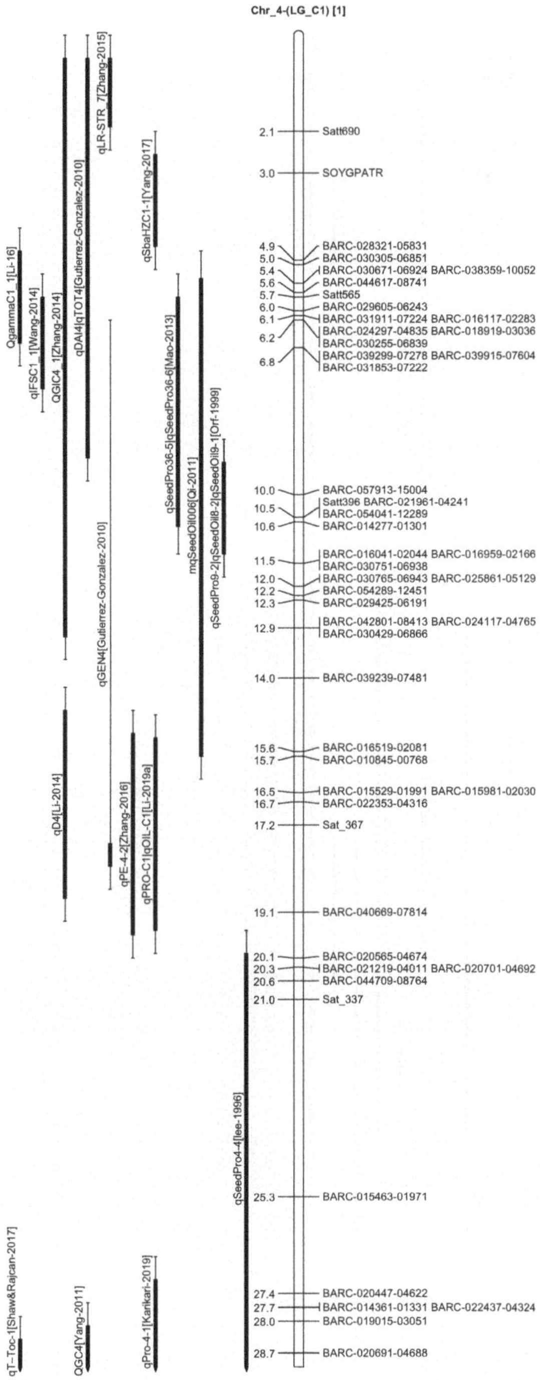

Fig 2.1 (continued)

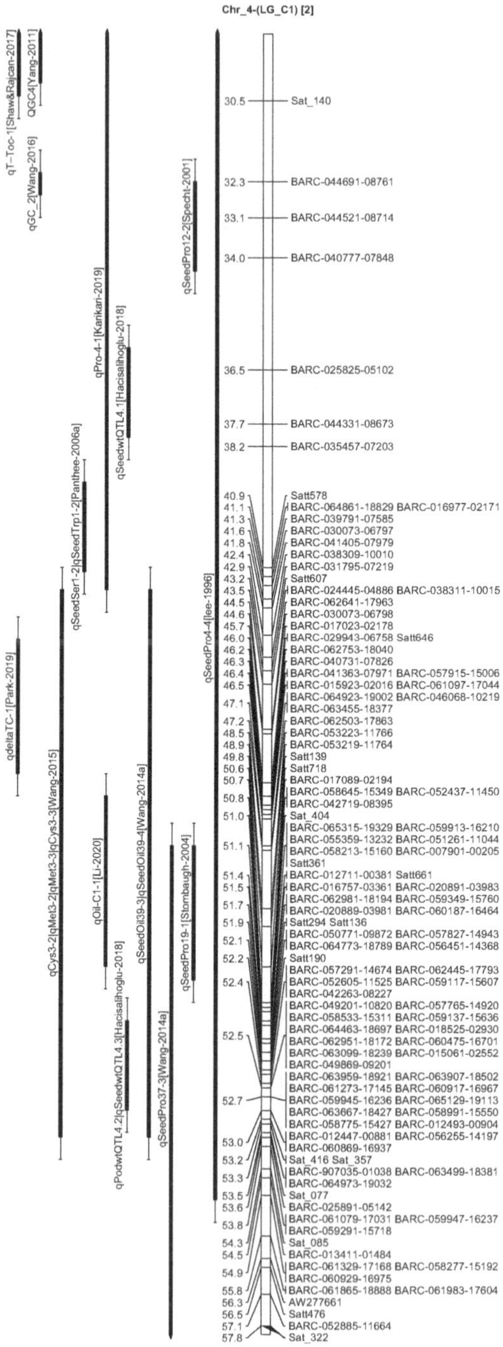

Fig 2.1 (continued)

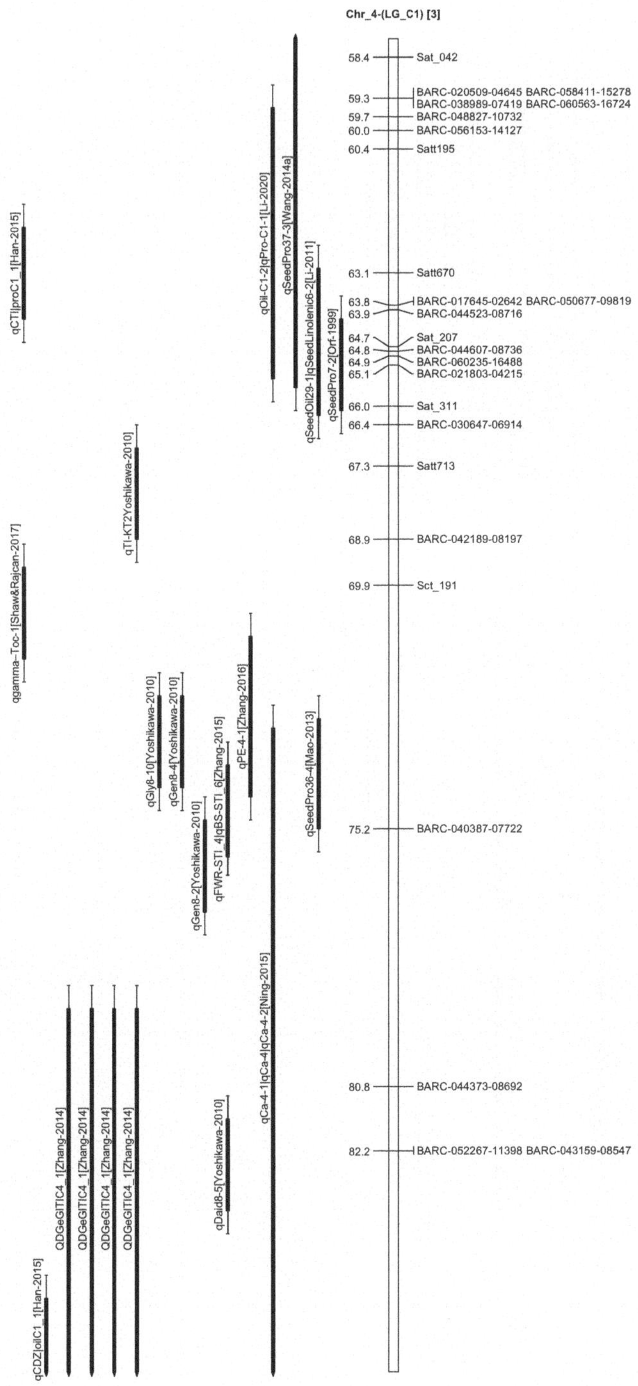

Fig 2.1 (continued)

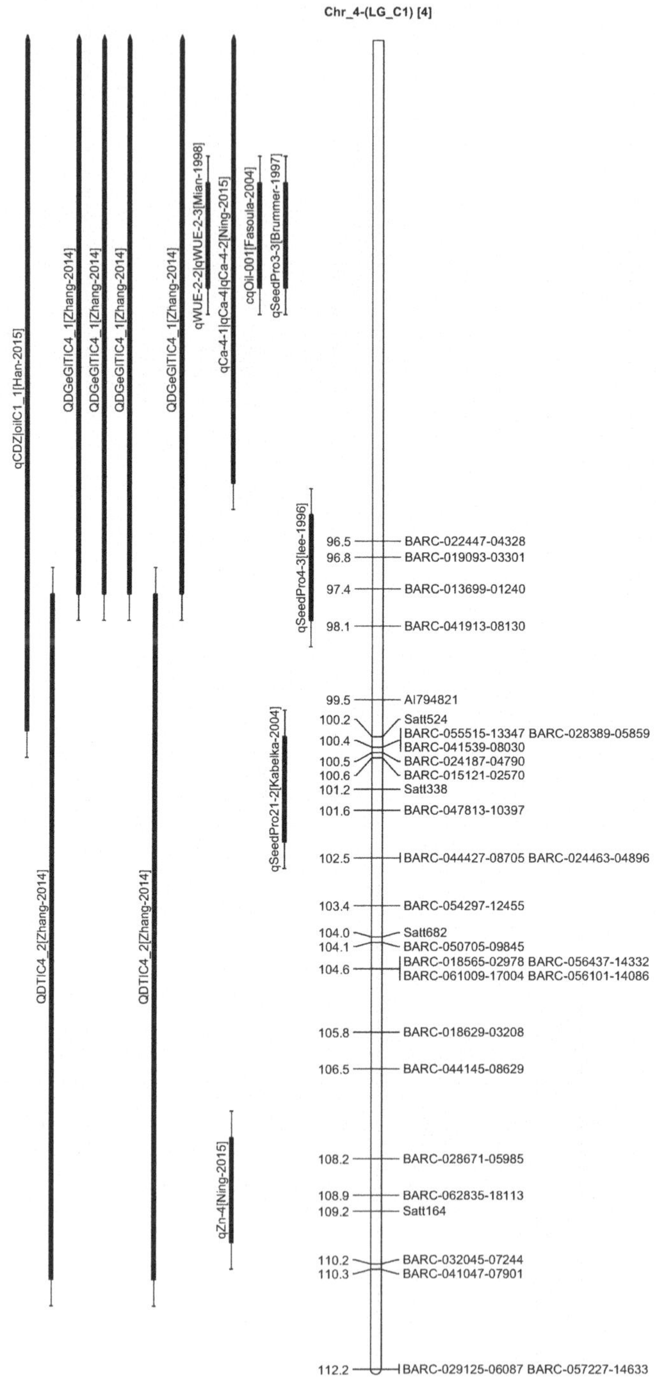

Fig 2.1 (continued)

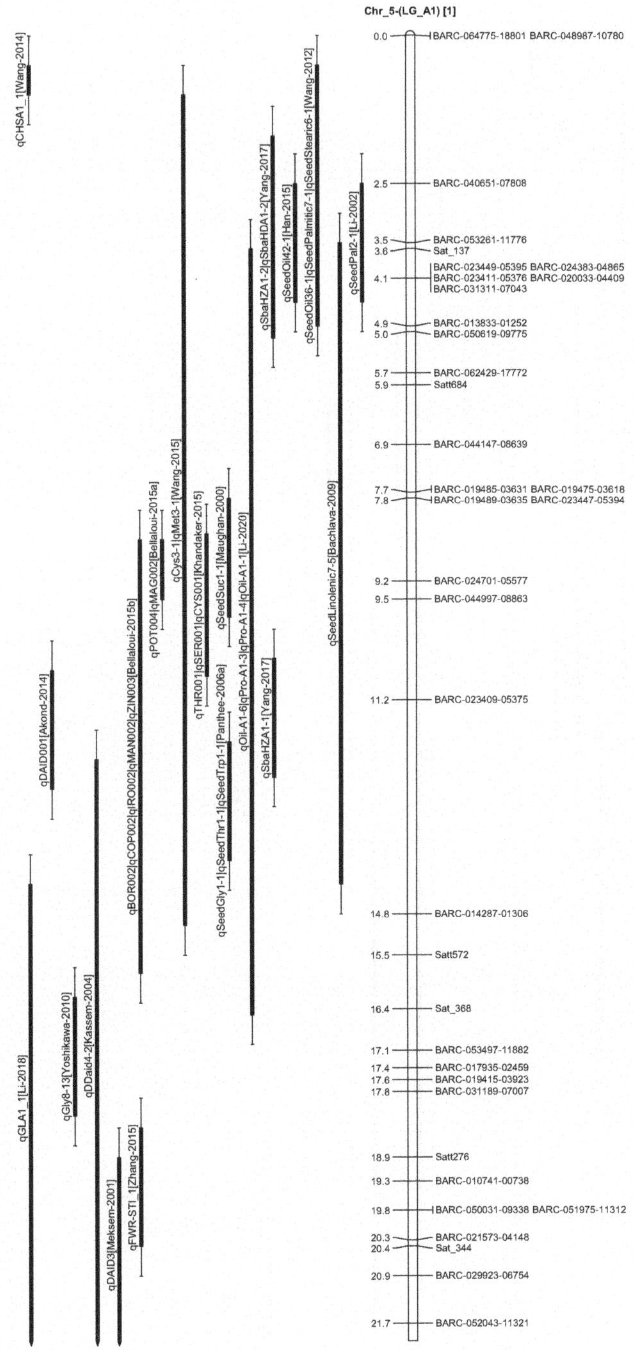

Fig 2.1 (continued)

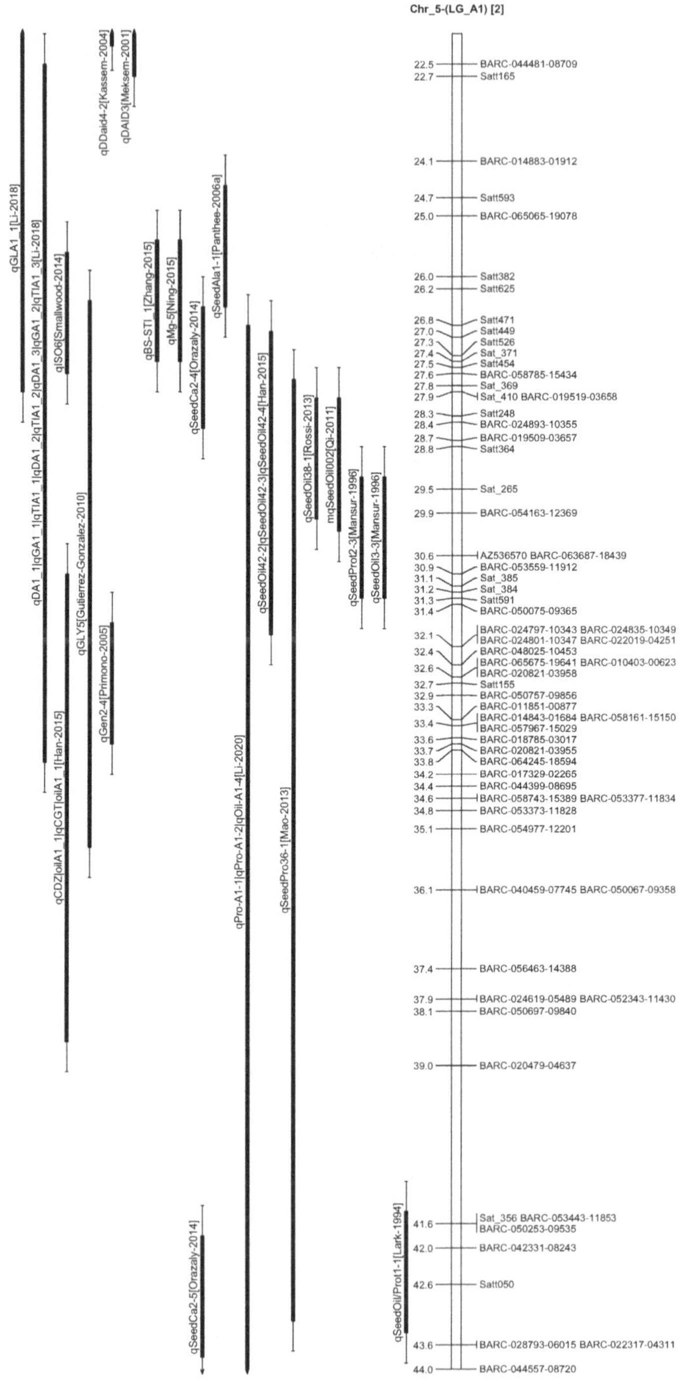

Fig 2.1 (continued)

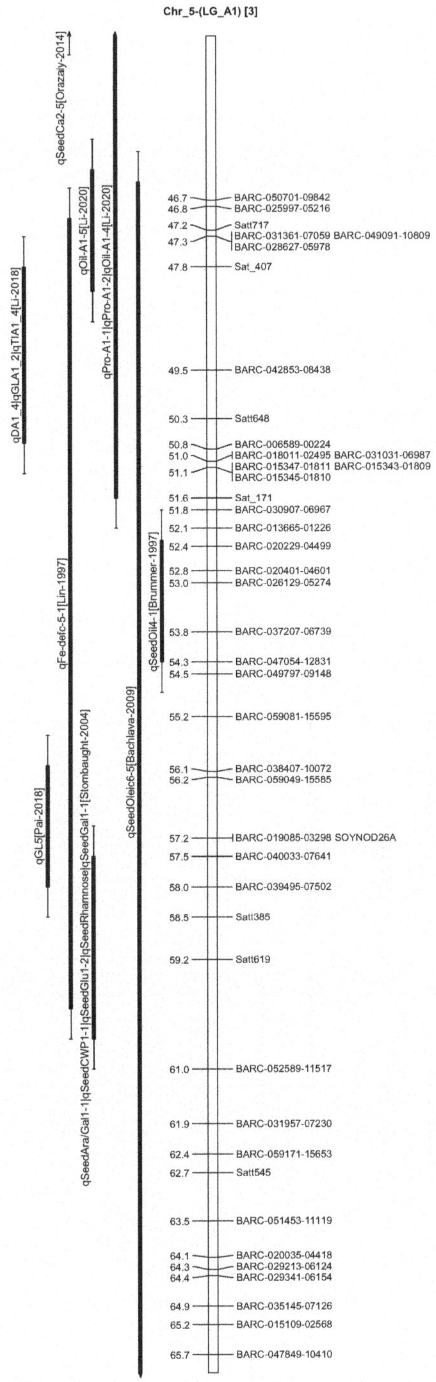

Fig 2.1 (continued)

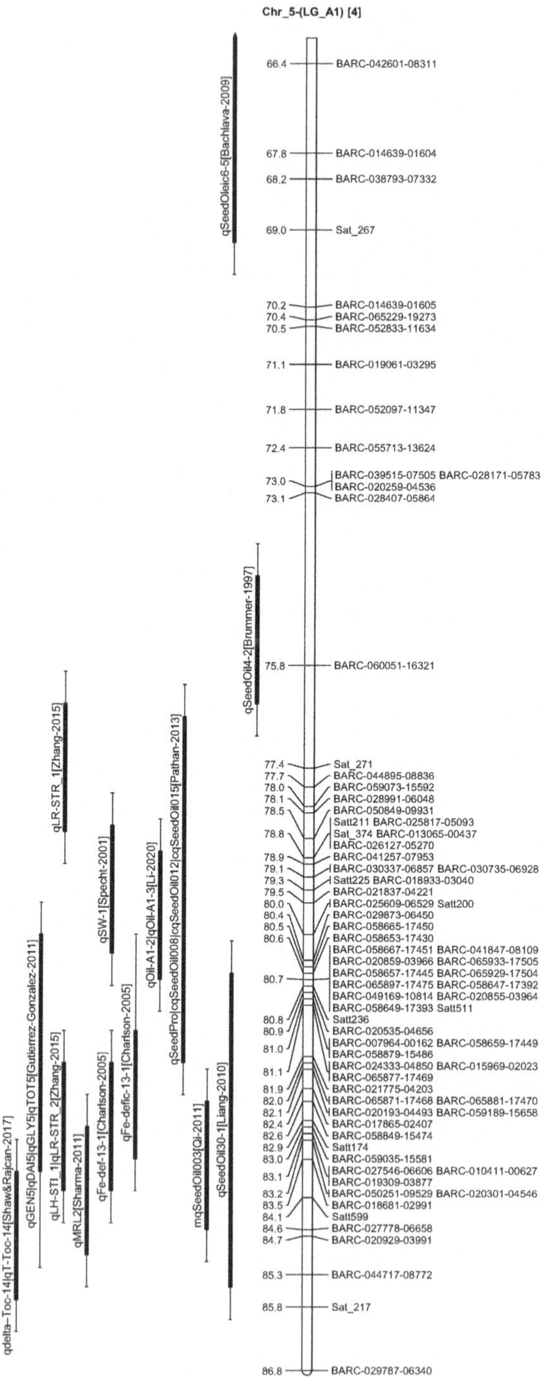

Fig 2.1 (continued)

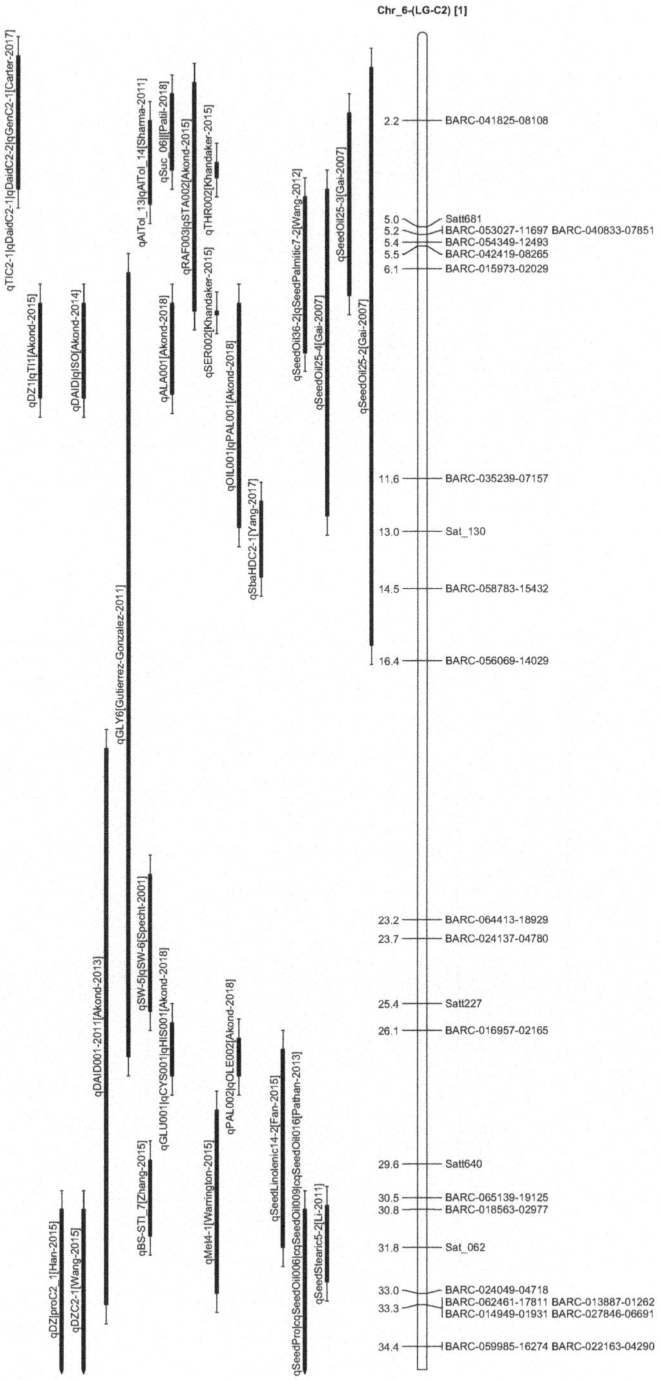

Fig 2.1 (continued)

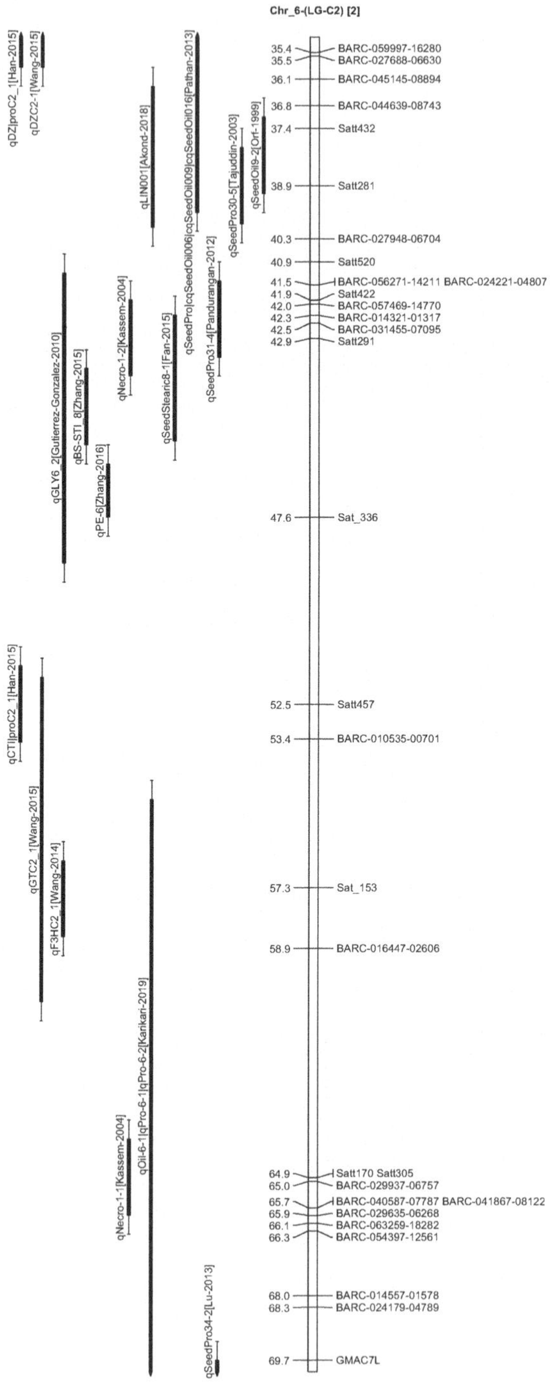

Fig 2.1 (continued)

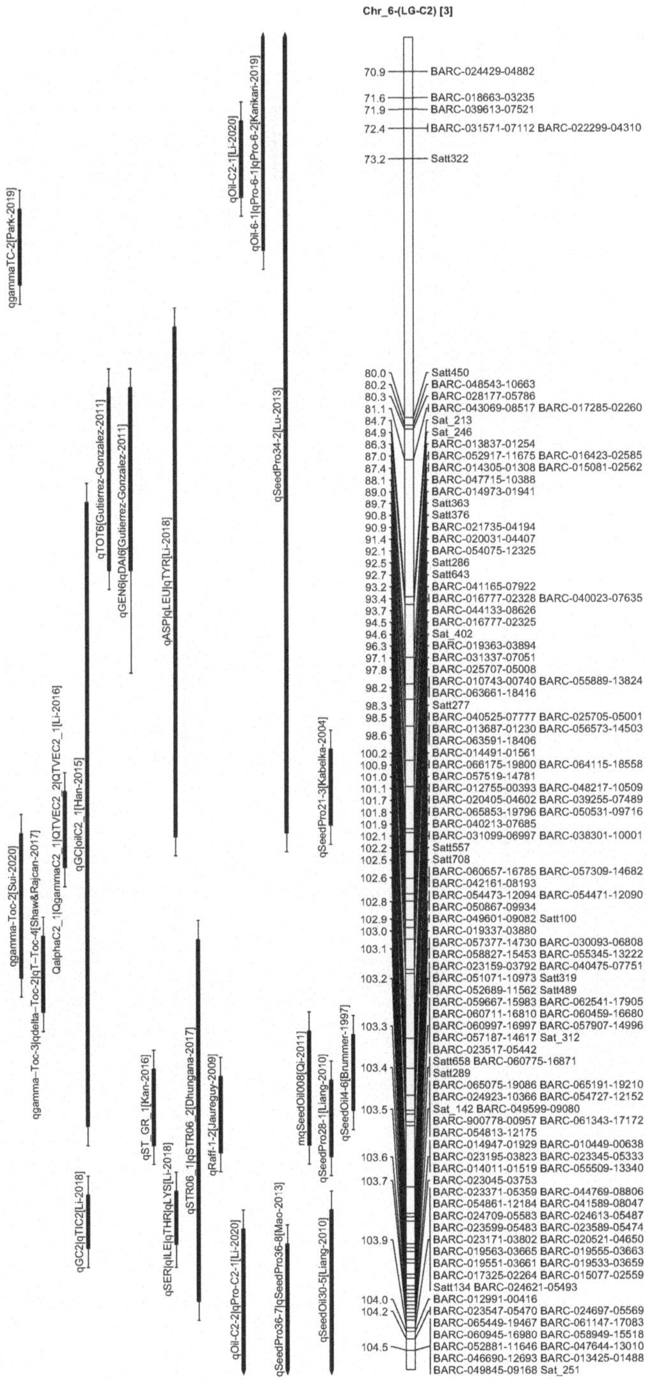

Fig 2.1 (continued)

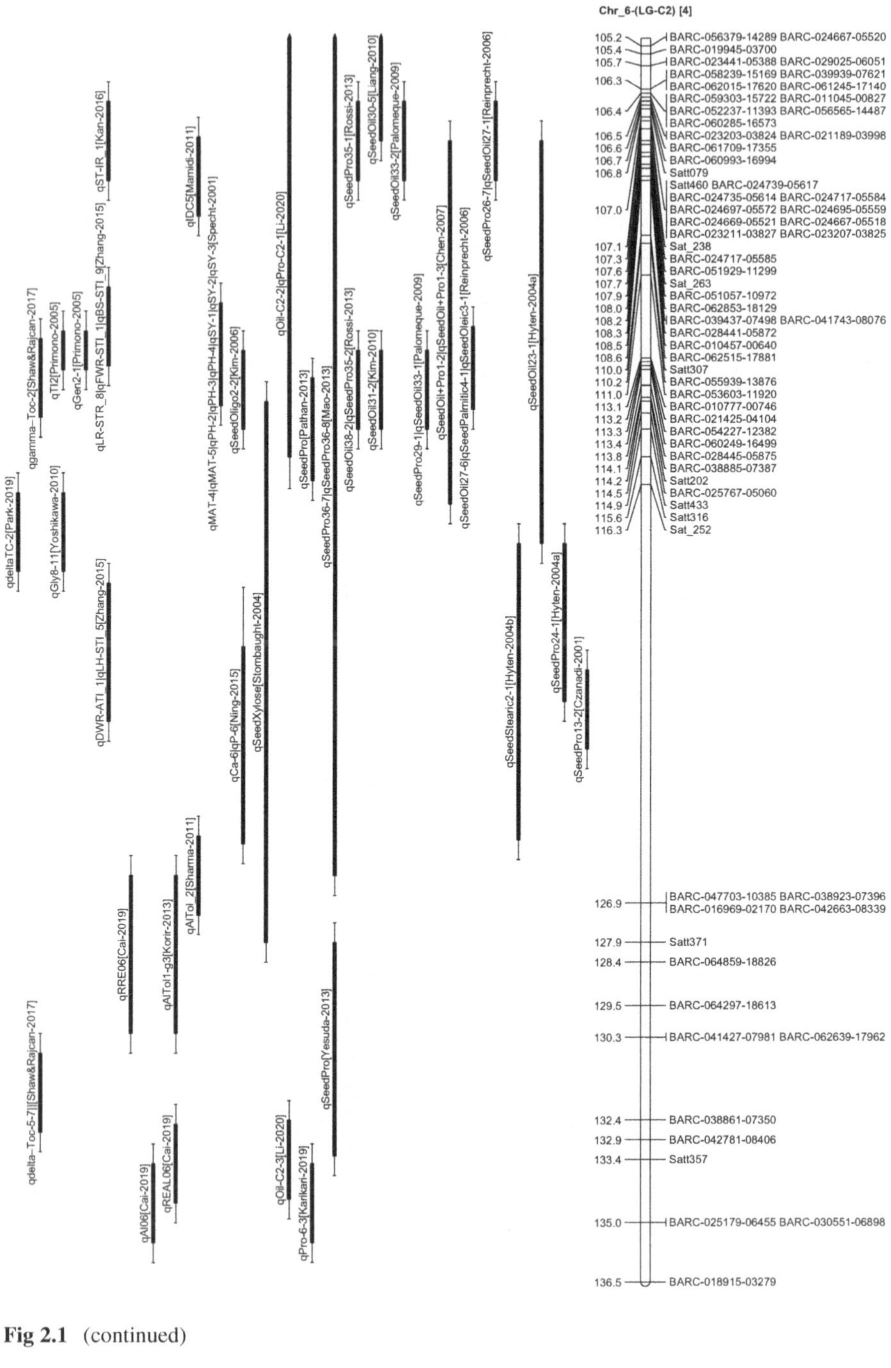

Fig 2.1 (continued)

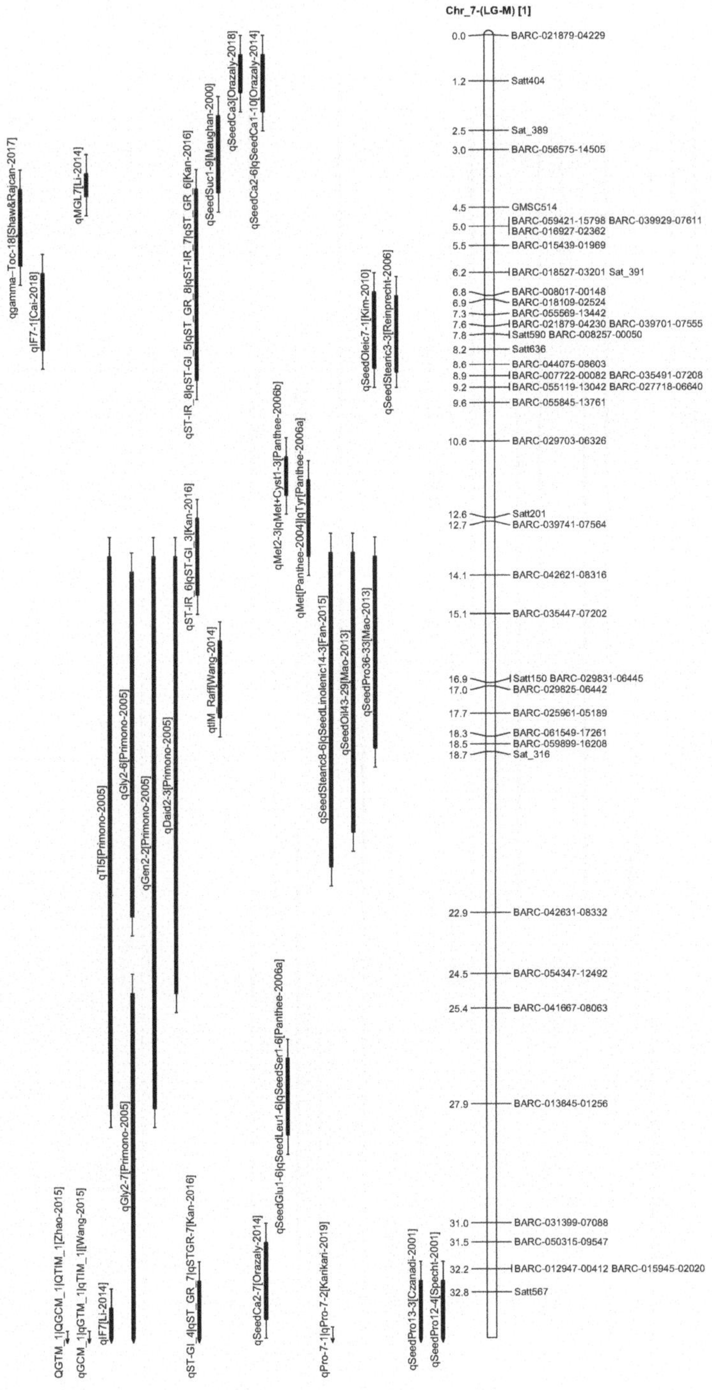

Fig 2.1 (continued)

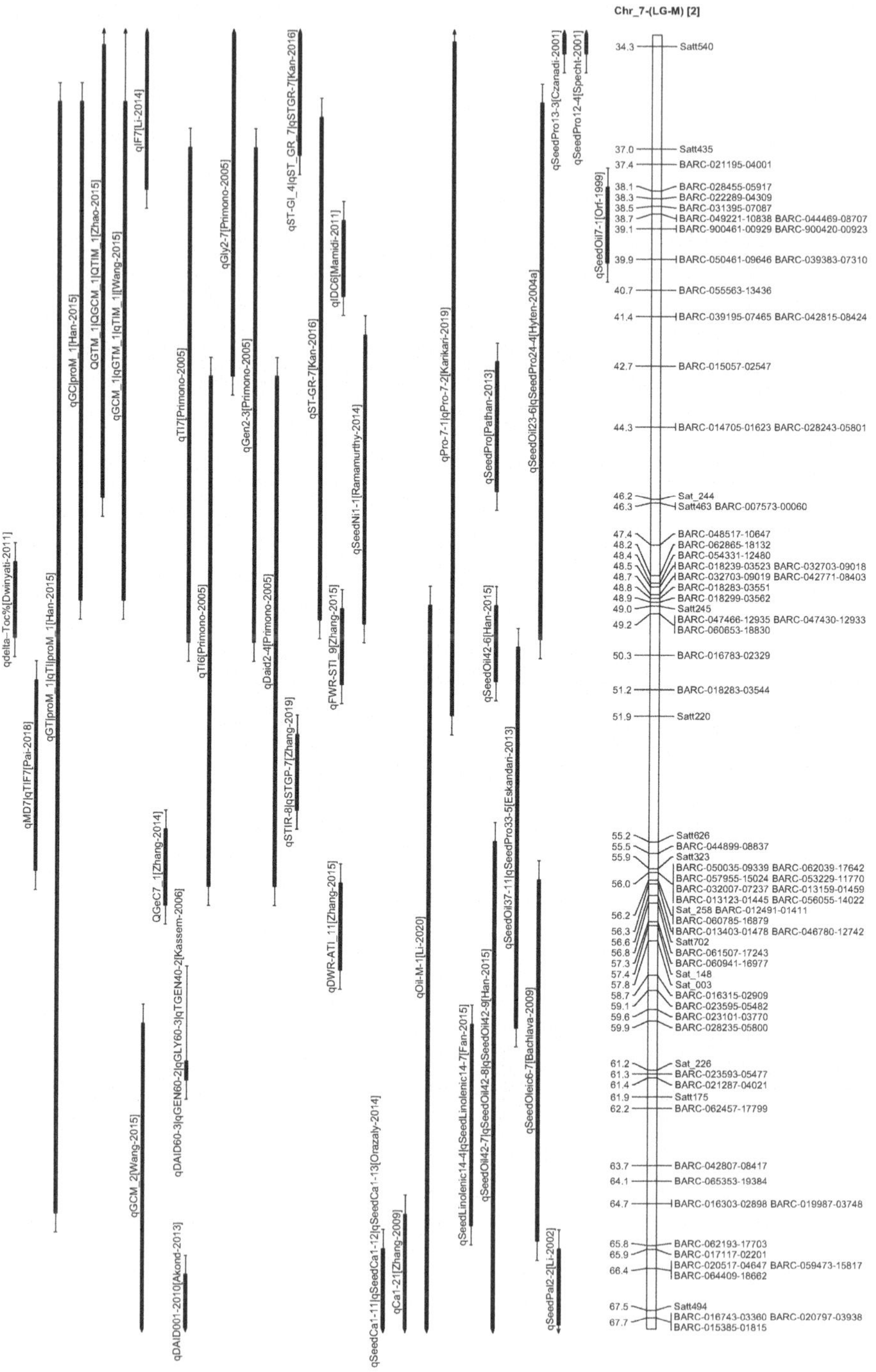

Fig 2.1 (continued)

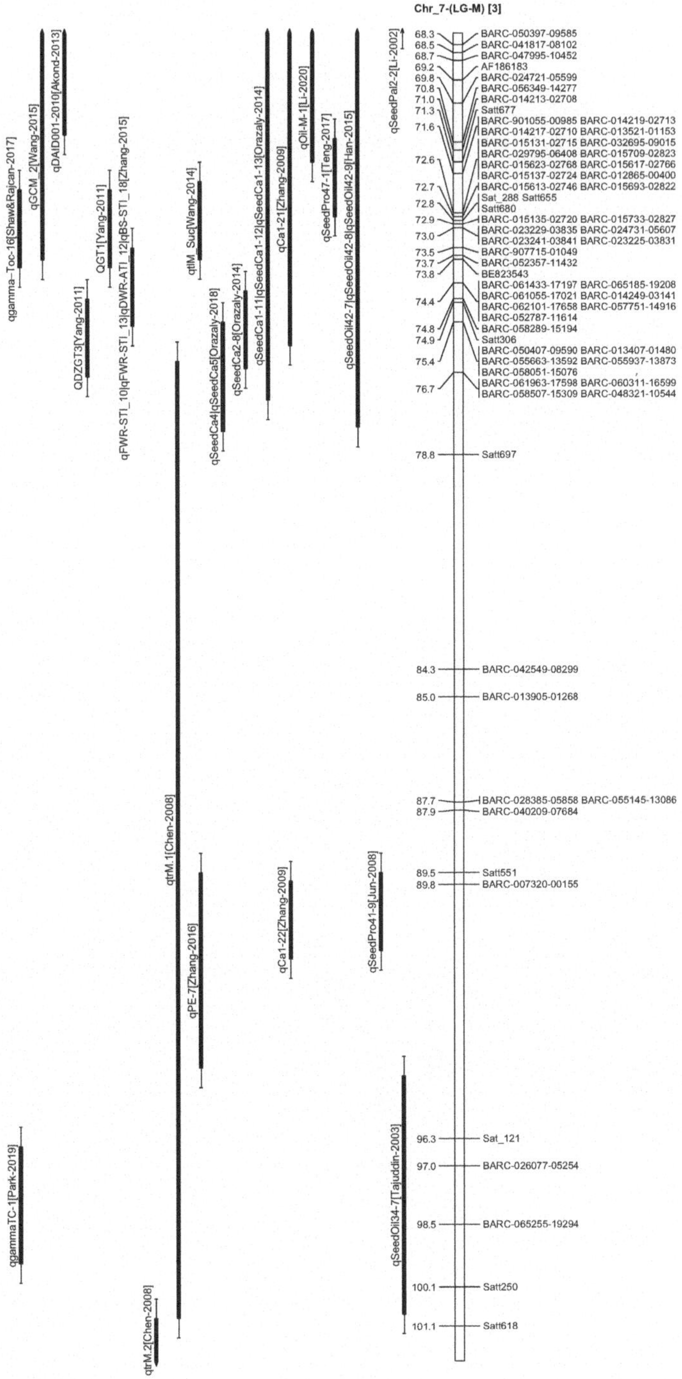

Fig 2.1 (continued)

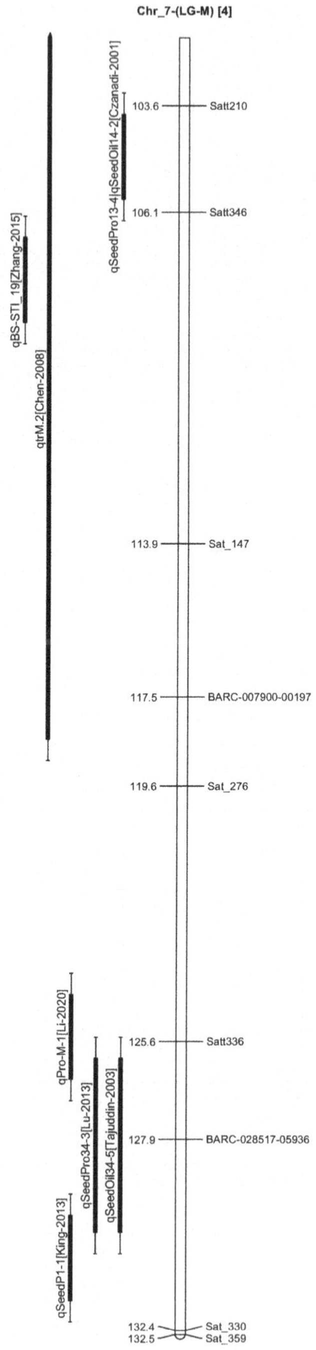

Fig 2.1 (continued)

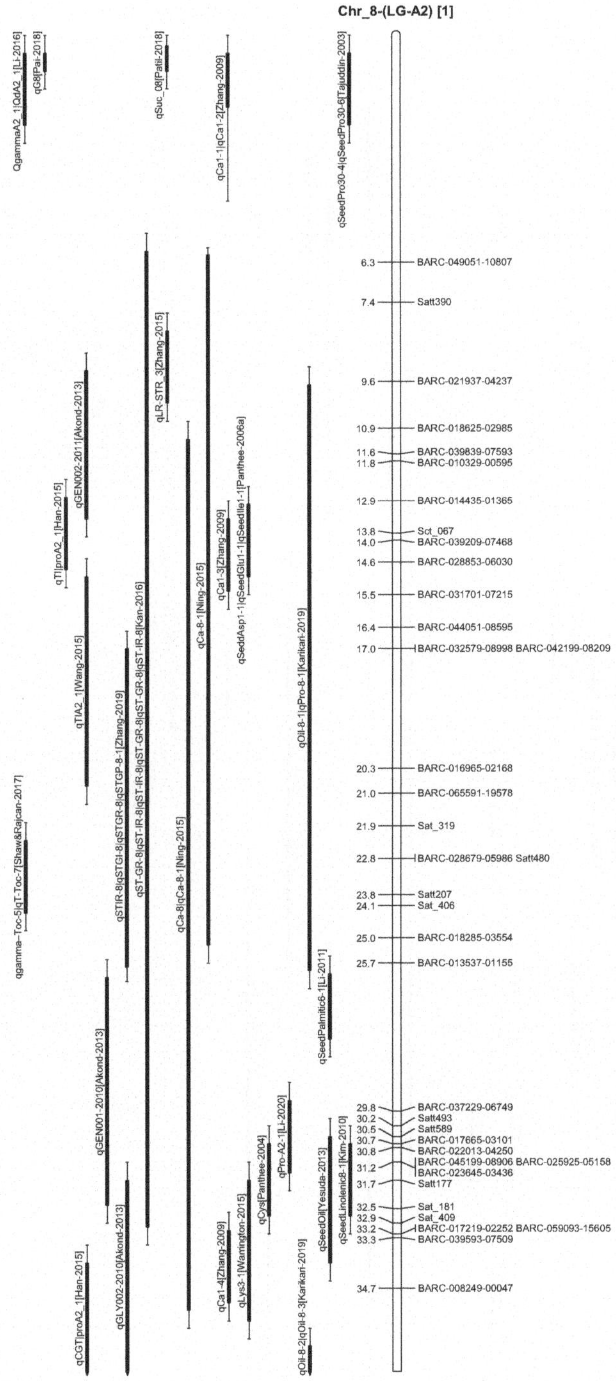

Fig 2.1 (continued)

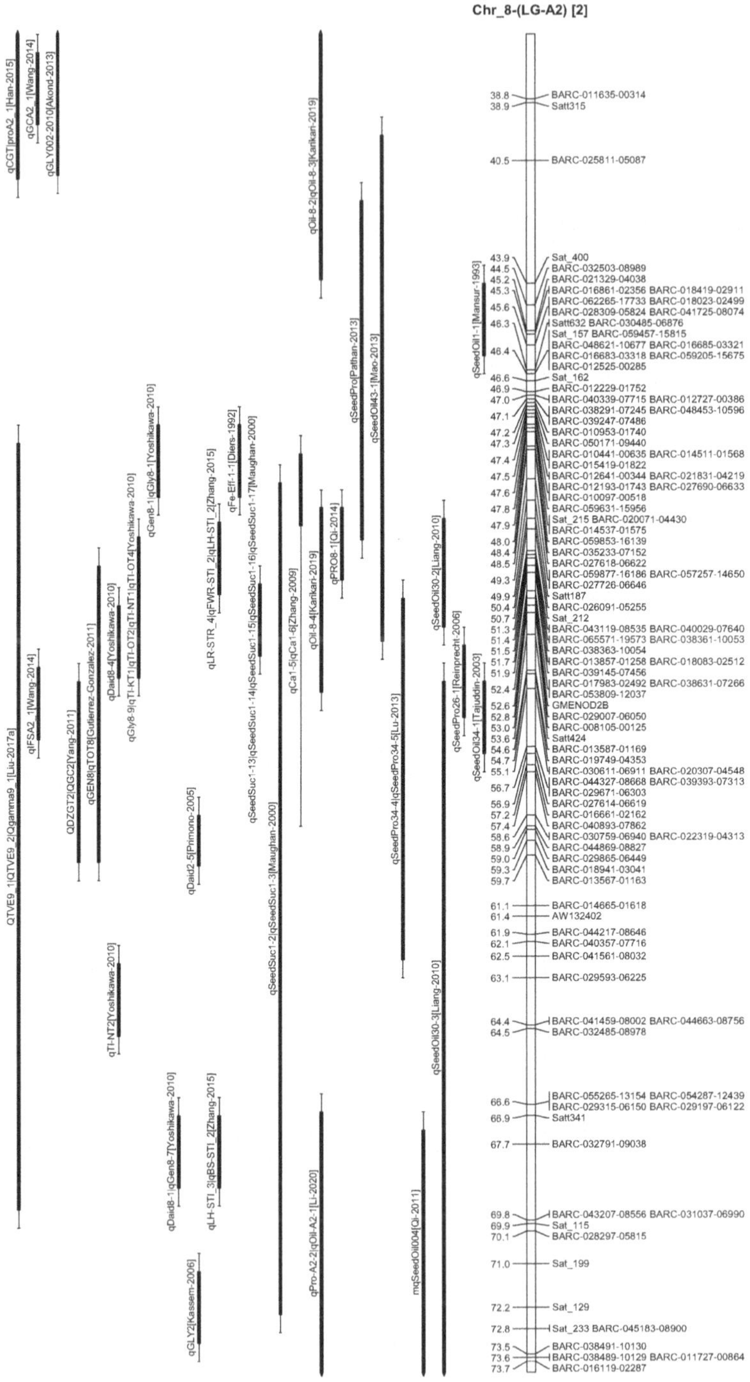

Fig 2.1 (continued)

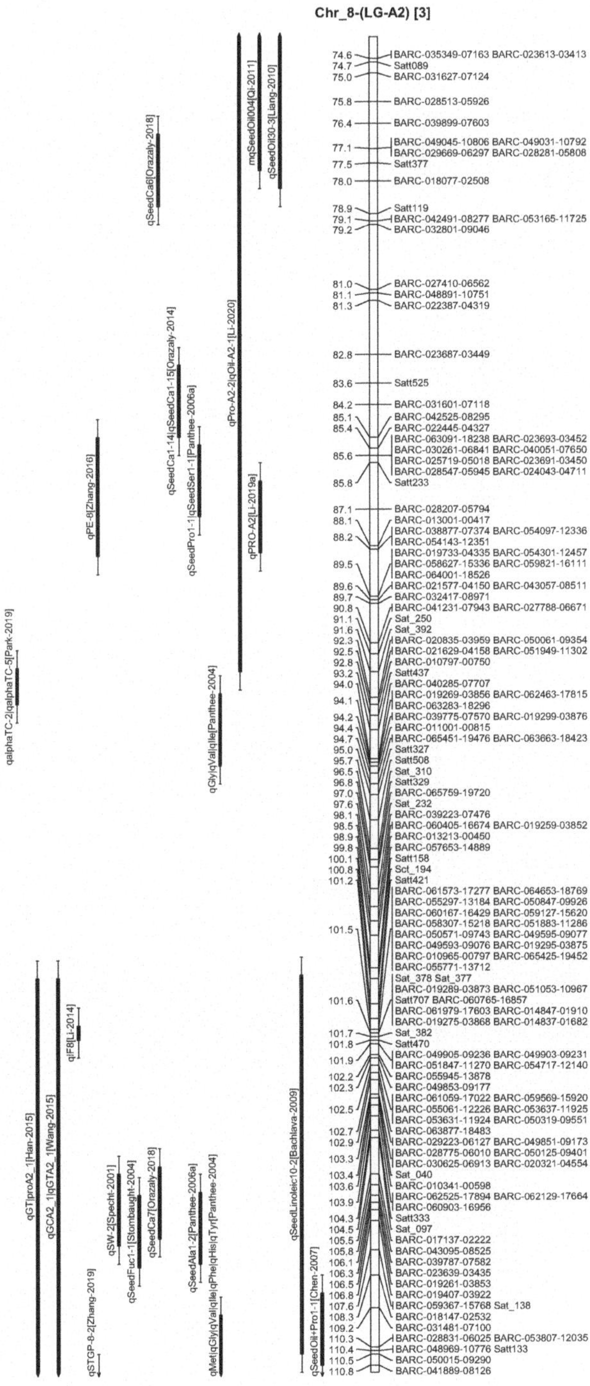

Fig 2.1 (continued)

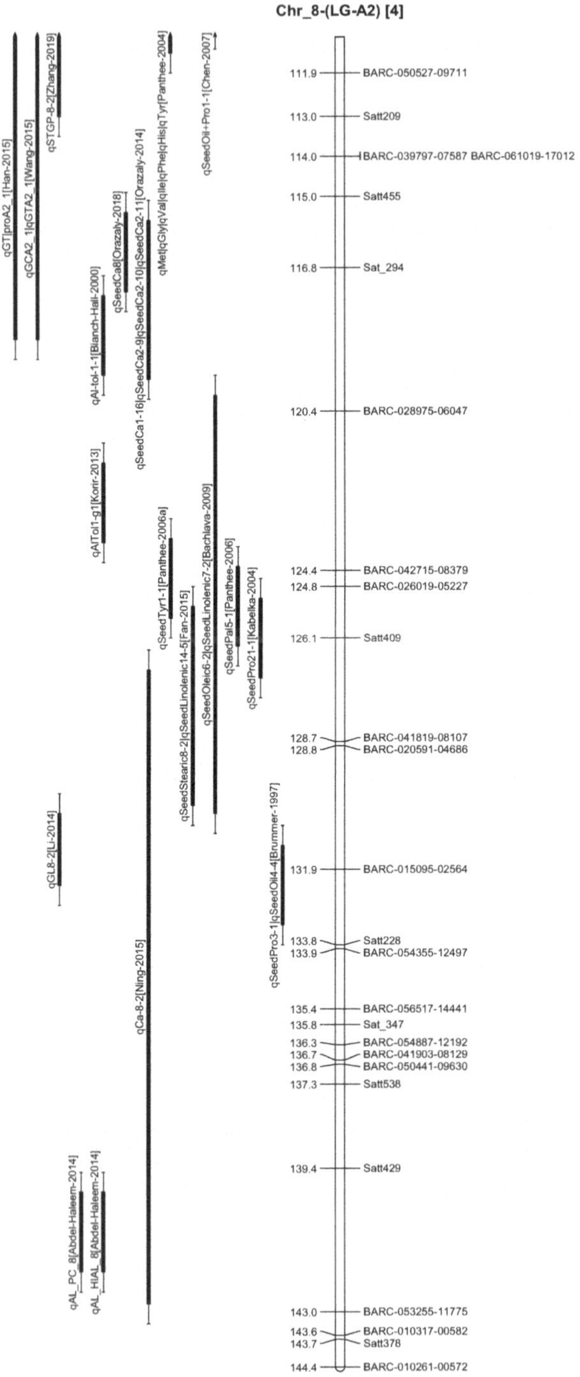

Fig 2.1 (continued)

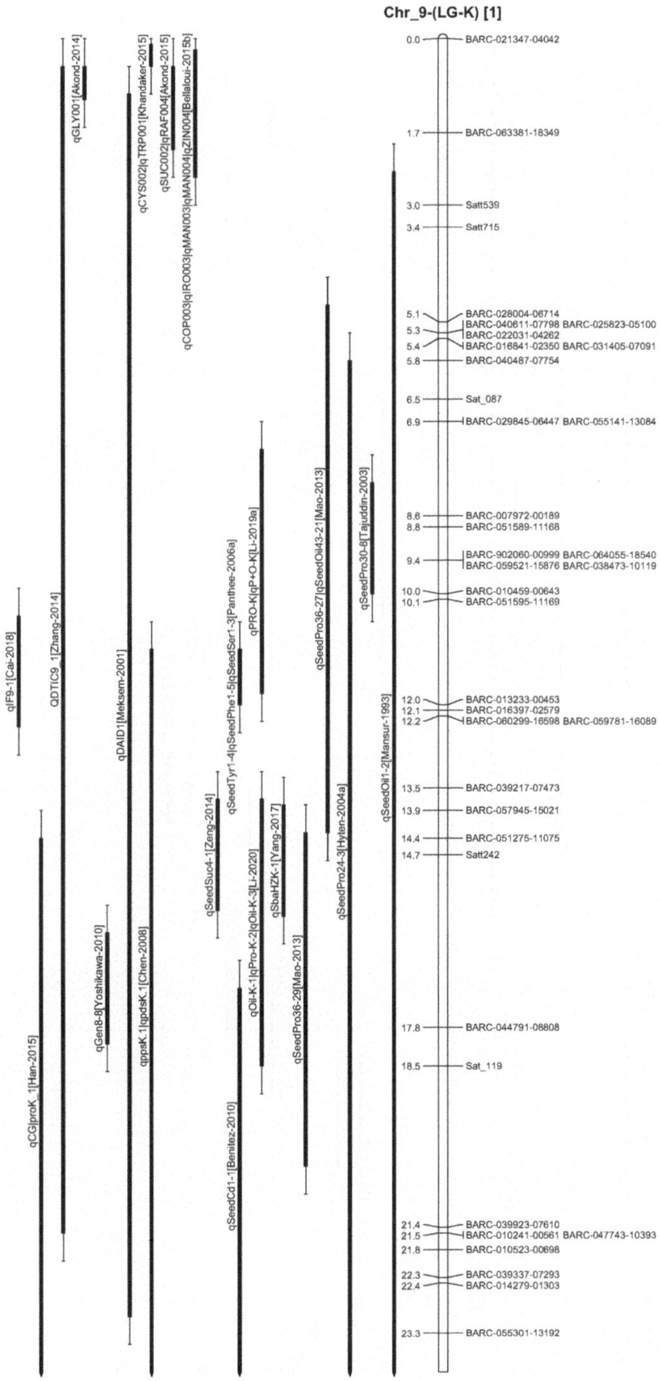

Fig 2.1 (continued)

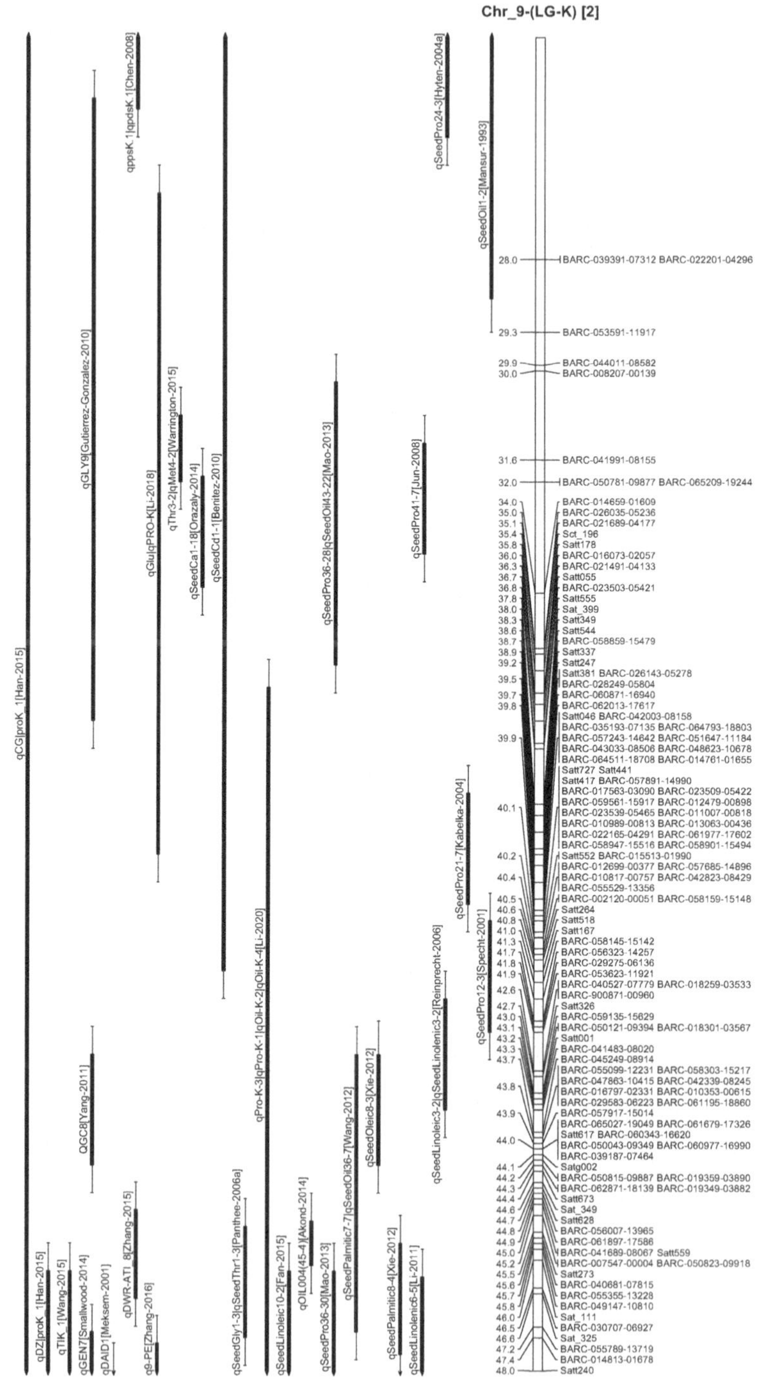

Fig 2.1 (continued)

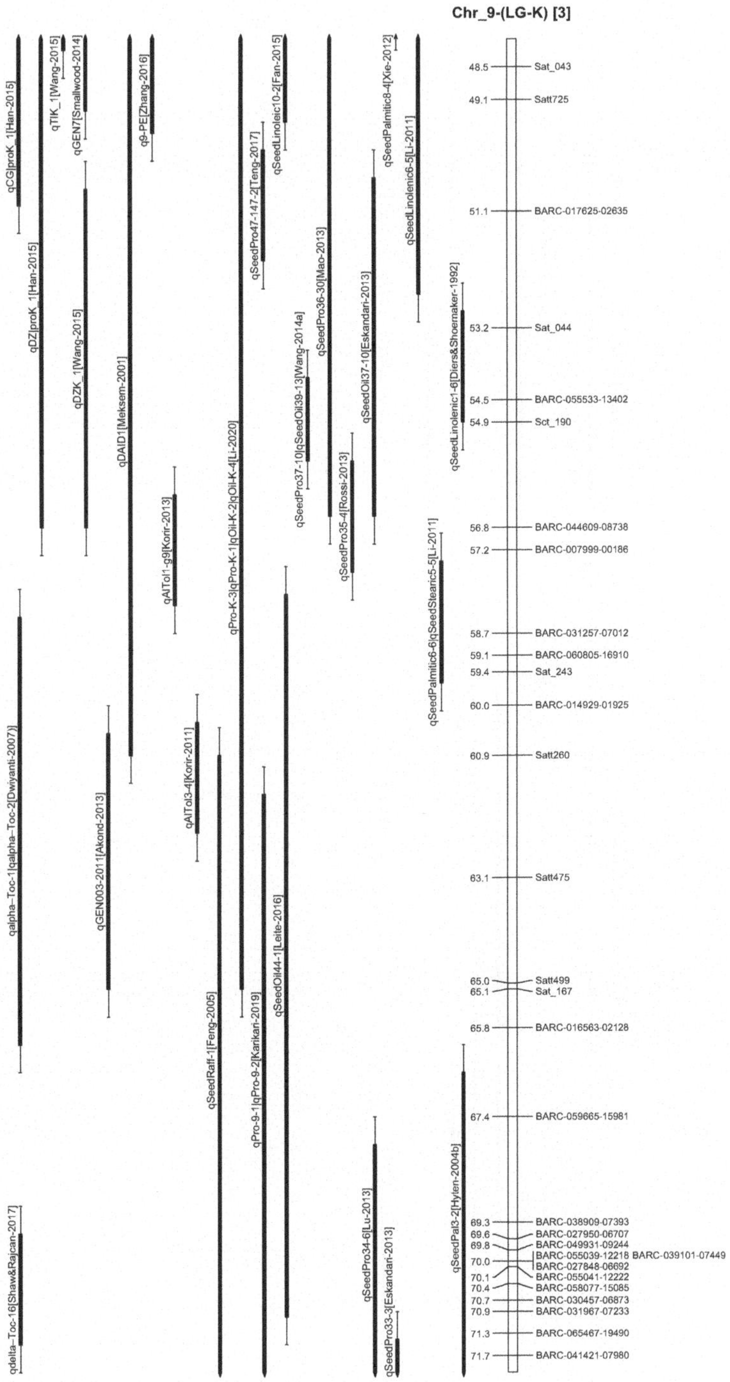

Fig 2.1 (continued)

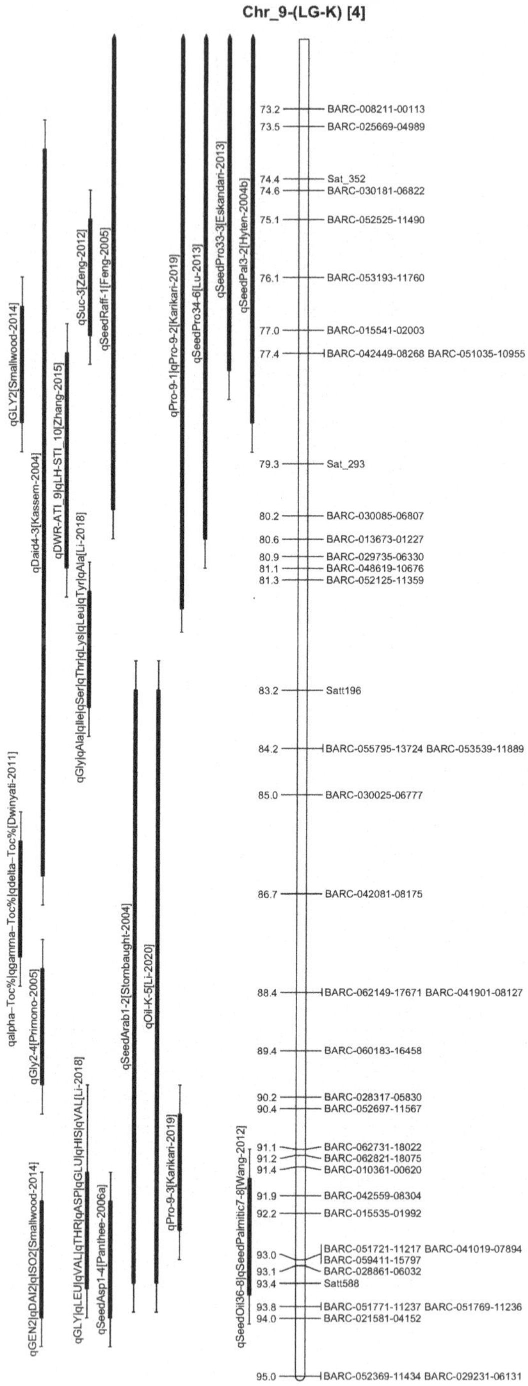

Fig 2.1 (continued)

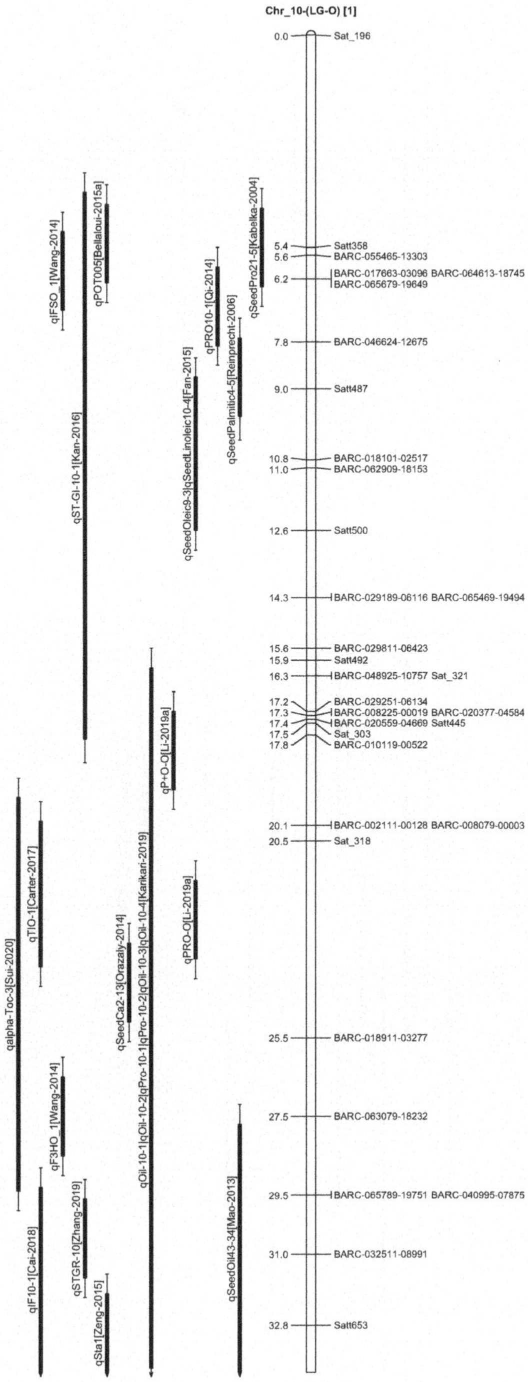

Fig 2.1 (continued)

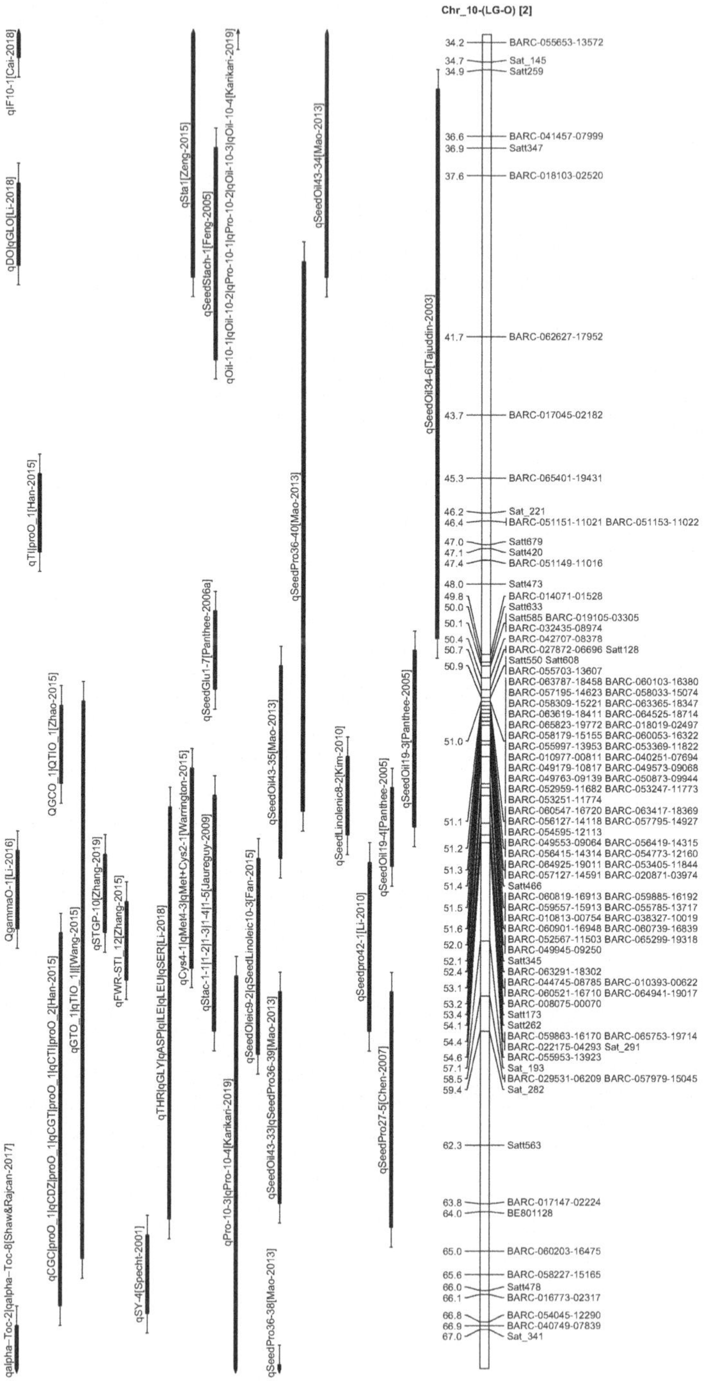

Fig 2.1 (continued)

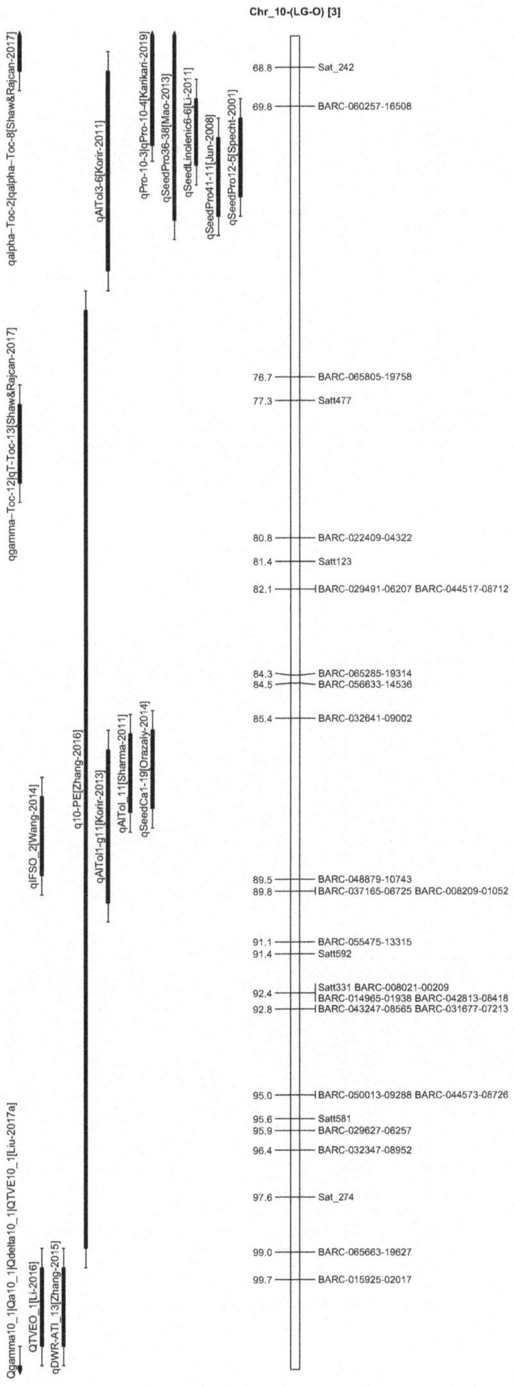

Fig 2.1 (continued)

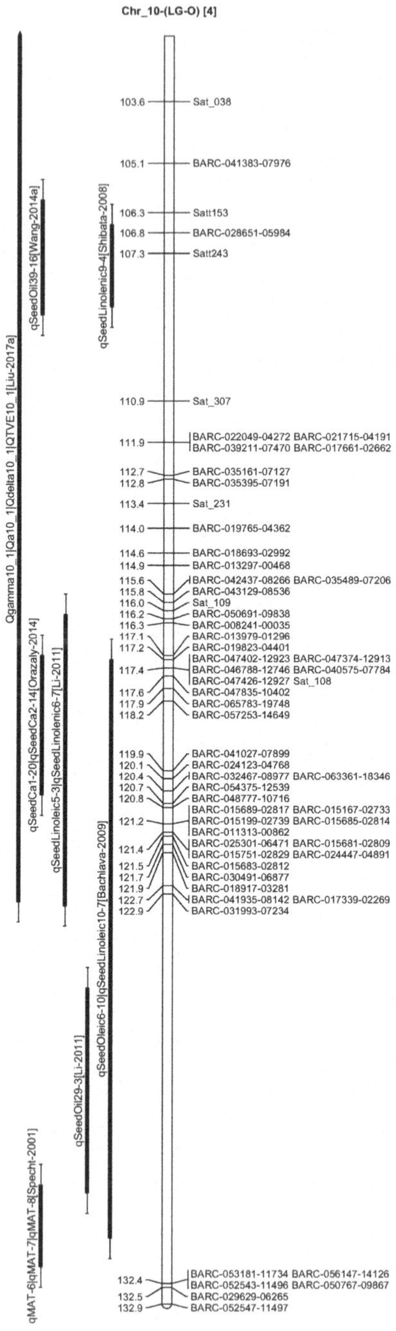

Fig 2.1 (continued)

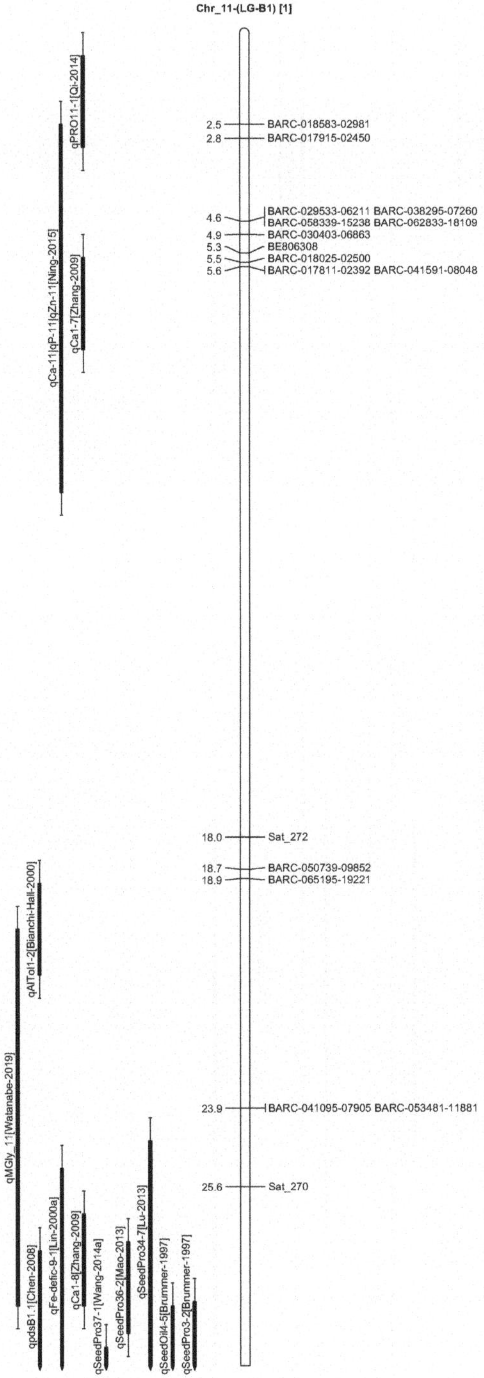

Fig 2.1 (continued)

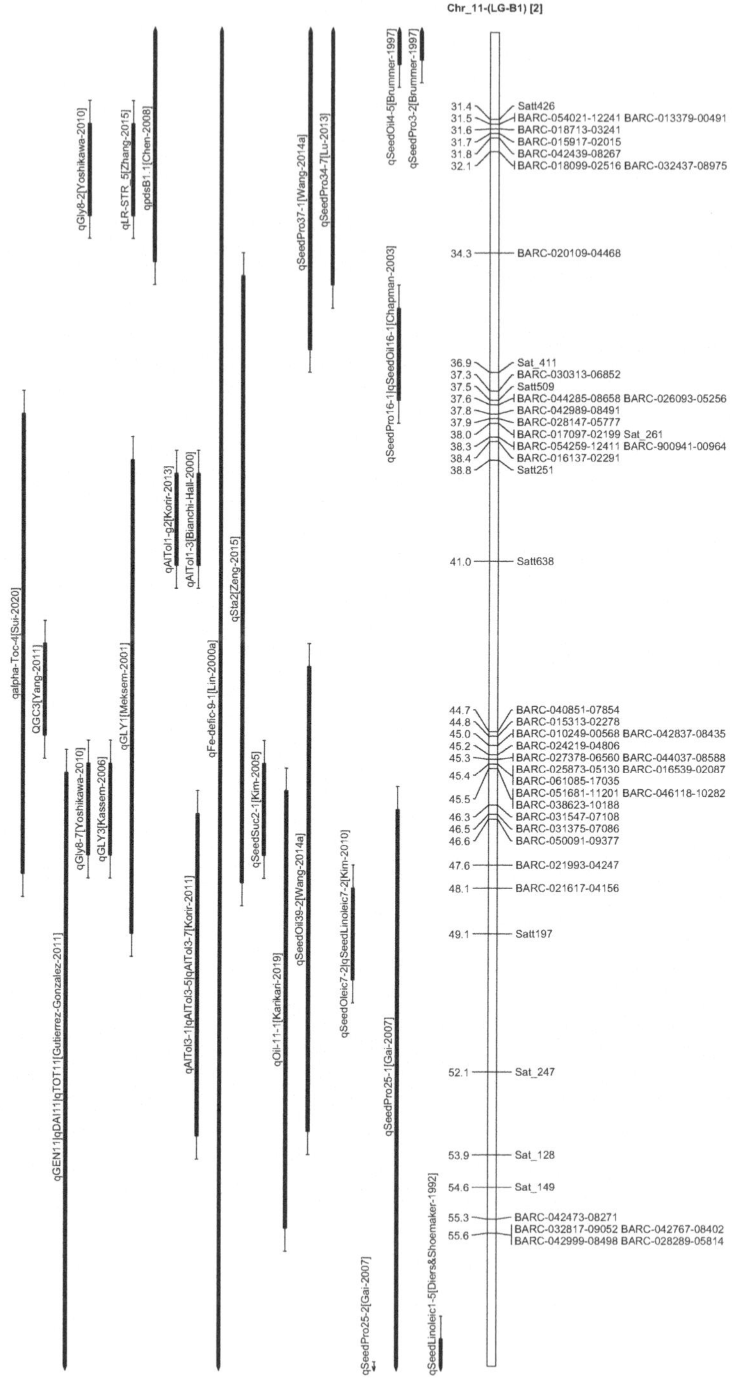

Fig 2.1 (continued)

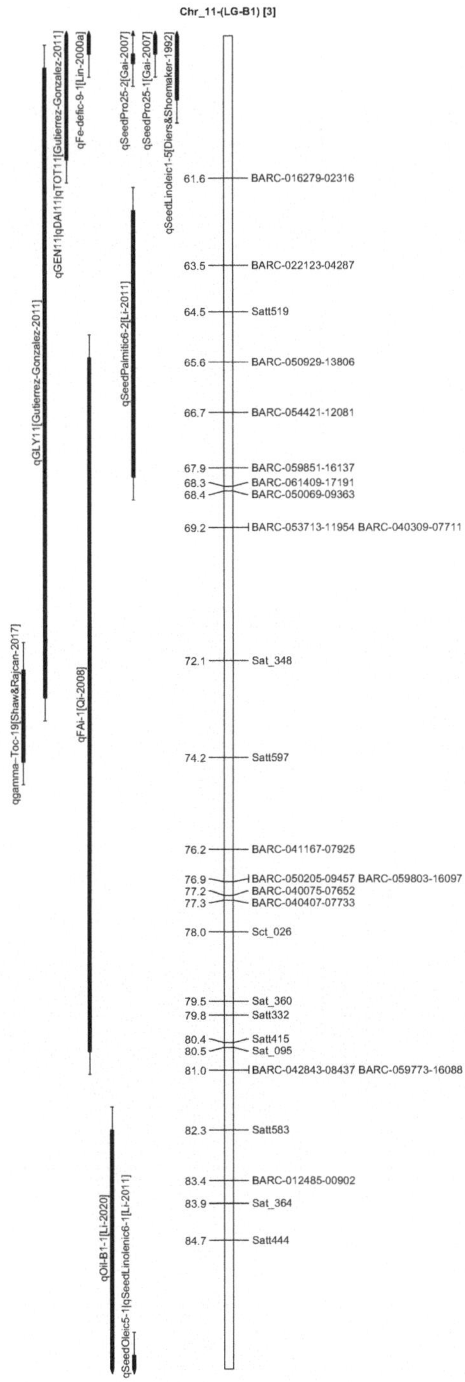

Fig 2.1 (continued)

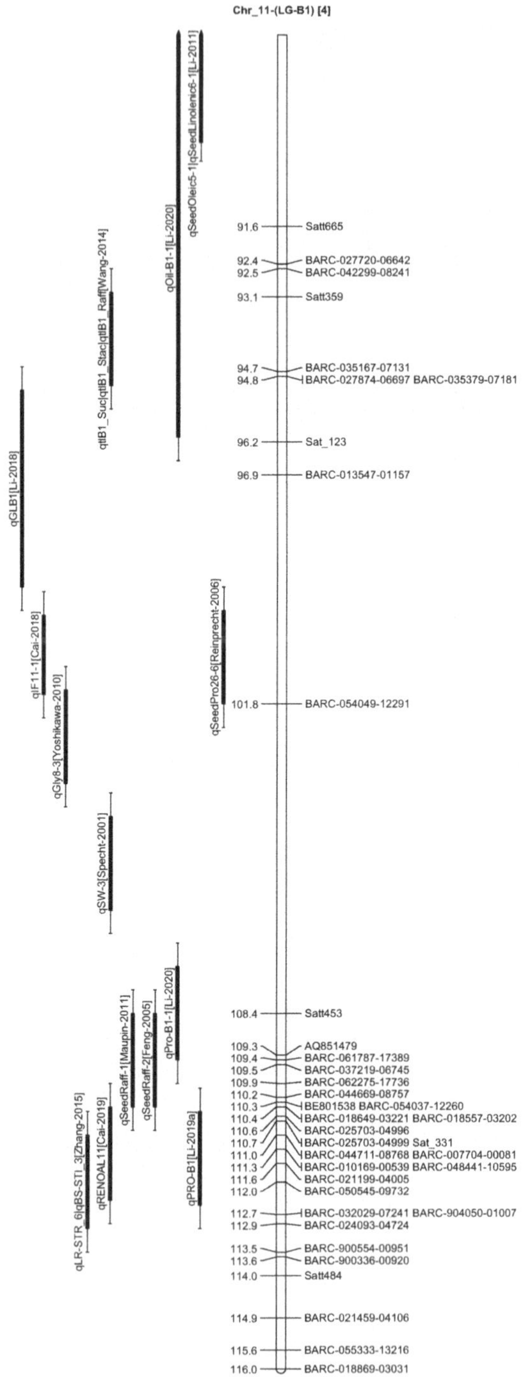

Fig 2.1 (continued)

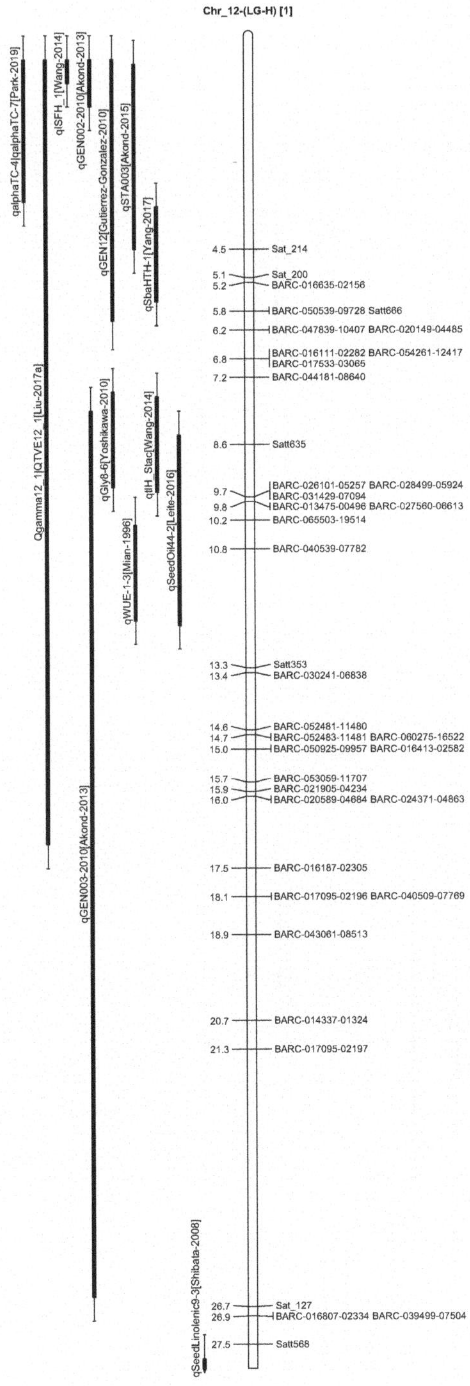

Fig 2.1 (continued)

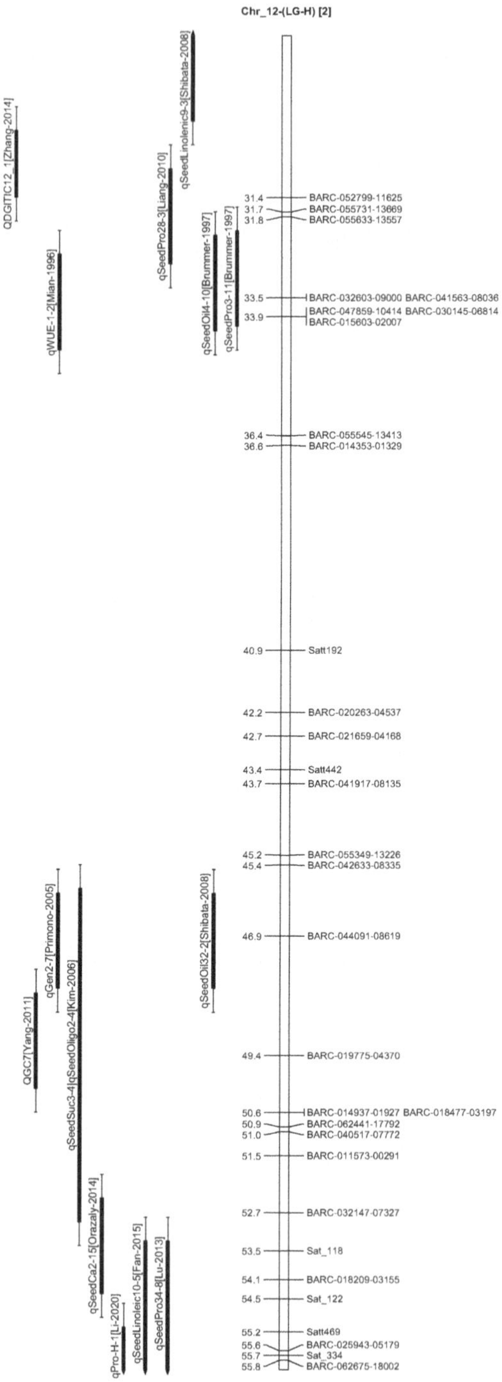

Fig 2.1 (continued)

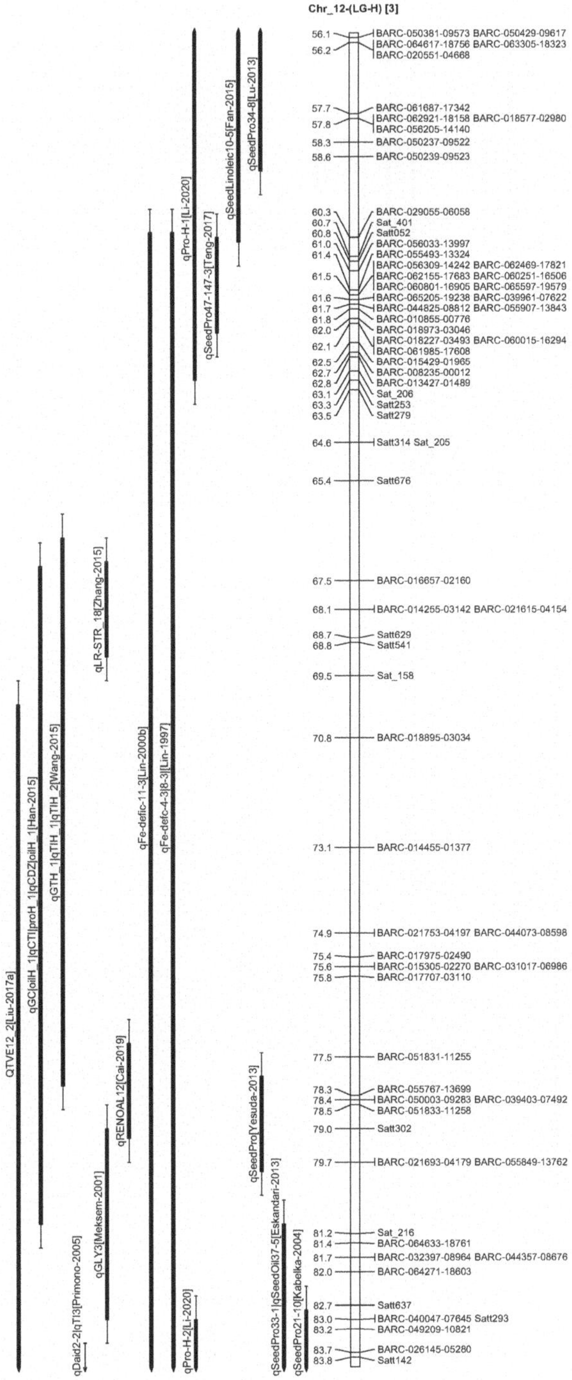

Fig 2.1 (continued)

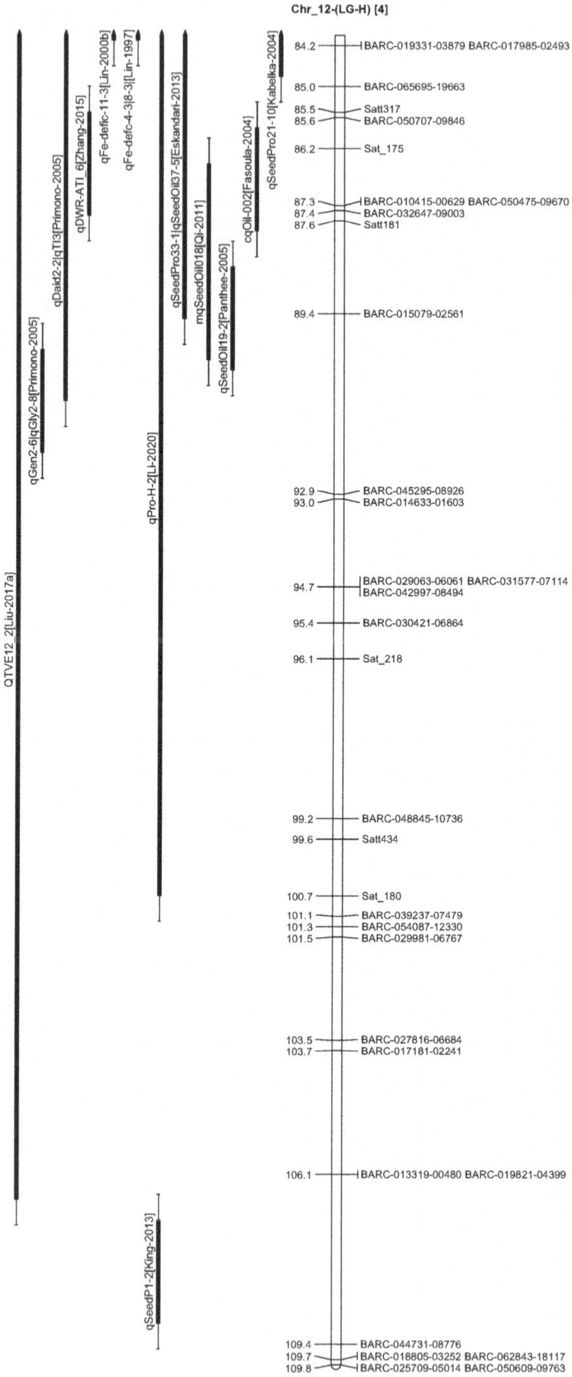

Fig 2.1 (continued)

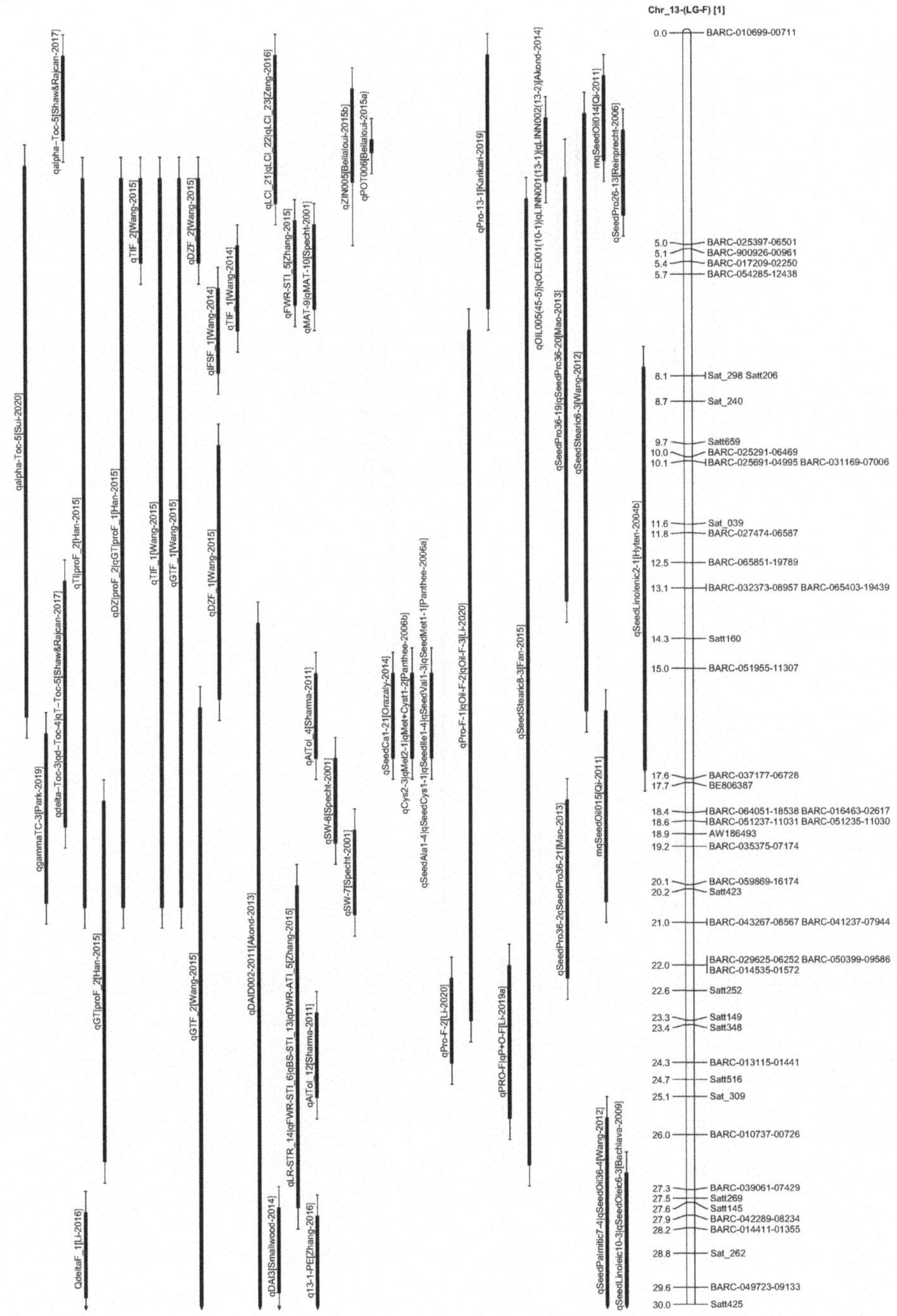

Fig 2.1 (continued)

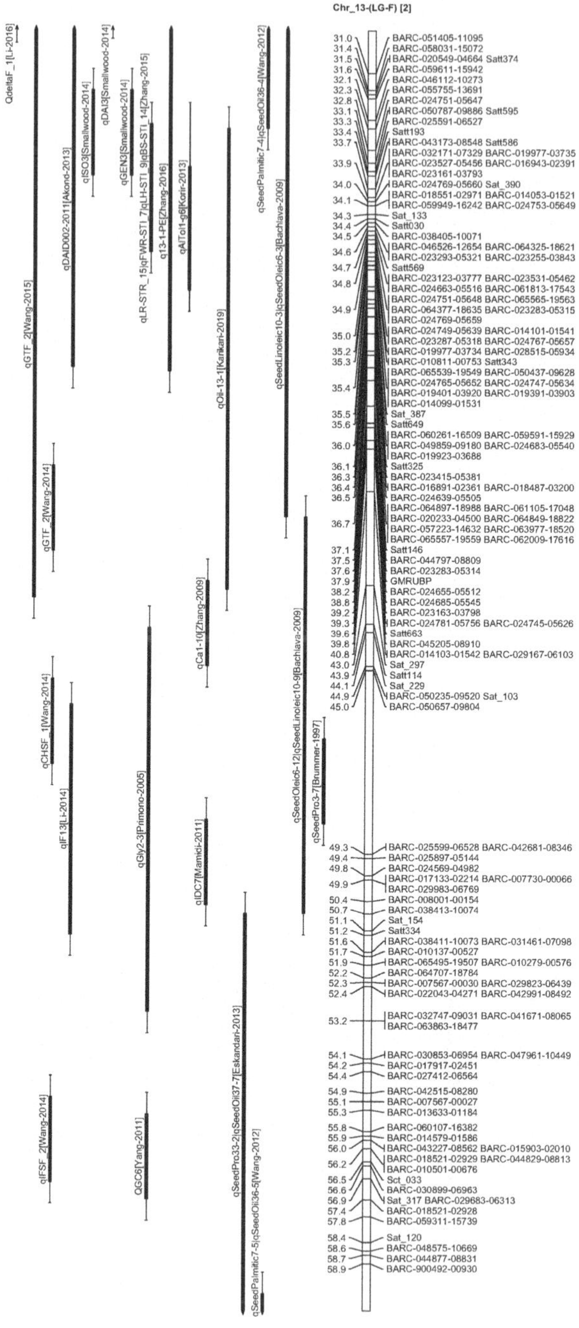

Fig 2.1 (continued)

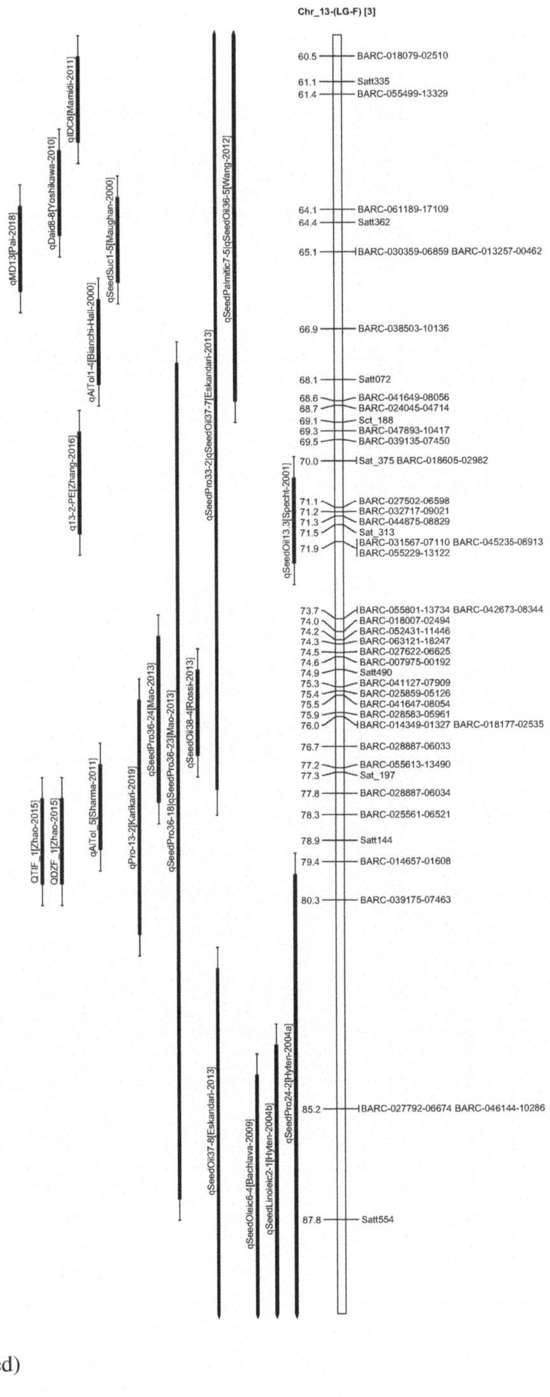

Fig 2.1 (continued)

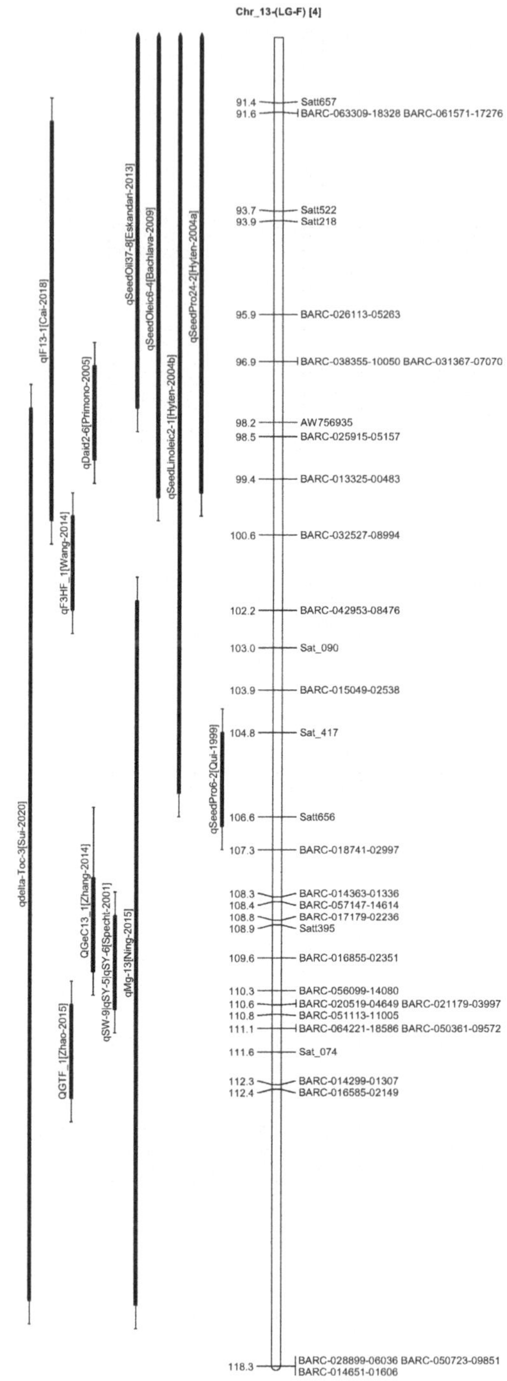

Fig 2.1 (continued)

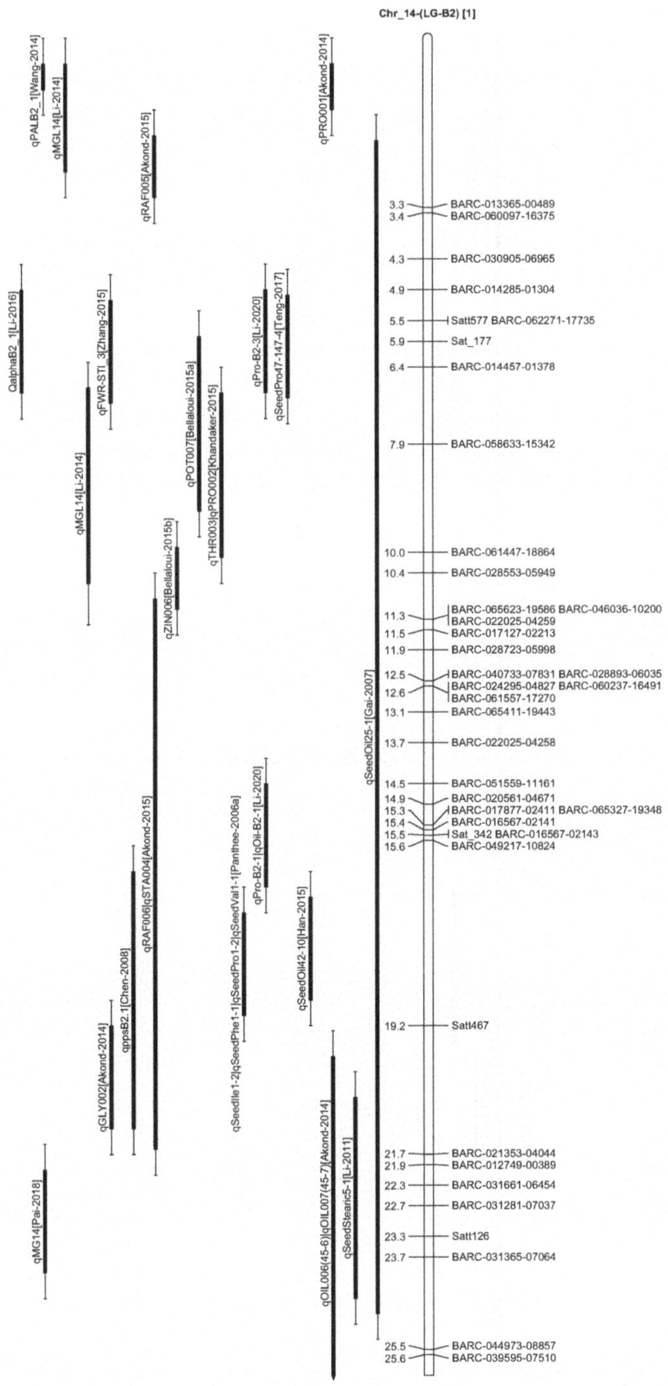

Fig 2.1 (continued)

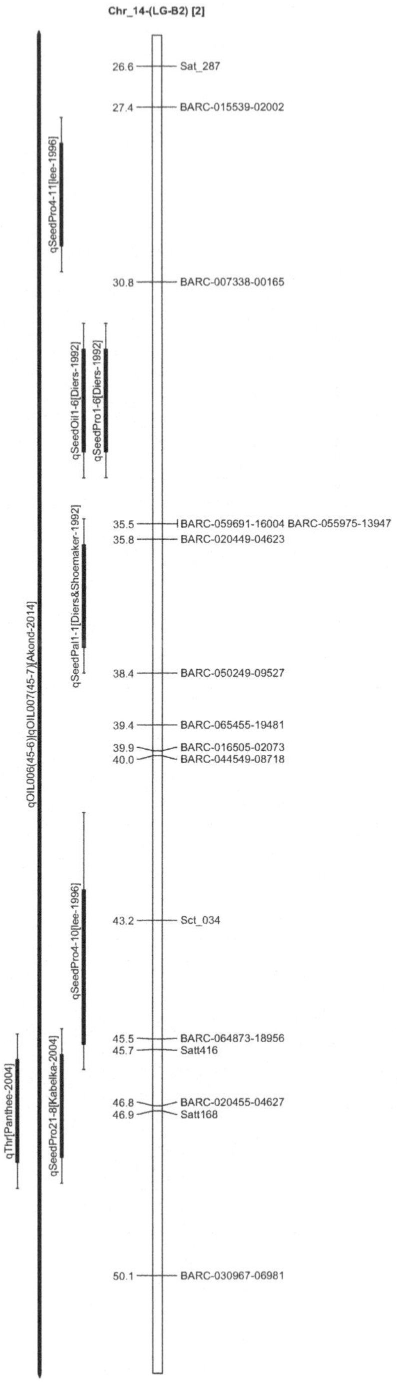

Fig 2.1 (continued)

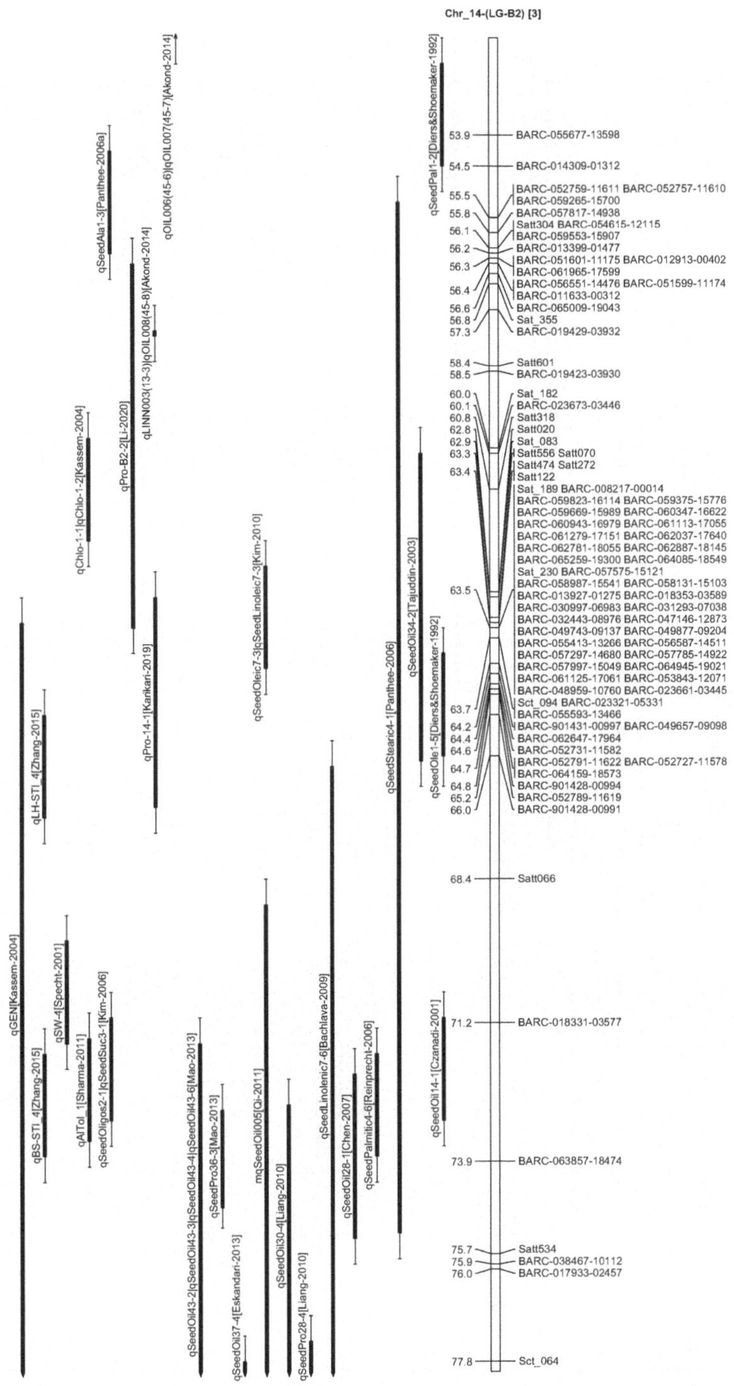

Fig 2.1 (continued)

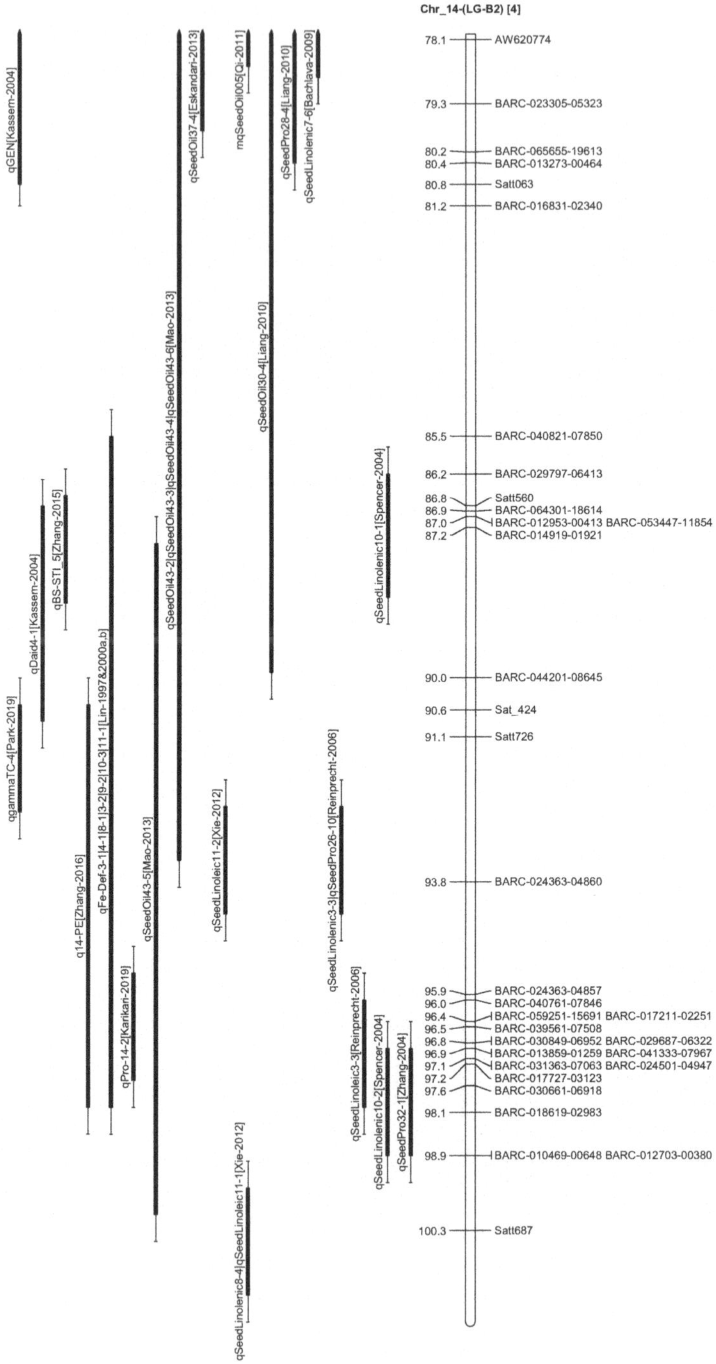

Fig 2.1 (continued)

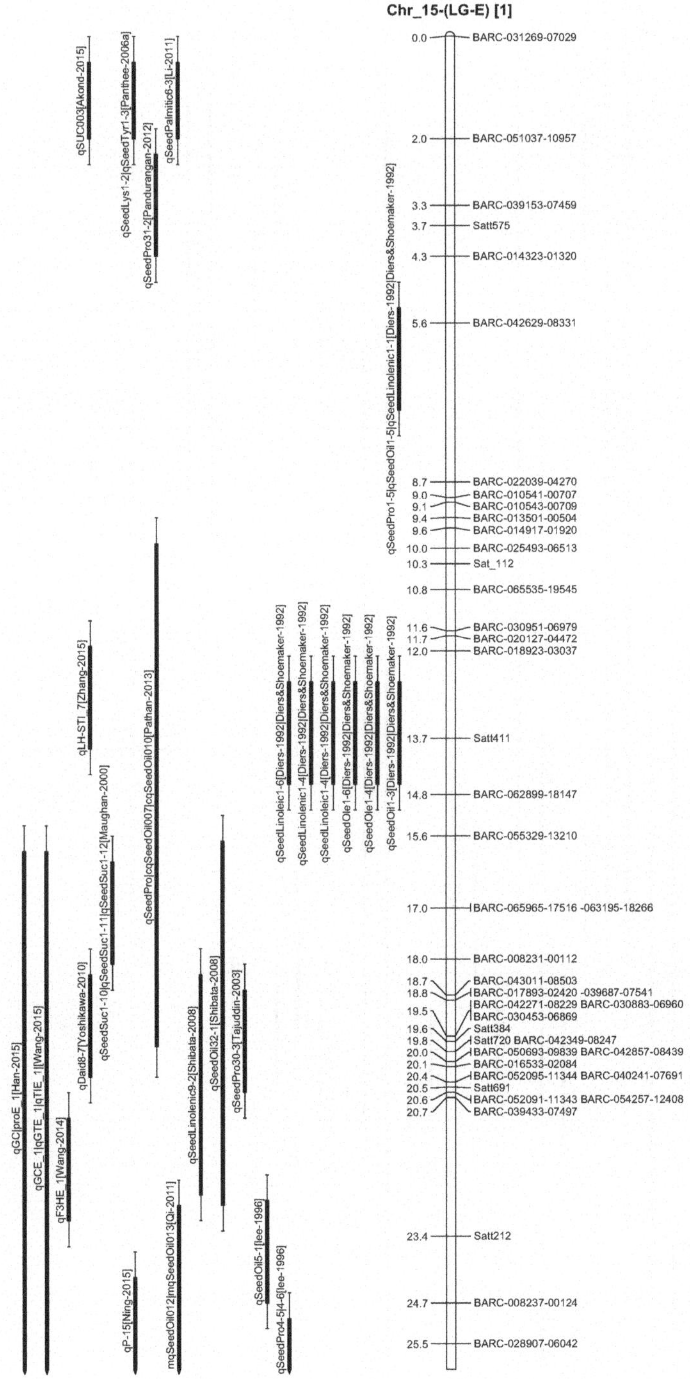

Fig 2.1 (continued)

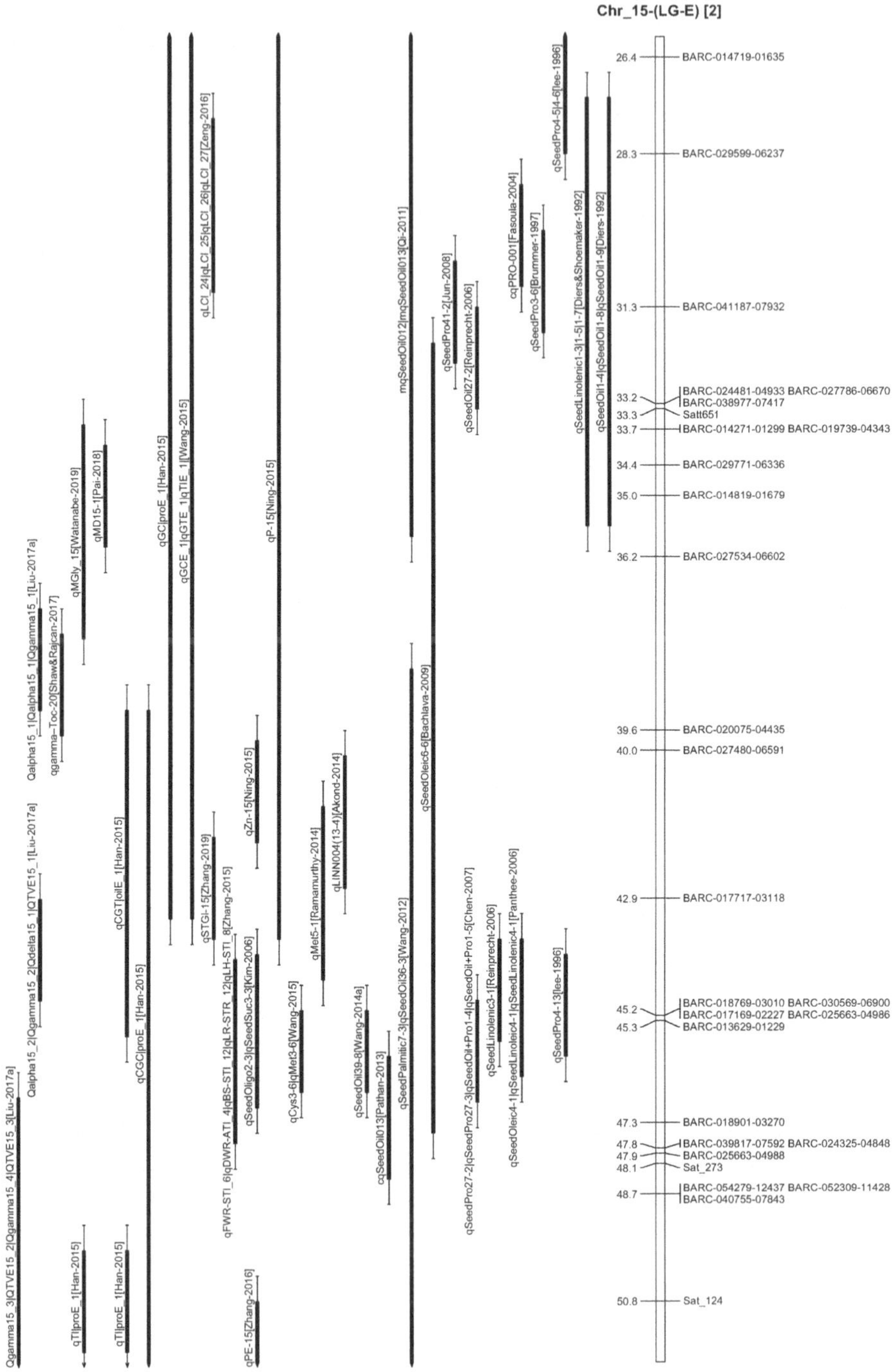

Fig 2.1 (continued)

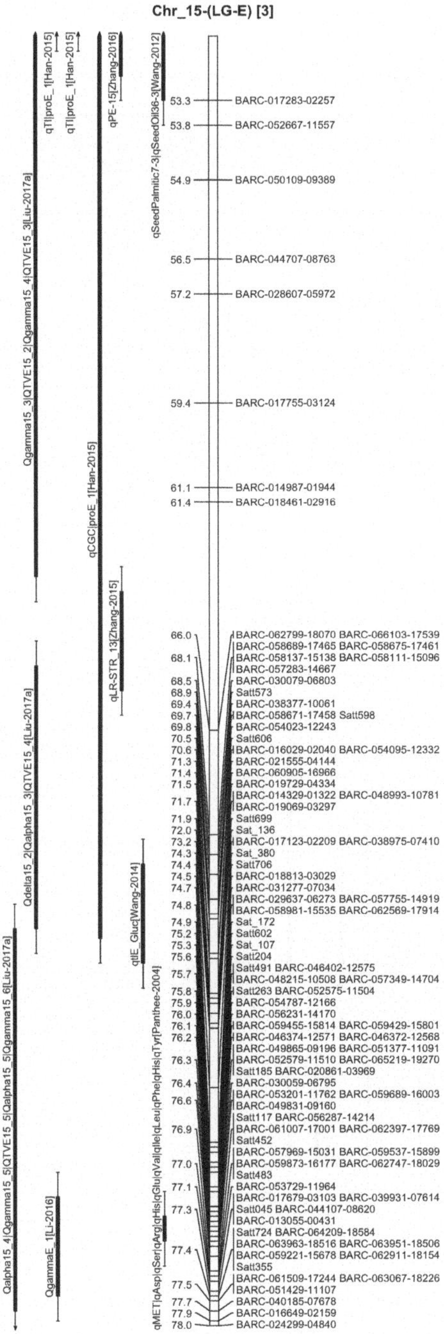

Fig 2.1 (continued)

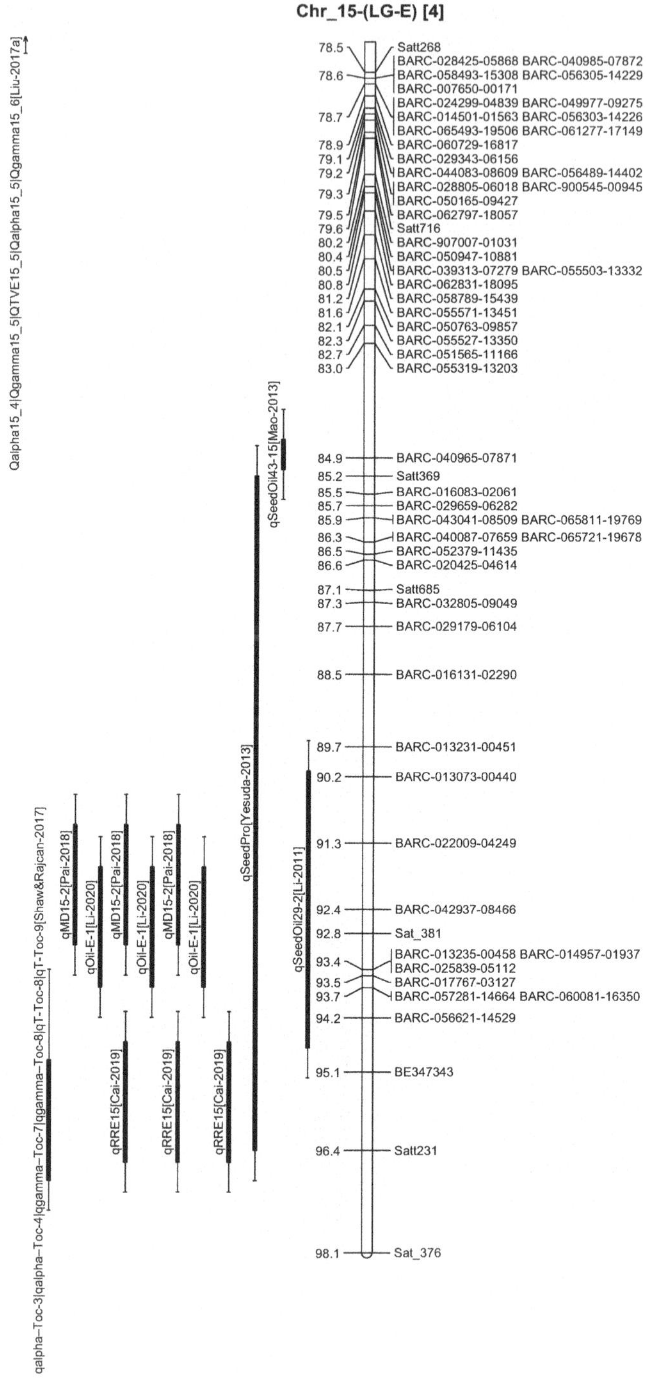

Fig 2.1 (continued)

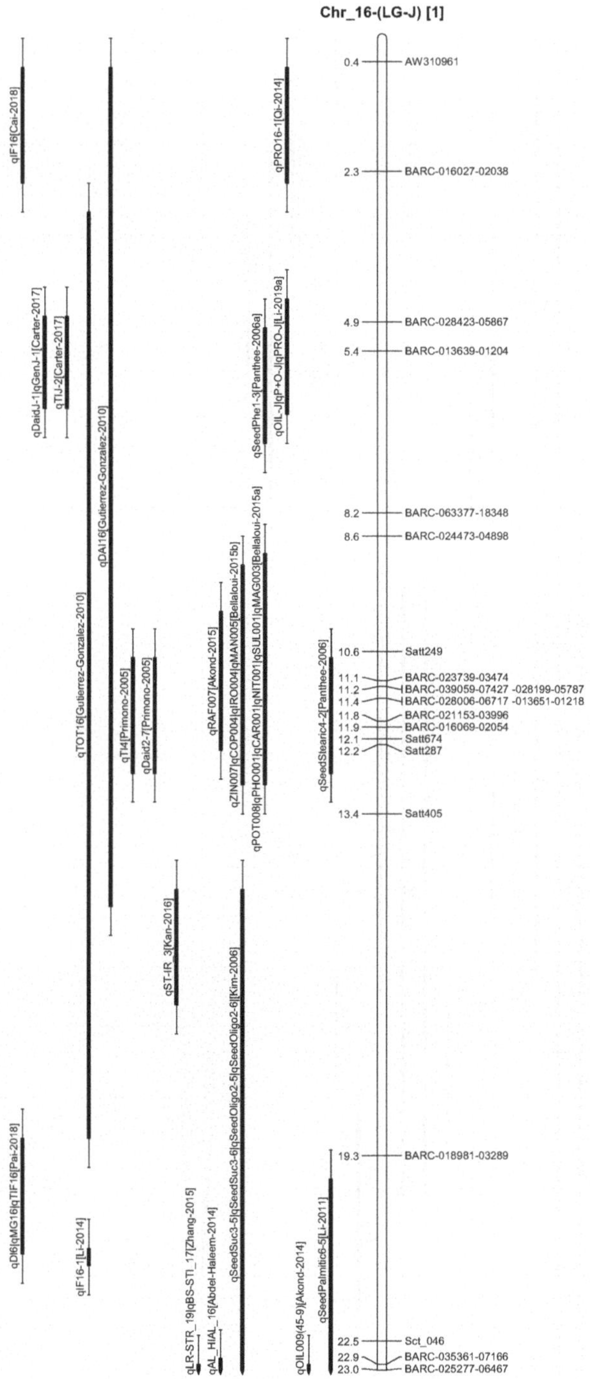

Fig 2.1 (continued)

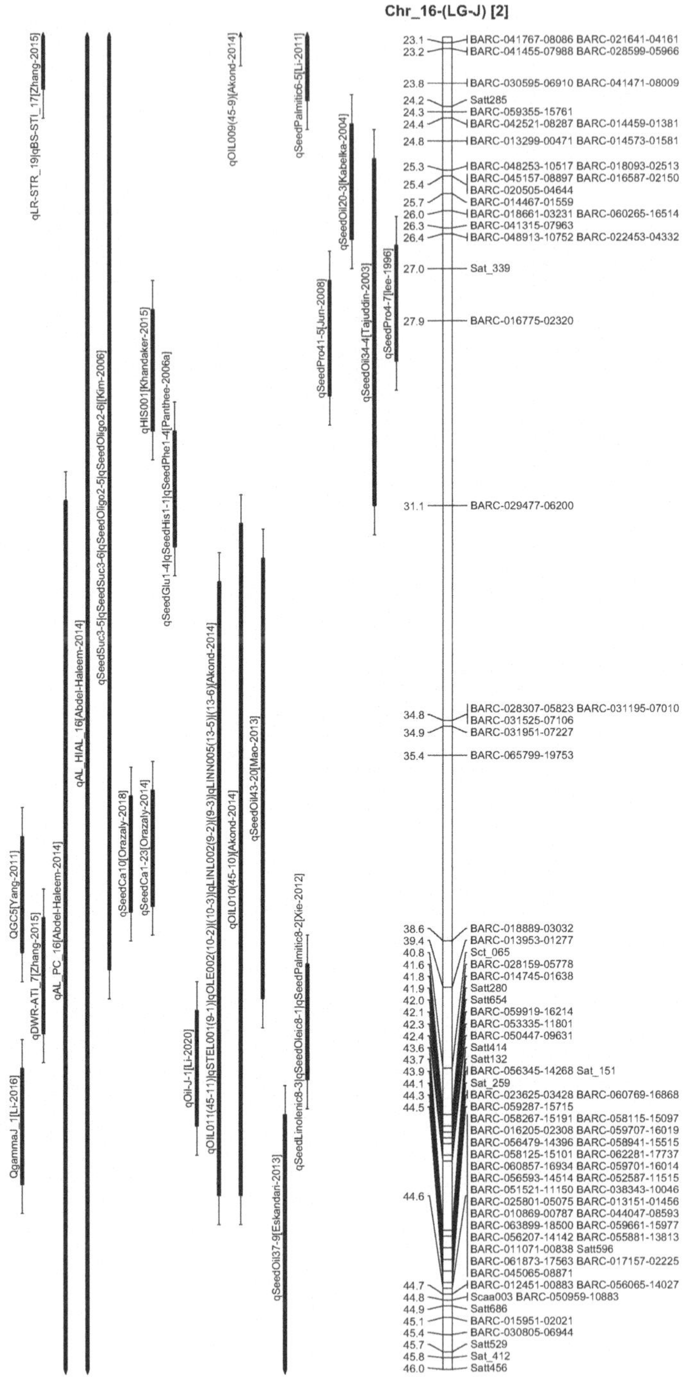

Fig 2.1 (continued)

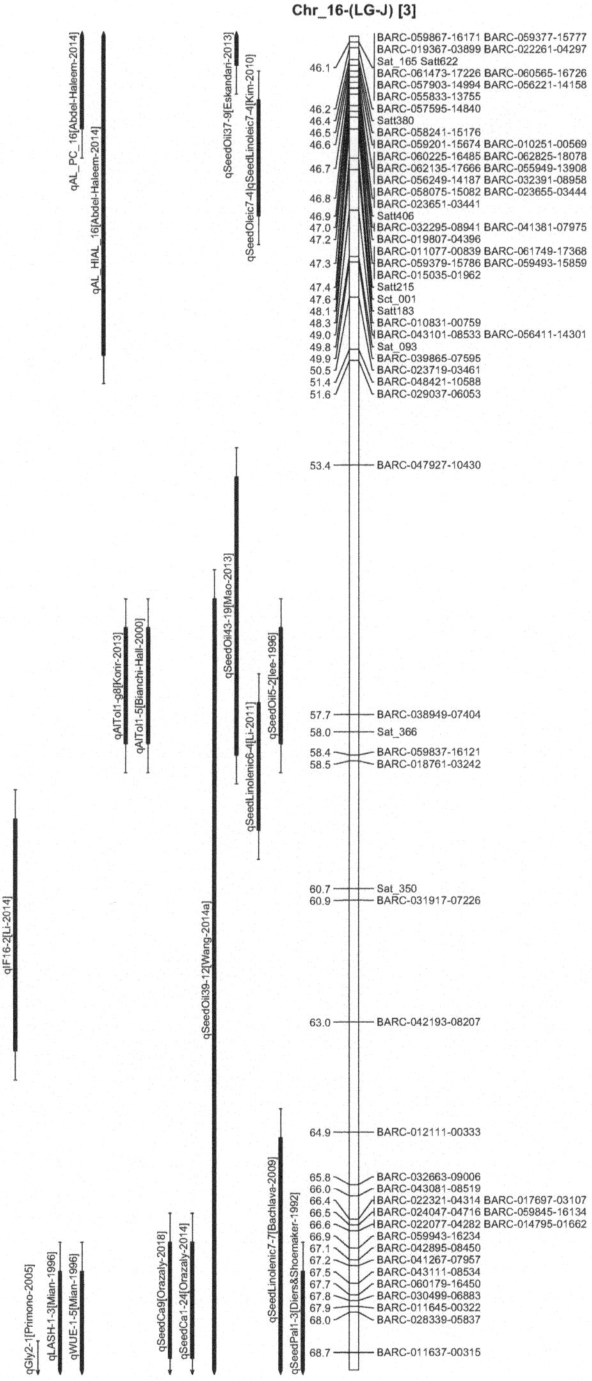

Fig 2.1 (continued)

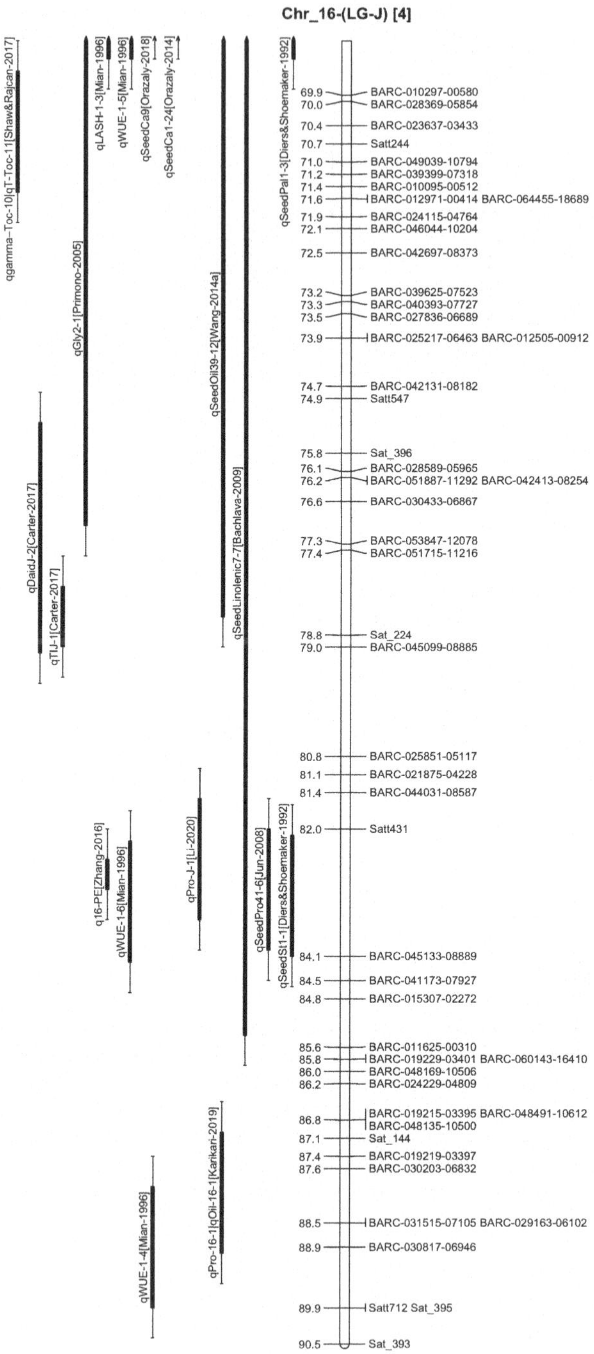

Fig 2.1 (continued)

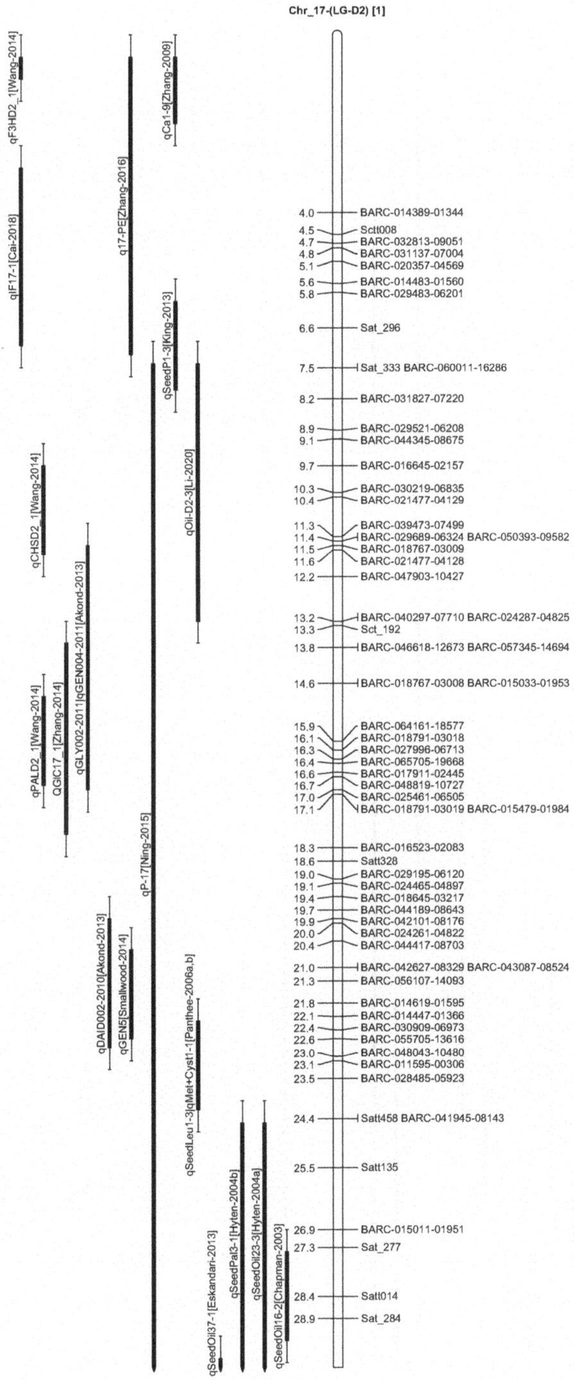

Fig 2.1 (continued)

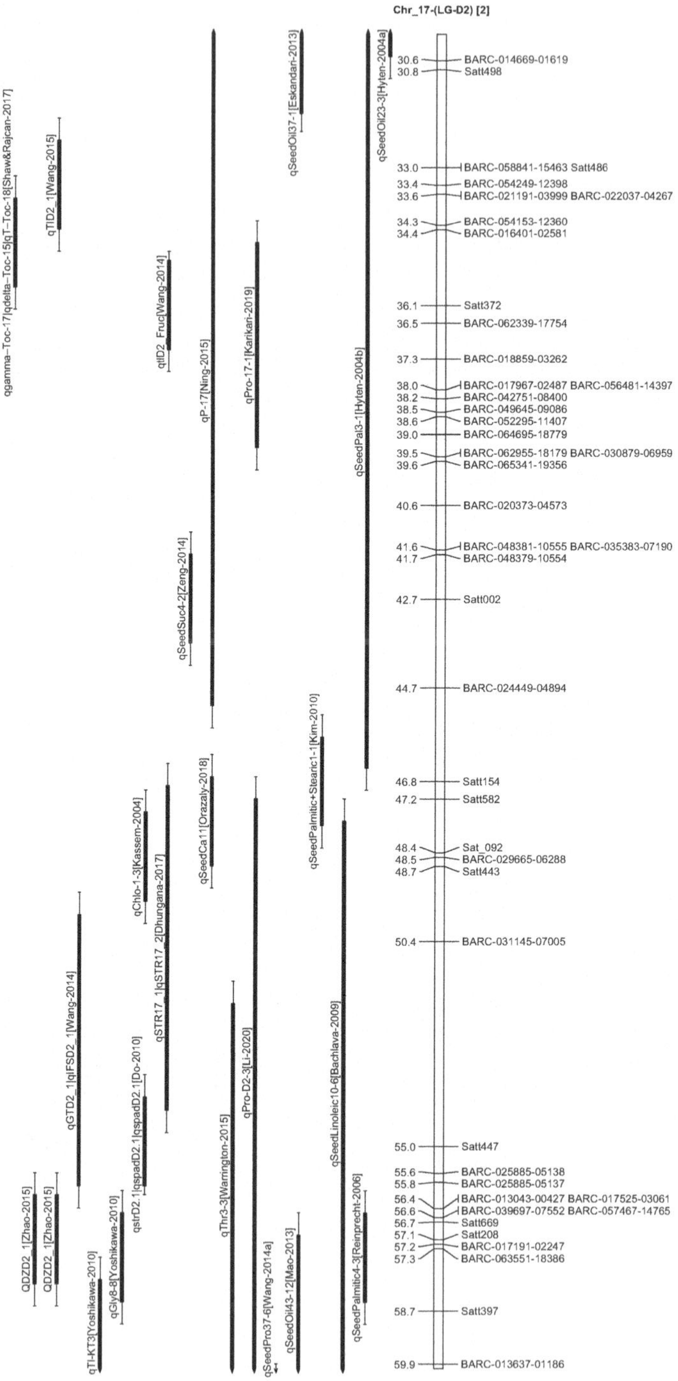

Fig 2.1 (continued)

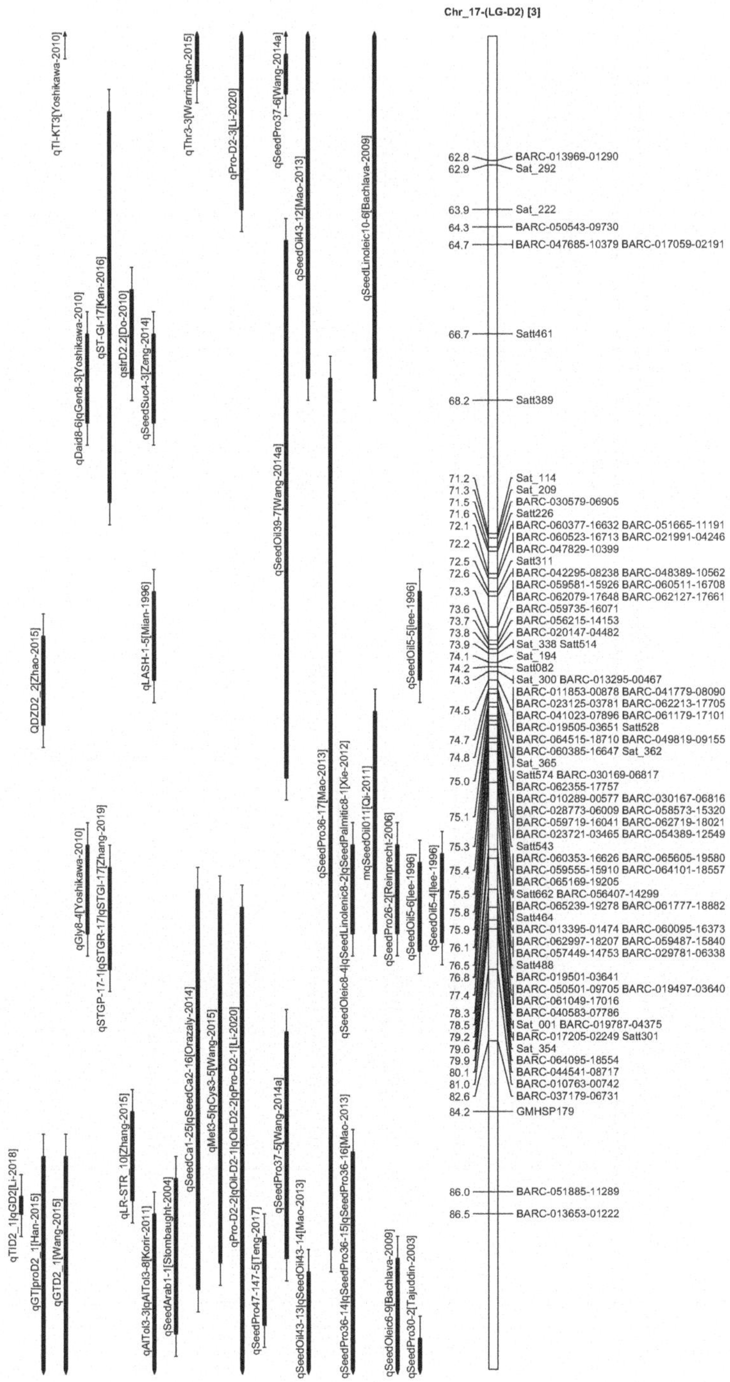

Fig 2.1 (continued)

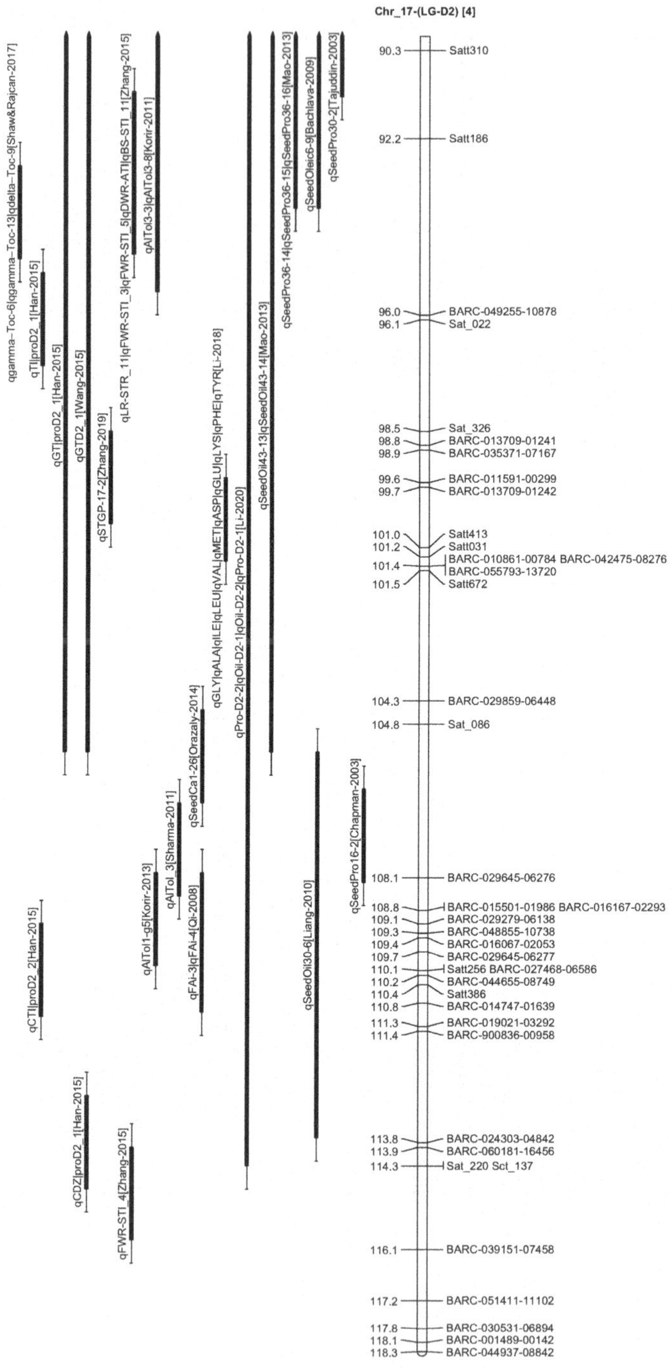

Fig 2.1 (continued)

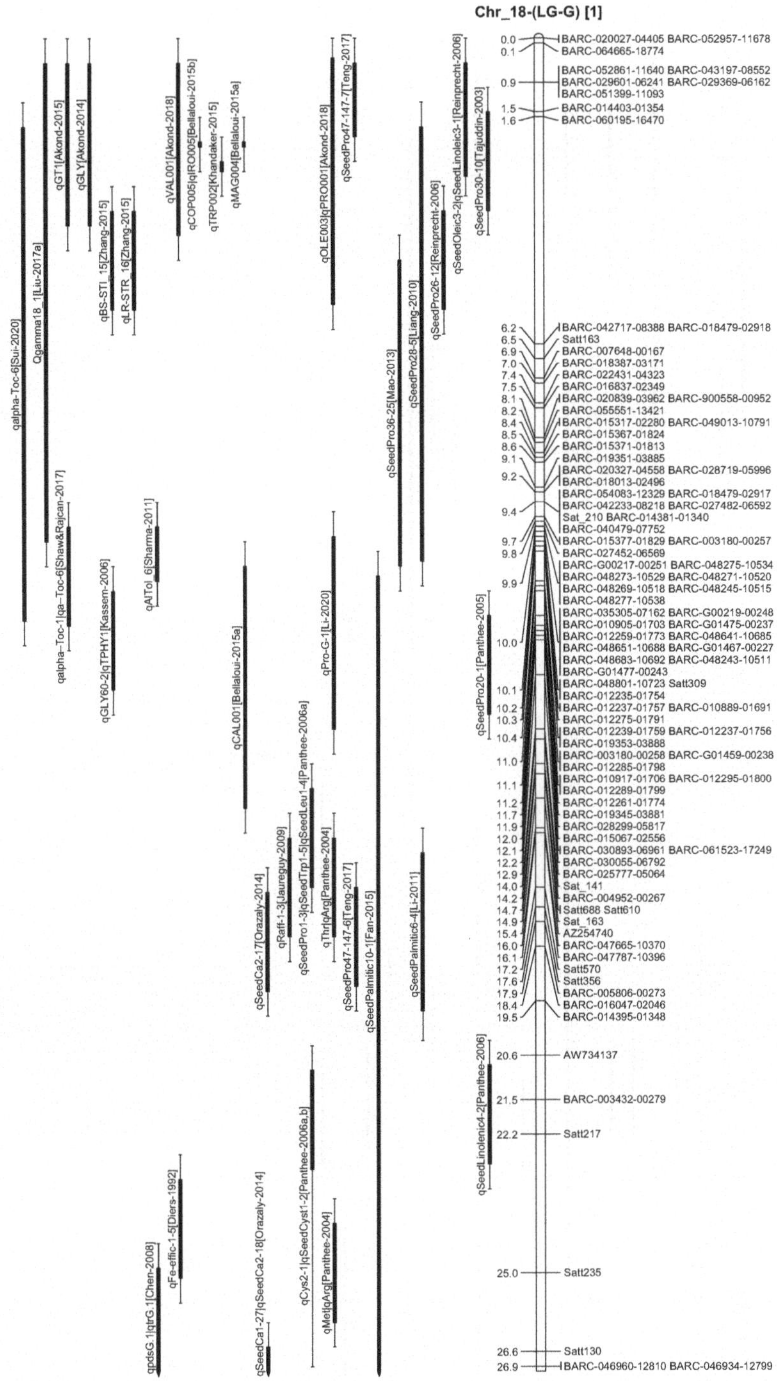

Fig 2.1 (continued)

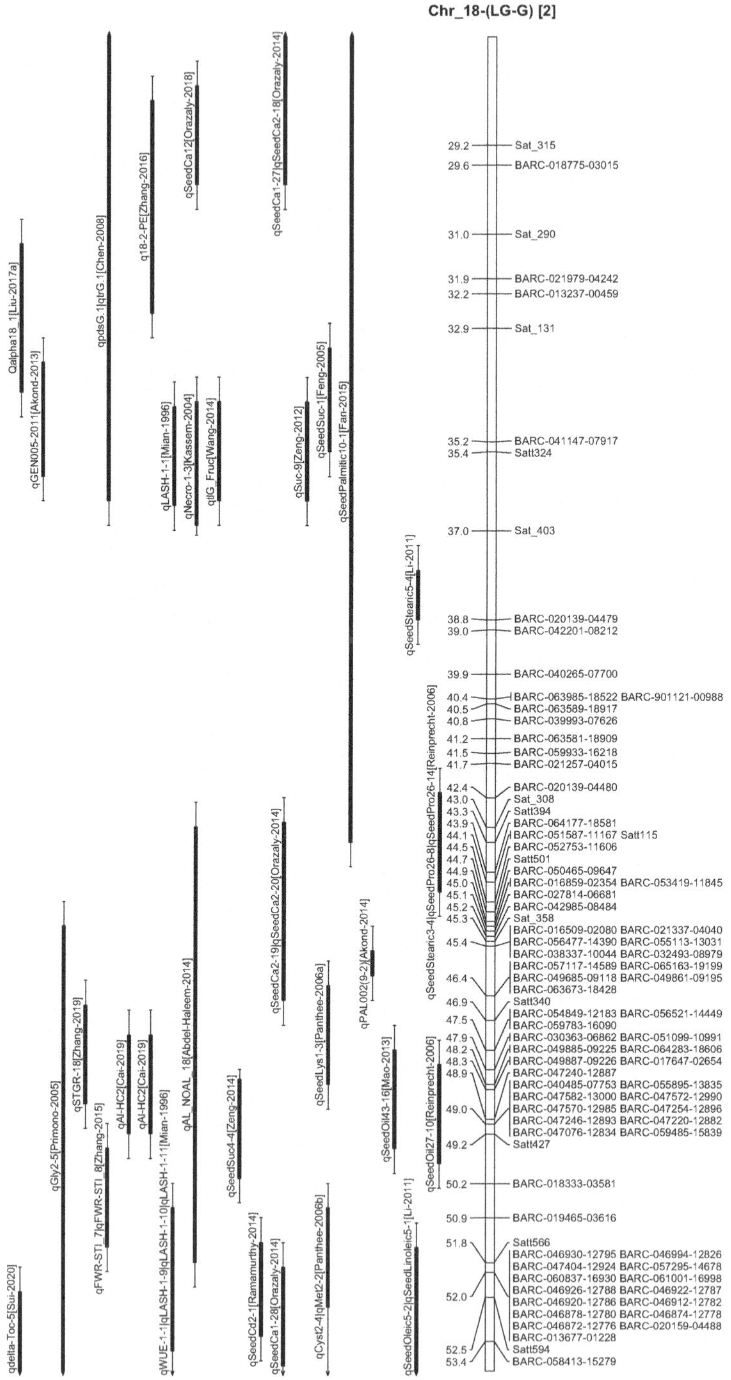

Fig 2.1 (continued)

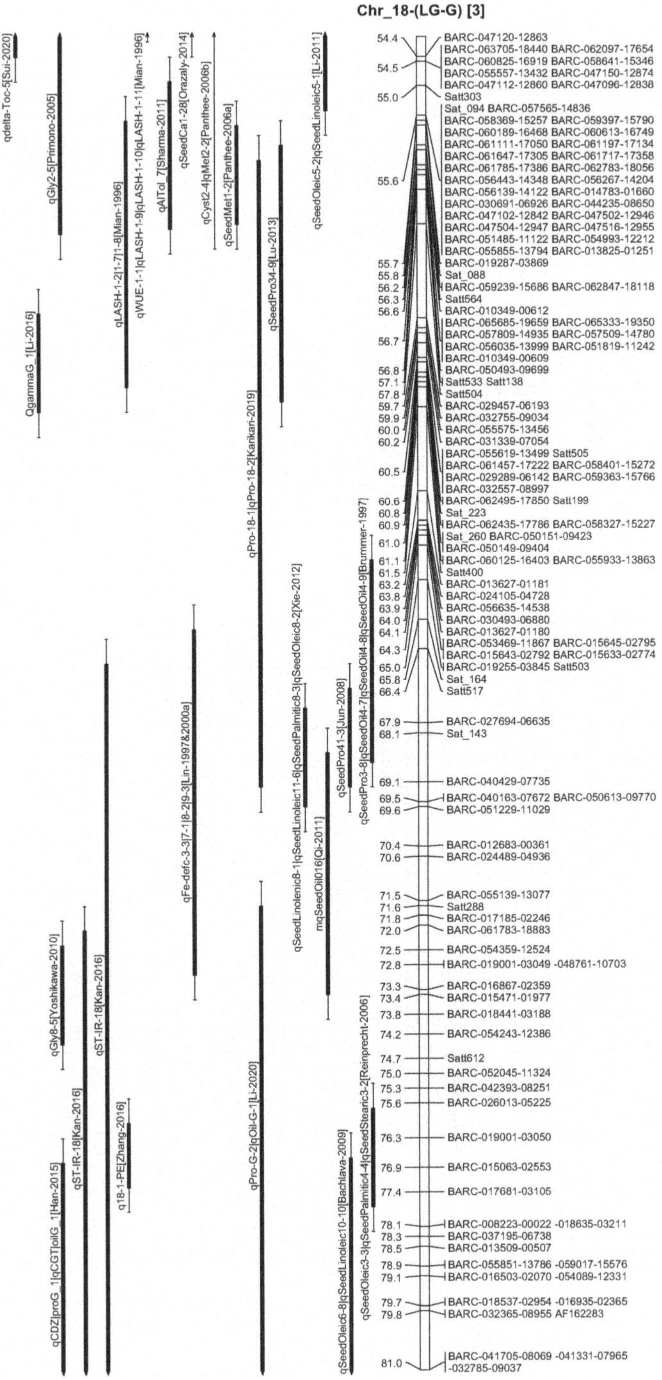

Fig 2.1 (continued)

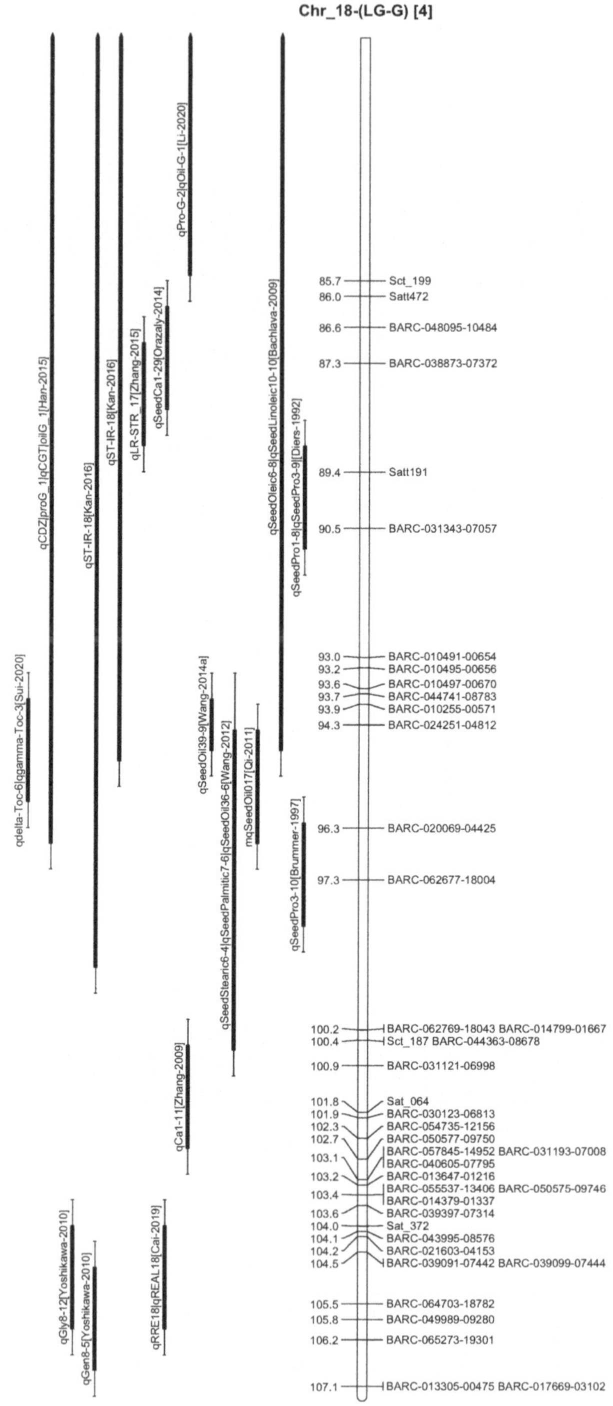

Fig 2.1 (continued)

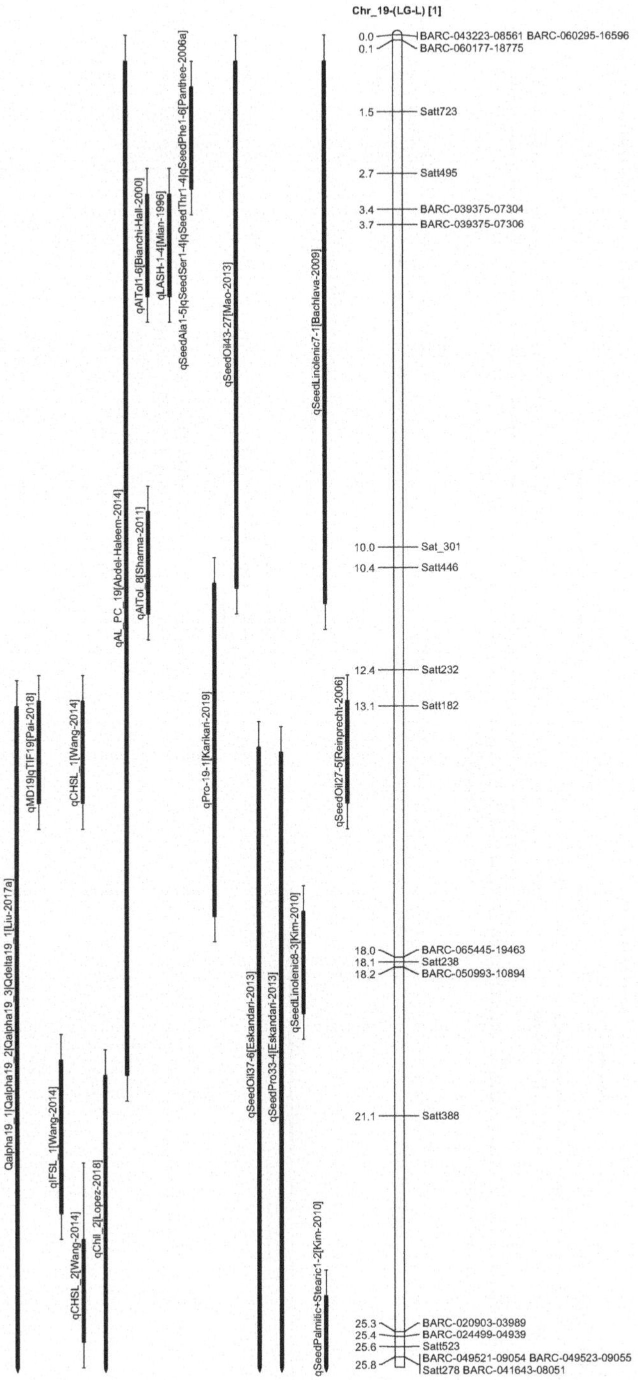

Fig 2.1 (continued)

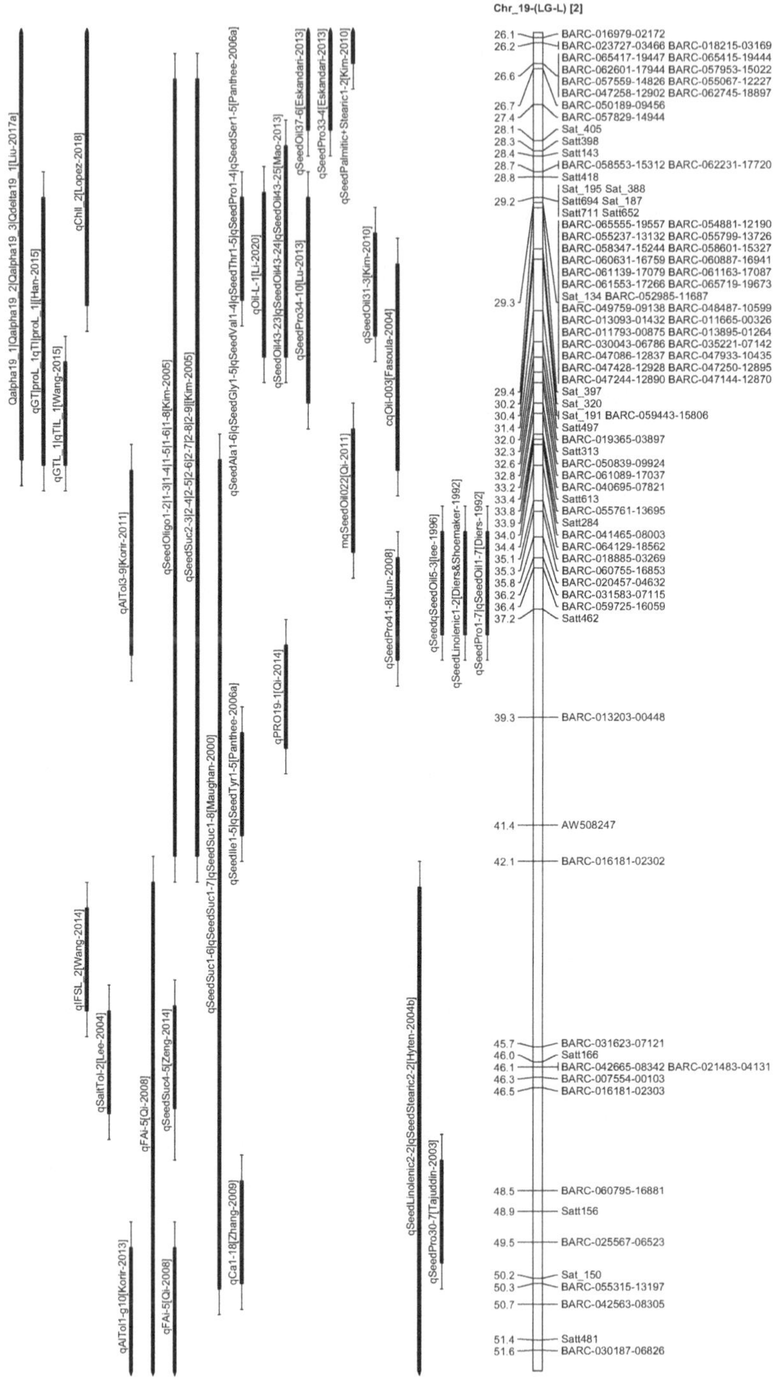

Fig 2.1 (continued)

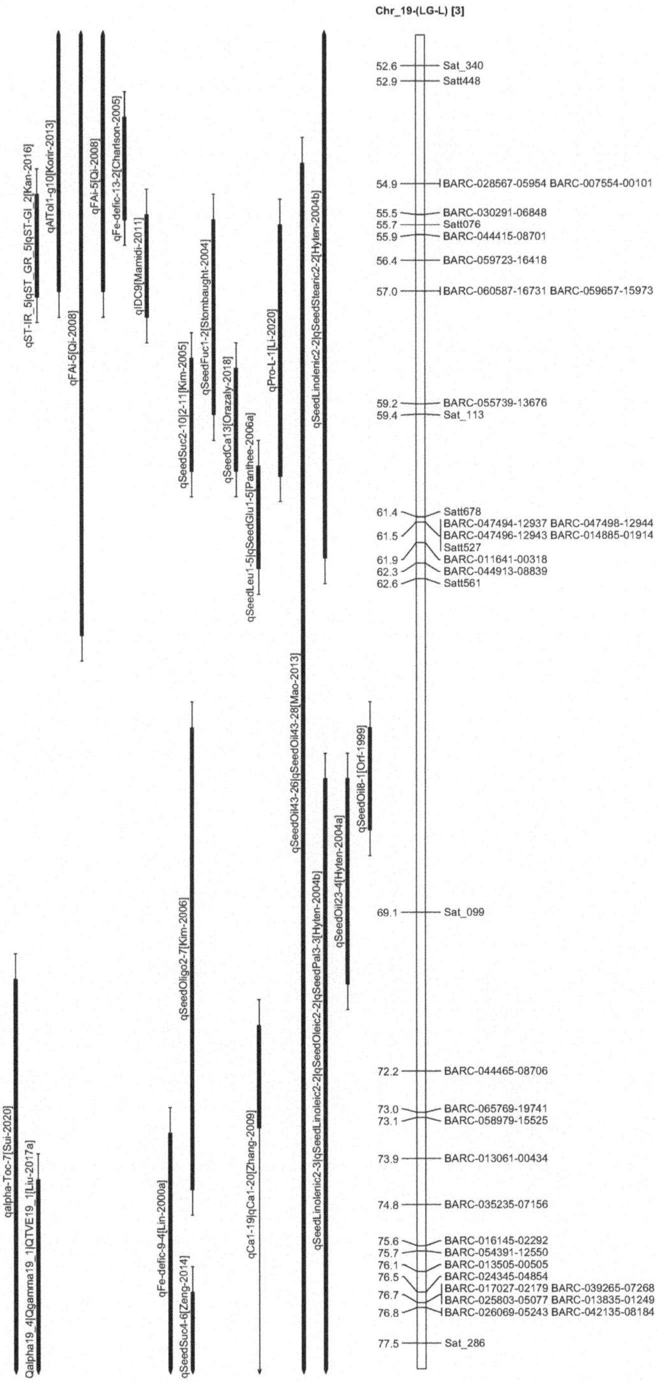

Fig 2.1 (continued)

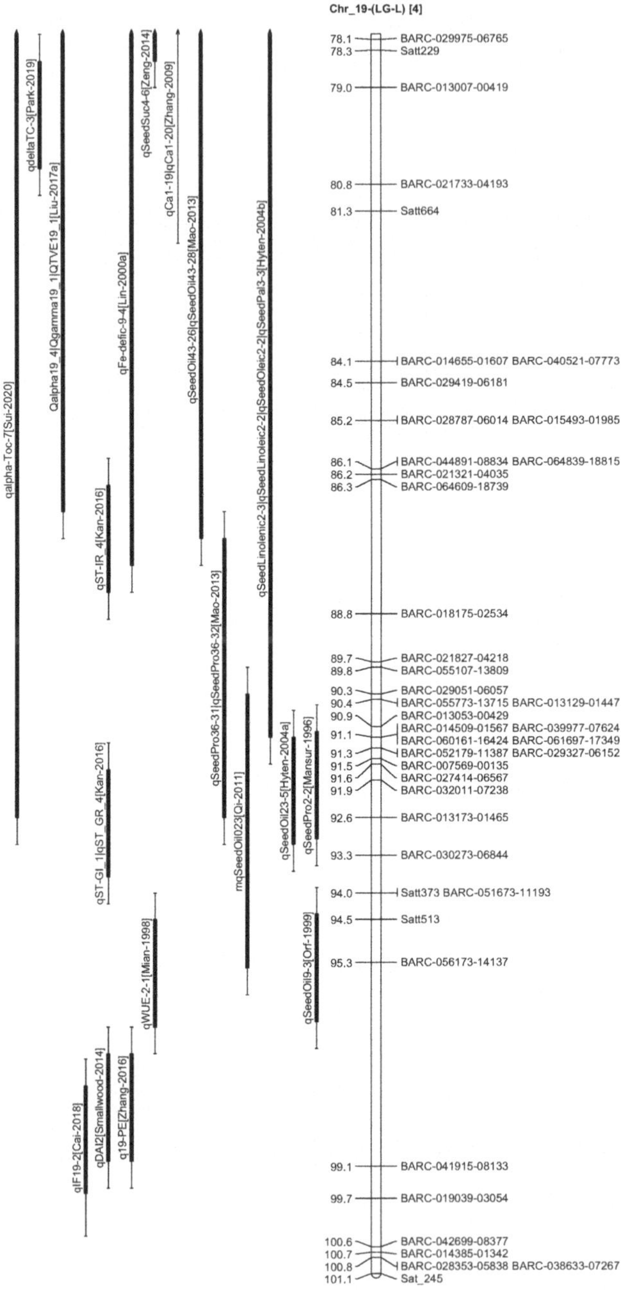

Fig 2.1 (continued)

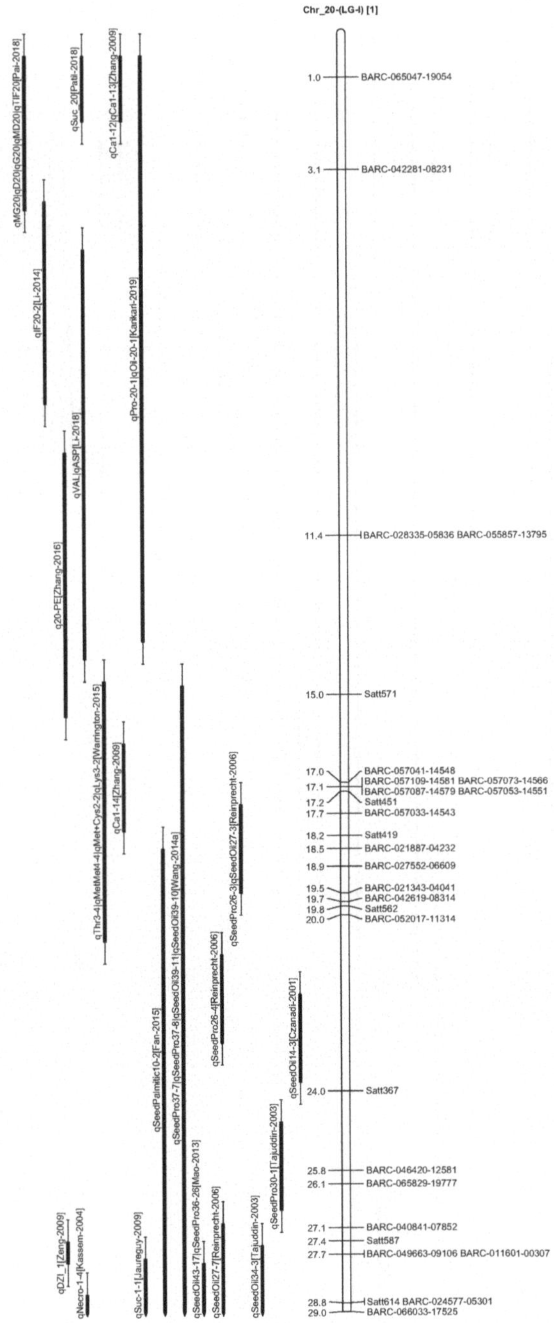

Fig 2.1 (continued)

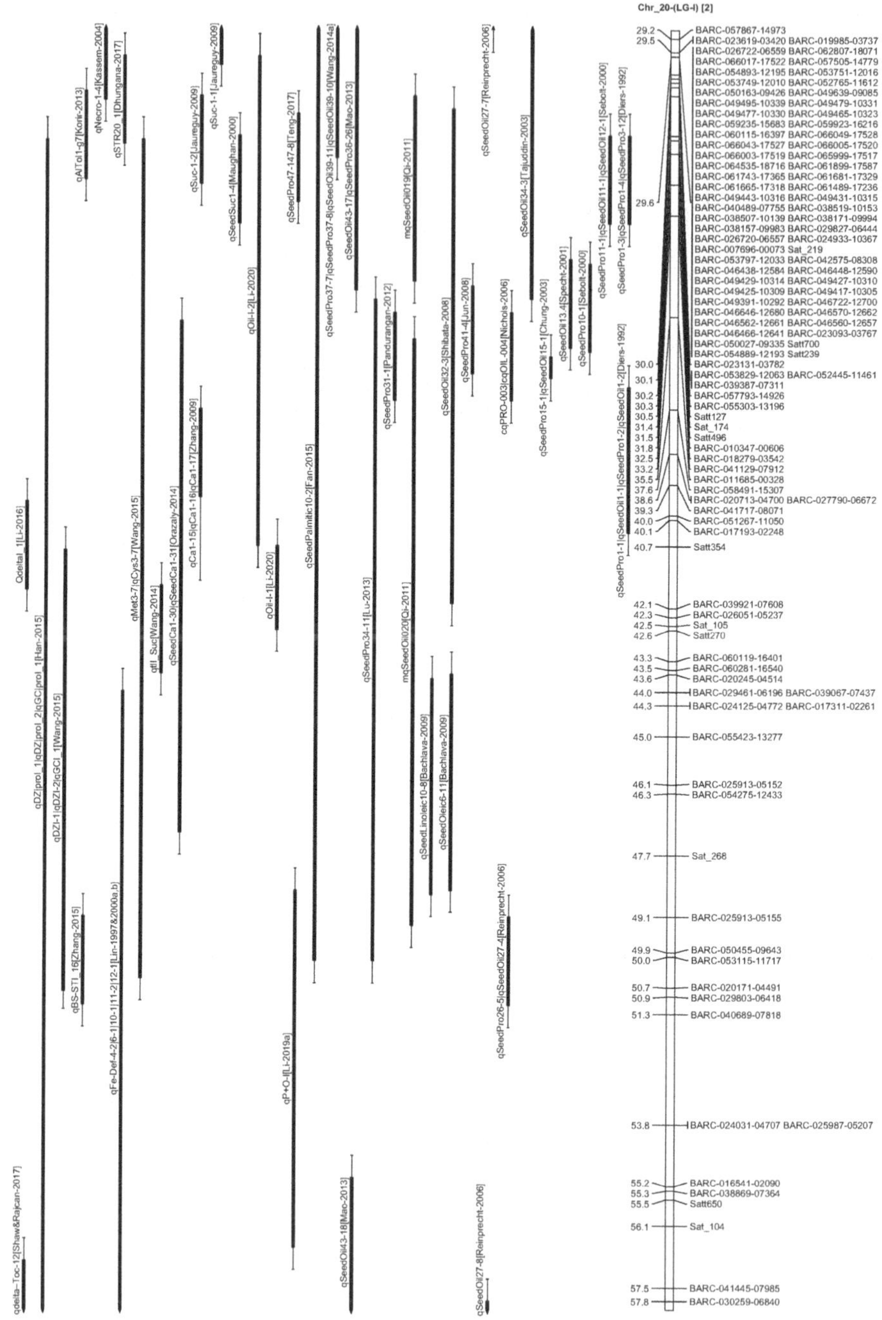

Fig 2.1 (continued)

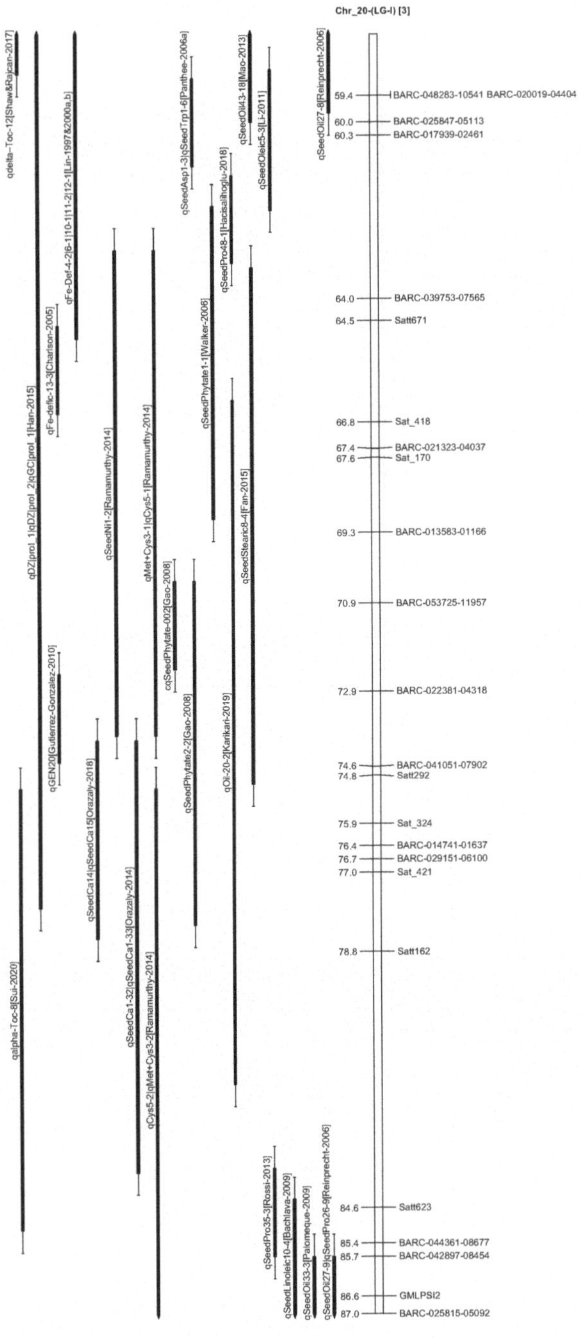

Fig 2.1 (continued)

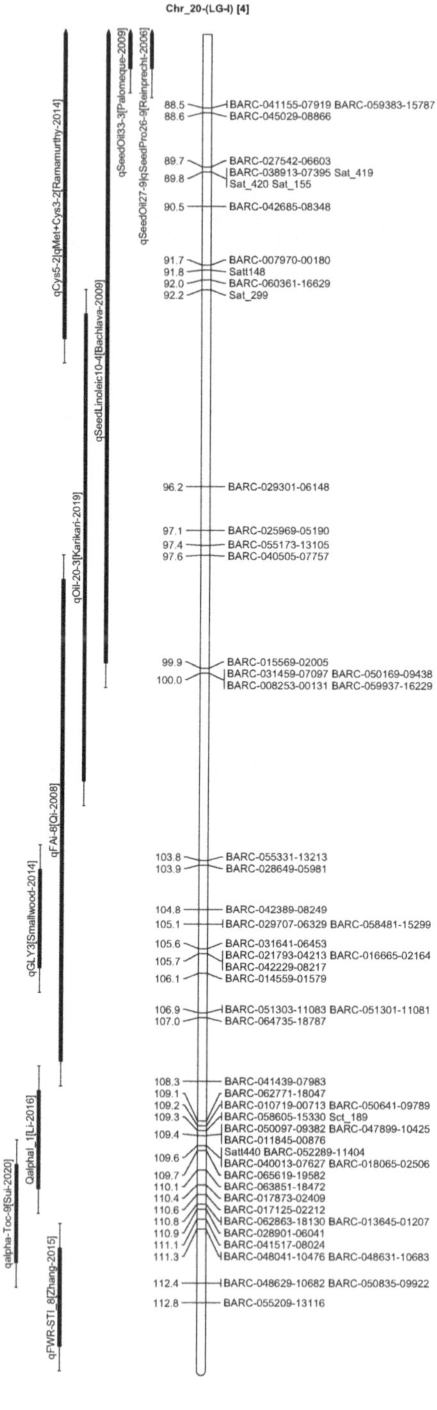

Fig 2.1 (continued)

Table 2.3 Major candidate genes identified with the QTL regions that control seed protein, oil, and fatty acids contents in soybean

Trait	QTL	Chr.	Marker	Candidate gene ID	Annotation/pathway	Reference
Oil	*Seed oil*	A1/5	A170_1-Satt511	AF243182	β-Ketoacyl-ACP synthase I	Qi et al. (2011)
	Seed oil	G/18	A073_1-Sat_143	AAF70509	β-Ketoacyl carrier protein synthase III	
	Seed oil	B2/14	Sat_083-Sat_009	ABM45911	Stearoyl-ACP desaturase	
	Seed oil	F/13	M8E6mr1-Satt343	EEF30167	Long-chain-fatty-acid-CoA ligase	
	Seed oil	I/20	Sat_219-Sat_105	P48630	Microsomal omega-6 fatty acid desaturase	
	Seed oil	B2/14	*	ACC91237	Long-chain-fatty-acid-CoA ligase	
Fatty acids	Oil 23-2	D1a/1	*	Glyma01g20720	LTA2, 2-oxoacid dehydrogenases acyltransferase family protein	Zhou et al. (2015)
	*	D1b/2	*	Glyma02g31320	LPAT2, lysophosphatidyl acyltransferase 2	
	Oil 31-1	D1b/2	*	Glyma02g38310	GLY1, NAD-dependent glycerol-3-phosphate dehydrogenase family protein	
	*	C2/6	*	Glyma06g11120	ACC1, acetyl-CoA carboxylase 1	
	*	M/7	*	Glyma07g05550	PDH-E1 alpha, pyruvate dehydrogenase E1 alpha	
	*	M/7	*	Glyma07g17720	GPAT6, glycerol-3-phosphate acyltransferase 6	
	Oil 34-5	M/7	*	Glyma07g39500	ATCES1, alkaline phytoceramidase (aPHC)	
	*	A2/8	*	Glyma08g01180	ACP4, acyl carrier protein 4	
	*	A2/8	*	Glyma08g45990	MOD1, NAD(P)-binding Rossmann-fold superfamily protein	
	*	O/10	*	Glyma10g42470	FAD2, fatty acid desaturase 2	
	*	F/13	*	Glyma13g31440	DES-1-like, fatty acid desaturase family protein	
	Oil 24-17, oil 30-4	B2/14	*	Glyma14g27990	Plant stearoyl-acyl-carrier-protein desaturase family protein	

(continued)

Table 2.3 (continued)

Trait	QTL	Chr.	Marker	Candidate gene ID	Annotation/pathway	Reference
		E/15	*	Glyma15g00550	KAS III, 3-ketoacyl-acyl carrier protein synthase III	
	Oil 32-1, oil 2-3	E/15	*	Glyma15g05800	Thioesterase superfamily protein	
	Oil 32-1, oil 2-3, Oil 24-23, oil 5-1, Oil 2-9	E/15	*	Glyma15g07900	DES-1-like, fatty acid desaturase family protein	
	Oil 2-3, oil 2-9, oil 2-8, oil 24-24, oil 27-2	E/15	*	Glyma15g11230	NAD(P)-binding Rossmann-fold superfamily protein	
	Oil 2-9, oil 24-24, oil 27-2	E/15	*	Glyma15g15310	LPD1, lipoamide dehydrogenase 1	
	Oil 23-3	D2/17	*	Glyma17g12940	FATB, fatty acyl-ACP thioesterases B	
	Oil 32-3, oil 11-1, oil 12-1, oil 24-29, oil 13-4, oil 16-3, cqOil-004	I/20	*	Glyma20g12290	Protein of unknown function (DUF630 and DUF632)	
Palmitic	*Seed palmitic*	A1/5	Map-0751	Glyma05g07630	Strictosidine synthase activity	Li et al. (2015)
	Seed palmitic	A1/5	Map-6064	Glyma05g07730	Translocase	
	Seed palmitic	M/7	Map-1323	Glyma07g18890	Carbohydrate-binding	
	Seed palmitic	K/9	Map-6356	Glyma09g09800	PPR repeat	
	Seed palmitic	H/12	Map-6395	Glyma12g01380	Triacylglycerol biosynthesis and Eukaryotic phospholipid synthesis & Editing	
	Seed palmitic	E/15	Map-6540	Glyma15g17090	Sequence-specific DNA-binding transcription factor activity	
	Seed palmitic	E/15	Map-6590	Glyma15g41680	Cell cycle control protein	
Stearic	*Seed stearic*	B2/14	BARC-013927-01275	Glyma14g27380	Transcription cofactor activity	
	Seed stearic	B2/14	Map-6506	Glyma14g27990	SACPD-C or FAB2C; fatty acid synthesis	

Trait	QTL	Chr.	Marker	Candidate gene ID	Annotation/pathway	Reference
	Seed stearic	E/15	Map-6568	Glyma15g37410	Serine/threonine protein kinase	
	Seed stearic	G/18	Map-3670	Glyma18g44690	MAC/perforin domain	
	Seed stearic	G/18	Map-6776	Glyma18g50790	Iron−/ion-binding	
Oleic	*Seed oleic*	N/3	Map-6056	Glyma03g41850	Phospholipid signaling	
	Seed oleic	C1/4	BARC-042719-08393	Glyma04g12490	Protein-binding	
	Seed oleic	C2/6	Map-6135	Glyma06g10830	DNA photolyase activity	
	Seed oleic	A2/8	Map-6325	Glyma08g17170	Motor activity	
	Seed oleic	G/18	Map-3670	Glyma18g44690	MAC/perforin domain	
Linoleic	*Seed linoleic*	N/3	BARC-028709-05992	Glyma03g32920	Oxidoreductase activity	
	Seed linoleic	C2/6	Map-6135	Glyma06g10830	DNA photolyase activity	
	Seed linoleic	C2/6	Map-6146	Glyma06g24830	Molecular chaperone	
	Seed linoleic	A2/8	Map-6283	Glyma08g14200	Protein-binding	
	Seed linoleic	A2/8	Map-6325	Glyma08g17170	Motor activity	
	Seed linoleic	A2/8	Map-6326	Glyma08g17190	Fatty acid elongation and wax biosynthesis, Lipid transfer proteins	
	Seed linoleic	O/10	Map-6383	Glyma10g36250	*	
	Seed linoleic	E/15	Map-6520	Glyma15g11810	Malonyl-CoA decarboxylase	
Linolenic	*Seed linolenic*	D1a/1	Map-0008	Glyma01g01830	Catalytic activity	
	Seed linolenic	D1a/1	BARC-017909-02439	Glyma01g07120	Ureide permease	
	Seed linolenic	D1a/1	Map-0076	Glyma01g24850	Galactosyltransferases	
	Seed linolenic	D1b/2	Map-6017	Glyma02g15600	SACPD-B or FAB2; fatty acid synthesis and fatty acid elongation	
	Seed linolenic	A1/5	Map-6077	Glyma05g34030	Phosphatidylethanolamine-binding proteins	
	Seed linolenic	A2/8	Map-6270	Glyma08g13060	Protein kinase activity	
	Seed linolenic	E/15	Map-6520	Glyma15g11810	Malonyl-CoA decarboxylase	
	Seed linolenic	G/18	Map-6738	Glyma18g05980	*	
	Seed linolenic	G/18	Map-3569	*	*	

(continued)

Table 2.3 (continued)

Trait	QTL	Chr.	Marker	Candidate gene ID	Annotation/pathway	Reference
	Seed linolenic	G/18	Map-3580	Glyma18g34290	*	
	Seed linolenic	G/18	Map-3665	Glyma18g43900	Dirigent-like protein	
	Seed linolenic	G/18	Map6782	Glyma18g51400	*	
Protein cont.	*GqPC3*	N/3	AX-93995056	Glyma.03G156100	Glycinin A2B1a precursor	Zhang et al. (2017)
PC	*GqPC4-1*	C1/4	AX-94274580	*	Family not named	
	GqPC4-2	C1/4	AX-93920058	Glyma.04G157500	Leucine-rich repeat protein	
	GqPC5	A1/5	AX-94012201	Glyma.05G070600	Nitrate transporter	
	GqPC6	C2/6	AX-93731783	Glyma.06G220300	Phospholipid acyltransferase	
	GqPC9	K/9	AX-93764734	Glyma.09G087200	Inositol transporter	
	GqPC10	O/10	AX-93933852	Glyma.10G177000	RmlC-like cupin superfamily protein	
	GqPC11-1	B1/11	AX-94089116	*	Family not named	
	GqPC11-2	B1/11	AX-93795201	Glyma.11G234600	Transmembrane amino acid transporter	
	GqPC12	H/12	AX-93804391	Glyma.12G179700	Serine carboxypeptidase S28 family protein	
	GqPC13	F/13	AX-94109235	Glyma.13G123500	Glycinin A3B4 subunits	
	GqPC14	B2/14	AX-93822697	Glyma.14G086300	Asparaginase	
	GqPC15-1	E/15	AX-93649790	Glyma.15G098100	Beta-amylase	
	GqPC15-2	E/15	AX-93837099	Glyma.15G138200	Rab5-interacting family protein	
	GqPC15-3	E/15	AX-94136114	Glyma.15G175200	Zinc finger WD40 repeat protein	
	GqPC19	L/19	AX-94196006	Glyma.19G236600	7S globulin precursor	
Water-soluble	*GqWSPC1*	D1a/1	AX-93961814	*	Family not named	
Protein cont.	*GqWSPC7*	M/7	AX-94283226	Glyma.07G041800	RNA polymerase III subunit RPC6	
WSPC	*GqWSPC8*	A2/8	AX-94048210	Glyma.08G112300	Seed storage 2S albumin protein	

Trait	QTL	Chr.	Marker	Candidate gene ID	Annotation/pathway	Reference
	GqWSPC10	O/10	AX-93933852	Glyma.10G177000	RmlC-like cupin superfamily protein	
	GqWSPC11-2	B1/11	AX-93795201	Glyma.11G234600	Transmembrane amino acid transporter	
	GqWSPC12-1	H/12	AX-94268787	Glyma.12G034900	Auxin-responsive protein	
	GqWSPC12-2	H/12	AX-93805526	Glyma.12G205100	Concanavalin A-like/legume lectin domain	
	GqWSPC13	F/13	AX-94112235	Glyma.13G194400	Albumin 1 gene	
	GqWSPC15	E/15	AX-93649393	Glyma.15G258100	Lob domain-containing protein 4	
	GqWSPC16	J/16	AX-94150758	Glyma.16G130900	Proline dipeptidase	
	GqWSPC17	D2/17	AX-94156641	Glyma.17G097800	Aldehyde dehydrogenase	
	GqWSPC18-1	G/18	AX-93870261	Glyma.18G071900	Amino acid permease	
	GqWSPC18-2	G/18	AX-94170844	Glyma.18G126900	GDSL-like lipase	
	GqWSPC18-3	G/18	AX-94182257	Glyma.18G276800	Amino acid transporter 6	
	GqWSPC19	L/19	AX-94196006	Glyma.19G236600	7S globulin precursor	
	GqWSPC20	I/20	AX-94198333	Glyma.20G031900	DNA replication licensing factor	
Protein	*Protein*	B2/14	ss715617976	Glyma.14G136800.1	ATP phosphoribosyl transferase 2	Leamy et al. (2017)
	Protein	B2/14		Glyma.14G136900.1	Cellulose synthase-like C5	
Oil	*Oil*	N/3	ss715585518	Glyma.03G131900.1	dehydration-induced protein (ERD15)	
	Oil	N/3		Glyma.03G132000.1	P-loop-containing nucleoside triphosphate hydrolases superfamily protein	
	Oil	N/3		Glyma.03G132200.1	N-acetyltransferases; N-acetyltransferases	
	Oil	N/3		Glyma.03G132300.1	Elongation-defective 1 protein/ELD1 protein	
	Oil	N/3		Glyma.03G132400.1	Nucleic acid-binding, OB-fold-like protein	

(continued)

Table 2.3 (continued)

Trait	QTL	Chr.	Marker	Candidate gene ID	Annotation/pathway	Reference
	Oil	N/3		Glyma.03G132500.1	Eukaryotic translation initiation factor eIF2A family protein	
	Oil	N/3		Glyma.03G132600.1	beta-1,3-glucanase 1	
	Oil	N/3		Glyma.03G132700.1	beta-1,3-glucanase 1	
	Oil	N/3		Glyma.03G132900.1	beta-1,3-glucanase 1	
	Oil	N/3		Glyma.03G133000.1	RECA homolog 3	
	Oil	N/3		Glyma.03G133100.1	Histone superfamily protein	
	Oil	N/3		Glyma.03G133200.1	Transmembrane proteins 14C	
	Oil	N/3		Glyma.03G133300.1	cytokinin oxidase/dehydrogenase 1	
	Oil	N/3		Glyma.03G133500.1	Heat shock protein DnaJ with tetratricopeptide repeat	
	Oil	N/3		Glyma.03G133600.1	Alpha carbonic anhydrase 4	
	Oil	N/3		Glyma.03G133700.1	INO80 ortholog	
	Oil	I/20	ss715637730	Glyma.20G124000.1	Endosomal targeting BRO1-like domain-containing protein	
	Oil	I/20		Glyma.20G124100.1	Tetratricopeptide repeat (TPR)-like superfamily protein	
	Oil	I/20		Glyma.20G124600.1	Transcription regulators	
	Oil	I/20		Glyma.20G124700.1	Quiescin-sulfhydryl oxidase 1	
	Oil	I/20		Glyma.20G124800.1	RING/FYVE/PHD zinc finger superfamily	
	Oil	I/20		Glyma.20G124900.1	Putative fasciclin-like arabinogalactan protein	
	Oil	I/20		Glyma.20G125200.1	ribose-5-phosphate isomerase 2	
Palmitic	*Palmitic*	C2/6	ss715595323	Glyma.06G068700.1	phospholipase D alpha 1	
	Palmitic	C2/6		Glyma.06G068800.1	phospholipase D alpha 1	

Trait	QTL	Chr.	Marker	Candidate gene ID	Annotation/pathway	Reference
	Palmitic	C2/6		Glyma.06G068900.1	Integrase-type DNA-binding superfamily protein	
	Palmitic	C2/6		Glyma.06G069000.1	RNA-dependent RNA polymerase 6	
	Palmitic	C2/6		Glyma.06G069100.1	SPX domain gene 2	
	Palmitic	C2/6		Glyma.06G069200.1	DnaJ/Hsp40 cysteine-rich domain superfamily protein	
	Palmitic	C2/6		Glyma.06G069300.1	DEAD/DEAH box RNA helicase family protein	
	Palmitic	C2/6		Glyma.06G069400.1	Nucleotide sugar transporter family protein	
	Palmitic	C2/6		Glyma.06G069500.1	Plant natriuretic peptide A	
	Palmitic	C2/6		Glyma.06G069600.1	Mitochondrial substrate carrier family protein	
	Palmitic	C2/6		Glyma.06G069700.1	Cellulose synthase 1	
	Palmitic	C2/6		Glyma.06G069800.1	Overexpressor of cationic peroxidase 3	
	Palmitic	M/7	ss715597684	Glyma.07G208900.1	Tetratricopeptide repeat (TPR)-like superfamily protein	
	Palmitic	M/7		Glyma.07G209000.1	2-oxoglutarate (2OG) and Fe (II)-dependent oxygenase superfamily protein	
	Palmitic	M/7		Glyma.07G209100.1	2-oxoglutarate (2OG) and Fe (II)-dependent oxygenase superfamily protein	
	Palmitic	M/7		Glyma.07G209200.1	2-oxoglutarate (2OG) and Fe (II)-dependent oxygenase superfamily protein	
	Palmitic	M/7		Glyma.07G209400.1	Phragmoplast-orienting kinesin 2	
	Palmitic	B2/14	ss715617910	Glyma.14G131800.1	Protein of unknown function (DUF1218)	
Stearic	*Stearic*	B2/14	ss715618427	Glyma.14G121200.1	Nucleic acid-binding, OB-fold-like protein	
	Stearic	B2/14		Glyma.14G121300.1	alcohol dehydrogenase 1	

(continued)

Table 2.3 (continued)

Trait	QTL	Chr.	Marker	Candidate gene ID	Annotation/pathway	Reference
	Stearic	B2/14		Glyma.14G121400.1	Chaperone DnaJ-domain superfamily protein	
	Stearic	B2/14		Glyma.14G121500.1	Plant stearoyl-acyl-carrier-protein desaturase family protein	
	Stearic	B2/14	ss715618430	Glyma.14G121700.1	Light-mediated development protein 1/ deetiolated1 (DET1)	
	Stearic	J/16	ss715625341	Glyma.16G068200.1	Cytochrome P450, family 724, subfamily A, polypeptide 1	
	Stearic	J/16		Glyma.16G068400.1	NHL domain-containing protein	
	Stearic	J/16		Glyma.16G068500.1	RING domain ligase2	
	Stearic	J/16		Glyma.16G068600.1	Surfeit locus protein 2 (SURF2)	
	Stearic	J/16		Glyma.16G068700.1	Plant invertase/pectin methylesterase inhibitor superfamily	
	Stearic	J/16		Glyma.16G068800.1	GDSL-like lipase/acylhydrolase superfamily protein	
Oleic	*Oleic*	M/7	ss715596070	Glyma.07G112000	Plant protein of unknown function (DUF247)	
	Oleic	M/7		Glyma.07G112100	photosystem II subunit Q-2	
	Oleic	M/7		Glyma.07G112200	Mitochondrial transcription termination factor family protein	
	Oleic	M/7		Glyma.07G112300	LMBR1-like membrane protein	
	Oleic	B2/14		Glyma.14G131800.1	Protein of unknown function (DUF1218)	
	Oleic	J/16		Glyma.16G014000.1	Alanine aminotransferase 2	
	Oleic	J/16		Glyma.16G014100.1	Pectin methylesterase 1	
	Oleic	J/16		Glyma.16G014200.1	Pectin methylesterase 3	
	Oleic	J/16		Glyma.16G014300.1	Exocyst complex component sec10	
	Oleic	J/16		Glyma.16G014500.1	Ubiquitin-conjugating enzyme 34	

Trait	QTL	Chr.	Marker	Candidate gene ID	Annotation/pathway	Reference
	Oleic	J/16		Glyma.16G014700.1	RING/U-box superfamily protein	
	Oleic	J/16		Glyma.16G014800.1	RING/U-box superfamily protein	
	Oleic	J/16		Glyma.16G014900.1	O-fucosyltransferase family protein	
	Oleic	J/16	ss715623400	Glyma.16G013900.1	Remorin family protein	
	Oleic	J/16		Glyma.16G014000.1	Alanine aminotransferase 2	
	Oleic	J/16		Glyma.16G014100.1	Pectin methylesterase 1	
	Oleic	J/16		Glyma.16G014200.1	Pectin methylesterase 3	
	Oleic	J/16		Glyma.16G014300.1	Exocyst complex component sec10	
	Oleic	J/16		Glyma.16G014500.1	Ubiquitin-conjugating enzyme 34	
	Oleic	J/16		Glyma.16G014700.1	RING/U-box superfamily protein	
	Oleic	J/16		Glyma.16G014800.1	RING/U-box superfamily protein	
	Oleic	J/16		Glyma.16G014900.1	O-fucosyltransferase family protein	
Linoleic	*Linoleic*	D1b/2	ss715582510	Glyma.02G213100.1	Eukaryotic aspartyl protease family protein	
	Linoleic	D1b/2		Glyma.02G213200.1	Adenine nucleotide alpha hydrolase-like superfamily protein	
	Linoleic	D1b/2		Glyma.02G213400.1	Conserved peptide upstream open reading frame 45	
	Linoleic	D1b/2		Glyma.02G213500.1	ubiquitin-specific protease 17	
	Linoleic	D1b/2		Glyma.02G213600.1	Predicted AT-hook DNA-binding family protein	
	Linoleic	D1b/2		Glyma.02G213700.1	Clathrin adaptor complexes medium subunit family protein	
	Linoleic	D1b/2		Glyma.02G212700.1	Beta-carbonic anhydrase 5	
	Linoleic	D1b/2	ss715582512	Glyma.02G213100.1	Eukaryotic aspartyl protease family protein	
	Linoleic	D1b/2		Glyma.02G213200.1	Adenine nucleotide alpha hydrolase-like superfamily protein	

(continued)

Table 2.3 (continued)

Trait	QTL	Chr.	Marker	Candidate gene ID	Annotation/pathway	Reference
	Linoleic	D1b/2		Glyma.02G213400.1	Conserved peptide upstream open reading frame 45	
	Linoleic	D1b/2		Glyma.02G213500.1	Ubiquitin-specific protease 17	
	Linoleic	D1b/2		Glyma.02G213600.1	Predicted AT-hook DNA-binding family protein	
	Linoleic	D1b/2		Glyma.02G213700.1	Clathrin adaptor complexes medium subunit family protein	
	Linoleic	D1b/2		Glyma.02G213800.1	Beta-carbonic anhydrase 5	
	Linoleic	M/7		Glyma.07G111800.1	Basic helix-loop-helix 32	
	Linoleic	M/7	ss715596075	Glyma.07G111900.1	AGAMOUS-like 44	
	Linoleic	M/7		Glyma.07G112000.1	Plant protein of unknown function (DUF247)	
	Linoleic	M/7		Glyma.07G112100.1	photosystem II subunit Q-2	
	Linoleic	M/7		Glyma.07G112200.1	Mitochondrial transcription termination factor family protein	
	Linoleic	M/7		Glyma.07G111800.1	LMBR1-like membrane protein	
	Linoleic	M/7	ss715596074	Glyma.07G111900.1	AGAMOUS-like 44	
	Linoleic	M/7		Glyma.07G112000.1	Plant protein of unknown function (DUF247)	
	Linoleic	M/7		Glyma.07G112100.1	Photosystem II subunit Q-2	
	Linoleic	M/7		Glyma.07G112200.1	Mitochondrial transcription termination factor family protein	
	Linoleic	M/7		Glyma.07G111900.1	LMBR1-like membrane protein	
	Linoleic	M/7	ss715596072	Glyma.07G112000.1	Plant protein of unknown function (DUF247)	
	Linoleic	M/7		Glyma.07G112100.1	Photosystem II subunit Q-2	

Trait	QTL	Chr.	Marker	Candidate gene ID	Annotation/pathway	Reference
	Linoleic	M/7		Glyma.07G112200.1	Mitochondrial transcription termination factor family protein	
	Linoleic	M/7		Glyma.07G112300.1	LMBR1-like membrane protein	
	Linoleic	M/7		Glyma.07G111900.1	LMBR1-like membrane protein	
	Linoleic	M/7	ss715596071	Glyma.07G112000.1	Plant protein of unknown function (DUF247)	
	Linoleic	M/7		Glyma.07G112100.1	photosystem II subunit Q-2	
	Linoleic	M/7		Glyma.07G112200.1	Mitochondrial transcription termination factor family protein	
	Linoleic	M/7		Glyma.07G112300.1	LMBR1-like membrane protein	
	Linoleic	M/7		Glyma.07G111900.1	LMBR1-like membrane protein	
	Linoleic	M/7	ss715596070	Glyma.07G112000.1	Plant protein of unknown function (DUF247)	
	Linoleic	M/7		Glyma.07G112100.1	Photosystem II subunit Q-2	
	Linoleic	M/7		Glyma.07G112200.1	Mitochondrial transcription termination factor family protein	
	Linoleic	M/7		Glyma.07G112300.1	LMBR1-like membrane protein	
	Linoleic	M/7		Glyma.07G112900.1	LMBR1-like membrane protein	
	Linoleic	M/7	ss715596058	Glyma.07G113200.1	Auxin efflux carrier family protein	
	Linoleic	K/9		Glyma.09G218500.1	Protein prenylyltransferase superfamily protein	
	Linoleic	K/9	ss715604488	Glyma.09G218600.1	Adaptin family protein	
	Linoleic	K/9		Glyma.09G218700.1	Cytochrome P450, family 707, subfamily A, polypeptide 1	
	Linoleic	K/9		Glyma.09G218800.1	High-affinity nickel-transport family protein	

(continued)

Table 2.3 (continued)

Trait	QTL	Chr.	Marker	Candidate gene ID	Annotation/pathway	Reference
	Linoleic	K/9		Glyma.09G218900.1	Pentatricopeptide repeat (PPR) superfamily protein	
	Linoleic	K/9		Glyma.09G219000.1	SAUR-like auxin-responsive protein family	
Linolenic	*Linolenic*	D1b/2	ss715583655	Glyma.02G005200.1	Plant protein of unknown function (DUF828) with plant pleckstrin homology-like region	
	Linolenic	D1b/2		Glyma.02G005300.1	NAD(P)-linked oxidoreductase superfamily protein	
	Linolenic	D1b/2		Glyma.02G005400.1	Regulator of chromosome condensation (RCC1) family protein	
	Linolenic	D1b/2		Glyma.02G005600.1	Lactate/malate dehydrogenase family protein	
	Linolenic	D1b/2	ss715583662	Glyma.02G005200.1	Plant protein of unknown function (DUF828) with plant pleckstrin homology-like region	
	Linolenic	D1b/2		Glyma.02G005300.1	NAD(P)-linked oxidoreductase superfamily protein	
	Linolenic	D1b/2		Glyma.02G005400.1	Regulator of chromosome condensation (RCC1) family protein	
	Linolenic	D1b/2		Glyma.02G005600.1	Lactate/malate dehydrogenase family protein	
	Linolenic	D1b/2		Glyma.02G005700.1	myb domain protein 15	
	Linolenic	B2/14	ss715617909	Glyma.14G131800.1	Protein of unknown function (DUF1218)	
	Linolenic	B2/14	ss715617910	Glyma.14G131800.1	Protein of unknown function (DUF1218)	
Palmitic	*Palmitic*	A1/5	Map-0751	Glyma05g07630	Strictosidine synthase activity	Li et al. (2019b)
	Palmitic	A1/5	Map-6064	Glyma05g07730	Translocase	

Trait	QTL	Chr.	Marker	Candidate gene ID	Annotation/pathway	Reference
	Palmitic	M/7	Map-1323	Glyma07g18890	Carbohydrate-binding	
	Palmitic	K/9	Map-6356	Glyma09g09800	PPR repeat	
	Palmitic	H/12	Map-6395	Glyma12g01380	Triacylglycerol biosynthesis, eukaryotic phospholipid synthesis and editing	
	Palmitic	E/15	Map-6540	Glyma15g17090	Sequence-specific DNA-binding transcription factor activity	
	Palmitic	E/15	Map-6590	Glyma15g41680	Cell cycle control protein	
Stearic	*Stearic*	B2/14	BARC-013927-01275	Glyma14g27380	Transcription cofactor activity	
	Stearic	B2/14	Map-6506	Glyma14g27990	SACPD-C or FAB2C; fatty acid synthesis	
	Stearic	E/15	Map-6568	Glyma15g37410	Serine/threonine protein kinase	
	Stearic	G/18	Map-3670	Glyma18g44690	MAC/perforin domain	
	Stearic	G/18	Map-6776	Glyma18g50790	Iron−/ion-binding	
Oleic	*Oleic*	N/3	Map-6056	Glyma03g41850	Phospholipid signaling	
	Oleic	C1/4	BARC-042719-08393	Glyma04g12490	Protein-binding	
	Oleic	C2/6	Map-6135	Glyma06g10830	DNA photolyase activity	
	Oleic	A2/8	Map-6325	Glyma08g17170	Motor activity	
	Oleic	G/18	Map-3670	Glyma18g44690	MAC/perforin domain	
Linoleic	*Linoleic*	N/3	BARC-028709-05992	Glyma03g32920	Oxidoreductase activity	
	Linoleic	C2/6	Map-6135	Glyma06g10830	DNA photolyase activity	
	Linoleic	C2/6	Map-6146	Glyma06g24830	Molecular chaperone	
	Linoleic	A2/8	Map-6283	Glyma08g14200	Protein-binding	
	Linoleic	A2/8	Map-6325	Glyma08g17170	Motor activity	
	Linoleic	A2/8	Map-6326	Glyma08g17190	Fatty acid elongation, wax biosynthesis, lipid transfer protein	
	Linoleic	O/10	Map-6383	Glyma10g36250	*	
	Linoleic	E/15	Map-6520	Glyma15g11810	Malonyl-CoA decarboxylase	
Linolenic	*Linolenic*	D1a/1	Map-0008	Glyma01g01830	Catalytic activity	

(continued)

Table 2.3 (continued)

Trait	QTL	Chr.	Marker	Candidate gene ID	Annotation/pathway	Reference
	Linolenic	D1a/1	BARC-017909-02439	Glyma01g07120	Ureide permease	
	Linolenic	D1a/1	Map-0076	Glyma01g24850	Galactosyltransferases	
	Linolenic	D1b/2	Map-6017	Glyma02g15600	SACPD-B or FAB2; Fatty acid synthesis, and Fatty acid elongation	
	Linolenic	A1/5	Map-6077	Glyma05g34030	Phosphatidylethanolamine-binding protein	
	Linolenic	A2/8	Map-6270	Glyma08g13060	Protein kinase activity	
	Linolenic	E/15	Map-6520	Glyma15g11810	Malonyl-CoA decarboxylase	
	Linolenic	G/18	Map-6738	Glyma18g05980	*	
	Linolenic	G/18	Map-3580	Glyma18g34290	*	
	Linolenic	G/18	Map-3665	Glyma18g43900	Dirigent-like protein	
	Linolenic	G/18	Map-6782	Glyma18g51400	*	
Protein	*qProt-2-1*	D1b/2	*	Glyma02g31440	NAD(P)-linked oxidoreductase superfamily protein	Li et al. (2019c)
	qProt-3-1	N/3	*	Glyma03g40780	Ankyrin repeat family protein	
	qProt-4-1	C1/4	*	Glyma04g02850	Basic leucine zipper (bZIP) transcription factor family protein	
		C1/4	*	Glyma04g02880	Glyoxal oxidase-related protein	
		C1/4	*	Glyma04g02980	K-box region and MADS-box transcription factor family protein	
		C1/4	*	Glyma04g03180	AAA-type ATPase family protein	
		C1/4	*	Glyma04g03230	Sec14p-like phosphatidylinositol transfer family protein	
	qProt-4-2	C1/4	*	Glyma04g10451	CRT (chloroquine-resistance transporter)-like transporter 2	
		C1/4	*	Glyma04g10481	Ubiquitin-protein ligase 1	

Trait	QTL	Chr.	Marker	Candidate gene ID	Annotation/pathway	Reference
		C1/4	*	Glyma04g10510	Target SNARE coiled-coil domain protein	
	qProt-6-1	C2/6	*	Glyma06g40780	Disease resistance protein (TIR-NBS-LRR class) family	
		C2/6	*	Glyma06g40820	Disease resistance protein (TIR-NBS class)	
	qProt-7-1	M/7	*	Glyma07g10621	Protein kinase superfamily protein	
	qProt-8-1	A2/8	*	Glyma08g21340	GDSL-like lipase/acylhydrolase superfamily protein	
		A2/8	*	Glyma08g21450	Hexokinase 2	
		A2/8	*	Glyma08g21530	NADP-malic enzyme 4	
	qProt-8-2	A2/8	*	Glyma08g42100	RNI-like superfamily protein	
		A2/8	*	Glyma08g42110	Glucan synthase-like 7	
		A2/8	*	Glyma08g42170	Protein kinase superfamily protein	
		A2/8	*	Glyma08g42230	MEI2-like protein 5	
		A2/8	*	Glyma08g42240	MAP kinase 20	
	qProt-9-1	K/9	*	Glyma09g24910	Hydroxyproline-rich glycoprotein family protein	
		K/9	*	Glyma09g24956	Divalent ion symporter	
	qProt-10-1	O/10	*	Glyma10g01180	Plant invertase/pectin methylesterase inhibitor superfamily	
		O/10	*	Glyma10g01450	WRKY DNA-binding protein 23	
	qProt-10-2	O/10	*	Glyma10g29000	Importin alpha isoform 2	
		O/10	*	Glyma10g29020	Syntaxin of plants 43	
		O/10	*	Glyma10g29190	Dormancy–/auxin-associated family protein	
		O/10	*	Glyma10g29280	Respiratory burst oxidase homolog B	

(continued)

Table 2.3 (continued)

Trait	QTL	Chr.	Marker	Candidate gene ID	Annotation/pathway	Reference
	qProt-10-3	O/10	*	Glyma10g29831	Mitochondrial inner membrane translocase complex, subunit Tim44-related protein	
		O/10	*	Glyma10g29851	ROP-binding protein kinases 2	
		O/10	*	Glyma10g29861	Protein of unknown function (DUF3049)	
		O/10	*	Glyma10g29900	Prefoldin chaperone subunit family protein	
		O/10	*	Glyma10g29970	Dehydroquinate dehydratase, putative / shikimate dehydrogenase, putative	
		O/10	*	Glyma10g30070	Protein kinase superfamily protein	
	qProt-15-2	E/15	*	Glyma15g28100	Guanyl-nucleotide exchange factors; GTPase binding; GTP-binding	
	qProt-17-1	D2/17	*	Glyma17g08880	Nonsense-mediated mRNA decay NMD3 family protein	
		D2/17	*	Glyma17g08890	AGAMOUS-like 8	
	qProt-19-1	L/19	*	Glyma19g29730	Exostosin family protein	
		L/19	*	Glyma19g29760	F-box family protein	
	qProt-19-2	L/19	*	Glyma19g31120	Glutamate synthase 1	
	qProt-20-3	I/20	*	Glyma20g21693	Subtilase family protein	
		I/20	*	Glyma20g21726	Aldehyde dehydrogenase 12A1	
	qProt-20-4	I/20	*	Glyma20g34880	Differentiation and greening-like 1	
		I/20	*	Glyma20g34910	Embryo sac development arrest 6	
		I/20	*	Glyma20g34975	ROP-interactive CRIB motif-containing protein 1	
		I/20	*	Glyma20g35020	COP1-interacting protein-related	
		I/20	*	Glyma20g35090	Calcium-binding EF-hand family protein	

Trait	QTL	Chr.	Marker	Candidate gene ID	Annotation/pathway	Reference
Oil	*qOil-1-1*	D1a/1	*	Glyma01g30670	PEBP (phosphatidylethanolamine-binding protein) family protein	
		D1a/1	*	Glyma01g30910	BCL-2-associated athanogene 4	
	qOil-3-1	N/3	*	Glyma03g04340	Protein kinase superfamily protein	
		N/3	*	Glyma03g04420	MATE efflux family protein	
		N/3	*	Glyma03g04470	Ovate family protein 1	
	qOil-3-2	N/3	*	Glyma03g09870	Protein kinase 1B	
		N/3	*	Glyma03g10002	Peroxisomal membrane 22 kDa (Mpv17/PMP22) family protein	
	qOil-4-2	C1/4	*	Glyma04g34745	FAR1-related sequence 5	
		C1/4	*	Glyma04g34790	ARM repeat superfamily protein	
	qOil-4-3	C1/4	*	Glyma04g42167	ACT domain repeat 3	
		C1/4	*	Glyma04g42174	Protein kinase protein with adenine nucleotide alpha hydrolase-like domain	
		C1/4	*	Glyma04g42246	Target of Myb protein 1	
	qOil-6-1	C2/6	*	Glyma06g40780	Disease resistance protein (TIR-NBS-LRR class) family	
		C2/6	*	Glyma06g40820	Disease resistance protein (TIR-NBS class)	
	qOil-8-2	A2/8	*	Glyma08g21340	GDSL-like lipase/acylhydrolase superfamily protein	
		A2/8	*	Glyma08g21450	Hexokinase 2	
		A2/8	*	Glyma08g21530	NADP-malic enzyme 4	
	qOil-8-3	A2/8	*	Glyma08g23660	Galactose oxidase/kelch repeat superfamily protein	
		A2/8	*	Glyma08g23760	Autoinhibited Ca(2+)-ATPase 9	

(continued)

Table 2.3 (continued)

Trait	QTL	Chr.	Marker	Candidate gene ID	Annotation/pathway	Reference
	qOil-10-1	O/10	*	Glyma10g06670	Protein phosphatase 2A regulatory B subunit family protein	
		O/10	*	Glyma10g06800	Phosphatidylinositol monophosphate 5 kinase	
		O/10	*	Glyma10g06820	Glutamate-ammonia ligases; catalytics; glutamate-ammonia ligases	
	qOil-10-3	O/10	*	Glyma10g29831	Mitochondrial inner membrane translocase complex, subunit Tim44-related protein	
		O/10	*	Glyma10g29851	ROP-binding protein kinases 2	
		O/10	*	Glyma10g29861	Protein of unknown function (DUF3049)	
		O/10	*	Glyma10g29900	Prefoldin chaperone subunit family protein	
		O/10	*	Glyma10g29970	Dehydroquinate dehydratase, putative/ shikimate dehydrogenase, putative	
		O/10	*	Glyma10g30070	Protein kinase superfamily protein	
	qOil-10-4	O/10	*	Glyma10g38890	NAD(P)-linked oxidoreductase superfamily protein	
		O/10	*	Glyma10g38930	Heat shock transcription factor A6B	
		O/10	*	Glyma10g38970	TIFY domain/divergent CCT motif family protein	
	qOil-13-1	F/13	*	Glyma13g27480	Starch synthase 3	
		F/13	*	Glyma13g27570	RNA-binding (RRM/RBD/RNP motifs) family protein	
		F/13	*	Glyma13g27610	Pentatricopeptide repeat 336	
	qOil-15-1	E/15	*	Glyma15g14700	Chaperone DnaJ-domain superfamily protein	

Trait	QTL	Chr.	Marker	Candidate gene ID	Annotation/pathway	Reference
	qOil-16-1	J/16	*	Glyma16g27980	WD-40 repeat family protein / notchless protein, putative	
		J/16	*	Glyma16g28080	Cellulose synthase 6	
	qOIl-17-1	D2/17	*	Glyma17g07341	UDP-glucosyl transferase 78D2	
	qOil-17-2	D2/17	*	Glyma17g11950	Tetratricopeptide repeat (TPR)-like superfamily protein	
		D2/17	*	Glyma17g12040	Calcium-binding EF-hand family protein	
		D2/17	*	Glyma17g12120	Zinc finger (ubiquitin-hydrolase) domain-containing protein	
	qOil-20-1	I/20	*	Glyma20g21693	Subtilase family protein	
		I/20	*	Glyma20g21726	Aldehyde dehydrogenase 12A1	
Protein	*qPro-7-1*	M/7	*	Glyma07g09060	Protein phosphorylation	Karikari et al. (2019)
		M/7	*	Glyma07g09100	Protein maturation	
		M/7	*	Glyma07g09230	Protein targeting to mitochondrion; protein maturation	
		M/7	*	Glyma07g09260	Regulation of cyclin-dependent protein kinase activity; stomatal lineage progression	
		M/7	*	Glyma07g09270	Transport; transmembrane transport	
		M/7	*	Glyma07g09370	Fatty acid metabolic process; jasmonic acid biosynthetic process	
		M/7	*	Glyma07g09410	Protein glycosylation; ER to Golgi vesicle-mediated transport	
		M/7	*	Glyma07g09420	Protein phosphorylation; response to wounding	
		M/7	*	Glyma07g09440	Protein metabolic process	

(continued)

Table 2.3 (continued)

Trait	QTL	Chr.	Marker	Candidate gene ID	Annotation/pathway	Reference
		M/7	*	Glyma07g09480	Amino acid import; transmembrane transport	
		M/7	*	Glyma07g09630	Systemic acquired resistance; phenylpropanoid biosynthetic process	
		M/7	*	Glyma07g09640	Response to endoplasmic reticulum stress; oxidation-reduction process	
		M/7	*	Glyma07g09650	Metabolic process; systemic acquired resistance	
		M/7	*	Glyma07g09780	Pyrimidine ribonucleotide biosynthetic process	
		M/7	*	Glyma07g09790	Fatty acid beta-oxidation; protein import into peroxisome matrix	
		M/7	*	Glyma07g09810	Tryptophan biosynthetic process	
		M/7	*	Glyma07g09820	Maltose metabolic process; rRNA processing; chloroplast relocation	
		M/7	*	Glyma07g09860	Glycerol biosynthetic process; lipid metabolic process	
		M/7	*	Glyma07g09880	Protein catabolic process	
		M/7	*	Glyma07g09980	N-terminal protein myristoylation; proteolysis	
		M/7	*	Glyma07g10050	Lipid transport	
		M/7	*	Glyma07g08950	Response to red or far red light; gibberellin biosynthetic process	
Oil	*qOil-8-3*	A2/8	*	Glyma08g14920	ABC1 family	
		A2/8	*	Glyma08g15750	Zinc finger, C3HC4 type (RING finger)	
		A2/8	*	Glyma08g15921	Protein kinase domain	

Trait	QTL	Chr.	Marker	Candidate gene ID	Annotation/pathway	Reference
		A2/8	*	Glyma08g16072	protein kinase activity; protein serine/threonine kinase activity	
		A2/8	*	Glyma08g16160	Protease inhibitor/seed storage/LTP family	
		A2/8	*	Glyma08g16620	Phosphatidylinositol 3- and 4-kinase	
		A2/8	*	Glyma08g17170	DIL domain; IQ calmodulin-binding motif	
		A2/8	*	Glyma08g17180	Arabidopsis proteins of unknown function	
		A2/8	*	Glyma08g17190	Protease inhibitor/seed storage/LTP family	
		A2/8	*	Glyma08g17555	Pyridoxal-phosphate dependent enzyme	
		A2/8	*	Glyma08g17600	Phosphoglycerate kinase	
		A2/8	*	Glyma08g17610	Phosphoglycerate kinase	
		A2/8	*	Glyma08g17640	Protein tyrosine kinase; "PB1 domain"	
		A2/8	*	Glyma08g17650	PB1 domain; "Protein tyrosine kinase"	
		A2/8	*	Glyma08g17760	Rhodanese-like domain	
		A2/8	*	Glyma08g17776	Methyltransferase small domain	
		A2/8	*	Glyma08g17793	Protein tyrosine kinase	
		A2/8	*	Glyma08g18110	Elongation factor Tu domain 2; Elongation factor G C-terminus	
		A2/8	*	Glyma08g18121	Elongation factor Tu GTP-binding domain	
		A2/8	*	Glyma08g18240	Elongation factor Tu GTP-binding domain; Initiation factor eIF2 gamma, C terminal	
	qOIl-10-2	O/10	*	Glyma10g06800	Phosphatidylinositol-4-phosphate 5-Kinase	

(continued)

Table 2.3 (continued)

Trait	QTL	Chr.	Marker	Candidate gene ID	Annotation/pathway	Reference
		O/10	*	Glyma10g06810	Amidohydrolase	
		O/10	*	Glyma10g06820	Amidohydrolase; Glutamine synthetase, catalytic domain	
		O/10	*	Glyma10g07160	UDP-glucoronosyl and UDP-glucosyl transferase	
		O/10	*	Glyma10g07270	Eukaryotic aspartyl protease	
		O/10	*	Glyma10g07290	FKBP-type peptidyl-prolyl cis-trans isomerase	
		O/10	*	Glyma10g07420	ATP-DEPENDENT PROTEASE (CEREBLON)	
		O/10	*	Glyma10g07430	MITOGEN-ACTIVATED PROTEIN KINASE	
		O/10	*	Glyma10g07500	Protein kinase domain; leucine-rich repeat	
		O/10	*	Glyma10g07580	Ergosterol biosynthesis ERG4/ERG24 family	
		O/10	*	Glyma10g07610	Protein tyrosine kinase	
		O/10	*	Glyma10g08210	GDSL-like lipase/acylhydrolase	
		O/10	*	Glyma10g08280	A20-like zinc finger; AN1-like zinc finger	
	qOil-10-4	O/10	*	Glyma10g23740	Zinc finger, C3HC4 type (RING finger)	
		O/10	*	Glyma10g23800	Protein kinase domain; legume lectin domain	
		O/10	*	Glyma10g23860	Clp amino terminal domain	
		O/10	*	Glyma10g24090	Protein of unknown function DUF89	
		O/10	*	Glyma10g24100	Clp amino terminal domain	
		O/10	*	Glyma10g24540	Phosphotransferase enzyme family; ABC1 family	
		O/10	*	Glyma10g24580	Zinc finger, C3HC4 type (RING finger)	

Trait	QTL	Chr.	Marker	Candidate gene ID	Annotation/pathway	Reference
		O/10	*	Glyma10g24590	Thiolase, N-terminal domain; thiolase, C-terminal domain	
		O/10	*	Glyma10g24620	Aldo/keto reductase family	
		O/10	*	Glyma10g25440	Protein kinase domain; leucine-rich repeat	
		O/10	*	Glyma10g25500	Core-2/I-branching enzyme	
		O/10	*	Glyma10g25550	Hydroxyethylthiazole kinase family	
		O/10	*	Glyma10g25630	TCP-1/cpn60 chaperonin family	
		O/10	*	Glyma10g26510	C2 domain; lipase (class 3)	
		O/10	*	Glyma10g26770	Tetratricopeptide repeat	
		O/10	*	Glyma10g26870	WD domain, G-beta repeat	
		O/10	*	Glyma10g26920	RNA recognition motif	
		O/10	*	Glyma10g27800	Aminotransferase class-V	
		O/10	*	Glyma10g27810	SecA DEAD-like domain; SecA preprotein cross-linking domain	
		O/10	*	Glyma10g27820	Ribosomal protein L7Ae/L30e/S12e/Gadd45 family	
		O/10	*	Glyma10g27850	DHHC zinc finger domain	
		O/10	*	Glyma10g27980	Haloacid dehalogenase-like hydrolase	
		O/10	*	Glyma10g27990	Cyclophilin type peptidyl-prolyl cis-trans isomerase/CLD	
		O/10	*	Glyma10g28180	tRNA synthetases class I (E and Q)	
		O/10	*	Glyma10g28210	Pumilio family RNA binding repeat	
		O/10	*	Glyma10g28370	Saposin-like type-B, region 2	
		O/10	*	Glyma10g28490	Protein kinase domain	
Oleic, 2018	*Oleic*	N/3	*	Glyma.03G054100.1	*Glycine max* TMV resistance protein N-like, transcript variant X3, mRNA	Wang et al. (2020)

(continued)

Table 2.3 (continued)

Trait	QTL	Chr.	Marker	Candidate gene ID	Annotation/pathway	Reference
	Oleic	N/3	*	Glyma.03G168200.3	*Glycine max* pleiotropic drug resistance protein 1-like, mRNA	
	Oleic	C1/4	*	Glyma.04G191100.1	*Glycine max* probable pectate lyase 18-like, mRNA	
	Oleic	C1/4	*	Glyma.04G203200.1	*Glycine max* respiratory burst oxidase homolog protein C-like (LOC100800248), mRNA	
	Oleic	A1/5	*	Glyma.05G155300.1	*Glycine max* ATP carrier protein 2, chloroplastic-like, mRNA	
	Oleic	M/7	*	Glyma.07G033100.1	*Glycine max* ADP,ATP carrier protein 1, chloroplastic-like (LOC100793284), mRNA	
	Oleic	M/7	*	Glyma.07G089000.1	*Glycine max* VIN3-like protein 1-like, transcript variant X2, mRNA	
	Oleic	A2/8	*	Glyma.08G019700.1	*Glycine max* calcium-dependent protein kinase 3-like, transcript variant 1, mRNA	
	Oleic	A2/8	*	Glyma.08G185000.2	*Glycine max* probable plastid-lipid-associated protein 4, chloroplastic-like, transcript variant 1, mRNA	
	Oleic	B1/11	*	Glyma.11G229600.1	*Glycine max* DNA replication complex BAG protein, transcript variant 2, mRNA	
	Oleic	F/13	*	Glyma.13G163400.1	*Glycine max* protein S-acyltransferase 24-like, misc_RNA	
	Oleic	B2/14	*	Glyma.14G045100.1	*Glycine max* abscisic-aldehyde oxidase-like, mRNA	
	Oleic	E/15	*	Glyma.15G117700.1	*Glycine max* uncharacterized LOC102666654, mRNA	

	qOil-D2-2	D2/17	98.5–100.9	Satt413-Sat_326	6.31	12.97	China	
	qOil-D2-3	D2/17	7.4–13.2	Sct_192-Sat_333	2.59	1.98	China	
	qOil-E-1	E/15	91.7–93.7	Sat_381-Satt553	2.52	4.70	China	
	qOil-F-1	F/13	8.7–10.7	AW186493-Satt659	3.80	6.36	China	
	qOil-F-2	F/13	7.1–9.1	Sat_298-AW186493	6.50	8.04	China	
	qOil-F-3	F/13	9.6–23.3	Satt659-Satt149	2.76	6.08	China	
	qOil-G-1	G/18	71.6–85.6	Satt288-Sct_199	3.65	8.21	China	
	qOil-I-1	I/20	40.7–42.6	Satt270-Satt354	3.10	4.56	China	
	qOil-I-2	I/20	29.6–40.7	Satt354-Satt239	3.96	5.15	China	
	qOil-J-1	J/16	39.8–41.8	BARCSOYSSR_16_0566-Sct_065	2.64	4.86	China	
	qOil-K-1	K/9	13.7–15.7	BARCSOYSSR_09_0183-Satt242	4.25	8.36	China	
	qOil-K-2	K/9	35.7–37.7	Satt055-BARCSOYSSR_09_0183	3.09	10.84	China	
	qOil-K-3	K/9	14.7–18.5	Satt242-Sat_119	2.75	2.93	China	
	qOil-K-4	K/9	36.7–40.1	Satt727-Satt055	2.63	9.64	China	
	qOil-K-5	K/9	83.2–93.4	Satt588-Satt196	2.58	3.06	China	
	qOil-L-1	L/19	29.1–32.3	Satt313-Sat_187	2.77	6.06	China	
	qOil-M-1	M/7	49–71.3	Satt245-Satt677	3.16	7.39	China	
	qOil-M-2	M/7	49–71.3	Satt677-Satt728	3.61	7.25	China	
	qOil-N-1	N/3	60.5–62.5	BARCSOYSSR_03_1671-Sat_304	2.79	3.49	China	
	qOil-N-2	N/3	57.2–61.5	Sat_304-Satt549	2.76	4.32	China	
	qOil-N-3	N/3	83.5–85.5	Satt584-Satt022	3.39	5.24	China	
	qOil-N-4	N/3	61.2–83.2	Satt237-Sat_295	3.00	4.48	China	

(continued)

Table 2.3 (continued)

Trait	QTL	Chr.	Marker	Candidate gene ID	Annotation/pathway	Reference
	Oleic	E/15	*	Glyma.15G120100.1	*Glycine max* tRNA methyltransferase 10 homolog A-like, mRNA	
	Oleic	E/15	*	Glyma.15G120200.2	*Glycine max* uncharacterized LOC102665381, mRNA	
	Oleic	E/15	*	Glyma.15G127500.1	*Glycine max* polygalacturonase-like, mRNA	
	Oleic	E/15	*	Glyma.15G201700.1	*Glycine max* uncharacterized LOC100814752, mRNA	
	Oleic	E/15	*	Glyma.15G210100.3	*Glycine max* alpha, alpha-trehalose-phosphate synthase [UDP-forming] 1-like, transcript variant X6, mRNA	
	Oleic	E/15	*	Glyma.15G244000.1	*Glycine max* uncharacterized, mRNA	
	Oleic	E/15	*	Glyma.15G261100.1	*Glycine max* uncharacterized, transcript variant X1, mRNA	
Oleic, 2019	*Oleic*	L/19	*	Glyma.19G110600.1	*Glycine max* uncharacterized, mRNA	
	Oleic	D1b/2	*	Glyma.02G220300.1	*Glycine max* ataxin-2-like, mRNA	
	Oleic	A2/8	*	Glyma.08G071600.1	*Glycine max* metacaspase-3-like, transcript variant X2, mRNA+F297	
	Oleic	B1/11	*	Glyma.11G229600.1	*Glycine max* DNA replication complex BAG protein, transcript variant 2	
	Oleic	H/12	*	Glyma.12G224000.1	*Glycine max* uncharacterized, mRNA	
	Oleic	H/12	*	Glyma.12G227300.1	*Glycine max* DNA ligase 1-like, mRNA	
	Oleic	I/20	*	Glyma.20G026100.1	*Glycine max* DNA ligase 1-like, mRNA+F297	

contents previously reported, narrowed the genomic regions of these reported QTL, and identified several candidate genes within these QTL regions (Hwang et al. 2014). Moreover, the major genomic region of 8.4 Mbp reported on chromosome 20 containing and about 12 candidate genes controlling both seed protein and oil contents (Bolon et al. 2010) is narrowed further to a region of 2.4 Mbp containing 6 candidate genes of the 12 reported earlier. These candidate genes are *Glyma.20 g19620*, *Glyma.20 g19630* [putative ammonium transporter AMT1], *Glyma.20 g19680* [Hsp22.5], *Glyma.20 g21030*, *Glyma.20 g21040*, and *Glyma.20 g21080* [ATP synthase D chain] (Hwang et al. 2014). Haun et al. (2014) introduced mutated two genes encoding for fatty acid desaturase (FAD2-1A; FAD2-1B). FAD converts oleic acid (C18:2) into linoleic acid (C18:3) in soybean seeds.

Transcription activator-like effector nucleases (TALENs) were engineered and used to introduce the mutations in *FAD2-1A* and *FAD2-1B* genes. The mutated seeds' oleic acid contents increased tremendously to up to 80% compared to the wild-type and non-mutated seeds. However, palmitic and linolenic acids have been lowered, and stearic acid was unchanged in the mutated seeds. The study demonstrated the efficiency of using TALENs mutagenesis method to engineer soybean seeds for high-quality beneficial fatty acids (Haun et al. 2014).

Zhou et al. (2015) re-sequenced 302 wild and cultivated soybean accessions, used a genome-wide association study (GWAS) and combined these sequences with QTL regions, and identified a total of 230 selected regions among which 96 fall within QTL controlling oil content and 21 fall within QTL controlling fatty acid contents. Twenty-one candidate genes involved in fatty acids biosynthetic pathways have been identified by this study, and 10 of them fall within previously identified QTL controlling oil content (Zhou et al. 2015).

Li et al. (2015) used a set of 421 accessions, genotyped them with 1536 SNP markers, and used GWAS to identify SNPs associated with the five major fatty acid (FAs = palmitic, stearic, oleic, linoleic, and linolenic acids) contents. The study identified 37 QTL (33 SNPs) associated with fatty acids contents, and a third of them mapped to previously reported QTL controlling fatty acid contents. Seven QTL controlling palmitic acid content have been identified on chromosomes 5, 7, 9, 12, and 15; 5 QTL controlling stearic acid content have been identified on chromosomes 14, 15, and 18; 5 QTL controlling oleic acid content have been identified on chromosomes 3, 4, 6, 8, and 18; 12 QTL controlling linolenic acid content have been identified on chromosomes 1, 2, 5, 8, 15, and 18; and 8 QTL controlling linoleic acid content have been identified on chromosomes 3, 6, 8, 10, and 15. Candidate genes within these QTL regions are shown in Table 2.3 (Li et al. 2015). Among these genes are genes involved in triacylglycerol biosynthesis; fatty acid elongation; wax biosynthesis; eukaryotic phospholipid synthesis and editing; fatty acid synthesis; phospholipid signaling, in addition to gene encoding for oxidoreductases; lipid transfer protein; SACPD-C or FAB2C; malonyl-CoA decarboxylase; translocase; and many others (Table 2.3) (Li et al. 2015).

Fang et al. (2017) used a genome-wide association study (GWAS) on 809 soybean accessions collected worldwide, grown in 2 environments in China in

2013–2014 and 2014–2015, and studied 84 traits among them seed lipid and fatty acid contents. One QTL that controls seed palmitic acid content has been identified and mapped on chromosome 5; four QTL that control seed stearic acid content have been identified and mapped on chromosomes 2, 3, and 14; two QTL that control seed linoleic acid content have been identified and mapped on chromosomes 4 and 14; three QTL that control seed palmitic acid ratio have been identified and mapped on chromosomes 5 and 13; four QTL that control seed stearic acid ratio have been identified and mapped on chromosomes 2, 3, and 14; one QTL that control seed linoleic acid ratio has been identified and mapped on chromosome 14; five QTL that control seed stearic acid to palmitic acid ratio (stearic/palmitic) have been identified and mapped on chromosomes 2, 3, 5, and 14; two QTL that control seed linolenic acid to linoleic acid ratio (linolenic/linoleic) have been identified and mapped on chromosomes 10 and 14; one QTL that control seed saturated fatty acids (SFA) have been identified and mapped on chromosome 5; and two QTL that control seed saturated fatty acids (SFA) to unsaturated fatty acids (USFA) ratio (SFA/USFA) have been identified and mapped on chromosome 5 (Table 7.1). In addition to the three main genes involved in fatty acids biosynthesis: FatB gene encoding by fatty acyl-ACP thioesterase, SAD gene encoding plant stearoyl-acyl-carrier protein desaturase, and FAD3 gene encoding fatty acid desaturase, the study identified five additional genes involved in fatty acid biosynthetic pathways and six genes involved in lipid biosynthetic pathways. Among these genes are genes encoding for phospholipase C2, phospholipase D alpha 1, reduced oleate desaturation 1, 3-ketoacyl-CoA synthase 11, diacylglycerol kinase family protein, fatty acid desaturase 5, O-acyltransferase (WSD1-like) family protein, allene oxide cyclase 3, glycolipid transfer protein 2, and several others (Table 2.3) (Fang et al. 2017).

Zhang et al. (2018) used a genome-wide association study (GWAS) on 313 soybean accessions and studied seed protein, oil, fatty acid, and amino acid contents. The study identified 3 QTL controlling protein content, 25 QTL controlling oil content, 18 QTL controlling fatty acids contents, and 39–53 QTL for amino acids contents (Fig. 2.2) (Zhang et al. 2018). Several candidate genes have been identified on the same regions or very close to these identified QTL. Among these candidate genes are *GmSAT1* encoding the *Glycine max* symbiotic ammonium transporter 1 involved in N_2 fixation identified on chromosome 15: *Glyma.15 g049900*, *Glyma.15 g049700*, and *Glyma.15 g049800* encoding the nodulin MtN21 protein family (Zhang et al. 2018).

Li et al. (2019c) identified 26 QTL that control seed protein content and 23 QTL that control seed oil content (Fig. 2.2; Table 2.2). Within the QTL regions that control seed protein content, they identified *Glyma02g31440* that encodes for a NAD(P)-linked oxidoreductase superfamily protein, *Glyma03g40780* that encodes for an ankyrin repeat family protein, *Glyma04g10451* that encodes for a CRT (chloroquine-resistance transporter)-like transporter 2, *Glyma06g40780* that encodes for a disease resistance protein (TIR-NBS-LRR class) family, *Glyma07g10621* that encodes for a protein kinase superfamily protein, *Glyma08g21340* that encodes for a GDSL-like lipase/acylhydrolase superfamily protein, *Glyma08g42100* that encodes for an RNI-like superfamily protein,

Glyma09g24910 that encodes for a hydroxyproline-rich glycoprotein family protein, *Glyma10g01180* that encodes for a plant invertase/pectin methylesterase inhibitor superfamily, *Glyma10g29000* that encodes for a importin alpha isoform 2, *Glyma10g29831* that encodes for a mitochondrial inner membrane translocase complex, *Glyma15g28100* that encodes for guanyl-nucleotide exchange factors, *Glyma17g08880* that encodes for a nonsense-mediated mRNA decay NMD3 family protein, *Glyma19g29730* that encodes for a exostosin family protein, *Glyma19g31120* that encodes for a glutamate synthase 1, *Glyma20g21693* that encodes for a subtilase family protein, *Glyma20g34880* that encodes for a differentiation and greening-like 1, and many other genes. Within the QTL regions that control seed oil content, they identified *Glyma01g30670* that encodes for a PEBP (phosphatidylethanolamine-binding protein) family protein, *Glyma03g04340* that encodes for a protein kinase superfamily protein, *Glyma03g09870* that encodes for a protein kinase 1B, *Glyma04g34745* that encodes for a FAR1-related sequence 5, *Glyma04g42167* that encodes for an ACT domain repeat 3, *Glyma06g40780* that encodes for a disease resistance protein (TIR-NBS-LRR class) family, *Glyma08g21340* that encodes for a GDSL-like lipase/acylhydrolase superfamily protein, *Glyma08g23660* that encodes for a galactose oxidase/kelch repeat superfamily protein, *Glyma10g06800* that encodes for a phosphatidylinositol monophosphate 5 kinase, *Glyma10g29851* that encodes for an ROP-binding protein kinases 2, *Glyma10g38890* that encodes for an NAD(P)-linked oxidoreductase superfamily protein, *Glyma13g27480* that encodes for a starch synthase 3, *Glyma15g14700* that encodes for a chaperone DNAJ-domain superfamily protein, *Glyma16g27980* that encodes for a WD-40 repeat family protein/notchless protein, *Glyma17g07341* that encodes for a UDP-glucosyl transferase 78D2, *Glyma17g11950* that encodes for a tetratricopeptide repeat (TPR)-like superfamily protein, *Glyma20g21693* that encodes for a subtilase family protein, *Glyma20g21726* that encodes for an aldehyde dehydrogenase 12A1, and many other genes (Table 2.3) (Li et al. 2019c).

Karikari et al. (2019) identified 25 QTL that control seed protein content and 19 QTL that control seed oil content. Within the QTL (*qPro-7-1*), authors identified the genes *Glyma07g09260* involved in regulation of cyclin-dependent protein kinase activity and stomatal lineage progression, *Glyma07g09270* involved in transmembrane transport, and *Glyma07g09630* involved in systemic acquired resistance and phenylpropanoid biosynthetic process. Within the QTL (*qOil-8-3*), authors identified the genes *Glyma08g14920* that encodes for an ABC1 family, *Glyma08g15750* that encodes for a zinc finger protein, *Glyma08g15921* that encodes for a protein kinase domain, and *Glyma08g16072* that encodes for a protein serine/threonine kinase. Within the QTL (*qOIl-10-2*), authors identified the genes *Glyma10g06800* that encodes for a phosphatidylinositol-4-phosphate 5-Kinase, *Glyma10g06810* that encodes for an amidohydrolase, and *Glyma10g06820* that encodes for a glutamine synthetase. Within the QTL (*qOil-10-4*), authors identified the genes *Glyma10g25630* that encodes for a TCP-1/cpn60 chaperonin family, *Glyma10g26510* that encodes for a lipase (class 3), *Glyma10g26770* that encodes for a tetratricopeptide repeat, *Glyma10g26870* that encodes for a WD domain,

Glyma10g26920 that encodes for an RNA recognition motif, and many other genes (Table 2.3) (Karikari et al. 2019).

In conclusion, more than a thousand of QTL that control seed protein, oil, and fatty acids (mainly palmitic, stearic, oleic, linoleic, and linolenic) have been identified and mapped in all 20 soybean chromosomes since 1992; however, only few studies identified candidate genes within these QTL regions (Bolon et al. 2010; Hwang et al. 2014; Zhou et al. 2015; Li et al. 2015, 2018a, 2019b, 2019c; Fang et al. 2017; Zhang et al. 2018). The uses of both biparental populations and GWAS (Atwell et al. 2010; Zhang et al. 2014) to map QTL for complex polygenic traits are shown to be efficient; however, decades of followed studies are needed to fine map the regions, identify candidate genes, and understand the full extent of the metabolic pathways involved in these traits.

References

Akond, M., S. Liu, M. Boney, S.K. Kantartzi, K. Meksem, N. Bellaloui, D.A. Lightfoot, and M.A. Kassem. 2014. Identification of Quantitative Trait Loci (QTL) underlying protein, oil, and five major fatty acids' contents in soybean. *American Journal of Plant Sciences* 5 (1): 158–167.

Arcade, A., A. Labourdette, M. Falque, B. Mangin, F. Chardon, A. Charcosset, and J. Joets. 2004. Biomercator: integrating genetic maps and QTL towards discovery of candidate genes. *Bioinformatics* 20: 2324–2326.

Atwell, S., Y.S. Huang, B.J. Vilhjálmsson, G. Willems, M. Horton, Y. Li, D. Meng, A. Platt, A.M. Tarone, T.T. Hu, R. Jiang, N.W. Muliyati, X. Zhang, M.A. Amer, I. Baxter, B. Brachi, J. Chory, C. Dean, M. Debieu, J. de Meaux, J.R. Ecker, N. Faure, J.M. Kniskern, J.D.G. Jones, T. Michael, A. Nemri, F. Roux, D.E. Salt, C. Tang, M. Todesco, M.B. Traw, D. Weigel, P. Marjoram, J.O. Borevitz, J. Bergelson, and M. Nordborg. 2010. Genome-wide association study of 107 phenotypes in Arabidopsis thaliana inbred lines. *Nature* 465: 627–631.

Bachlava, E., R.E. Dewey, J.W. Burton, and A.J. Cardinal. 2009. Mapping and comparison of quantitative trait loci for oleic acid seed content in two segregating soybean populations. *Crop Science* 49: 433–442. https://doi.org/10.2135/cropsci2008.06.0324.

Bandillo, N., D. Jarquin, Q.J. Song, R. Nelson, P.B. Cregan, J. Specht, and A. Lorenz. 2015. A population structure and genome-wide association analysis on the USDA soybean germplasm collection. *Plant Genome* 8 (3): 1–13. https://doi.org/10.3835/plantgenome2015.04.0024.

Bolon, Y.T., B. Joseph, S.B. Cannon, M.A. Graham, B.W. Diers, A.D. Farmer, G.D. May, G.J. Muehlbauer, J.E. Specht, Z.J. Tu, N. Weeks, W.W. Xu, R.C. Shoemaker, and C.P. Vance. 2010. Complementary genetic and genomic approaches help characterize the linkage group I seed protein QTL in soybean. *BMC Plant Biology* 10: 41. https://doi.org/10.1186/1471-2229-10-41.

Bolon, Y.T., D.L. Hyten, J.H. Orf, C.P. Vance, and G.J. Muehlbauer. 2014. eQTL networks reveal complex genetic architecture in the immature soybean seed. *Plant Genome* 7: 1–14. https://doi.org/10.3835/plantgenome2013.08.0027.

Brown, A.V., S.I. Conners, W. Huang, A.P. Wilkey, D. Grant, N.T. Weeks, S.B. Cannon, M.A. Graham, and R.T. Nelson. 2020. A new decade and new data at SoyBase, the USDA-ARS soybean genetics and genomics database. *Nucleic Acids Research* 2020: 1–6. https://doi.org/10.1093/nar/gkaa1107.

Brummer, E.C., G.L. Graef, J. Orf, J.R. Wilcox, and R.C. Shoemaker. 1997. Mapping QTL for Seed Protein and Oil Content in Eight Soybean Populations. *Crop Science* 37: 370–378.

Chapman, A., V.R. Pantalone, A. Ustun, F.L. Allen, D. Landau-Ellis, R.N. Trigiano, and M. Gresshoff. 2003. Quantitative trait loci for agronomic and seed quality traits in an F-2 and F-4:6 soybean population. *Euphytica* 129 (3): 387–393.

Chen, Q.S., Z.C. Zhang, C.Y. Liu, W.Q. Wang, and W.B. Li. 2005. Construction and analysis of soybean genetic map using recombinant inbred line of Charleston 9 Dongnong 594. *Agricultural Sciences in China* 38: 1312–1316.

Chen, Q.S., Z.C. Zhang, C.Y. Liu, D.W. Xin, H.M. Qiu, D.P. Shan, and G.H. Hu. 2007. QTL analysis of major agronomic traits in soybean. *Agricultural Sciences in China* 6 (4): 399–405. https://doi.org/10.1016/S1671-2927(07)60062-5.

Chung J, H. L Babka, GL Graef, PE Staswick, DJ Lee, PB Cregan, RC Shoemaker, and JE Specht (2003) The Seed Protein, Oil, and Yield QTL on Soybean Linkage Group I. Crop Science 43 (3): 1053-1067.

Csanadi, G., J. Vollmann, G. Stift, and T. Lelley. 2001. Seed quality QTLs identified in a molecular map of early maturing soybean. *Theoretical and Applied Genetics* 103: 912–919.

Diers, B.W., and R.C. Shoemaker. 1992. Restriction fragment length polymorphism analysis of soybean fatty acid content. *Journal of the American Oil Chemists Society* 69 (12): 1242–1244.

Diers, B.W., P. Keim, W.R. Fehr, and R.C. Shoemaker. 1992. RFLP analysis of soybean seed protein and oil content. *Theoretical and Applied Genetics* 83 (5): 608–612.

Eskandari, M., E.R. Cober, and I. Rajcan. 2013. Genetic control of soybean seed oil: II. QTL and genes that increase oil concentration without decreasing protein or with increased seed yield. *Theoretical and Applied Genetics* 126: 1677–1687.

Fan, S., B. Li, F. Yu, F. Han, S. Yan, L. Wang, and J. Sun. 2015. Analysis of additive and epistatic quantitative trait loci underlying fatty acid concentrations in soybean seeds across multiple environments. *Euphytica* 206: 689–700. https://doi.org/10.1007/s10681-015-1491-3.

Fang, C., Y. Ma, S. Wu, L.C. Fang, Y. Ma, S. Wu, Z. Liu, Z. Wang, R. Yang, G. Hu, Z. Zhou, H. Yu, M. Zhang, Y. Pan, G. Zhou, H. Ren, W. Du, H. Yan, Y. Wang, D. Han, Y. Shen, S. Liu, T. Liu, J. Zhang, H. Qin, J. Yuan, X. Yuan, F. Kong, B. Liu, J. Li, Z. Zhang, G. Wang, B. Zhu, and Z. Tian. 2017. Genome-wide association studies dissect the genetic networks underlying agronomical traits in soybean. *Genome Biology* 18: 161. https://doi.org/10.1186/s13059-017-1289-9.

Fasoula, V.A., D.K. Harris, and H.R. Boerma. 2004. Validation and designation of quantitative trait loci for seed protein, seed oil, and seed weight from two soybean populations. *Crop Science* 44 (4): 1218–1225. https://doi.org/10.2135/cropsci2004.1218.

Fehr, W.R. 2007. Breeding for modified fatty acid composition in soybean. *Crop Science* 47 (S3): S72–S87.

Gai, J., W. Yongjun, W. Xiaolei, and C. Shouyi. 2007. A comparative study on segregation analysis and QTL mapping of quantitative traits in plants—with a case in soybean. *Frontiers of Agriculture in China* 1 (1): 1–7.

Graef, G.L., W.R. Fehr, L.A. Miller, E.G. Hammond, and S.R. Cianzo. 1988. Inheritance of fatty acid composition in a soybean mutant with low linolenic acid. *Crops Science* 28 (1): 55–58.

Ha, B.K., H.J. Kim, V. Velusamy, T.D. Vuong, H.T. Nguyen, J.G. Shannon, and J.D. Lee. 2014. Identification of quantitative trait loci controlling linolenic acid concentration in PI483463 (*Glycine soja*). *Theoretical and Applied Genetics* 127: 1501–1512.

Hacisalihoglu, G., A.L. Burton, J.L. Gustin, S. Eker, S. Asikli, E.H. Heybet, L. Ozturk, I. Cakmak, A. Yazici, K.O. Burkey, J. Orf, and A.M. Settles. 2018. Quantitative trait loci associated with soybean seed weight and composition under different phosphorus levels. *Journal of Integrative Plant Biology* 60 (3): 232–241.

Han, Y., W. Teng, Y. Wang, X. Zhao, L. Wu, D. Li, and W. Li. 2015. Unconditional and conditional QTL underlying the genetic interrelationships between soybean seed isoflavone, and protein or oil contents. *Plant Breeding* 134: 300–309. https://doi.org/10.1111/pbr.12259.

Haun, W., A. Coffman, B.M. Clasen, Z.L. Demorest, A. Lowy, E. Ray, A. Retterath, T. Stoddard, A. Juillerat, F. Cedrone, L. Mathis, D.F. Voytas, and F. Zhang. 2014. Improved soybean oil quality by targeted mutagenesis of the fatty acid desaturase 2 gene family. *Plant Biotechnology Journal* 12: 934–940.

Hwang, E.Y., Q. Song, G. Jia, J.E. Specht, D.L. Hyten, J. Costa, and P.B. Cregan. 2014. A genome-wide association study of seed protein and oil content in soybean. *BMC Genomics* 15: 1. https://doi.org/10.1186/1471-2164-15-1.

Hyten, D.L., V.R. Pantalone, A.M. Saxton, M.E. Schmidt, and C.E. Sams. 2004a. Molecular mapping and identification of soybean fatty acid modifier quantitative trait loci. *Journal of the American Oil Chemists' Society* 81 (12): 1115–1118. https://doi.org/10.1007/s11746-004-1027-z.

Hyten, D.L., V.R. Pantalone, C.E. Sams, A.M. Saxton, D. Landau-Ellis, T.R. Stefaniak, and M.E. Schmidt. 2004b. Seed quality QTL in a prominent soybean population. *Theoretical and Applied Genetics* 109: 552–561. https://doi.org/10.1007/s00122-004-1661-5.

Jun, T.H., K. Van, M.Y. Kim, S.H. Lee, and D.R. Walker. 2008. Association analysis using SSR markers to find QTL for seed protein content in soybean. *Euphytica* 162 (2): 179–191.

Kabelka, E.A., B.W. Diers, W.R. Fehr, A.R. LeRoy, I.C. Baianu, T. You, D.J. Neece, and R.L. Nelson. 2004. Putative alleles for increased yield from soybean plant introductions. *Crop Science* 44: 784–791. https://doi.org/10.2135/cropsci2004.0784.

Karikari, B., S. Li, J.A. Bhat, Y. Cao, J. Kong, J. Yang, J. Gai, and T. Zhao. 2019. Genome-wide detection of major and epistatic effect qtls for seed protein and oil content in soybean under multiple environments using high-density bin map. *International Journal of Molecular Sciences* 2019 (20): 979. https://doi.org/10.3390/ijms20040979.

Kim, H.K., Y.C. Kim, S.T. Kim, B.G. Son, Y.W. Choi, J.S. Kang, Y.H. Park, Y.S. Cho, and I.S. Choi. 2010. Analysis of quantitative trait loci (QTLs) for seed size and fatty acid composition using recombinant inbred lines in soybean. *Journal of Life Science* 20 (8): 1186–1192.

Kim, H.J., B.K. Ha, K.S. Ha, J.H. Chae, J.H. Park, M.S. Kim, S. Asekova, J.G. Shannon, C.K. Son, and J.D. Lee. 2015. Comparison of a high oleic acid soybean line to cultivated cultivars for seed yield, protein and oil concentrations. *Euphytica* 201: 285–292.

La, T., E. Large, E. Taliercio, Q. Song, J.D. Gillman, D. Xu, H.T. Nguyen, G. Shannon, and A. Scaboo. 2019. Characterization of select wild soybean accessions in the USDA germplasm collection for seed composition and agronomic traits. *Crop Science* 58 (1): 1–19.

Lalles, J.P., and H.M. Tukur. 1996. Analytical criteria for predicting apparent digestibility of soybean protein in preruminant calves. *Dairy Science* 79: 475–482. https://doi.org/10.3168/jds.S0022-0302(96)76388-9.

Lark, K.G., J. Orf, and L.M. Mansur. 1994. Epistatic expression of quantitative trait loci (QTL) in soybean [*Glycine max* (L.) Merr.] determined by QTL association with RFLP alleles. *Theoretical and Applied Genetics* 88: 486–489.

Leamy, L.J., H. Zhang, C. Li, C.Y. Chen, and B.H. Song. 2017. A genome-wide association study of seed composition traits in wild soybean (Glycine soja). *BMC Genomics* 18: 18. https://doi.org/10.1186/s12864-016-3397-4.

Lee, S.H., M.A. Bailey, M.A.R. Mian, T.E. Carter, E.R. Shipe, D.A. Ashley, W.A. Parrott, R.S. Hussey, and H.R. Boerma. 1996. RFLP loci associated with soybean seed protein and oil content across populations and locations. *Theoretical and Applied Genetics* 93 (5-6): 649–657.

Leite, D.C., J.B. Pinheiro, J.B. Campos, A.O.D. Mauro, and S.H. Uneda-Trevisoli. 2016. QTL mapping of soybean oil content for marker-assisted selection in plant breeding program. *Genetics and Molecular Research* 15 (1): 1–11. https://doi.org/10.4238/gmr.15017685.

Li, Z., R.F. Wilson, W.E. Rayford, and H.R. Boerma. 2002. Molecular mapping genes conditioning reduced palmitic acid content in N87-2122-4 soybean. *Crops Science* 42 (2): 373–378. https://doi.org/10.2135/cropsci2002.3730.

Li, D., M. Sun, Y. Han, W. Teng, and W. Li. 2010. Identification of QTL underlying soluble pigment content in soybean stems related to resistance to soybean white mold (*Sclerotinia sclerotiorum*). *Euphytica* 172: 49–57. https://doi.org/10.1007/s10681-009-0036-z.

Li, H., T. Zhao, Y. Wang, D. Yu, S. Chen, R. Zhou, and J. Gai. 2011. Genetic structure composed of additive QTL, epistatic QTL pairs and collective unmapped minor QTL conferring oil content and fatty acid components of soybeans. *Euphytica* 182: 117–132. https://doi.org/10.1007/s10681-011-0524-9.

Li, Y.H., J.C. Reif, Y.S. Ma, H.L. Hong, Z.X. Liu, R.Z. Chang, and L.J. Qiu. 2015. Targeted association mapping demonstrating the complex molecular genetics of fatty acid formation in soybean. *BMC Genomics* 16: 841. https://doi.org/10.1186/s12864-015-2049-4.

Li, Y., J.C. Reif, H. Hong, H. Lia, Z. Liu, Y. Ma, J. Li, Y. Tian, Y. Li, and L. Qiua. 2018a. Genome-wide association mapping of QTL underlying seed oil and protein contents of a diverse panel of soybean accessions. *Plant Science* 266: 95–101. https://doi.org/10.1016/j.plantsci.2017.04.013.

Li, X., R. Tian, S. Kamala, H. Du, W. Li, Y. Kong, and C. Zhang. 2018b. Identification and verification of pleiotropic QTL controlling multiple amino acid contents in soybean seed. *Euphytica* 214: 93. https://doi.org/10.1007/s10681-018-2170-y.

Li, W., Z. Shao, R. Tian, H. Zhang, H. Du, Y. Kong, W. Li, and C. Zhang. 2019a. Mining QTLs and candidate genes for seed protein and oil contents across multiple environments and backgrounds in soybean. *Molecular Breeding* 39: 139. https://doi.org/10.1007/s11032-019-1055-7.

Li, D., X. Zhao, Y. Han, W. Li, and F. Xie. 2019b. Genome-wide association mapping for seed protein and oil contents using a large panel of soybean accessions. *Genomics* 111: 90–95. https://doi.org/10.1016/j.ygeno.2018.01.004.

Li, S., H. Xu, J. Yang, and T. Zhao. 2019c. Dissecting the genetic architecture of seed protein and oil content in soybean from the Yangtze and Huaihe River Valleys using multi-locus genome-wide association studies. *International Journal of Molecular Science* 2019 (20): 3041. https://doi.org/10.3390/ijms20123041.

Li, X., H. Xue, K. Zhang, W. Li, Y. Fang, Z. Qi, Y. Wang, X. Tian, J. Song, W. Li, and H. Ning. 2020. Mapping QTLs for protein and oil content in soybean by removing the influence of related traits in a four-way recombinant inbred line population. *The Journal of Agricultural Science* 157: 659–675. https://doi.org/10.1017/S0021859620000040.

Liang, H.Z., S.F. Wang, Y.L. Yu, Y. Lian, T.F. Wang, Y.L. Wei, P.T. Geong, X.Y. Liu, and X.J. Fang. 2010. QTL mapping of isoflavone, oil and protein content in soybean. *Agricultural Sciences in China* 9 (8): 1108–1116.

Liener, I.E. 1996. Effects of processing on antinutritional factors in legumes: the soybean case. *Archivos Latinoamericanos de Nutrición* 44: 48S–54S. PMID:9137638.

Lu, W., Z. Wen, H. Li, D. Yuan, J. Li, H. Zhang, Huang Zn, S. Cui, and W. Du. 2013. Identification of the quantitative trait loci (QTL) underlying water-soluble protein content in soybean. *Theoretical and Applied Genetics* 126: 425–433.

Mansur, L.M., K.G. Lark, H. Kross, and A. Oliveira. 1993. Interval mapping of quantitative trait loci for reproductive, morphological, and seed traits of soybean (Glycine max L.). *Theoretical and Applied Genetics* 86: 907–913.

Mansur, L.M., J.H. Orf, K. Chase, T. Jarvik, P.B. Cregan, and K.G. Lark. 1996. Genetic mapping of agronomic traits using recombinant inbred lines of soybean. *Crop Science* 36 (5): 1327–1336.

Mao, T., Z. Jiang, Y. Han, W. Teng, X. Zhao, and W. Li. 2013. Identification of quantitative trait loci underlying seed protein and oil contents of soybean across multi-genetic backgrounds and environments. *Plant Breeding* 132 (6): 630–641. https://doi.org/10.1111/pbr.12091.

Meksem, K., T. Doubler, K. Chancharoenchai, N. Nijti, S.J. Chang, A.P. Rao-Arelli, P.E. Cregan, L.E. Gray, P.T. Gibson, and D.A. Lightfoot. 1999. Clustering among loci underlying soybean resistance to Fusarium solani, SDS and SCN in near-isogenic lines. *Theoretical and Applied Genetics* 99: 1131–1142. https://doi.org/10.1007/s001220051317.

Moongkanna, J., S. Nakasathien, W.P. Novitzky, P. Kwanyuen, P. Sinchaisri, and P. Srinives. 2011. SSR Markers Linking to Seed Traits and Total Oil Content in Soybean. *Thai Journal of Agricultural Science* 44 (4): 233–241.

Nichols, D.M., K.D. Glover, S.R. Carlson, J.E. Specht, and B.W. Diers. 2006. Fine mapping of a seed protein QTL on soybean linkage group I and its correlated effects on agronomic traits. *Crop Science* 46: 834–839. https://doi.org/10.2135/cropsci2005.05-0168.

Orf, J.H., K. Chase, T. Jarvik, L.M. Mansur, P.B. Cregan, F.R. Adler, and K.G. Lark. 1999. Genetics of soybean agronomic traits: I. comparison of three related recombinant inbred populations. *Crop Science* 39 (6): 1642–1651.

Palomeque, L., L. Li-Jun, W. Li, B. Hedges, E.R. Cober, and I. Rajcan. 2009a. QTL in mega-environments: I. Universal and specific seed yield QTL detected in a population derived from a cross of high-yielding adapted high-yielding exotic soybean lines. *Theoretical and Applied Genetics* 119: 417–427. https://doi.org/10.1007/s00122-009-1049-7.

———. 2009b. QTL in mega-environments: II. Agronomic trait QTL co-localized with seed yield QTL detected in a population derived from a cross of high-yielding adapted £ high-yielding exotic soybean lines. *Theoretical and Applied Genetics* 119: 429–436. https://doi.org/10.1007/s00122-009-1048-8.

Pandurangan, S., A. Pajak, S.J. Molnar, E.R. Cober, S. Dhaubhadel, C. Hernandez-Sebastia, W.M. Kaiser, R.L. Nelson, S.C. Huber, and F. Marsolais. 2012. Relationship between asparagine metabolism and protein concentration in soybean seed. *Journal of Experimental Botany* 63 (8): 3173–3184. https://doi.org/10.1093/jxb/ers039.

Panthee, D.R., V.R. Pantalone, D.R. Westa, A.M. Saxtonb, and C.E. Samsa. 2005. Quantitative trait loci for seed protein and oil concentration, and seed size in soybean. *Crop Science* 45 (5): 2015–2022.

Panthee, D.R., V.R. Pantalone, and A.M. Saxton. 2006. Modifier QTL for fatty acid composition in soybean oil. *Euphytica* 152: 67–73.

Pathan, S.M., T. Vuong, K. Clark, J.D. Lee, J.G. Shannon, C.A. Roberts, M.R. Ellersieck, J.W. Burton, P.B. Cregan, D.L. Hyten, H.T. Nguyen, and D.A. Sleper. 2013. Genetic mapping and confirmation of quantitative trait loci for seed protein and oil contents and seed weight in soybean. *Crop Science* 53 (3): 765–774. https://doi.org/10.2135/cropsci2012.03.0153.

Patil, G., T.D. Vuong, S. Kale, B. Valliyodan, R. Deshmukh, C. Zhu, X. Wu, Y. Bai, D. Yungbluth, F. Lu, S. Kumpatla, J.G. Shannon, R.K. Varshney, and H.T. Nguyen. 2018. Dissecting genomic hotspots underlying seed protein, oil, and sucrose content in an interspecific mapping population of soybean using high-density linkage mapping. *Plant Biotechnology Journal* 16: 1939–1953. https://doi.org/10.1111/pbi.12929.

Priolli, R.H.G., J.B. Campos, N.S. Stabellini, J.B. Pinheiro, and N.A. Vello. 2015. Association mapping of oil content and fatty acid components in soybean. *Euphytica* 203: 83–96. https://doi.org/10.1007/s10681-014-1264-4.

Qi, Z.M., X. Han, Y.N. Sun, Q. Wu, D.P. Shan, X.Y. Du, C.Y. Liu, H.W. Jiang, G.H. Hu, and Q.S. Chen. 2011. An integrated quantitative trait locus map of oil content in soybean, Glycine max (L.) Merr., generated using a meta-analysis method for mining genes. *Agricultural Sciences in China* 10 (11): 1681–1692. https://doi.org/10.1016/S1671-2927(11)60166-1.

Qi, Z.M., M. Hou, X. Han, C. Liu, H. Jiang, D. Xin, G. Hu, and Q. Chen. 2014. Identification of quantitative trait loci (QTLs) for seed protein concentration in soybean and analysis for additive effects and epistatic effects of QTLs under multiple environments. *Plant Breeding* 133 (4): 499–507. https://doi.org/10.1111/pbr.12179.

Qiu, B.X., P.R. Arelli, and D.A. Sleper. 1999. RFLP markers associated with soybean cyst nematode resistance and seed composition in a 'Peking'×'Essex' population. *Theoretical and Applied Genetics* 98: 356–364.

Reinprecht, Y., V.W. Poysa, K. Yu, I. Rajcan, G.R. Ablett, and K.P. Paulsa. 2006. Seed and agronomic QTL in low linolenic acid, lipoxygenase-free soybean (*Glycine max* (L.) Merrill) germplasm. *Genome* 49 (12): 1510–1527. https://doi.org/10.1139/g06-112.

Rossi, M.E., J.H. Orf, L.J. Liu, Z. Dong, and I. Rajcan. 2013. Genetic basis of soybean adaptation to North American vs. Asian mega-environments in two independent populations from Canadian × Chinese crosses. *Theor and Appl Genet* 126: 1809–1823.

Sebolt, A.M., R.C. Shoemaker, and B.W. Diers. 2000. Analysis of a Quantitative Trait Locus Allele from Wild Soybean That Increases Seed Protein Concentration in Soybean. *Crop Science* 40 (95): 1438–1444.

Shibata, M., K. Takayama, A. Ujiie, T. Yamada, J. Abe, and K. Kitamura. 2008. Genetic relationship between lipid content and linolenic acid concentration in soybean seeds. *Breeding Science* 58: 361–366.

Sonah, H., L. O'Donoughue, E. Cober, I. Rajcan, and F. Belzile. 2015. Identification of loci governing eight agronomic traits using a GBS-GWAS approach and validation by QTL mapping in soya bean. *Plant Biotech Journal* 13: 211–221. https://doi.org/10.1111/pbi.12249.

Specht, J.E., K. Chase, M. Macrander, G.L. Graef, J. Chung, J.P. Markwell, M. Germann, J.H. Orf, and K.G. Lark. 2001. Soybean Response to water: A QTL analysis of drought tolerance. *Crop Science* 41 (2): 493–509.

Spencer, M.M., D. Landau-Ellis, E.J. Meyer, and V.R. Pantalone. 2004. Molecular markers associated with linolenic acid content in soybean. *JAOCS* 81: 559–562.

Stevenson, D.G., J.L. Jane, and G.E. Ingletta. 2007. Structures and physicochemical properties of starch from immature seeds of soybean varieties (*Glycine max* (L.) Merr.) exhibiting normal, low-linolenic or low-saturated fatty acid oil profiles at maturity-Carbohydrate. *Polymers* 70 (2): 149–159.

Stombaugh, S.K., J.H. Orf, H.G. Jung, K. Chase, K.G. Lark, and D.A. Somers. 2004. Quantitative trait loci associated with cell wall polysaccharides in soybean seed. *Crop Science* 44: 2101–2106.

Tajuddin, T., S. Watanabe, N. Yamanaka, and K. Harada. 2003. Analysis of quantitative trait loci for protein and lipid contents in soybean seeds using recombinant inbred lines. *Breeding Science* 53: 133–140.

Teng, W., W. Li, Q. Zhang, D. Wu, X. Zhao, H. Li, Y. Han, and W. Li. 2017. Identification of quantitative trait loci underlying seed protein content of soybean including main, epistatic, and QTL × environment effects in different regions of Northeast China. *Genome* 60 (8): 649–655. https://doi.org/10.1139/gen-2016-0189.

Van Ooijen JM (2004) Software for the mapping of quantitative trait loci in experimental populations. MapQTL5 Kyazma BV, Wageningen. The Netherlands.

Vollmann, J., H. Schausberger, H. Bistrich, and T. Lelley. 2002. The presence or absence of the soybean Kunitz trypsin inhibitor as a quantitative trait locus for seed protein content. *Plant Breeding* 121 (3): 272–274.

Voorrips, R.E. 2002. MapChart: software for the graphical presentation of the linkage maps and QTLs. *Journal of Heredity* 93: 77–78.

Vuong, T.D., D.A. Sleper, J.G. Shannon, X. Wu, and H.T. Nguyen. 2010. Novel quantitative trait loci for broad-based resistance to soybean cyst nematode (*Heterodera glycines* I.) in soybean PI 567516C. *Theoretical and Applied Genetics* 121: 1253–1266. https://doi.org/10.1007/s00122-010-1385-7.

———. 2011. Confirmation of quantitative trait loci for resistance to multiple-HG types of soybean cyst nematode (*Heterodera glycines* I.). *Euphytica* 181: 101–113. https://doi.org/10.1007/s10681-011-0430-1.

Wang, X., G.L. Jiang, M. Green, R.A. Scott, D.L. Hyten, and P.B. Cregan. 2012. Quantitative trait locus analysis of saturated fatty acids in a population of recombinant inbred lines of soybean. *Molecular Breeding* 30 (2): 1163–1179.

Wang, X., G.L. Jiang, M. Green, R.A. Scott, Q. Song, D.L. Hyten, and P.B. Cregan. 2014a. Identification and validation of quantitative trait loci for seed yield, oil, and protein contents in two recombinant inbred line populations of soybean. *Molecular Genetics and Genomics* 289 (5): 935–949.

Wang, X., G.L. Jiang, M. Green, R.A. Scott, D.L. Hyten, and P.B. Cregan. 2014b. Quantitative trait locus analysis of unsaturated fatty acids in a recombinant inbred population of soybean. *Molecular Breeding* 33: 281–296.

Xie, D., W. Chang, Y. Han, Y. Zeng, W. Teng, and W. Li. 2012. SSR- and SNP-related QTL underlying linolenic acid and other fatty acid contents in soybean seeds across multiple environments. *Molecular Breeding* 30: 169–179. https://doi.org/10.1007/s11032-011-9607-5.

Yang, M.L., J.J. Ding, H.Y. Li, M.N. Sui, and J.A. Wang. 2017. Identification of QTL underlying soybean agglutinin content in soybean seeds and analysis for epistatic effects among multiple genetic backgrounds and environments. *Canadian Journal of Plant Science* 97: 674–683. https://doi.org/10.1139/cjps-2016-0270.

Yesudas, C.R., R. Bashir, M.B. Geisler, and D.A. Lightfoot. 2013. Identification of germplasm with stacked QTL underlying seed traits in an inbred soybean population from cultivars Essex and Forrest. *Plant Breeding* 31: 693–703. https://doi.org/10.1007/s11032-012-9827-3.

Zhang, K.W., Y.J. Wang, G.Z. Luo, J.S. Zhang, C.Y. He, X.L. Wu, J.Y. Gai, and S.Y. Chen. 2004. QTL mapping of ten agronomic traits on the soybean (*Glycine max* L. Merr.) genetic map and their association with EST markers. *Theoretical and Applied Genetics* 108: 1131–1139.

Zhang, D., H. Song, H. Cheng, D. Hao, H. Wang, G. Kan, H. Jin, and D. Yu. 2014. The acid phosphatase-encoding gene GmACP1 contributes to soybean tolerance to low-phosphorus stress. *PLoS Genetics* 10 (1): e1004061. https://doi.org/10.1371/journal.pgen.1004061.

Zhang, D., H. Lu, S. Chu, H. Zhang, H. Zhang, Y. Yang, H. Li, and D. Yu. 2017. The genetic architecture of water-soluble protein content and its genetic relationship to total protein content in soybean. *Scientific Reports* 7: 5053. https://doi.org/10.1038/s41598-017-04685-7.

Zhang, J., X. Wang, Y. Lu, S.J. Bhusal, Q. Song, P.B. Cregan, Y. Yen, M. Brown, and G.L. Jiang. 2018. Genome-wide scan for seed composition provides insights into soybean quality improvement and the impacts of domestication and breeding. *Molecular Plant* 11 (3): 460–472. https://doi.org/10.1016/j.molp.2017.12.016.

Zhou, Z., Y. Jiang, Z. Wang, Z. Gou, J. Lyu, W. Li, Y. Yu, L. Shu, Y. Zhao, Y. Ma, C. Fang, Y. Shen, T. Liu, C. Li, Q. Li, M. Wu, M. Wang, Y. Wu, Y. Dong, W. Wan, X. Wang, Z. Ding, Y. Gao, H. Xiang, B. Zhu, S.H. Lee, W. Wang, and Z. Tian. 2015. Resequencing 302 wild and cultivated accessions identifies genes related to domestication and improvement in soybean. *Nature Biotechnology* 33: 408–414.

Chapter 3
Seed Amino Acids, Macronutrients, Micronutrients, Sugars, and Other Compounds

Moulay Abdelmajid Kassem

3.1 Introduction

Through their root system, plants take up water and mineral nutrients as inorganic ions from the soil, and they are transported into various plant parts where they are needed for plant growth and development (Taiz and Zeiger 1998). Plant mineral nutrition studies how plants acquire and process these mineral nutrients called *essential nutrients or essential elements* which are divided into two groups: macronutrients or macroelements and micronutrients or macroelements (Taiz and Zeiger 1998). Macronutrients required by plants in large amounts are nitrogen (N), potassium (K), calcium (Ca), magnesium (Mg), phosphorus (P), sulfur (S), and silicon (Si), and micronutrients required by plants in low amounts are chlorine (Cl), iron (Fe), boron (B), manganese (Mn), sodium (Na), zinc (Zn), copper (Cu), nickel (Ni), and molybdenum (Mo) (Taiz and Zeiger 1998). Moreover, plants require carbon (C), hydrogen (H), and oxygen (O) that they obtain from water and carbon dioxide (CO_2) during the process of photosynthesis. When the sunlight and these essential elements are provided to the plant, it can synthesize all other compounds that it needs for its growth and development; however, any absence or scarceness of one or more of these essential elements will cause symptoms of mineral deficiencies that can be chlorosis (yellow spots on leaves and stems), necrosis (browning and local death of tissues), dwarfism (reduced stem and internodes elongation), impairment in flowers and fruit development, sterility, etc. (Taiz and Zeiger 1998; Welch 1999).

Like many other seed composition traits such as protein, oil, fatty acids, isoflavones, sugars, and organic and amino acids, macronutrients and macronutrients' contents vary considerably in soybean seeds (Tables 3.1 and 3.2) depending on the

M. A. Kassem (✉)
Plant Genomics and Biotechnology Lab, Department of Biological and Forensic Sciences, Fayetteville State University, Fayetteville, NC, USA
e-mail: mkassem@uncfsu.edu

M. A. Kassem (ed.), *Soybean Seed Composition*,
https://doi.org/10.1007/978-3-030-82906-3_3

Table 3.1 Soybean seed amino acids, sugars (sucrose, raffinose, stachyose), phosphorus (P), boron (B), iron (Fe), and ureide contents

Seed composition trait	Seed content	Reference
Amino Acids		
Asparagine (Asn)	16.9–21.7μmol g^{-1}	Rainbird et al. (1984)
Glutamine (Gln)	1.4–20.9μmol g^{-1}	
Asparagine (Asn)	0.056–0.111 mg g^{-1}	Takahashi et al. (2003)
Glutamine (Gln)	0.005–0.007 mg g^{-1}	
Serine (Ser)	0.024–0.025 mg g^{-1}	
Histidine (His)	0.024–0.068 mg g^{-1}	
Proline (Pro)	0.026–0.042 mg g^{-1}	
Threonine (Thr)	0.021–0.024 mg g^{-1}	
Arginine (Arg)	0.172–0.523 mg g^{-1}	
Valine (Val)	0.038–0.048 mg g^{-1}	
Alanine (Ala)	0.054–0.087 mg g^{-1}	
Aspartate (Asp)	0.293–0.297 mg g^{-1}	
Leucine (Leu)	0.023–0.031 mg g^{-1}	
Glutamate (Glu)	0.238–0.400 mg g^{-1}	
Isoleucine (Ile)	0.022–0.029 mg g^{-1}	
Glycine (Gly)	0.015–0.023 mg g^{-1}	
Phenylalanine (Phe)	0.055–0.061 mg g^{-1}	
Tyrosine (Tyr)	0.094–0.099 mg g^{-1}	
Methionine (Met)	0.038–0.051 mg g^{-1}	
Sugars		
Sucrose	165.2–269μmol g^{-1}	Hitz et al. (2002)
	30.1–79.4 mg g^{-1}	Bellaloui et al. (2014a, b)
Raffinose	3.5–24.2μmol g^{-1}	Hitz et al. (2002)
	2.7–5.5 mg g^{-1}	Bellaloui et al. (2014a, b)
Stachyose	0.7–70.5μmol g^{-1}	Hitz et al. (2002)
	44.3–54.5 mg g^{-1}	Bellaloui et al. (2014a, b)
Galactinol	0–59.5μmol g^{-1}	Hitz et al. (2002)
Other compounds		
Inorganic phosphate	1.5–76μmol phosphate g^{-1}	Hitz et al. (2002)
Phytic acid phosphate	62.5–155μmol phosphate g^{-1}	
Phosphorus (P)	3.9–5.9 g kg^{-1}	Bellaloui et al. (2014a, b)
Boron (B)	33.3–54.8 mg kg^{-1}	
Iron (Fe)	34.0–54.8 mg kg^{-1}	
`Ureides	4.9–37.3μmol g^{-1}	Rainbird et al. (1984)

genotype, biotic, and abiotic factors such as temperature extremes, water supply or availability, elevated or low CO_2 levels, growing season, presence or absence of pathogens, mineral deficiencies, salinity, soil pH, mineral toxicity, etc. (Yazdi-Samadi et al. 1977; Wolf et al. 1982; Mazur et al. 1999; Welch 1999; Takahashi et al. 2003; Welch and Graham 2004; Jaureguy et al. 2011; Bellaloui et al. 2014a, b, Bellaloui et al. 2015a, b, 2016, 2017). These fluctuations in seed composition traits

Table 3.2 Soybean seed protein amino acids, sugars (sucrose, raffinose, stachyose, fructose, glucose, and starch), and organic acids' contents

Seed composition trait	Seed content	Reference
Proteins and amino acids		Yazdi-Samadi et al. (1977)
Protein	1.4–2.1 mg 100 g^{-1}	Yazdi-Samadi et al. (1977)
Crude protein	392.6–481.7 g kg^{-1}	La et al. (2019)
Cysteine (Cys)	14.9–19.2 g kg^{-1}	
Proline (Pro)	46.2–52.1 g kg^{-1}	
Threonine (Thr)	35.1–39.6 g kg^{-1}	
Valine (Val)	43.8–48.9 g kg^{-1}	
Alanine (Ala)	38.9–42.9 g kg^{-1}	
Aspartate (Asp)	104.7–116.6 g kg^{-1}	
Leucine (Leu)	68.6–75.6 g kg^{-1}	
Glutamate (Glu)	158.3–181.6 g kg^{-1}	
Isoleucine (Ile)	41.1–46.0 g kg^{-1}	
Glycine (Gly)	39.3–44.4 g kg^{-1}	
Lysine (Lys)	60.9–67.3 g kg^{-1}	
Methionine (Met)	14.0–16.9 g kg^{-1}	
Sugars		
Sucrose	0.8–1.0 mg 100 g^{-1}	Yazdi-Samadi et al. (1977)
Glucose	0.1–0.2 mg 100 g^{-1}	
Fructose	0.1–0.3 mg 100 g^{-1}	
Stachyose	0.2–0.4 mg 100 g^{-1}	
Starch	1.1–1.3 mg 100 g^{-1}	
Sucrose	14.6–39.5 g kg^{-1}	La et al. (2019)
Raffinose	6.6–9.3 g kg^{-1}	
Stachyose	37.2–58.9 g kg^{-1}	
Fructose	5.1–11.6 g kg^{-1}	
Glucose	4.3–6.5 g kg^{-1}	
Organic acids		
Citric	0.1–0.1 mg 100 g^{-1}	Yazdi-Samadi et al. (1977)
Malic	0–0.1 mg 100 g^{-1}	
Succinic	0–0.1 mg 100 g^{-1}	
Glycolic	0–0.2 mg 100 g^{-1}	

could be due to fluctuations in photosynthetic rates, interplant competition, carbon metabolism, availability of nutrients, access to sunlight, crop rotation, etc. (Spaeth and Sinclair 1984; Ball et al. 2000; Bellaloui and Gillen 2010; Bellaloui et al. 2012b, Bellaloui et al. 2014a, b, 2016, 2017).

For our normal health and development, we as humans must consume about 49 nutrients and compounds including water, carbohydrates, proteins (amino acids), lipids, macroelements, microelements, and vitamins that come primarily from agricultural products including seeds, vegetables, and fruits, among other products, and any inadequate supply of these nutrients can lead to deficiencies, impaired development, and poor health (Welch and Graham 2004). Several studies estimate that 2–3 billion people experience some kind of mineral malnutrition worldwide (Ramakrishnan 2002; Welch and Graham 2004; WHO 2019). Malnutrition symptoms include underweight, stunting, obesity, overweight, and vitamin and/or macronutrient or micronutrient deficiencies (WHO 2019). Approximately, 262 million people are underweight, 2 billion are overweight, and more than 250 million children under the age of 17 experience some kind of malnutrition symptoms, and most of them live in developing countries (WHO 2019).

Poor people including children, adolescents, and women living in these developing and even in industrial countries are exposed to high risks of malnutrition (Cabarello 2002; Welch and Graham 2004; Kaidar-Person et al. 2008; WHO 2019). The first 2–3 years of a child life is critical to his/her life, and any malnutrition during this period will affect his/her overall growth and development which occur in one third of malnourished children (Cabarello 2002; Kaidar-Person et al. 2008; WHO 2019). Healthcare costs are rising worldwide because of malnutrition, poverty, and wealth and income inequalities.

In this chapter, we will discuss the roles of amino acids, macronutrients, micronutrients, and sugars in seed development and summarize genetic and quantitative trait loci (QTL) mapping of these important seed composition traits during the past three decades. We will also summarize briefly several biotic and abiotic factors affecting seed composition.

3.2 The Role of Organic and Amino Acids in Seed Development

Essential amino acids for human and animal consumption are Leu, Ile, Met, Lys, Val, Trp, Phe, and Thr; His is semi-essential; and Pro, Ser, Ala, Asp, Arg, Asn, Cys, Glu, Gln, Gly, and Tyr are nonessential amino acids (National Research Council 1998; Ufaz and Galili 2008). There must be adequate amounts of these amino acids in human and animal diets because of their pivotal role in proteins and enzymes formation, stability, and function (Ufaz and Galili 2008; Khandaker et al. 2015). This classification of amino acids into essential or indispensable and nonessential or dispensable dated goes back to Womack and Rose (1947); however, others divide amino acids into three categories: indispensable, dispensable, and conditionally essential amino acids (Jackson et al. 1981; Reeds 2000).

It is well established that proteins, amino acids, and organic acids play a major role in seed development, embryo growth, and development (Abdul-Baki and Anderson 1973; Yazdi-Samadi et al. 1977; Rainbird et al. 1984; Chrispeels and Raikhelb 1991; Takahashi et al. 2003). They are also essential for animal and human consumption (Berk 1992; Welch and Graham 2004; Ufaz and Galili 2008). Abdul-Baki and Anderson (1973) showed that excised developing embryos take up high amounts of amino acids and sugars during their early stages of development and convert them into proteins and polysaccharides (Abdul-Baki and Anderson 1973).

Yazdi-Samadi et al. (1977) analyzed organic and amino acids, starch, sugars, oil, and proteins in cultivars "Harosoy 63" and "Steele" 10 days after flowering and found that they represented 75% of the seed dry weight. They also found that protein and oil were low in Steele, but they were in high contents in Harosoy 63. It took 50 days after flowering to detect stachyose, raffinose, and starch in both cultivars and at least 17 of the 20 amino acids were detected in these cultivars (Yazdi-Samadi et al. 1977). Rainbird et al. (1984) showed that the amino acids glutamine (Gln) and asparagine (Asn) were the most abundant in soybean seed embryos and contributed 55% and 20%, respectively, of the total nitrogen content of the seed embryos (Rainbird et al. 1984). For example, soybean seed agglutinin, the lectin phytohemagglutinin, and concanavalin A are synthesized in embryos axes and cotyledons during seed development, and these lectins provide amino acids for the biosynthesis of seed storage proteins and contribute to seedling and embryo development (Chrispeels and Raikhelb 1991). Many roles have been attributed to lectins; however, their prominent role is in plant defense, and certain lectins in roots are involved in recognizing *Rhizobia* bacteria and play a role in the formation of symbiotic relationships (Chrispeels and Raikhelb 1991). It is well-known that the amino acid proline (Pro) plays a major role in plant defense against biotic and abiotic stresses, mainly by the means of osmotic regulation or adjustment, but also in seed germination and development (Hare and Cress 1997; Thakur and Sharma 2005; Hayat et al. 2012). The high proline levels during stress are also linked to providing precursors for secondary metabolites which also increase drastically during periods of stress (Hare and Cress 1997). Proline content increased drastically under salt stress (>700%) and ab acid (ABA) stress (>300%), respectively, and seed germination decreased by 35% in salt-stressed and 65% in ABA-stressed sorghum plants (Thakur and Sharma 2005). Under stress, plants accumulate reactive oxygen species (ROS) that can cause damage to important molecules; therefore, plants increase the biosynthesis of a wide range of solutes including proline, polyols, sucrose, glycine betaine, and trehalose to protect them against these ROS (Rhodes and Hanson 1993; Yancey 1994; Bohnert and Jensen 1996; Ashraf and Harris 2004; Ashraf and Foolad 2007). For a thorough review of the role of proline in coping with stress in plants, see Hayat et al. (2012).

Since soybean seeds are low in sulfur-containing amino acids cysteine (Cys) and methionine (Met) which limit their nutritional value for humans and animals, several strategies have been applied to increase the contents of these two amino acids including genetic engineering, mutagenesis, and traditional breeding especially for poultry and swine (Shewry et al. 1995; Imsande 2001) but also for human consumption (Shewry et al. 1995; Krishnan 2005, 2008; Amir and Hacham 2008; Hitsuda et al. 2008). With the increase of soybean transformation methods (Olhoft et al. 2003; Somers et al. 2003), attempts to improve seed composition and quality especially increasing protein, methionine, cysteine, and lysine contents have been made in recent years (Mazur et al. 1999; Wang et al. 2003; Haun et al. 2014). The 2S albumins from the Brazil nut (*Bertholletia excelsa* H.B.K.) (BNA) and sunflower (SFA8) are rich in methionine and cysteine [BNA, Met (18%), and Cys (8%); SFA8, Met, and Cyst (23%)] (Altenbach et al. 1987; Kortt et al. 1991). A study transformed soybean with the gene encoding for 2S albumin and demonstrated that methionine increased 15–40% in the obtained transgenic lines (Townsend and Thomas 1994; Streit et al. 2001) and sulfur amino acid contents of some of these transgenic lines surpassed the contents of wild-type lines (Dinkins et al. 2001). A study showed that one single mutation in amino acid residue 396's codon changing it from lysine (Lys) to asparagine (Asn) caused an increase in seed inorganic phosphate content and a decrease in seed phytic acid, sucrose, raffinose, and stachyose contents triggered by a 90% decrease in *myo*-inositol 1-phosphate synthase activity (Hitz et al. 2002). Another study found that a mutant seeds contained 4.5–8.2% of nitrogen content, while wild-type varieties contained 0.3–0.8% of nitrogen content contributed by free amino acids, and arginine (Arg) was a prominent amino acid in the mutant seeds; thus, nitrogen accumulates in the form of free amino acids or proteins glycinin and β-conglycinin in soybean seeds (Takahashi et al. 2003). Synthetic genes encoding for proteins with high contents of sulfur-containing amino acids such as *ASP1* gene have been introduced and expressed in sweet potato and corn, but not yet in soybean, and the results showed the production of high levels of essential amino acids (Zhang et al. 2002; Krishnan 2005; Potrykus 2003).

3.3 The Role of Macronutrients and Micronutrients in Seed Development

Macronutrients (N, P, K, S, Ca, Mg, Si) are required by plants in large amounts, and micronutrients (B, Cl, Fe, Mn, Na, Zn, Cu, Ni, Mo) are required by plants in low amounts (Table 3.3; Taiz and Zeiger 1998). Plants require C, H, and O in high quantities that are obtained from H_2O and CO_2 during photosynthesis. These nutrients are necessary for the seed embryos at early stages of growth and development. The major roles and functions of macroelements and microelements are detailed in Table 3.3.

Carbohydrates which are sugars and their polymers (monosaccharides, disaccharides, and polysaccharides) contain only C, H, and O, while proteins contain C, H, O, and N in addition to S which is part of sulfur-containing amino acids (Cys and Met). Nucleic acids (DNA and RNA) are made of nucleotides that contain mainly C, H, O, N and P. K is required as a cofactor for more than 40 enzymes; Ca plays a role in cell wall and is required as a cofactor for several enzymes including calmodulins; Mg is required as a cofactor for several enzymes; S is part of sulfur-containing amino acids (Cys and Met) and thus proteins, coenzyme A, and lipoic acid; Si plays a role in cell walls; Cl plays a role in photosynthesis; Fe is part of the cytochromes and iron proteins and plays a role in N_2 fixation (Bellaloui et al. 2014a, b) and cell respiration; B plays a role in cell walls and cell elongation and is part of polymannuronic acid; Mn is required as a cofactor for some enzymes especially the water-splitting enzyme attached to the photosystem II in the thylakoid membranes of the chloroplast; Na helps in phosphoenol pyruvate (PEP) regeneration and can

Table 3.3 Contents of macroelements and microelements needed for plant growth and development in plant tissues

Nutrients	Content (μmol g^{-1})	Functions in plants	Reference
1. Obtained from H_2O and CO_2			
Hydrogen (H)	60,000	Carbohydrates, proteins, nucleic acids, and lipids	Taiz and Zeiger (1998)
Carbon (C)	40,000	Carbohydrates, proteins, nucleic acids, and lipids	
Oxygen (O)	30,000	Carbohydrates, proteins, nucleic acids, and lipids	
2. Obtained from soil			
Macroelements			
Nitrogen (N)	1000	Amino acids, proteins, nucleic acids, amides, coenzymes, etc.	
Potassium (K)	250	Required as a cofactor for >40 enzymes	
Calcium (Ca)	125	Cell wall, required as a cofactor for some enzymes	
Magnesium (Mg)	80	Required as a cofactor for some enzymes	
Phosphorus (P)	60	Nucleic acids, required as a cofactor for some enzymes	
Sulfur (S)	30	Cys, Met, proteins, coenzyme A, lipoic acid	
Silicon (Si)	30	Cell walls	
Microelements			
Chlorine (Cl)	3.00	Photosynthesis	
Iron (Fe)	2.00	Cytochromes, iron proteins, N_2 fixation, cell respiration	
Boron (B)	2.00	Cell walls, polymannuronic acid, cell elongation	
Manganese (Mn)	1.00	Required as a cofactor for some enzymes	
Sodium (Na)	0.40	Phosphoenol pyruvate (PEP) regeneration, substitutes K	
Zinc (Zn)	0.30	Required as a cofactor for some enzymes (alcohol dehydrogenase)	
Copper (Cu)	0.10	Required as a cofactor for some enzymes (ascorbic acid oxidase)	
Nickel (Ni)	0.002	Required as a cofactor for some enzymes (urease, hydrogenases)	
Molybdenum (Mo)	0.001	Required as a cofactor for some enzymes (nitrogenase, nitrate reductase)	

From Taiz and Zeiger (1998)

substitutes K; Zn is required as a cofactor for some enzymes (alcohol dehydrogenase); Cu is required as a cofactor for some enzymes (ascorbic acid oxidase); Ni is required as a cofactor for some enzymes (urease, hydrogenases); and Mo is required as a cofactor for some enzymes (nitrogenase, nitrate reductase) (Table 3.3).

These compounds or molecules of life are essential for all living organisms including plants and play major roles in seed and embryo development. Seed mineral composition is also influenced by genotype, environmental factors, crop rotation, row spacing, and biotic and abiotic stresses which affect the plants' ability to withstand these stresses (Welch 1999; Bellaloui et al. 2010a, b, c, 2015a, b). Low concentrations of Mn, Zn, or Cu can make plants more vulnerable to soil pathogens, cold temperatures, soil acidity, and water deficit. Moreover, the size and the number of seeds produced by a plant are widely determined by the availability of nutrients to the plant mainly during the flowering time, floral buds' emergence, and seed development (Welch 1999). Other studies developed enhanced value soybeans by modifying sugars, protein, and oil contents of seeds and commercialized these high-yielding germplasm combined with high-quality seed nutrients (Kwanyuen et al. 1997; Mazur et al. 1999).

For example, Boron (B) is an essential element for plant growth and development and plays a major role in seed setting, flowering, and nitrogen fixation (Dell et al. 2002; Brown et al. 2002; Dordas 2006; Bolanos et al. 1994, 1996; Bellaloui et al. 2010a, 2014a, b) in addition to seed quality and seed yield in alfalfa (Dordas 2006). Bellaloui et al. (2010a) studied the effect of foliar boron (FB) application on seed composition of soybean by measuring the activities of nitrate reductase and nitrogenase, two key enzymes in nitrogen fixation and metabolism, and observed high activities of these two enzymes in FB-treated plants compared to control plants. Moreover, seed oleic acid and proteins increased by 13.7% and 30.9%, respectively, in FB-treated plants compared to control plants and leaves of treated plants contained high levels of boron (Bellaloui et al. 2010a). FB application caused an increase in nodule numbers, nodule masses, nitrogen fixation, and seed weight in soybean plants under moderate water stress (Bellaloui et al. 2014a, b). Another study investigated the effect of soybean-corn rotation on soybean seed composition. The study showed that 3 successive years of rotation with corn was sufficient to increase seed oleic acid content from 61% to 68%, phosphorus (P) content from 60% to 75%, iron (Fe) content from 70% to 72%, and boron (B) content from 34% to 69% or more. The overall 3-year (2006–2008) percent increases of P, Fe, and B were 88.9%, 82.9%, and 64.2%, respectively. The study demonstrated also an increase of 22% in seed yield (Table 3.4, Bellaloui et al. 2010b).

The effect of row spacing (RS) and seeding rate (SDR) on seed composition was investigated in four soybean cultivars (P 93M90, AG 3906, P 94B73, and V 52N3) over 2 years (2006–2007) in a field in Stoneville, MS (Bellaloui et al. 2015a, b). The study showed that SDR, cultivar, and year affected seed Fe, P, B, sucrose, protein, and oil contents significantly but not stachyose and linolenic acid contents. Likewise, RS affected Fe, B, raffinose, and sucrose contents significantly but not protein and oil contents (Bellaloui et al. 2015a, b).

Table 3.4 Increases of seed yield and several seed composition traits (phosphorus (P), iron (Fe), and boron (B), protein, oil, oleic acid, linoleic acid, palmitic acid, and stearic acid) due to soybean-corn crop rotation

Component	Contents				Reference
	2006	2007	2008	% Increase	
P (%)	2.7	4.1	5.1	88.9	Bellaloui et al. (2010b)
Fe (mg kg^{-1})	54.3	91.8	99.3	82.9	
B (mg kg^{-1})	41.9	59.5	66.8	64.2	
Yield (kg h^{-1})	4549	3824	5024	10.44	
Protein (%)	1860	1774	2338	25.7	
Oil (kg h^{-1})	984	833	1099	11.7	
Palmitic (kg h^{-1})	476	380	549	15.34	
Stearic (kg h^{-1})	225	128	176	–	
Oleic (kg h^{-1})	1085	1125	1521	40.2	
Linoleic (kg h^{-1})	2520	2258	3173	25.9	

From Bellaloui et al. (2010b)

3.4 The Role of Sugars and Other Compounds in Seed Development

Sugars play major and different roles in plants, and certain sugars such as raffinose, stachyose, and galactinol and maturation proteins or late embryogenesis accumulating proteins (Dure et al. 1989) play a major role in seed desiccation tolerance (Blackman et al. 1992; Sun et al. 1994; Weber et al. 1997; Koch 2004). For example, there was a drastic increase in stachyose content from traces to high levels reaching 483μg/axis in soybean seed axes during the period of slow drying (desiccation). Sucrose contents also increased from 25 to 128μg/axis during the same period of desiccation. Likewise, maturation proteins also increased (Blackman et al. 1992). Sucrose breakdown provides glucose, other sugars such as hexoses, and carbon in general for seed and fruit development. Sucrose stimulates cell division among other things and is mainly cleaved by invertases and sucrose synthases. In developing seeds and grains, invertases from the cell walls convert sucrose into hexoses (Wobus and Webber 1999; Patrick and Offler 2001; Weschke et al. 2003; Koch 2004). Sucrose accumulates in epidermis which makes the embryo independent of maternal control during seed development and upregulates the expression of genes encoding seed storage proteins (Borisjuk et al. 2004). Sucrose, raffinose, and stachyose are involved in seed desiccation tolerance (Castillo et al. 1990; Blackman et al. 1992; Sun et al. 1994; Obendorf 1997; Hitz et al. 2002). Castillo et al. (1990) investigated galactinol synthase (GS) and raffinose synthase (RS) activities in soybean and kidney bean seeds. These enzymes are involved in the raffinose family oligosaccharide (RFO) biosynthesis. They showed that RS was active in early stages of soybean seeds development from 5 to 60 days after flowering (DAF), while GS was active in late stages of kidney bean seeds development (Castillo et al. 1990). Saravitz et al. (1987) showed that sucrose content increased during seed development and maturation; however, the increase was higher, and the galactinol synthase activity was higher during late stages of seed development; therefore, galactinol synthase is a key enzyme in carbon metabolism and portioning during seed development (Saravitz et al. 1987). Other studies showed that phytic acid is abundant in seeds (up to 60%); however, it may not be required for seed viability and germination (Sebastian et al. 2000; Wilcox et al. 2000; Hitz et al. 2002).

Jaureguy et al. (2011) investigated the heritability and correlations among several traits including sucrose, raffinose, stachyose, seed size, and protein in a RIL population derived from "R95-1705" and "MFL-552" (*n* = 98). They found that R95-1705 seeds had lower sucrose, higher stachyose, and higher protein content and are smaller in size compared to MFL-552; however, the RILs were similar in these traits across all tested environments. Their results also showed a high variability among lines rather than genotype by environment interactions for these traits (Jaureguy et al. 2011).

3.5 Biotic, Abiotic, and Other Factors Affecting Seed Composition

As mentioned earlier, many seed composition traits are affected by abiotic and biotic stresses including elevated or low temperatures, water deficit, flooding, rainfall, growing season, presence or absence of pathogens, mineral deficiencies, elevated or low CO_2 levels, soil salinity, soil pH, mineral toxicity, irrigation regimes, glyphosate application, maturity, boron application, seeding rates, row spacing, etc. (Yazdi-Samadi et al. 1977; Wolf et al. 1982; Takahashi et al. 2003; Bellaloui and Mengistu 2008; Bellaloui et al. 2008, 2009a, b, c, 2010a, b, c, 2011, 2012a, b, c, 2014a, b, 2015a, b, 2016, 2017, 2020a, b; Wijewardana et al. 2019).

Wolf et al. (1982) investigated the effect of temperature on seed composition by growing soybean plants 18 °C, 24 °C, 27 °C, 30 °C, and 33 °C. They found that oleic acid contents in mature seeds increased with the temperature increase, while linoleic and linolenic acid contents decreased, and stearic and palmitic acids contents did not change. Protein and oil contents also increased with the temperature increase. However, a 56% decrease in sucrose content was observed with an increase of only 15 °C in temperature. Other sugars and most amino acids remained stable, but stachyose content decreased slightly and methionine content increased with the temperature increase (Wolf et al. 1982). Another study found that a mutant seeds contained 4.5–8.2% of nitrogen content, while wild-type varieties contained 0.3–0.8% of nitrogen content contributed by free amino acids, and arginine (Arg) was a prominent amino acid in the mutant seeds; thus, nitrogen accumulates in the form of free amino acids or proteins glycinin and β-conglycinin in soybean seeds (Takahashi et al. 2003).

Several studies reported that crop rotation affects seed yield and seed composition including protein, oil, fatty acids, and nutrient contents (Temperly and Borges

2006; Bellaloui et al. 2009a, b, 2010b). It was shown that soybean seed protein content decreased from 357 mg kg^{-1} to 351 mg kg^{-1} from the first to fifth year of soybean consecutive growth after 5 years of corn consecutive growth (Temperly and Borges 2006).

It is well-known that elevated temperatures and CO_2 levels affect plant growth and development including seed composition (Thomas et al. 2003; Long et al. 2004; Prasad et al. 2005; Hay and Porter 2006; Taub et al. 2008; Bellaloui et al. 2016). In a study, authors exposed soybean plants to increased temperatures [28 °C, 32 °C, 36 °C, 40 °C, 44 °C] and CO_2 levels [350, 700μmol mol^{-1}] to investigate their effects on seed composition and found that seed oil, linolenic acid, and carbohydrate contents decreased when temperatures increased; however, seed oleic acid, N, and P contents increased when temperatures increased, while CO_2 increase had a minimal effect on the alteration of seed composition (Thomas et al. 2003). Other studies however predict alteration of seed yield and composition due to global climate changes including high CO_2 levels that will affect stomatal conductance and photosynthetic rates (Long et al. 2004; Prasad et al. 2005; Hay and Porter 2006; Taub et al. 2008). In another recent study, authors exposed soybean plants to increased temperatures [26 °C and 45 °C] and CO_2 levels [360 and 700μmol mol^{-1}] to investigate their effects on seed composition and found that seed protein, linolenic acid, and mineral nutrient (P, K, N, Fe, Zn, and B) contents decreased when temperature and CO_2 levels increased; however, seed oil and oleic acid contents increased when temperatures and CO_2 levels increased. Moreover, they found an increase in seed fructose, glucose, and sucrose contents with increased temperature and CO_2 levels, while seed raffinose and stachyose contents remain unchanged (Bellaloui et al. 2016, 2017).

Drought, water stress, water deficit, and CO_2 levels affect plant growth, development, root architecture, seed yield, and composition (Mayoral et al. 1981; Farquhar and Sharkey 1982; Cu et al. 1987; Heagle et al. 1998; Thomas et al. 2003; Gregory et al. 2005; Rogers et al. 2009; Ambrocio et al. 2016; Bellaloui et al. 2017; Wijewardana et al. 2019). Heagle et al. (1998) treated three soybean cultivars Essex, Holladay, and NK 6955 with increasing levels of CO_2 (1.3X, 1.6X, and 2X ambient) and ozone O_3 (0.4X, 0.9X, and 1.5X ambient) and found that CO_2 increase had a positive effect on yield and growth which increased, while O_3 increase caused a decrease in plant growth and yield. For example, Essex yield increased by up to 81% at 1.5X CO_2 ambient level. Fluctuation of seed oil content caused by increased CO_2 and O_3 levels depended on the cultivar; however, seed protein contents remain unchanged (Heagle et al. 1998). Root length (RL), root surface area (RSA), average root diameter (ARD), and average root volume (ARV) and their correlation with seed protein and oil contents have been studied in water-stressed and non-stressed soybean plants. The authors found a positive correlation between RL, RSA, ARD, and ARV and seed protein contents and a negative correlation with seed oil content in control plants. However, RSA, RL, and ARV were negatively correlated with seed protein and oil contents, and ARD was positively correlated with seed protein and oil contents in water-stressed plants (Ambrocio et al. 2016). A recent study demonstrated that drought and elevated CO_2 levels caused a *decrease* in seed sucrose, fructose, glucose, linolenic acid, oil, K, P, Mn, Zn, Cu, B, Fe, and Mg contents and an *increase* in seed stachyose, raffinose, oleic acid, protein, and N contents. A difference in response to drought and elevated CO_2 levels was also observed between the two studied soybean cultivars Freedom and Hutcheson (Bellaloui et al. 2016, 2017).

Other factors such as seed size and seed color also affect seed composition (Redondo-Cuenca et al. 2007; Guleria et al. 2008; Saldivar et al. 2011; Lee et al. 2017). For example, it has been reported that seed coat color and seed weight affect seed composition. The fluctuations in the contents of metabolites of black soybean seeds are more influenced, by seed coat color, than those of yellow soybean seeds. Seed contents of oligosaccharides, sucrose, total carbohydrates, genistein (an isoflavone), and fatty acids (stearic, linoleic, linolenic, and oleic) were significantly affected by seed coat color. Likewise, 100 seed weight (seed size) also affected significantly the seed contents of the same compounds in addition to stachyose, protein, oil, and two other isoflavones (genistein and daidzein) (Lee et al. 2017).

Wijewardana et al. (2019) investigated the effect of soil moisture on several seed composition traits such as protein, oil, fatty acids, and amino acids as well as sugars (sucrose, raffinose, and stachyose) and several mineral nutrients (N, P, and K) in two maturity group V cultivars, Asgrow AG5332 and Progeny 5333RY, using five irrigation regimes of 20% ET, 40% ET, 60% ET, 80% ET, and 100% ET (percent of evapotranspiration – ET). For seed sugars' contents, both cultivars' seeds had higher levels of sucrose, then stachyose, and then raffinose. Seed sucrose contents increased with increased soil moisture for both cultivars; seed raffinose contents increased with increased soil moisture for both cultivars; however, seed stachyose contents decreased with increased soil moisture for both cultivars. It was reported that water deficit (WD) is negatively correlated with seed stachyose content and sucrose seed content is positively correlated with raffinose seed content (Bellaloui et al. 2012a, b, c). Under WD, sucrose synthase activity decreases drastically which leads to a decrease in sucrose content and in the activities of other enzymes involved in stachyose and raffinose biosynthesis (Gonzalez et al. 1995). Sucrose biosynthesis and accumulation is highly sensitive to WD than the other sugars since sucrose translocation from sources (leaves) to sinks (developing seeds and fruits) is much affected by WD (Taiz and Zeiger 1998; Bellaloui et al. 2012b). Although stachyose and raffinose play a major role during seed germination, they are not desirable for animal and human consumption due to the production of phytate complexes, tannins, and trypsin inhibitors that cause severely reduced gastrointestinal performance (Arendt and Zannini 2013; Avilles-Gaxiola et al. 2017).

A recent study showed that seed protein concentration had an effect on seed amino acids composition especially those essential for animal growth such as Thr, Lys, Met, Trp, and Cys. They used different stress treatments that created seed protein levels from 371 to 430 g of protein per kg of seed. These stress treatments are growing plants in open environment, defoliation of plants at the R5 and R6 growth stages, removal of 40% and 70% of pods from plants, and growing plants at 80% shade. With these levels of protein concentrations (371–430 g kg^{-1}), seed amino acids abundance was affected. For example, lower seed amino acids abundance was caused by higher seed protein concentrations achieved by pod removal and growth in open environment. Likewise, increased seed amino acids abundance was caused by lower seed protein concentrations caused by defoliation of plants at R5 and R6 growth stages (Pfarr et al. 2018).

In a recent review, Krishnan and Jez (2018) summarized several approaches used to enhance essential seed amino acids abundance especially of sulfur-containing amino acids (Cys and Met) that play a major role in animal and human health and nutrition. These approaches encompass the manipulation of enzymes involved in these amino acids' biosynthetic pathway and sulfur metabolism manipulation in transgenic soybean plants; however, to date, transgenic plants expressing Met-rich heterologous proteins showed very little success in increasing seed sulfur-containing amino acids. Moreover, QTL for seed amino acids including sulfur-containing amino acids have been identified and mapped on all soybean chromosomes (see paragraph below); however, no soybean lines with increased amino acids have been developed to date; therefore, both approaches need to be fine-tuned to reach the much suited goal of developing soybean lines with increased amino acids especially sulfur-containing amino acids (Ingenbleek and Kimura 2013; Krishnan and Jez 2018).

Another recent study showed that the use of harvest aids such as paraquat, carfentrazone-ethyl (AIM), glyphosate, and sodium chlorate affects seed composition especially seed protein, fructose, oleic acid, oil, and fructose contents; however, they had small effects on seed amino acid contents during development stages R6 and R7 depending on the year of growth (Bellaloui et al. 2020a). For example, seed palmitic acid, stearic acid, and linolenic acid have not been affected by the addition of harvest aids at the R7 growth stage; however, the addition of paraquat plus AIM or paraquat caused a decrease in seed oleic acid content and a high increase in seed oil content. The addition of glyphosate caused an increase of seed protein content, while the addition of $NaClO_3$ caused its decrease. The addition of both AIM and glyphosate caused a decrease in seed fructose content; however, a small nonsignificant effect of the addition of these harvest aids was observed for seed amino acid contents (Bellaloui et al. 2020a). In another study, Bellaloui et al. (2020b) showed that row spacing, seeding rate, and Cobra herbicide and N application influence seed amino acid, sugar, oil, and protein contents in two soybean cultivars (Pioneer 47T46 and Asgrow 4632). For example, they found that seed sucrose, protein, and linoleic acid contents were higher in 100,000 seeds/ac (40,000 seeds ha^{-1}; 40K) rate compared to 140,000 seeds/ac (56,000 seeds ha^{-1}; 56K) rate; however, the amino acids Ser, Gly, Asp, Arg, Lys, and Tyr have been higher at 56K rate compared to 40K rate. At 76 cm row spacing, seed linoleic acid, stearic acid, and oil contents have been higher at 40K rate than 56K rate (Bellaloui et al. 2020b).

3.6 Quantitative Trait Loci (QTL) Mapping

Likewise, different types of molecular markers and mapping populations have been used to genetically map quantitative trait loci (QTL) for seed amino acids, sugars, and mineral nutrients in soybean. The main approaches are the use of biparental recombinant inbred line populations and the use of genome-wide association studies (GWAS). There are 265+ QTL controlling seed amino acids contents that have been identified and mapped on 19 chromosomes reported in SoyBase (Table 3.5) (Grant et al. 2010; http://www.soybase.org). Other studies identified and mapped QTL for seed sugars, macronutrients, micronutrients, and other compounds' contents in soybean (Maughan et al. 2000; Stombaugh et al. 2004; Kim et al. 2005, 2006; Panthee et al. 2006a, b; Bellaloui et al. 2015a, b; Zhang et al. 2018a, b; Zeng et al. 2014; Vaughn et al. 2014; Qiu et al. 2015; Kaler et al. 2017; Fang et al. 2017; Li et al. 2018; Krishnan and Jez 2018). Here, we will summarize the most important QTL mapping studies for these compounds reported on SoyBase and others not reported on SoyBase (Khandaker et al. 2015; Akond et al. 2018; Gong et al. 2018). As stated before, soybean seeds contain high levels of minerals, vitamins, amino acids, tocopherols, sugars, anthocyanins, saponins, isoflavones, etc.; however, they have low levels of sulfur-containing amino acids, Cys, and Met. Therefore, it is the goal of plant breeders and farmers to improve the contents of these sulfur-containing amino acids (Cys and Met) and other lower compounds in soybean seeds (Cooper 1983; Mazur et al. 1999; Yu et al. 2004; Amir et al. 2012).

3.6.1 QTL That Control Seed Amino Acids' Contents

As stated earlier, there are 265+ QTL controlling amino acids contents that have been identified and mapped using various mapping populations, accessions, and mapping methods such as gnome-wide association studies (GWAS), single-marker analysis (SMA), composite interval mapping (CIM), and multiple-interval mapping (MIM) (Table 3.5) (Panthee et al. 2004, 2006a, b; Ramamurthy et al. 2014; Vaughn et al. 2014; Khandaker et al. 2015; Wang et al. 2015; Warrington et al. 2015; Qiu et al. 2015; Fang et al. 2017; Zhang et al. 2018a, b; Akond et al. 2018; Li et al. 2018; Gong et al. 2018). In this section, we will summarize the most important QTL controlling seed amino acid contents (Fig. 2.1; Table 3.8).

Panthee et al. (2004) grew a RIL population derived from a cross between "N87-984-16" and "TN93-99" lines ($n = 101$) in a field in Knoxville, TN, and quantified seed nitrogen (N) content, as amino acids, of all RILs and parents at R_5, R_6, and R_7 stages of growth. The study screened the parents with 568 SSR markers; however, the genetic linkage map was constructed with 86 polymorphic SSR markers. A total of 17 QTL that control seed nitrogen (amino acids) contents have been identified. QTL for seed Cys, Met, Gly, Pro, Val, Ile, His, Tyr, Asp, Ser, Arg, and Leu have been identified and mapped on chromosomes 2, 7, 8, 14, 15, and 18 at R_5,

Table 3.5 Numbers of reported QTL that control individual amino acids contents reported in SoyBase (Grant et al. 2010; http://www.soybase.org). As of August 2020

Chromosome#/linkage group (LG)											
	1	*2*	*3*	*4*	*5*	*6*	*7*	*8*	*9*	*10*	
Trait	*D1a*	*D1b*	*N*	*C1*	*A1*	*C2*	*M*	*A2*	*K*	*O*	*Total*
Ala	0	0	0	0	1	0	0	2	2	0	5
Arg	0	1	0	0	0	1	1	3	0	1	7
Asn	0	0	0	0	0	0	0	0	0	0	0
Asp	1	0	0	0	0	2	1	3	2	1	10
Cys	2	0	1	2	1	1	1	0	0	1	9
Gln	0	0	0	0	0	0	0	0	0	0	0
Glu	1	1	0	0	0	1	1	1	2	1	8
Gly	0	1	0	0	1	0	0	3	3	2	10
His	0	0	0	0	0	0	0	1	2	0	3
Ile	1	0	0	0	0	1	0	1	1	1	5
Leu	1	1	1	0	0	1	1	0	2	2	9
Lys	1	0	0	0	0	3	0	3	1	0	8
Met	1	0	1	2	1	2	3	0	1	1	12
Phe	0	1	0	0	0	1	0	1	1	0	4
Pro	0	0	0	0	0	0	0	1	0	0	1
Ser	0	0	0	1	0	1	1	3	2	1	9
Thr	1	1	0	0	1	1	0	1	4	1	10
Trp	2	4	0	2	1	3	1	5	0	1	19
Tyr	0	1	0	0	0	1	0	2	2	0	6
Val	0	1	0	0	0	0	0	0	2	0	3
Total	*11*	*12*	*3*	*7*	*6*	*19*	*10*	*30*	*27*	*13*	*138*
–	*11*	*12*	*13*	*14*	*15*	*16*	*17*	*18*	*19*	*20*	
Trait	*B1*	*H*	*F*	*B2*	*E*	*J*	*D2*	*G*	*L*	*I*	*Total*
Aal	0	0	1	1	1	0	1	0	2	0	6
Arg	0	0	1	0	4	3	2	1	0	0	11
Asn	0	0	0	0	0	0	0	0	0	0	0
Asp	1	0	3	0	1	0	3	1	1	1	11
Cys	0	0	3	0	2	0	2	3	0	2	12
Gln	0	0	0	0	0	0	0	0	0	0	0
Glu	0	0	1	0	2	1	2	0	1	0	7
Gly	0	0	0	0	0	0	1	0	3	0	4
His	1	0	0	0	1	1	0	0	0	1	4
Ile	0	0	2	1	1	0	2	0	1	0	7
Leu	0	0	1	0	2	0	2	1	0	0	6
Lys	0	0	2	0	3	0	2	1	0	0	8
Met	0	0	3	0	2	0	3	2	0	2	12
Phe	0	0	0	1	1	2	1	0	1	0	6
Pro	0	0	0	1	1	0	0	0	1	0	3
Ser	0	0	0	0	0	0	0	0	2	0	2
Thr	0	0	0	0	1	0	1	0	2	1	5
Trp	0	0	0	1	5	0	0	2	1	1	10
Tyr	0	0	1	0	1	0	2	1	1	0	6
Val	0	0	2	1	1	0	1	0	1	1	7
Total	*2*	*0*	*20*	*6*	*29*	*7*	*25*	*12*	*17*	*9*	*127*
Grand total											*265*

R_6, and R_7 stages of growth (Fig. 2.1; Table 3.8) (Panthee et al. 2004). In a subsequent study, Panthee et al. (2004) used the same RIL population derived from the cross of "N87-984-16" and "TN93-99" (NxTN, $n = 101$) and 94 SSR markers and single-marker analysis method by ANOVA (analysis of variance) and found 80 QTL controlling various amino acid contents, and these QTL were mapped on different chromosomes of the soybean genome (Fig. 2.1, Table 3.6) (Panthee et al. 2006a). Six QTL that underlie seed Ala contents (*Ala 1-1* to *Ala 1-6*) have been identified and mapped on chromosomes 5, 8, 13, 14, and 19, respectively. One QTL that underlies seed Arg content (*Arg 1-1*) has been identified and mapped on chromosome 2. Four QTL that underlie Asp content (*Asp 1-1* to *Asp 1-4*) have been identified and mapped on chromosomes 1, 8, 9, and 20, respectively. Two QTL that underlie seed Cys content (*Cyst 1-1* and *Cys 1-2*) have been identified and mapped on chromosomes 13 and 18, respectively. Seven QTL controlling seed Glu content (*Glu 1-1* to *Glu 1-7*) have been identified and mapped on chromosomes 1, 2, 7, 8, 10, 16, and 19, respectively. Five QTL controlling seed Gly content (*Gly 1-1* to *Gly 1-5*) have been identified and mapped on chromosomes 2, 5, 9, and 19, respectively. One QTL for seed His content (*His 1-1*) has been identified and mapped on chromosome 16. Five QTL controlling seed Ile content (*Ile 1-1* to *Ile 1-5*) have been identified and mapped on chromosomes 1, 8, 13, 14, and 19, respectively. Seven QTL that control seed Leu content (*Leu 1-1* to *Leu 1-7*) have been identified and mapped on chromosomes 1, 2, 3, 7, 18, and 19 (Fig. 2.1; Table 3.6). Three QTL controlling seed Lys content (*Lys 1-1* to *Lys 1-3*) have been identified and mapped on chromosomes 1, 15, and 18, respectively. Three QTL controlling seed Met content (*Met 1-1* to *Met 1-3*) have been identified and mapped on chromosomes 7, 13, and 18, respectively. Six QTL controlling seed Phe content (*Phe 1-1* to *Phe 1-6*) have been identified and mapped on chromosomes 2, 9, 14, 16, and 19, respectively. Four QTL controlling seed Pro content (*Pro 1-1* to *Pro 1-4*) have been identified and mapped on chromosomes 8, 14, 18, and 19, respectively. Six QTL controlling seed Ser content (*Ser 1-1* to *Ser 1-6*) have been identified and mapped on chromosomes 4, 7, 8, 9, and 19, respectively. Five QTL controlling seed Thr content (*Thr 1-1* to *Thr 1-5*) have been identified and mapped on chromosomes 2, 5, 9, and 19, respectively. Six QTL controlling seed Trp content (*Trp 1-1* to *Trp 1-6*) have been identified and mapped on chromosomes 1, 2, 4, 5, 18, and 19, respectively. Five QTL controlling seed Tyr content (*Tyr 1-1* to *Tyr 1-5*) have been identified and mapped on chromosomes 2, 8, 9, 15, and 19, respectively. Four QTL controlling seed Val content (*Val 1-1* to *Val 1-4*) have been identified and mapped on chromosomes 2, 13, 14, and 19, respectively. Using the same RIL population "N87-984-16" by "TN93-99" (NxTN, $n = 101$) and the 94 SSRs, Panthee et al. (2006b) used composite interval mapping (CIM) method of QTL Cartographer and confirmed 10 QTL for amino acids contents in three environments (Holston, Knoxville, and Ames, TN). Four QTL that control seed Cys content (*Cys 2-1* to *Cys 2-4*) have been identified and mapped on chromosomes 1, 13, and 18; three QTL that control seed Met content (*Met 2-1* to *Met 2-3*) have been identified and mapped on chromosomes 7, 13, and 18; and three QTL that control seed Met + Cys combined (*Met+Cys 1-1* to *Met+Cys 1-3*) have

Table 3.6 Detailed information of QTL that control seed amino acid contents reported and not reported in SoyBase (Grant et al. 2010; http://www.soybase.org)

Trait	QTL	LG/Chr	Position (cM)	Marker/interval	LOD, *P*, or *F*	R^2 (%)	Environment	Reference
Seed amino acid contents								
Cys	*Cys*	A2/8	30.7–32.7	Satt177	0.0270	6.10	Knoxville, TN	Panthee et al. (2004)
Met, Gly, Pro, Val, Ile, Phe, His, Tyr	*Met, Gly, Pro, Val, Ile, Phe, His, Tyr*	A2/8	109.4–111.4	Satt133	0.0270	6.60	Knoxville, TN	
Cys, Thr	*Cys, Thr*	D1b/2	117.6–119.6	Satt274	0.0004	10.70	Knoxville, TN	
Asp, Ser, Arg, His	*Asp, Ser, Arg, His*	E/15	75.8–76.3	Satt185	0.0110	9.00	Knoxville, TN	
Asp, Glu, Sar, Val, Ile, Leu, Phe, His, Tyr	*Asp, Glu, Sar, Val, Ile, Leu, Phe, His, Tyr*	E/15	75.8–76.3	Satt263	0.0260	7.00	Knoxville, TN	
Met, Arg	*Met, Arg*	G/18	24–26	Satt235	0.0190	7.10	Knoxville, TN	
Gly, Val, Ile	*Gly, Val, Ile*	A2/8	92.2–94.2	Satt437	0.0150	8.20	Knoxville, TN	
Met, Glu, Gly, Ile	*Met, Glu, Gly, Ile*	D1b/2	118.5–120.5	Satt459	0.1000	9.30	Knoxville, TN	
Met, Tyr	*Met, Tyr*	M/7	11.6–13.6	Satt201	0.0280	6.50	Knoxville, TN	
Thr	*Thr*	B2/14	45.9–47.9	Satt168	0.0490	5.00	Knoxville, TN	
Met, Arg	*Met, Arg*	E/15	75.8–78.5	Satt268	0.0190	7.00	Knoxville, TN	
Asp, Glu, Sar, Val, Ile, Leu, Phe, His, Tyr	*Asp, Glu, Sar, Val, Ile, Leu, Phe, His, Tyr*	E/15	75.8–78.5	Satt263	0.0040	11.60	Knoxville, TN	
Met, Asp, Glu, Arg	*Met, Asp, Glu, Arg*	E/15	75.8–78.5	Satt045	0.0040	9.90	Knoxville, TN	
Thr, Arg, Pro	*Thr, Arg, Pro*	G/18	16.2–18.2	Satt570	0.0280	6.30	Knoxville, TN	

(continued)

Table 3.6 (continued)

Trait	QTL	LG/Chr	Position (cM)	Marker/interval	LOD, *P*, or *F*	R^2 (%)	Environment	Reference
Ala	*Seed Ala 1-1*	A1/5	24.55–26.55	Satt593	2.75		Knoxville, TN	Panthee et al. (2006a)
	Seed Ala 1-2	A2/8	106.05–108.05	Satt437	3.28		Knoxville, TN	
	Seed Ala 1-3	B2/14	54.20–56.20	Satt168	3.00		Knoxville, TN	
	Seed Ala 1-4	F/13	15.08–17.08	Satt252	4.11		Knoxville, TN	
	Seed Ala 1-5	L/19	0.00–2.00	Satt495	3.21		Knoxville, TN	
	Seed Ala 1-6	L/19	29.19–31.19	Satt143	5.21		Knoxville, TN	
Arg	*Seed Arg 1-1*	D1b/2	57.8–59.8	Satt274	2.68		Knoxville, TN	
Asp	*Seed Asp 1-1*	A2/8	13–15	Satt177	3.37		Knoxville, TN	
	Seed Asp 1-2	D1a/1	40.5–42.5	Satt203	2.71		Knoxville, TN	
	Seed Asp 1-3	I/20	59–61	Satt292	2.75		Knoxville, TN	
	Seed Asp 1-4	K/9	95–97	Satt196	2.97		Knoxville, TN	
Cys	*Seed Cys 1-1*	F/13	15–17	Satt252	4.11		Knoxville, TN	
	Seed Cyst 1-2	G/18	24.4–26.4	Satt235	2.78		Knoxville, TN	
Glu	*Seed Glu 1-1*	A2/8	13–15	Satt177	3.13		Knoxville, TN	
	Seed Glu 1-2	D1a/1	40.5–42.5	Satt203	2.55		Knoxville, TN	
	Seed Glu 1-3	D1b/2	57.8–59.8	Satt274	3.73		Knoxville, TN	
	Seed Glu 1-4	J/16	29.8–31.8	Satt249	2.97		Knoxville, TN	
	Seed Glu 1-5	L/19	60.4–62.4	Satt076	4.62		Knoxville, TN	
	Seed Glu 1-6	M/7	26.7–28.7	Satt567	3.11		Knoxville, TN	
	Seed Glu 1-7	O/10	48.7–50.7	Satt420	3.57		Knoxville, TN	
Gly	*Seed Gly 1-1*	A1/5	11.9–13.9	Satt593	2.91		Knoxville, TN	
	Seed Gly 1-2	D1b/2	57.8–59.8	Satt274	3.11		Knoxville, TN	
	Seed Gly 1-3	K/9	45.4–47.4	Satt518	4.15		Knoxville, TN	
	Seed Gly 1-4	L/19	43833.00	Satt495	3.83		Knoxville, TN	
	Seed Gly 1-5	L/19	29–31	Satt143	2.80		Knoxville, TN	

Trait	QTL	LG/Chr	Position (cM)	Marker/interval	LOD, *P*, or *F*	R^2 (%)	Environment	Reference
His	*Seed His 1-1*	J/16	29.8–31.8	Satt380	2.80		Knoxville, TN	
Ile	*Seed Ile 1-1*	A2/8	13–15	Satt177	2.52		Knoxville, TN	
	Seed Ile 1-2	B2/14	17–19	Satt168	2.72		Knoxville, TN	
	Seed Ile 1-3	D1a/1	40.5–42.5	Satt203	2.98		Knoxville, TN	
	Seed Ile 1-4	F/13	15–17	Satt252	3.25		Knoxville, TN	
	Seed Ile 1-5	L/19	39.6–41.6	Satt143	2.64		Knoxville, TN	
Leu	*Seed Leu 1-1*	D1a/1	40.5–42.5	Satt203	3.39		Knoxville, TN	
	Seed Leu 1-2	D1b/2	57.8–59.8	Satt274	3.15		Knoxville, TN	
	Seed Leu 1-3	D2/17	22.2–24.2	Satt002	2.89		Knoxville, TN	
	Seed Leu 1-4	G/18	15.2–17.2	Satt570	2.83		Knoxville, TN	
	Seed Leu 1-5	L/19	60.4–62.4	Satt076	2.95		Knoxville, TN	
	Seed Leu 1-6	M/7	26.7–28.7	Satt567	2.64		Knoxville, TN	
	Seed Leu 1-7	N/3	6.00–9.00	Satt255	3.14		Knoxville, TN	
Lys	*Seed Lys 1-1*	D1a/1	23–24	Satt184	3.11		Knoxville, TN	
	Seed Lys 1-2	E/15	0–2	Satt185	3.93		Knoxville, TN	
	Seed Lys 1-3	G/18	46.2–48.2	Satt427	3.62		Knoxville, TN	
Met	*Seed Met 1-1*	F/13	15–17	Satt252	2.84		Knoxville, TN	
	Seed Met 1-2	G/18	55.8–57.8	Satt564	2.59		Knoxville, TN	
	Seed Met 1-3	M/7	11.00–12.00	Satt590	2.42		Knoxville, TN	
Phe	*Seed Phe 1-1*	B2/14	17–19	Satt168	3.39		Knoxville, TN	
	Seed Phe 1-2	D1b/2	57.8–59.8	Satt274	3.56		Knoxville, TN	
	Seed Phe 1-3	J/16	5.00–7.00	Satt249	3.17		Knoxville, TN	
	Seed Phe 1-4	J/16	29.8–31.8	Satt380	3.08		Knoxville, TN	
	Seed Phe 1-5	K/9	11.00–12.00	Satt102	2.74		Knoxville, TN	
	Seed Phe 1-6	L/19	1.00–3.00	Satt495	3.36		Knoxville, TN	

(continued)

Table 3.6 (Continued)

Trait	QTL	LG/Chr	Position (cM)	Marker/interval	LOD, *P*, or *F*	R^2 (%)	Environment	Reference
Pro	*Seed Pro 1-1*	A2/8	85.3–87.3	Satt437	3.26		Knoxville, TN	
	Seed Pro 1-2	B2/14	17–19	Satt168	3.04		Knoxville, TN	
	Seed Pro 1-3	G/18	15.2–17.2	Satt570	2.66		Knoxville, TN	
	Seed Pro 1-4	L/19	29–31	Satt143	3.64		Knoxville, TN	
Ser	*Seed Ser 1-1*	A2/8	85.3–87.3	Satt437	2.84		Knoxville, TN	
	Seed Ser 1-2	C1/4	39–41	Satt139	3.04		Knoxville, TN	
	Seed Ser 1-3	K/9	11.00–13.00	Satt102	3.19		Knoxville, TN	
	Seed Ser 1-4	L/19	1.00–3.00	Satt495	4.92		Knoxville, TN	
	Seed Ser 1-5	L/19	29–31	Satt143	5.41		Knoxville, TN	
	Seed Ser 1-6	M/7	26.7–28.7	Satt567	4.09		Knoxville, TN	
Thr	*Seed Thr 1-1*	A1/5	11.9–13.9	Satt593	2.99		Knoxville, TN	
	Seed Thr 1-2	D1b/2	57.8–59.8	Satt274	2.82		Knoxville, TN	
	Seed Thr 1-3	K/9	45.4–47.4	Satt518	2.83		Knoxville, TN	
	Seed Thr 1-4	L/19	1.00–3.00	Satt495	4.00		Knoxville, TN	
	Seed Thr 1-5	L/19	29–31	Satt143	3.37		Knoxville, TN	
Trp	*Seed Trp 1-1*	A1/5	11.9–13.9	Satt593	2.54		Knoxville, TN	
	Seed Trp 1-2	C1/4	39–41	Satt139	2.67		Knoxville, TN	
	Seed Trp 1-3	D1a/1	40.5–42.5	Satt203	3.20		Knoxville, TN	
	Seed Trp 1-4	D1b/2	15.1–17.1	Satt537	2.83		Knoxville, TN	
	Seed Trp 1-5	G/18	15.2–17.2	Satt570	2.91		Knoxville, TN	
	Seed Trp 1-6	I/20	59–61	Satt292	3.10		Knoxville, TN	
Tyr	*Seed Tyr 1-1*	A2/8	123.6–125.6	Satt133	3.67		Knoxville, TN	
	Seed Tyr 1-2	D1b/2	57.8–59.8	Satt274	3.51		Knoxville, TN	
	Seed Tyr 1-3	E/15	0–2	Satt185	7.44		Knoxville, TN	
	Seed Tyr 1-4	K/9	11.00–12.00	Satt102	3.49		Knoxville, TN	
	Seed Tyr 1-5	L/19	39.6–41.6	Satt313	3.23		Knoxville, TN	

Trait	QTL	LG/Chr	Position (cM)	Marker/interval	LOD, *P*, or *F*	R^2 (%)	Environment	Reference
Val	*Seed Val 1-1*	B2/14	17–19	Satt168	2.58		Knoxville, TN	
	Seed Val 1-2	D1b/2	15.1–17.1	Satt537	2.71		Knoxville, TN	
	Seed Val 1-3	F/13	15–17	Satt252	4.45		Knoxville, TN	
	Seed Val 1-4	L/19	29–31	Satt143	3.37		Knoxville, TN	
Cys	*Cys 2-1*	G/18	20.9–22.9	Satt235	2.78		Ames & Holston	Panthee et al. (2006b)
	Cys 2-2	D1a/1	69.7–71.7	Satt436	2.46		Holston	
	Cys 2-3	F/13	15.1–17.1	Satt252	2.22		Combined	
	Cyst 2-4	G/18	50.7–52.7	Satt427	2.29		Ames	
Met	*Met 2-1*	F/13	15–17	Satt252	2.84		KPSF	
	Met 2-2	G/18	55.8–57.8	Satt564	2.59		Combined	
	Met 2-3	M/7	11.00–12.00	Satt590	2.42		Combined	
Met + Cys	*Met+Cyst 1-1*	D2/17	22.2–24.2	Satt002	2.10		Holston	
	Met+Cyst 1-2	F/13	15–17	Satt252	2.45		Combined	
	Met+Cyst 1-3	M/7	11.00–12.00	Satt590	2.43		Ames	
Cys	*Cys 5-1*	I/20	62.9–73.9	BARC-038869-07364	5.35	16.00	Lincoln, NE	Ramamurthy et al. (2014)
	Cys 5-2	I/20	75.1–93.2	BARC-013583-01166	3.66	9.00	Lincoln, NE	
Met	*Met 5-1*	E/15	41.1–44.5	BARC-018461-02916	3.23	6.00	Lincoln, NE	
Met + Cys	*Met+Cys 3-1*	I/20	62.9–73.9	BARC-038869-07364	5.35	16.00	Lincoln, NE	
	Met+Cys 3-2	I/20	75.1–93.2	BARC-013583-01166	3.66	9.00	Lincoln, NE	
Pro	*qPRO001*	D1a/1	3.30–4.50	ss244519430-ss244503611	3.64	0.10	Fayetteville, NC	Khandaker et al. (2015)
Arg	*qARG001*	N/3	3.20–4.10	ss244936977-ss244942217	3.74	0.60	Fayetteville, NC	
Thr	*qTHR001*	A1/5	8.40–9.60	ss245747167-ss245786667	30.57	0.06	Fayetteville, NC	
Ser	*qSER001*	A1/5	8.50–10.80	ss245747167-ss245786667	7.24	0.09	Fayetteville, NC	
Cys	*qCYS001*	A1/5	8.50–10.20	ss245747167-ss245786667	2.80	0.06	Fayetteville, NC	
Thr	*qTHR002*	C2/6	3.30–3.70	ss246091245-ss246092064	2.82	0.12	Fayetteville, NC	

(continued)

Table 3.6 (continued)

Trait	QTL	LG/Chr	Position (cM)	Marker/interval	LOD, *P*, or *F*	R^2 (%)	Environment	Reference
Ser	*qSER002*	C2/6	7.20–7.30	ss246087580-ss246091735	5.00	0.01	Fayetteville, NC	
Cys	*qCYS002*	K/9	0.05–0.06	ss246894176-ss246871129	3.98	0.05	Fayetteville, NC	
Trp	*qTRP001*	K/9	0.40–0.50	ss246871129-ss246860974	11.98	0.03	Fayetteville, NC	
Thr	*qTHR003*	B2/14	7.90–10.10	ss24829340-ss248275088	29.52	0.26	Fayetteville, NC	
Pro	*qPRO002*	B2/14	6.90–11.10	ss24829340-ss248275088	7.32	0.08	Fayetteville, NC	
His	*qHIS001*	J/16	27.70–29.80	ss248982430-ss249037210	3.00	0.05	Fayetteville, NC	
Trp	*qTRP002*	G/18	2.50–2.70	ss249599552-ss249581714	10.63	0.04	Fayetteville, NC	
Cys	*Cys 3-1*	A1/5	1.00–15.00	BARC_04_0366-BARC-07848	2.1–3.7	10.8–19.9	E10AU, E11BF	Wang et al. (2015)
	Cys 3-2	C1/4	41.4–53.6	Sat_140-BARC_04_0366	2.4–3.5	15.7–21.9	E10AU, E11BF	
	Cys 3-3	C1/4	48.6–49.1	BARC_04_0366--07848	2.9–4.0	14.7–19.6	E10AU, E11BF	
	Cyst 3-4	D1a/1	57.5–62.4	Satt580-BARC-054393-12560	2.3–3.7	11.6–17.9	E10AU, E11BF	
	Cys 3-5	D2/17	79.4–87.6	BARC-17_0844-Satt300	3.5–4.9	17.8–24.5	E10AU, E11BF	
	Cys 3-6	E/15	45.1–46.7	Satt452-Satt045	2.7–4.3	13.9–21.1	E10AU, E11BF	
	Cys 3-7	I/20	31.5–50.7	Satt587-BARC-20_0819	4.9–7.8	25.8–37.3	E10AU, E11BF	
	Cyst 3-8	N/3	37.9–38.9	Sat_084-Satt584	2.5–4.8	12.7–22.5	E10AU, E11VG	

Trait	QTL	LG/Chr	Position (cM)	Marker/interval	LOD, *P*, or *F*	R^2 (%)	Environment	Reference
Met	*Met 3-1*	A1/5	1.00–15.00	BARC-053261-11776	2.0–3.5	10.2–17.8	E10AU, E11VG	
	Met 3-2	C1/4	41.4–53.6	Sat_140-BARC_04_0366	2.5–3.5	15.4–21.6	E10AU, E11BF	
	Met 3-3	C1/4	48.6–49.1	BARC_04_0366--07848	2.9–4.1	14.9–19.9	E10AU, E11VG	
	Met 3-4	D1a/1	57.5–62.4	Satt580-BARC-054393-12560	2.1–3.8	10.7–19.9	E10AU, E11VG	
	Met 3-5	D2/17	79.4–87.6	BARC-17_0844-Satt300	3.3–4.7	16.9–24.9	E10AU, E11VG	
	Met 3-6	E/15	45.1–46.7	Satt452-Satt045	2.7–4.1	13.4–20.1	E10AU, E11VG	
	Met 3-7	I/20	31.5–50.7	Satt587-BARC-20_0819	5.2–7.7	27.9–33.8	E10AU, E11VG	
	Met 3-8	N/3	37.9–38.9	Sat_084-Satt584	2.5–4.9	12.7–22.9	E10AU, E11VG	
Cys	*qCys_Gm10 (Cys 4-1)*	O/10	52.7–57.8	Satt592-BARC-043247	3.50	0.10	Athens, GA	Warrington et al. (2015)
Lys	*qLys_Gm08 (Lys 3-1)*	A2/8	31.7–35.6	BARC-055265-Satt089	3.40	0.06	Athens, GA	
	qLys_Gm20 (Lys 3-2)	I/20	14.7–20.6	GSM0012-Satt354	23.30	0.48	Athens, GA	
Met	*qMet_Gm06 (Met 4-1)*	C2/6	28.2–33.0	BARC-055889-BARC-048217	3.60	0.08	Athens, GA	
	qMet_Gm09 (Met 4-2)	K/9	30.8–32.0	BARC-31967-BARC-042449	4.10	0.09	Athens, GA	
	qMet_Gm10 (Met 4-3)	O/10	52.9–57.8	Satt592-BARC-043247	5.90	0.14	Athens, GA	
	qMet_Gm20 (Met 4-4)	I/20	14.6–17.0	GSM0012-BARC-020713	7.70	0.19	Athens, GA	

(continued)

Table 3.6 (continued)

Trait	QTL	LG/Chr	Position (cM)	Marker/interval	LOD, *P*, or *F*	R^2 (%)	Environment	Reference
Met + Cyst	*qMet+Cys_Gm10 (2-1)*	O/10	52.0–57.8	Satt592-BARC-043247	4.30	0.11	Athens, GA	
	qMet+Cys_Gm20 (2-2)	I/20	14.6–19.5	GSM0012-BARC-020713	4.40	0.12	Athens, GA	
Thr	*qThr_Gm01 (Thr 3-1)*	D1a/1	53.1–64.4	BARC-035219-BARC-018211	3.30	0.06	Athens, GA	
	qThr_Gm09 (Thr 3-2)	K/9	30.8–32.0	BARC-31967-BARC-042449	4.00	0.05	Athens, GA	
	qThr_Gm17 (Thr 3-3)	D2/17	51.8–61.0	Satt256-BARC-039151	3.70	0.06	Athens, GA	
	qThr_Gm20 (Thr 3-4)	I/20	14.6–16.1	GSM0012-BARC-020713	26.20	0.52	Athens, GA	
Gln	L_Glu (Gln) content	D1a/1		Chr01_53486467	3.16E-08		Beijing, China	Fang et al. (2017)
	L_Glu (Gln) content	D1a/1		Chr01_53486650	3.16E-08		Beijing, China	
	L_Glu (Gln) content	A2/8		Chr08_35266140	2.19E-08		Beijing, China	
	L_Glu (Gln) content	A2/8		Chr08_35266158	2.19E-08		Beijing, China	
	L_Glu (Gln) content	A2/8		Chr08_35266324	2.19E-08		Beijing, China	
	L_Glu (Gln) content	K/9		Chr09_33309127	6.03E-08		Beijing, China	
Gln	L_Glu (Gln) content	F/13		Chr13_2546775	1.12E-08		Beijing, China	
	L_Glu (Gln) content	F/13		Chr13_2546862	1.12E-08		Beijing, China	
	L_Glu (Gln) content	B2/14		Chr14_5055466	5.62E-08		Beijing, China	
	L_Glu (Gln) content	B2/14		Chr14_5055474	5.62E-08		Beijing, China	
Gly	L_Gly content	O/10		Chr10_2422267	1.05E-07		Beijing, China	
	L_Gly content	O/10		Chr10_2428789	1.05E-07		Beijing, China	
	L_Gly content	O/10		Chr10_2436611	1.05E-07		Beijing, China	
Phe	L_Phe content	F/13		Chr13_2546862	1.82E-07		Beijing, China	

Trait	QTL	LG/Chr	Position (cM)	Marker/interval	LOD, *P*, or *F*	R^2 (%)	Environment	Reference
Ser	L_Ser content	D1a/1		Chr01_989187	7.94E-10		Beijing, China	
	L_Ser content	D1a/1		Chr01_53486467	1.82E-10		Beijing, China	
	L_Ser content	D1a/1		Chr01_53486490	1.82E-10		Beijing, China	
	L_Ser content	D1a/1		Chr01_53486686	1.82E-10		Beijing, China	
	L_Ser content	A2/8		Chr08_35266140	7.94E-11		Beijing, China	
	L_Ser content	A2/8		Chr08_35266158	7.94E-11		Beijing, China	
	L_Ser content	A2/8		Chr08_35266324	7.94E-11		Beijing, China	
	L_Ser content	K/9		Chr09_33309127	1.74E-09		Beijing, China	
	L_Ser content	F/13		Chr13_2546775	1.02E-11		Beijing, China	
	L_Ser content	F/13		Chr13_2546862	1.02E-11		Beijing, China	
	L_Ser content	B2/14		Chr14_5055466	9.77E-10		Beijing, China	
	L_Ser content	B2/14		Chr14_5055474	9.77E-10		Beijing, China	
Thr	L_Thr content	D1a/1		Chr01_53486490	1.91E-07		Beijing, China	
	L_Thr content	O/10		Chr10_2436611	1.07E-07		Beijing, China	
	L_Thr content	F/13		Chr13_2546862	4.57E-08		Beijing, China	
	L_Thr content	I/20		Chr20_38215861	2.45E-07		Beijing, China	

(continued)

Table 3.6 (continued)

Trait	QTL	LG/Chr	Position (cM)	Marker/interval	LOD, *P*, or *F*	R^2 (%)	Environment	Reference
Tyr	L_Tyr content	D1a/1		Chr01_989187	1.66E-12		Beijing, China	
	L_Tyr content	D1a/1		Chr01_989204	1.66E-12		Beijing, China	
	L_Tyr content	D1a/1		Chr01_2181558	5.37E-08		Beijing, China	
	L_Tyr content	D1a/1		Chr01_53486467	1.35E-16		Beijing, China	
	L_Tyr content	D1a/1		Chr01_53486686	1.35E-16		Beijing, China	
	L_Tyr content	A2/8		Chr08_35266140	1.15E-16		Beijing, China	
	L_Tyr content	A2/8		Chr08_35266324	1.15E-16		Beijing, China	
	L_Tyr content	K/9		Chr09_33309127	1.41E-16		Beijing, China	
	L_Tyr content	F/13		Chr13_2546775	9.12E-17		Beijing, China	
	L_Tyr content	F/13		Chr13_2546862	9.12E-17		Beijing, China	
	L_Tyr content	B2/14		Chr14_5055466	2.09E-15		Beijing, China	
	L_Tyr content	B2/14		Chr14_5055474	2.09E-15		Beijing, China	
	L_Tyr content	L/19		Chr19_44806855	1.29E-07		Beijing, China	
	L_Tyr content	L/19		Chr19_44912660	1.29E-07		Beijing, China	
	L_Tyr content	L/19		Chr19_44922846	1.29E-07		Beijing, China	
Total AA	Total AA content	O/10		Chr10_2400625	1.17E-07		Beijing, China	
	Total AA content	O/10		Chr10_2415581	1.17E-07		Beijing, China	
	Total AA content	D1a/1		Chr01_989187	1.45E-08		Beijing, China	
	Total AA content	D1a/1		Chr01_53486467	7.76E-11		Beijing, China	
	Total AA content	D1a/1		Chr01_53486686	7.76E-11		Beijing, China	
	Total AA content	A2/8		Chr08_35266140	7.94E-11		Beijing, China	
	Total AA content	A2/8		Chr08_35266324	7.94E-11		Beijing, China	
	Total AA content	K/9		Chr09_33309127	1.10E-10		Beijing, China	
	Total AA content	F/13		Chr13_2546775	2.19E-11		Beijing, China	
	Total AA content	F/13		Chr13_2546862	2.19E-11		Beijing, China	
	Total AA content	B2/14		Chr14_5055466	2.82E-10		Beijing, China	
	Total AA content	B2/14		Chr14_5055474	2.82E-10		Beijing, China	

Trait	QTL	LG/Chr	Position (cM)	Marker/interval	LOD, *P*, or *F*	R^2 (%)	Environment	Reference
Ala	*Seed alanine*	A2/8	9323205	ss715602851	7.54E-10	4.70	South Dakota	Zhang et al. (2018a)
Ala_P	*Seed alanine_P*	E/15	3846538	ss715621777	4.34E-07	4.80	South Dakota	
Arg	*Seed arginine*	A2/8	9220290	ss715602843	5.29E-09	2.60	South Dakota	
	Seed arginine	O/10	36801502	ss715606358	2.96E-07	1.90	South Dakota	
	Seed arginine	E/15	3846538	ss715621777	5.93E-14	4.00	South Dakota	
	Seed arginine	E/15	10102255	ss715623324	1.37E-06	2.60	South Dakota	
	Seed arginine	J/16	7122876	ss715625391	8.72E-07	2.70	South Dakota	
	Seed arginine	J/16	35721993	ss715624781	7.56E-08	2.30	South Dakota	
Arg_P	*Seed arginine_P*	N/3	42377595	ss715586293	7.60E-07	2.20	South Dakota	
	Seed arginine_P	C2/6	18230296	ss715593784	9.09E-08	2.60	South Dakota	
	Seed arginine_P	M/7	3713267	ss715597475	1.52E-06	2.10	South Dakota	
	Seed arginine_P	A2/8	7792993	ss715602707	1.91E-11	6.10	South Dakota	
	Seed arginine_P	A2/8	9323205	ss715602851	2.58E-07	4.30	South Dakota	
	Seed arginine_P	F/13	43777586	ss715616488	1.55E-06	2.10	South Dakota	
	Seed arginine_P	E/15	3846538	ss715621777	8.64E-09	2.40	South Dakota	
	Seed arginine_P	E/15	6853445	ss715622952	3.71E-07	2.30	South Dakota	
	Seed arginine_P	J/16	7122876	ss715625391	3.22E-07	2.40	South Dakota	
	Seed arginine_P	G/18	5543903	ss715631635	1.33E-06	2.10	South Dakota	
Asp	*Seed aspartic*	C2/6	6079473	ss715595388	3.77E-07	2.80	South Dakota	
	Seed aspartic	E/15	3846538	ss715621777	1.15E-11	4.90	South Dakota	

(continued)

Table 3.6 (continued)

Trait	QTL	LG/Chr	Position (cM)	Marker/interval	LOD, *P*, or *F*	R^2 (%)	Environment	Reference
Asp_P	*Seed aspartic_P*	M/7	3713267	ss715597475	5.87E-07	3.50	South Dakota	
	Seed aspartic_P	A2/8	7792993	ss715602707	3.39E-15	9.30	South Dakota	
	Seed aspartic_P	A2/8	8276381	ss715602750	2.68E-12	7.10	South Dakota	
	Seed aspartic_P	B1/11	33670325	ss715610611	5.66E-08	4.20	South Dakota	
	Seed aspartic_P	F/13	16286313	ss715616790	1.15E-06	3.30	South Dakota	
	Seed aspartic_P	D2/17	5761052	ss715628060	9.57E-08	4.00	South Dakota	
	Seed aspartic_P	G/18	5597675	ss715631691	1.48E-06	3.20	South Dakota	
	Seed aspartic_P	L/19	5149450	ss715636094	1.33E-06	3.30	South Dakota	
Cys	*Seed cysteine*	C2/6	6079473	ss715595388	2.79E-07	3.80	South Dakota	
	Seed cysteine	E/15	3846538	ss715621777	2.16E-11	6.20	South Dakota	
Glu	*Seed glutamic*	C2/6	6079473	ss715595388	9.49E-07	2.80	South Dakota	
	Seed glutamic	E/15	3846538	ss715621777	2.67E-12	5.70	South Dakota	
Glu_P	*Seed glutamic_P*	E/15	4778258	ss715622337	1.19E-06	5.20	South Dakota	
Gly	*Seed glycine*	A2/8	8276381	ss715602750	4.59E-07	5.30	South Dakota	
	Seed glycine	A2/8	9323205	ss715602851	3.67E-09	6.20	South Dakota	
Gly_P	*Seed glycine_P*	A2/8	8276381	ss715602750	5.18E-07	6.20	South Dakota	
	Seed glycine_P	O/10	40818239	ss715607012	6.11E-07	4.60	South Dakota	
	Seed glycine_P	L/19	37868145	ss715634642	4.90E-11	4.00	South Dakota	
His	*Seed histidine*	A2/8	20249542	ss715600491	6.50E-07	3.30	South Dakota	
	Seed histidine	K/9	3502638	ss715603692	4.94E-07	3.40	South Dakota	
	Seed histidine	B1/11	1530847	ss715609163	6.25E-08	4.00	South Dakota	
	Seed histidine	E/15	3846538	ss715621777	1.02E-07	3.60	South Dakota	
	Seed histidine	I/20	25164461	ss715637040	1.29E-06	3.10	South Dakota	
Ile	*Seed isoleucine*	E/15	4685574	ss715622170	1.05E-07	3.80	South Dakota	

Trait	QTL	LG/Chr	Position (cM)	Marker/interval	LOD, *P*, or *F*	R^2 (%)	Environment	Reference
Leu	*Seed leucine*	O/10	1829221	ss715605762	1.37E-06	2.30	South Dakota	
	Seed leucine	E/15	4685574	ss715622170	1.21E-08	4.80	South Dakota	
	Seed leucine	E/15	16673056	ss715621082	1.32E-07	3.10	South Dakota	
Lys	*Seed lysine*	C2/6	6079473	ss715595388	1.81E-07	3.70	South Dakota	
	Seed lysine	E/15	4685574	ss715622170	1.57E-07	4.70	South Dakota	
Lys_P	*Seed lysine_P*	C2/6	5631766	ss715595354	2.83E-07	3.30	South Dakota	
	Seed lysine_P	A2/8	8276381	ss715602750	7.49E-14	7.20	South Dakota	
	Seed lysine_P	A2/8	9323205	ss715602851	2.30E-09	6.20	South Dakota	
	Seed lysine_P	E/15	12150934	ss715620399	1.26E-06	3.00	South Dakota	
	Seed lysine_P	D2/17	5761052	ss715628060	6.75E-07	4.20	South Dakota	
Met	*Seed methionine*	C2/6	10277748	ss715592689	5.23E-07	2.80	South Dakota	
	Seed methionine	E/15	3846538	ss715621777	5.19E-07	2.70	South Dakota	
Phe	*Seed phenylalanine*	C2/6	6079473	ss715595388	1.63E-07	3.10	South Dakota	
	Seed phenylalanine	E/15	3846538	ss715621777	6.19E-12	5.20	South Dakota	
Phe_P	*Seed phenylalanine_P*	A2/8	8276381	ss715602750	3.12E-08	5.70	South Dakota	
Pro_P	*Seed proline_P*	E/15	3846538	ss715621777	7.53E-08	5.40	South Dakota	
Ser	*Seed serine*	A2/8	8276381	ss715602750	7.80E-07	4.00	South Dakota	
	Seed serine	A2/8	9323205	ss715602851	2.24E-08	4.60	South Dakota	
Thr	*Seed threonine*	A2/8	9323205	ss715602851	2.24E-07	3.40	South Dakota	
Thr_P	*Seed threonine_P*	E/15	4783644	ss715622343	5.89E-07	5.00	South Dakota	
Trp	*Seed tryptophan*	D1a/1	51679468	ss715580150	7.68E-07	3.80	South Dakota	
	Seed tryptophan	A2/8	8276381	ss715602750	4.72E-10	6.10	South Dakota	
	Seed tryptophan	O/10	1829221	ss715605762	4.15E-08	2.30	South Dakota	
	Seed tryptophan	E/15	3846538	ss715621777	5.70E-09	2.60	South Dakota	

(continued)

Table 3.6 (continued)

Trait	QTL	LG/Chr	Position (cM)	Marker/interval	LOD, *P*, or *F*	R^2 (%)	Environment	Reference
Trp_P	*Seed tryptophan_P*	D1b/2	11045403	ss715580968	4.85E-07	2.80	South Dakota	
	Seed tryptophan_P	D1b/2	13098947	ss715581136	2.98E-07	2.90	South Dakota	
	Seed tryptophan_P	D1b/2	39504882	ss715582418	6.19E-07	2.80	South Dakota	
	Seed tryptophan_P	C1/4	7646745	ss715589288	1.49E-06	0.80	South Dakota	
	Seed tryptophan_P	C2/6	1446085	ss715593213	1.12E-08	3.70	South Dakota	
	Seed tryptophan_P	C2/6	39693731	ss715594345	1.32E-19	3.00	South Dakota	
	Seed tryptophan_P	C2/6	40410331	ss715594365	1.67E-10	1.70	South Dakota	
	Seed tryptophan_P	M/7	3713267	ss715597475	1.95E-08	3.60	South Dakota	
	Seed tryptophan_P	A2/8	8276381	ss715602750	1.13E-17	8.90	South Dakota	
	Seed tryptophan_P	A2/8	8390974	ss715602764	3.63E-17	8.60	South Dakota	
	Seed tryptophan_P	A2/8	12227484	ss715599384	4.58E-08	3.40	South Dakota	
	Seed tryptophan_P	A2/8	14477724	ss715599575	1.23E-07	3.10	South Dakota	
	Seed tryptophan_P	B2/14	10057919	ss715617366	1.30E-07	1.50	South Dakota	
	Seed tryptophan_P	E/15	3032666	ss715621537	2.59E-08	3.50	South Dakota	
	Seed tryptophan_P	E/15	4006986	ss715621822	4.87E-08	3.40	South Dakota	
	Seed tryptophan_P	E/15	43631120	ss715621917	3.97E-07	2.90	South Dakota	
	Seed tryptophan_P	E/15	49442375	ss715622474	7.35E-07	2.70	South Dakota	
	Seed tryptophan_P	G/18	5543903	ss715631635	5.01E-07	2.80	South Dakota	
	Seed tryptophan_P	L/19	5251007	ss715636111	7.39E-09	3.80	South Dakota	
Tyr	*Seed tyrosine*	A2/8	9323205	ss715602851	8.19E-07	3.30	South Dakota	
	Seed tyrosine	G/18	18333296	ss715629450	2.32E-07	3.40	South Dakota	
Tyr_P	*Seed tyrosine_P*	F/13	16254965	ss715616797	1.45E-06	4.80	South Dakota	
Val	*Seed valine*	A2/8	9323205	ss715602851	2.01E-08	4.50	South Dakota	
Val_P	*Seed valine_P*	E/15	3846538	ss715621777	1.51E-06	5.40	South Dakota	

Trait	QTL	LG/Chr	Position (cM)	Marker/interval	LOD, *P*, or *F*	R^2 (%)	Environment	Reference
Amino acids	*Seed Gly, Ser, Trp, Arg, Asp, Gly, Lys, Phe, and Trp*	A2/8	8276381 (pb)	ss715602750				Zhang et al. (2018b)
Amino acids	*Seed Arg, Asp, Cys, Glu, His, Ile, Leu, Lys, Met, Phe, Trp, Ala, Arg, Glu, Pro, Thr, and Val*	E/15	4685574 (bp)	ss715622170				
Glu	*qGLU001*	C2/6	26.30–27.30	ss245914593-ss245908292	2.51	14.00	Fayetteville, NC	Akond et al. (2018)
Ala	*qALA001*	C2/6	7.00–9.40	ss246100375-ss245879277	50.24	74.00	Fayetteville, NC	
Cys	*qCYS001*	C2/6	25.90–27.30	ss245898080-ss245908292	2.53	11.00	Fayetteville, NC	
Val	*qVAL001*	G/18	0.00–4.00	ss249909538-ss249919445	2.62	11.00	Fayetteville, NC	
His	*qHIS001*	C2/6	26.00–27.30	ss245898080-ss245908292	2.70	12.00	Fayetteville, NC	
Ser	*qSER-C2*	C2/6	100.35–101.75	ss715593740-ss715593717	3.95	5.48	China	Li et al. (2018)
Ile	*qILE-C2*	C2/6	100.35–101.75	ss715593740-ss715593717	3.27	3.36	China	
Thr	*qTHR-C2*	C2/6	100.35–101.75	ss715593740-ss715593717	2.58	3.67	China	
Lys	*qLYS-C2*	C2/6	101.65–101.75	ss715593740-ss715593717	2.53	3.58	China	
Asp	*qASP-C2*	C2/6	77.65–91.05	ss715594534-ss715594314	2.68	2.57	China	
Leu	*qLEU-C2*	C2/6	77.45–91.05	ss715594534-ss715594314	3.07	3.27	China	
Tyr	*qTYR-C2*	C2/6	83.75–91.05	ss715594534-ss715594314	4.03	3.88	China	
Glu	*qGLU-K-1*	K/9	28.85–38.75	ss715603464-ss715603594	2.89	6.67	China	
Pro	*qPRO-K*	K/9	26.65–38.75	ss715603181-ss715603464	2.6	7.28	China	
Gly	*qGly-K-1*	K/9	122.05–122.35	ss715603840-ss715603826	2.94	4.91	China	
Ala	*qAla-K-2*	K/9	122.05–122.35	ss715603840-ss715603826	2.88	4.57	China	
Ile	*qIle-K*	K/9	122.05–122.35	ss715603840-ss715603826	4.02	4.18	China	
Ser	*qSer-K*	K/9	122.05–122.35	ss715603840-ss715603826	4.01	5.57	China	

(continued)

Table 3.6 (continued)

Trait	QTL	LG/Chr	Position (cM)	Marker/interval	LOD, *P*, or *F*	R^2 (%)	Environment	Reference
Thr	*qThr-K-2*	K/9	122.05–122.35	ss715603840-ss715603826	3.72	5.36	China	
Lys	*qLys-K*	K/9	122.05–122.35	ss715603840-ss715603826	3.12	4.45	China	
Leu	*qLeu-K-2*	K/9	122.05–122.35	Sct_196-ss715603840	3.45	6.25	China	
Tyr	*qTyr-K*	K/9	121.95–122.35	Sct_196-ss715603840	3.71	3.74	China	
Ala	*qAla-K-1*	K/9	120.25–121.85	Satt349-Sct_196	2.79	5.99	China	
Asp	*qASP-K*	K/9	162.85–185.25	ss715604355-ss715604767	3.08	5.53	China	
Glu	*qGLU-K-2*	K/9	163.85–187.25	ss715604355-ss715604767	3.53	5.07	China	
His	*qHIS-K*	K/9	162.95–182.55	ss715604355-ss715604767	2.99	4.33	China	
Val	*qVAL-K-2*	K/9	162.65–189.55	ss715604355-ss715604767	2.93	6.39	China	
Gly	*qGLY-K-2*	K/9	154.65–160.65	ss715604149-ss715604297	2.65	2.59	China	
Leu	*qLEU-K-1*	K/9	153.75–160.65	ss715604149-ss715604297	3.01	2.7	China	
Val	*qVAL-K-1*	K/9	153.75–160.65	ss715604149-ss715604297	2.85	2.58	China	
Thr	*qTHR-K-1*	K/9	153.75–160.65	ss715604149-ss715604297	2.55	2.64	China	
Thr	*qTHR-O*	O/10	53.75–56.35	ss715607711-ss715607716	3.16	4.28	China	
Gly	*qGLY-O*	O/10	53.75–56.35	ss715607711-ss715607716	3.19	4.02	China	
Asp	*qASP-O*	O/10	53.75–56.35	ss715607711-ss715607716	5.07	5.11	China	
Ile	*qILE-O*	O/10	53.75–56.35	ss715607711-ss715607716	4.62	4.27	China	
Leu	*qLEU-O*	O/10	63.5–64.05	ss715607872-ss715607213	2.55	2.4	China	
Ser	*qSER-O*	O/10	64.05–64.15	ss715607213-ss715607993	2.69	4.02	China	
Ile	*qILE-F*	F/13	212.65–216.65	ss715615780-ss715615827	2.86	2.93	China	
Leu	*qLEU-F*	F/13	213.15–216.65	ss715615780-ss715615827	2.77	4.2	China	
Val	*qVAL-F*	F/13	212.15–216.65	ss715615780-ss715615827	2.79	3.04	China	
Asp	*qASP-F-1*	F/13	212.45–216.65	ss715615780-ss715615827	3.52	3.89	China	
Glu	*qGLU-F*	F/13	213.15–216.65	ss715615780-ss715615827	3.72	2.85	China	
Lys	*qLYS-F-1*	F/13	212.05–216.65	ss715615780-ss715615827	3.27	4.66	China	
Lys	*qLYS-F-2*	F/13	216.65–219.25	ss715615837-ss715615876	3.15	3.14	China	

Trait	QTL	LG/Chr	Position (cM)	Marker/interval	LOD, *P*, or *F*	R^2 (%)	Environment	Reference
Asp	*qASP-F-2*	F/13	216.45–219.25	ss715615837-ss715615876	2.71	2.38	China	
Gly	*qGLY-D2*	D2/17	100.75–101.35	ss715625893-Satt389	8.79	11.71	China	
Ala	*qALA-D2*	D2/17	100.05–101.35	ss715625893-Satt389	7.49	9.45	China	
Ile	*qILE-D2-1*	D2/17	100.75–101.35	ss715625893-Satt389	11.22	12.88	China	
Leu	*qLEU-D2*	D2/17	100.75–101.35	ss715625893-Satt389	11.22	13.51	China	
Val	*qVAL-D2*	D2/17	100.35–101.25	ss715625893-Satt389	10.28	11.61	China	
Met	*qMET-D2*	D2/17	100.05–101.15	ss715625893-Satt389	6.93	8.84	China	
Asp	*qASP-D2-1*	D2/17	100.75–101.35	ss715625893-Satt389	12.52	14.29	China	
Glu	*qGLU-D2-1*	D2/17	100.45–101.25	ss715625893-Satt389	11.24	12.81	China	
Lys	*qLYS-D2*	D2/17	100.05–101.35	ss715625893-Satt389	8.33	10.27	China	
Phe	*qPHE-D2*	D2/17	100.95–101.45	ss715625893-Satt389	14.8	16.79	China	
Tyr	*qTYR-D2-1*	D2/17	99.45–100.65	ss715625974-ss715625893	10.32	10.41	China	
Arg	*qARG-D2-1*	D2/17	06.95–111.85	Satt461-ss715625820 1	9.75	11.27	China	
Ile	*qILE-D2-2*	D2/17	125.35–145.75	Sat_222-ss715628342	3.19	7.35	China	
Glu	*qGLU-D2-2*	D2/17	127.85–148.85	Sat_222-ss715628342	2.71	4.66	China	
Tyr	*qTYR-D2-2*	D2/17	127.756–146.45	Sat_222-ss715628342	3.48	6.69	China	
Asp	*qASP-D2-2*	D2/17	148.85–152.55	ss715628342-ss715628332	3.22	3.62	China	
Arg	*qARG-D2-2*	D2/17	148.85–152.55	ss715628342-ss715628332	3.42	4.31	China	
Val	*qVAL-I*	I/20	4.95–14.15	ss715638718-Sat_420	2.51	3.16	China	
Asp	*qASP-I*	I/20	6.65–13.65	ss715638718-Sat_420	2.73	3.04	China	

been identified and mapped on chromosomes 7, 13, and 17 (Fig. 2.1; Table 3.6) (Panthee et al. 2006b).

In another study, other authors used three RIL populations derived from the crosses of "Williams 82" by "NKS19-90" (*n* = 100), "Williams 82" by "Vinton 81" (*n* = 95), and "Williams 82" by 'DSR-173' (*n* = 93) to identify QTL that control seed methionine, cysteine, and minerals contents. "Williams 82" by "NKS19-90" (*n* = 100) RIL population was genotyped by 391 polymorphic markers, "Williams 82" by "Vinton 81" (*n* = 95) RIL population was genotyped by 400 polymorphic markers, and "Williams 82" by 'DSR-173' (*n* = 93) RIL population was genotyped by 350 polymorphic markers; and the joint genetic linkage map consisted of 766 markers. Two QTL that control seed cysteine (*Cys 5-1* and *Cys 5-2*), one QTL that controls seed Met content has been identified and mapped on chromosome 15, and two QTL that control Met and Cys (*Met+Cys 3-1* and *Met+Cys 3-2*) contents have been identified on chromosome 20 (Fig. 2.1; Table 3.6) (Ramamurthy et al. 2014).

Vaughn et al. (2014) employed GWAS in three populations to investigate QTL for seed protein, oil, and amino acids. The three populations were IL-1964 (*n* = 619 accessions), IL-1996 (*n* = 728 accessions), and MS-2000 (*n* = 934 accessions). Like in other studies, they found negative correlations between seed protein and oil contents. A genomic region narrowed earlier to 8.5 Mbp (Bolon et al. 2010) was narrowed further by this study to 1 Mbp. The study identified also several SNP markers associated with seed amino acids Thr, Cys, and Lys contents on chromosome 8. The study also showed a high resolution-specific QTL for seed Met content, and those seed amino acids contents are minimally influenced by the plant's maturity group. The study also quantified several sugars (sucrose and stachyose) in the seeds of these populations; however, it failed to identify any QTL for these sugars (Vaughn et al. 2014).

In a recent study, we used the "MD96-5722" by "Spencer" RIL population (MDxS, *n* = 50) and its SNP-based genetic linkage map (Akond et al. 2013), composed of 550 SNP markers, to map QTL for amino acids (Khandaker et al. 2015). We identified and mapped 13 QTL for different amino acids using the CIM method of QTL Cartographer (Fig. 2.1; Table 3.6). One QTL controlling seed Arg content (*qARG001*) has been identified and mapped on chromosome 3. One QTL controlling seed His content (*qHIS001*) has been identified and mapped on chromosome 16. Two QTL controlling seed Pro content (*qPRO001* and *qPRO002*) have been identified and mapped on chromosomes 1 and 14, respectively. Three QTL controlling seed Thr content (*qTHR001 to qTHR003*) have been identified and mapped on chromosomes 5, 6, and 14, respectively. Two QTL controlling seed Ser content (*qSER001* and *qSER002*) have been identified and mapped on chromosomes 5 and 6, respectively. Two QTL controlling seed Cys content (*qCYS001* and *qCYS002*) have been identified and mapped on chromosomes 5 and 9, respectively. Two QTL controlling seed Trp content (*qTRP001* and *qTRP002*) have been identified and mapped on chromosomes 9 and 18, respectively (Fig. 2.1; Table 3.6) (Khandaker et al. 2015).

In another study, Wang et al. (2015) used two RIL populations: SD02-4-59 by A02-381100 (*n* = 85) (Wang et al. 2014a) and SD02-911 by SD00-1501 (*n* = 196)

(Fehr 1987). They identified and mapped 16 QTL, 8 QTL for each of seed Cys (*Cys 1-8*) and Met (*Met 1-8*) contents. The LOD scores of these QTL ranged from 2.0 for *Met 5* on chromosome 5 to 7.8 for *Cys 8* on chromosome 20 (Fig. 2.1; Table 3.6) (Wang et al. 2015).

In another study, Warrington et al. (2015) used the Benning (PI595645) by Danbaekkong (PI619083) RIL population (n = 140) that was grown in Athens, GA, in order to genetically map QTL for protein and amino acids (Warrington et al. 2015). They identified and mapped 13 QTL for different amino acids on different chromosomes. Two QTL for Lys (*qLys_Hm08* and qLys_Gm20) have been identified and mapped on chromosomes 8 and 20, respectively; four QTL for Thr (*qThr_Gm01*, *qThr_Gm09*, *qThr_Gm17*, and *qThr_Gm20*) have been identified and mapped on chromosomes 1, 9, 17, and 20, respectively; one QTL for Cys (qCys_Gm10) on chromosome 10; and two QTL for Cys + Met combined (*qMet* + *Cys_Gm10* and *qMet* + *Cys_Gm20*) have been identified and mapped on chromosomes 10 and 20, respectively (Fig. 2.1; Table 3.6) (Warrington et al. 2015).

Qiu et al. (2015) used QTL meta-analysis to identify QTL that underlie sulfur-containing amino acids in soybean. They identified 33 QTL related to seed sulfur-containing amino acids, 8 meta-QTL regions have been estimated, and 7 sequences from these regions have been used to identify candidate genes. Sixteen candidate genes involved in amino acids biosynthetic pathways have been identified and mapped on chromosomes 6, 7, 9, 10, and 20. Among these are genes that encode for PPR protein, transcription factors (TFs), and F-Box protein (Qiu et al. 2015).

Fang et al. (2017) used a genome-wide association study (GWAS) on 809 soybean accessions collected worldwide and studied 84 traits among them seed lipids and fatty acids contents. In addition to the three main genes involved in fatty acids biosynthesis: FatB gene encoding by fatty acyl-ACP thioesterase, SAD gene encoding plant stearoyl-acyl-carrier protein desaturase, and FAD3 gene encoding fatty acid desaturase, the study identified five additional genes involved in fatty acid biosynthetic pathways and six genes involved in lipid biosynthetic pathways. Among these genes are genes encoding for phospholipase C2, phospholipase D alpha 1, reduced oleate desaturation 1, 3-ketoacyl-CoA synthase 11, diacylglycerol kinase family protein, fatty acid desaturase 5, O-acyltransferase (WSD1-like) family protein, allene oxide cyclase 3, glycolipid transfer protein 2, and several others (Fig. 2.1; Table 3.6) (Fang et al. 2017).

Zhang et al. (2018a) used a genome-wide association study (GWAS) on 313 soybean accessions and studied seed protein, oil, fatty acid, and amino acid contents. Three QTL that control seed protein content have been identified and mapped on chromosomes 15 and 16; 25 QTL that control seed oil content have been identified and mapped on chromosomes 3, 5, 6, 7, 8, 13, 14, 15, 16, 17, and 18; 18 QTL that control seed palmitic acid content have been identified and mapped on chromosomes 5, 6, 8, 14, and 15; 3 QTL that control seed stearic acid content have been identified and mapped on chromosomes 5 and 14; 4 QTL that control seed oleic acid content have been identified and mapped on chromosomes 5, 8, 14, and 19; 1 QTL that controls seed linoleic acid content have been identified and mapped on chromosome 17; 3 QTL that control seed linolenic acid content have been identified and mapped on chromosomes 14 and 15 (Table 2.2); and 92 QTL that control seed amino acid contents have been identified and mapped on all 20 soybean chromosomes (Zhang et al. 2018a). Several candidate genes have been identified on the same regions or very close to these identified QTL. Among these candidate genes are *GmSAT1* encoding the *Glycine max* symbiotic ammonium transporter 1 involved in N_2 fixation identified on chromosome 15 and *Glyma.15g049900*, *Glyma.15g049700*, and *Glyma.15g049800* encoding the nodulin MtN21 protein family. QTL that control seed Gly, Ser, Trp, Arg, Lys, Asp, Cys, Glu, His, Ile, Leu, Met, Ala, Pro, Phe, Thr, and Val have been identified and mapped on chromosomes 8 and 15 (Fig. 2.1; Table 2.2) (Zhang et al. 2018a).

In another study, we grew the RIL population derived from "Hamilton" and "Spencer" ($n = 93$) in a field in Fayetteville, NC, in 2012 and genotyped it with 1510 SNP markers. The method of CIM of QTL Cartographer identified a total of 13 QTL that control seed protein, oil, fatty acid, and amino acid contents. One QTL that controls protein content (*qPRO001*) has been identified and mapped on chromosome 18; one QTL that controls oil content (*qOIL001*) has been identified and mapped on chromosome 6; two QTL that control palmitic acid content (*qPAL001* and *qPAL002*) have been identified and mapped on chromosome 6; three QTL that control oleic acid have been identified and mapped on chromosome 3 (*qOLE001*), chromosome 6 (*qOLE002*), and chromosome 18 (*qOLE003*); and one QTL that controls linolenic acid content (*qLIN001*) has been identified and mapped on chromosome 6. For amino acids, four QTL have been identified in this RIL population. One QTL that controls seed glutamic acid content (*qGLU001*), one QTL that controls seed alanine contents (*qALA001*), and one QTL that controls seed histidine content (*qHIS001*) have been both identified and mapped on chromosome 6; and one QTL that controls seed valine content (*qVAL001*) has been identified and mapped on chromosome (Fig. 2.1; Table 3.6) (Akond et al. 2018).

Li et al. (2018) grew a RIL population derived from a cross between "Zheng 92116" (high protein and amino acid contents (46%)) and "Liaodou 14" (low protein and amino acids contents (38%)) ($n = 345$) in four different environments (E1–E4) in China. They genotyped the RIL population with 1739 SNP and 127 SSR markers and constructed a genetic linkage map covering 4265.76 cM. The study identified 12 clusters containing 59 QTL controlling seed amino acids contents on chromosomes 6, 9, 10, 13, 17, and 20. One major QTL cluster containing ten QTL (*qALA-D2*, *qASP-D2-1*, *qGLU-D2-1*, *qGLY-D2*, *qILE-D2-1*, *qLEU-D2*, *qLYS-D2*, *qMET-D2*, *qPHE-D2*, and *qVAL-D2*) has been identified and mapped on chromosome 17 (Fig. 2.1; Table 3.6) (Li et al. 2018).

Gong et al. (2018) conducted a meta-analysis and overview analysis methods (Chardon et al. 2004; Qi et al. 2011; Thilakarathna and Raizada 2017) on 140 QTL that control seed amino acids contents in soybean reported in SoyBase (Grant et al. 2010. http://www.soybase.org) using the software BioMercator 2.1. The results showed 33 consensus QTL with confidence intervals (CIs) of 0.07 Mb to 19.85 Mb. The overview method analysis identified 57 "real" QTL for amino acids that have been mapped in all soybean chromosomes except chromosomes 10 and 12. Using both methods (meta-analysis and overview analysis), 27 "real" QTL have been identified and mapped on 16 chromosomes among which 2 major "real" QTL have been mapped on chromosomes 3 and 9. Twenty candidate genes have been identified with these 27 "real" QTL regions among which 4 have been published earlier [von Groll et al. 2002 (Glyma.03G167600), Catoni et al. 2003 (Glyma.04G143500), Shah et al. 2001 (Glyma.05G083100), and Lewis et al. 2010 (Glyma.05G083100)] which left 16 new candidate genes identified by the study. Among these candidate genes are genes that encode serine-type endopeptidase activity (Glyma.06G236800), protein SER/THR kinases (Glyma.08G237200, Glyma.09G066600, GO:0004674, and Glyma.09G099200), amino acid binding (GO:0016597), serine-type endopeptidase (Glyma.16G008900), aspartic-type endopeptidase (Glyma.17G157500), lysine-tRNA ligase (Glyma.17G158700), and several others (Fig. 2.1; Table 3.6) (Gong et al. 2018).

3.6.2 QTL That Control Seed Macronutrients' and Micronutrients' Contents

As stated earlier, plants take up both macronutrients (C, O, H, N, K, Ca, Mg, P, S, and Si) and micronutrients (Cl, Fe, B, Mn, Na, Zn, Cu, Ni, and Mo) or essential elements from the soil through their root systems, combine them with sunlight, and synthesize all other compounds that they need for their growth and development; however, any absence or scarceness of one or more of these essential elements will cause symptoms of mineral deficiencies that can be chlorosis (yellow spots on leaves and stems), necrosis (browning and local death of tissues), dwarfism (reduced stem and internodes elongation), impairment in flowers and fruit development, sterility, etc. (Taiz and Zeiger 1998; Welch 1999). In acidic soils around the world, Al toxicity causes root chlorosis and necrosis, reduced root development, and increased root thickness and poses a threat to seed yield and productivity (Reicht et al. 1981; Carter et al. 1999; Bianchi-Hall et al. 2000; Silva et al. 2008), while Mn toxicity causes leaf chlorosis and root necrosis (Nable et al. 1988; Kilo and Lightfoot 1996).

These nutrients are also essential for human and animal growth and development, and any imbalance in their availability will cause malnutrition, diseases, and health issues (Fletcher et al. 2004; Lu et al. 2008); therefore, it is imperative to develop crop cultivars with high seed amounts of these nutrients (Heinemann et al. 2005; Philip and Martin 2005; Bellaloui et al. 2015a, b). Very few studies identified and mapped QTL that control seed macronutrient and micronutrient contents (Bellaloui et al. 2015a, b); however, many studies identified and mapped QTL that control nutrients toxicity, water-use efficiency (WUE), and other traits related to soil nutrients (SoyBase: https://www.soybase.org; Grant et al. 2010) using biparental populations and GWAS. Here, we will summarize only studies dealing with mapping QTL that control seed nutrient contents (Zhang et al. 2009; King et al. 2013; Benitez et al. 2010; Bellaloui et al. 2015a, b; Ning et al. 2015; Vollmann et al. 2015; Orazaly et al. 2014, 2018; Malle et al. 2020).

Zhang et al. (2009) grew a RIL population derived from a cross between “SS-516” (low calcium (Ca)) and “Camp” (high Ca) lines (n = 178 and n = 157) in a field in Fayetteville, AR, and quantified seed Ca contents. The SS-516 X Camp genetic linkage map consisted of 148 SSR markers and covered approximately 1363.7 cM (Zhang et al. 2008). Using single-marker data analysis (ANOVA) for QTL detection, 22 QTL that control seed Ca content have been identified and mapped on chromosomes 7, 8, 13, 14, 17, 18, 19, and 20. Using CIM and MIM methods, only four QTL that control seed Ca content have been identified and mapped on chromosomes 7, 8, and 20 (Fig. 2.1; Table 3.7) (Zhang et al. 2009).

Benitez et al. (2010) grew a RIL population derived from a cross between “Harosoy” and “Fukuyutaka” (n = 93) in Yawara, Japan (2006–2008), genotyped it with 220 SSR markers, and used ANOVA for QTL analysis of seed cadmium (Cd) content. The linkage map used consisted of 22 linkage groups and covered approximately 2400 cM. A major QTL for Cd content (cd1) has been identified and mapped on chromosome 9 and confirmed using NILs. The study identified two SSR markers on chromosomes 9 (Gm09:4770663 and Gm09:4790483) that could be used to select for genotypes with low harmful Cd contents I soybean seeds through marker-assisted selection (MAS) (Fig. 2.1; Table 3.7) (Benitez et al. 2010).

Jegadeesan et al. (2010) grew a RIL population derived from a cross between “AC Hime” (high seed Cd) and “WEstag-97” (low seed Cd) lines (n = 166) in Ontario, CA, in 2005, and genotyped it with 177 SSR markers. Another population derived from a cross between “Leo” and “Westag-97” lines (n = 95) was also grown in the same location and included in the QTL validation analysis. Single-marker analysis (ANOVA) identified nine SSR markers associated with seed Cd content. Three of these SSR markers have been validated and were close to the *Cda1* gene controlling seed Cd content on chromosome 9. *Cda1* gene is located between SSR markers SatK147 and SacK149 on chromosome 9. Between the SSR markers SatK140 and SaatK155 flanking *Cda1* gene, three other genes have been identified, one coding for a plant type serine-threonine protein kinase and two coding for a cation-transporting ATPase (Jegadeesan et al. 2010).

Maupin et al. (2011) grew a RIL population derived from a cross between “V99-5089” (low raffinose and phytate and high sucrose and Pi) and “Essex” lines (n = 178) in Blacksburg and Warsaw, VA, in 2008–2009, and used single-marker analysis to identify QTL for seed sucrose, raffinose, stachyose, and Pi. A major QTL that controls seed stachyose content has been identified by the SSR marker Satt453 and mapped on chromosome 11 (Fig. 2.1; Table 3.7). They further demonstrated that the mutation in the D-myo-inositol 3-phosphate synthase 1 gene (MIPS1) carried by the line V99-5089 enhances both sugar and phosphate (P) contents in soybean (Maupin et al. 2011).

King et al. (2013) grew a RIL population derived from a cross between “Anoka” and “A7” lines (n = 92) in a field in Ames, IA, and quantified seed total phosphorus (TP) contents. Three QTL that control seed TP have been identified and mapped on chromosomes 10, 18, and 20 (Fig. 2.1; Table 3.7). Within these QTL regions, they identified many candidate genes including *Glyma07g34870* that encodes for an inorganic phosphate transporter; *Glyma07g34030* that encodes for a

Table 3.7 Detailed information of QTL that control seed macronutrient and micronutrient contents in soybean reported and not reported in SoyBase (Grant et al. 2010; http://www.soybase.org)

Trait	*QTL*	LG/Chr	Interval (cM)	Marker/interval	LOD, *P*, or *F*	R^2 (%)	Environment	Reference
Macronutrients, macroelements, and related traits								
Seed calcium (Ca)	*Ca 1-1*	A2/8	0–2	Sat_382	p = 0.112	*	Fayetteville, AR	Zhang et al. (2009)
	Ca 1-2	A2/8	2.1–4.1	satt429	p = 0.11	*	Fayetteville, AR	
	Ca 1-3	A2/8	13.4–15.4	Satt228	4.3000	10.70	Fayetteville, AR	
	Ca 1-4	A2/8	33.1–35.1	Sat_392	p = 0.004	*	Fayetteville, AR	
	Ca 1-5	A2/8	48.6–50.6	Satt089	p = 0.005	*	Fayetteville, AR	
	Ca 1-6	A2/8	56.3–58.4	Satt133	p = 0.034	*	Fayetteville, AR	
	Ca 1-7	B1/11	5.4–7.4	Satt359	p = 0.1	*	Fayetteville, AR	
	Ca 1-8	B1/11	26.2–28.2	Sat_095	p = 0.088	*	Fayetteville, AR	
	Ca 1-9	D2/17	0–2	Sctt_008	p = 0.001	*	Fayetteville, AR	
	Ca 1-10	F/13	42.9–44.9	Satt252	p = 0.045	*	Fayetteville, AR	
	Ca 1-11	G/18	100.5–102.5	Satt610	p = 0.026	*	Fayetteville, AR	
	Ca 1-12	I/20	0–2	Satt496	p = 0.023	*	Fayetteville, AR	
	Ca 1-13	I/20	2.1–4.1	Sat_174	6.8000	16.30	Fayetteville, AR	
	Ca 1-14	I/20	16.1–18.1	Satt354	p = 0.004	*	Fayetteville, AR	
	Ca 1-15	I/20	35.7–37.7	Sat_170	p = 0.03	*	Fayetteville, AR	
	Ca 1-16	I/20	37.6–39.6	Satt330	p = 0.01	*	Fayetteville, AR	
	Ca 1-17	I/20	39–41	Satt292	p = 0.011	*	Fayetteville, AR	
	Ca 1-18	L/19	48.3–50.3	Satt462	p = 0.028	*	Fayetteville, AR	
	Ca 1-19	L/19	71.3–73.3	Satt229	p = 0.005	*	Fayetteville, AR	
	Ca 1-20	L/19	79.4–81.4	Satt373	p = 0.042	*	Fayetteville, AR	
	Ca 1-21	M/7	65–76	Satt150	4.9000	9.70	Fayetteville, AR	
	Ca 1-22	M/7	89.7–91.7	Satt677	4.5000	14.90	Fayetteville, AR	

(continued)

Table 3.7 (Continued)

Trait	*QTL*	LG/Chr	Interval (cM)	Marker/interval	LOD, *P*, or *F*	R^2 (%)	Environment	Reference
Seed cadmium (Cd)	*Seed Cd 1-1*	K/9	17.1–40.8	Satt178	42.9000		Yawara, Japan	Benitez et al. (2010)
Seed phosphorus (P)	*Seed P 1-1*	M/7	132.7–135.7	pk_70T-Satt308	2.3000	18.40	Ames, IA	King et al. (2013)
	Seed P 1-2	H/12	117.9–119.9	S12_0711-S12_1103	2.2000	29.90	Ames, IA	
	Seed P 1-3	D2/17	6.0–8.0	pa_257-Sat_277	2.3000	16.30	Ames, IA	
Seed calcium (Ca)	*Seed Ca 1-1*	D1a/1	16.5–18.5	Satt184	p = 0.0001	15.00	Fayetteville, AR	Orazaly et al. (2014)
	Seed Ca 1-2	D1a/1	35.2–37.2	Sat_353	p = 0.01	10.00	Fayetteville, AR	
	Seed Ca 1-3	D1a/1	47.1–49.1	Satt342	p = 0.001	13.00	Fayetteville, AR	
	Seed Ca 1-4	D1a/1	55.4–57.4	Satt254	p = 0.05	14.00	Fayetteville, AR	
	Seed Ca 1-5	D1a/1	107.9–109.9	Satt147	p = 0.001	8.00	Fayetteville, AR	
	Seed Ca 1-6	D1b/2	24.6–26.6	Satt095	p = 0.001	14.00	Fayetteville, AR	
	Seed Ca 1-7	D1b/2	71.9–73.9	Satt141	p = 0.001	5.00	Fayetteville, AR	
	Seed Ca 1-8	D1b/2	74.4–76.4	Satt600	p = 0.0001	5.00	Fayetteville, AR	
	Seed Ca 1-9	D1b/2	83–85	Satt041	p = 0.0001	7.00	Fayetteville, AR	
	Seed Ca 1-10	M/7	0–2	Sat_391	p = 0.001	*	Fayetteville, AR	
	Seed Ca 1-11	M/7	65.9–67.9	Satt175	p = 0.001	34.00	Fayetteville, AR	
	Seed Ca 1-12	M/7	74.6–76.6	Satt677	p = 0.001	10.00	Fayetteville, AR	
	Seed Ca 1-13	M/7	75.4–77.4	Sat_288	p = 0.01	8.00	Fayetteville, AR	
	Seed Ca 1-14	A2/8	83.1–85.1	Sat_199	p = 0.05	11.00	Fayetteville, AR	
	Seed Ca 1-15	A2/8	89.8–91.8	Satt377	p = 0.01	13.00	Fayetteville, AR	
	Seed Ca 1-16	A2/8	115.6–117.6	Sat_377	p = 0.001	10.00	Fayetteville, AR	
	Seed Ca 1-17	A2/8	153.1–155.1	Satt220	p = 0.001	*	Fayetteville, AR	
	Seed Ca 1-18	K/9	31.9–33.9	Satt055	p = 0.001	9.00	Fayetteville, AR	
	Seed Ca 1-19	O/10	85.7–87.7	Satt123	p = 0.001	16.00	Fayetteville, AR	
	Seed Ca 1-20	O/10	117.1–119.1	Satt153	p = 0.001	7.00	Fayetteville, AR	
	Seed Ca 1-21	F/13	15.1–17.1	Satt252	p = 0.01	10.00	Fayetteville, AR	
	Seed Ca 1-22	F/13	134.1–135.1	Satt656	p = 0.01	7.00	Fayetteville, AR	

Trait	*QTL*	LG/Chr	Interval (cM)	Marker/interval	LOD, *P*, or *F*	R^2 (%)	Environment	Reference
	Seed Ca 1-23	J/16	36–38	Satt414	p = 0.01	5.00	Fayetteville, AR	
	Seed Ca 1-24	J/16	66.8–68.8	Satt547	p = 0.01	14.00	Fayetteville, AR	
	Seed Ca 1-25	D2/17	79.2–81.2	Satt461	p = 0.01	9.00	Fayetteville, AR	
	Seed Ca 1-26	D2/17	104.5–106.5	Satt186	p = 0.01	7.00	Fayetteville, AR	
	Seed Ca 1-27	G/18	26.5–28.5	Sat_315	p = 0.01	8.00	Fayetteville, AR	
	Seed Ca 1-28	G/18	51.9–53.9	Satt594	p = 0.05	6.00	Fayetteville, AR	
	Seed Ca 1-29	G/18	86.2–88.2	Satt564	p = 0.05	7.00	Fayetteville, AR	
	Seed Ca 1-30	I/20	35.6–37.6	Sat_174	p = 0.05	12.00	Fayetteville, AR	
	Seed Ca 1-31	I/20	45.2–47.2	Satt354	p = 0.05	*	Fayetteville, AR	
	Seed Ca 1-32	I/20	74–76	Sat_170	p = 0.01	5.00	Fayetteville, AR	
	Seed Ca 1-33	I/20	81.8–83.8	Satt292	p = 0.001	7.00	Fayetteville, AR	
Seed calcium (Ca)	*Seed Ca 2-1*	D1a/1	47.1–49.1	Satt342	p = 0.001	5.00	Fayetteville, AR	
	Seed Ca 2-2	D1a/1	76.5–78.5	Satt077	p = 0.0001	7.00	Fayetteville, AR	
	Seed Ca 2-3	D1a/1	105.3–107.3	Sat_408	p = 0.0001	*	Fayetteville, AR	
	Seed Ca 2-4	A1/5	26.5–28.5	Satt042	p = 0.0001	5.00	Fayetteville, AR	
	Seed Ca 2-5	A1/5	41.8–43.8	Sat_356	p = 0.0001	12.00	Fayetteville, AR	
	Seed Ca 2-6	M/7	0–2	Sat_391	p = 0.001	*	Fayetteville, AR	
	Seed Ca 2-7	M/7	31.5–33.5	Satt567	p = 0.0001	7.00	Fayetteville, AR	
	Seed Ca 2-8	M/7	74.6–76.6	Satt677	p = 0.0001	*	Fayetteville, AR	
	Seed Ca 2-9	A2/8	115.6–117.6	Sat_377	p = 0.0001	6.00	Fayetteville, AR	
	Seed Ca 2-10	A2/8	115.7–117.7	Satt470	p = 0.001	6.00	Fayetteville, AR	
	Seed Ca 2-11	A2/8	117.6–119.6	Sat_040	p = 0.001	7.00	Fayetteville, AR	
	Seed Ca 2-12	A2/8	153.1–155.1	Satt220	p = 0.001	*	Fayetteville, AR	
	Seed Ca 2-13	O/10	23.1–25.1	Sat_318	p = 0.001	5.00	Fayetteville, AR	
	Seed Ca 2-14	O/10	118.5–120.5	Satt243	p = 0.01	7.00	Fayetteville, AR	
	Seed Ca 2-15	H/12	52.4–54.4	Satt541	p = 0.01	8.00	Fayetteville, AR	

(continued)

Table 3.7 (Continued)

Trait	*QTL*	LG/Chr	Interval (cM)	Marker/interval	LOD, *P*, or *F*	R^2 (%)	Environment	Reference
	Seed Ca 2-16	D2/17	86.2–88.2	Sat_338	p = 0.001	5.00	Fayetteville, AR	
	Seed Ca 2-17	G/18	17.3–19.3	Satt217	p = 0.01	8.00	Fayetteville, AR	
	Seed Ca 2-18	G/18	28–30	Sat_290	p = 0.01	10.00	Fayetteville, AR	
	Seed Ca 2-19	G/18	42.9–44.9	Satt115	p = 0.01	13.00	Fayetteville, AR	
	Seed Ca 2-20	G/18	44.5–46.5	Sat_358	p = 0.01	8.00	Fayetteville, AR	
Seed nickel	*Seed Ni 1-1*	M/7	41.9–49.5	BARC-042815-08424	3.27	5.00	Lincoln, NE	Ramamurthy et al. (2014)
	Seed Ni 1-2	I/20	62.9–73.9	BARC-038869-07364	3.70	5.00	Lincoln, NE	
Seed cadmium (Cd)	*Seed Cd 2-1*	G/18	51.4–53.3	BARC-059485-15839	3.21	5.00	Lincoln, NE	
Seed potassium (K)	*qPOT001*	D1b/2	4.40–4.60	ss244503611-ss244513054	2.63	0.02	Fayetteville, NC	Bellaloui et al. (2015a)
	qPOT002	D1b/2	11.10–13.10	ss244525143-ss244552583	2.81	0.02	Fayetteville, NC	
	qPOT003	N/3	16.00–16.40	ss245025187-ss245021991	9.88	0.04	Fayetteville, NC	
	qPOT004	A1/5	8.50–9.50	ss245747167-ss245786667	3.43	0.69	Fayetteville, NC	
	qPOT005	O/10	4.30–6.30	ss247085505-ss247098566	11.82	0.03	Fayetteville, NC	
	qPOT006	F/13	2.50–2.80	ss247942156-ss247937719	8.55	0.01	Fayetteville, NC	
	qPOT007	B2/14	5.80–9.20	ss248293401-ss248275088	7.25	0.57	Fayetteville, NC	
	qPOT008	J/16	9.90–12.90	ss248983974-ss248977568	3.24	0.05	Fayetteville, NC	
Seed magnesium (Mg)	*qMAG001*	N/3	8.10–9.10	ss244947608-ss244967332	8.26	0.91	Fayetteville, NC	
	qMAG002	A1/5	8.50–9.50	ss245747167-ss245786667	8.68	0.02	Fayetteville, NC	
	qMAG003	J/16	11.00–12.30	ss248983974-ss248977568	8.49	0.94	Fayetteville, NC	
	qMAG004	G/18	2.10–2.20	ss249623816-ss249632893	8.59	0.95	Fayetteville, NC	
Seed phosphorus (P)	*qPHO001*	J/16	9.90–11.00	ss248983974-ss248977568	9.49	0.94	Fayetteville, NC	
Seed carbon (C)	*qCAR001*	J/16	9.00–10.00	ss248983974-ss248977568	10.01	0.94	Fayetteville, NC	
Seed nitrogen (N)	*qNIT001*	J/16	8.90–10.00	ss248983974-ss248977568	9.85	0.94	Fayetteville, NC	
Seed sulfur (S)	*qSUL001*	J/16	9.90–11.10	ss248983974-ss248977568	9.73	0.94	Fayetteville, NC	
Seed calcium (Ca)	*qCAL001*	G/18	10.70–15.60	ss249715368-ss249830195	2.72	0.66	Fayetteville, NC	

Trait	QTL	LG/Chr	Interval (cM)	Marker/interval	LOD, *P*, or *F*	R^2 (%)	Environment	Reference
Seed boron (B)	*qBOR001*	N/3	15.50–15.80	ss245030179-ss245023395	8.86	0.96	Fayetteville, NC	Bellaloui et al. (2015b)
	qBOR002	A1/5	10.30–11.50	ss245747167-ss245786667	8.02	0.79	Fayetteville, NC	
Seed copper (Cu)	*qCOP001*	N/3	15.50–15.80	ss245030179-ss245023395	8.91	0.94	Fayetteville, NC	
	qCOP002	A1/5	8.50–9.50	ss245747167-ss245786667	8.97	0.06	Fayetteville, NC	
	qCOP003	K/9	0.20–0.30	ss246894176-ss246871129	8.94	0.94	Fayetteville, NC	
	qCOP004	J/16	9.90–12.00	ss248983974-ss248977568	8.78	0.94	Fayetteville, NC	
	qCOP005	G/18	2.10–2.20	ss249623816-ss249632893	8.69	0.95	Fayetteville, NC	
Seed iron (Fe)	*qIRO001*	N/3	15.60–15.80	ss245030179-ss245023395	8.61	0.94	Fayetteville, NC	
	qIRO002	A1/5	8.50–9.50	ss245747167-ss245786667	8.71	0.05	Fayetteville, NC	
	qIRO003	K/9	0.40–0.50	ss246870684-ss246877864	8.77	0.94	Fayetteville, NC	
	qIRO004	J/16	10.90–12.20	ss248983974-ss248977568	8.58	0.94	Fayetteville, NC	
	qIRO005	G/18	2.10–2.20	ss249623816-ss249632893	8.69	0.94	Fayetteville, NC	
Seed manganese (Mn)	*qMAN001*	N/3	15.60–15.80	ss245030179-ss245023395	8.37	0.94	Fayetteville, NC	
	qMAN002	A1/5	8.50–9.50	ss245747167-ss245786667	8.38	0.05	Fayetteville, NC	
	qMAN003	K/9	0.40–0.50	ss246870684-ss246877864	8.64	0.94	Fayetteville, NC	
	qMAN004	K/9	0.40–0.50	ss246870684-ss246877864	8.4	0.94	Fayetteville, NC	
	qMAN005	J/16	10.90–12.20	ss248983974-ss248977568	8.23	0.94	Fayetteville, NC	
Seed zinc (Zn)	*qZIN001*	D1b/1	15.60–15.70	ss244552583-ss244554797	9.14	0.83	Fayetteville, NC	
	qZIN002	N/3	15.60–15.80	ss245030179-ss245023395	9.63	0.96	Fayetteville, NC	
	qZIN003	A1/5	8.50–9.50	ss245747167-ss245786667	9.52	0.03	Fayetteville, NC	
	qZIN004	K/9	1.40–1.50	ss246893538-ss246895099	4.16	0.02	Fayetteville, NC	
	qZIN005	F/13	1.30–1.50	ss247940172-ss247965852	8.72	0.02	Fayetteville, NC	
	qZIN006	B2/14	9.90–11.10	ss248293401-ss248275088	9.09	0.94	Fayetteville, NC	
	qZIN007	J/16	9.10–12.90	ss248983974-ss248977568	9.44	0.94	Fayetteville, NC	
Biparental RILs analysis:								

(continued)

Table 3.7 (Continued)

Trait	*QTL*	LG/Chr	Interval (cM)	Marker/interval	LOD, *P*, or *F*	R^2 (%)	Environment	Reference
Seed calcium (Ca)	*12-qCa-1*	D1a/1	46.5–86.8	Satt468-sat_160	3.64	17.94	Nanjing, China	Ning et al. (2015)
	12-qCa-2	D1b/2	270.8–290.3	Sat_351	2.53	4.34	Nanjing, China	
	12-qCa-4-1	C1/4	73–82.8	Satt294	3.64	6.61	Nanjing, China	
	12-qCa-4-2	C1/4	84.3–95.4	Sat_311-sct_191	2.72	6.21	Nanjing, China	
	12-qCa-6	C2/6	126.2–147.7	Sat_312-satt316	2.58	6.5	Nanjing, China	
	12-qCa-8-1	A2/8	6.1–25.2	BE820148-sat_162	5.5	13.36	Nanjing, China	
	12-qCa-8-2	A2/8	126.9–142.8	Satt333-satt133	4.36	8.86	Nanjing, China	
	12-qCa-11	B1/11	35–39.5	Sat_261	3.21	5.77	Nanjing, China	
	13-qCa-2	D1b/2	275.9–290.5	Sat_351	3.4	7.14	Nanjing, China	
	13-qCa-8	A2/8	11.6–35.3	Sat_162-AW132402	2.52	8.37	Nanjing, China	
	13-qCa-11	B1/11	0–5.8	Sat_149	2.6	5.39	Nanjing, China	
	qCa-1	D1a/1	52.7–88.6	Satt468-sat_160	3.46	8.48	Nanjing, China	
	qCa-2	D1b/2	275.1–290.1	Sat_351	3.42	6.36	Nanjing, China	
	qCa-4	C1/4	72.9–82.5	Satt294	2.83	4.75	Nanjing, China	
	qCa-8-1	A2/8	11.2–30.2	Sat_162-AW132402	5.88	12.35	Nanjing, China	
	qCa-8-2	A2/8	128.1–143.6	Satt333-satt133	3.71	7.41	Nanjing, China	
	qCa-11	B1/11	30.5–40.4	Sat_261	2.65	4.73	Nanjing, China	
Seed phosphorus (P)	*12-qP-2*	D1b/2	121.8–146.2	Satt428	2.73	6.82	Nanjing, China	
	12-qP-17	D2/17	15.5–45.1	Sat_277-satt135	3.14	9.5	Nanjing, China	
	13-qP-6	C2/6	119.4–139.5	Sat_312-satt316	2.56	6.53	Nanjing, China	
	13-qP-11	B1/11	6.8–23.1	Satt197-satt251	4.15	11.99	Nanjing, China	
	qP-11	B1/11	1.8–17.9	Satt197	3.52	8.21	Nanjing, China	
	qP-15	E/15	24.2–43.7	Satt606-satt706	3.17	8.11	Nanjing, China	
	qP-17	D2/17	7.4–34.7	Sat_296-sat_277	3.47	14.96	Nanjing, China	
Seed iron (Fe)	*12-qFe-3*	N/3	32.4–56.9	Satt009-satt675	2.55	6.21	Nanjing, China	
	13-qFe-3	N/3	76.3–82.1	Satt675-satt237	2.91	7.11	Nanjing, China	
	qFe-3	N/3	72.0–80.6	Satt675-satt237	2.8	6.88	Nanjing, China	

Trait	*QTL*	LG/Chr	Interval (cM)	Marker/interval	LOD, *P*, or *F*	R^2 (%)	Environment	Reference
Seed magnesium (Mg)	*12-qMg-5*	A1/5	25.4–51.6	Sat_171	2.91	7.39	Nanjing, China	
	13-qMg-13	F/13	99–134.9	Satt335satt522	2.99	28.18	Nanjing, China	
	qMg-13	F/13	92.9–134.8	Satt335-satt522	3.95	21.9	Nanjing, China	
Seed zinc (Zn)	*12-qZn-4*	C1/4	107.8–119.3	AI794821	5.55	15.64	Nanjing, China	
	13-qZn-11	B1/11	18.8–31.2	Satt197-satt251	4.94	12.44	Nanjing, China	
	qZn-11-1	B1/11	20.5–29.6	Satt251	4.62	12.55	Nanjing, China	
	qZn-11-2	B1/11	35.4–46.4	Sat_261-sat_156	3.69	12.06	Nanjing, China	
	qZn-15	E/15	197.7–41.8	Satt606	3.04	7.41	Nanjing, China	
GWAS analysis:								
Seed calcium (Ca)	*Seed Ca*	C2/6	19,071,254	BARC-041345-07968	ns	*	Nanjing, China	
	Seed Ca	D2/17	4,966,375	BARC-014619-01595	4.38	*	Nanjing, China	
	Seed Ca	G/18	871,120	BARC-044817-08810	3.23	*	Nanjing, China	
Seed magnesium (Mg)	*Seed Mg*	D1a/1	2,840,247	BARC-029293-06144	3.21	*	Nanjing, China	
Seed zinc (Zn)	*Seed Zn*	D1a/1	2,699,256	BARC-024477-04900	3.05	*	Nanjing, China	
	Seed Zn	D1a/1	52,543,750	BARC-039805-07589	ns	*	Nanjing, China	
	Seed Zn	C1/4	6,862,890	BARC-044331-08669	ns	*	Nanjing, China	
	Seed Zn	A1/5	9,097,069	BARC-044481-08709	ns	*	Nanjing, China	
	Seed Zn	M/7	8,426,497	BARC-012945-00406	3.43	*	Nanjing, China	
	Seed Zn	B1/11	456,273	BARC-018025-02500	3.89	*	Nanjing, China	
	Seed Zn	B1/11	5,265,730	BARC-018099-02516	3.88	*	Nanjing, China	

(continued)

Table 3.7 (Continued)

Trait	*QTL*	LG/Chr	Interval (cM)	Marker/interval	LOD, *P*, or *F*	R^2 (%)	Environment	Reference
Seed iron (Fe)	*Seed Fe*	N/3	46,719,894	BARC-032309-08945	3.82	*	Nanjing, China	
	Seed Fe	N/3	46,941,746	BARC-044595-08729	3.12	*	Nanjing, China	
	Seed Fe	C2/6	17,423,963	BARC-014491-01561	3.55	*	Nanjing, China	
	Seed Fe	O/10	3,519,980	BARC-018911-03272	ns	*	Nanjing, China	
	Seed Fe	B1/11	11,571,689	BARC-042819-08425	3.55	*	Nanjing, China	
	Seed Fe	B1/11	15,576,626	BARC-040309-07711	4.07	*	Nanjing, China	
	Seed Fe	H/12	35,019,428	BARC-039403-07490	3.92	*	Nanjing, China	
Seed phosphorus (P)	*Seed P*	N/3	40,585,010	BARC-016485-02069	ns	*	Nanjing, China	
	Seed P	A1/5	31,137,034	BARC-044399-08694	ns	*	Nanjing, China	
	Seed P	C2/6	13,550,805	BARC-028177-05786	3.99	*	Nanjing, China	
	Seed P	M/7	8,426,497	BARC-012945-00406	3.18	*	Nanjing, China	
	Seed P	A2/8	3,056,231	BARC-031701-07215	4.23	*	Nanjing, China	
	Seed P	B1/11	456,273	BARC-018025-02500	3.11	*	Nanjing, China	
	Seed P	B1/11	7,619,324	BARC-040851-07853	ns	*	Nanjing, China	
	Seed P	B1/11	1,656,367	BARC-018583-02981	3.07	*	Nanjing, China	
	Seed P	B1/11	38,631,243	BARC-024093-04724	ns	*	Nanjing, China	
	Seed P	B2/14	46,713,866	BARC-013273-00466	3.07	*	Nanjing, China	
Seed calcium (Ca)	*Seed Ca 1*	D1a/1	56.3–58.4	Satt267	p = 0.001	34.00	Fayetteville, AR	Orazaly et al. (2018)
	Seed Ca 2	D1a/1	64.6–66.6	Sat_345	p = 0.05	21.00	Fayetteville, AR	
	Seed Ca 3	M/7	0.5–1.5	Satt391	p = 0.001	34.00	Fayetteville, AR	
	Seed Ca 4	M/7	75.4–77.4	Satt288	p = 0.001	20.00	Fayetteville, AR	
	Seed Ca 5	M/7	76.2–78.2	Satt680	p = 0.001	22.00	Fayetteville, AR	
	Seed Ca 6	A2/8	76.7–78.7	Satt341	p = 0.0001	40.00	Fayetteville, AR	
	Seed Ca 7	A2/8	105.3–107.3	Sat_392	p = 0.0001	38.00	Fayetteville, AR	
	Seed Ca 8	A2/8	115.4–117.4	Sat_382	p = 0.05	23.00	Fayetteville, AR	
	Seed Ca 9	J/16	66.8–68.8	Satt547	p = 0.05	20.00	Fayetteville, AR	
	Seed Ca 10	J/16	36.1–38.1	Satt414	p = 0.05	22.00	Fayetteville, AR	

Trait	*QTL*	LG/Chr	Interval (cM)	Marker/interval	LOD, *P*, or *F*	R^2 (%)	Environment	Reference
	Seed Ca 11	D2/17	46.7–48.7	Satt002	p = 0.001	32.00	Fayetteville, AR	
	Seed Ca 12	G/18	28–30	Sat_290	p = 0.001	42.00	Fayetteville, AR	
	Seed Ca 13	L/19	58.5–60.5	Sct_010	p = 0.05	32.00	Fayetteville, AR	
	Seed Ca 14	I/20	74–76	Sat_170	p = 0.05	23.00	Fayetteville, AR	
	Seed Ca 15	I/20	76.5–78.5	Satt330	p = 0.001	30.00	Fayetteville, AR	
Seed calcium (Ca)	*Seed Ca 3*	C2/6	3,354,869	SNP	2.94E-08	20.00	ON and QC, Canada	Malle et al. (2020)
	Seed Ca 4	K/9	6,092,970	SNP	3.70E-08	21.00	ON and QC, Canada	
Seed potassium (K)	*Seed K 1*	C1/4	49,071,552	SNP	1.75E-06	17.00	ON and QC, Canada	
	Seed K 3	O/10	1,925,709	SNP	4.31E-10	31.00	ON and QC, Canada	
Seed phosphorus (P)	*Seed P 1*	C1/4	49,071,286	SNP	6.12E-08	22.00	ON and QC, Canada	
Seed sulfur (S)	*Seed S 4*	O/10	1,602,998	SNP	2.84E-08	23.00	ON and QC, Canada	
	Seed S 7	E/15	3,986,243	SNP	2.80E-07	19.00	ON and QC, Canada	
	Seed S 10	I/20	39,076,484	SNP	9.13E-07	18.00	ON and QC, Canada	

phosphatidylinositol-4-phosphate 5-kinase (PIP5K); *Glyma07g32190* which is a solute carrier of glucose-6-phosphate/phosphate; *Glyma07g34930* that encodes for a ZIP Zinc transporter; *Glyma12g16410*, *Glyma12g30100*, and *Glyma12g30070* that encode for a protein members of the ABC transporter family; *Glyma12g23920* that encodes for a triose-phosphate transporter gene; Glyma12g22680 that encodes for an inositol monophosphatase family; *Glyma12g29790* that encodes for a triose-phosphate transporter gene; *Glyma20g34610* that encodes for an Pi transporter (Wu et al. 2011); *Glyma20g34620* that encodes for an P/H$^+$ symporter (PHS) transporter (Pao et al. 1998); and many other genes (Table 3.9) (King et al. 2013).

Orazaly et al. (2014) grew two RIL populations in two fields in Arkansas. The first RIL population was derived from the cross of "KS4303sp" and "PI 407818B" (*n* = 120), and the second RIL population was derived from the cross of "PI 408052C" and "PI 407818B" (*n* = 124). The two RIL populations were genotyped with 486 SSR markers; however, 105 and 101 SSR markers were polymorphic in these populations, respectively. Both single-marker analysis (ANOVA), IM and CIM of WinQTL Cartographer, have been used for QTL analysis of seed Ca content. In "KS4303sp" X "PI 407818B" RIL population, 33 QTL that control seed Ca content have been identified and mapped on chromosomes 1, 2, 7, 8, 9, 10, 13, 16, 17, 18, and 20 (Fig. 2.1; Table 3.7). In "PI 408052C" X "PI 407818B" RIL population, 20 QTL that control seed Ca content have been identified and mapped on chromosomes 1, 5, 7, 8, 10, 12, 17, and 18 (Fig. 2.1; Table 3.7) (Orazaly et al. 2014).

In another study, Ramamurthy et al. (2014) used three RIL populations derived from the crosses of "Williams 82" by "NKS19-90" (*n* = 100), "Williams 82" by "Vinton 81" (*n* = 95), and "Williams 82" by "DSR-173" (*n* = 93) to identify QTL that control seed methionine, cysteine, and minerals contents. "Williams 82" by "NKS19-90" (*n* = 100) RIL population was genotyped by 391 polymorphic markers, "Williams 82" by "Vinton 81" (*n* = 95) RIL population was genotyped by 400 polymorphic markers, and "Williams 82" by "DSR-173" (*n* = 93) RIL population was genotyped by 350 polymorphic markers; and the joint genetic linkage map consisted of 766 markers. In addition to the QTL that control seed amino acids (*Cys 5-1*, *Cys 5-2*, *Met+Cys 3-1*, and *Met+Cys 3-2*) mentioned above, two QTL that control seed nickel (Ni) contents and one QTL that controls seed cadmium (Cd) content have been identified and mapped on chromosomes 7, 18, and 20 (Fig. 2.1; Table 3.7) (Ramamurthy et al. 2014).

Vollmann et al. (2015) analyzed seed cadmium content in 48 soybean cultivars (*n* = 48) using single-maker ANOVA at the SSR marker SacK149. They found that 29 cultivars showed high seed Cd content and 19 showed low seed Cd content. The SSR marker SacK149 is associated with *Cda1* gene controlling seed Cd content (Jegadeesan et al. 2010; Vollmann et al. 2015).

In two recent studies, we grew a RIL population derived from a cross between "MD96-5722" and "Spencer" lines (*n* = 94) in a field in Fayetteville, NC. The genetic linkage map used in these two studies was constructed by 550 SNP markers and covered 201.6 cM (Akond et al. 2013). In Bellaloui et al. (2015a), 17 QTL that control seed C, Ca, K, Mg, N, P, and S contents have been identified and mapped on chromosomes 2, 3, 5, 10, 13, 14, 16, and 18 (Fig. 2.1; Table 3.7) (Bellaloui et al.

2015a). In Bellaloui et al. (2015b), 24 QTL that control seed B, Cu, Fe, Mn, and Zn contents have been identified and mapped on chromosomes 1, 3, 5, 9, 13, 14, 16, and 18 (Fig. 2.1; Table 3.7) (Bellaloui et al. 2015b).

Ning et al. (2015) used a RIL population and a natural population. They grew a RIL population derived from a cross between "Kefeng 1" and "Nanong 1138-2" lines ($n = 184$). The natural population ($n = 219$) was composed of various landraces from diverse geographical regions and used for GWAS analysis. Both populations were grown in Nanjing, China, and evaluated for seed mineral (P, Ca, Mg, Zn, and Fe) contents. In the RIL population, 17 QTL that control seed Ca content have identified and mapped on chromosomes 1, 2, 4, 6, 8, and 11; 7 QTL that control seed P content have identified and mapped on chromosomes 2, 6, 11, 15, and 17; 3 QTL that control seed Fe content have identified and mapped on chromosomes 3; 3 QTL that control seed Mg content have identified and mapped on chromosomes 5 and 13; and 5 QTL that control seed Zn content have identified and mapped on chromosomes 4, 11, and 15 (Fig. 2.1; Table 3.7). In the natural population and using GWAS, three QTL (SNPs) that control seed Ca content have identified and mapped on chromosomes 6, 17, and 18; one QTL (SNP) that controls seed Mg content has identified and mapped on chromosome 1; seven QTL (SNPs) that control seed Zn content have identified and mapped on chromosomes 1, 4, 5, 7, and 11; seven QTL (SNPs) that control seed Fe content have identified and mapped on chromosomes 3, 6, 10, 11, and 12; and ten QTL (SNPs) that control seed P content have identified and mapped on chromosomes 3, 5, 6, 7, 8, 11, and 14 (Table 3.7) (Ning et al. 2015).

Orazaly et al. (2018) grew six RIL populations ($n = 164$), derived from six crosses between cultivars contrasting for their seed calcium (Ca) contents, in two fields in Arkansas in 2007 (Kibler and Fayetteville) and 2008 (Keiser and Fayetteville): a cross between "Hutcheson" and "Camp" ($n = 31$), a cross between "SS-516" and "Camp" ($n = 30$), a cross between "SS-516" and "V96-4486" ($n = 20$), a cross between "Hutcheson" and "V96-4486" ($n = 20$), a cross between "MFS-591" and "Camp" ($n = 20$), and a cross between "MFS-591" and "V96-4486" ($n = 43$). They also used the KS4303sp by PI 408078B RIL population ($n = 120$) and single-marker analysis (ANOVA) and CIM of WinQTL Cartographer to identify QTL that control seed Ca content. Fifteen QTL that control seed Ca content have been identified and mapped on chromosomes 1, 7, 8, 16, 17, 18, 19, and 20 (Fig. 2.1; Table 3.7) (Orazaly et al. 2018).

Malle et al. (2020) grew a population of 137 soybean accessions from a wider population of 304 accessions (Sonah et al. 2015) in 2 environments in Canada (St-Mathieu-de-Beloeil, QC and Woodstock, ON) and used GWAS analysis for seed mineral contents (P, S, K, and Ca) analysis. The study identified five QTL that control seed P content, ten QTL that control seed S content, ten QTL that control seed K content, and seven QTL that control seed Ca content. Among these 32 QTL, 8 robust QTL have been identified and mapped on chromosomes 4, 6, 9, 10, 15, and 20 (Table 3.7) (Malle et al. 2020).

3.6.3 QTL That Control Seed Sugars and Other Compounds Contents

There are 53+ QTL controlling seed sugars contents (mainly sucrose, raffinose, glucose, galactose, rhamnose, fucose, stachyose, and starch) that have been identified and mapped using various mapping populations, accessions, and mapping methods such as genome-wide association studies (GWAS), single-marker analysis (SMA), and composite interval mapping (CIM) (Maughan et al. 2000; Cicek 2001; Stombaugh et al. 2004; Feng et al. 2005; Kim et al. 2005, 2006; Jaureguy 2009; Zeng et al. 2012; Zeng et al. 2014; Akond et al. 2015; Dhungana et al. 2017). In general, breeding programs aim at developing soybean lines with low stachyose and raffinose contents, and high sucrose contents. In this section, we will summarize the most important QTL controlling seed sugars (sucrose, raffinose, glucose, galactose, rhamnose, fucose, stachyose, and starch) contents (Fig. 2.1; Table 3.8).

Maughan et al. (2000) grew a RIL population derived from a cross between "V71-370" and "PI407.162" ($n = 149$) in Mt. Holly, VA, and genotyped it with 178 RFLP, RAPD, and SSR markers. Seventeen markers mapped into 7 QTL that control seed sucrose content have been identified and mapped on chromosome 5, 7, 8, 13, 15, 19, and 20 among which several are also associated with protein and oil contents (Brummer et al. 1997) (Fig. 2.1; Table 3.8) (Maughan et al. 2000).

Cicek (2001) grew a RIL population derived from a cross between "V71-370" and "PI 407162" ($n = 309$) in Blacksburg and Warsaw, VA, in 1998, genotyped it with 54 RFLP and SSR markers, and used single-marker analysis (ANOVA) and CIM of WinQTL Cartographer for QTL data analysis; however, the genetic linkage map used covered only 860 cM of the soybean genome. Five QTL that control seed sucrose content have been identified and mapped on chromosomes 5, 7, 8, 13, and 18, one QTL that controls seed raffinose content has been identified and mapped on chromosome 18, and one QTL that controls seed stachyose sucrose content has been identified and mapped on chromosome 5 (Fig. 2.1; Table 3.8) (Cicek 2001).

Stombaugh et al. (2004) grew a RIL population derived from a cross between "Minsoy" and "Archer" ($n = 108$) in two Waseca and Rosemount, MN, in 1997 and 2000, respectively and investigated seed sugars contents. In combined environments, two QTL that control seed arabinose content have been identified and mapped on chromosomes 9 and 17; one QTL that controls seed arabinose/galactose ratio has been identified and mapped on chromosome 5; one QTL that controls seed cell wall polysaccharides (CWP: fructose, arabinose, and galactose are incorporated into CWP) content has been identified and mapped on chromosome 5; three QTL that control seed fucose content have been identified and mapped on chromosomes 2, 8, and 19; one QTL that control seed galactose content has been identified and mapped on chromosome 5; two QTL that control seed glucose content have been identified and mapped on chromosomes 2 and 5; one QTL that control seed rhamnose content has been identified and mapped on chromosome 5; and one QTL that control seed xylose content has been identified and mapped on chromosome 6 (Fig. 2.1; Table 3.8) (Stombaugh et al. 2004).

Table 3.8 Detailed information of QTL that control seed sugars (sucrose, glucose, raffinose, stachyose, and oligosaccharide) and other compounds' contents reported and not reported in SoyBase (Grant et al. 2010; http://www.soybase.org)

Trait	QTL	LG/Chr	Interval (cM)	Marker	LOD/*P*/*F*	R^2 (%)	Environment	Reference
Sugars								
Seed sucrose	*Seed sucrose 1-1*	A1/5	7.82–9.82	sg:laSU-A487V	0.003	0.081	Virginia	Maughan et al. (2000)
	Seed sucrose 1-2	A2/8	70.38–72.38	sg:laSU-T169H	0.007	0.07	Virginia	
	Seed sucrose 1-3	A2/8	49.40–51.40	sg:laSU-A458-1H	0.001	0.077	Virginia	
	Seed sucrose 1-4	I/20	31.42–33.42	sg:laSU-A136V	0.002	0.087	Virginia	
	Seed sucrose 1-5	F/13	63.85–65.85	sg:laSU-A486I	0.004	0.084	Virginia	
	Seed sucrose 1-6	L19	35.70–37.70	sg:laSU-T153-1I	0.005	0.076	Virginia	
	Seed sucrose 1-7	L/19	34.30–36.30	sg:laSU-A144H	0.000	0.124	Virginia	
	Seed sucrose 1-8	L/19	48.40–50.40	sg:laSU-K227-1I	0.000	0.114	Virginia	
	Seed sucrose 1-9	M/7	2.05–4.05	sg:VT-NBS37	0.000	0.113	Virginia	
	Seed sucrose 1-10	E/15	16.12–18.12	sg:laSU-A186I	0.001	0.096	Virginia	
	Seed sucrose 1-11			OO20850e	0.006	0.061	Virginia	
	Seed sucrose 1-12			sg:laSU-A23I	0.001	0.097	Virginia	
	Seed sucrose 1-13	A2/8	52.16–54.16	sg:laSU-B164D	0.001	0.096	Virginia	
	Seed sucrose 1-14			sg:laSU-B162V	0.008	0.07	Virginia	
	Seed sucrose 1-15			OB41600	0.006	0.086	Virginia	
	Seed sucrose 1-16			BARC-SC514	0.009	0.073	Virginia	
	Seed sucrose 1-17			sg:laSU-A963H	0.009	0.069	Virginia	
Seed sucrose	*qSeedSuc1*	A1/5	99.1–101.1	Satt042-Satt276	2.20	3.80	Virginia	Cicek (2001)
	qSeedSuc2	A2/8	75.8–77.8	Satt424-Satt089	3.80	12.70	Virginia	
	qSeedSuc3	F/13	51.7–53.7	Satt133-Satt114	2.80	4.90	Virginia	
	qSeedSuc4	G/18	89.7–91.7	Satt115-Satt517	3.00	4.70	Virginia	
	qSeedSuc5	M/7	50.3–52.3	Satt540-Satt551	3.00	6.70	Virginia	
Seed raffinose	*qSeedRaff1*	G/18	65.4–67.4	Satt324-Satt115	2.60	4.80	Virginia	
Seed stachyose	*qSeedStach1*	A1/5	15–17	Satt511-Satt200	2.80	4.90	Virginia	

(continued)

Table 3.8 (Continued)

Trait	QTL	LG/Chr	Interval (cM)	Marker	LOD/*P*/*F*	R^2 (%)	Environment	Reference
Arabinose	*Seed arabinose 1-1*	D2/17	85.7–89.2	Satt514-Satt488	3.20	23.30	Minnesota	Stombaugh et al. (2004)
	Seed arabinose 1-2	K/9	83.2–93.4	Satt196-Satt588	8.90	54.60	Minnesota	
Arabinose/ galactose	*Seed Ara/Gal 1-1*	A1/5	57.5–59.5	Satt385	3.30	24.00	Minnesota	
CWP	*Seed CWP 1-1*	A1/5	62.7–84.1	Satt545-Satt599	4.80	43.30	Minnesota	
Fucose	*Seed fucose 1-1*	A2/8	106.1–108.1	Satt437	6.70	42.80	Minnesota	
	Seed fucose 1-2	L/19	55.6–59.4	Satt076-Sat_113	3.30	24.10	Minnesota	
	Seed fucose 1-3	D1b/2	46.2–53	Satt157-Satt542	3.40	30.00	Minnesota	
Galactose	*Seed galactose 1-1*	A1/5	62.7–84.1	Satt545-Satt599	5.60	38.30	Minnesota	
Glucose	*Seed glucose 1-1*	D1b/2	53–73.3	Satt542-Satt290	3.40	24.50	Minnesota	
Glucose	*Seed glucose 1-2*	A1/5	82.8–95.9	Satt174-Satt211	3.20	24.30	Minnesota	
Rhamnose	*Seed rhamnose*	A1/5	62.7–84.1	Satt545-Satt599	3.50	26.00	Minnesota	
Xylose	*Seed xylose*	C2/6	114.2–127.9	Satt202-Satt371	3.30	24.60	Minnesota	
Seed sucrose	*qSeedSuc-1*	G/18	33.3–35.4	Satt324	p = 0.0001	12.90	China	Feng et al. (2005)
Seed raffinose	*qSeedRaff-1*	K/9	60.9–80.1	Satt260	p = 0.0001	11.60	China	
	qSeedRaff-2	B1/11	108.4–124	Satt453	p = 0.0001	25.20	China	
Seed stachyose	*qSeedStach-1*	O/10	36.9–42.3	Satt347	p = 0.0001	18.80	China	
Seed sucrose	*Seed sucrose 2-1*	B1/11	45.38–47.38	Satt197	6.50	*	Korea	Kim et al. (2005)
	Seed sucrose 2-2	D1b/2	86.19–88.19	Satt546	8.47	*	Korea	
	Seed sucrose 2-3	L/19	26.92–28.92	Satt523	17.80	*	Korea	
	Seed sucrose 2-4	L/19	30.21–32.21	Satt278	18.40	*	Korea	
	Seed sucrose 2-5	L/19	29.93–31.93	Satt418	19.80	*	Korea	
	Seed sucrose 2-6	L/19	29.58–31.58	Satt398	18.00	*	Korea	
	Seed sucrose 2-7	L/19	40.00–42.00	Satt462	9.20	*	Korea	
	Seed sucrose 2-8	L/19	32.70–34.70	Satt497	18.80	*	Korea	
	Seed sucrose 2-9	L/19	33.54–35.54	Satt313	16.80	*	Korea	
	Seed sucrose 2-10	L/19	58.52–60.52	Sct_010	4.40	*	Korea	
	Seed sucrose 2-11	L/19	58.34–60.34	Satt076	4.00	*	Korea	

Trait	QTL	LG/Chr	Interval (cM)	Marker	LOD/*P*/*F*	R^2 (%)	Environment	Reference
Seed oligosaccharide	*Oligosaccharides 1-1*	D1b/2	86.19–88.19	Satt546	7.70	*	Korea	
	Oligosaccharides 1-2	L/19	26.92–28.92	Satt523	14.40	*	Korea	
	Oligosaccharides 1-3	L/19	30.21–32.21	Satt278	15.50	*	Korea	
	Oligosaccharides 1-4	L/19	29.93–31.93	Satt418	16.60	*	Korea	
	Oligosaccharides 1-5	L/19	29.58–31.58	Satt398	15.00	*	Korea	
	Oligosaccharides 1-6	L/19	40.00–42.00	Satt462	6.20	*	Korea	
	Oligosaccharides 1-7	L/19	32.70–34.70	Satt497	15.50	*	Korea	
	Oligosaccharides 1-8	L/19	33.54–35.54	Satt313	13.50	*	Korea	
Seed sucrose	*Seed sucrose 3-1*	B2/14	72.12–73.12	Satt020	4.68	*	Korea	Kim et al. (2006)
	Seed sucrose 3-2	D1b/2	74.66–76.66	Satt537	p = 0.0481	*	Korea	
	Seed sucrose 3-3	E/15	43.98–45.98	Satt483	p = 0.0001	*	Korea	
	Seed sucrose 3-4	H/12	45.95–47.95	Satt442	p = 0.0111	*	Korea	
	Seed sucrose 3-5	J/16	14.69–24.69	Satt287	13.50	*	Korea	
	Seed sucrose 3-6	J/16	31.09–33.09	Sct_065	p = 0.0472	*	Korea	
Seed oligosaccharide	*Oligosaccharides 2-1*	B2/14	71.12–73.12	Satt020	p = 0.0012	*	Korea	
	Oligosaccharides 2-2	C2/6	112.95–114.95	Satt100	p = 0.0002	*	Korea	
	Oligosaccharides 2-3	E/15	43.98–45.98	Satt483	p = 0.0008	*	Korea	
	Oligosaccharides 2-4	H/12	45.95–52.95	Satt442	p = 0.0191	*	Korea	
	Oligosaccharides 2-5	J/16	14.69–26.69	Satt287	p = 0.0025	*	Korea	
	Oligosaccharides 2-6	J/16	31.09–39.09	Sct_065	p = 0.0076	*	Korea	
	Oligosaccharides 2-7	L/19	65.51–74.51	Satt166	p = 0.0105	*	Korea	
Seed sucrose	*qSucrose-1-1*	I/20	27.8–29.8	Satt614	p = 0.0001	*	Arkansas	Jaureguy (2009)
	qSucrose-1-2	I/20	30.5–32.5	Satt496	p = 0.0004	*	Arkansas	

(continued)

Table 3.8 (Continued)

Trait	QTL	LG/Chr	Interval (cM)	Marker	LOD/*P*/*F*	R^2 (%)	Environment	Reference
Seed raffinose	*qRaffinose-1-1*	M/7	*	GMSC514	p = 0.0005	*	Arkansas	
	qRaffinose-1-2	C2/6	97.3–99.3	Satt277	p = 0.0001	*	Arkansas	
	qRaffinose-1-3	G/18	16.2–18.2	Satt570	p = 0.0001	*	Arkansas	
Seed stachyose	*qStachyose-1-1*	O/10	53.4–59.4	Sat_282	p = 0.0001	*	Arkansas	
	qStachyose-1-2	O/10	53.4–59.4	Satt173	p = 0.0001	*	Arkansas	
	qStachyose-1-3	O/10	53.4–59.4	Satt241	p = 0.0001	*	Arkansas	
	qStachyose-1-4	O/10	53.4–59.4	Satt262	p = 0.0001	*	Arkansas	
	qStachyose-1-5	O/10	33.9–35.9	Satt259	p = 0.0001	*	Arkansas	
Seed raffinose	*qSeedRaff-1*	B1/11	108.4–124	Satt453	p = 0.0001	25.20	Virginia	Maupin et al. (2011)
Seed sucrose	*qSuc-1*	A1/5	124.1–126.1	ss245668753	p = 0.0102	*	Arkansas	Zeng et al. (2012)
	qSuc-2	A1/5	153.9–155.9	ss245672731	p = 0.0433	*	Arkansas	
	qSuc-3	K/9	75.1–77.1	ss246796276	p = 0.0238	*	Arkansas	
	qSuc-4	F/13	178.2–180.2	ss248014959	p = 0.096	*	Arkansas	
	qSuc-5	F/13	186.7–188.7	ss248007511	p = 0.0188	*	Arkansas	
	qSuc-6	J/16	276.3–278.2	ss249186914	p = 0.0282	*	Arkansas	
	qSuc-7	J/16	280.8–282.8	ss248953718	p = 0.0319	*	Arkansas	
	qSuc-8	D2/17	300–302	ss249412632	p = 0.0016	*	Arkansas	
	qSuc-9	G/18	34.4–36.4	ss249519978	p = 0.0813	*	Arkansas	
	qSuc-10	L/19	301.1–303.1	ss250098318	p = 0.1015	*	Arkansas	
Seed sucrose	*Seed sucrose 4-1*	K/9	13.7–15.7	Satt242	p = 0.0246	3.00	Arkansas	Zeng et al. (2014)
	Seed sucrose 4-2	D2/17	41.7–43.7	Satt002	p = 0.0415	3.00	Arkansas	
	Seed sucrose 4-3	D2/17	66.7–68.7	Satt669	p = 0.0280	3.00	Arkansas	
	Seed sucrose 4-4	G/18	48.1–50.1	Satt427	p = 0.0259	3.00	Arkansas	
	Seed sucrose 4-5	L/19	44.9–46.9	Satt166	p = 0.0211	4.00	Arkansas	
	Seed sucrose 4-6	L/19	76.5–78.5	Sat_286	p = 0.0139	4.00	Arkansas	

Trait	QTL	LG/Chr	Interval (cM)	Marker	LOD/*P*/*F*	R^2 (%)	Environment	Reference
Seed glucose	*qtlE_glucose*	E/15	68.7–70.7	Satt598	2.29	6.20	Arkansas	Wang et al. (2014b)
	qtlE_glucose	E/15	68.7–70.7	Satt598	2.13	5.70	Arkansas	
Seed fructose	*qtlD2_fructose*	D2/17	35.1–37.1	Satt372	1.91	5.10	Arkansas	
	qtlG_fructose	G/18	34.4–36.4	Satt324	2.74	7.30	Arkansas	
Seed sucrose	*qtlB1_sucrose*	B1/11	92.1–94.4	Satt359	5.19	13.40	Arkansas	
	qtlM_sucrose	M/7	71.8–73.8	Satt680	1.72	4.60	Arkansas	
	qtlI_sucrose	I/20	41.6–43.6	Satt270	1.91	5.20	Arkansas	
Seed raffinose	*qtlB1_raffinose*	B1/11	92.1–94.4	Satt359	3.95	10.40	Arkansas	
	qtlM_raffinose	M/7	15.8–17.8	Satt150	1.96	5.30	Arkansas	
Seed stachyose	*qtlB1_stachyose*	B1/11	92.1–94.4	Satt359	13.57	31.40	Arkansas	
	qtlH_stachyose	H/12	7.6–9.6	Satt635	3.61	10.50	Arkansas	
Seed sucrose	*qSUC001*	N/3	20.80–21.10	ss244980861-ss245114421	3.36	8.00	North Carolina	Akond et al. (2015)
	qSUC002	K/9	0.06–0.07	ss246864598-ss246876159	3.24	8.00	North Carolina	
	qSUC003	E/15	0.40–1.90	ss248604753-ss248616287	3.24	68.00	North Carolina	
Seed raffinose	*qRAF001*	D1a/1	14.70–15.80	ss244562583-ss244554797	26.28	73.00	North Carolina	
	qRAF002	N/3	17.50–17.70	ss245003089-ss245010765	25.64	73.00	North Carolina	
	qRAF003	C2/6	7.10–7.20	ss246088874-ss246091245	20.08	73.00	North Carolina	
	qRAF004	K/9	0.05–0.06	ss246872868-ss246868120	32.36	76.00	North Carolina	
	qRAF005	B2/14	1.90–3.10	ss248293401-ss248275088	26.06	74.00	North Carolina	
	qRAF006	B2/14	10.90–16.7	ss248293401-ss248275088	27.78	74.00	North Carolina	
	qRAF007	J/16	9.90–12.30	ss248983974-ss248977568	28.58	73.00	North Carolina	
Seed stachyose	*qSTA001*	D1a/1	17.00–17.40	ss244552971-ss244555288	3.62	27.00	North Carolina	
	qSTA002	C2/6	1.20–1.90	ss246024749-ss246041195	3.25	8.00	North Carolina	
	qSTA003	H/12	0.60–4.5	ss247690345-ss247704473	4.68	8.00	North Carolina	
	qSTA004	B2/14	20.60–21.60	ss248278947-ss248330608	3.47	28.00	North Carolina	

(continued)

Table 3.8 (Continued)

Trait	QTL	LG/Chr	Interval (cM)	Marker	LOD/*P*/*F*	R^2 (%)	Environment	Reference
Seed stachyose	*sta1*	O/10	32–40.2	Satt173-Sat_282	p < 0.0001	14.00	North Carolina	Zeng et al. (2015)
	sta2	B1/11	34.8–48	SSR_11_1373-453	p < 0.0001	50.00	North Carolina	
Seed sucrose	*qtl_SUC1*	M/7	142,962	AX90425348-AX90424508	6.04	17.18	Korea	Lee et al. (2016)
	qtl_SUC2	A2/8	43,670,249	AX90497857-AX90396735	5.48	15.94	Korea	
	qtl_SUC3	L/19	34,327,636	AX90508115-AX90509468	11.73	36.87	Korea	
Seed oligosaccharide	*qtl_Oli1*	A2/8	41,836,991	AX90412478-A-90446465	5.70	13.38	Korea	
	qtl_Oli2	B1/11	10,922,426	AX90359608-AX90415216	3.29	9.52	Korea	
Seed starch	*qSTR06_1*	C2/6	93.7–95.7	BARC-054471-12090	3.49	7.4	Korea	Dhungana et al. (2017)
	qSTR06_2	C2/6	101.2–103.2	BARC-062515-17881	4.54	9.6	Korea	
	qSTR17_1	D2/17	46.9–48.9	BARC-044417-08703	3.04	6.6	Korea	
	qSTR17_2	D2/17	52.2–54.2	BARC-028485-05923	4.37	9.2	Korea	
	qSTR20_1	I/20	29.6–31.6	BARC-057087-14579	2.82	9.4	Korea	
Sucrose	*qSuc_06*	C2/6	1.4–3.4	bin-6-37707727	3.30	7.40	BREC12	Patil et al. (2018)
	qSuc_08	A2/8	0–0.30	bin-8-9130096-bin-8-9319041	9.70	24.00	BREC12	
	qSuc_06	C2/6	1.4–3.4	bin-6-40104066	7.10	13.00	BREC13	
	qSuc_08	A2/8	0–0.30	bin-8-8561674-bin-8-8602440	8.30	16.00	BREC13	
	qSuc_20	I/20	0–0.95	bin-20-2558940	3.80	6.50	BREC13	
	qSuc_06	C2/6	2.0–4.0	bin-6-18870607	5.40	10.00	BREC14	
	qSuc_08	A2/8	0–0.30	bin-8-7951579-bin-8-8095904	13.00	27.00	CR13	
	qSuc_08	A2/8	0–0.30	bin-8-8561674-bin-8-8602440	12.00	29.00	CR13	

Trait	QTL	LG/Chr	Interval (cM)	Marker	LOD/*P*/*F*	R^2 (%)	Environment	Reference
Sucrose	*qSucrose-1-1*	D1a/1	43,979,136	*	*	10.00	IL, IN, NE	Salari et al. (2020)
	qSucrose-1-2	N/3	192,792	*	*	22.00	IL, IN, NE	
	qSucrose-1-3	N/3	1,061,417	*	*	22.00	IL, IN, NE	
Raffinose	*qRaffinose-1-1*	C2/6	17,204,660	*	*	6.00	IL, IN, NE	
Other compounds								
Seed antioxidants	*qAox19_1*	L/19	37457038	*	*	62.14	China	Li et al. (2016)
Seed phenolics	*qPhen19_1*	L/19	37457038	*	*	64.27	China	
Seed flavonoids	*qFlav18_1*	G/18	52716510	*	*	19.27	China	
	qFlav19_1	L/19	37315018	*	*	31.69	China	

Feng et al. (2005) grew a RIL population derived from a cross between "MFS-591" (high seed sugars) and "Camp" (low seed sugars) lines (*n* = 156) in China, genotyped it with 65 SSR markers, and used single-marker analysis to identify QTL that control seed sucrose, stachyose, glucose, fructose, and raffinose. The study identified one QTL that controls each of seed stachyose, glucose, and sucrose and two QTL that control seed raffinose content. These QTL explained 11–25% of the total variation of seed sugars in this MFS-591 by Camp RIL population (Fig. 2.1; Table 3.8) (Feng et al. 2005).

Kim et al. (2005) grew a RIL population derived from a cross between "Keunolkong" and "Iksan10" (*n* = 115) in Korea and genotyped it with 99 SSR and two morphological markers. The "Keunolkong" X "Iksan10" genetic linkage map covered approximately 1590 cM. Eleven QTL that control seed sucrose contents have been identified and mapped on chromosomes 2, 11, and 19. Eight QTL that control seed oligosaccharide contents have been identified and mapped on chromosomes 2 and 19 (Fig. 2.1; Table 3.8) (Kim et al. 2005).

Kim et al. (2006) grew a RIL population derived from a cross between "Keunolkong" and "Shinpaldalkong" (*n* = 117) in Korea and genotyped it with 108 SSR and 2 morphological markers. The "Keunolkong" X "Shinpaldalkong" genetic linkage map covered approximately 1890 cM. Single-marker analysis (ANOVA) identified 25 markers associated with seed sucrose content and 24 markers associated with seed oligosaccharide content; however, 2 major QTL that control seed sucrose contents have been identified and mapped on chromosomes 12 and 16; and 5 major QTL that control seed oligosaccharide contents have been identified and mapped on chromosomes 6, 12, 16, and 19 (Fig. 2.1; Table 3.8) (Kim et al. 2006).

Jaureguy (2009) grew a RIL population derived from a cross between "R95-1705" (high seed protein content) and "MFL-552" (moderate seed sucrose content) lines (*n* = 98) in Fayetteville, AR, in 2007–2008, genotyped it with 186 SSR markers, and used single-marker analysis (ANOVA) to identify QTL that control seed size, sugars, and protein contents. Two SSRs linked to QTL that control seed sucrose content have been identified and mapped on chromosome 20. These two QTL explained about 59% of the total variation of seed sucrose content. Two SSRs linked to QTL associated with seed protein content have also been identified and mapped on chromosome 20 (Table 3.8) (Jaureguy 2009).

Zeng et al. (2012) grew a RIL population derived from a cross between "MFS-553" (low seed sucrose content) and "PI 243545" (high seed sucrose content) lines (*n* = 220) in Arkansas, genotyped it with 209 SSRs and 2016 SNPs, and used CIM of WinQTL Cartographer to identify QTL that control seed sucrose and stachyose contents. Three QTL that control seed sucrose content have been identified and mapped on chromosomes 5, 9, and 16 (Table 3.7) (Zeng et al. 2012).

Zeng et al. (2014) used a RIL population derived from a cross between "MFS-553" and "PI 243545" (*n* = 220) in Fayetteville, AR, in 2010, and genotyped it with 209 SSR markers. Later on, a subpopulation of RILs derived from the same cross (*n* = 94) was genotyped with 2016 single-nucleotide polymorphism (SNP) markers. Using both composite interval mapping (CIM) and multiple-interval mapping (MIM) methods, six QTL that control seed sucrose content have been identified and mapped on chromosomes 9, 17, 18, and 19 (Fig. 2.1; Table 3.8) (Zeng et al. 2014).

Wang et al. (2014b) aimed at identifying QTL for seed soluble sugar (sucrose, stachyose, raffinose, glucose, and fructose) contents in soybean. They grew an F_2 population derived from a cross between "V97-3000" and "V99-5089" lines (n = 170) in a field in Fayetteville, AR, in 2009, genotyped with 106 SSR markers, and used single-marker analysis, IM, CIM, and MIM of QTL Cartographer for QTL data analysis. Two QTL that control seed glucose content have been identified and mapped on chromosome 15; two QTL that control seed fructose content have been identified and mapped on chromosomes 17 and 18; three QTL that control seed sucrose content have been identified and mapped on chromosomes 7, 11, and 20; two QTL that control seed raffinose content have been identified and mapped on chromosomes 7 and 11; and two QTL that control seed stachyose content have been identified and mapped on chromosomes 11 and 12 (Fig. 2.1; Table 3.8) (Wang et al. 2014b).

In Akond et al. (2015), we grew the RIL population derived from the cross of "MD 96-5722" and "Spencer" (n = 92) in Fayetteville, NC, in 2012, genotyped it with 5376 SNP markers, and used CIM of WinQTL Cartographer for QTL data analysis. Three QTL that control seed sucrose content have been identified and mapped on chromosomes 3, 9, and 15; seven QTL that control seed raffinose content have been identified and mapped on chromosomes 1, 3, 6, 9, 14, and 16; and four QTL that control seed stachyose content have been identified and mapped on chromosomes 1, 6, 12, and 14 (Fig. 2.1; Table 3.8) (Akond et al. 2015).

Zeng et al. (2015) grew a RIL population derived from a cross between "Osage" and "V99-5089" lines (n = 129) in Fayetteville, AR, Costa Rica, and Argentina in 2010–2011 and genotyped it with 33 SSR and 37 SNP markers and 20 SSR and 19 SNP markers distributed on chromosomes 10 and 11, respectively. They used one-way ANOVA using PROC GLM of SAS program for QTL data analysis. Two QTL that control seed stachyose content have been identified and mapped on chromosomes 10 and 11 across the three environments in 2 years (Fig. 2.1; Table 3.8) (Zeng et al. 2015).

Lee et al. (2016) grew a RIL population derived from a cross between "Keunolkong" and "Iksan 10" lines (n = 116) in Korea, genotyped it with the Axiom® 180 K SoyaSNP array, used single-factor ANOVA and inclusive composite interval mapping (ICIM) for QTL data analysis. Three QTL that control seed sucrose content have been identified and mapped on chromosomes 7, 8$ and 19, and two QTL that control seed oligosaccharide content have been identified and mapped on chromosomes 8 and 11 (Table 3.8). Within these QTL regions, 15, 10, 18, 42, and 190 candidate genes have been identified, respectively. The candidate genes include *Glyma.07g002300* that encodes for a V-ATPase subunit involved in glycolysis (Hanitzsch et al. 2007), *Glyma.08g302900* that encodes for a BAM receptor kinase-like protein involved in polysaccharide biosynthesis (DeYoung et al. 2006) and genes involved in glycolysis and gluconeogenesis (Table 3.9) (Lee et al. 2016).

Dhungana et al. (2017) grew a RIL population derived from a cross between "Williams 82" and "PI 366121" lines (n = 169) in Korea in 2013–2015, genotyped it with 501 SNP and 2 SSR markers, used a genetic linkage map of 2555.78 cM, and used CIM of WinQTL Cartographer for QTL data analysis of seed starch content. Three QTL that control seed starch content have been identified and mapped on chromosomes 6, 17, and 20 (Fig. 2.1; Table 3.8) (Dhungana et al. 2017).

Patil et al. (2018) grew an interspecific RIL population derived from a cross between “Williams 82” (*Glycine max*) and “PI 483460B” (*Glycine soja*) ($n = 188$) in Portageville, MO, in 2012–2014, genotyped it with 6000+ SNP markers, used a genetic linkage map of 2925 cM, and used CIM of WinQTL Cartographer for QTL data analysis of seed protein, oil, and sucrose contents. Twelve QTL that control seed protein content have been identified and mapped on chromosomes 6, 8, 13, 19, and 20; 16 QTL that control seed oil content have been identified and mapped on chromosomes 2, 8, 9, 15, 17, and 20; and 8 QTL that control seed sucrose content have been identified and mapped on chromosomes 6, 8, and 20 (Fig. 2.1; Table 3.8) (Patil et al. 2018).

Several other studies investigated QTL analysis for other compounds in soybean seeds such as phytic acid or phytate (Walker et al. 2006; Gao et al. 2008; Scaboo et al. 2009), carbon and oxygen isotope ratios (Kaler et al. 2017), antioxidants, phenolics, flavonoids (Li et al. 2016), etc. For example, Walker et al. (2006) identified and mapped two QTL that control seed phytate content on chromosomes 3 (close to the SSR marker Satt237) and 19 (close to Satt527). The authors think that these loci correspond to the previously described *pha1* and *pha2* loci (Oltmans et al. 2004). Gao et al. (2008) grew a RIL population derived from a cross between “CX1834” and “V99-3337” lines ($n = 208$) in Blacksburg, VA, in 2003 and aimed at validating seed low-phytate (LP) QTL. The two QTL for seed LP content that were previously identified on chromosomes 3 and 19 (Walker et al. 2006) have been validated in this population. The two LP QTL have been mapped to chromosomes 3 (Satt387–Satt257) and 19 (Satt156–Satt099) (Fig. 2.1; Table 3.8) (Gao et al. 2008). Scaboo et al. (2009) grew two RIL populations in Knoxville, TN, in 2004. The first RIL population was derived from a cross between “5601T” and “CX1834-1-2” lines ($n = 187$), and the second RIL population was derived from a cross between “S97-1688” and “CX1834-1-2” lines ($n = 186$). The aim was also to validate the seed low-phytate (LP) QTL identified previously (Walker et al. 2006; Gao et al. 2008) using different environment and genetic backgrounds. The study validated the two seed LP QTL previously identified on the same chromosomes 3 (Satt237) and 19 (Satt561) (Fig. 2.1; Table 3.8) (Scaboo et al. 2009).

Li et al. (2016) grew a RIL population derived from a cross between the wild-type line “W05” and cultivated line “C08” ($n = 96$) in the greenhouse and fields from 2010 to 2012 in China, genotyped it with SNP markers, and investigated QTL for salt tolerance (Qi et al. 2014), antioxidants, phenolics, and flavonoids (Li et al. 2016) using IM and CIM of WinQTL Cartographer. One QTL that controls both seed antioxidants and total phenolics contents has been identified and mapped on chromosome 19, and two QTL that control seed flavonoid contents have been identified and mapped on chromosomes 18 and 19. The QTL cluster on chromosome 19 controls seed antioxidants, total phenolics, and flavonoids with high phenotypic variations and peak LOD scores of 3.97 and 62.14%, 4.66 and 64.27%, and 3.81 and

Table 3.9 Candidate genes identified within the QTL regions that control seed sugars (sucrose, glucose, raffinose, stachyose, and oligosaccharide) and phosphorus contents from 2013 to 2020

Trait	Candidate gene	LG/Chr	Annotation/pathway	References
Seed phosphorus	Glyma07g34870	M/7	Inorganic phosphate transporter	King et al. (2013)
	Glyma07g34030	M/7	Phosphatidylinositol-4-phosphate 5-kinase (PIP5K)	
	Glyma07g32190	M/7	Solute carrier of glucose-6-phosphate/phosphate	
	Glyma07g34930	M/7	ZIP zinc transporter	
	Glyma07g33480	M/7	Leucine-rich repeat N-terminal domain	
	Glyma07g33510	M/7	AP2 domain	
	Glyma07g33560	M/7	Cytochrome P450	
	Glyma07g33580	M/7	Glycosyltransferase family 10 (fucosyltransferase)	
	Glyma07g33640	M/7	Thioredoxin	
	Glyma07g33770	M/7	Zinc finger, C3HC4 type (RING finger)	
	Glyma12g16410	H/12	ABC transporter family member	
	Glyma12g30100	H/12	ABC transporter family member	
	Glyma12g30070	H/12	ABC transporter family member	
	Glyma12g23920	H/12	Triose-phosphate transporter	
	Glyma12g22680	H/12	Inositol monophosphatase family	
	Glyma12g29790	H/12	Triose-phosphate transporter	
	Glyma12g17720	H/12	SRF-type transcription factor (DNA-binding domain)	
	Glyma12g18230	H/12	Peptide N-acetyl-beta-D-glucosaminyl asparaginase amidase A	
	Glyma12g18250	H/12	Poly (ADP-ribose) polymerase and DNA-ligase Zn-finger region	
	Glyma12g19520	H/12	Lactate/malate dehydrogenase, NAD-binding domain	
	Glyma17g03830	D2/17	Protein phosphatase 2C	
	Glyma17g03950	D2/17	WRKY DNA-binding domain	
	Glyma17g03960	D2/17	RNA recognition motif	
	Glyma17g03990	D2/17	Galactokinase galactose-binding signature	

(continued)

Table 3.9 (Continued)

Trait	Candidate gene	LG/Chr	Annotation/pathway	References
	Glyma17g04020	D2/17	Plant invertase/pectin methylesterase inhibitor	
	Glyma17g04030	D2/17	Peroxidase	
	Glyma17g04070	D2/17	Cas1p-like protein	
	Glyma17g04080	D2/17	Rab5-interacting protein (Rab5ip)	
	Glyma17g04110	D2/17	Serine carboxypeptidase	
	Glyma17g04140	D2/17	Transcription initiation factor IID, 31kD subunit	
	Glyma18g41340	G/18	Glutathione S-transferase, N-terminal domain	
	Glyma18g41450	G/18	NB-ARC domain	
	Glyma18g41520	G/18	Myb-like DNA-binding domain	
	Glyma18g41530	G/18	Ctr copper transporter family	
	Glyma18g41540	G/18	Trypsin and protease inhibitor	
	Glyma18g41590	G/18	Protease inhibitor/seed storage/LTP family	
	Glyma18g41600	G/18	Leucine-rich repeat	
	Glyma18g41650	G/18	Hsp20/alpha crystallin family	
	Glyma18g41860	G/18	Multicopper oxidase	
	Glyma18g41940	G/18	Fructose-1-6-bisphosphatase	
	Glyma20g34610	I/20	Pi transporter	
	Glyma20g34620	I/20	P/H+ symporter (PHS) transporter	
	Glyma20g28400	I/20	Aminotransferase class IV	
	Glyma20g28420	I/20	Imidazoleglycerol-phosphate dehydratase	
	Glyma20g28460	I/20	Cupin	
	Glyma20g28490	I/20	Protease inhibitor/seed storage/LTP family	
	Glyma20g28700	I/20	Carbamoyl-phosphate synthetase large chain	
	Glyma20g28720	I/20	Homocysteine S-methyltransferase	
	Glyma20g28740	I/20	Tesmin/TSO1-like CXC domain	
	Glyma20g28780	I/20	Ribosomal S3Ae family	

Trait	Candidate gene	LG/Chr	Annotation/pathway	References
Seed sucrose	Glyma07g002300	M/7	V-ATPase subunits	Lee et al. (2016)
	Glyma.08G302900	A2/8	BAM receptor kinase-like proteins	
Seed sucrose	Glyma.06G236800	C2/6	Serine-type endopeptidase activity	Gong et al. (2018)
	Glyma.08G237200	A2/8	Protein SER/THR kinase	
	Glyma.09G066600	K/9	Protein SER/THR kinase	
	Glyma.09G099200	K/9	Protein SER/THR kinase	
	GO:0004674	*	Protein SER/THR kinase	
	GO:0016597	*	Amino acid binding	
	Glyma.16G008900	J/16	Serine-type endopeptidase	
	Glyma.17G157500	D2/17	Aspartic-type endopeptidase	
	Glyma.17G158700	D2/17	Lysine-tRNA ligase	
Seed sucrose	Glyma.01g127600	D1a/1	Protein phosphatase 2C family protein	Salari et al. (2020)
	Glyma.03g019300	N/3	MADS box protein domain	
	Glyma.03g064700	N/3	Phosphatidylinositol monophosphate 5 kinase	
Seed raffinose	Glyma.06g194200	C2/6	Gibberellin-regulated family protein	

31.69%, respectively (Table 3.8). Based on Williams 82 reference genome, 14 genes have been identified within this cluster including 4 putative MATE transporter genes (*GmMATE1-4*) (Li et al. 2016).

Kaler et al. (2017) grew a population of various genotypes ($n = 373$) in two fields in Columbia, MO, and Stuttgart, AR, in 2009 and 2010, genotyped it with 31,260 SNPs, and used GWAS to identify SNPs associated with $^{\delta 13}C$ and $^{\delta 18}O$ contents. Forty six SNPs associated with $^{\delta 13}C$ have been identified and mapped on chromosomes 1, 2, 3, 4, 5, 6, 7, 9, 10, 11, 13, 14, 15, 16, 17, 18, 19, and 20; and 21 SNPs associated with $^{\delta 18}O$ content have been identified and mapped on chromosomes 1, 2, 3, 4, 5, 6, 7, 9, 15, 17, 18, 19, and 20 (Fig. 2.1; Table 3.8) (Kaler et al. 2017).

Salari et al. (2020) grew a SoyNAM (Nested Association Mapping) RIL population derived from a cross between "IA3023" and "LD02-4485" lines ($n = 140$) in four environments in Nebraska, Illinois, and Indiana in 2012–2013 and genotyped it with 3038 SNPs to identify QTL that control seed sucrose, stachyose, and raffinose contents. They used CIM of WinQTL Cartographer for QTL data analysis. Three QTL that control seed sucrose content have been identified and mapped on chromosomes 1 and 3. Both QTL explained approximately 32% of the total variation of sucrose. One QTL that controls seed raffinose content has been identified and mapped on chromosome 6 which explained only 7% of the total variation raffinose. All the four QTL are novel (Table 3.8). Within these QTL regions, authors identified the candidate genes *Glyma.01g127600* (*qSucrose-1-1*, Chr 1) that encode for a protein phosphatase 2C family protein, *Glyma.03g019300* (*qSucrose-1-2*, Chr 3) that encodes for a MADS box protein, *Glyma.03g064700* (*qSucrose-1-3*, Chr 3) that encodes for a phosphatidylinositol monophosphate 5 kinase, and *Glyma.06g194200* (*qRaffinose-1-1*, Chu 6) that encodes for a gibberellin-regulated family protein (Salari et al. 2020).

3.7 Conclusion

Since the first article on genetic mapping (Lander and Botstein 1989), thousands of QTL for different complex polygenic traits have been identified and mapped in humans, plants, and other species (Altshuler et al. 2008; Donnelly 2008; Frazer et al. 2009; Mackay et al. 2009; Grant et al. 2010) including hundreds of QTL for seed amino acids, sugars, macronutrients, and micronutrients that have been identified and mapped on the soybean genome, very little is known about genes within these QTL positions (Altshuler et al. 2008; Donnelly 2008; Frazer et al. 2009; Mackay et al. 2009), and decades of research are needed before candidate genes are discovered.

In soybean, very few studies identified candidate genes for these seed composition traits (Duressa et al. 2010; Lee et al. 2016; Zhang et al. 2018; Gong et al. 2018). Duressa et al. (2010) performed a root transcriptomics analysis of PI416937 (Al tolerant line) treated with 10μM of Al solution (group 1) and untreated control plants (group 2) using microarray analysis. Their results showed that 38 genes have been differentially expressed between the 2 groups of plants after only 2 h of treating plants in group 1 with 10μM of Al solution; only 1 gene was differentially expressed between the 2 groups of plants after 12 h and 72 h of treatment; and 542 genes have been differentially expressed between the 2 groups of plants after 48 h of treatment (Duressa et al. 2010). These genes regulated by Al include genes involved in different metabolic pathways (15%), transcription factors (15%), genes involved in different types of molecular transport (5%), genes involved in stress responses (7%), genes involved in signaling responses (6%), structural genes (2%), and unclassified genes (50%). Among the most promising genes are a cell wall-modifying enzyme, Cys2His2 zinc finger protein (Iuchi et al. 2007), phytosulfokines growth factor, and *ADR6* transcription activator (Iuchi et al. 2007) that may play major roles in Al tolerance (Duressa et al. 2010). Lee et al. (2016) grew a RIL population derived from a cross between "Keunolkong" and "Iksan 10" lines ($n = 116$) in Korea, genotyped it with the Axiom® 180 K SoyaSNP array, and used single-factor ANOVA and inclusive composite interval mapping (ICIM) for QTL data analysis. Three QTL that control seed sucrose content have been identified and mapped on chromosomes 7, 8, and 19, and two QTL that control seed oligosaccharide content have been identified and mapped on chromosomes 8 and 11 (Table 3.8). Within these QTL regions, 15, 10, 18, 42, and 190 candidate genes have been identified, respectively. The candidate genes include Glyma.07G002300 involved in glycolysis, Glyma.08G302900 involved in polysaccharide biosynthesis, and genes involved in glycolysis and gluconeogenesis (Lee et al. 2016). Li et al. (2016) a major QTL that controls both seed antioxidants, total phenolics, and flavonoids contents on chromosome 19. The QTL cluster controls seed antioxidants, total phenolics, and flavonoids with high phenotypic variations and peak LOD scores of 3.97 and 62.14%, 4.66 and 64.27%, and 3.81 and 31.69%, respectively. Based on Williams 82 reference genome, 14 genes have been identified within this cluster including four putative MATE transporter genes (*GmMATE1-4*). According to a phylogenetic analysis, the authors showed that these genes are not related to known MATE transporters of flavonoids such as S1MTP77 or AtTT12 (Li et al. 2002; Li et al. 2016; Chen et al. 2020).

Zhang et al. (2018a) used a genome-wide association study (GWAS) on 313 soybean accessions and studied seed protein, oil, fatty acid, and amino acid contents. The study identified 3 QTL controlling protein content, 25 QTL controlling oil content, 18 QTL controlling fatty acids contents, and 39–53 QTL for amino acid contents (Fig. 2.1). Several candidate genes have been identified on the same regions or very close to these identified QTL. Among these candidate genes are *GmSAT1* encoding the *Glycine max* symbiotic ammonium transporter 1 involved in N_2 fixation identified on chromosome 15: *Glyma.15g049900*, *Glyma.15g049700*, and *Glyma.15g049800* encoding the nodulin MtN21 protein family (Zhang et al. 2018a).

Gong et al. (2018) conducted a meta-analysis and overview analysis methods (Chardon et al. 2004; Qi et al. 2011; Thilakarathna and Raizada 2017) on 140 QTL that control seed amino acids contents in soybean reported in SoyBase (Grant et al. 2010. http://www.soybase.org) using the software BioMercator 2.1. The results showed 33 consensus QTL with confidence intervals (CIs) of 0.07 Mb to 19.85 Mb.

The overview method analysis identified 57 "real" QTL for amino acids that have been mapped in all soybean chromosomes except chromosomes 10 and 12. Using both methods (meta-analysis and overview analysis), 27 "real" QTL have been identified and mapped on 16 chromosomes among which 2 major "real" QTL have been mapped on chromosomes 3 and 9. Twenty candidate genes have been identified with these 27 "real" QTL regions among which 4 have been published earlier [von Groll et al. 2002 (*Glyma.03G167600*), Catoni et al. 2003 (*Glyma.04G143500*), Shah et al. 2001 (*Glyma.05G083100*), and Lewis et al. 2010 (*Glyma.05G083100*)] which left 16 new candidate genes identified by the study. Among these candidate genes are genes that encode serine-type endopeptidase activity (*Glyma.06G236800*), protein SER/THR kinases (*Glyma.08G237200*, *Glyma.09G066600*, *GO:0004674*, and *Glyma.09G099200*), amino acid binding (GO:0016597), serine-type endopeptidase (*Glyma.16G008900*), aspartic-type endopeptidase (Glyma.17G157500), lysine-tRNA ligase (*Glyma.17G158700*), and several others (Table 3.9) (Gong et al. 2018).

Salari et al. (2020) identified three QTL that control seed sucrose content on chromosomes 1 and 3 and one QTL that controls seed raffinose content chromosome 6, and, within these QTL regions, the candidate genes *Glyma.01g127600* that encodes for a protein phosphatase 2C family protein, *Glyma.03g019300* that encodes for a MADS box protein, *Glyma.03g064700* that encodes for a phosphatidylinositol monophosphate 5 kinase, and *Glyma.06g194200* that encodes for a gibberellin-regulated family protein have been identified (Table 3.9) (Salari et al. 2020).

References

Abdul-Baki, A.A., and J.D. Anderson. 1973. Vigor determination in soybean seed by multiple criteria. *Crop Science* 13 (6): 630–633.

Akond, A., S. Liu, L. Schoener, J.A. Anderson, S.K. Kantartzi, K. Meksem, Q. Song, D. Wang, Z. Wen, D.A. Lightfoot, and M.A. Kassem. 2013. A SNP-Based Genetic Linkage Map of Soybean Using the SoySNP6K Illumina Infinium BeadChip Genotyping Array. *Plant Genetics, Genomics, and Biotechnology 1*(3): 80–89. https://doi.org/10.5147/jpgs.2013.0090.

Akond, M., S. Liu, S.K. Kantartzi, K. Meksem, N. Bellaloui, D.A. Lightfoot, J. Yuan, D. Wang, J. Anderson, D.A. Lightfoot, and M.A. Kassem. 2015. A SNP genetic linkage map based on the 'Hamilton' by 'Spencer' recombinant inbred line (RIL) population of soybean [Glycine max (L.) Merr.] identified QTL for seed isoflavone contents. *Plant Breeding 134*(5): 580–588. https://doi.org/10.1111/pbr.12298.

Akond, M., J. Yuan, S. Liu, S.K. Kantartzi, K. Meksem, N. Bellaloui, D.A. Lightfoot, and M.A. Kassem. 2018. Detection of QTL underlying seed quality components in soybean [*Glycine max* (L.) Merr.]. *Canadian Journal of Plant Science* 98: 881–888. https://doi.org/10.1139/cjps-2017-0204.

Altenbach, S.B., K.W. Pearson, F.W. Leung, and S.S.M. Sun. 1987. Cloning and sequence analysis of a cDNA encoding a Brazil nut protein exceptionally rich in methionine. *Plant Mol. Biol.* 8: 239–250.

Altshuler, D., M.J. Daly, and E.S. Lander. 2008. Genetic mapping in human disease. *Science* 322: 881–888.

Ambrocio, Z., S.K. Kantartzi, K. Meksem, and M.A. Kasseml. 2016. Influence of Drought Stress on Several Root Traits and their Correlation with Seed Protein and Oil Contents in Soybean. *Atlas Journal of Biology* 2016: 267–273. https://doi.org/10.5147/ajb.2016.0140.

Amir, R., and Y. Hacham. 2008. Methionine metabolism in plants. In A missing link between soils, crops, and nutrition. Joseph Jez (ed.). pp. 235–249. Published by American Society of Agronomy, Crops Science Society of America, and Soil Science Society of America. ISBN 9780891181866.

Amir, R., T. Han, and F. Ma. 2012. Bioengineering approaches to improve the nutritional values of seeds by increasing their methionine content. *Molecular Breeding* 29: 915–924.

Arendt, E.K., and E. Zannini. 2013. Cereal grains for the food and beverage industries. *Food Science Technology and Nutrition* 409–438.

Ashraf, M., and M.R. Foolad. 2007. Roles of glycine betaine and proline in improving plant abiotic stress resistance. *Env. and Exp. Botany* 59: 206–216. https://doi.org/10.1016/j.envexpbot.2005.12.006.

Ashraf, M., and P.J.C. Harris. 2004. Potential biochemical indicators of salinity tolerance in plants. *Plant Sci* 166: 3–16. http://dx.doi.org/10.1016/j.plantsci.2003.10.024.

Aviles-Gaxiola, S., C. Chuck-Hernandez, and S.O. Saldivar. 2017. Inactivation methods of trypsin inhibitor in legumes: A review. *Journal of Food Science* 83: 17–29.

Ball, R.A., L.C. Purcell, and E.D. Vories. 2000. Optimizing soybean plant population for a short-season production system in the Southern USA. *Crop Sci.* 40: 757–764. https://doi.org/10.2135/cropsci2000.403757x.

Bellaloui, N., and A.M. Gillen. 2010. Soybean seed protein, oil, fatty acids, N, and S partitioning as affected by node position and cultivar differences. *Agricultural Sciences* 1: 110–118.

Bellaloui, N., and A. Mengistu. 2008. Seed composition is influenced by irrigation regimes and cultivar differences in soybean. *Irrigation Science* 26: 261–268. https://doi.org/10.1007/s00271-007-0091-y.

Bellaloui, N., R.M. Zablotowicz, K.N. Reddy, and C.A. Abel. 2008. Nitrogen metabolism and seed composition as influenced by glyphosate application in glyphosate-resistant soybean. *Journal of Agricultural and Food Chemistry* 56: 2765–2772. https://doi.org/10.1021/jf703615m.

Bellaloui, N., H.K. Abbas, A.M. Gillen, and C.A. Abel. 2009a. Effect of glyphosate-boron application on seed composition and nitrogen metabolism in glyphosate-resistant soybean. *Journal of Agricultural and Food Chemistry* 57: 9050–9056. https://doi.org/10.1021/jf901801z.

Bellaloui, N., J.E. Hanks, D.K. Fisher, and A. Mengistu. 2009b. Soybean seed composition is influenced by within-field variability in soil nutrients. *Crop Management* 8 (1): 1–12. https://doi.org/10.1094/CM-2009-1203-01-RS.

Bellaloui N., K.N. Reddy, R.M. Zablotowicz, H.K. Abbas, and C.A. Abel. 2009c. Effects of glyphosate application on seed iron and root ferric (III) reductase in soybean cultivars. *Agricultural and Food Chemistry* 57 (20): 9569–9574. https://doi.org/10.1021/jf902175y.

Bellaloui, N., J.R. Smith, J.D. Ray, and A.M. Gillen. 2009d. Effect of maturity on seed composition in the early soybean production system as measured on near-isogenic soybean lines. *Crop Science* 49: 608–620. https://doi.org/10.2135/cropsci2008.04.0192.

Bellaloui, N., H.A. Bruns, A.M. Gillen, H.K. Abbas, R.M. Zablotowicz, A. Mengistu, and R.L. Paris. 2010a. Soybean seed protein, oil, fatty acids, and mineral composition as influenced by soybean-corn rotation. *Agricultural Sciences* 1 (3): 102–109.

Bellaloui, N., J.R. Smith, A.M. Gillen, and J.D. Ray. 2010b. Effect of maturity on seed sugars as measured on near-isogenic soybean (*Glycine max*) lines. *Crop Science* 50: 1978–1987. https://doi.org/10.2135/cropsci2009.10.0596.

Bellaloui, N., K.N. Reddy, A.M. Gillen, and C.A. Abel. 2010c. Nitrogen metabolism and seed composition as influenced by foliar B application in soybean. *Plant and Soil* 336:143–155. https://doi.org/10.1007/s11104-010-0455-6.

Bellaloui, N., J.R. Smith, A.M. Gillen, and J.D. Ray. 2011. Effects of maturity, genotypic background, and temperature on seed mineral composition in near-isogenic soybean lines in the early soybean production system. *Crop Science* 51: 1161–1171. https://doi.org/10.2135/cropsci2010.04.0187.

Bellaloui, N., A. Mengistu, and L.H.S. Zobiole. 2012a. Phomopsis seed infection effects on soybean seed phenol, lignin, and isoflavones in maturity group V genotypes differing in phomopsis resistance. *Journal of Crop Improvement* 26: 693–710. https://doi.org/10.1080/15427528.2012.671236.

Bellaloui, N., A. Mengistu, D.K. Fisher, and C.A. Abel. 2012b. Soybean seed composition constituents as affected by drought and Phomopsis in phomopsis susceptible and resistant genotypes. *Journal of Crop Improvement* 26: 428–453. https://doi.org/10.1080/15427528.2011.651774.

Bellaloui, N., J.R. Smith, A.M. Gillen, D.K. Fisher, and A. Mengistu. 2012c. Effect of shade on seed protein, oil, fatty acids, and minerals in soybean lines varying in seed germinability in the Early Soybean Production System. *American Journal of Plant Sciences* 3: 84–95. https://doi.org/10.4236/ajps.2012.31008.

Bellaloui, N., A. Mengistu, E.R. Walker, and L.D. Young. 2014a. Soybean seed composition as affected by seeding rates and row spacing. *Crop Science* 54: 1782–1795.

Bellaloui, N., A. Mengistu, M.A. Kassem, C.A. Abel, and L.H.S. Zobiole. 2014b. Role of boron nutrient in nodule growth and nitrogen fixation in soybean genotypes under water stress conditions. In *Advances in biology and ecology of nitrogen fixation*, ed. T. Ohyama. IntechOpen. https://doi.org/10.5772/56994.

Bellaloui, N., L. Khandaker, M. Akond, S.K. Kantartzi, K. Meksem, A. Mengistu, and M.A. Kassem. 2015a. Identification of QTLs underlying seed macronutrients accumulation in 'MD96-5722' by 'Spencer' recombinant inbred lines of soybean. *Atlas Journal of Biology* 3 (2): 224–235. https://doi.org/10.5147/ajb.v3i2.34.

Bellaloui, N., L. Khandaker, M. Akond, S.K. Kantartzi, K. Meksem, A. Mengistu, D.A. Lightfoot, and M.A. Kassem. 2015b. Identification of QTL underlying seed micronutrients accumulation in 'MD 96-5722' by 'Spencer' recombinant inbred lines of soybean. *Atlas Journal of Plant Biology* 1 (3): 39–49. https://doi.org/10.5147/ajpb.v1i3.112.

Bellaloui, N., Y. Hu, A. Mengistu, H.K. Abbas, M.A. Kassem, and M. Tigabu. 2016. Elevated atmospheric carbon dioxide and temperature affect seed composition, mineral nutrition, and ^{15}N and ^{13}C dynamics in soybean genotypes under controlled environments. *Atlas Journal of Plant Biology* 2016: 56–65. https://doi.org/10.5147/ajpb.2016.0157.

Bellaloui, N., A. Mengistu, H.K. Abbas, and M.A. Kassem. 2017. Effects of drought and elevated atmospheric carbon dioxide on seed nutrition and 15N and 13C natural abundance isotopes in soybean under controlled environments. In *Soybean – The basis of yield, biomass and productivity*, ed. M. Kasai. IntechOpen. https://doi.org/10.5772/67511.

Bellaloui, N., H.A. Bruns, H.K. Abbas, D.K. Fisher, and A. Mengistu. 2020a. Effects of harvest-aids on seed nutrition in soybean under midsouth USA conditions. *Plants* 9: 1007. https://doi.org/10.3390/plants9081007.

Bellaloui, N., A.M. McClure, A. Mengistu, and H.K. Abbas. 2020b. The influence of agricultural practices, the environment, and cultivar differences on soybean seed protein, oil, sugars, and amino acids. *Plants* 9: 378. https://doi.org/10.3390/plants9030378.

Benitez, E.R., M. Hajika, T. Yamada, K. Takahashi, N. Oki, N. Yamada, T. Nakamura, and K. Kanamaru. 2010. A major QTL controlling seed cadmium accumulation in soybean. *Crop Science* 50: 1728–1734. https://doi.org/10.2135/cropsci2009.11.0664.

Berk, Z. 1992. Technology of Production of Edible Flours and Protein Products from Soybeans. *FAO Agricultural Services Bulletin* No. 97, Rome, 178.

Bianchi-Hall, C.M., T.E. Carter, M.A. Baileyd, M.A.R. Miane, T.W. Ruftya, D.A. Ashleye, H.R. Boermae, C. Arellanoc, R.S. Husseye, and W.A. Parrotte. 2000. Aluminum Tolerance Associated with Quantitative Trait Loci Derived from Soybean PI 416937 in Hydroponics. *Crop Science 40*(2): 538–545. http://doi.org/10.2135/cropsci2000.402538x.

Blackman, S.A., R.L. Obendorf, and A.C. Leopold. 1992. Maturation proteins and sugars in desiccation tolerance of developing soybean seeds. *Plant Physiology* 100: 225–230.

Bohnert, H.J., and R.G. Jensen. 1996. Strategies for engineering water-stress tolerance in plants. *Trends in Biotechnology* 14: 89–97; http://dx.doi.org/10.1016/0167-7799(96)80929-2.

Bolanos, L., E. Esteban, C. de Lorenzo, M. Fernández-Pascual, M.R. deFelipe, A. Garate, and I. Bonilla. 1994. Essentiality of boron for symbiotic dinitrogen fixation in pea (Pisum sativum) rhizobium nodules. *Plant Physiology* 104: 85–90.

Bolanos, L., N.J. Brewin, and I. Bonilla. 1996. Effects of boron on rhizobium-legume cell—surface interactions and node development. *Plant Physiology* 110: 1249–1256.

Bolon, Y.T., B. Joseph, S.B. Cannon, M.A. Graham, B.W. Diers, A.D. Farmer, G.D. May, G.J. Muehlbauer, J.E. Specht, Z.J. Tu, N. Weeks, W.W. Xu, R.C. Shoemaker, and C.P. Vance. 2010. Complementary genetic and genomic approaches help characterize the linkage group I seed protein QTL in soybean. *BMC Plant Biology* 10: 41. https://doi.org/10.1186/1471-2229-10-41.

Borisjuk, L., H. Rolletschek, R. Radchuk, W. Weschke, U. Wobus, and H. Weber. 2004. Seed development and differentiation: a role for metabolic regulation. *Plant Biology* 6 (4): 375–386. https://doi.org/10.1055/s-2004-817908.

Brown, P.H., N. Bellaloui, M.A. Wimmer, E.S. Bassil, J. Ruiz, H. Hu, H. Pfeffer, F. Dannel, and V. Romheld. 2002. Boron in plant biology. *Plant Biology* 4: 205–223.

Brummer, E.C., G.L. Graef, J. Orf, J.R. Wilcox and R.C. Shoemaker. 1997. Mapping QTL for seed protein and oil content in eight soybean populations. *Crop Science* 37 (2): 370–378.

Cabarello, B. 2002. Global patterns of child health: the role of nutrition. *Annals of Nutrition and Metabolism* 46: 3–7.

Carter, T.E., P.I. DeSouza, and L.C. Purcell. 1999. Recent advances in breeding for drought and aluminum resistance in soybean p. 106–125. In H. Kauffman (ed.) Proceedings of the World Soybean Research Conference, VI, Chicago. 4–7 Aug. 1999. National Soybean Research Lab., Urbana, IL.

Castillo, E.M., B.O. de Lumen, P.S. Reyes, and H.Z. de Lumen. 1990. Raffinose synthase and galactinol synthase in developing seeds and leaves of legumes. *Journal of Agricultural and Food Chemistry* 38: 351–355.

Catoni, E., M. Desimone, M. Hilpert, D. Wipf, R. Kunze, A. Schneider, U. Flügge, K. Schumacher, and W.B. Frommer. 2003. Expression pattern of a nuclear encoded mitochondrial arginine ornithine translocator gene from Arabidopsis. *BMC Plant Biology* 3: 1. https://doi.org/10.1186/1471-2229-3-1.

Chardon, F., B. Virlon, L. Moreau, M. Falque, J. Joets, L. Decousset, A. Murigneux, and A. Charcosset. 2004. Genetic architecture of flowering time in maize as inferred from quantitative trait loci meta-analysis and synteny conservation with the rice genome. *Genetics* 168: 2169–2185.

Chen, G., H. Liang, Q. Zhao, A.M. Wu, and B. Wang. 2020. Exploiting MATE efflux proteins to improve flavonoid accumulation in Camellia sinensis in silico. *International Journal of Biological Macromolecules* 143 (15): 732–743. https://doi.org/10.1016/j.ijbiomac.2019.10.028.

Chrispeels, M.J., and N.V. Raikhelb. 1991. Lectins, lectin genes, and their role in plant defense. *The Plant Cell* 3: 1–9.

Cicek, M. 2001. Genetic marker analysis of three major carbohydrates in soybean seeds. PhD Dissertation, Virginia Polytechnic Institute and State University, Blacksburg, VA, USA.

Cooper, A.J. 1983. Biochemistry of sulfur-containing amino acids. *Annual Review of Biochemistry* 52: 187–222.

Cu, J., V. Vu, L.H. Allen, and G. Bowes. 1987. Drought stress and elevated CO2 effects on soybean ribulose bisphosphate carboxylase activity and canopy photosynthetic rates. *Plant Physiology* 83: 573–578.

Dell, B., L. Huang, and R.W. Bell. 2002. Boron in plant reproduction. In *Boron in plant and animal nutrition*, eds. Golbrach, H.E., B. Rerkasem, M.A. Wimmer, P.H. Brown, M. Thellier, and R.W. Bell, 103–117. New York: Kluwer Academic/Pledum.

DeYoung, B.J., K.L. Bickle, K.J. Schrage, P. Muskett, K. Patel, and S.E. Clark. 2006. The CLAVATA1-related BAM1, BAM2 and BAM3 receptor kinase-like proteins are required for meristem function in Arabidopsis. *The Plant Journal* 45 (1): 1–16. https://doi.org/10.1111/j.1365-313X.2005.02592.x.

Dhungana, S.K., K.P. Kulkarni, C.W. Park, H. Jo, J.T. Song, D.H. Shin, and J.D. Lee. 2017. Mapping quantitative trait loci controlling soybean seed starch content in an interspecific cross of 'Williams 82' (*Glycine max*) and 'PI 366121' (*Glycine soja*). *Plant Breeding* 136: 379–385. https://doi.org/10.1111/pbr.12480.

Dinkins, R.D., M.S.S. Reddy, C.A. Meurer, B. Yan, H. Trick, F. Thibaud-Nissen, J.J. Finer, W.A. Parrott, and G.B. Collins. 2001. Increased sulfur amino acids in soybean plants overexpressing the maize 15 kDa zein protein. *In Vitro Cellular and Developmental Biology - Plant* 37: 742–747.

Donnelly, P. 2008. Progress and challenges in genome-wide association studies in humans. *Nature* 456: 728–731.

Dordas, C. 2006. Foliar boron application improves seed set, seed yield, and seed quality of alfalfa. *Agronomy Journal* 98: 907–913.

Dure, L., M. Crouch, J. Harada, T.D. Ho, J. Mundy, R. Quatrano, T. Thomas, and Z.R. Sung. 1989. Common amino acid sequence domains among the Lea proteins of higher plants. *Plant Molecular Biology* 12: 475–486.

Duressa, D., K. Soliman, and D. Chen. 2010. Identification of aluminum responsive genes in Al-tolerant soybean line PI 416937. *International Journal of Plant Genomics*. Article ID 164862, 13 pages. https://doi.org/10.1155/2010/164862.

Fang, C., Y. Ma, S. Wu, Z. Liu, Z. Wang, R. Yang, G. Hu, Z. Zhou, H. Yu, M. Zhang, Y. Pan, G. Zhou, H. Ren, W. Du, H. Yan, Y. Wang, D. Han, Y. Shen, S. Liu, T. Liu, J. Zhang, H. Qin, J. Yuan, X. Yuan, F. Kong, B. Liu, J. Li, Z. Zhang, Gg Wang, B. Zhu, and Z. Tian. 2017. Genome-wide association studies dissect the genetic networks underlying agronomical traits in soybean. *Genome Biology* 18: 161. https://doi.org/10.1186/s13059-017-1289-9.

Farquhar, G.D., and T.D. Sharkey. 1982. Stomatal conductance and photosynthesis. *Annual Review of Plant Physiology* 33: 317–345.

Fehr, W.R. 1987. *Principles of cultivar development—Volume 1, Theory and Technique*. New York: Macmillan Publishing Company.

Feng, C.B., M. Morsy, E. Giannocaro, B. Zhang, and P.Y. Chen. 2005. Soybean seed sugar content and quantitative trait loci mapping. In *Plant nutrition for food security, human health and environmental protection*, ed. C.J. Li, 438–439. Beijing: Tsinghua University Press.

Fletcher, R.J., I.P. Bell, and J.P. Lambert. 2004. Public health aspects of food fortification: A question of balance. *The Proceedings of the Nutrition Society* 63: 605–614.

Frazer, K.A., S.S. Murray, N.J. Schork, and E.J. Topol. 2009. Human genetic variation and its contribution to complex traits. *Nature Reviews Genetics* 10: 241–251.

Gao, Y., R.M. Biyashev, M.A. Saghai Maroof, N.M. Glover, D.M. Tucker, and G.R. Buss. 2008. Validation of low-phytate QTLs and evaluation of seedling emergence of low-phytate soybeans. *Crop Science* 48: 1355–1364. https://doi.org/10.2135/cropsci2007.11.0633.

Gong, Q.C., H.X. Yu, X.R. Mao, H.D. Qi, Y. Shi, W. Xiang, Q.S. Chen, and Z.M. Qi. 2018. Meta-analysis of soybean amino acid QTLs and candidate gene mining. *Journal of Integrative Agriculture* 17 (5): 1074–1084. https://doi.org/10.1016/S2095-3119(17)61783-0.

Gonzalez, E.M., A.J. Gordon, C.L. James, and C. Arrese-lgor. 1995. The role of sucrose synthase in the response of soybean nodulesto drought. *Journal of Experiment Botany* 46: 1515–1523.

Grant, D., R.T. Nelson, S.B. Cannon, and R.C. Shoemaker. 2010. SoyBase, the USDA-ARS soybean genetics and genomics database. *Nucleic Acids Research* 38 (suppl 1): D843–D846. https://doi.org/10.1093/nar/gkp798.

Gregory, P.J., J.S.I. Ingram, and M. Brklacich. 2005. Climate change and food security. *Philosophical Transactions of the Royal Society B* 360: 2139–2148. https://doi.org/10.1098/rstb.2005.1745.

von Groll, U., D. Berger, and T. Altmann. 2002. The subtilisin-like serine protease SDD1 mediates cell-to-cell signaling during Arabidopsis stomatal development. *The Plant Cell* 14: 1527–1539.

Guleria, S., S. Sharma, B.S. Gill, and S.K. Munshi. 2008. Distribution and biochemical composition of large and small seeds of soybean (Glycine max L.). *Journal of the Science of Food and Agriculture* 88: 269–272.

Hanitzsch, M., D. Schnitzer, T. Seidel, D. Golldack, and K.J. Dietz. 2007. Transcript level regulation of the vacuolar H+-ATPase subunit isoforms VHA-a, VHA-E and VHA-G in Arabidopsis thaliana. *Molecular Membrane Biology* 24 (5–6): 507–518. https://doi.org/10.1080/09687680701447393.

Hare, P.D., and W.A. Cress. 1997. Implications of stress-induced proline accumulation in plants. *Plant Growth Regulation* 21 (2): 79–102.

Haun, W., A. Coffman, B.M. Clasen, Z.L. Demorest, A. Lowy, E. Ray, A. Retterath, T. Stoddard, A. Juillerat, F. Cedrone, L. Mathis, D.F. Voytas, and F. Zhang. 2014. Improved soybean oil quality by targeted mutagenesis of the fatty acid desaturase 2 gene family. *Plant Biotechnology Journal* 12: 934–940.

Hay R., and J. Porter. 2006. *The physiology of crop yield (2nd ed.)*. Oxford: Blackwell.

Hayat, S., Q. Hayat, M.N. Alyemeni, A.S. Wani, J. Pichtel, and A. Ahmad. 2012. Role of proline under changing environments. *Plant Signaling and Behavior* 7 (11): 1456–1466. https://doi.org/10.4161/psb.21949.

Heagle, A.S., J.E. Miller, and W.A. Pursley. 1998. Influence of ozone stress on soybean response to carbon dioxide enrichment: III. Yield and seed quality. *Crop Science* 38: 128–134.

Heinemann, R.J.B., P.L. Fagundes, E.A. Pinto, M.V.C. Penteado, and U.M. Lanfer-Marquez. 2005. Comparative study of nutrient composition of commercial brown, parboiled and milled rice from Brazil. *Journal of Food Composition and Analysis* 18: 287–296.

Hitsuda, K., K. Toriyama, G.V. Subbarao, and O. Ito. 2008. Sulfur management for soybean production. In *A missing link between soils, crops, and nutrition*. ed. Joseph Jez, 235–249. Published by American Society of Agronomy, Crops Science Society of America, and Soil Science Society of America. ISBN 9780891181866.

Hitz, W.D., T.J. Carlson, P.S. Kerr, and S.A. Sebastian. 2002. Biochemical and molecular characterization of a mutation that confers a decreased raffinosaccharide and phytic acid phenotype on soybean seeds. *Plant Physiology* 128: 650–660.

Imsande, J. 2001. Selection of soybean mutants with increased concentrations of seed methionine and cysteine. *Crop Science* 41: 510–515.

Ingenbleek, Y., and H. Kimura. 2013. Nutritional essentiality of sulfur in health and disease. *Nutrition Reviews* 71 (7): 413–432. https://doi.org/10.1111/nure.12050.

Iuchi, S., H. Koyama, A. Iuchi, Y. Kobayashi, S. Kitabayashi, Y. Kobayashi, T. Ikka, T. Hirayama, K. Shinozaki, and M. Kobayashi. 2007. Zinc finger protein STOP1 is critical for proton tolerance in Arabidopsis and coregulates a key gene in aluminum tolerance. *PNAS* 104 (23): 9900–9905. https://doi.org/10.1073/pnas.0700117104.

Jackson, A.A., J.C. Shaw, A. Barber, and M.H. Golden. 1981. Nitrogen metabolism in preterm infants fed human donor breast milk: the possible essentiality of glycine. *Pediatric Research* 15: 1454–1461.

Jaureguy, L.M. 2009. *Identification of molecular markers associated with seed size, protein and sugar content in soybean*. MS Thesis, University of Arkansas, Fayetteville, AR, USA.

Jaureguy, L.M., P. Chen, and A.M. Scaboo. 2011. Heritability and correlations among food-grade traits in soybean. *Plant Breeding* 130: 647–652. https://doi.org/10.1111/j.1439-0523.2011.01887.x.

Jegadeesan, S., K. Yu, V. Poysa, E. Gawalko, M.J. Morrison, C. Shi, and E. Cober. 2010. Mapping and validation of simple sequence repeat markers linked to a major gene controlling seed cadmium accumulation in soybean [Glycine max (L.) Merr]. *Theoretical and Applied Genetics* 121: 283–294.

Kaidar-Person, O., B. Person, S. Szomstein, and R.J. Rosenthal. 2008. Nutritional deficiencies in morbidly obese patients: a new form of malnutrition? Part A: Vitamins. *Obesity Surgery* 18 (7): 870–876. https://doi.org/10.1007/s11695-007-9349-y.

Kaler, A.S., A.P. Dhanapal, J.D. Ray, C.A. King, F.B. Fritschi, and L.C. Purcell. 2017. Genome-wide association mapping of carbon isotope and oxygen isotope ratios in diverse soybean genotypes. *Crop Science* 57 (6): 3085–3100. https://doi.org/10.2135/cropsci2017.03.0160.

Khandaker, L., M. Akond, S. Liu, S.K. Kantartzi, K. Meksem, N. Bellaloui, D.A. Lightfoot, and M.A. Kassem. 2015. Mapping of QTL associated with seed amino acids content in 'MD 96-5722' by 'Spencer' RIL population of soybean using SNP markers. *Food and Nutrition Sciences* 6: 974–984. https://doi.org/10.4236/fns.2015.611101.

Kilo, V., and D.A. Lightfoot. 1996. Genetic analysis of resistance to manganese toxicity in soybean using molecular markers Soybean. *Genetics Newsletter* 23: 155–157.

Kim, H.K., S.T. Kang, J.H. Cho, M.G. Choung, and D.Y. Suh. 2005. Quantitative trait loci associated with oligosaccharide and sucrose contents on soybean (*Glycine max* L.). *Journal of Plant Bioloy* 48 (1): 106–112.

Kim, H., S. Kang, and K. Oh. 2006. Mapping of putative quantitative trait loci controlling the total oligosaccharide and sucrose content of Glycine max seeds. *Journal of Plant Research* 119 (5): 533–538.

King, K.E., N. Lauter, SFu Lin, M.P. Scott, and R.C. Shoemaker. 2013. Evaluation and QTL mapping of phosphorus concentration in soybean seed. *Euphytica* 189: 261–269. https://doi.org/10.1007/s10681-012-0797-7.

Koch, K. 2004. Sucrose metabolism: regulatory mechanisms and pivotal roles in sugar sensing and plant development. *Current Opinion in Plant Biology* 7: 235–246.

Kortt, A.A., J.B. Caldwell, G.G. Lilley, and T.J.V. Higgins. 1991. Amino acid and cDNA sequence of a methionine-rich 2S protein from sunflower seed (Helianthus annuus L.). *European Journal of Biochemistry* 195: 329–334.

Krishnan, H.B. 2005. Engineering Soybean for Enhanced Sulfur Amino Acid Content. *Crop Science* 45: 454–461.

Krishnan, H.B. 2008. Improving the sulfur-containing amino acids of soybean to enhance its nutritional value in animal feed. In A missing link between soils, crops, and nutrition. Joseph Jez (ed.). pp. 235–249. Published by American Society of Agronomy, *Crops Science Society of America, and Soil Science Society of America*. ISBN 9780891181866.

Krishnan, H.B., and J.M. Jez. 2018. Review: The promise and limits for enhancing sulfur-containing amino acid content of soybean seed. *Plant Science* 272: 14–21. https://doi.org/10.1016/j.plantsci.2018.03.030.

Kwanyuen, P., V.R. Pantalone, J.W. Burton, and R.F. Wilson. 1997. A new approach to genetic alteration of soybean protein composition and quality. *Journal of the American Oil Chemists Society* 74: 983–987. https://doi.org/10.1007/s11746-997-0015-2.

La, T., E. Large, E. Taliercio, Q. Song, J.D. Gillman, D. Xu, H.T. Nguyen, G. Shannon, and A. Scaboo. 2019. Characterization of Select Wild Soybean Accessions in the USDA Germplasm Collection for Seed Composition and Agronomic Traits. *Crop Science 59* (1): 233–251. https://doi.org/10.2135/cropsci2017.08.0514.

Lander, E.S., and D. Botstein. 1989. Mapping Mendelian factors underlying quantitative traits using RFLP linkage maps. *Genetics* 121: 185–199. https://doi.org/10.1093/genetics/121.1.185.

Lee, J.S., S.M. Kim, and S. Kang. 2016. Fine mapping of quantitative trait loci for sucrose and oligosaccharide contents in soybean [*Glycine max* (L.) Merr.] using 180 K Axiom® SoyaSNP genotyping platform. *Euphytica* 208: 195–203. https://doi.org/10.1007/s10681-015-1622-x.

Lee. J., Y.S. Hwang, S.T. Kim, W.B. Yoon, W.Y. Han, I.K. Kang, and M.G. Choung. 2017. Seed coat color and seed weight contribute differential responses of targeted metabolites in soybean seeds. *Food Chemistry* 214: 248–258.

Lewis, M.W., M.E. Leslie, E.H. Fulcher, L. Darnielle, P.N. Healy, J.Y. Youn, and S.J. Liljegren. 2010. The SERK1 receptorlike kinase regulates organ separation in Arabidopsis flowers. *The Plant Journal* 62: 817–828.

Li, L.G., Z.Y. He, G.K. Pandey, T. Tsuchiya, and S. Luan. 2002. Functional cloning and characterization of a plant efflux carrier for multi-drug and heavy metal detoxification. *The Journal of Biological Chemistry* 277: 5360–5368.

Li, M.W., N.B. Muñoz, C.F. Wong, F.L. Wong, K.S. Wong, J.W.H. Wong, X. Qi, K.P. Li, M.S. Ng, and H.M. Lam. 2016. QTLs regulating the contents of antioxidants, phenolics, and flavonoids in soybean seeds share a common genomic region. *Frontiers in Plant Science*. https://doi.org/10.3389/fpls.2016.00854.

Li, X., R. Tian, S. Kamala, H. Du, W. Li, Y. Kong, and C. Zhang. 2018. Identification and verification of pleiotropic QTL controlling multiple amino acid contents in soybean seed. *Euphytica* 214: 93. https://doi.org/10.1007/s10681-018-2170-y.

Long, S.P., E.A. Ainsworth, A. Rogers, and D.R. Ort. 2004. Rising atmospheric carbon dioxide: Plants FACE the future. *Annu. Rev. Plant Biol.* 55: 591–628.

Lu, K., L. Li, X. Zheng, Z. Zhang, T. Mou, and Z. Hu. 2008. Quantitative trait loci controlling Cu, Ca, Zn, Mn and Fe content in rice grains. *Journal of Genetics* 87: 305–310.

Mackay, T.F.C., E.A. Stone, and J.F. Ayroles. 2009. The genetics of quantitative traits: challenges and prospects. *Nature Reviews Genetics* 10: 565–577. https://doi.org/10.1038/nrg2612.

Malle, S., M. Morrison, and F. Belzile. 2020. Identification of loci controlling mineral element concentration in soybean seeds. Preprint available at Research Square. Under Review in *BMC Plant Biology*. https://doi.org/10.21203/rs.3.rs-22007/v2.

Maughan, P.J., M.A.S. Maroof, and G.R. Buss. 2000. Identification of quantitative trait loci controlling sucrose content in soybean (Glycine max). *Molecular Breeding* 6 (1): 105–111.

Maupin, L.M., M.L. Rosso, and K.M. Rainey. 2011. Environmental Effects on Soybean with Modified Phosphorus and Sugar Composition. *Crop Science* 51 (2): 642–650. https://doi.org/10.2135/cropsci2010.07.0396.

Mayoral, M.L., D. Atsmon, D. Shimshi, and Z. GRoMET-Elhanan. 1981. Effect of water stress on enzyme activities in wheat and related wild species: carboxylase activity, electron transport and photophosphorylation in isolated chloroplasts. *Aust J Plant Physiol.* 8: 385–393.

Mazur, B., E. Krebbers, S. Tingey. 1999. Gene Discovery and Product Development for Grain Quality Traits. *Science* 285 (5426): 372–375. https://doi.org/10.1126/science.285.5426.372.

Nable, R.O., R.L. Houtz, and G.M. Cheniae. 1988. Early inhibition of photosynthesis during development of Mn toxicity in tobacco. *Plant Physiol.* 86: 1136–1142.

Ning, L., P. Sun, Q. Wang, D. Ma, Z. Hu, D. Zhang, G. Zhang, H. Cheng, and D. Yu. 2015. Genetic architecture of biofortification traits in soybean (Glycine max L. Merr.) revealed through association analysis and linkage mapping. *Euphytica* 204: 353–369. https://doi.org/10.1007/s10681-014-1340-9.

Obendorf, R.L. 1997. Oligosaccharides and galactosyl cyclitols in seed desiccation tolerance. *Seed Sci. Res* 7: 63–74.

Olhoft, P.M., L.E. Flagel, C.M. Donovan, and D.A. Somers. 2003. Efficient soybean transformation using hygromycin B selection in the cotyledonary-node method. *Planta* 216 (5): 723–735.

Oltmans, S.E., W.R. Fehr, G.A. Welke, and S.R. Cianzio. 2004. Inheritance of low-phytate phosphorus in soybean. *Crop Science* 44: 433–435. https://doi.org/10.2135/cropsci2004.4330.

Orazaly, M., P. Chen, B. Zhang, and A. Zeng. 2014. Quantitative Trait Loci Mapping for Seed Calcium Content of Soybean. *Crop Science* 54 (2): 500–506. https://doi.org/10.2135/cropsci2013.01.0026.

Orazaly, M., P. Chen, B. Zhang, L. Florez-Palacios, and A. Zeng. 2018. Confirmation of SSR markers and QTL for seed calcium content and hardness of soybean. *Journal of Crop Improvement* 32 (1): 71–89. https://doi.org/10.1080/15427528.2017.1398701.

Panthee, D.R., P. Kwanyuen, C.E. Sams, D.R. West, A.M. Saxton, and V.R. Pantalone. 2004. Quantitative trait loci for β-conglycinin (7S) and glycinin (11S) fractions of soybean storage protein. J Am. *Oil Chem Society* 81 (11): 1005–1012. https://doi.org/10.1007/s11746-004-1014-4.

Panthee, D.R., V.R. Pantalone, A.M. Saxton, D.R. West, and C.E. Sams. 2006a. Genomic regions associated with amino acid composition in soybean. *Molecular Breeding* 17: 79–89.

Panthee, D.R., V.R. Pantalone, C.E. Sams, A.M. Saxton, D.R. West, J.H. Orf, and A.S. Killam. 2006b. Quantitative trait loci controlling sulfur containing amino acids, methionine and cysteine, in soybean seeds. *Theoretical and Applied Genetics* 112: 546–553.

Pao, S.S., I.A.N.T. Paulsen, and M.H. Saier. 1998. Major facilitator superfamily. *Microbiology* 62: 1–34.

Patil, G., T.D. Vuong, S. Kale, B. Valliyodan, R. Deshmukh, C. Zhu, X. Wu, Y. Bai, D. Yungbluth, F. Lu, S. Kumpatla, J.G. Shannon, R.K. Varshney, and H.T. Nguyen. 2018. Dissecting genomic hotspots underlying seed protein, oil, and sucrose content in an interspecific mapping population of soybean using high-density linkage mapping. *Plant Biotechnology Journal* 16: 1939–1953. https://doi.org/10.1111/pbi.12929.

Patrick, J.W., and C.E. Offler. 2001. Compartmentation of transport and transfer events in developing seeds. *J Exp Bot* 52: 551–564.

Pfarr, M.D., M.J. Kazula, J.E. Miller-Garvin, and S.L. Naeve. 2018. Amino acids balance is affected by protein concentration in soybean. *Crop Science* 58: 2050–2062. https://doi.org/10.2135/cropsci2017.11.0703.

Philip, J.W., and R.B. Martin. 2005. Biofortifying crops with essential mineral elements. *Trends in Plant Science* 10: 586–593.

Potrykus, I. 2003. Nutritional improvement of rice to reduce malnutrition in developing countries. p. 401–406. In I.K. Vasil (ed.) Plant biotechnology 2002 and beyond. Kluwer Academic Publishers, Dordrecht, The Netherlands.

Prasad, P., V. Vara, L.H. Allen, and K.J. Boote. 2005. Crop responses to elevated carbon dioxide and interaction with temperature: grain legumes. *J. Crop Improv*. 13: 113–155.

Qi, Z.M., Q. Wu, X. Han, Y.N. Sun, X.Y. Du, C.Y. Liu, H.W. Jiang, G.H. Hu, and Q.S. Chen. 2011. Soybean oil content QTL mapping and integratingwith meta-analysis method for mining genes. *Euphytica* 179: 499–514. https://doi.org/10.1007/s10681-011-0386-1.

Qi, X.P., M.W. Li, M. Xie, X. Liu, M. Ni, G. Shao, C. Song, A.K.Y. Yim, Y. Tao, F.L. Wong, S. Isobe, C.F. Wong, K.S. Wong, C. Xu, C. Li, Y. Wang, R. Guan, F. Sun, G. Fan, Z. Xiao, F. Zhou, T.H. Phang, X. Liu, S.W. Tong, T.F. Chan, S.M. Yiu, S. Tabata, J. Wang, X. Xu, and H. Lam. 2014. Identification of a novel salt tolerance gene in wild soybean by whole-genome sequencing. *Nature Communications* 5: 4340. https://doi.org/10.1038/Ncomms5340.

Qiu, H., Z. Li, Y. Yu, S. Gao, X. Ma, Y. Zheng, F. Meng, Y. Hou, Y. Wang, and S. Wang. 2015. Mining and analysis of genes related to sulfur–containing amino acids in soybean based on meta–QTL. *Chinese Journal of Oil Crop Sciences* 37 (2): 141–147. https://doi.org/10.7505/j.issn.1007-9084.2015.02.003.

Rainbird, R.M., J.H. Thorne, and R.W.F. Hardy. 1984. Role of amides, amino acids, and ureides in the nutrition of developing soybean seeds. *Plant Physiology* 74: 329–334.

Ramakrishnan, U. 2002. Prevalence of micronutrient malnutrition worldwide. *Nutrition Reviews* 60 (S5): S46–S52. https://doi.org/10.1301/00296640260130731.

Ramamurthy, R.K., J. Jedlicka, G.L. Graef, and B.M. Waters. 2014. Identification of new QTLs for seed mineral, cysteine, and methionine concentrations in soybean [*Glycine max* (L.) Merr.]. *MolecularBreeding* 34: 431–445. https://doi.org/10.1007/s11032-014-0045-z.

Redondo-Cuenca, A., M.J. Villanueva-Suárez, M.D. Rodríguez-Sevilla, and I. Mateos-Aparicio. 2007. Chemical composition and dietary fibre of yellow and green commercial soybeans (Glycine max). *Food Chemistry* 101: 1216–1222.

Reeds, P.J. 2000. Dispensable and indispensable amino acids for humans. *Journal of Nutrition* 130: 1835S–1840S.

Reich, R.C., E.J. Kamprath, and L.A. Nelson. 1981. Soil and management factors correlated with soybean yields in the southeastern U.S. *Coastal Plain. Agron. J*. 73: 90–95.

Rhodes, D., and A.D. Hanson. 1993. Quaternary ammonium and tertiary sulfonium compounds in higher plants. *Annual Review of Plant Physiology and Plant Molecular Biology* 44: 357–384. https://dx.doi.org/10.1146/annurev.pp.44.060193.002041.

Rogers, A., E.A. Ainsworth, and A.D.B. Leakey. 2009. Update on legumes and elevated CO2: Will elevated carbon dioxide concentration amplify the benefits of nitrogen fixation in legumes? *Plant Physiol*. 151: 1009–1016.

Salari, M.W., P.O. Ongom, R. Thapa, H.T. Nguyen, T.D. Vuong, and K.M. Rainey. 2020. Mapping QTL controlling soybean seed sucrose and oligosaccharides in a single family of soybean nested association mapping (SoyNAM) population. *Plant Breeding* 00: 1–13. https://doi.org/10.1111/pbr.12883.

Saldivar, X., Y.J. Wang, P. Chen, and A. Hou. 2011. Changes in chemical composition during soybean seed development. *Food Chemistry* 124: 1369–1375.

Saravitz, D.M., M.D. Pharr, and T.E. Carter. 1987. Galactinol Synthase Activity and Soluble Sugars in Developing Seeds of Four Soybean Genotypes. *Plant Physiology* 83: 185–189.

Scaboo, A.M., V.R. Pantalone, D.R. Walker, H.R. Boerma, D.R. West, F.R. Walker, and C.E. Sams. 2009. Confirmation of molecular markers and agronomic traits associated with seed phytate content in two soybean RIL populations. *Crop Science* 49: 426–432. https://doi.org/10.2135/cropsci2007.11.0614.

Sebastian, S.A., P.S. Kerr, R.W. Pearlstein, and W.D. Hitz. 2000. Soybean germplasm with novel genes for improved digestibility. In JK Drackley, ed, Soy in Animal Nutrition. *Federation of Animal Science Societies*, Savoy, IL, pp 56–74.

Shah, K., J. Vervoort, and S.C. de Vries. 2001. Role of threonines in the Arabidopsis thaliana somatic embryogenesis receptor kinase 1 activation loop in phosphorylation. *Journal of Biological Chemistry* 276: 41263–41269.

Shewry, P.R., J.A. Napier, and A.S. Tatham. 1995. Seed storage proteins: structure and biosynthesis. *The Plant Cell* 7: 945–956.

Silva, D.C.G., N. Yamanaka, R.L. Brogin, C.A.A. Arias, A.L. Nepomuceno, A.O. Di Mauro, S.S. Pereira, L.M. Nogueira, A.L.L. Passianotto, and R.V. Abdelnoor. 2008. Molecular mapping of two loci that confer resistance to Asian rust in soybean. *Theor Appl Genet* 117:57–63. https://doi.org/10.1007/s00122-008-0752-0.

Somers, D.A., D.A. Samac, and P.M. Olhoft. 2003. Recent advances in legume transformation. *Plant Physiol.* 131: 892–899.

Sonah, H., L. O'Donoughue, E. Cober, I. Rajcan, and F. Belzile. 2015. Indentification of loci governing eight agronomic traits using a GBS-GWAS approach and validation by QTL mapping in soya bean. *Plant Biotech Journal* 13: 211–221. https://doi.org/10.1111/pbi.12249.

Spaeth, S.C., and T.R. Sinclair. 1984. Soybean seed growth. I. Timing of growth of individual seeds. *Agronomy Journal* 76:123–217. https://doi.org/10.2134/agronj1984.00021962007600010030x
.

Stombaugh, S.K., J.H. Orf, H.G. Jung, K. Chase, K.G. Lark, and D.A. Somers. 2004. Quantitative trait loci associated with cell wall polysaccharides in soybean seed. *Crop Science* 44: 2101–2106.

Streit, L.G., L.R. Beach, J.C. Register, R. Jung, and W.R. Fehr. 2001. Association of the Brazil nut protein gene and Kunitz trypsin inhibitor alleles with soybean protease inhibitor activity and agronomic traits. *Crop Sci.* 41: 1757–1760.

Sun, W.Q., T.C. Irving, and A.C. Leopold. 1994. The role of sugar, vitrification and membrane phase transition in seed desiccation tolerance. *Physiologia Plantarum* 90: 621–628.

Taiz, L. and E. Zeiger. 1998. Mineral nutrition. In Plant Physiology (ed.). Sinauer Associates Inc., P.O. Box 407, Sunderland, MA 01375, USA.

Takahashi, M., Y. Uematsu, K. Kashiwaba, K. Yagasaki, M. Hajika, R. Matsunaga, K Komatsu, and M. Ishimoto. 2003. Accumulation of high levels of free amino acids in soybean seeds through integration of mutations conferring seed protein deficiency. *Planta* 217 (4): 577–586.

Taub, D., B. Miller, and H. Allen. 2008. Effects of elevated CO2 on the protein concentration of food crops: A meta-analysis. *Global Ch. Biol.* 14: 565–575.

Temperly, R.T., and R. Borges. 2006. Tillage and crop rotation impact on soybean grain yield and composition. *Agronomy Journal* 98: 999–1004.

Thakur, M., and A.D. Sharma. 2005. Salt-stress-induced proline accumulation in germinating embryos: Evidence suggesting a role of proline in seed germination. *Journal of Arid Environments* 62 (3): 517–523.

Thomas, J.M.G., K.J. Boote, L.H. Allen, M. Gallo-Meagher, and J.M. Davis. 2003. Elevated temperature and carbon dioxide effects on soybean seed composition and transcript abundance. *Crop Sci.* 43: 1548–1557.

Thilakarathna, M.S., and M.N. Raizada. 2017. A meta-analysis of the effectiveness of diverse rhizobia inoculants on soybean traits under field conditions. *Soil Biology and Biochemistry* 105: 177–196.

Townsend, J.A., and L.A. Thomas. 1994. Factors which influence the Agrobacterium-mediated transformation of soybean. J. Cell. Biochem. Suppl. 18A: 78.

Ufaz, S., and G. Galili. 2008. Improving the Content of Essential Amino Acids in Crop Plants: Goals and Opportunities. *Plant Phys*. 147: 954–961. https://dx.doi.org/10.1104/pp.108.118091.

Vaughn, J.N., R.L. Nelson, Q. Song, P.B. Cregan, and Z. Li. 2014. The genetic architecture of seed composition in soybean is refined by genome-wide association scans across multiple populations. *G3-Genes Genomes Genetics* 4: 2282–2294. https://doi.org/10.1534/g3.114.013433.

Vollmann, J., T. Losak, M. Pachner, D. Watanabe, L. Musilova, and J. Hlusek. 2015. Soybean cadmium concentration: Validation of a QTL affecting seed cadmium accumulation for improved food safety. *Euphytica* 203: 177–184. https://doi.org/10.1007/s10681-014-1297-8.

Walker, D.R., A.M. Scaboo, V.R. Pantalone, J.R. Wilcox, and H.R. Boerma. 2006. Genetic mapping of loci associated with seed phytic acid content in CX1834-1-2 soybean. *Crop Science* 46 (1): 390–397. https://doi.org/10.2135/cropsci2005.0245.

Wang, T.L., C. Domoney, C.L. Hedley, R. Casey, and M.A. Grusak. 2003. Can we improve the nutritional quality of legume seeds? *Plant Physiol.* 131: 886–891.

Wang, X., G.L. Jiang, M. Green, R.A. Scott, Q. Song, D.L. Hyten, and P.B. Cregan. 2014a. Identification and validation of quantitative trait loci for seed yield, oil and protein contents in two recombinant inbred line populations of soybean. *Molecular Genetics and Genomics* 289 (5): 935–949.

Wang, Y., P. Chen, and B. Zhang. 2014b. Quantitative trait loci analysis of soluble sugar contents in soybean. *Plant Breeding* 133: 493–498. https://doi.org/10.1111/pbr.12178.

Wang, X., G.L. Jiang, Q. Song, P.B. Cregan, R.A. Scott, J. Zhang, Y. Yen, and M. Brown. 2015. Quantitative trait locus analysis of seed sulfur-containing amino acids in two recombinant inbred line populations of soybean. *Euphytica* 201 (2): 293–305. https://doi.org/10.1007/s10681-014-1223-0.

Warrington, C.V., H. Abdel-Haleem, D.L. Hyten, P.B. Cregan, J.H. Orf, A.S. Killam, N. Bajjalieh, Z. Li, and H.R. Boerma. 2015. QTL for seed protein and amino acids in the Benning × Danbaekkong soybean population. *Theoretical and Applied Genetics* 128 (5): 839–850. https://doi.org/10.1007/s00122-015-2474-4.

Weber, H., L. Borisjuk, and U. Wobus. 1997. Sugar import and metabolism during seed development. *Trends in Plant Science* 2 (5): 169–174.

Welch, R.M. 1999. Importance of seed mineral nutrient reserves in crop growth and development. In: Rengel Z, ed. Mineral nutrition of crops. Fundamental mechanisms and implications. New York: *Food Products Press*, 205–226.

Welch, R.M., and R.D. Graham. 2004. Breeding for micronutrients in staple food crops from a human nutrition perspective. *Journal of Experimental Botany* 55 (396: 353–364. https://doi.org/10.1093/jxb/erh064.

Weschke, W., R. Panitz, S. Gubatz, Q. Wang, R. Radchuk, H. Weber, and U. Wobus. 2003. The role of invertases and hexose transporters in controlling sugar ratios in maternal and filial tissues of barley caryopses during early development. *Plant Journal* 33: 395–411.

WHO. 2019. https://www.who.int/mediacentre/factsheets/fs311/en/. Accessed on December 22, 2019.

Wijewardana, C., K.R. Reddy, and N. Bellaloui. 2019. Soybean seed physiology, quality, and chemical composition under soil moisture stress. *Food Chemistry* 278: 92–100. https://doi.org/10.1016/j.foodchem.2018.11.0.35.

Wilcox, J., G. Premachandra, K. Yound, and V. Raboy. 2000. Isolation of high seed inorganic P, low-phytate soybean mutants. *Crop Sci* 40: 1601–1605.

Wobus, U., and H. Weber. 1999. Sugars as signal molecules in plant seed development. *Biol Chem* 390: 937–944.

Wolf, R.B., J.F. Cavins, R. Kleiman, and L.T. Black. 1982. Effect of temperature on soybean seed constituents: Oil, protein, moisture, fatty acids, amino acids and sugars. *Journal of the American Oil Chemists' Society* 59 (5): 230–232.

Womack, M., and W.C. Rose. 1947. The role of proline, hydroxyproline, and glutamic acid in growth. *Journal of Biological Chemistry* 171: 37–50.

Wu, Z., J. Zhao, R. Gao, G. Hu, J. Gai, G. Xu, and H. Xing. 2011. Molecular cloning, characterization and expression analysis of two members of the Pht1 family of phosphate transporters in *Glycine max*. *PLoS One* 6: 1–12.

Yancey, P.H. 1994. Compatible and counteracting solutes In: Strange K ed. Cellular and Molecular Physiology of Cell Volume Regulation. Boca Raton, FL: CRC Press 1994: 81–109.

Yazdi-Samadi, B., R.W. Rinne, and R.D. Seif. 1977. Components of developing soybean seeds: Oil, protein, sugars, starch, organic acids, and amino acids. *Agronomy Journal* 69: 481–486.

Yu, C.H., Lee, J.S., Lee, L., Kim, S.H., Lee, S.S., and Kang, S.A. 2004. Nutritional Factors Related to Bone Mineral Density in the Different Age Groups of Korean Men. *Korean J Nutr* 37 (2): 132–142.

Zeng, A., P. Chen, A. Shi, D. Wang, B. Zhang, M. Orazaly, L. Florez-Palacios, K. Brye, Q. Song, and P. Cregan. 2012. Identification of Quantitative Trait Loci for Sucrose Content in Soybean Seed. *Crop Science* 54 (2): 554–564. https://doi.org/10.2135/cropsci2013.01.0036.

Zeng, P., P. Chen, A. Shi, D. Wang, B. Zhang, M. Orazaly, L. Florez-Palacios, K. Brye, Q. Song, and P. Cregan. 2014. Identification of quantitative trait loci for sucrose content in soybean seed. *Crop Science* 54 (2): 554–564.

Zeng, A., P. Chen, B. Zhang, M. Orazaly, L. Florez-Palacios, and K.R. Brye. 2015. Identification and confirmation of quantitative trait loci for stachyose content in soybean seed. *Plant Breeding* 134 (2): 178–185. https://doi.org/10.1111/pbr.12246.

Zhang, P., J.M. Jaynes, I. Potrykus, W. Gruissem, and J. Puonti-Kaerlas. 2002. Transfer and expression of an artificial storage protein (ASP1) gene in cassava: Towards improving nutritive value of storage proteins. p. 425–427. In I.K. Vasil (ed.) Plant biotechnology 2002 and beyond. Kluwer Academic Publishing.

Zhang, B., P. Chen, C.Y. Chen, D. Wang, A. Shi, A. Hou, and T. Ishibashi. 2008. Quantitative trait loci mapping of seed hardness in soybean. *Crop Science* 48: 1341–1349.

Zhang, B., P. Chen, A. Shi, A. Hou, T. Ishibashi, and D. Wang. 2009. Putative quantitative trait loci associated with calcium content in soybean seed. *Journal of Heredity* 100 (2): 263–269. https://doi.org/10.1093/jhered/esn096.

Zhang, J., X. Wang, Y. Lu, S.J. Bhusal, Q. Song, P.B. Cregan, Y. Yen, M. Brown, and G.L. Jiang. 2018. Genome-wide scan for seed composition provides insights into soybean quality improvement and the impacts of domestication and breeding. *Molecular Plant* 11 (3): 460–472. https://doi.org/10.1016/j.molp.2017.12.016.

Zhang, J., X. Wang, Y. Lu, S.J. Bhusal, Q. Song, P.B. Cregan, Y. Yen, M. Brown, and G.L. Jiang. 2018a. Genome-wide scan for seed composition provides insights into soybean quality improvement and the impacts of domestication and breeding. *Molecular Plant* 11 (3): 460–472. https://doi.org/10.1016/j.molp.2017.12.016.

Zhang, X., W. Wang, N. Guo, Y. Zhang, Y. Bu, J. Zhao, and H. Xing. 2018b. Combining QTL-seq and linkage mapping to fine map a wild soybean allele characteristic of greater plant height. *BMC Genomics* 19: 226. https://doi.org/10.1186/s12864-018-4582-4.

Chapter 4
Two Decades of QTL Mapping of Mineral Nutrient Deficiencies in Soybean

Moulay Abdelmajid Kassem

4.1 Introduction

Likewise, different types of molecular markers and mapping populations have been used to genetically map quantitative trait loci (QTL) for mineral nutrients efficiencies, deficiencies, and tolerances in soybean. The main approaches are the use of biparental recombinant inbred line populations and the use of genome-wide association studies (GWAS). There are many QTL-controlling mineral nutrients efficiencies, deficiencies, and tolerances that have been identified and mapped on the soybean genome (Diers et al. 1992; Kilo and Lightfoot 1996; Lin et al. 1997, 2000a, 2000b; Bianchi-Hall et al. 2000; Kassem et al. 2004; Korir et al. 2011, 2013; Peiffer et al. 2012; Mamidi et al. 2011, 2014; Zhang et al. 2014; Dhanapal et al. 2015, 2018; Kumar and Lal 2015; Kaler et al. 2017).

As stated earlier, plants take up both essential elements (macronutrients and micronutrients) from the soil through their root systems, combine them with sunlight, and can synthesize all other compounds that they need for their growth and development; however, any absence or scarceness of one or more of these essential elements will cause symptoms of mineral deficiencies that can be chlorosis (yellow spots on leaves and stems), necrosis (browning and local death of tissues), dwarfism (reduced stem and internodes elongation), impairment in flowers and fruit development, sterility, etc. (Taiz and Zeiger 1998; Welch 1995). For example, in acidic soils around the world, Al toxicity causes root chlorosis and necrosis, reduced root development, and increased root thickness and poses a threat to seed yield and productivity (Carter et al. 1999; Bianchi-Hall et al. 2000; Silva et al. 2001). Mn toxicity causes leaf chlorosis and root necrosis (Nable et al. 1988; Kilo and Lightfoot 1996).

M. A. Kassem (✉)
Plant Genomics and Biotechnology Lab, Department of Biological and Forensic Sciences, Fayetteville State University, Fayetteville, NC, USA
e-mail: mkassem@uncfsu.edu

M. A. Kassem (ed.), *Soybean Seed Composition*,
https://doi.org/10.1007/978-3-030-82906-3_4

Many studies identified and mapped QTL that nutrients' toxicities, efficiencies, deficiencies, tolerances, water use efficiency (WUE), and other traits related to soil nutrients using biparental populations and GWAS. Here, we will summarize the most important QTL mapping studies for these traits reported and not reported on SoyBase (Diers et al. 1992; Kilo and Lightfoot 1996; Lin et al. 1997, 2000a, 2000b; Bianchi-Hall et al. 2000; Kassem et al. 2004; Qi et al. 2008; Korir et al. 2011, 2013; Peiffer et al. 2012; Mamidi et al. 2011, 2014; Grant et al. 2010; Zhang et al. 2014; Dhanapal et al. 2015, 2018; Kumar and Lal 2015; Kaler et al. 2017; Cai et al. 2019).

4.1.1 QTL for Iron Deficiency and Mn Toxicity

Iron deficiency chlorosis (IDC) reduces seed yield severely in many parts of the USA and the world especially in calcareous soils, and these losses are estimated to $260 million per year (Waters et al. 2018). The development of iron-efficient soybean lines is the best approach to fight against IDC-caused yield losses (Diers et al. 1992; Hansen et al. 2004; Peiffer et al. 2012; Waters et al. 2018). Mn toxicity rises in acidic soils especially in Midwestern USA and is manifested in soybean plants via leaf chlorosis and root necrosis (Kilo and Lightfoot 1996; Kassem et al. 2004).

Diers et al. (1992) grew a RIL population derived from a cross between "A86–356022" (Fe-inefficient) and PI468916 (Fe-efficient) (n = 110) in 1989 and 1990 in Ames and Humboldt Iowa, USA, and genotyped it with 272 markers mostly RFLPs (263). Five markers linked to five QTL that control Fe efficiency have been identified and mapped on chromosomes 1, 8, and 18 (Fig. 2.1; Table 4.1) (Diers et al. 1992).

To identify QTL for iron deficiency, Lin et al. (1997) grew two RIL populations in Ames, Iowa, USA. The first RIL population was derived from a cross between "Pride B216" (Fe-inefficient) and "A15" (Fe-efficient) (n = 120), and the second RIL population was derived from a cross between "Anoka" and "A7" (n = 92). "Pride B216" by "A15" RIL population was genotyped with 99 markers (89 RFLPs and 10 SSRs), and "Anoka" by "A7" RIL population was genotyped with 97 markers (82 RFLPs, 14 SSRs, and 1 morphologic marker). In "Pride B216" by "A15" RIL population, six markers linked to four major QTL that control Fe deficiency have been identified and mapped on chromosomes 3, 14, and 18. In "Anoka" by "A7" RIL population, one major gene/QTL that controls iron deficiency has been mapped to the same position on chromosome 3 (Fig. 2.1; Table 4.1) (Lin et al. 1997). In two subsequent studies using the same two RIL populations, 13 additional QTL that control iron deficiency chlorosis have been identified and mapped on chromosomes 3, 11, 12, 14, 18, 19, and 20 (Fig. 2.1; Table 4.1) (Lin et al. 2000a, 2000b).

Kilo and Lightfoot (1996) grew the "Essex" by "Forrest" RIL population (ExF) in a field in Carbondale, IL, and genotyped it with several RAPD markers and identified four QTL that control Mn toxicity (Kilo and Lightfoot 1996). In a subsequent study, we grew the same RIL population derived from a cross between "Essex" and

Table 4.1 Detailed information on QTL that control iron deficiency chlorosis (IDC) and manganese (Mn) toxicity reported and not reported in SoyBase (http://www.soybase.org) from 1992 to 2020

Trait	*QTL*	LG/Chr	Interval (cM)	Marker/interval	LOD, *P*, or *F*	R^2 (%)	Environment	Reference
Iron efficiency	*qFe-effc 1–1*	A2/8	47.8–49.8	*i*	$p = 0.029$	0.17	Ames, Iowa	Diers et al. (1992)
	qFe-effc 1–2	D1a/1	113.9–115.9	pc-63	$p = 0.0001$	0.33	Ames, Iowa	
	qFe-effc 1–3	D1a/1	101.3–103.3	pK-417b	$p = 0.0002$	0.28	Ames, Iowa	
	qFe-effc 1–4	D1a/1	93.9–95.9	pK-647	$p = 0.031$	0.12	Ames, Iowa	
	qFe-effc 1–5	G/18	23.1–25.1	pK-69a	$p = 0.035$	0.12	Ames, Iowa	
Iron deficiency	*qFe-defc 3–1*	B2/14	72.8–98.5	Satt070	2.5	10.80	Ames, Iowa	Lin et al. (1997)
	qFe-defc 3–2	B2/14	93.5–104.7	Satt063	2.4	7.70	Ames, Iowa	
	qFe-defc 3–3	G/18	73.0–76.0	T036–2	2.2	9.20	Ames, Iowa	
	qFe-defc 3–4	N/3	28.5–30.4	Satt009-K418_1	2.1	8.00	Ames, Iowa	
	qFe-defc 4–1	B2/14	71.10–73.80	Satt020-Satt070	2.4	11.30	Ames, Iowa	
	qFe-defc 4–2	I/20	44.0–64.9	A515_1-K644_2	2.6	19.30	Ames, Iowa	
	qFe-defc 4–3	H/12	60.20–84.10	A404_1-B069–1	2.4	21.70	Ames, Iowa	
	qFe-defc 5–1	A1/5	42.0–60.0	A256–2-A258_1	2.8	35.20	Ames, Iowa	
	qFe-defc 5–2	N/3	58.40–74.50	Satt_033	13.1	72.70	Ames, Iowa	
	qFe-defc 6–1	I/20	44.0–64.90	A515_1	3.5	80.00	Ames, Iowa	
	qFe-defc 6–2	N/3	58.40–74.50	Satt_033	7.3	68.80	Ames, Iowa	
	qFe-defc 7–1	G/18	66.60–73.0	Satt012-T036_2	4.0	15.40	Ames, Iowa	
	qFe-defc 7–2	N/3	30.40–42.0	K418_1	2.4	23.40	Ames, Iowa	
	qFe-defc 8–1	B2/14	72.13–98.5	Satt020	2.1	7.70	Ames, Iowa	
	qFe-defc 8–2	G/18	66.0–73.0	Satt012	4.2	16.00	Ames, Iowa	
	qFe-defc 8–3	H/12	60.2–84.1	A404_1	2.4	25.00	Ames, Iowa	
	qFe-defc 8–4	N/3	30.40–42.0	K418_1	2.1	11.50	Ames, Iowa	
Iron deficiency	*qFe defic 9–1*	B1/11	25.20–58.90	A118_1	2.3	19.60	Ames, Iowa	Lin et al. (2000a)
	qFe defic 9–2	B2/14	93.50–104.70	Satt063	3.0	12.10	Ames, Iowa	

(continued)

Table 4.1 (continued)

Trait	*QTL*	LG/Chr	Interval (cM)	Marker/interval	LOD, *P*, or *F*	R^2 (%)	Environment	Reference
	qFe defic 9–3	G/18	66.60–73.00	Satt012	2.6	9.60	Ames, Iowa	
	qFe defic 9–4	L/19	73.40–87.90	A132_2	2.7	16.30	Ames, Iowa	
	qFe defic 9–5	N/3	30.40–42.0	K418–1	2.7	21.70	Ames, Iowa	
	qFe defic 10–1	I/20	44.00–64.90	A515_1	2.9	29.40	Ames, Iowa	
	qFe defic 10–2	N/3	58.40–74.50	Sat_033	5.2	39.50	Ames, Iowa	
	qFe defic 10–3	B2/14	73.00–104.70	A148_2	2.3	11.70	Ames, Iowa	
Iron deficiency	*qFe defic 11–1*	B2/14	71.10–73.80	Satt020	2.4	11.30	Ames, Iowa	Lin et al. (2000b)
	qFe defic 11–2	I/20	44.00–64.90	A515_1	2.6	19.30	Ames, Iowa	
	qFe defic 11–3	H/12	60.20–84.10	A404_1	2.4	21.70	Ames, Iowa	
	qFe defic 12–1	I/20	44.00–64.90	A515_1	3.5	80.00	Ames, Iowa	
	qFe defic 12–2	N/3	58.40–74.50	Sat_033	7.3	68.80	Ames, Iowa	
Mn toxicity (leaf chl.)	*qChlo 1–1*	B2/14	60.77	Satt318	2.30	0.14	Carbondale, IL	Kassem et al. (2004)
	qChlo 1–2	B2/14	60.77	Satt318	2.60	0.18	Carbondale, IL	
	qChlo 1–3	D2/17	48.44	Sat_092	2.20	0.11	Carbondale, IL	
Mn toxicity (root nec.)	*qNecro 1–1*	C2/6	64.91	Satt305	2.70	0.27	Carbondale, IL	
	qNecro 1–2	C2/6	42.94	Satt291	2.10	0.23	Carbondale, IL	
	qNecro 1–3	G/18	35.43	$OE02_{1000}$	2.00	0.59	Carbondale, IL	
	qNecro 1–4	I/20	29.61	Satt239	2.10	0.30	Carbondale, IL	
Iron deficiency	*qFe defic 13–1*	A1/5	94.95–96–95	Satt211	–	–		Charlson et al. (2005)
	qFe defic 13–2	L/19	53.60–55.60	Satt481	–	0.12		
	qFe defic 13–3	I/20	64.60–66.60	Sat_104	–	–		
Iron def. Chl. (IDC)	*qIDC1*	D1b/2	17.8–19.8	ss107918691	(–log10p = 7.531)	24.30		Mamidi et al. (2011)
	qIDC2	N/3	85.8–88.2	ss107920575	(–log10p = 4.898)	15.70		

Trait	*QTL*	LG/Chr	Interval (cM)	Marker/interval	LOD, *P*, or *F*	R^2 (%)	Environment	Reference
	qIDC3	N/3	85.8–88.2	ss107927657	(–log10*p* = 3.781)	16.30		
	qIDC4	N/3	85.8–88.2	ss107913086	(–log10*p* = 3.781)	15.30		
	qIDC5	C2/6	107.5–109.5	ss107912865	(–log10*p* = 4.018)	2.60		
	qIDC6	M/7	38.9–40.9	ss107913778	(–log10*p* = 4.548)	17.80		
	qIDC7	F/13	48.4–50.4	ss107913443	(–log10*p* = 4.008)	15.90		
	qIDC8	F/13	60.4–62.4	ss107924542	(–log10*p* = 2.601)	13.40		
	qIDC9	L/19	55.4–57.4	ss107927321	(–log10*p* = 4.124)	15.40		
	qIDC3-g1	N/3	45,391,018 (bp)	ss107927657	*	16.30		
	qIDC3-g2	A1/5	937,302 (bp)	ss107923476	*	7.20		
	qIDC3-g3	A1/5	41,114,078 (bp)	ss107916955	*	7.00		
	qIDC3-g4	M/7	8,652,831 (bp)	ss107924009	*	4.78		
	qIDC3-g6	B1/11	38,020,165 (bp)	ss107928908	*	1.64		
	qIDC3-g7	F/13	30,457,599 (bp)	ss107915800	*	7.83		
	qIDC3-g8	E/15	48,694,193 (bp)	ss107920392	*	12.66		
	qIDC3-g9	J/16	3,039,691 (bp)	ss107918800	*	4.55		
	qIDC3-g10	J/16	36,544,010 (bp)	ss107914365	*	0.47		
	qIDC3-g11	D27	4,899,023 (bp)	ss107920404	*	0.25		
	qIDC3-g12	G/18	1,957,710 (bp)	ss107914461	*	2.07		
	qIDC3-g13	G/18	56,429,447 (bp)	ss107913107	*	2.77		

(continued)

Table 4.1 (continued)

Trait	*QTL*	LG/Chr	Interval (cM)	Marker/interval	LOD, *P*, or *F*	R^2 (%)	Environment	Reference
	qIDC3-g14	L/19	40,357,687 (bp)	ss107927321	*	15.41		
	qIDC3-g15	I/20	343,106 (bp)	ss107920153	*	2.33		
Iron def. Chlorosis (IDC)	*qIDC4-g1*	D1b/2	46,939,135	Gm02:43850877	(−log10 = 3.784)	5.20		Mamidi et al. (2014)
	qIDC4-g2	N/3	44,927,455	Gm03:42920376	(−log10 = 4.165)	15.20		
	qIDC4-g3	N/3	44,930,187	Gm03:42923228	(−log10 = 4.165)	15.20		
	qIDC4-g4	N/3	44,970,860	Gm03:42963950	(−log10 = 4.275)	14.42		
	qIDC4-g5	N/3	44,985,022	Gm03:42977988	(−log10 = 4.165)	15.20		
	qIDC4-g6.1	N/3	45,031,929	Gm03:43104890	(−log10 = 5.127)	15.77		
	qIDC4-g6.2	N/3	45,038,912	ss715586372	(−log10 = 3.684)	14.71		
	qIDC4-g7	N/3	45,107,141	ss715586373	(−log10 = 3.905)	13.87		
	qIDC4-g8	N/3	45,343,317	Gm03:43338975	(−log10 = 3.689)	17.28		
	qIDC4-g9	N/3	45,343,612	Gm03:43339270	(−log10 = 4.231)	18.85		
	qIDC4-g10	A1/5	8,817,714	Gm05:300244	(−log10 = 3.912)	10.81		
	qIDC4-g11	A1/5	8,823,135	Gm05:294856	(−log10 = 4.617)	11.43		
	qIDC4-g12	A1/5	8,877,264	Gm05:240477	(−log10 = 3.855)	8.30		
	qIDC4-g13.1	A1/5	17,397,118	Gm05:15885333	(−log10 = 3.704)	0.08		
	qIDC4-g13.2	A1/5	17,397,130	Gm05:15885321	(−log10 = 3.704)	0.08		
	qIDC4-g13.3	A1/5	17,397,135	Gm05:15885316	(−log10 = 3.704)	0.08		
	qIDC4-g14	M/7	6,397,319	Gm07:6435967	(−log10 = 3.961)	8.80		
	qIDC4-g15	M/7	6,405,421	Gm07:6444069	(−log10 = 3.961)	8.80		
	qIDC4-g16	M/7	6,540,541	ss715598504	(−log10 = 4.128)	8.10		
	qIDC4-g17	M/7	6,575,306	Gm07:6613954	(−log10 = 3.961)	8.80		
	qIDC4-g18.1	B1/11	530,116	Gm11:539091	(−log10 = 4.401)	4.73		
	qIDC4-g18.2	B1/11	530,117	Gm11:539092	(−log10 = 4.401)	4.73		
	qIDC4-g18.3	B1/11	530,118	Gm11:539093	(−log10 = 4.401)	4.73		

Trait	*QTL*	LG/Chr	Interval (cM)	Marker/interval	LOD, *P*, or *F*	R^2 (%)	Environment	Reference
	qIDC4-g19	H/12	20,406,669	Gm12:27690435	(−log10 = 3.697)	0.02		
	qIDC4-g20	E/15	21,422,057	Gm15:21479453	(−log10 = 4.152)	2.10		
	qIDC4-g21	J/16	27,300,116	Gm16:27643739	(−log10 = 3.779)	4.80		
	qIDC4-g22	D2/17	25,859,992	Gm17:25482971	(−log10 = 4.322)	0.66		
	qIDC4-g23.1	D2/17	25,859,995	Gm17:25482974	(−log10 = 4.322)	0.66		
	qIDC4-g23.2	G/18	15,905,372	Gm18:15646012	(−log10 = 3.781)	3.25		
	qIDC4-g24	G/18	28,141,888	Gm18:30878841	(−log10 = 3.786)	3.78		
	qIDC4-g25	G/18	59,284,729	Gm18:55012843	(−log10 = 4.006)	0.70		
	qIDC4-g26	L/19	34,914,546	Gm19:35099001	(−log10 = 3.769)	0.81		
	qIDC4-g27	L/19	40,193,564	Gm19:40386502	(−log10 = 3.995)	15.62		

"Forrest" ($n = 100$) in Carbondale, IL, and genotyped it with 201 SSR markers, and the genetic linkage map (ExF) used in this study covered 2823.1 cM and was constructed using MapMaker-Exp (Lander et al. 1987). We used single-marker analysis (ANOVA) to detect QTL that control resistance to manganese toxicity. Our results identified three additional QTL that control Mn toxicity. The QTL that controls root necrosis 7 days after treatment (NECRO) has been mapped on chromosomes 6 (*qNECRO1* linked to the SSR markers Satt291 and Satt305), 18 (*qNECRO2* linked to the RAPD marker OEO2), and 20 (*qNECRO3* linked to the SSR marker Satt231) (Kassem et al. 2004).

Charlson et al. (2005) grew a RIL population derived from a cross between "Pioneer 9254" and "A97–770012" lines in Iowa and genotyped it with 108 SSR markers. Three QTL that control iron deficiency chlorosis (Fe-def 1–1 to Fe-def 1–3) have been identified on chromosomes 5, 19, and 20 (Fig. 2.1; Table 4.1) (Charlson et al. 2005).

Mamidi et al. (2011) grew two populations in several locations in the USA over 2 years (2005 and 2006). The first population ($n = 143$) was grown in 2005, and the second population ($n = 141$) was grown in 2006. They were genotyped with 858 SNPs and 868 SNPS, respectively, and used GWAS analysis to identify markers associated with IDC. The study identified 42 loci and 88 loci associated with IDC in 2005 and 2006, respectively. Nine major QTL that control IDC have been identified and mapped on chromosomes 2, 3, 6, 7, 13, and 19. These loci explained 74.5% and 93.8 and of the total IDC variation in 2005 and 2006, respectively (Fig. 2.1; Table 4.1) (Mamidi et al. 2011). In a subsequent study using the same populations ($n = 132$) and ($n = 138$) in 2005 and 2006, respectively, Mamidi et al. (2014) identified five major QTL regions associated with IDC on chromosomes 3, 5, 7, 16, and 19. Within the major QTL on chromosome 3, they identified four candidate genes such as Glyma03g38620 (*FRE1*) encoding for the phytochrome A, Glyma03g38640 (*NRT1:2*) encoding for the nitrate transporter 1:2, Glyma03g39050 (*NAS3*) encoding for the nicotianamine synthase 3, and Glyma03g40700 (*AMT2:1*) encoding for the ammonium transporter 2. With the major QTL on chromosome 5, they identified three candidate genes including Glyma05g01460 (*AHA11*) encoding for the H^+–ATPase 11, Glyma05g08170 (*AGO10*) encoding for a stabilizer of iron transporter and SufD/polynucleotidyl transferase, and Glyma05g08260 (*RAN1*) encoding for the RAS-related nuclear protein-1. Within the major QTL on chromosome 7; they identified two candidate genes including Glyma07g07380 (*FRO2*) encoding for a ferric reduction oxidase 2 and Glyma07g08240 (*FRD4*) encoding for the signal recognition particle receptor protein. Within the major QTL on chromosome 16, they identified the candidate gene Glyma16g23710 (*FNR1*) encoding for a ferredoxin $NADP^+$ oxidase 1. Within the major QTL on chromosome 19, they identified the two candidate genes Glyma19g32190 (HMAS) encoding for a heavy metal ATPase 5 and Glyma19g32401 (*OPT5*) encoding for the oligopeptide transporter 5 (Mamidi et al. 2014).

Peiffer et al. (2012) attempted to identify candidate genes involved in iron deficiency chlorosis within the QTL region previously identified an mapped on chromosome 3 (Lin et al. 1997, 200a,b). They used transcriptome analysis and qRT-PCR, to analyze two near-isogenic lines (NILs) generated by crossing the two cultivars "Clark" and "Harosy" with T203 (PI 54619). The study identified two candidate genes within this QTL region (Satt387-Satt339) narrowed to 250 kb. They identified 18 candidate genes in the region; however, further analysis revealed 2 differentially expressed genes (Glyma03g28610 and Glyma03g28630) that exclusively expressed in the roots of control and stressed plants. These two genes encode for transcription factors (TFs) (Peiffer et al. 2012). It has been reported that regulatory elements of some genes involved in Fe metabolism may be outside of the QTL region (O'Rourke et al. 2009; Peiffer et al. 2012). In *A. thaliana*, the two genes encode for TFs AtbHLH038 and AtbHLH039 which are known enhance the expression of AtIRT1 and AtFRO2 involved in Fe acquisition (Yuan et al. 2008; Peiffer et al. 2012).

4.1.2 QTL for Aluminum Tolerance

Al toxicity limits plant growth and development especially in acidic soils and causes severe crop losses in agriculture land worldwide (Von Uexkull and Mutert 1995; Samac and Tesfaye 2003; Tasma et al. 2008; Simoes et al. 2012). It is caused by a heavy plant roots' uptake of Al^{3+} and H^{+} ions which leads to a reduction in the overall root system's growth and plant productivity (Stephens et al. 1990; Lazof et al. 1996; Kochian 1995; Von Uexkull and Mutert 1995). The mechanisms by which plants cope with Al toxicity include pumping Al^{3+} ions outside of the root cells or their internal detoxification (Simoes et al. 2012).

The development of Al tolerant soybean cultivars is desirable by plant scientists, breeders, and farmers (Foy 1988; Tasma et al. 2008; Qi et al. 2008; Bianchi-Hall et al. 2000; Korir et al. 2010); therefore, several studies have been conducted to identify QTL for Al tolerance in soybean (Bianchi-Hall et al. 2000; Qi et al. 2008; Sharma et al. 2011; Korir et al. 2013; Abdel-Haleem et al. 2014; Cai et al. 2019; Wang et al. 2019).

Bianchi-Hall et al. (2000) grew a RIL population derived from a cross between "Young" and "PI416937" lines (n = 120) in hydroponics, genotyped it with 155 RFLP markers, and used single-marker and pairwise analyses for QTL data analysis. The identified 14 RFLP markers associated with 6 QTL-controlling Al tolerance (root tap extension) that have been identified and mapped on chromosomes 8, 11, 13, 16, and 19 (Fig. 2.1; Tables 4.2 and 4.3) (Bianchi-Hall et al. 2000).

Qi et al. (2008) grew a RIL population derived from a cross between "Kefeng No.1" and "Nannong 1138–2" lines (n = 184) on sand supplied with 28 mg/L of Al or no Al (control) in the greenhouse and genotyped it with 219 SSR markers, 189 RFLP markers, 40 ESTs, 1 phenotypic marker, and 3 resistance genes. The genetic linkage map used covered 4151.2 cM; they used average membership values (FAi) to estimate Al tolerance using four traits (root dry weight (RDW), number of leaves (NL), shoot dry weight (RDW), and plant height (PH) at the seedling stage. For QTL data analysis, they used CIM and MIM of WinQTL Cartographer. Eight QTL

Table 4.2 Detailed information on QTL that control aluminum (Al^{3+}) tolerance in soybean reported and not reported in SoyBase (http://www.soybase.org) from 2000 to 2020

Trait	*QTL*	LG/ Chr	Interval (cM)	Marker/interval	LOD, *P*, or *F*	R^2 (%)	Environment	Reference
Aluminum tolerance QTL								
Aluminum tolerance	*qAlTol1–1*	A2/8	117.5–119.5	O103–1n	$p = 0.003$	32.00	Hydroponics	Bianchi-Hall et al. (2000)
(Root tap extension)	*qAlTol1–2*	B1/11	19–21	Cr207–1	$p = 0.0976$	32.00	Hydroponics	
	qAlTol1–3	B1/11	39.1–41.1	A847	$p = 0.0460$	32.00	Hydroponics	
	qAlTol1–4	F/13	66.2–68.2	K644–1	$p = 0.0193$	32.00	Hydroponics	
	qAlTol1–5	J/16	56.2–58.2	B122–1	$p = 0.0041$	32.00	Hydroponics	
	qAlTol1–6	L/19	3.1–5.1	EV2–1	$p = 0.0038$	24.00	Hydroponics	
Al-Tol -CIM	*qFAi-1*	B1/11	65.4–78	GMKF046-GMKF080	2.60	6.14	China	Qi et al. (2008)
	qFAi-2	D1b/2	75.8–93.8	Satt172-Satt703	2.50	5.20	China	
	qFAi-3	D2/17	108–124	GMKF058-Satt397	3.80	9.07	China	
	qFAi-4	D2/17	120.5–127.8	Satt397-Satt669	3.10	6.83	China	
	qFAi-5	L/19	49.6–57	Satt278-Sat_195	3.10	6.01	China	
Al-Tol -MIM	*qFAi-1*	B1/11	65.1–80.6	GMKF046-GMKF080	2.10	8.90	China	
	qFAi-5	L/19	42.5–63.7	Satt278-sat_195	2.50	5.70	China	
	qFAi-6	D1b/2	87.5–98.1	Satt703- LE45T	3.70	24.60	China	
	qFAi-7	D1b/2	101.7–109.3	A516_3H-A953_3H	2.90	1.90	China	
	qFAi-8	I/20	97.1–107.9	B39V-Sat_419	2.50	10.50	China	
Aluminum tolerance	*qMRL1-(AlTol_1)*	B2/14	72.5	Satt122	$p = 0.011$	18.00	Hydroponics	Sharma et al. (2011)
Mean root length (MRL)	*qMRL2-(AlTol_2)*	C2/6	126.2	Satt202	$p = 0.013$	17.00	Hydroponics	
	qMRL3-(AlTol_3)	D2/17	107.5	Satt310	$p = 0.045$	10.00	Hydroponics	

Trait	*QTL*	LG/ Chr	Interval (cM)	Marker/interval	LOD, *P*, or *F*	R^2 (%)	Environment	Reference
	qMRL4-(AlTol_4)	F/13	16.1	Satt252	$p = 0.011$	16.00	Hydroponics	
	qMRL5-(AlTol_5)	F/13	78.1	Satt334	$p = 0.024$	15.00	Hydroponics	
	qMRL6-(AlTol_16	G/18	10.9	Satt610	$p = 0.008$	24.00	Hydroponics	
	qMRL7-(AlTol_7)	G/18	55.9	Satt138	$p = 0.04$	18.00	Hydroponics	
	qMRL8-(AlTol_8)	L/19	10.3	Satt232	$p = 0.028$	11.00	Hydroponics	
	qMRL9-(AlTol_9)	N/3	22.7	Satt152	$p = 0.039$	10.00	Hydroponics	
	qMRL10-(AlTol_10)	N/3	27.1	Satt159	$p = 0.039$	15.00	Hydroponics	
	qMRL11-(AlTol_11)	O/10	86.8	Satt123	$p = 0.05$	13.00	Hydroponics	
	qMRL12-(AlTol_12)	F/13	24.1	Satt160-Satt252	3.300	31.00	Hydroponics	
	qMRL13-(AlTol_13)	C2/6	3.2	Satt202-Satt371	14.700	34.00	Hydroponics	
Relative growth (RMG)	*qRMG1-(AlTol_14)*	C2/6	3.4	Satt202-Satt371	14.100	26.00	Hydroponics	
Aluminum tolerance	*qAlTol3–1*	B1/11	46.4–53.4	GMKF046-Sat_128	0.246	9.23	China	Korir et al. (2011)
	qAlTol3–2	D1b/2	38–42	GNT007-Satt698	0.103	5.06	China	
	qAlTol3–3	D2/17	86.4–95.4	LI10TSatt311	0.099	3.10	China	
	qAlTol3–4	K/9	60.3–62.3	Sat_043-GNE054	0.114	4.91	China	
	qAlTol3–5	B1/11	46.4–53.4	GMKF046-Sat_128	0.124	7.05	China	
	qAlTol3–6	O/10	68.9–74	GNB054-BE801128	0.113	3.82	China	
	qAlTol3–7	B1/11	46.4–53.4	GMKF046-Sat_128	0.165	8.85	China	
	qAlTol3–8	D2/17	86.4–95.5	Satt311-Satt464	0.158	6.64	China	
	qAlTol3–9	L/19	34.5–38.1	Satt313-GNE029A	0.138	7.53	China	

(continued)

Table 4.2 (continued)

Trait	*QTL*	LG/ Chr	Interval (cM)	Marker/interval	LOD, *P*, or *F*	R^2 (%)	Environment	Reference
Aluminum tolerance	*qAlTol1-g1*	A2/8	121.7–123.7	Satt209	$p < 0.01$	14.50	China	Korir et al. (2013)
	qAlTol1-g2	B1/11	39.1–41.1	Sat_364	$p < 0.01$	8.20	China	
	qAlTol1-g3	C2/6	126.2–130.2	Satt286	$p < 0.01$	5.20	China	
	qAlTol1-g4	D1b/2	75.8–93.8	Satt005	$p < 0.01$	8.60	China	
	qAlTol1-g5	D2/17	108–110	Satt186	$p < 0.01$	10.40	China	
	qAlTol1-g6	F/13	33.2–36.1	Sat_240	$p < 0.01$	8.50	China	
	qAlTol1-g7	I/20	30.4–32.4	Sat_174	$p < 0.01$	6.70	China	
	qAlTol1-g8	J/16	56.2–58.2	GMES0069	$p < 0.01$	5.20	China	
	qAlTol1-g9	K/9	56.2–58.2	Sct_190	$p < 0.01$	4.30	China	
	qAlTol1-g10	L/19	49.6–57	Satt284	$p < 0.01$	7.70	China	
	qAlTol1-g11	O/10	86.2–90.1	GMES1703	$p < 0.01$	5.90	China	
Aluminum tolerance	*qAL_NOAL_03*	N/3	80.9–102.3	Satt237-K494_1	2.53	10.30	Hydroponics	Abdel-Haleem et al. (2014)
	qAL_NOAL_18	G/18	43–51.8	Satt570-Satt501	2.61	9.40	Hydroponics	
	qAL_HIAL_08	A2/8	147.9–156.8	Satt409	11.22	44.80	Hydroponics	
	qAL_HIAL_16	J/16	22.8–51.5	Sat_093-Satt431	2.25	8.80	Hydroponics	
	qAL_PC_08	A2/8	147.7–155.7	Satt409	11.42	40.70	Hydroponics	
	qAL_PC_16	J/16	31–47.6	B122_1-Satt431	3.41	12.80	Hydroponics	
	qAL_PC_19	L/19	0–20.3	A169_1-A106_1	2.58	5.60	Hydroponics	
Aluminum tolerance	*qRENOAL11*	B1/11	110.4–112.4	Bin1831	2.86	5.75	Hydroponics	Cai et al. (2019)
	qRENOAL12	H/12	77.2–79.2	Bin1979	2.94	6.08	Hydroponics	
	qREAL02	D1b/2	5.9–7.9	Bin182	2.59	5.56	Hydroponics	
	qREAL06	C2/6	140–142	Bin954	4.66	8.99	Hydroponics	
	qREAL18	G/18	169.9–171.9	Bin3057	2.70	5.02	Hydroponics	
	qRRE06	C2/6	150.3–152.3	Bin964	3.66	6.89	Hydroponics	

Trait	*QTL*	LG/ Chr	Interval (cM)	Marker/interval	LOD, *P*, or *F*	R^2 (%)	Environment	Reference
	qRRE15	E/15	109.2–111.2	Bin2492	2.56	4.80	Hydroponics	
	qAl-HC2	G/18	47.2–49.2	Bin2960	3.76	7.04	Hydroponics	
	qRRE18	G/18	216.8–218.8	Bin3087	2.78	5.15	Hydroponics	
	qAl06	C2/6	132.5–134.5	Bin951	3.10	5.89	Hydroponics	
	qAl-HC2	G/18	47.2–49.2	Bin2960	3.75	7.13	Hydroponics	
Aluminum tolerance	*qRRE_04*	C1/4	726,277 bp	bin93-bin94	2.76	7.09	Hydroponics	Wang et al. (2019)
	qRRE_16	J/16	60,558 bp	bin8	2.73	7.80	Hydroponics	
	qRRE_17	D2/17	32,282 bp	bin93	3.32	8.52	Hydroponics	
	qACC_04	C1/4	726,277 bp	bin93-bin94	3.27	8.98	Hydroponics	
	qACC_19	L/19	302,930 bp	bin1979	2.66	7.26	Hydroponics	

Table 4.3 Identified QTL that control various indicators of Al^{3+} tolerance reported and not reported in SoyBase (www.soybase.org) as of December 31, 2020

Identified QTL	LG/Chr.	Al^{3+} tolerance indicators	In SoyBase	Reference
qAlTol1–1, qAlTol1–2, qAlTol1–3, qAlTol1–4, qAlTol1–5, qAlTol1–6	A2/8, B1/11, B1/11, F/13 J/16, L/19	Tap root extension (TRE)	Yes	Bianchi-Hall et al. (2000)
qFAi-1, qFAi-2, qFAi-3, qFAi-4, qFAi-5, qFAi-1, qFAi-5, qFAi-6, qFAi-7, qFAi-8	B1/11, D1b/2, D2/17, D2/17, L/19, B1/11, L/19, D1b/2, D1b/2, I/20	Plant height (PH), leaf number (LN), shoot dry weight (SDW), root dry weight (RDW)	No	Qi et al. (2008)
qMRL1, qMRL2, qMRL3, qMRL4, qMRL5, qMRL6, qMRL7, qMRL8, qMRL9, qMRL10, qMRL11, qMRL12, qMRL13, qRMG1	B2/14, C2/6, D2/17, F/13, F/13, G/18, G/18, L/19, N/3, N/3, O/10, F/13, C2/6, C2/6	Root tolerance index (RTI), Root relative growth (RRG)	Yes	Sharma et al. (2011)
qAlTol3–1, qAlTol3–2, qAlTol3–3, qAlTol3–4, qAlTol3–5, qAlTol3–6, qAlTol3–7, qAlTol3–8, qAlTol3–9	B1/11, D1b/2, D2/17, K/9, B1/11, O/10, B1/11, D2/17, L/19	Plant dry weight (PDW), shoot dry weight (SDW), root dry weight (RDW)	Yes	Korir et al. (2011)
qAlTol1-g1, qAlTol1-g2, qAlTol1-g3, qAlTol1-g4, qAlTol1-g5, qAlTol1-g6, qAlTol1-g7, qAlTol1-g8, qAlTol1-g9, qAlTol1-g10, qAlTol1-g11	A2/8, B1/11, C2/6, D1b/2, D2/17, F/13, I/20, J/16, K/9, L/19, O/10	Relative root elongation (RRE)	No	Korir et al. (2013)
qAL_NOAL_03, qAL_NOAL_18, qAL_HIAL_08, qAL_HIAL_16, qAL_PC_08, qAL_PC_16, qAL_PC_19	N/3, G/18, A2/8, J/16, A2/8, J/16, L/19	Tap root extension at HIAL, Tap root extension at NOAL, Relative root elongation (RRE)	No	Abdel-Haleem et al. (2014)
qRENOAL11, qRENOAL12, qREAL02, qREAL06, qREAL18, qRRE06, qRRE15, qAl-HC2, qRRE18, qAl06, qAl-HC2	B1/11, H/12, D1b/2, C2/6, G/18, C2/6, E/15, G/18, G/18, C2/6, G/18	Relative root elongation (RRE)	No	Cai et al. (2019)
qRRE_04, qRRE_16, qRRE_17, qACC_04, qACC_19	C1/4, J/16, D2/17, C1/4, L/19	Relative root elongation (RRE), Apical (root) Al^{3+} content	No	Wang et al. (2019)

that control Al toxin tolerance (*qFAi1–8*) have been identified and mapped on chromosomes 2, 11, 17, 19, and 20 (Fig. 2.1; Tables 4.2 and 4.3) (Qi et al. 2008).

In a subsequent study, Korir et al. (2011) grew a RIL population derived from a cross between "KF No. 1" and "NN 1138–2" lines (n = 184) on hydroponics and in a field in China and genotyped it with 834 markers. The genetic linkage map used in this study covered 2307.8 cM, and they used CIM for QTL analysis. Six major QTL that control relative total plant dry weight (RTDW), relative shoot dry weight (RSDW), and relative root dry weight (RRDW) have been identified and mapped on chromosomes 2, 9, 10, 11, 17, and 19 (Fig. 2.1; Table 4.2) (Korir et al. 2011).

Sharma et al. (2011) grew the same "Essex" by "Forrest" RIL population (ExF, $n = 100$) (Kassem et al. 2004, 2006) on hydroponics, genotyped it with 201 SSR markers, and used one-way ANOVA. They identified 11 SSR markers associated with 11 QTL-controlling Al tolerance (root length mean, RLMN, and relative mean growth, RMG) that have been identified and mapped on chromosomes 3, 6, 10, 13, 14, 17, 18, and 19. CIM identified three QTL for RLM and RMG on chromosomes 6 and 13 (Fig. 2.1; Tables 4.2 and 4.3) (Sharma et al. 2011).

Korir et al. (2013) grew a RIL population derived from a cross between "KF No. 1" and "NN1138–2" lines ($n = 188$) on hydroponics and used relative root elongation (RRE) as a measure in order to identify QTL for Al tolerance. They used single-marker analysis for QTL data analysis. Eleven QTL that control Al tolerance (measured as RRE) have been identified and mapped on chromosomes 2, 6, 8, 9, 10, 11, 13, 16, 17, 19, and 20 (Fig. 2.1; Tables 4.2 and 4.3) (Korir et al. 2013).

Abdel-Haleem et al. (2014) grew a RIL population derived from a cross between "Young" and "PI 416937" lines ($n = 120$) in hydroponics, genotyped it with 279 RFLP and SSR markers, and used MIM for QTL data analysis of Al tolerance estimated as root tap extension in non-Al stress (NOAL), Al stress (HIAL), and percent of control (PC) traits. Two QTL that control root tap extension in NOAL conditions have been identified and mapped on chromosomes 3 and 18; two QTL that control root tap extension in HIAL conditions have been identified and mapped on chromosomes 8 and 16; and three QTL that control root tap extension in PC conditions have been identified and mapped on chromosomes 8, 16, and 19 (Fig. 2.1; Tables 4.2 and 4.3). Moreover, one candidate gene that control Al tolerance has been identified and located on the QTL region on chromosome 8. This candidate gene, *Glyma08g42400*, encodes for citrate synthase (Table 4.4) (Abdel-Haleem et al. 2014).

Cai et al. (2019) grew a RIL population derived from a cross between "Huachun2" (Al tolerant) and "Wayao" (Al-sensitive) lines ($n = 196$) in hydroponics, genotyped it with 3469 bin markers, and used CIM of WinQTL cartographer for QTL data analysis of Al tolerance estimated as relative root elongation (RRE). Eleven QTL that control Al tolerance (RRE) have been identified and mapped on chromosomes 2, 6, 11, 12, 15, and 18; however, 5 of them are novel and the others have been previously identified (Korir et al. 2013; Abdel-Haleem et al. 2014) (Fig. 2.1; Tables 4.2 and 4.3). Moreover, 11 candidate genes are within these QTL regions. The candidate genes are *Glyma06g24500* encoding for the histone H2A; *Glyma06g24520* encoding for a protein disulfide isomerase; *Glyma15g21890* encoding for S-adenosylmethionine synthase; *Glyma18g06220*, *Glyma18g06230*, and *Glyma18g06250* encoding for a peroxidase; *Glyma18g48730* and *Glyma18g48740* encoding for ethylene-responsive transcription factor 13; *Glyma18g49140* encoding for WRKY DNA-binding domain; *Glyma06g20710* encoding for a putative S-adenosyl-L-methionine-dependent methyltransferase; and *Glyma06g20730* encoding for a glutathione S-transferase U9 (Table 4.4) (Cai et al. 2019).

Table 4.4 Candidate genes identified within the genomic regions harboring QTL that control various indicators of Al^{3+} tolerance from 2000 to 2020

Gene	LG/ Chr	Position	Annotation	References
Glyma02g45790	D1b/2	49.93 Mbp	Citrate synthase (CS)	Abdel-Haleem et al. (2014)
Glyma08g42400	A2/8	12.29 Mbp	Citrate synthase (CS)	
Glyma08g16770	A2/8	42.39 Mbp	Citrate synthase (CS)	
Glyma14g03000	B2/14	1.90 Mbp	Citrate synthase (CS)	
Glyma15g42300	E/15	49.69 Mbp	Citrate synthase (CS)	
Glyma18g12410	G/18	11.46 Mbp	Citrate synthase (CS)	
Glyma06g24500	C2/6	21,904,185–21,906,364	Histone H2A	Cai et al. (2019)
Glyma06g24520	C2/6	21,981,184–21,984,389	Protein disulfide isomerase	
Glyma06g20710	C2/6	17,109,334–17,113,973	Putative S-adenosyl-L-methionine-dependent methyltransferase	
Glyma15g21890	E/15	20,191,634–20,202,861	S-adenosylmethionine synthetase	
Glyma18g06220	G/18	4,808,002–4,811,827	Peroxidase	
Glyma18g06230	G/18	4,815,244–4,817,920	Peroxidase	
Glyma18g06250	G/18	4,825,091–4,827,712	Peroxidase	
Glyma18g48730	G/18	58,176,828–58,177,740	Ethylene-responsive TF 13	
Glyma18g48740	G/18	58,181,838–58,183,134	Ethylene-responsive TF 13	
Glyma18g49140	G/18	58,523,898–58,526,841	WRKY DNA-binding domain	
Glyma.04G212400	C1/4	48419972..48430337	Rab family protein	Wang et al. (2019)
Glyma.04G212800	C1/4	48446746..48452104	Nucleotide-sugar transporter	
Glyma.04G213000	C1/4	48461372..48467197	Catalytic domain of the serine/threonine kinase	
Glyma.04G213100	C1/4	48485249..48499658	1,3-beta-glucan synthase component	
Glyma.04G213300	C1/4	48507581..48512003	NAC family transcription factor	
Glyma.04G213400	C1/4	48513546..48526035	Catalytic domain of type III phosphoinositide 4-kinase beta	
Glyma.04G213600	C1/4	48531249..48534528	Flavoprotein CzcO-associated	
Glyma.04G213700	C1/4	48536823..48545911	GDP-fucose protein O fucosyltransferase	

(continued)

Table 4.4 (continued)

Gene	LG/ Chr	Position	Annotation	References
Glyma.04G213800	C1/4	48547107..48556098	ATP-dependent metalloprotease	
Glyma.04G213900	C1/4	48558152..48562318	Intercellular signaling	
Glyma.04G214000	C1/4	48562842..48563882	PME inhibitor	
Glyma.04G214100	C1/4	48576689..48577420	HLH (helix loop helix domain)	
Glyma.04G214200	C1/4	48588818..48595244	Zinc finger	
Glyma.04G214400	C1/4	48603113..48610147	Plant alcohol dehydrogenase	
Glyma.04G214600	C1/4	48619635..48623696	PHD finger	
Glyma.04G214700	C1/4	48626650..48627708	σ factors for regulation of plastid	
Glyma.04G214800	C1/4	48632734..48637158	PHD zinc finger	
Glyma.04G214900	C1/4	48638122..48642786	WD40/YVTN repeat-like-containing domain superfamily	
Glyma.04G215200	C1/4	48682262..48684713	Catalytic domain of the serine/threonine kinase	
Glyma.04G215400	C1/4	48688559..48693977	Domain in TBC and LysM domain-containing proteins	
Glyma.04G215500	C1/4	48694724..48697753	Serine/threonine protein kinases	
Glyma.04G215600	C1/4	48701372..48703052	Ring finger	
Glyma.04G215800	C1/4	48709355..48711106	Photosystem I reaction center subunit N	
Glyma.04G215900	C1/4	48716688..48719585	Actin-related protein	
Glyma.04G216000	C1/4	48723367..48726816	Carbamoyl phosphate synthase	
Glyma.04G216100	C1/4	48744386..48746847	GT-2A trihelix transcription factor: myb-like protein	
Glyma.04g216300	C1/4	45626281..45636778	DnaJ molecular chaperone	
Glyma.04g216400	C1/4	45639837..45646610	N-terminal LabA-like	
Glyma.04g216500	C1/4	45648605..45652897	RNA-binding domain	
Glyma.04g216700	C1/4	45656880..45660390	Transcription elongation factor S-II	
Glyma.04G216900	C1/4	48794680..48795762	Polysaccharide lyase family 7 protein	
Glyma.04G217100	C1/4	48812524..48816184	Uncharacterized protein At2g34160-like	
Glyma.04G217200	C1/4	48818011..48822613	Membrane protein	
Glyma.04G217300	C1/4	48823535..48828533	Ribonucleoside-diphosphate reductase large subunit	
Glyma.04g217400	C1/4	48838002..48839592	Transcription elongation factor S-II	

(continued)

Table 4.4 (continued)

Gene	LG/ Chr	Position	Annotation	References
Glyma.04G217500	C1/4	48841683..48847602	26S proteasome regulatory complex component	
Glyma.04G218000	C1/4	48871360..48877837	Serine/threonine protein kinases	
Glyma.04G218200	C1/4	48898190..48912726	Triose-phosphate transporter family	
Glyma.04G218300	C1/4	48926008..48930251	BRI1B brassinosteroid receptor	
Glyma.04G218400	C1/4	48944031..48947321	WRKY transcription factor	
Glyma.04G218500	C1/4	48960783..48965306	Putative PEP-CTERM system TPR-repeat lipoprotein	
Glyma.04G218600	C1/4	48965931..48966949	Glycosyl hydrolases family 17	
Glyma.04G218700	C1/4	48971780..48974496	WRKY transcription factor	
Glyma.04G218800	C1/4	48983780..48987348	Programmed cell death protein 2	
Glyma.04G218900	C1/4	48988587..48993895	ABC transporter	
Glyma.04G219100	C1/4	49003316..49010609	Ribosome-interacting GTPase 1	
Glyma.04G219400	C1/4	49025185..49026960	Kelch-like protein	
Glyma.04G219500	C1/4	49027221..49032753	Disproportionating enzyme; 4-alpha-glucanotransferase	
Glyma.04G219600	C1/4	49034941..49037744	The Toll/interleukin-1 receptor	
Glyma.04G220000	C1/4	49047151..49058197	Zinc finger, C3HC4 type	
Glyma.04G220200	C1/4	49075208..49089579	Unusual protein kinase regulating ubiquinone biosynthesis	
Glyma.04G220400	C1/4	49102263..49108229	Catalytic domain of the serine/threonine kinases	
Glyma.04G220500	C1/4	49114993..49117068	Unknown function	
Glyma.04G220600	C1/4	49133080..49135000	Peroxidase	
Glyma.16g043300	J/16	4059536..4066000	Nucleoside phosphatase	
Glyma.16g043400	J/16	4075555..4080995	Nucleoside phosphatase	
Glyma.16g043500	J/16	4088719..4095606	Nucleoside phosphatase	
Glyma.16g043600	J/16	4104900..4107633	Unknown function	
Glyma.16g043700	J/16	4109008..4114624	Nucleoside phosphatase	
Glyma.16g043800	J/16	4117014..4123335	Aquaporin	
Glyma.17g228100	D2/17	38305771..38313230	ATP-dependent metalloprotease	
Glyma.17g228300	D2/17	38316355..38326030	Serine aminopeptidase	

(continued)

Table 4.4 (continued)

Gene	LG/ Chr	Position	Annotation	References
Glyma.19G056500	L/19	10193428..10196404	Glycosyltransferase family 17	
Glyma.19G056600	L/19	10297831..10300181	Unknown function	
Glyma.19G056700	L/19	10302499..10320179	Vesicle tethering protein	
Glyma.19G056800	L/19	10351816..10355739	Hypothetical methyltransferase	

Wang et al. (2019) grew a RIL population derived from a cross between "Zhonghuang 24" and "Huaxia 3" lines ($n = 160$) in hydroponics, genotyped it with 4991 SNPs and 874 SSRs, and used highly dense genetic linkage map covering 2638.24 cM and CIM of WinQTL Cartographer for QTL data analysis of Al tolerance estimated as relative root elongation (RRE) and apical Al^{3+} content (AAC). Three QTL that control RRE (*qRRE_04*, *qRRE_16*, and *qRRE_17*) have been identified and mapped on chromosomes 4, 16, and 17, and two QTL that control AAC (*qAAC_04* and *qAAC_19*) have been identified and mapped on chromosomes 4 and 19 (Fig. 2.1; Tables 4.2 and 4.3). The study identified 66 differentially expressed genes (DEGs) within these QTL regions; however, only 6 highly DEGs under Al^{3+} stress have been identified. These genes were *Glyma.04g218700* which encodes for transcription factor (TF) called WRKY21 (Zhou et al. 2008), *Glyma.04g217400* which encodes for the ethylene-responsive TF ABR1 (Pandey et al. 2005), *Glyma.04g216100* and *Glyma.04g213300* which encode for TFs involved in coping with abiotic stresses (Xie et al. 2018; Hussain et al. 2017), *Glyma.04g220600* which encodes a peroxidase which may be involved in Al tolerance (Inostroza-Blancheteau et al. 2012), and *Glyma.04g212800* which encodes for a GDP-mannose transporter also involved in Al tolerance. The authors elected *Glyma.04g218700* to be a strong candidate gene involved in Al stress (Table 4.4) (Wang et al. 2019).

4.2 QTL for Water Use Efficiency (WUE) and Phosphorus Efficiency (PE)

Several studies identified QTL that control phosphorus efficiency (PE) and water use efficiency (WUE) using different RIL and other populations (Mian et al. 1996, 1998; Specht et al. 2001; Dhanapal et al. 2015; Zhang et al. 2014, 2016; Dhanapal et al. 2015; Kumar and Lal 2015; Li et al. 2016; Kaler et al. 2017).

To identify QTL that control WUE, Mian et al. (1996) grew a RIL population derived from a cross between "Young" and "PI 416937" lines ($n = 120$) in the greenhouse in 1994 and genotyped it with 155 RFLP markers. To estimate WUE, they calculated leaf dry weight (LDW), stem dry weight (SDW), root dry weight (RDW), and total dry weight (TDW). TDW = (LDW + SDW + RDW) and WUE = TDW/

amount of water used in liters. They also estimated leaf ash (LASH) which is related to WUE. They used single-marker analysis (ANOVA) for QTL data analysis of SAS program. Four RFLP markers associated with QTL that control WUE have been identified and mapped on chromosomes 12, 16, and 18, and six RFLP markers associated with QTL that control LASH have been identified and mapped on chromosomes 16, 18, and 19. One marker was unlinked (Fig. 2.1; Table 4.5) (Mian et al. 1998).

Mian et al. (1998) grew an F_2 population derived from a cross between "S100" and "Tokyo" lines ($n = 116$) in the greenhouse in 1996, genotyped it with 142 RFLP markers, used single-marker analysis (ANOVA) for QTL data analysis, and estimated WUE as described earlier (Mian et al. 1996). Three new markers associated with QTL that control WUE have been identified and mapped on chromosomes (Fig. 2.1; Table 4.5) (Mian et al. 1998).

Specht et al. (2001) grew a RIL population derived from a cross between "Minsoy" and "Noir 1" cultivars ($n = 236$) and genotyped it with 665 markers (SSRs, RFLPs, and classical markers) in Mead, NE, in 1994–1995 under water stress (0%, 20%, 40%, 60%, 80%, and 100% replenishment of evapotranspiration (ET)) on a weekly basis. They used single-marker analysis (ANOVA) for QTL data analysis. *Five QTL that control seed protein content have been identified and mapped on chromosomes 4, 5, 7, 9, and 10, and five QTL that control seed oil content have been identified and mapped on chromosomes 1, 5, 13, and 20* (Fig. 2.2; Table 4.5). Nine *QTL that control seed weight (SW) have been identified and mapped on chromosomes 5, 6, 8, 11, 13, and 14; six QTL that control seed yield (SY) have been identified and mapped on chromosomes 6, 10, and 13; ten QTL that control maturity date (MAT) have been identified and mapped on chromosomes 1, 2, 6, 10, and 13; and four QTL that control plant height (PH) have been identified and mapped on chromosomes 2 and 6 (*Fig. 2.1; Table *4.5).* A major QTL that controls seed yield, explained 23–29% of the total variation in seed yield, and had a positive effect on seed protein content has been identified and mapped on chromosome 7. A second major QTL that controls seed yield and maturity, explained 7–13% of the total variation in seed yield and 13–15% of the total variation in maturity, and had a positive effect on yield-to-water response (WUE) has been identified and mapped on chromosome 6. A third major QTL that controls seed yield and plant height, explained 20–40% of the total variation in plant height, and had a positive effect on yield-to-water response (WUE) has been identified and mapped on chromosome 19. This QTL may be the *S* gene responsible for short plant height (Orf et al. 1999). All these QTL are related to WUE and how soybeans cope with water stress deficit (Specht et al. 2001).

Dhanapal et al. (2015) grew a population of different genotypes ($n = 385$) of soybean germplasm in Columbia, MO, in 2009–2010 and genotyped it with 12,347 SNPs to identify QTL for WUE estimated as carbon isotope ratio ($\delta^{13}C$). They used GWAS analysis to identify SNPs associated with WUE (carbon isotope ratio, $\delta^{13}C$). Fifty two SNPs associated with QTL that control WUE (carbon isotope ratio, $\delta^{13}C$) has been identified and mapped on chromosomes 2, 4, 5, 6, 7, 9, 13, 15, 16, 19, and 20 (Table 4.5). The authors also identified many genes within these QTL regions

Table 4.5 Detailed information of QTL that control water use efficiency (WUE) and phosphorus efficiency (PE) reported and *not reported* in SoyBase (Grant et al. 2010; http://www.soybase.org)

Trait	*QTL*	LG/ Chr	Interval (cM)	Marker/interval	LOD, *P*, or *F*	R^2 (%)	Environment	Reference
WUE	*qWUE-1-1*	G/18	51.6–53.6	B031–1	$p = 0.0100$	8.50	Greenhouse	Mian et al. (1996)
	qWUE-1-2	H/12	32.6–34.6	A089–1	$p = 0.0033$	8.70	Greenhouse	
	qWUE-1-3	H/12	10.3–12.3	A381–1	$p = 0.0100$	8.00	Greenhouse	
	qWUE-1-4	J/16	87.9–89.9	cr497–1	$p = 0.0002$	13.20	Greenhouse	
	qWUE-1-5	J/16	67.3–69.3	K375–1	$p = 0.0029$	7.50	Greenhouse	
	qWUE-1-6	J/16	82.2–84.2	A233–1	$p = 0.0100$	8.00	Greenhouse	
LASH	*qLASH-1-1*	G/18	34.5–36.5	A112–1	p = 0.0001	13.80	Greenhouse	
	qLASH-1-2	G/18	57.7–59.7	A458–2	$p = 0.0003$	12.70	Greenhouse	
	qLASH-1-3	J/16	67.3–69.3	K375–1	$p = 0.0036$	7.20	Greenhouse	
	qLASH-1-4	L/19	3.1–5.1	EV2–1	$p = 0.0150$	6.20	Greenhouse	
	qLASH-1-5	D2/17	72.5–74.5	K258–2	$p = 0.0098$	6.30	Greenhouse	
	qLASH-1-6	*	*	cr107–1	$p = 0.0083$	6.60	Greenhouse	
	qLASH-1-7	G/18	59.1–61.1	cr509–1	p = 0.0100	6.00	Greenhouse	
	qLASH-1-8	G/18	55.7–57.7	A638–2	$p = 0.0100$	10.00	Greenhouse	
	qLASH-1-9	G/18	53.5–55.5	K443–1	$p = 0.0100$	7.00	Greenhouse	
	qLASH-1-10	G/18	51.6–53.6	B031–2	$p = 0.0100$	6.00	Greenhouse	
	qLASH-1-11	G/18	50.7–52.7	EV1–2	$p = 0.0100$	6.00	Greenhouse	
WUE	*qWUE-2-1*	L/19	94.4–96.4	A489–1	3.4	14.00	Greenhouse	Mian et al. (1998)
	qWUE-2-2	C1/4	89.7–91.7	A063–1	$p = 0.0090$	8.00	Greenhouse	
	qWUE-2-3	C1/4	89.7–91.7	A063–1	$p = 0.03300$	5.00	Greenhouse	

(continued)

Table 4.5 (continued)

Trait	*QTL*	LG/ Chr	Interval (cM)	Marker/interval	LOD, *P*, or *F*	R^2 (%)	Environment	Reference
WUE/seed weight	*qSW-1*	A1/5	78.3–80.3	Satt200	3.55	3.40	Nebraska	Specht et al. (2001)
	qSW-2	A2/8	105.4–107.4	Satt470	6.17	6.20	Nebraska	
	qSW-3	B1/11	104.2–106.2	Sat_095	2.65	4.40	Nebraska	
	qSW-4	B2/14	69.6–71.6	BLT057	4.83	4.70	Nebraska	
	qSW-5	C2/6	22–24	A059	4.14	4.20	Nebraska	
	qSW-6	C2/6	23.6–25.6	L199a	4.61	4.60	Nebraska	
	qSW-7	F/13	18.8–20.8	W1	4.08	4.50	Nebraska	
	qSW-8	F/13	17.1–19.1	Satt343	3.3	3.20	Nebraska	
	qSW-9	F/13	108.7–110.7	L195	3.07	4.10	Nebraska	
WUE/seed yield	*qSY-1*	C2/6	112.3–114.3	Satt489	9.12	7.00	Nebraska	
	qSY-2	C2/6	111.7–113.7	Satt281	5.25	12.80	Nebraska	
	qSY-3	C2/6	112.3–114.3	Satt489	4.73	6.40	Nebraska	
	qSY-4	O/10	64.6–66.6	Satt420	2.54	1.80	Nebraska	
	qSY-5	F/13	173.8–175.8	Sat_074	3.39	2.40	Nebraska	
	qSY-6	F/13	173.8–175.8	Sat_074	2.65	3.90	Nebraska	

Trait	*QTL*	LG/ Chr	Interval (cM)	Marker/interval	LOD, *P*, or *F*	R^2 (%)	Environment	Reference
WUE/ maturity date	*qMAT-1*	D1a/1	0–2.5	Tel1075A	4.12	3.30	Nebraska	
	qMAT-2	D1a/1	59.8–61.8	Satt360	3.92	2.40	Nebraska	
	qMAT-3	D1b/2	15.2–17.2	L216b	2.67	1.80	Nebraska	
	qMAT-4	C2/6	111.7–113.7	Satt205	20.52	12.70	Nebraska	
	qMAT-5	C2/6	112.3–114.3	Satt291	4.61	15.40	Nebraska	
	qMAT-6	O/10	151.4–153.4	Sat_109	5.8	4.00	Nebraska	
	qMAT-7	O/10	149.3–151.3	Sat_108	4.28	2.60	Nebraska	
	qMAT-8	O/10	148.1–150.1	G214p	2.9	2.90	Nebraska	
	qMAT-9	F/13	4.4–4.6	BL030	4.85	2.60	Nebraska	
	qMAT-10	F/13	4.4–4.6	BL030	4.13	2.40	Nebraska	
WUE/plant height	*qPH-1*	D1b/2	25.7–27.7	Wc	8.18	4.50	Nebraska	
	qPH-2	C2/6	111.7–113.7	Satt205	10.55	5.40	Nebraska	
	qPH-3	C2/6	112.3–114.3	Satt489	26.7	15.20	Nebraska	
	qPH-4	C2/6	112.7–114.7	Satt365	9.52	11.10	Nebraska	

(continued)

Table 4.5 (continued)

Trait	*QTL*	LG/ Chr	Interval (cM)	Marker/interval	LOD, *P*, or *F*	R^2 (%)	Environment	Reference
Phosphorus efficiency (PE)	*qPE2-g1.1*	A2/8	*	Satt089	*	41.05	China	Zhang et al. (2014)
	qPE2-g1.2	A2/8	*	Sat_310	*	41.05	China	
	qPE2-g2.1	F/13	*	Sat_262	*	7.72	China	
	qPE2-g2.2	F/13	*	Satt030	*	7.72	China	
	qPE2-g3.1	A1/5	*	Satt276	*	5.67	China	
	qPE2-g3.2	A1/5	*	Sat_368	*	5.67	China	
	qPE2-g4.1	B2/14	*	Sat_424	*	21.28	China	
	qPE2-g4.2	B2/14	*	Satt063	*	21.28	China	
	qPE2-g5.1	D1a/1	*	Satt436	*	8.20	China	
	qPE2-g5.2	D1a/1	*	Sat_343	*	8.20	China	
	qPE2-g5.3	D1a/1	*	Sat_346	*	6.75	China	
	qPE2-g6.1	L/19	*	Sat_191	*	7.05	China	
	qPE2-g6.2	L/19	*	Satt313	*	7.05	China	
	qPE2-g7.1	G/18	*	Sat_185	*	26.06	China	
	qPE2-g7.2	G/18	*	Satt012	*	26.06	China	
	qPE2-g8.1	K/9	*	Sat_352	*	7.54	China	
	qPE2-g8.2	G/18	*	Satt138	*	7.54	China	
Water use eff. (WUE)	*qWUE3-g1*	D1b/2	*	ss715581017	*	*	Missouri	Dhanapal et al. (2015)
Carbon isotope ratio, δ13C	*qWUE3-g2*	D1b/2	*	ss715582880	*	*	Missouri	
	qWUE3-g3	D1b/2	*	ss715582885	*	*	Missouri	

Trait	*QTL*	LG/ Chr	Interval (cM)	Marker/interval	LOD, *P*, or *F*	R^2 (%)	Environment	Reference
	qWUE3-g4	D1b/2	*	ss715582890	*	*	Missouri	
	qWUE3-g5	D1b/2	*	ss715582892	*	*	Missouri	
	qWUE3-g6	D1b/2	*	ss715582906	*	*	Missouri	
	qWUE3-g7	D1b/2	*	ss715582962	*	*	Missouri	
	qWUE3-g8	N/3	*	ss715586406	*	*	Missouri	
	qWUE3-g9.1	C1/4	*	ss715588676	*	*	Missouri	
	qWUE3-g9.2	C1/4	*	ss715588677	*	*	Missouri	
	qWUE3-g10	C1/4	*	ss715588831	*	*	Missouri	
	qWUE3-g11	C1/4	*	ss715588846	*	*	Missouri	
	qWUE3-g12	C1/4	*	ss715588861	*	*	Missouri	
	qWUE3-g13	C1/4	*	ss715588873	*	*	Missouri	
	qWUE3-g14	C2/6	*	ss715594602	*	*	Missouri	
	qWUE3-g15	M/7	*	ss715596465	*	*	Missouri	
	qWUE3-g16	A2/8	*	ss715602516	*	*	Missouri	
	qWUE3-g17	A2/8	*	ss715602521	*	*	Missouri	
	qWUE3-g18	K/9	*	ss715603285	*	*	Missouri	
	qWUE3-g19	O/10	*	ss715606377	*	*	Missouri	
	qWUE3-g20.1	O/10	*	ss715608656	*	*	Missouri	
	qWUE3-g20.2	O/10	*	ss715608657	*	*	Missouri	
	qWUE3-g21	B1/11	*	ss715610958	*	*	Missouri	
	qWUE3-g22	H/12	*	ss715612946	*	*	Missouri	
	qWUE3-g23	F/13	*	ss715613850	*	*	Missouri	
	qWUE3-g24	F/13	*	ss715614732	*	*	Missouri	
	qWUE3-g25	E/15	*	ss715621062	*	*	Missouri	

(continued)

Table 4.5 (continued)

Trait	*QTL*	LG/ Chr	Interval (cM)	Marker/interval	LOD, *P*, or *F*	R^2 (%)	Environment	Reference
	qWUE3-g26	E/15	*	ss715623203	*	*	Missouri	
	qWUE3-g27.1	E/15	*	ss715623206	*	*	Missouri	
	qWUE3-g27.2	E/15	*	ss715623207	*	*	Missouri	
	qWUE3-g28.1	E/15	*	ss715623209	*	*	Missouri	
	qWUE3-g28.2	E/15	*	ss715623210	*	*	Missouri	
	qWUE3-g29	D2/17	*	ss715627734	*	*	Missouri	
	qWUE3-g30	G/18	*	ss715631698	*	*	Missouri	
	qWUE3-g31	G/18	*	ss715632228	*	*	Missouri	
	qWUE3-g32	G/18	*	ss715632407	*	*	Missouri	
	qWUE3-g33	G/18	*	ss715632431	*	*	Missouri	
	qWUE3-g34	L/19	*	ss715634542	*	*	Missouri	
	qWUE3-g35	I/20	*	ss715637230	*	*	Missouri	
Water use Eff. (WUE)	*qWUE-13C-1*	D1b/2	*	Gm02_45509202_T_C	(– Log$p$ = 5.32)	6.00	Missouri	
Carbon isotope ratio, δ13C	*qWUE-13C-2*	D1b/2	*	Gm02_45509202_T_C	(– Log$p$ = 2.56)	3.00	Missouri	
	qWUE-13C-3	D1b/2	*	Gm02_45542400_A_C	(– Log$p$ = 5.05)	7.00	Missouri	
	qWUE-13C-4	D1b/2	*	Gm02_45542400_A_C	(– Log$p$ = 2.64)	4.00	Missouri	
	qWUE-13C-5	D1b/2	*	Gm02_45573752_A_G	(– Log$p$ = 4.98)	6.00	Missouri	
	qWUE-13C-6	D1b/2	*	Gm02_45573752_A_G	(– Log$p$ = 2.62)	3.00	Missouri	

Trait	*QTL*	LG/ Chr	Interval (cM)	Marker/interval	LOD, *P*, or *F*	R^2 (%)	Environment	Reference
	qWUE-13C-7	D1b/2	*	Gm02_45595304_A_G	(–Log*p* = 4.98)	6.00	Missouri	
	qWUE-13C-8	D1b/2	*	Gm02_45595304_A_G	(–Log*p* = 2.62)	3.00	Missouri	
	qWUE-13C-9	D1b/2	*	Gm02_45691551_A_G	(–Log*p* = 5.04)	6.00	Missouri	
	qWUE-13C-10	D1b/2	*	Gm02_45691551_A_G	(–Log*p* = 2.99)	4.00	Missouri	
	qWUE-13C-11	D1b/2	*	Gm02_46187810_C_T	(–Log*p* = 9.70)	11.00	Missouri	
	qWUE-13C-12	D1b/2	*	Gm02_46187810_C_T	(–Log*p* = 2.67)	3.00	Missouri	
	qWUE-13C-13	C1/4	*	Gm04_46049853_A_G	(–Log*p* = 7.40)	9.00	Missouri	
	qWUE-13C-14	C1/4	*	Gm04_46049853_A_G	(–Log*p* = 2.24)	3.00	Missouri	
	qWUE-13C-15	A1/5	*	Gm05_33458251_G_A	(–Log*p* = 5.09)	5.00	Missouri	
	qWUE-13C-16	A1/5	*	Gm05_33458251_G_A	(–Log*p* = 2.99)	3.00	Missouri	
	qWUE-13C-17	C2/6	*	Gm06_46350377_T_C	(–Log*p* = 5.30)	7.00	Missouri	
	qWUE-13C-18	C2/6	*	Gm06_46350377_T_C	(–Log*p* = 2.11)	3.00	Missouri	
	qWUE-13C-19	M/7	*	Gm07_30573150_A_G	(–Log*p* = 3.06)	4.00	Missouri	

(continued)

Table 4.5 (continued)

Trait	*QTL*	LG/ Chr	Interval (cM)	Marker/interval	LOD, *P*, or *F*	R^2 (%)	Environment	Reference
	qWUE-13C-20	M/7	*	Gm07_30573150_A_G	(–Log*p* = 2.71)	3.00	Missouri	
	qWUE-13C-21	K/9	*	Gm09_942620_T_G	(–Log*p* = 6.50)	8.00	Missouri	
	qWUE-13C-22	K/9	*	Gm09_942620_T_G	(–Log*p* = 2.17)	3.00	Missouri	
	qWUE-13C-23	K/9	*	Gm09_948977_T_C	(–Log*p* = 7.85)	9.00	Missouri	
	qWUE-13C-24	K/9	*	Gm09_948977_T_C	(–Log*p* = 2.48)	3.00	Missouri	
	qWUE-13C-25	K/9	*	Gm09_964605_A_G	(–Log*p* = 6.01)	8.00	Missouri	
	qWUE-13C-26	K/9	*	Gm09_964605_A_G	(–Log*p* = 2.07)	3.00	Missouri	
	qWUE-13C-27	K/9	*	Gm09_967930_C_T	(–Log*p* = 6.50)	8.00	Missouri	
	qWUE-13C-28	K/9	*	Gm09_967930_C_T	(–Log*p* = 2.17)	3.00	Missouri	
	qWUE-13C-29	F/13	*	Gm13_14458750_G_T	(–Log*p* = 7.31)	8.00	Missouri	
	qWUE-13C-30	F/13	*	Gm13_14458750_G_T	(–Log*p* = 4.00)	4.00	Missouri	
	qWUE-13C-31	F/13	*	Gm13_19554349_C_A	(–Log*p* = 4.69)	6.00	Missouri	
	qWUE-13C-32	F/13	*	Gm13_19554349_C_A	(–Log*p* = 2.08)	3.00	Missouri	
	qWUE-13C-33	E/15	*	Gm15_1609497_C_T	(–Log*p* = 5.95)	7.00	Missouri	

Trait	*QTL*	LG/ Chr	Interval (cM)	Marker/interval	LOD, *P*, or *F*	R^2 (%)	Environment	Reference
	qWUE-13C-34	E/15	*	Gm15_1609497_C_T	(–Logp = 4.00)	5.00	Missouri	
	qWUE-13C-35	E/15	*	Gm15_2427922_C_T	(–Log*p* = 5.26)	5.00	Missouri	
	qWUE-13C-36	E/15	*	Gm15_2427922_C_T	(–Log*p* = 2.23)	2.00	Missouri	
	qWUE-13C-37	E/15	*	Gm15_9117659_T_C	(–Log*p* = 3.64)	4.00	Missouri	
	qWUE-13C-38	E/15	*	Gm15_9117659_T_C	(–Log*p* = 3.21)	4.00	Missouri	
	qWUE-13C-39	E/15	*	Gm15_9127493_G_A	(–Log*p* = 4.05)	4.00	Missouri	
	qWUE-13C-40	E/15	*	Gm15_9127493_G_A	(–Logp = 3.44)	3.00	Missouri	
	qWUE-13C-41	E/15	*	Gm15_9145025_A_G	(–Log*p* = 4.05)	4.00	Missouri	
	qWUE-13C-42	E/15	*	Gm15_9145025_A_G	(–Log*p* = 3.44)	3.00	Missouri	
	qWUE-13C-43	E/15	*	Gm15_9151938_A_C	(–Log*p* = 4.05)	4.00	Missouri	
	qWUE-13C-44	E/15	*	Gm15_9151938_A_C	(–Log*p* = 3.44)	3.00	Missouri	
	qWUE-13C-45	E/15	*	Gm15_9162160_A_G	(–Logp = 3.89)	5.00	Missouri	
	qWUE-13C-46	E/15	*	Gm15_9162160_A_G	(–Log*p* = 3.16)	5.00	Missouri	

(continued)

Table 4.5 (continued)

Trait	*QTL*	LG/ Chr	Interval (cM)	Marker/interval	LOD, *P*, or *F*	R^2 (%)	Environment	Reference
	qWUE-13C-47	J/16	*	Gm16_28368515_C_T	(–Log*p* = 5.19)	6.00	Missouri	
	qWUE-13C-48	J/16	*	Gm16_28368515_C_T	(–Log*p* = 2.18)	3.00	Missouri	
	qWUE-13C-49	L/19	*	Gm19_48054733_G_A	(–Log*p* = 3.69)	5.00	Missouri	
	qWUE-13C-50	L/19	*	Gm19_48054733_G_A	(–Log*p* = 2.48)	4.00	Missouri	
	qWUE-13C-51	I/20	*	Gm20_2972792_A_G	(–Log*p* = 9.15)	12.00	Missouri	
	qWUE-13C-52	I/20	*	Gm20_2972792_A_G	(–Log*p* = 2.26)	3.00	Missouri	
Water use efficiency	*qWUE1-g1*	D1b/2	*	Satt157	*	*	India	Kumar and Lal (2015)
	qWUE1-g2	H/12	*	Satt434	*	*	India	
	qWUE1-g3	A1/5	*	Satt545	*	*	India	
	qWUE1-g4	L/19	*	Satt664	*	*	India	
	qWUE1-g5	E/15	*	Satt685	*	*	India	
	qWUE1-g6	K/9	*	Satt260	*	*	India	
	qWUE1-g7	I/20	*	Satt330	*	*	India	
	qWUE1-g8	O/13	*	Satt331	*	*	India	
	qWUE1-g9	C2/6	*	Satt460	*	*	India	
	qWUE1-g10	F/13	*	Satt522	*	*	India	
Phosphorus efficiency (PE)	*q1*	D1a/1	72.21–75.01	Mrk338587–Mrk397044	3.39	8.4	China	Zhang et al. (2016)

Trait	*QTL*	LG/ Chr	Interval (cM)	Marker/interval	LOD, *P*, or *F*	R^2 (%)	Environment	Reference
	q2	D1b/2	19.19–20.81	Mrk2373489–Mrk2377055	3.71	8.63	China	
	q3	N/3	58.11–59.27	Mrk962011–Mrk923093	3.57	7.37	China	
	q4–1	C1/4	71.02–84.42	Mrk41910–Mrk131379	4.14	11.11	China	
	q4–2	C1/4	15.21–19.56	Mrk53079–Mrk63513	5.14	12.59	China	
	q6	C2/6	46.19–47.59	Mrk2151148–Mrk2105391	3.41	7.59	China	
	q7	M/7	89.41–94.55	Mrk2525208–Mrk2596654	3.89	9.38	China	
	q8	A2/8	85.14–88.42	Mrk2634911–Mrk2645532	4.51	11.95	China	
	q9	K/9	47.53–49.68	Mrk1603422–Mrk1540438	3.20	7.21	China	
	q10	O/10	75.05–98.87	Mark641231–Mrk703395	4.24	9.76	China	
	q13–1	F/13	27.92–38.03	Mrk1786250–Mrk1796581	3.84	9.20	China	
	q13–2	F/13	69.28–71.74	Mrk1719580–Mrk1684017	3.43	7.31	China	
	q14	B2/14	144.06–148.67	Mrk473565–Mrk461735	3.80	8.44	China	
	q15	E/15	50.34–51.40	Mrk1381067–Mrk1370961	4.89	12.17	China	

(continued)

Table 4.5 (continued)

Trait	*QTL*	LG/ Chr	Interval (cM)	Marker/interval	LOD, *P*, or *F*	R^2 (%)	Environment	Reference
	q16	J/16	82.51–83.05	Mrk1134679–Mrk1148700	3.66	8.67	China	
	q17	D2/17	0–7.21	Mrk2031528–Mrk1984881	4.24	10.14	China	
	q18–1	G/18	75.99–77.28	Mrk2194216–Mrk2328462	3.33	10.52	China	
	q18–2	G/18	28.25–32.59	Mrk2329091–Mrk2270259	3.51	7.06	China	
	q19	L/19	155.48–158.99	Mrk1014646–Mrk1106196	2.76	5.14	China	
Phosphorus efficiency (PE)	*q1 [CiLP-E1,CiNP-E2]*	D1a/1	984,367	Mrk399740–Mrk405817	2.86	12.28	Hydroponics	Li et al. (2016)
	q2–1 [CoLP-E1,PCLP-E2,PupLP-E2, PCLP-E1,PupLP-E1,PCLP-E3]	D1b/2	13,942,698	Mrk2426304–Mrk2355122	3.76	14.04	Hydroponics	
	q2–2 [RSNP-E2,TDWNP-E2,CCNP-E1, CCLP-E5]	D1b/2	41,077,137	Mrk2359119–Mrk2407344	4.02	9.16	Hydroponics	
	q3–1 SDWLP-E1,PCNP-E1,PupNP-E1, CCNP-E1,TDWLP-E1]	N/3	2,384,930	Mrk873205–Mrk923093	3.19	8.39	Hydroponics	
	q3–2 [TDWLP-E2,CCNP-E2,CCLP-E1, CCLP-E2]	N/3	15,406,533	Mrk968996–Mrk945189	3.70	8.68	Hydroponics	
	q3–3 [CCNP-E1]	N/3	38,757,448	Mrk953521–Mrk863963	3.09	5.56	Hydroponics	
	q4–1 [PCNP-E2,PupNP-E2]	C1/4	2,244,572	Mrk70489–Mrk94279	3.52	11.16	Hydroponics	
	q4–2 [PCNP-E4,PupNP-E2]	C1/4	32,704,336	Mrk67935–Mrk129026	3.16	8.62	Hydroponics	

Trait	*QTL*	LG/ Chr	Interval (cM)	Marker/interval	LOD, *P*, or *F*	R^2 (%)	Environment	Reference
	q4–3 [RDWLP-E1,SDWNP-E2,SDWLP-E1, SDWNP-E1,RDWNP-E1,RDWNP-E2, TDWNP-E1,TDWNP-E2,SDWLP-E2,TDWLP-E2,CCNP-E1]	C1/4	46,024,793	Mrk101824–Mrk58600	4.44	10.86	Hydroponics	
	q5 [CCLP-E3,SDWLP-E2]	A1/5	5,023,300	Mrk1911126–Mrk1840642	4.30	9.16	Hydroponics	
	q6 [PCNP-E1,PupNP-E1,PupNP-E2]	C2/6	15,623,936	Mrk2071221–Mrk2174814	4.02	11.22	Hydroponics	
	q8–1 [CCLP-E1,CCLP-E2,PCLP-E3]	A2/8	1,796,491	Mrk2655981–Mrk2726667	4.12	8.89	Hydroponics	
	q8–2 [BYLP-E5,PUELP-E4,PCLP-E5, PupLP-E5,PUELP-E5]	A2/8	11,520,732	Mrk2655900–Mrk2651744	3.55	8.01	Hydroponics	
	q9 [PnNP-E1]	K/9	43,233,332	Mrk1658126–Mrk1641616	3.54	9.82	Hydroponics	
	q10 [CCLP–E4]	O/10	36,602,559	Mrk713711–Mrk700268	3.31	9.50	Hydroponics	
	q11–1 [PCNP-E1,PupNP-E1,PCNP-E3]	B1/11	7,615,924	Mrk799502–Mrk844658	5.33	15.96	Hydroponics	
	q11–2 [PCNP-E5,BYNP-E3,BYNP-E4, BYNP-E5,PupNP-E5,PUENP-E5, BYLP-E3,BYLP-E5,PupLP-E5]	B1/11	24,450,424	Mrk789315–Mrk822756	3.22	7.18	Hydroponics	
	q12–1 [CCNP-E1,PnNP-E1,CoNP-E2, PnLP-E1,PnLP-E2,CoLP-E2, CCNP-E2,CCLP-E1,CCLP-E2]	H/12	3,008,087	Mrk249870–Mrk262276	7.37	23.74	Hydroponics	
	q12–2 [PnNP-E2,CCLP-E3]	H/12	31,303,690	Mrk200457–Mrk203972	3.05	13.78	Hydroponics	

(continued)

Table 4.5 (continued)

Trait	*QTL*	LG/ Chr	Interval (cM)	Marker/interval	LOD, *P*, or *F*	R^2 (%)	Environment	Reference
	q13 [CCNP-E2,SDWNP-E1,SDWNP-E2, RDWNP-E2,TDWNP-E1,TDWNP-E2]	F/13	35,339,389	Mrk1729776–Mrk1778051	4.22	10.41	Hydroponics	
	q14–1 [TDWLP-E1]	B2/14	33,522,076	Mrk489485–Mrk557885	3.83	9.26	Hydroponics	
	q14–2 [PCLP-E1,PupLP-E1,PCNP-E2, PupNP-E2,PCNP-E4,PupNP-E4, PUENP-E4,PupNP-E1,PCNP-E1, CiNP-E2,PCNP-E3,CiLP-E1,PCLP-E4, PupLP-E4,PUELP-E4,PupLP-E5, PnLP-E2,PCLP-E5,PUELP-E5, PCNP-E5,BYNP-E3,BYNP-E4, BYNP-E5,PupNP-E5,PUENP-E5, BYLP-E3,BYLP-E4,BYLP-E5]	B2/14	45,573,911	Mrk505100–Mrk509340	4.59	14.66	Hydroponics	
	q15–1 [CoNP-E2,RSLP-E2]	E/15	2,943,851	Mrk1309756–Mrk1323793	3.54	10.18	Hydroponics	
	q15–2 [SDWLP-E1,SDWNP-E2,RDWNP-E2, TDWLP-E1,TDWNP-E1,TDWNP-E2, CoLP-E1,SDWNP-E1,RDWNP-E1, TrNP-E1,TrNP-E2,CiNP-E1, SDWLP-E2,TDWLP-2,PUELP-E2, PCLP-E3,PupLP-E3,PUELP-E3]	E/15	9,586,975	Mrk1333127–Mrk1381067	6.04	20.00	Hydroponics	
	q16 [PupNP-E3,PupNP-E4,PupLP-E4, PUELP-E4	J/16	31,778,047	Mrk1146835–Mrk1164294	3.01	12.21	Hydroponics	
	q18–1 [CCLP-E5,CCLP-E3,CCLP-E3,CCLP-E4,CCLP-E5]	G/18	2,037,678	Mrk2217116–Mrk2237156	8.36	22.00	Hydroponics	
	q18–2 [RDWLP-E1]	G/18	53,977,091	Mrk2270259–Mrk2325242	4.16	12.73	Hydroponics	

Trait	*QTL*	LG/ Chr	Interval (cM)	Marker/interval	LOD, *P*, or *F*	R^2 (%)	Environment	Reference
	q19–1 [PCLP-E2,PupLP-E2]	L/19	36,624,688	Mrk1039368–Mrk1043830	3.43	11.47	Hydroponics	
	q19–2 [PUELP-E1,PUELP-E2,BYNP-E3, BYNP-E4,PCLP-E5,BYLP-E3, BYLP-E4,PupLP-E5,CCNP-E3, CCNP-E4,CCNP-E5,PCNP-E5, BYNP-E5,PupNP-E5,BYLP-E5, PUELP-E5,PCLP-E1,PCLP-E2, PupLP-E1,PupLP-E2,PCLP-E3, PCLP-E4,PUELP-E3,PUELP-E4, CiLP-E2,PUENP-E5,TrLP-E1,TrLP-E2]	L/19	44,865,752	Mrk1094659–Mrk1066605	6.51	20.03	Hydroponics	
Water use efficiency	*qWUE2-g1*	D1a/1	*	ss715579430	3.90E+04	*	Missouri	Kaler et al. (2017)
	qWUE2-g2	D1b/2	*	ss715582801	1.00E+06	*	Missouri	
	qWUE2-g3	D1b/2	*	ss715580894	6.60E+05	*	Missouri	
	qWUE2-g4	D1b/2	*	ss715582359	5.30E+04	*	Missouri	
	qWUE2-g5	D1b/2	*	ss715582706	3.80E+04	*	Missouri	
	qWUE2-g6	D1b/2	*	ss715582890	4.60E+04	*	Missouri	
	qWUE2-g7	D1b/2	*	ss715582944	4.80E+08	*	Missouri	
	qWUE2-g8	N/3	*	ss715584746	5.70E+04	*	Missouri	
	qWUE2-g9	C1/4	*	ss715587339	1.10E+04	*	Missouri	
	qWUE2-g10	C1/4	*	ss715589508	3.90E+03	*	Missouri	
	qWUE2-g11	C1/4	*	ss715588669	1.90E+06	*	Missouri	
	qWUE2-g12	C1/4	*	ss715588823	1.00E+06	*	Missouri	
	qWUE2-g13	A1/5	*	ss715591885	1.20E+04	*	Missouri	
	qWUE2-g14	A1/5	*	ss715592204	1.40E+04	*	Missouri	

(continued)

Table 4.5 (continued)

Trait	*QTL*	LG/ Chr	Interval (cM)	Marker/interval	LOD, *P*, or *F*	R^2 (%)	Environment	Reference
	qWUE2-g15	C2/6	*	ss715593478	6.60E+05	*	Missouri	
	qWUE2-g16	C2/6	*	ss715595695	4.70E+06	*	Missouri	
	qWUE2-g17	M/7	*	ss715598752	7.20E+04	*	Missouri	
	qWUE2-g18	M/7	*	ss715599042	1.80E+05	*	Missouri	
	qWUE2-g19	M/7	*	ss715597447	1.10E+04	*	Missouri	
	qWUE2-g20	K//9	*	ss715605464	4.30E+03	*	Missouri	
	qWUE2-g21	K//9	*	ss715603235	3.10E+03	*	Missouri	
	qWUE2-g22	K//9	*	ss715603246	3.10E+03	*	Missouri	
	qWUE2-g23	K//9	*	ss715603692	2.80E+06	*	Missouri	
	qWUE2-g24	K//9	*	ss715605385	1.70E+04	*	Missouri	
	qWUE2-g25	O/10	*	ss715608065	3.80E+04	*	Missouri	
	qWUE2-g26	B1/11	*	ss715611189	5.40E+05	*	Missouri	
	qWUE2-g27	B1/11	*	ss715611203	5.80E+04	*	Missouri	
	qWUE2-g28	F/13	*	ss715614983	2.80E+06	*	Missouri	
	qWUE2-g29	F/13	*	ss715615278	2.40E+07	*	Missouri	
	qWUE2-g30	B2/14	*	ss715618209	2.40E+04	*	Missouri	
	qWUE2-g31	E/15	*	ss715621121	9.70E+06	*	Missouri	
	qWUE2-g32	E/15	*	ss715621159	1.60E+04	*	Missouri	
	qWUE2-g33	E/15	*	ss715621444	1.10E+05	*	Missouri	
	qWUE2-g34	E/15	*	ss715623062	3.50E+05	*	Missouri	
	qWUE2-g35	E/15	*	ss715623207	2.30E+05	*	Missouri	
	qWUE2-g36	J/16	*	ss715623427	3.40E+03	*	Missouri	
	qWUE2-g37	J/16	*	ss715624450	7.50E+05	*	Missouri	
	qWUE2-g38	J/16	*	ss715624931	3.60E+03	*	Missouri	
	qWUE2-g39	D2/17	*	ss715628013	4.70E+06	*	Missouri	

Trait	*QTL*	LG/ Chr	Interval (cM)	Marker/interval	LOD, *P*, or *F*	R^2 (%)	Environment	Reference
Water use efficiency	*qWUE2-g40*	D2/17	*	ss715628319	9.70E+03	*	Missouri	
	qWUE2-g41	D2/17	*	ss715625747	2.10E+04	*	Missouri	
	qWUE2-g42	D2/17	*	ss715626097	1.20E+04	*	Missouri	
	qWUE2-g43	D2/17	*	ss715626210	1.60E+04	*	Missouri	
	qWUE2-g44	D2/17	*	ss715626630	3.50E+04	*	Missouri	
	qWUE 2-g45	D2/17	*	ss715626831	3.30E+04	*	Missouri	
	qWUE2-g46	D2/17	*	ss715627051	1.30E+04	*	Missouri	
	qWUE2-g47	G/18	*	ss715632594	2.70E+04	*	Missouri	
	qWUE2-g48	G/18	*	ss715628612	7.20E+03	*	Missouri	
	qWUE2-g49	G/18	*	ss715631898	5.30E+06	*	Missouri	
	qWUE2-g50	L/19	*	ss715633033	1.20E+06	*	Missouri	
	qWUE2-g51	L/19	*	ss715636187	7.50E+03	*	Missouri	
	qWUE2-g52	I/20	*	ss715637331	3.00E+05	*	Missouri	
	qWUE2-g53	I/20	*	ss715637883	1.40E+05	*	Missouri	
	qWUE2-g54	I/20	*	ss715638590	3.80E+03	*	Missouri	
	qWUE2-g55	D1a/1	*	ss715579733	2.90E+03	*	Missouri	
	qWUE2-g56.1	D1b/2	*	ss715582161	1.20E+04	*	Missouri	
	qWUE2-g56.2	D1b/2	*	ss715582162	1.60E+04	*	Missouri	
	qWUE2-g56.3	D1b/2	*	ss715582164	8.30E+03	*	Missouri	
	qWUE2-g57	D1b/2	*	ss715582166	5.80E+03	*	Missouri	
	qWUE2-g58	D1b/2	*	ss715582171	3.80E+03	*	Missouri	
	qWUE2-g59.1	D1b/2	*	ss715583378	5.40E+03	*	Missouri	
	qWUE2-g59.2	D1b/2	*	ss715583381	5.80E+03	*	Missouri	

(continued)

Table 4.5 (continued)

Trait	*QTL*	LG/ Chr	Interval (cM)	Marker/interval	LOD, *P*, or *F*	R^2 (%)	Environment	Reference
	qWUE2-g60	N/3	*	ss715584347	3.90E+03	*	Missouri	
	qWUE2-g61	N/3	*	ss715584350	3.50E+03	*	Missouri	
	qWUE2-g62	C1/4	*	ss715588095	9.10E+03	*	Missouri	
	qWUE2-g63	A1/5	*	ss715592509	3.40E+03	*	Missouri	
	qWUE2-g64	A1/5	*	ss715590791	5.00E+03	*	Missouri	
	qWUE2-g65.1	C2/6	*	ss715593777	5.00E+04	*	Missouri	
	qWUE2-g65.2	C2/6	*	ss715593778	5.00E+04	*	Missouri	
	qWUE2-g65.3	C2/6	*	ss715593779	1.40E+04	*	Missouri	
	qWUE2-g65.4	C2/6	*	ss715593780	1.40E+04	*	Missouri	
	qWUE2-g65.5	C2/6	*	ss715593781	3.30E+04	*	Missouri	
	qWUE2-g66	C2/6	*	ss715593786	1.90E+04	*	Missouri	
	qWUE2-g67	C2/6	*	ss715593793	1.60E+04	*	Missouri	
	qWUE2-g68	C2/6	*	ss715593810	5.20E+03	*	Missouri	
	qWUE2-g69	C2/6	*	ss715593814	2.50E+04	*	Missouri	
	qWUE2-g70.1	C2/6	*	ss715593820	1.60E+04	*	Missouri	
	qWUE2-g70.2	C2/6	*	ss715593822	1.90E+04	*	Missouri	
	qWUE2-g71	C2/6	*	ss715593829	4.50E+03	*	Missouri	
	qWUE2-g72	C2/6	*	ss715593831	2.50E+04	*	Missouri	
	qWUE2-g73	M/7	*	ss715598280	3.00E+04	*	Missouri	
	qWUE2-g74	M/7	*	ss715596496	3.00E+04	*	Missouri	
	qWUE2-g75	M/7	*	ss715596547	8.30E+03	*	Missouri	
	qWUE2-g76	M/7	*	ss715597538	4.50E+03	*	Missouri	
	qWUE2-g77.1	K//9	*	ss715605414	3.30E+04	*	Missouri	
	qWUE2-g77.2	K//9	*	ss715605417	3.30E+04	*	Missouri	
	qWUE2-g78	K//9	*	ss715605426	3.30E+04	*	Missouri	

Trait	*QTL*	LG/ Chr	Interval (cM)	Marker/interval	LOD, *P*, or *F*	R^2 (%)	Environment	Reference
	qWUE2-g79	K//9	*	ss715605451	1.00E+04	*	Missouri	
	qWUE2-g80	K//9	*	ss715605482	5.20E+03	*	Missouri	
	qWUE2-g81	K//9	*	ss715603084	6.60E+03	*	Missouri	
	qWUE2-g82	E/15	*	ss715620587	4.10E+03	*	Missouri	
Water use efficiency	*qWUE2-g83.1*	D2/17	*	ss715628212	3.30E+03	*	Missouri	
	qWUE2-g83.2	D2/17	*	ss715628213	8.30E+03	*	Missouri	
	qWUE2-g83.3	D2/17	*	ss715628214	3.80E+03	*	Missouri	
	qWUE2-g84	G/18	*	ss715631240	5.00E+04	*	Missouri	
	qWUE2-g85	G/18	*	ss715631406	2.80E+03	*	Missouri	
	qWUE2-g86	L/19	*	ss715636181	3.30E+03	*	Missouri	
	qWUE2-g87	L/19	*	ss715636462	3.90E+03	*	Missouri	
	qWUE2-g88	I/20	*	ss715636819	2.90E+03	*	Missouri	
	qWUE2-g89	I/20	*	ss715637488	1.10E+04	*	Missouri	
	qWUE2-g90	C2/6	*	ss715593775	5.00E+04	*	Missouri	

including *Glyma02g13240* that encodes for YhbC family, *Glyma02g40340* that encodes for a protein kinase, Glyma03g04410 that encodes for a protein kinase/phosphorylation activity, *Glyma04g39910* that encodes for a PPR repeat, *Glyma04g41540* that encodes for glutamate synthase activity, *Glyma08g48050* that encodes for a structural constituent of ribosome, Glyma09g02960 that encodes for a histidinol dehydrogenase, *Glyma10g09720* that encodes for a plant syntaxin, *Glyma10g27660* that encodes for an aconitase, Glyma13g11700 that encodes for an AMP-binding enzyme, *Glyma15g02390* that encodes for a mitochondrial processing peptidase, *Glyma15g12330* that encodes for the NADH-ubiquinone oxidoreductase complex I, *Glyma18g46050* that encodes for a leucine-rich repeat (LRR)-containing protein, *Glyma18g53170* that encodes for enzymes involved in the glutamine metabolic process, *Glyma19g29316* that encodes for a peptide deformylase, *Glyma20g03275* that encodes for an MSS1–/TRME-related GTP-binding protein, and many other genes (Table 4.6) (Dhanapal et al. 2015) (Dhanapal et al. 2015).

Kumar and Lal (2015) grew a population of different soybean genotypes ($n = 91$) from across the world, genotyped it with 26 SSR markers, and characterized it for WUE estimated as carbon isotope discrimination (CID). The genotypes were classified as having high (CID < 20.5‰), medium (20.6‰ < CID < 22.4‰), and low (CID > 22.4‰) WUE according to the CID values which were negatively correlated with WUE. The study reported low allelic variation in the population, less genetic variability, and no QTL for CID (WUE) (Kumar and Lal 2015).

Kaler et al. (2017) grew a population of different genotypes ($n = 373$) of soybean germplasm in Columbia, MO, in 2009–2010 and genotyped it with 31,260 SNPs to identify QTL for WUE estimated as carbon isotope ratio ($\delta^{13}C$). They used GWAS analysis to identify SNPs associated with WUE (carbon isotope ratio, $\delta^{13}C$) and oxygen isotope ratio ($\delta^{18}O$). Fifty four SNPs associated with QTL that control carbon isotope ratio ($\delta^{13}C$) have been identified and mapped on chromosomes 1, 2, 3, 4, 5, 6, 7, 9, 10, 11, 13, 14, 15, 16, 17, 18, 19, and 20, and 47 SNPs associated with QTL that control oxygen isotope ratio ($\delta^{18}O$) have been identified and mapped on chromosomes 1, 2, 3, 4, 5, 6, 7, 9, 15, 17, 18, 19, and 20 (Table 4.5). Within the $\delta^{13}C$ QTL regions, the authors identified many genes including *Glyma02g05595* that encodes for a heterogeneous nuclear ribonucleoprotein R, *Glyma02g36753* that encodes for a mitochondrial calcineurin B, *Glyma02g40340* that encodes for a aerine/threonine protein kinase, *Glyma04g39850* that encodes for a nodulin Mtn21-like transporter family protein, *Glyma06g02400* that encodes for a thioredoxin superfamily protein, *Glyma09g02800* that encodes for an oxidoreductase, *Glyma09g04640* that encodes for a polygalacturonase, *Glyma15g02690* that encodes for a 3′-5′ exonuclease (zinc ion binding), *Glyma16g01780* that encodes for an RNA-binding family protein, *Glyma18g48220* that encodes for an acyl-CoA thioesterase, and many other genes. Within the $\delta^{18}O$ QTL regions, the authors identified many genes including *Glyma01g00820* that encodes for a CCCH-type zinc finger family protein, *Glyma02g35400* that encodes for a COBRA-like protein, *Glyma02g46330* that encodes for an O-glycosyl hydrolases family 17 protein, *Glyma03g09050* that encodes for a galactosyltransferase activity, *Glyma04g35190*

Table 4.6 Candidate genes identified within the genomic regions harboring QTL that control water use efficiency (WUE) and phosphorus efficiency (PE)

Gene	LG/ Chr	Position	Annotation	References
Glyma08g20560	A2/8	*	Mitochondrial chaperonin	Zhang et al. (2014)
Glyma08g20570	A2/8	*	Uncharacterized protein	
Glyma08g20580	A2/8	*	Leucine-rich repeat	
Glyma08g20590	A2/8	*	Protein tyrosine kinase	
Glyma08g20600	A2/8	*	Glucose methanol choline oxidoreductase	
Glyma08g20610	A2/8	*	GTP-binding protein	
Glyma08g20620	A2/8	*	Protein kinase family protein	
Glyma08g20630	A2/8	*	Uncharacterized protein	
Glyma08g20640	A2/8	*	Endomembrane protein	
Glyma08g20650	A2/8	*	Beta-galactosidase	
Glyma08g20660	A2/8	*	uncharacterized protein	
Glyma08g20670	A2/8	*	ATP-dependent RNA helicase	
Glyma08g20680	A2/8	*	Uncharacterized protein	
Glyma08g20690	A2/8	*	Cytochrome P450	
Glyma08g20700	A2/8	*	Calcineurin B	
Glyma08g20710	A2/8	*	Phospholipase D	
Glyma08g20720	A2/8	*	Prenylated rab acceptor 1 family protein	
Glyma08g20730	A2/8	*	RNA-binding protein	
Glyma08g20740	A2/8	*	Alpha-helical protein	
Glyma08g20750	A2/8	*	Serine/threonine protein kinase	
Glyma08g20760	A2/8	*	Uncharacterized protein	
Glyma08g20770	A2/8	*	ABC transporter	
Glyma08g20780	A2/8	*	ABC transporter	
Glyma08g20790	A2/8	*	Uncharacterized protein	
Glyma08g20800	A2/8	*	Putative phosphatase	
Glyma08g20810	A2/8	*	Haloacid dehalogenase	
Glyma08g20820	A2/8	*	Putative phosphatase	
Glyma08g20830	A2/8	*	Protein phosphatase	
Glyma02g13240	D1b/2	*	Uncharacterized BCR, YhbC family COG0779	Dhanapal et al. (2015)
Glyma02g40340	D1b/2	*	Protein kinase/phosphorylation activity	
Glyma02g40310	D1b/2	*	Translation initiation factor activity	
Glyma02g40340	D1b/2	*	Protein kinase/phosphorylation activity	
Glyma02g40341	D1b/2	*	Protein kinase/phosphorylation activity	
Glyma02g40460	D1b/2	*	Protein of unknown function (DUF1191)	
Glyma02g40981	D1b/2	*	Protein kinase/phosphorylation activity	
Glyma03g04410	N/3	*	Protein kinase/phosphorylation activity	
Glyma04g39890	C1/4	*	Dual specificity phosphatase, catalytic domain	
Glyma04g39910	C1/4	*	PPR repeat	

Table 4.6 (continued)

Gene	LG/ Chr	Position	Annotation	References
Glyma04g41210	C1/4	*	PPR repeat	
Glyma04g41420	C1/4	*	PPR repeat	
Glyma04g41490	C1/4	*	Protein-binding	
Glyma04g41540	C1/4	*	Glutamate synthase activity	
Glyma06g43040	C2/6	*	Protein of unknown function (DUF640)	
Glyma07g16030	M/7	*	No annotation available	
Glyma08g48050	A2/8	*	Structural constituent of ribosome (intracellular)	
Glyma08g48160	A2/8	*	No annotation available	
Glyma09g02960	K/9	*	Oxidation-reduction process (histidinol dehydrogenase)	
Glyma10g09720	O/10	*	Syntaxin, plant	
Glyma10g09720	O/10	*	Syntaxin, plant	
Glyma10g27660	O/10	*	Metabolic process (aconitase)	
Glyma11g08590	B1/11	*	No annotation available	
Glyma12g36060	H/12	*	Protein of unknown function (DUF822)	
Glyma13g11700	F/13	*	AMP-binding enzyme	
Glyma13g24380	F/13	*	No annotation available	
Glyma15g02390	E/15	*	Mitochondrial processing peptidase, alpha subunit	
Glyma15g12330	E/15	*	NADH-ubiquinone oxidoreductase complex I, 21 kDa subunit	
Glyma15g20631	E/15	*	No annotation available	
Glyma15g12370	E/15	*	Domain of unknown function (DUF588)	
Glyma15g12380	E/15	*	Stress-associated endoplasmic reticulum protein (Serp1/Ramp4)	
Glyma15g12400	E/15	*	Protein/DNA-binding	
Glyma17g36700	D2/17	*	Protein of unknown function, DUF607	
Glyma18g46050	G/18	*	Leucine-rich repeat-containing protein	
Glyma18g41760	G/18	*	No annotation available	
Glyma18g52720	G/18	*	Regulation of sequence-specific DNA-binding TF	
Glyma18g53170	G/18	*	Glutamine metabolic process	
Glyma19g29316	L/19	*	Peptide deformylase.	
Glyma20g03275	I/20	*	MSS1–/TRME-related GTP-binding protein	

that encodes for a trehalose-phosphatase, *Glyma05g08200* that encodes for a serine/threonine kinase receptor, *Glyma06g22160* that encodes for a exocyst complex component 7 (salicylic acid biosynthetic process), *Glyma06g22771* that encodes for an 18S pre-ribosomal assembly protein Gar2-related, *Glyma07g32283* that encodes for a cellulose synthase (polysaccharide biosynthetic process), *Glyma09g13480* v 2-hydroxyacid dehydrogenase, and may other genes (Table 4.6) (Kaler et al. 2017).

To identify QTL for phosphorus efficiency (PE), Zhang et al. (2014) grew a RIL population derived from a cross between "Nannong 94–156" and "Bogao" lines ($n = 152$) in different fields in China in 2006 and 2007 and genotyped with 1536 SNP markers. They also used GWAS using a natural population of soybean accessions ($n = 192$) that was genotyped with 82 SSR markers. They used IM and CIM of WinQTL Cartographer for QTL data analysis. They identified QTL for RPC, RPPC, RAPA, RPHt, and 100-RSW as measures for PE (RPC, relative P concentration in the plant at two P conditions; RPPC, relative P concentration in the pod at the filling stage at two P conditions; RAPA, relative acid phosphatase activity at two P conditions; RPHt, relative plant height at two P conditions; 100-RSW, relative value of 100-seed weight at two P conditions). Three QTL for RPC have been identified and mapped on chromosomes 8 and 13; three QTL for RPPC have been identified and mapped on chromosomes 5 and 8; two QTL for RAPA have been identified and mapped on chromosomes 8; five QTL for RPHt have been identified and mapped on chromosomes 1, 14, and 19; and three QTL for 100-RSW have been identified and mapped on chromosomes 18 (Fig. 2.1; Table 4.5). Within the major QTL region on chromosomes 8 (*qPE8*), authors identified 28 genes including *Glyma08g20560* that encodes for a mitochondrial chaperonin, *Glyma08g20580* that encodes for a leucine-rich repeat, *Glyma08g20590* that encodes for a protein tyrosine kinase, *Glyma08g20600* that encodes for a glucose methanol choline oxidoreductase, *Glyma08g20610* that encodes for a GTP-binding protein, *Glyma08g20640* that encodes for an endomembrane protein, *Glyma08g20650* that encodes for a beta-galactosidase, *Glyma08g20690* that encodes for a cytochrome P450 (Barlier et al. 2000), *Glyma08g20700* that encodes for a calcineurin B, *Glyma08g20710* that encodes for a phospholipase D, and many other genes. The gene *GmACP1* that encodes for an acid phosphatase was a prominent gene within this region (Zhang et al. 2014). In a subsequent study using the same RIL population (Nannong 94–156 by Bogao; $n = 146$) grown in China in 2014 and the derived high-density genetic linkage map constructed by 6159 specific-locus amplified fragment sequencing (SLAF-seq) markers and covering 3020.59 cM, Zhang et al. (2016) identified 20 loci associated with QTL that control phosphorus efficiency (PE) across multiple environments, and 8 of these loci were novel (*q1*, *q3*, *q4–1*, *q4–2*, *q7*, *q13–1*, *q13–2*, and *q16*) including the major QTL on chromosome 4 (*q4–2*). Within *q8* genomic region, *GmACP1* gene was identified. Within *q2* genomic region, 18 candidate genes have been identified including 3 candidate genes (*Glyma.02G266700*, *Glyma.02G267500*, and *Glyma.02G267900*) that were related to P efficiency and/or plant stress. Within *q4–2* genomic region, 38 candidate genes have been identified including *Glyma.04G214000* that encodes for a pectin methylesterase. Within *q13–2* genomic region, 146 candidate genes have been identified including

Table 4.7 Detailed information of QTL that control shoot macronutrients' (P, K, Mg, S, and Ca) and micronutrients' (Zn, Mn, B, Cu, and Fe) contents reported and *not reported* in SoyBase (Grant et al., 2010; http://www.soybase.org)

Trait	*QTL*	LG/Chr	Interval (cM)	Marker/interval		R^2 (%)	Environment	Reference
Shoot potassium (K)	*qShootK1-g1*	D1a/1	*	ss715579931		12.00	Missouri	Dhanapal et al. (2018)
	qShootK1-g2	D1b/2	*	ss715581242		12.00	Missouri	
	qShootK1-g3	D1b/2	*	ss715581244		17.00	Missouri	
	qShootK1-g4	D1b/2	*	ss715581329		8.00	Missouri	
	qShootK1-g5	D1b/2	*	ss715582594		11.00	Missouri	
	qShootK1-g6	D1b/2	*	ss715583151		11.00	Missouri	
	qShootK1-g7	C2/6	*	ss715593829		10.00	Missouri	
	qShootK1-g8	C2/6	*	ss715593841		9.00	Missouri	
	qShootK1-g9	C2/6	*	ss715593852		8.00	Missouri	
	qShootK1-g10	C2/6	*	ss715593868		9.00	Missouri	
	qShootK1-g11	M/7	*	ss715597932		8.00	Missouri	
	qShootK1-g12.1	M/7	*	ss715598065		12.00	Missouri	
	qShootK1-g12.2	M/7	*	ss715598067		9.00	Missouri	
	qShootK1-g12.3	M/7	*	ss715598070		13.00	Missouri	
	qShootK1-g12.4	M/7	*	ss715598081		8.00	Missouri	
	qShootK1-g13	M/7	*	ss715598093		13.00	Missouri	
	qShootK1-g14	M/7	*	ss715598106		8.00	Missouri	
	qShootK1-g15	M/7	*	ss715598117		11.00	Missouri	
	qShootK1-g16	A2/8	*	ss715600019		9.00	Missouri	
	qShootK1-g17	A2/8	*	ss715600022		14.00	Missouri	
	qShootK1-g18	K/9	*	ss715603535		8.00	Missouri	
	qShootK1-g19	K/9	*	ss715604244		8.00	Missouri	
	qShootK1-g20	K/9	*	ss715604256		8.00	Missouri	
	qShootK1-g21	K/9	*	ss715604274		8.00	Missouri	

Trait	*QTL*	LG/Chr	Interval (cM)	Marker/interval		R^2 (%)	Environment	Reference
	qShootK1-g22	O/10	*	ss715607037		9.00	Missouri	
	qShootK1-g23.1	O/10	*	ss715607053		9.00	Missouri	
	qShootK1-g23.2	O/10	*	ss715607054		9.00	Missouri	
	qShootK1-g23.3	O/10	*	ss715607056		9.00	Missouri	
	qShootK1-g24	O/10	*	ss715607072		7.00	Missouri	
	qShootK1-g25	O/10	*	ss715608927		9.00	Missouri	
	qShootK1-g26.1	O/10	*	ss715608951		10.00	Missouri	
	qShootK1-g26.2	O/10	*	ss715608952		11.00	Missouri	
	qShootK1-g27	O/10	*	ss715609030		7.00	Missouri	
	qShootK1-g28	O/10	*	ss715609051		12.00	Missouri	
	qShootK1-g29	O/10	*	ss715609059		7.00	Missouri	
	qShootK1-g30	F/13	*	ss715615849		7.00	Missouri	
	qShootK1-g31	B2/14	*	ss715618815		8.00	Missouri	
	qShootK1-g32	B2/14	*	ss715619792		11.00	Missouri	
	qShootK1-g33	J/16	*	ss715624781		12.00	Missouri	
	qShootK1-g34	G/18	*	ss715629801		14.00	Missouri	
	qShootK1-g35	G/18	*	ss715629913		12.00	Missouri	
	qShootK1-g36	G/18	*	ss715629921		12.00	Missouri	
	qShootK1-g37	G/18	*	ss715630092		8.00	Missouri	
	qShootK1-g38.1	G/18	*	ss715630099		10.00	Missouri	
	qShootK1-g38.2	G/18	*	ss715630101		8.00	Missouri	
	qShootK1-g39	G/18	*	ss715632404		12.00	Missouri	
	qShootK1-g40	G/18	*	ss715632710		10.00	Missouri	
	qShootK1-g41	G/18	*	ss715632713		10.00	Missouri	

(continued)

Table 4.7 (continued)

Trait	*QTL*	LG/Chr	Interval (cM)	Marker/interval		R^2 (%)	Environment	Reference
Shoot calcium (Ca)	*qShootCa1-g1*	D1a/1	*	ss715579848		8.00	Missouri	
	qShootCa1-g2	D1a/1	*	ss715579861		7.00	Missouri	
	qShootCa1-g3	D1a/1	*	ss715579863		7.00	Missouri	
	qShootCa1-g4	D1a/1	*	ss715579872		9.00	Missouri	
	qShootCa1-g5.1	D1a/1	*	ss715579877		9.00	Missouri	
	qShootCa1-g5.2	D1a/1	*	ss715579878		9.00	Missouri	
	qShootCa1-g6	D1a/1	*	ss715579881		9.00	Missouri	
	qShootCa1-g7	D1b/2	*	ss715582357		12.00	Missouri	
	qShootCa1-g8	D1b/2	*	ss715582360		7.00	Missouri	
	qShootCa1-g9.1	D1b/2	*	ss715582365		11.00	Missouri	
	qShootCa1-g9.2	D1b/2	*	ss715582366		12.00	Missouri	
	qShootCa1-g9.3	D1b/2	*	ss715582368		10.00	Missouri	
	qShootCa1-g10	C1/4	*	ss715587071		7.00	Missouri	
	qShootCa1-g11.1	C1/4	*	ss715588329		8.00	Missouri	
	qShootCa1-g11.2	C1/4	*	ss715588330		8.00	Missouri	
	qShootCa1-g12	M/7	*	ss715596467		7.00	Missouri	
	qShootCa1-g13	A2/8	*	ss715599197		7.00	Missouri	
	qShootCa1-g14	A2/8	*	ss715599203		7.00	Missouri	
	qShootCa1-g15	A2/8	*	ss715600504		7.00	Missouri	
	qShootCa1-g16	A2/8	*	ss715601326		10.00	Missouri	
	qShootCa1-g17	A2/8	*	ss715601328		9.00	Missouri	
	qShootCa1-g18	B1/11	*	ss715609081		9.00	Missouri	
	qShootCa1-g19	B1/11	*	ss715611254		9.00	Missouri	
	qShootCa1-g20	D2/17	*	ss715625751		10.00	Missouri	
	qShootCa1-g21	D2/17	*	ss715625753		9.00	Missouri	

Trait	*QTL*	LG/Chr	Interval (cM)	Marker/interval		R^2 (%)	Environment	Reference
	qShootCa1-g22	D2/17	*	ss715625757		9.00	Missouri	
	qShootCa1-g23	D2/17	*	ss715625764		10.00	Missouri	
	qShootCa1-g24.1	D2/17	*	ss715628353		7.00	Missouri	
	qShootCa1-g24.2	D2/17	*	ss715628355		7.00	Missouri	
	qShootCa1-g25	D2/17	*	ss715628358		8.00	Missouri	
	qShootCa1-g26.1	D2/17	*	ss715628360		8.00	Missouri	
	qShootCa1-g26.2	D2/17	*	ss715628361		8.00	Missouri	
	qShootCa1-g26.3	D2/17	*	ss715628362		9.00	Missouri	
	qShootCa1-g26.4	D2/17	*	ss715628363		8.00	Missouri	
Shoot magnesium (Mg)	*qShootMg1-g1*	N/3	*	ss715585605		9.00	Missouri	
	qShootMg1-g2	N/3	*	ss715586435		8.00	Missouri	
	qShootMg1-g3	A1/5	*	ss715590854		8.00	Missouri	
	qShootMg1-g4	A1/5	*	ss715591448		9.00	Missouri	
	qShootMg1-g5	C2/6	*	ss715594812		9.00	Missouri	
	qShootMg1-g6	A2/8	*	ss715599184		8.00	Missouri	
	qShootMg1-g7	A2/8	*	ss715602899		9.00	Missouri	
	qShootMg1-g8	O/10	*	ss715606679		9.00	Missouri	
	qShootMg1-g9	B1/11	*	ss715608796		11.00	Missouri	
	qShootMg1-g10	H/12	*	ss715612181		11.00	Missouri	
	qShootMg1-g11	F/13	*	ss715615305		10.00	Missouri	
	qShootMg1-g12	F/13	*	ss715616287		8.00	Missouri	
	qShootMg1-g13	F/13	*	ss715616755		8.00	Missouri	
	qShootMg1-g14	F/13	*	ss715616917		9.00	Missouri	

(continued)

Table 4.7 (continued)

Trait	*QTL*	LG/Chr	Interval (cM)	Marker/interval		R^2 (%)	Environment	Reference
	qShootMg1-g15	F/13	*	ss715616945		12.00	Missouri	
	qShootMg1-g16	E/15	*	ss715622570		9.00	Missouri	
	qShootMg1-g17	G/18	*	ss715630257		9.00	Missouri	
	qShootMg1-g18	L/19	*	ss715635824		9.00	Missouri	
Shoot phosphorus (P)	*qShootP1-g1*	D1b/2	*	ss715583151		12.00	Missouri	
	qShootP1-g2	N/3	*	ss715585521		10.00	Missouri	
	qShootP1-g3	N/3	*	ss715585596		11.00	Missouri	
	qShootP1-g4	C1/4	*	ss715588318		9.00	Missouri	
	qShootP1-g5	C1/4	*	ss715588329		9.00	Missouri	
	qShootP1-g6	A1/5	*	ss715590418		9.00	Missouri	
	qShootP1-g7	A1/5	*	ss715591409		12.00	Missouri	
	qShootP1-g8.1	A1/5	*	ss715591411		11.00	Missouri	
	qShootP1-g8.2	A1/5	*	ss715591412		11.00	Missouri	
	qShootP1-g9	A1/5	*	ss715591419		9.00	Missouri	
	qShootP1-g10.1	A1/5	*	ss715591882		10.00	Missouri	
	qShootP1-g10.2	A1/5	*	ss715591883		11.00	Missouri	
	qShootP1-g11.1	C2/6	*	ss715594973		10.00	Missouri	
	qShootP1-g11.2	C2/6	*	ss715594979		10.00	Missouri	
	qShootP1-g11.3	C2/6	*	ss715594980		10.00	Missouri	
	qShootP1-g12	M/7	*	ss715597409		12.00	Missouri	
	qShootP1-g13.1	M/7	*	ss715597410		12.00	Missouri	
	qShootP1-g13.2	M/7	*	ss715597411		10.00	Missouri	
	qShootP1-g14	M/7	*	ss715597413		11.00	Missouri	
	qShootP1-g15.1	M/7	*	ss715597421		12.00	Missouri	
	qShootP1-g15.2	M/7	*	ss715597422		16.00	Missouri	

Trait	*QTL*	LG/Chr	Interval (cM)	Marker/interval		R^2 (%)	Environment	Reference
	qShootP1-g16	M/7	*	ss715597431		10.00	Missouri	
	qShootP1-g17.1	M/7	*	ss715597434		10.00	Missouri	
	qShootP1-g17.2	M/7	*	ss715597435		10.00	Missouri	
	qShootP1-g18	M/7	*	ss715597441		11.00	Missouri	
	qShootP1-g19	M/7	*	ss715597868		12.00	Missouri	
	qShootP1-g20	A2/8	*	ss715599260		9.00	Missouri	
	qShootP1-g21	B1/11	*	ss715608726		13.00	Missouri	
	qShootP1-g22	B2/14	*	ss715619393		10.00	Missouri	
	qShootP1-g23	G/18	*	ss715630156		11.00	Missouri	
	qShootP1-g24	G/18	*	ss715630720		10.00	Missouri	
	qShootP1-g25	G/18	*	ss715630844		10.00	Missouri	
	qShootP1-g26	G/18	*	ss715630868		11.00	Missouri	
	qShootP1-g27	G/18	*	ss715630904		10.00	Missouri	
	qShootP1-g28	G/18	*	ss715630905		10.00	Missouri	
	qShootP1-g29	G/18	*	ss715630917		10.00	Missouri	
	qShootP1-g30	G/18	*	ss715632236		11.00	Missouri	
	qShootP1-g31	L/19	*	ss715633156		10.00	Missouri	
	qShootP1-g32	L/19	*	ss715633177		12.00	Missouri	
	qShootP1-g33	I/20	*	ss715636663		9.00	Missouri	
	qShootP1-g34	I/20	*	ss715637157		10.00	Missouri	
	qShootP1-g35	I/20	*	ss715637621		10.00	Missouri	
Shoot sulfur (S)	*qShootS1-g1*	D1b/2	*	ss715583445		8.00	Missouri	
	qShootS1-g2	A1/5	*	ss715590151		7.00	Missouri	
	qShootS1-g3	A1/5	*	ss715590415		9.00	Missouri	

(continued)

Table 4.7 (continued)

Trait	*QTL*	LG/Chr	Interval (cM)	Marker/interval		R^2 (%)	Environment	Reference
	qShootS1-g4	A1/5	*	ss715590418		11.00	Missouri	
	qShootS1-g5	A1/5	*	ss715591901		9.00	Missouri	
	qShootS1-g6	A2/8	*	ss715601069		7.00	Missouri	
	qShootS1-g7	A2/8	*	ss715601994		7.00	Missouri	
	qShootS1-g8	K/9	*	ss715604112		7.00	Missouri	
	qShootS1-g9	O/10	*	ss715607072		7.00	Missouri	
	qShootS1-g10	O/10	*	ss715607651		10.00	Missouri	
	qShootS1-g11	B1/11	*	ss715609121		9.00	Missouri	
	qShootS1-g12	B1/11	*	ss715609125		10.00	Missouri	
	qShootS1-g13.1	B1/11	*	ss715609135		9.00	Missouri	
	qShootS1-g13.2	B1/11	*	ss715609136		9.00	Missouri	
	qShootS1-g14	B1/11	*	ss715609143		9.00	Missouri	
	qShootS1-g15	B1/11	*	ss715609154		9.00	Missouri	
	qShootS1-g16.1	B1/11	*	ss715609165		9.00	Missouri	
	qShootS1-g16.2	B1/11	*	ss715609166		9.00	Missouri	
	qShootS1-g17	B1/11	*	ss715609171		9.00	Missouri	
	qShootS1-g18.1	B1/11	*	ss715609173		9.00	Missouri	
	qShootS1-g18.2	B1/11	*	ss715609174		9.00	Missouri	
	qShootS1-g19	B1/11	*	ss715609179		9.00	Missouri	
	qShootS1-g20.1	B1/11	*	ss715609182		9.00	Missouri	
	qShootS1-g20.2	B1/11	*	ss715609183		9.00	Missouri	
	qShootS1-g21	H/12	*	ss715612181		7.00	Missouri	
	qShootS1-g22	F/13	*	ss715616829		7.00	Missouri	
	qShootS1-g23	B2/14	*	ss715619407		9.00	Missouri	
	qShootS1-g24	E/15	*	ss715622004		8.00	Missouri	

Trait	*QTL*	LG/Chr	Interval (cM)	Marker/interval		R^2 (%)	Environment	Reference
	qShootS1-g25	D2/17	*	ss715626508		8.00	Missouri	
	qShootS1-g26	I/20	*	ss715636663		12.00	Missouri	
	qShootS1-g27	I/20	*	ss715636768		7.00	Missouri	
	qShootS1-g28	I/20	*	ss715636773		7.00	Missouri	
	qShootS1-g29.1	I/20	*	ss715636780		7.00	Missouri	
	qShootS1-g29.2	I/20	*	ss715636781		7.00	Missouri	
Shoot iron (Fe)	*qShootFe1-g1*	D1a/1	*	ss715579327		12.00	Missouri	
	qShootFe1-g2.1	D1b/2	*	ss715581093		12.00	Missouri	
	qShootFe1-g2.2	D1b/2	*	ss715581094		12.00	Missouri	
	qShootFe1-g3.1	D1b/2	*	ss715581096		12.00	Missouri	
	qShootFe1-g3.2	D1b/2	*	ss715581097		12.00	Missouri	
	qShootFe1-g4	D1b/2	*	ss715581523		12.00	Missouri	
	qShootFe1-g5	A1/5	*	ss715590845		12.00	Missouri	
	qShootFe1-g6	M/7	*	ss715597822		12.00	Missouri	
	qShootFe1-g7.1	K/9	*	ss715603713		12.00	Missouri	
	qShootFe1-g7.2	K/9	*	ss715603714		12.00	Missouri	
	qShootFe1-g8.1	K/9	*	ss715603939		15.00	Missouri	
	qShootFe1-g8.2	K/9	*	ss715603940		20.00	Missouri	
	qShootFe1-g8.3	K/9	*	ss715603941		15.00	Missouri	
	qShootFe1-g9	K/9	*	ss715603946		15.00	Missouri	
	qShootFe1-g10	K/9	*	ss715603949		14.00	Missouri	
	qShootFe1-g11	K/9	*	ss715603953		15.00	Missouri	
	qShootFe1-g12.1	K/9	*	ss715604767		12.00	Missouri	
	qShootFe1-g12.2	K/9	*	ss715604768		14.00	Missouri	

(continued)

Table 4.7 (continued)

Trait	QTL	LG/Chr	Interval (cM)	Marker/interval		R^2 (%)	Environment	Reference
	qShootFe1-g13.1	K/9	*	ss715604773		14.00	Missouri	
	qShootFe1-g13.2	K/9	*	ss715604774		14.00	Missouri	
	qShootFe1-g14	K/9	*	ss715604775		16.00	Missouri	
	qShootFe1-g15	K/9	*	ss715604779		12.00	Missouri	
	qShootFe1-g16	K/9	*	ss715604853		13.00	Missouri	
	qShootFe1-g17.1	O/10	*	ss715606682		12.00	Missouri	
	qShootFe1-g17.2	O/10	*	ss715606683		13.00	Missouri	
	qShootFe1-g18	H/12	*	ss715611801		14.00	Missouri	
	qShootFe1-g19.1	H/12	*	ss715612489		12.00	Missouri	
	qShootFe1-g19.2	H/12	*	ss715612490		12.00	Missouri	
	qShootFe1-g20	H/12	*	ss715612979		12.00	Missouri	
	qShootFe1-g21	H/12	*	ss715613122		12.00	Missouri	
	qShootFe1-g22	B1/11	*	ss715614291		17.00	Missouri	
	qShootFe1-g23	B1/11	*	ss715614293		17.00	Missouri	
	qShootFe1-g24	B1/11	*	ss715614298		12.00	Missouri	
	qShootFe1-g25	B1/11	*	ss715614300		17.00	Missouri	
	qShootFe1-g26	B1/11	*	ss715614306		15.00	Missouri	
	qShootFe1-g27.1	B1/11	*	ss715614311		14.00	Missouri	
	qShootFe1-g27.2	B1/11	*	ss715614312		12.00	Missouri	
	qShootFe1-g28.1	B1/11	*	ss715614315		15.00	Missouri	
	qShootFe1-g28.2	B1/11	*	ss715614316		15.00	Missouri	
	qShootFe1-g28.3	B1/11	*	ss715614317		12.00	Missouri	
	qShootFe1-g29	B1/11	*	ss715614320		16.00	Missouri	
	qShootFe1-g30.1	B1/11	*	ss715614322		15.00	Missouri	

Trait	QTL	LG/Chr	Interval (cM)	Marker/interval	R^2 (%)	Environment	Reference
	qShootFe1-g30.2	B1/11	*	ss715614323	16.00	Missouri	
	qShootFe1-g31.1	B1/11	*	ss715614325	12.00	Missouri	
	qShootFe1-g31.2	B1/11	*	ss715614327	15.00	Missouri	
	qShootFe1-g32.1	B1/11	*	ss715614329	16.00	Missouri	
	qShootFe1-g32.2	B1/11	*	ss715614330	12.00	Missouri	
	qShootFe1-g32.3	B1/11	*	ss715614331	16.00	Missouri	
	qShootFe1-g32.4	B1/11	*	ss715614332	16.00	Missouri	
	qShootFe1-g32.5	B1/11	*	ss715614333	12.00	Missouri	
	qShootFe1-g32.6	B1/11	*	ss715614334	16.00	Missouri	
	qShootFe1-g33	F/13	*	ss715614441	15.00	Missouri	
	qShootFe1-g34	F/13	*	ss715614444	16.00	Missouri	
	qShootFe1-g35	F/13	*	ss715614495	12.00	Missouri	
	qShootFe1-g36	F/13	*	ss715614506	12.00	Missouri	
	qShootFe1-g37	F/13	*	ss715615493	12.00	Missouri	
	qShootFe1-g38	F/13	*	ss715617055	13.00	Missouri	
	qShootFe1-g39	B2/14	*	ss715617366	11.00	Missouri	
	qShootFe1-g40	B2/14	*	ss715619587	13.00	Missouri	
	qShootFe1-g41.1	E/15	*	ss715620329	12.00	Missouri	
	qShootFe1-g41.2	E/15	*	ss715620330	13.00	Missouri	
	qShootFe1-g41.3	E/15	*	ss715620331	13.00	Missouri	
	qShootFe1-g42	E/15	*	ss715622954	15.00	Missouri	
	qShootFe1-g43	E/15	*	ss715622982	12.00	Missouri	
	qShootFe1-g44.1	J/16	*	ss715623847	12.00	Missouri	
	qShootFe1-g44.2	J/16	*	ss715623848	12.00	Missouri	

(continued)

Table 4.7 (continued)

Trait	*QTL*	LG/Chr	Interval (cM)	Marker/interval		R^2 (%)	Environment	Reference
	qShootFe1-g45.1	J/16	*	ss715623856		15.00	Missouri	
	qShootFe1-g45.2	J/16	*	ss715623859		15.00	Missouri	
	qShootFe1-g46	J/16	*	ss715623861		16.00	Missouri	
	qShootFe1-g47	G/18	*	ss715629958		13.00	Missouri	
	qShootFe1-g48	G/18	*	ss715632057		13.00	Missouri	
	qShootFe1-g49	L/19	*	ss715635930		16.00	Missouri	
	qShootFe1-g50	L/19	*	ss715635934		17.00	Missouri	
	qShootFe1-g51	L/19	*	ss715635948		14.00	Missouri	
	qShootFe1-g52.1	L/19	*	ss715635950		17.00	Missouri	
	qShootFe1-g52.2	L/19	*	ss715635951		17.00	Missouri	
Shoot boron content	*qShootB1-g1*	D1a/1	*	ss715579327		10.00	Missouri	
	qShootB1-g2	C1/4	*	ss715588329		9.00	Missouri	
	qShootB1-g3	C1/4	*	ss715588347		11.00	Missouri	
	qShootB1-g4	C1/4	*	ss715589522		9.00	Missouri	
	qShootB1-g5	A1/5	*	ss715590333		10.00	Missouri	
	qShootB1-g6	A1/5	*	ss715591896		9.00	Missouri	
	qShootB1-g7	C2/6	*	ss715593124		9.00	Missouri	
	qShootB1-g8.1	C2/6	*	ss715593150		9.00	Missouri	
	qShootB1-g8.2	C2/6	*	ss715593152		10.00	Missouri	
	qShootB1-g8.3	C2/6	*	ss715593154		9.00	Missouri	
	qShootB1-g9	C2/6	*	ss715594345		10.00	Missouri	
	qShootB1-g10	C2/6	*	ss715594354		12.00	Missouri	
	qShootB1-g11	O/10	*	ss715607072		10.00	Missouri	
	qShootB1-g12	F/13	*	ss715614515		10.00	Missouri	
	qShootB1-g13	F/13	*	ss715614784		9.00	Missouri	

Trait	*QTL*	LG/Chr	Interval (cM)	Marker/interval		R^2 (%)	Environment	Reference
	qShootB1-g14	B2/14	*	ss715619127		10.00	Missouri	
	qShootB1-g15	E/15	*	ss715622093		9.00	Missouri	
	qShootB1-g16	E/15	*	ss715622711		9.00	Missouri	
	qShootB1-g17	L/19	*	ss715633104		9.00	Missouri	
	qShootB1-g18.1	L/19	*	ss715633123		9.00	Missouri	
	qShootB1-g18.2	L/19	*	ss715633124		9.00	Missouri	
	qShootB1-g19	L/19	*	ss715633156		11.00	Missouri	
	qShootB1-g20	L/19	*	ss715633177		12.00	Missouri	
	qShootB1-g21	L/19	*	ss715633221		11.00	Missouri	
	qShootB1-g22.1	L/19	*	ss715633232		14.00	Missouri	
	qShootB1-g22.2	L/19	*	ss715633233		11.00	Missouri	
	qShootB1-g22.3	L/19	*	ss715633236		13.00	Missouri	
	qShootB1-g22.4	L/19	*	ss715633238		10.00	Missouri	
	qShootB1-g22.5	L/19	*	ss715633239		14.00	Missouri	
	qShootB1-g22.6	L/19	*	ss715633243		11.00	Missouri	
	qShootB1-g22.7	L/19	*	ss715633245		14.00	Missouri	
	qShootB1-g22.8	L/19	*	ss715633247		12.00	Missouri	
	qShootB1-g23	L/19	*	ss715633252		9.00	Missouri	
	qShootB1-g24	L/19	*	ss715634527		10.00	Missouri	
Shoot manganese (Mn)	*qShootMn1-g1*	N/3	*	ss715584556		12.00	Missouri	
	qShootMn1-g2	N/3	*	ss715584595		12.00	Missouri	
	qShootMn1-g3	N/3	*	ss715584908		12.00	Missouri	
	qShootMn1-g4	A1/5	*	ss715589897		12.00	Missouri	
	qShootMn1-g5	C2/6	*	ss715594401		14.00	Missouri	

(continued)

Table 4.7 (continued)

Trait	*QTL*	LG/Chr	Interval (cM)	Marker/interval		R^2 (%)	Environment	Reference
	qShootMn1-g6	C2/6	*	ss715595056		13.00	Missouri	
	qShootMn1-g7	M/7	*	ss715598099		12.00	Missouri	
	qShootMn1-g8	A2/8	*	ss715599191		12.00	Missouri	
	qShootMn1-g9	A2/8	*	ss715599198		12.00	Missouri	
	qShootMn1-g10	A2/8	*	ss715599203		12.00	Missouri	
	qShootMn1-g11.1	D2/17	*	ss715626070		12.00	Missouri	
	qShootMn1-g11.2	D2/17	*	ss715626071		12.00	Missouri	
	qShootMn1-g11.3	D2/17	*	ss715626072		13.00	Missouri	
	qShootMn1-g12	D2/17	*	ss715627655		12.00	Missouri	
	qShootMn1-g13	L/19	*	ss715633238		14.00	Missouri	
Shoot zinc (Zn)	*qShootZn1-g1*	D1a/1	*	ss715580413		12.00	Missouri	
	qShootZn1-g2.1	N/3	*	ss715584971		11.00	Missouri	
	qShootZn1-g2.2	N/3	*	ss715584973		12.00	Missouri	
	qShootZn1-g3	N/3	*	ss715585611		14.00	Missouri	
	qShootZn1-g4	N/3	*	ss715586810		12.00	Missouri	
	qShootZn1-g5	N/3	*	ss715586848		14.00	Missouri	
	qShootZn1-g6	C1/4	*	ss715587241		11.00	Missouri	
	qShootZn1-g7.1	C1/4	*	ss715587257		12.00	Missouri	
	qShootZn1-g7.2	C1/4	*	ss715587261		17.00	Missouri	
	qShootZn1-g8	A1/5	*	ss715589793		12.00	Missouri	
	qShootZn1-g9	A1/5	*	ss715591409		16.00	Missouri	
	qShootZn1-g10.1	A1/5	*	ss715591411		12.00	Missouri	
	qShootZn1-g10.2	A1/5	*	ss715591412		15.00	Missouri	
	qShootZn1-g11	A1/5	*	ss715591419		14.00	Missouri	
	qShootZn1-g12.1	M/7	*	ss715596871		12.00	Missouri	

Trait	*QTL*	LG/Chr	Interval (cM)	Marker/interval		R^2 (%)	Environment	Reference
	qShootZn1-g12.2	M/7	*	ss715596872		12.00	Missouri	
	qShootZn1-g13	M/7	*	ss715596877		12.00	Missouri	
	qShootZn1-g14.1	M/7	*	ss715596896		13.00	Missouri	
	qShootZn1-g14.2	M/7	*	ss715596897		13.00	Missouri	
	qShootZn1-g15	M/7	*	ss715598480		12.00	Missouri	
	qShootZn1-g16	M/7	*	ss715598544		11.00	Missouri	
	qShootZn1-g17	A2/8	*	ss715600434		12.00	Missouri	
	qShootZn1-g18	K/9	*	ss715604421		12.00	Missouri	
	qShootZn1-g19	K/9	*	ss715604574		14.00	Missouri	
	qShootZn1-g20	K/9	*	ss715604600		12.00	Missouri	
	qShootZn1-g21	H/12	*	ss715612166		12.00	Missouri	
	qShootZn1-g22.1	H/12	*	ss715612168		12.00	Missouri	
	qShootZn1-g22.2	H/12	*	ss715612169		13.00	Missouri	
	qShootZn1-g23	E/15	*	ss715623110		15.00	Missouri	
	qShootZn1-g24	E/15	*	ss715623122		13.00	Missouri	
	qShootZn1-g25	E/15	*	ss715623134		12.00	Missouri	
	qShootZn1-g26	E/15	*	ss715623142		12.00	Missouri	
	qShootZn1-g27	E/15	*	ss715623149		12.00	Missouri	
	qShootZn1-g28.1	L/19	*	ss715635116		13.00	Missouri	
	qShootZn1-g28.2	L/19	*	ss715635117		12.00	Missouri	
	qShootZn1-g29.1	L/19	*	ss715635119		13.00	Missouri	
	qShootZn1-g29.2	L/19	*	ss715635120		12.00	Missouri	
	qShootZn1-g29.3	L/19	*	ss715635123		12.00	Missouri	
Shoot copper (Cu)	*qShootCu1-g1*	D1a/1	*	ss715580217		10.00	Missouri	

(continued)

Table 4.7 (continued)

Trait	QTL	LG/Chr	Interval (cM)	Marker/interval		R^2 (%)	Environment	Reference
	qShootCu1-g2.1	D1b/2	*	ss715583130		13.00	Missouri	
	qShootCu1-g2.2	D1b/2	*	ss715583131		17.00	Missouri	
	qShootCu1-g3	C1/4	*	ss715587241		11.00	Missouri	
	qShootCu1-g4.1	C1/4	*	ss715587257		12.00	Missouri	
	qShootCu1-g4.2	C1/4	*	ss715587261		12.00	Missouri	
	qShootCu1-g5	C1/4	*	ss715588403		12.00	Missouri	
	qShootCu1-g6	A1/5	*	ss715591438		10.00	Missouri	
	qShootCu1-g7	K/9	*	ss715604600		10.00	Missouri	
	qShootCu1-g8	O/10	*	ss715605639		10.00	Missouri	
	qShootCu1-g9	H/12	*	ss715612166		12.00	Missouri	
	qShootCu1-g10	H/12	*	ss715612168		12.00	Missouri	
	qShootCu1-g11	F/13	*	ss715614831		9.00	Missouri	
	qShootCu1-g12	B2/14	*	ss715617719		10.00	Missouri	
	qShootCu1-g13.1	B2/14	*	ss715618596		10.00	Missouri	
	qShootCu1-g13.2	B2/14	*	ss715618601		10.00	Missouri	
	qShootCu1-g13.3	B2/14	*	ss715618608		10.00	Missouri	
	qShootCu1-g13.4	B2/14	*	ss715618612		10.00	Missouri	
	qShootCu1-g13.5	B2/14	*	ss715618620		10.00	Missouri	
	qShootCu1-g14.1	E/15	*	ss715622632		10.00	Missouri	
	qShootCu1-g14.2	E/15	*	ss715622633		10.00	Missouri	
	qShootCu1-g15	J/16	*	ss715624945		11.00	Missouri	
	qShootCu1-g16	J/16	*	ss715625324		11.00	Missouri	
	qShootCu1-g17	J/16	*	ss715625328		11.00	Missouri	

Glyma.13G161900, *Glyma.13G161800*, *Glyma.13G167600*, *Glyma.13G162800*, and *Glyma.13G172100* that were related to P efficiency and/or plant stress. *Glyma.13G161900* encodes for a protein kinase, and *Glyma.13G280500* also encoding for a protein kinase was identified within *q13–1* genomic region which is very close to *q13–2* genomic region (Zhang et al. 2016).

Li et al. (2016) used the same RIL population ("Nannong 94–156" by "Bogao," *n* = 152) and genetic linkage map used previously (Zhang et al. 2016) to investigate the relationship between PE and photosynthesis and to identify QTL for PE. The RIL population was grown in hydroponics in 2014–2014 (E1, E2). They studied several traits including shoot dry weight (SDW), root dry weight (RDW), root-to-shoot ratio (R/S), phosphorus (P) concentration (PC), phosphorus use efficiency (PUE), and phosphorus uptake (Pup) and used QTL IciMapping software program (Meng et al. 2015) for QTL data analysis. One hundred seventy two QTL controlling these traits have been identified; however, they were grouped into 29 QTL that were mapped on chromosomes 1, 2, 3, 4, 5, 6, 8, 9, 10, 11, 12, 13, 14, 15, 16, and 18 (Table 4.5). Within these QTL regions, several candidate genes have been identified including *Glyma.19G193900* that encodes for a purple acid phosphatase, *Glyma.19G207900* that encodes for an NPH3 protein, *Glyma.19G254700* that encodes for an NADPH dehydrogenase, *Glyma.19G260600* that encodes for a photosystem I P subunit, Glyma.19G261400 that encodes for a chlorophyll a binding protein, *Glyma.19G227400* that encodes for a photosystem II reaction center PsbP family protein, and many other candidate genes (Li et al. 2016) (Li et al. 2016).

4.3 QTL for Shoot Minerals Contents (B, Ca, Cu, Fe, K, Mg, Mn, P, S, and Zn)

Mineral nutrients are essential for human and animal growth and development and any imbalance in their availability will cause malnutrition, diseases, and health issues (Fletcher et al. 2004); therefore, it is imperative to develop soybean cultivars with high amounts of these nutrients (Heinemann et al. 2005; Philip and Martin 2005). Few studies identified QTL that control seed macronutrients and micronutrients contents (Bellaloui et al. 2015a, 2015b), and few others identified QTL that control these mineral nutrients in other plant parts such as roots and shoots (Dhanapal et al. (2018).

As reported earlier, we grew a RIL population derived from a cross between "MD96–5722" and "Spencer" lines (*n* = 94) in a field in Fayetteville, NC, and used it in two studies. The genetic linkage map used in these two studies was constructed by 550 SNP markers and covered 201.6 cM (Akond et al. 2013). In the first study, 17 QTL that control seed C, Ca, K, Mg, N, P, and S contents have been identified and mapped on chromosomes 2, 3, 5, 10, 13, 14, 16, and 18 (Bellaloui et al. 2015a). In the second study, 24 QTL that control seed B, Cu, Fe, Mn, and Zn contents have been identified and mapped on chromosomes 1, 3, 5, 9, 13, 14, 16, and 18 (Bellaloui et al. 2015b).

Dhanapal et al. (2018) grew a population of different genotypes (n = 385) of soybean germplasm in four environments in 009–2010 in Columbia, MO, and genotyped it with 31,748 SNPs to identify QTL that control macronutrients (P, K, Mg, S, and Ca) and micronutrients' (Zn, Mn, B, Cu, and Fe) concentrations in plant shoots. They used GWAS analysis to identify SNPs associated with these macronutrients and micronutrient concentrations in shoots. Thirty four SNPs associated with QTL that control shoot Ca content have been identified and mapped on chromosomes 1, 2, 4, 7, 8, 11, and 17; 48 SNPs associated with QTL that control shoot K content have been identified and mapped on chromosomes 1, 2, 6, 7, 8, 9, 10, 13, 14, 16, and 18; 18 SNPs associated with QTL that control shoot Mg content have been identified and mapped on chromosomes 3, 5, 6, 8, 10, 11, 12, 13, 15, 18, and 19; 42 SNPs associated with QTL that control shoot P content have been identified and mapped on chromosomes 2, 3, 4, 5, 6, 7, 8, 11, 14, 18, 19, and 20; 34 SNPs associated with QTL that control shoot S content have been identified and mapped on chromosomes 2, 5, 8, 9, 10, 11, 12, 13, 14, 15, 17, and 20; 77 SNPs associated with QTL that control shoot Fe content have been identified and mapped on chromosomes 1, 2, 5, 7, 9, 10, 11, 12, 13, 14, 15, 16, 18, and 19; 34 SNPs associated with QTL that control shoot B content have been identified and mapped on chromosomes 1, 4, 5, 6, 10, 13, 14, 15, and 19; 15 SNPs associated with QTL that control shoot Mn content have been identified and mapped on chromosomes 3, 5, 6, 7, 8, 17, and 19; 38 SNPs associated with QTL that control shoot Zn content have been identified and mapped on chromosomes 1, 3, 4, 5, 7, 8, 9, 12, 15, and 19; and 24 SNPs associated with QTL that control shoot Cu content have been identified and mapped on chromosomes 1, 2, 4, 5, 9, 10, 12, 13, 14, 15, and 16 (Table 4.7) (Dhanapal et al. 2018).

4.4 Conclusion

Since the first QTL for mineral deficiencies, efficiencies, and tolerances such as iron deficiency chlorosis (IDC) and necrosis, Al tolerance, Mn toxicity, phosphorus efficiency (PE), water use efficiency (WUE), etc. have been identified and mapped on the soybean genome (Diers et al. 1992; Lin et al. 1997, 2000a, 2000b; Bianchi-Hall et al. 2000; Kassem et al. 2004; Charlson et al. 2005; Mamidi et al. 2011; Sharma et al. 2011; Korir et al. 2011, 2013; Zhang et al. 2014; Dhanapal et al. 2015, 2018; Kumar and Lal 2015; Kaler et al. 2017), only few studies identified candidate genes within these QTL regions (Duressa et al. 2010; Peiffer et al. 2012; Mamidi et al. 2014; Lauter et al. 2014; Waters et al. 2018).

4.4.1 Candidate Genes Involved in Iron Deficiency Chlorosis

Peiffer et al. (2012) attempted to identify candidate genes involved in iron deficiency chlorosis within the QTL region previously identified an mapped on chromosome 3 (Lin et al. 1997, 200a,b). They used transcriptome analysis and qRT-PCR, to analyze two near-isogenic lines (NILs) generated by crossing the two cultivars "Clark" and "Harosy" with T203 (PI 54619). The study identified two candidate genes within this QTL region (Satt387–Satt339) narrowed to 250 kb. They identified 18 candidate genes in the region; however, further analysis revealed 2 differentially expressed genes (*Glyma03g28610* and *Glyma03g28630*) that exclusively expressed in the roots of control and stressed plants. These two genes encode for transcription factors (TFs) (Peiffer et al. 2012). It has been reported that regulatory elements of some genes involved in Fe metabolism may be outside of the QTL region (O'Rourke et al. 2009; Peiffer et al. 2012). In *A. thaliana*, the two genes encode for TFs AtbHLH038 and AtbHLH039 which are known enhance the expression of AtIRT1 and AtFRO2 involved in Fe acquisition (Yuan et al. 2008; Peiffer et al. 2012).

Mamidi et al. (2011, 2014) identified QTL that control iron deficit chlorosis (IDC) including five major QTL regions associated with IDC on chromosomes 3, 5, 7, 16, and 19. Within the major QTL on chromosome 3, they identified four candidate genes such as *Glyma03g38620* (*FRE1*) encoding for the phytochrome A, *Glyma03g38640* (*NRT1:2*) encoding for the nitrate transporter 1:2, Glyma03g39050 (*NAS3*) encoding for the nicotianamine synthase 3, and *Glyma03g40700* (*AMT2:1*) encoding for the ammonium transporter 2. With the major QTL on chromosome 5, they identified three candidate genes including *Glyma05g01460* (*AHA11*) encoding for the H^+–ATPase 11, Glyma05g08170 (*AGO10*) encoding for a stabilizer of iron transporter and SufD/polynucleotidyl transferase, and *Glyma05g08260* (*RAN1*) encoding for the RAS-related nuclear protein-1. Within the major QTL on chromosome 7, they identified two candidate genes including *Glyma07g07380* (*FRO2*) encoding for a ferric reduction oxidase 2 and *Glyma07g08240* (*FRD4*) encoding for the signal recognition particle receptor protein. Within the major QTL on chromosome 16, they identified the candidate gene *Glyma16g23710* (*FNR1*) encoding for a ferredoxin $NADP^+$ oxidase 1. Within the major QTL on chromosome 19, they identified the two candidate genes Glyma19g32190 (HMAS) encoding for a heavy metal ATPase 5 and *Glyma19g32401* (*OPT5*) encoding for the oligopeptide transporter 5 (Mamidi et al. 2011; Mamidi et al. 2014).

Likewise, Lauter et al. (2014) identified several genes induced by IDC including *Glyma07g13790* (AtPLP1 encoding for pupation related phospholipase (Rietz et al. 2010)), *Glyma14g40400* (AtGASA1 encoding for a gibberellic acid-regulated protein (Herzog et al. 1995)), *Glyma05g38351* (AtSWEET12 encoding for a sucrose transporter (Chen et al. 2012)), *Glyma11g33040* and Glyma18g05160 (AtOXS3 involved in higher tolerance to heavy metals and oxidative stress), and *Glyma01g07860* (copper amine oxidase involved in wound healing and nematode resistance (Rea et al. 2002; Moller et al. 1998) (Lauter et al. 2014).

To identify differentially expressed genes related to IDC, Waters et al. (2018) grew the IDC tolerant ("U06–105454") and the IDC susceptible ("U06–625083") lines on hydroponics using RNA-seq. They found about 5288 differentially expressed genes (DEGs) in treated plant roots compared to 489 DEGs in previous studies (O'Rourke et al. 2009; Peiffer et al. 2012). Among these 5288 DEGs, 175 DEGs such as bHLH38, OPT3, FIT, IRT1, AHA2, NRAMP6, and FR02 have been upregulated; and 474 DEGs such as ZIP2, FSD2, OPT7, ZIP11, and YSL3 have been downregulated by iron deficiency (ID) and alkaline stress (Waters et al. 2018).

4.4.2 Candidate Genes Involved in Al Tolerance

Duressa et al. (2010) performed a root transcriptomics analysis of PI416937 (Al tolerant line) treated with 10 μM of Al solution (group 1) and untreated control plants (group 2) using microarray analysis. Their results showed that 38 genes have been differentially expressed between the two groups of plants after only 2 hours of treating plants in group 1 with 10 μM of Al solution; only one gene was differentially expressed between the two groups of plants after 12 hours and 72 hours of treatment; and 542 genes have been differentially expressed between the two groups of plants after 48 hours of treatment (Duressa et al. 2010). These genes regulated by Al include genes involved in different metabolic pathways (15%), transcription factors (15%), genes involved in different types of molecular transport (5%), genes involved in stress responses (7%), genes involved in signaling responses (6%), structural genes (2%), and unclassified genes (50%). Among the most promising genes are a cell wall-modifying enzyme, Cys2His2 zinc finger protein (Iuchi et al. 2007), phytosulfokines growth factor, and *ADR6* transcription activator (Iuchi et al. 2007) that may play major roles in Al tolerance (Duressa et al. 2010).

To identify candidate genes involved in Al tolerance, Abdel-Haleem et al. (2014) grew a RIL population derived from a cross between "Young" and "PI 416937" lines ($n = 120$) in hydroponics, genotyped it with 279 RFLP and SSR markers, and used MIM for QTL data analysis of Al tolerance estimated as root tap extension in non-Al stress (NOAL), Al stress (HIAL), and percent of control (PC) traits. Two QTL that control root tap extension in NOAL conditions have been identified and mapped on chromosomes 3 and 18; two QTL that control root tap extension in HIAL conditions have been identified and mapped on chromosomes 8 and 16; and three QTL that control root tap extension in PC conditions have been identified and mapped on chromosomes 8, 16, and 19. Within the QTL region on chromosome 8, a candidate gene (*Glyma08g42400*) encoding for citrate synthase was identified (Abdel-Haleem et al. 2014).

Cai et al. (2019) grew a RIL population derived from a cross between "Huachun2" (Al tolerant) and "Wayao" (Al-sensitive) lines ($n = 196$) in hydroponics, genotyped it with 3469 bin markers, and used CIM of WinQTL cartographer for QTL data analysis of Al tolerance estimated as relative root elongation (RRE). Eleven QTL that control Al tolerance (RRE) have been identified and mapped on chromosomes

2, 6, 11, 12, 15, and 18; however, 5 of them are novel and the others have been previously identified (Korir et al. 2013; Abdel-Haleem et al. 2014) (Fig. 2.1; Table 4.1). Moreover, 11 candidate genes are within these QTL regions. The candidate genes are *Glyma06g24500* encoding for the histone H2A; *Glyma06g24520* encoding for a protein disulfide isomerase; *Glyma15g21890* encoding for S-adenosylmethionine synthase; *Glyma18g06220, Glyma18g06230, and Glyma18g06250* encoding for a peroxidase; *Glyma18g48730* and *Glyma18g48740* encoding for ethylene-responsive transcription factor 13; *Glyma18g49140* encoding for WRKY DNA-binding domain; *Glyma06g20710* encoding for a putative S-adenosyl-L-methionine-dependent methyltransferase; and *Glyma06g20730* encoding for a glutathione S-transferase U9 (Table 4.2) (Cai et al. 2019).

Wang et al. (2019) grew a RIL population derived from a cross between "Zhonghuang 24" and "Huaxia 3" lines ($n = 160$) in hydroponics, genotyped it with 4991 SNPs and 874 SSRs, and used highly dense genetic linkage map covering 2638.24 cM and CIM of WinQTL Cartographer for QTL data analysis of Al tolerance estimated as relative root elongation (RRE) and apical Al^{3+} content (AAC). Three QTL that control RRE (*qRRE_04*, *qRRE_16*, and *qRRE_17*) have been identified and mapped on chromosomes 4, 16, and 17, and two QTL that control AAC (*qAAC_04* and *qAAC_19*) have been identified and mapped on chromosomes 4 and 19 (Fig. 2.1; Table 4.1). The study identified 66 differentially expressed genes (DEGs) within these QTL regions; however, only 6 highly DEGs under Al^{3+} stress have been identified. These genes were *Glyma.04g218700* which encodes for transcription factor (TF) called WRKY21 (Zhou et al. 2008), *Glyma.04g217400* which encodes for the ethylene-responsive TF ABR1 (Pandey et al. 2005), *Glyma.04g216100* and *Glyma.04g213300* which encode for TFs involved in coping with abiotic stresses (Xi et al. 2009; Hussain et al. 2017), *Glyma.04g220600* which encodes a peroxidase which may be involved in Al tolerance (Inostroza-Blancheteau et al. 2012), and *Glyma.04g212800* which encodes for a GDP-mannose transporter also involved in Al tolerance. The authors elected *Glyma.04g218700* to be a strong candidate gene involved in Al stress (Table 4.2) (Wang et al. 2019).

4.4.3 Candidate Genes Involved in WUE and PE

As stated above, several studies identified QTL that control water use efficiency (WUE) and phosphorus efficiency (PE) have been identified using different mapping populations and study approaches; however, very few identified candidate genes within these QTL regions (Zhang et al. 2014, 2016; Dhanapal et al. 2015; Kaler et al. 2017).

Zhang et al. (2014) identified a major QTL that controls PE on chromosome 8 (*qPE8*) among other QTL, and within this region, they identified 28 genes including *Glyma08g20560* that encodes for a mitochondrial chaperonin, *Glyma08g20580* that encodes for a leucine-rich repeat, *Glyma08g20590* that encodes for a protein tyrosine kinase, *Glyma08g20600* that encodes for a glucose methanol choline oxidoreductase, *Glyma08g20610* that encodes for a GTP-binding protein, *Glyma08g20640* that encodes for an endomembrane protein, *Glyma08g20650* that encodes for a beta-galactosidase, *Glyma08g20690* that encodes for a cytochrome P450, *Glyma08g20700* that encodes for a calcineurin B, *Glyma08g20710* that encodes for a phospholipase D, and many other genes. The gene *GmACP1* that encodes for an acid phosphatase was a prominent gene within this region (Zhang et al. 2014). In a subsequent study using the same RIL population (Nannong 94–156 by Bogao; n = 146) grown in China in 2014 and the derived high-density genetic linkage map constructed by 6159 specific-locus amplified fragment sequencing (SLAF-seq) markers and covering 3020.59 cM, Zhang et al. (2016) identified 20 loci associated with QTL that control phosphorus efficiency (PE) across multiple environments, and 8 of these loci were novel (*q1*, *q3*, *q4–1*, *q4–2*, *q7*, *q13–1*, *q13–2*, and *q16*) including the major QTL on chromosome 4 (*q4–2*). Within *q8* genomic region, *GmACP1* gene was identified. Within *q2* genomic region, 18 candidate genes have been identified including 3 candidate genes (*Glyma.02G266700*, *Glyma.02G267500*, and *Glyma.02G267900*) that were related to P efficiency and/or plant stress. Within *q4–2* genomic region, 38 candidate genes have been identified including *Glyma.04G214000* that encodes for a pectin methylesterase. Within *q13–2* genomic region, 146 candidate genes have been identified including *Glyma.13G161900*, *Glyma.13G161800*, *Glyma.13G167600*, *Glyma.13G162800*, and *Glyma.13G172100* that were related to P efficiency and/or plant stress. *Glyma.13G161900* encodes for a protein kinase, and *Glyma.13G280500* also encoding for a protein kinase was identified within *q13–1* genomic region which is very close to *q13–2* genomic region (Zhang et al. 2016).

Dhanapal et al. (2015) identified several QTL that control WUE (carbon isotope ratio, $\delta^{13}C$) and several genes within these QTL regions including *Glyma02g13240* that encodes for YhbC family, *Glyma02g40340* that encodes for a protein kinase, Glyma03g04410 that encodes for a protein kinase/phosphorylation activity, *Glyma04g39910* that encodes for a PPR repeat, *Glyma04g41540* that encodes for glutamate synthase activity, *Glyma08g48050* that encodes for a structural constituent of ribosome, Glyma09g02960 that encodes for a histidinol dehydrogenase, *Glyma10g09720* that encodes for a plant syntaxin, *Glyma10g27660* that encodes for an aconitase, Glyma13g11700 that encodes for an AMP-binding enzyme, *Glyma15g02390* that encodes for a mitochondrial processing peptidase, *Glyma15g12330* that encodes for the NADH-ubiquinone oxidoreductase complex I, *Glyma18g46050* that encodes for a leucine-rich repeat (LRR) containing protein, *Glyma18g53170* that encodes for enzymes involved in the glutamine metabolic process, *Glyma19g29316* that encodes for a peptide deformylase, *Glyma20g03275* that encodes for an MSS1–/TRME-related GTP-binding protein, and many other genes (Dhanapal et al. 2015).

Li et al. (2016) identified 29 QTL that control many traits related to PE and photosynthesis such as SDW, RDW, root-to-shoot ratio (R/S), phosphorus (P) concentration (PC), phosphorus use efficiency (PUE), and phosphorus uptake (Pup) (Table 4.5). Within these QTL regions, several candidate genes have been identified including *Glyma.19G193900* that encodes for a purple acid phosphatase,

Glyma.19G207900 that encodes for an NPH3 protein, *Glyma.19G254700* that encodes for an NADPH dehydrogenase, *Glyma.19G260600* that encodes for a photosystem I P subunit, Glyma.19G261400 that encodes for a chlorophyll a binding protein, *Glyma.19G227400* that encodes for a photosystem II reaction center PsbP family protein, and many other candidate genes (Li et al. 2016). (Li et al. 2016).

Kaler et al. (2017) identified several QTL that control WUE (carbon isotope ratio, $\delta^{13}C$) and oxygen isotope ratio ($\delta^{18}O$). Within the $\delta^{13}C$ QTL regions, the authors identified many genes including *Glyma02g05595* that encodes for a heterogeneous nuclear ribonucleoprotein R, *Glyma02g36753* that encodes for a mitochondrial calcineurin B, *Glyma02g40340* that encodes for an aerine/threonine protein kinase, *Glyma04g39850* that encodes for a nodulin Mtn21-like transporter family protein, *Glyma06g02400* that encodes for a thioredoxin superfamily protein, *Glyma09g02800* that encodes for an oxidoreductase, *Glyma09g04640* that encodes for a polygalacturonase, *Glyma15g02690* that encodes for a 3′–5′ exonuclease (zinc ion binding), *Glyma16g01780* that encodes for an RNA-binding family protein, *Glyma18g48220* that encodes for an acyl-CoA thioesterase, and many other genes. Within the $\delta^{18}O$ QTL regions, the authors identified many genes including *Glyma01g00820* that encodes for a CCCH-type zinc finger family protein, *Glyma02g35400* that encodes for a COBRA-like protein, *Glyma02g46330* that encodes for an O-glycosyl hydrolases family 17 protein, *Glyma03g09050* that encodes for a galactosyltransferase activity, *Glyma04g35190* that encodes for a trehalose-phosphatase, *Glyma05g08200* that encodes for a serine/threonine kinase receptor, *Glyma06g22160* that encodes for a exocyst complex component 7 (salicylic acid biosynthetic process), *Glyma06g22771* that encodes for an 18S preribosomal assembly protein Gar2-related, *Glyma07g32283* that encodes for a cellulose synthase (polysaccharide biosynthetic process), *Glyma09g13480* v 2-hydroxyacid dehydrogenase, and may other genes (Kaler et al. 2017).

References

Abdel-Haleem, H., T.E. Carter, T.W. Rufty, H.R. Boerma, and Z. Li. 2014. Quantitative trait loci controlling aluminum tolerance in soybean: candidate gene and single nucleotide polymorphism marker discovery. *Molecular Breeding* 33: 851–862. https://doi.org/10.1007/s11032-013-9999-5.

Akond, A., S. Liu, L. Schoener, J.A. Anderson, S.K. Kantartzi, K. Meksem, Q. Song, D. Wang, Z. Wen, D.A. Lightfoot, and M.A. Kassem. 2013. A SNP-based genetic linkage map of soybean using the SoySNP6K Illumina Infinium BeadChip genotyping array. *Plant Genetics, Genomics, and Biotechnology* 1 (3): 80–89. https://doi.org/10.5147/pggb.v1i3.154.

Barlier, I., M. Kowalczyk, A. Marchant, K. Ljung, R. Bhalerao, M. Bennett, G. Sandberg, and C. Bellini. 2000. The SUR2 gene of Arabidopsis thaliana encodes the cytochrome P450 CYP83B1, a modulator of auxin homeostasis. *PNAS* 97 (26): 14819–14824. https://doi.org/10.1073/pnas.260502697.

Bellaloui, N., L. Khandaker, M. Akond, S.K. Kantartzi, K. Meksem, A. Mengistu, and M.A. Kassem. 2015a. Identification of QTLs underlying seed macronutrients accumulation in

'MD96-5722' by 'Spencer' recombinant inbred lines of soybean. *Atlas Journal of Biology* 3 (2): 224–235. https://doi.org/10.5147/ajb.v3i2.34.

Bellaloui, N., L. Khandaker, M. Akond, S.K. Kantartzi, K. Meksem, A. Mengistu, D.A. Lightfoot, and M.A. Kassem. 2015b. Identification of QTL Underlying seed micronutrients accumulation in 'MD 96-5722' by 'Spencer' recombinant inbred lines of soybean. *Atlas Journal of Plant Biology* 1 (3): 39–49. https://doi.org/10.5147/ajpb.v1i3.112.

Bianchi-Hall, C.M., T.E. Carter, M.A. Bailey, M.A.R. Mian, T.W. Rufty, D.A. Ashley, H.R. Boerma, C. Arellano, R.S. Hussey, and W.A. Parrott. 2000. Aluminum Tolerance Associated with Quantitative Trait Loci Derived from Soybean PI 416937 in hydroponics. *Crop Science* 40: 538–545. http://doi.org/10.2135/cropsci2000.402538x.

Cai, Z., Y. Cheng, P. Xian, R. Lin, Q. Xia, X. He, Q. Liang, T. Lian, Q. Ma, and H. Nian. 2019. Fine-mapping QTLs and the validation of candidate genes for Aluminum tolerance using a high-density genetic map. *Plant and Soil* 444: 119–137.

Carter, T.E., P.I. DeSouza, and L.C. Purcell. 1999. Recent Advances in Breeding for Drought and Aluminum Resistance in Soybean p. 106–125. In H. Kauffman (ed.) Proceedings of the World Soybean Research Conference, VI, Chicago. 4–7 Aug. 1999. National Soybean Research Lab., Urbana, IL.

Charlson, D.V., T.B. Bailey, S.R. Cianzio, and R.C. Shoemaker. 2005. Molecular Marker Satt481 is Associated with Iron-Deficiency Chlorosis Resistance in a Soybean Breeding Population. *Crop Science* 45 (6): 2394–2399. https://doi.org/10.2135/cropsci2004.0510.

Chen, L.Q., X.Q. Qu, B.H. Hou, D. Sosso, S. Osorio, A.R. Fernie, and W.B. Frommer. 2012. Sucrose efflux mediated by SWEET proteins as a key step for phloem transport. *Science* 335 (6065): 207–211.

Dhanapal, A.P., J.D. Ray, S.K. Singh, V. Hoyos-Villegas, J.R. Smith, L.C. Purcell, C.A. King, P.B. Cregan, Q. Song, and F.B. Fritschi. 2015. Genome-wide association study (GWAS) of carbon isotope ratio ($\delta^{13}C$) in diverse soybean [*Glycine max* (L.) Merr.] genotypes. *Theoretical and Applied Genetics* 128 (1): 73–91.

Dhanapal, A.P., J.D. Ray, J.R. Smith, L.C. Purcell, and F.B. Fritschi. 2018. Identification of novel genomic loci associated with soybean shoot tissue macro- and micronutrient concentrations. *Plant Genome* 11: 170066. https://doi.org/10.3835/plantgenome2017.07.0066.

Diers, B.W., S.R. Cianzio, and R.C. Shoemaker. 1992. Possible identification of quantitative trait loci affecting iron efficiency in soybean. *Journal of Plant Nutrition* 15 (10): 2127–2136. https://doi.org/10.1080/01904169209364462.

Duressa, D., K. Soliman, and D. Chen. 2010. Identification of aluminum responsive genes in al-tolerant soybean line PI 416937. *International Journal of Plant Genomics* 2010: 164862, 13 pages. https://doi.org/10.1155/2010/164862.

Fletcher, R.J., I.P. Bell, and J.P. Lambert. 2004. Public health aspects of food fortification: a question of balance. Proceedings of the Nutrition Society (2004), 63, 605–614. https://doi.org/10.1079/PNS2004391.

Foy, C.D. 1988. Plant adaptation to acid aluminum toxic soils. *Common. Soil Sci. Plant Anal.* 19: 959–987.

Grant, D., R.T. Nelson, S.B. Cannon, and R.C. Shoemaker. 2010. SoyBase, the USDA-ARS soybean genetics and genomics database. *Nucleic Acids Research* 38 (suppl 1): D843–D846. http://doi.org/10.1093/nar/gkp798.

Heinemann, R.J.B., P.L. Fagundes, E.A. Pinto, M.V.C. Penteado, and U.M. Lanfer-Marquez. 2005. Comparative study of nutrient composition of commercial brown, parboiled and milled rice from Brazil. *Journal of Food Composition and Analysis* 18: 287–296.

Herzog, M., Am Dorne, and F. Grellet. 1995. GASA, a gibberellin-regulated gene family from Arabidopsis thaliana related to the tomato GAST1 gene. *Plant Molecular Biology* 27 (4): 743–752.

Hussain, R.M., M. Ali, X. Feng, and X. Li. 2017. The essence of NAC gene family to the cultivation of drought-resistant soybean (Glycine max L. Merr.) cultivars. *BMC Plant Biology* 17 (1): 55. https://doi.org/10.1186/s12870-017-1001-y.

Inostroza-Blancheteau, C., Z. Rengel, M. Alberdi, M.D.L.L. Mora, F. Aquea, P. Arce-Johnson, and M. Reyes-Diaz. 2012. Molecular and physiological strategies to increase aluminum resistance in plants. *Molecular Biology Reports* 39 (3): 2069–2079.

Iuchi, S., H. Koyama, A. Iuchi, Y. Kobayashi, S. Kitabayashi, Y. Kobayashi, T. Ikka, T. Hirayama, K. Shinozaki, and M. Kobayashi. 2007. Zinc finger protein STOP1 is critical for proton tolerance in Arabidopsis and coregulates a key gene in aluminum tolerance. *PNAS* 104 (23): 9900–9905. https://doi.org/10.1073/pnas.0700117104.

Kaler, S.A., A.P. Dhanapal, J.D. Ray, C.A. King, F.B. Fritschi, and L.C. Purcell. 2017. Genome-wide association mapping of carbon isotope and oxygen isotope ratios in diverse soybean genotypes. *Crop Science* 57: 1–16. https://doi.org/10.2135/cropsci2017.03.0160.

Kassem, M.A., K. Meksem, C.H. Kang, V.N. Njiti, V.Y. Kilo, A.J. Wood, and D.A. Lightfoot. 2004. Loci underlying resistance to manganese toxicity mapped in a soybean recombinant inbred line population of 'Essex' x 'Forrest'. *Plant and Soil* 260: 197–200.

Kassem, M. A., J. Shultz, K. Meksem, Y. Cho, A. J. Wood, M. J. Iqbal, and D. A. Lightfoot, 2006. An updated 'Essex' by 'Forrest' linkage map and first composite map of QTL underlying six soybean traits. *Theoretical and Applied Genetics* 113: 1015—1026.

Kilo, V., and D.A. Lightfoot. 1996. Genetic analysis of resistance to manganese toxicity in soybean using molecular markers Soy-bean. *Genetics Newsletter* 23: 155–157.

Kochian, L.V. 1995. Cellular mechanisms of aluminum toxicity and resistance in plants. *Annual Review of Plant Physiology and Plant Molecular Biology* 46: 237–260.

Korir, P.C., T. Zhao, and J. Gai. 2010. A study on indicators and evaluation stages of aluminum tolerance in soybean. *Frontiers of Agriculture in China* 4 (3): 280–286. https://doi.org/10.1007/s11703-010-1019-5.

Korir, P.C., B. Qi, Y. Wang, T. Zhao, D. Yu, S. Chen, and J. Gai. 2011. A study on relative importance of additive, epistasis and unmapped QTL for Aluminum tolerance at seedling stage in soybean. *Plant Breeding* 130 (5): 551–562.

Korir, P.C., J. Zhang, K. Wu, T. Zhao, and J. Gai. 2013. Association mapping combined with linkage analysis for aluminum tolerance among soybean cultivars released in Yellow and Changjiang River Valleys in China. *Theoretical and Applied Genetics* 126 (6): 1659–1675.

Kumar, M., and S.K. Lal. 2015. Molecular analysis of soybean varying in water use efficiency using SSR markers. *Journal of Environmental Biology* 36: 1011–1016.

Lander, E.S., P. Green, J. Abrahamson, A. Barlow, M.J. Daly, S.E. Lincoln, and L. Newburg. 1987. MAPMAKER: An interactive computer package for constructing primary genetic linkage maps of experimental and natural populations. *Genomics* 1 (2): 174–181. https://doi.org/10.1016/0888-7543(87)90010-3.

Lauter, A.N.M., G.A. Peiffer, T. Yin, S.A. Whitham, D. Cook, R.C. Shoemaker, and M.A. Graham. 2014. Identification of candidate genes involved in early iron deficiency chlorosis signaling in soybean (*Glycine max*) roots and leaves. *BMC Genomics* 15: 702. https://doi.org/10.1186/1471-2164-15-702.

Lazof, D.B., J.G. Goldsmith, T.W. Rufty, and R.W. Linton. 1996. The early entry of Al into cells of intact soybean roots. A comparison of three developmental root regions using secondary ion mass spectrometry imaging. *Plant Physiology* 112: 1289–1300.

Li H, Y Yang, H Zhang, S Chu, X Zhang, D Yin, D Yu, and D Zhang. 2016. A genetic relationship between phosphorus efficiency and photosynthetic traits in soybean as revealed by QTL analysis using a high-density genetic map. *Frontiers in Plant Science*. https://doi.org/10.3389/fpls.2016.00924.

Lin, A., S. Cianzio, and R.C. Shoemaker. 1997. Mapping genetic loci for iron deficiency chlorosis in soybean. *Molecular Breeding* 3: 219–229.

Lin, S.F., J. Baumer, D. Ivers, S. Cianzio, and R.C. Shoemaker. 2000a. Nutrient solution screening of Fe chlorosis resistance in soybean evaluated by molecular characterization. *Journal of Plant Nutrition* 23 (11–12): 1915–1928. https://doi.org/10.1080/01904160009382153.

Lin, S.F., D. Grant, S. Cianzio, and R.C. Shoemaker. 2000b. Molecular characterization of iron deficiency chlorosis in soybean. *Journal of Plant Nutrition* 23 (11–12): 1929–1939. https://doi.org/10.1080/01904160009382154.

Mamidi, S., S. Chikara, R.J. Goos, D.L. Hyten, D. Annam, S.M. Moghaddam, R.K. Lee, P.B. Cregan, and P.E. McClean. 2011. Genome-wide association analysis identifies candidate genes associated with iron deficiency chlorosis in soybean. *The Plant Genome* 4 (3): 154–164.

Mamidi, S., R.K. Lee, J.R. Goos, and P.E. McClean. 2014. Genome-wide association studies identifies seven major regions responsible for iron deficiency chlorosis in soybean (*Glycine max*). *PLoS One* 9 (9): e107469. https://doi.org/10.1371/journal.pone.0107469.

Meng, L., H. Li, L. Zhang, and J. Wang. 2015. QTL IciMapping: Integrated software for genetic linkage map construction and quantitative trait locus mapping in biparental populations. *The Crop Journal* 3: 269–283. https://doi.org/10.1016/j.cj.2015.01.001.

Mian, M.A.R., M.A. Bailey, D.A. Ashley, R. Wells, T.E. Carter, W.A. Parrott, and H.R. Boerma. 1996. Molecular markers associated with water use efficiency and leaf ash in soybean. *Crop Science* 36 (5): 1252–1257.

Mian, M.A.R., D.A. Ashley, and H.R. Boerma. 1998. An additional QTL for water use efficiency in soybean. *Crop Science* 38 (2): 390–393.

Moller, S.G., P.E. Urwin, H.J. Atkinson, and M.J. McPherson. 1998. Nematode-induced expression of atao1, a gene encoding an extracellular diamine oxidase associated with developing vascular tissue. *Physiol Mol Plant Phatol* 53 (2): 73–79.

Nable, R.O., R.L. Houtz, and G.M. Cheniae. (1988). Early inhibition of photosynthesis during development of Mn toxicity in tobacco. *Plant Physiology* 86: 1136–1142.

Orf, J.H., K. Chase, T. Jarvik, L.M. Mansur, P.B. Cregan, F.R. Adler, and K.G. Lark. 1999. Genetics of soybean agronomic traits: I. *Comparison of three related recombinant inbred populations. Crop Sci.* 39: 1642–1651.

O'Rourke, J.A., R.T. Nelson, D. Grant, J. Schmutz, J. Grimwood, S. Cannon, C.P. Vance, M.A. Graham, and R.C. Shoemaker. 2009. Integrating microarray analysis and the soybean genome to understand the soybeans iron deficiency response. *BMC Genomics* 10: 376. https://doi.org/10.1186/1471-2164-10-376.

Pandey, G.K., J.J. Grant, Y.H. Cheong, B.G. Kim, L. Li, and S. Luan. 2005. ABR1, an APETALA2-domain transcription factor that functions as a repressor of ABA response in Arabidopsis. *Plant Physiology* 139 (3): 1185–1193. https://doi.org/10.1104/pp.105.066324.

Peiffer, G.A., K.E. King, A.J. Severin, G.D. May, S.R. Cianzio, S.F. Lin, N.C. Lauter, and R.C. Shoemaker. 2012. Identification of candidate genes underlying an iron efficiency quantitative trait locus in soybean. *Plant Physiology* 158: 1745–1754.

Philip, J.W., and R.B. Martin. 2005. Biofortifying crops with essential mineral elements. *Trends Plant Science* 10: 586–593.

Qi, B., P. Korir, T. Zhao, D. Yu, S. Chen, and J. Gai. 2008. Mapping quantitative trait loci associated with aluminum toxin tolerance in NJRIKY recombinant inbred line population of soybean (*Glycine max*). *Journal of Integrative Plant Biology* 50 (9): 1089–1095.

Rea, G., O. Metoui, A. Infantino, R. Federico, and R. Angelini. 2002. Copper amine oxidase expression in defense responses to wounding and Ascochyta rabiei invasion. *Plant Physiology* 128 (3): 865–875.

Rietz, S., G. Dermendjiev, E. Oppermann, F.G. Tafesse, Y. Effendi, A. Holk, J.E. Parker, M. Teige, and G.F. Scherer. 2010. Roles of Arabidopsis patatin-related phospholipases a in root development are related to auxin responses and phosphate deficiency. *Molecular Plant* 3 (3): 524–538.

Samac, D.A., and M. Tesfaye. 2003. Plant improvement for tolerance to aluminum in acid soils – a review. *Plant Cell, Tissue and Organ Culture* 75: 189–207.

Sharma, A.D., H. Sharma, and D.A. Lightfoot. 2011. The genetic control of tolerance to aluminum toxicity in the 'Essex' by 'Forrest' recombinant inbred line population. *Theoretical and Applied Genetics* 122: 687–694. https://doi.org/10.1007/s00122-010-1478-3.

Silva, I.R., T.J. Smytha., C.D. Rapera, T.E. Carter, and T.W. Rufty. 2001. Differential aluminum tolerance in soybean: An evaluation of the role of organic acids. *Physiologia Plantarum* 112: 200–210.

Simoes, C.C., J.O. Melo, J.V. Magalhaes, and C.T. Guimaraes. 2012. Genetic and molecular mechanisms of aluminum tolerance in plants. *Genetics and Molecular Research* 11 (3): 1949–1957. https://doi.org/10.4238/2012.July.19.14.

Specht, J.E., K. Chase, M. Macrander, G.L. Graef, J. Chung, J.P. Markwell, M. Germann, J.H. Orf, and K.G. Lark. 2001. Soybean Response to water: A QTL analysis of drought tolerance. *Crop Science* 41 (2): 493–509.

Stephens, P.A., J.M. Widholm, and C.D. Nickell. 1990. Iron-deficiency chlorosis evaluation of soybean with tissue culture. *Theoretical and Applied Genetics* 80: 417–420.

Taiz, L., and E. Zeiger. 1998. Mineral nutrition. In Plant Physiology (ed.). Sinauer Associates Inc., P.O. Box 407, Sunderland, MA 01375, USA.

Tasma, I.M., A. Warsun, and Asadi. 2008. Development and characterization of F_2 population for molecular mapping of aluminum-toxicity tolerant QTL in soybean. *Journal AgroBiogen* 4 (1): 1–8.

Von Uexkull, H.R., and E. Mutert. 1995. Global extent, development and economic impact of acid soils. *Plant and Soil* 171: 1–15.

Wang, X., Y. Cheng, C. Yang, C. Yang, Y. Mu, Q. Xia, and Q. Ma. 2019. QTL mapping for aluminum tolerance in RIL population of soybean (*Glycine max* L.) by RAD sequencing. *PLoS One* 14 (10): e0223674. https://doi.org/10.1371/journal.pone.0223674.

Waters, B.M., K. Amundsen, and G. Graef. 2018. Gene expression profiling of iron deficiency chlorosis sensitive and tolerant soybean indicates key roles for phenylpropanoids under alkalinity stress. *Frontiers in Plant Science*. https://doi.org/10.3389/fpls.2018.00010.

Welch, R.M. 1995. Micronutrient nutrition of plants. *Critical Reviews in Plant Sciences* 14: 49–82.

Xie Y., P. Chen, Y. Yan, C. Bao, X. Li, L. Wang, X. Shen, H. Li, X. Liu, C. Niu, C. Zhu, N. Fang, Y. Shao, T. Zhao, J. Yu, J. Zhu, L. Xu, S. van Nocker, F. Ma, and Qi. Guan. 2018. An atypical R2R3 MYB transcription factor increases cold hardiness by CBF-dependent and CBF-independent pathways in apple. *New Phytologist* 218 (1): 201–218. https://doi.org/10.1371/journal.pone.0006898.

Yuan, Y., H. Wu, N. Wang, J. Li, W. Zhao, J. Du, D. Wang, and H.Q. Ling. 2008. FIT interacts with AtbHLH38 and AtbHLH39 in regulating iron uptake gene expression for iron homeostasis in *Arabidopsis*. *Cell Research* 18: 385–397.

Zhang, D., H. Song, H. Cheng, D. Hao, H. Wang, G. Kan, H. Jin, and D. Yu. 2014. The acid phosphatase-encoding gene *GmACP1* contributes to soybean tolerance to low-phosphorus stress. *PLoS Genetics* 10 (1): e1004061. https://doi.org/10.1371/journal.pgen.1004061.

Zhang, D., H. Li, J. Wang, H. Zhang, Z. Hu, S. Chu, H. Lv, and D. Yu. 2016. High-density genetic mapping identifies new major loci for tolerance to low-phosphorus stress in soybean. *Soybean. Frontiers in Plant Science* 7: 372. https://doi.org/10.3389/fpls.2016.00372.

Zhou, Q., A. Tian, H. Zou, Z. Xie, G. Lei, J. Huang, C. Wang, H. Wang, J. Zhang, and S. Chen. 2008. Soybean WRKY-type transcription factor genes, GmWRKY13, GmWRKY21, and GmWRKY54, confer differential tolerance to abiotic stresses in transgenic Arabidopsis plants. *Plant Biotechnology Journal* 6: 486–503. https://doi.org/10.1111/j.1467-7652.2008.00336.x.

Chapter 5
Salt Tolerance QTL Mapping in Soybean: 2004–2020

Moulay Abdelmajid Kassem

5.1 Introduction

Different types of molecular markers and mapping populations have been used to genetically map quantitative trait loci (QTL) that control salt tolerance (ST) in soybean using biparental populations and genome-wide association mapping studies (GWAS). Here, we will summarize the most important QTL mapping studies for salt stress tolerance or ST reported and not reported on SoyBase (Lee et al. 2004, 2009, 2018; Chen et al. 2008, 2018a, b; Hamwieh and Xu 2008; Huang 2013; Do et al. 2010, 2013, 2018, 2019; Hamwieh et al. 2011; Xu and Do 2012; Ha et al. 2013; Guan et al. 2014a, b; Qi et al. 2014; Zhang et al. 2014; Kan et al. 2015, 2016; Zeng et al. 2017; Shi et al. 2018a, b; Cao et al. 2018; Huang et al. 2018).

5.1.1 QTL for ST (Na^+ and Cl^-)

Salt stress also called salinity causes major crop losses worldwide and affects about 20% of agricultural land (Flowers and Yeo 1995; Pathan et al. 2007; Phang et al. 2008). Saline and sodic (alkaline) soils are the two main types of soils that cause the two types of slat stress in plants (Do et al. 2010; Xu and Do 2012). A plant or crop that can grow and complete its life cycle with good yield is said to be salt tolerant and is called a halophyte; however, this ST depends on the plant species and plant genotypes with the same plant species. Glycophytes like soybean and most crop species cannot tolerate high salinity conditions, and salt stress affects their growth

M. A. Kassem (✉)
Plant Genomics and Biotechnology Lab, Department of Biological and Forensic Sciences, Fayetteville State University, Fayetteville, NC, USA
e-mail: mkassem@uncfsu.edu

M. A. Kassem (ed.), *Soybean Seed Composition*,
https://doi.org/10.1007/978-3-030-82906-3_5

and productivity (Aoki et al. 2005; Pathan et al. 2007; Phang et al. 2008; Deshmukh et al. 2014; Liu et al. 2016; Cao et al. 2018).

In soybean, salt stress caused many symptoms including a reduction in plant height, root growth, photosynthesis rate and biomass, high concentrations of Cl^- ions in leaves and stems, leaf necrosis, and increased mortality (An et al. 2001, 2002; Essa 2002; Pathan et al. 2007; Phang et al. 2008; Cao et al. 2018). Among ST mechanisms in soybean are maintenance of iron homeostasis, osmotic adjustment and accumulation of osmoprotectants, restoration of oxidative balance, and a possible role of late embryogenesis (LEA) proteins (Shao et al. 2005; Pathan et al. 2007; Phang et al. 2008; Cao et al. 2018).

Soybean cultivars (*Glycine max*) respond differently to concentrations of NaCl ranging from 5.2 to 8 g/L; however, the perennial *Glycine* accessions can tolerate NaCl concentrations ranging from 3 to 17.5 g/L (Kao et al. 2006; Pathan et al. 2007). This made it possible to develop recombinant inbred line (RIL) and other populations to investigate the genetic basis of ST in soybean; therefore, several studies identified QTL that control ST using different RIL and other populations (Lee et al. 2004, 2009, 2018; Chen et al. 2008, 2018a, b; Hamwieh and Xu 2008; Huang 2013; Do et al. 2010, 2013, 2018, 2019; Hamwieh et al. 2011; Xu and Do 2012; Ha et al. 2013; Guan et al. 2014a, b; Qi et al. 2014; Zhang et al. 2014; Kan et al. 2015, 2016; Zeng et al. 2017; Shi et al. 2018a, b; Cao et al. 2018; Huang et al. 2018).

Lee et al. (2004) grew an F_2 population derived from a cross between 'S-100' (salt-tolerant) and 'Tokyo' (salt-sensitive) lines ($n = 106$) in the greenhouse and a saline field in NC in 1999, genotyped it with 32 SSRs and an unknown number of RFLP markers, and used single marker analysis (ANOVA) for QTL data analysis. Six markers (QTL) that control ST have been identified and mapped on chromosomes 3 and 19 (Fig. 2.1; Table 5.1) (Lee et al. 2004).

Chen et al. (2008) grew a RIL population derived from a cross between 'Kefeng No. 1' and 'Nannong 1138-2' lines ($n = 184$) both in the greenhouse and field trials and genotyped it with 221 SSR markers and 1 disease resistance gene (*Rsc-7*). To quantify ST under salt stress, they used percentage of plant survival (PPS) in the field trials and plant survival days (PSD) in the greenhouse trials and visual STR. For QTL data analysis, they used CIM of WinQTL cartographer. Four QTL that control PPS have been identified and mapped on chromosomes 2, 3, 9, and 14; three QTL that control PSD have been identified and mapped on chromosomes 9, 11, and 18; and three QTL that control TR have been identified and mapped on chromosomes 7 and 18 (Fig. 2.1; Table 5.1) (Chen et al. 2008).

Hamwieh and Xu (2008) grew an F_2 population derived from a cross between 'Jackson' (salt-sensitive) and the wild accession 'JWS156-1' (salt-tolerant) lines ($n = 225$) on hydroponics and genotyped it with SSR markers around the previously reported major QTL for ST on chromosome 3 (Lee et al. 2004). To estimate ST, they used STR and leaf chlorophyll content estimated as SPAD value. For QTL data analysis, they used CIM of IciMapping software (Li et al. 2007). One major QTL that controls both STR and SPAD was identified and mapped on chromosome 3 at the SSR marker Satt339. This major QTL alone explained 43.4% and 33.4% of the

Table 5.1 Detailed information of QTL that control ST reported and *not reported* in SoyBase (Grant et al. 2010; http://www.soybase.org) from 2004 to 2020

Trait	*QTL*	LG/Chr	Interval (cM)	Marker/interval	LOD, *P, or F*	R^2 (%)	Environment	References
ST	*qSaltTol-1*	L/19	*	Cr354a	p < 0.01	23.00	N. Carolina	Lee et al. (2004)
	qSaltTol-2	L/19	45–47	Satt166	p < 0.01	9.00	N. Carolina	
	qSaltTol-3	N/3	59.2–61.2	Satt339	p < 0.01	29.00	N. Carolina	
	qSaltTol-4	N/3	60.2–62.2	Satt237	p < 0.01	32.00	N. Carolina	
	qSaltTol-5	N/3	63.9–65.9	Sat_091	p < 0.01	35.00	N. Carolina	
	qSaltTol-6	N/3	*	Cr354b	p < 0.01	25.00	N. Carolina	
PPS	*qppsB2.1*	B2/14	16.2–21.2	Sat_182-Satt318	4.60	10.90	China	Chen et al. (2008)
	qppsD1b.1	D1b/2	239.1–268.4	Satt558-Satt157	3.20	12.50	China	
	qppsK.1	K/9	12.8–21.9	Sat_363-Satt710	5.00	19.20	China	
	qppsN.1	N/3	72.4–99.8	Satt237-Satt312	2.80	7.80	China	
PDS	*qpdsB1.1*	B1/11	27.0–34.5	Sat_151-Sat_261	2.80	8.40	China	
	qpdsG.1	G/18	24.9–46.4	Sat_164 -Sat_358	4.40	17.90	China	
	qpdsK.1	K/9	11.0–25.3	Sat_363-Satt710	3.70	11.30	China	
TR	*qtrG.1*	G/18	28.8–46.9	Sat_164 -Sat_358	4.20	10.80	China	
	qtrM.1	M/7	76.4–100.9	Satt702-Satt728	3.50	7.10	China	
	qtrM.2	M/7	100.9–128.5	Satt655-Satt210	3.60	19.70	China	
STR	*qstrN.1*	N/3	59.2–61.2	Satt339	43.40	68.70	Greenhouse	Hamwieh and Xu (2008)
SPAD	*qspadN.1*	N/3	59.2–61.2	Satt339	33.40	49.60	Greenhouse	
STR (F6-RIL)	*qstrD2.1*	D2/17	53.9–55.9	Satt447	14.50	50.20	Greenhouse	Do et al. (2010)
SPAD (F6-RIL)	*qspadD2.1*	D2/17	53.9–55.9	Satt447	11.80	43.00	Greenhouse	
STR (F2)	*qstrD2.2*	D2/17	65.7–67.7	Satt461	4.00	13.00	Greenhouse	
SPAD (F2)	*qspadD2.1*	D2/17	53.9–55.9	Satt447	4.60	13.90	Greenhouse	

(continued)

Table 5.1 (Continued)

Trait	*QTL*	LG/Chr	Interval (cM)	Marker/interval	LOD, *P, or F*	R^2 (%)	Environment	References
STR	*qSTR*	N/3	63.9–65.9	Sat_091	12.10	44.00	Greenhouse	Hamwieh et al. (2011)
	qSTR	N/3	63.9–65.11	Sat_091	11.20	47.10	Greenhouse	
SPAD	*qSPAD*	N/3	63.9–65.10	Sat_091	10.00	38.20	Greenhouse	
	qSPAD	N/3	63.9–65.12	Sat_091	8.60	40.90	Greenhouse	
LS	*qLS.1*	N/3	27.4–29.4	Satt255-BARC-038333-10036	18.80	56.50	Greenhouse	Ha et al. (2013)
	qLS.1	N/3	0.7–8.0	Satt255-Satt312	30.80	94.80	Greenhouse	
LS	*qLS.1*	D1b/2	11,494,870 bp	Gm02_11494870_T_C	(–Log10(p) = 3.7)	8.40	Greenhouse	Huang (2013)
	qLS.2	D1b/2	11,495,939	Gm02_11495939_C_T	(–Log10(p) = 3.7)	8.40	Greenhouse	
	qLS.3	D1b/2	11,698,384	Gm02_11698384_C_A	(–Log10(p) = 3.81)	8.70	Greenhouse	
	qLS.4	N/3	38,172,118	Gm03_38172118_A_G	(–Log10(p) = 4.25)	11.00	Greenhouse	
	qLS.5	N/3	39,868,800	Gm03_39868800_T_C	(–Log10(p) = 3.73)	9.00	Greenhouse	
	qLS.6	N/3	39,912,307	Gm03_39912307_T_G	(–Log10(p) = 3.73)	9.00	Greenhouse	
	qLS.7	N/3	39,915,523	Gm03_39915523_C_T	(–Log10(p) = 3.97)	9.80	Greenhouse	
	qLS.8	N/3	40,018,337	Gm03_40018337_C_T	(–Log10(p) = 4.12)	10.10	Greenhouse	
	qLS.9	N/3	40,019,280	Gm03_40019280_A_G	(–Log10(p) = 4.11)	10.20	Greenhouse	
	qLS.10	N/3	40,052,612	Gm03_40052612_T_C	(–Log10(p) = 3.98)	9.70	Greenhouse	
	qLS.11	N/3	40,060,562	Gm03_40060562_C_T	(–Log10(p) = 3.9)	9.70	Greenhouse	
	qLS.12	N/3	40,062,964	Gm03_40062964_C_T	(–Log10(p) = 3.88)	9.60	Greenhouse	
	qLS.13	N/3	40,153,205	Gm03_40153205_T_C	(–Log10(p) = 8.74)	25.60	Greenhouse	
	qLS.14	N/3	40,154,304	Gm03_40154304_A_G	(–Log10(p) = 8.74)	25.60	Greenhouse	
	qLS.15	N/3	40,155,554	Gm03_40155554_T_G	(–Log10(p) = 8.67)	28.20	Greenhouse	
	qLS.16	N/3	40,197,155	Gm03_40197155_A_C	(–Log10(p) = 7.29)	20.90	Greenhouse	
	qLS.17	N/3	40,278,033	Gm03_40278033_G_A	(–Log10(p) = 3.81)	9.30	Greenhouse	
	qLS.18	N/3	40,421,296	Gm03_40421296_A_G	(–Log10(p) = 4.15)	10.00	Greenhouse	
	qLS.19	N/3	40,452,899	Gm03_40452899_C_A	(–Log10(p) = 7.19)	20.20	Greenhouse	
	qLS.20	N/3	40,466,433	Gm03_40466433_C_T	(–Log10(p) = 8.68)	27.20	Greenhouse	

	qLS.21	N/3	40,467,180	Gm03_40467180_G_A	(–Log10(p) = 9.08)	27.50	Greenhouse	
	qLS.22	N/3	40,494,556	Gm03_40494556_A_G	(–Log10(p) = 13.35)	43.50	Greenhouse	
	qLS.23	N/3	40,516,071	Gm03_40516071_A_G	(–Log10(p) = 11.25)	35.20	Greenhouse	
	qLS.24	N/3	40,546,463	Gm03_40546463_C_A	(–Log10(p) = 12.87)	41.50	Greenhouse	
	qLS.25	N/3	40,557,941	Gm03_40557941_A_G	(–Log10(p) = 12.79)	40.50	Greenhouse	
	qLS.26	N/3	40,565,405	Gm03_40565405_G_A	(–Log10(p) = 15.34)	50.90	Greenhouse	
	qLS.27	N/3	40,584,028	Gm03_40584028_A_G	(–Log10(p) = 13.37)	43.50	Greenhouse	
	qLS.28	N/3	40,585,266	Gm03_40585266_G_A	(–Log10(p) = 13.14)	41.60	Greenhouse	
	qLS.29	N/3	40,593,882	Gm03_40593882_G_A	(–Log10(p) = 15.82)	54.10	Greenhouse	
	qLS.30	N/3	40,597,392	Gm03_40597392_C_T	(–Log10(p) = 5.44)	14.60	Greenhouse	
	qLS.31	N/3	40,600,088	Gm03_40600088_A_G	(–Log10(p) = 5.03)	13.50	Greenhouse	
	qLS.32	N/3	40,602,759	Gm03_40602759_G_A	(–Log10(p) = 4.06)	10.00	Greenhouse	
	qLS.33	N/3	40,606,894	Gm03_40606894_T_C	(–Log10(p) = 15.82)	54.10	Greenhouse	
	qLS.34	N/3	40,613,405	Gm03_40613405_T_C	(–Log10(p) = 6.7)	18.30	Greenhouse	
	qLS.35	N/3	40,656,449	Gm03_40656449_A_G	(–Log10(p) = 9.45)	29.50	Greenhouse	
	qLS.36	N/3	40,661,459	Gm03_40661459_A_G	(–Log10(p) = 11.3)	35.90	Greenhouse	
	qLS.37	N/3	40,663,609	Gm03_40663609_G_A	(–Log10(p) = 9.23)	28.40	Greenhouse	
	qLS.38	N/3	40,677,040	Gm03_40677040_G_A	(–Log10(p) = 4.58)	11.60	Greenhouse	
	qLS.39	N/3	40,699,200	Gm03_40699200_G_T	(–Log10(p) = 4.36)	11.00	Greenhouse	
	qLS.40	N/3	40,700,660	Gm03_40700660_T_C	(–Log10(p) = 4.48)	11.50	Greenhouse	
	qLS.41	N/3	40,740,328	Gm03_40740328_T_G	(–Log10(p) = 6.32)	16.80	Greenhouse	
	qLS.42	N/3	40,768,058	Gm03_40768058_C_A	(–Log10(p) = 6.58)	18.30	Greenhouse	
	qLS.43	N/3	40,800,808	Gm03_40800808_A_G	(–Log10(p) = 4.36)	10.60	Greenhouse	
	qLS.44	N/3	40,895,159	Gm03_40895159_T_C	(–Log10(p) = 4.18)	10.70	Greenhouse	
	qLS.45	N/3	41,100,337	Gm03_41100337_C_T	(–Log10(p) = 4.35)	10.40	Greenhouse	
	qLS.46	N/3	41,194,560	Gm03_41194560_G_A	(–Log10(p) = 4.33)	10.80	Greenhouse	
	qLS.47	N/3	41,205,784	Gm03_41205784_A_G	(–Log10(p) = 4.22)	10.10	Greenhouse	

(continued)

Table 5.1 (Continued)

Trait	*QTL*	LG/Chr	Interval (cM)	Marker/interval	LOD, *P, or F*	R^2 (%)	Environment	References
	qLS.48	N/3	41,206,640	Gm03_41206640_A_G	(−Log10(p) = 4.37)	10.80	Greenhouse	
	qLS.49	N/3	41,212,195	Gm03_41212195_C_T	(−Log10(p) = 4.1)	9.80	Greenhouse	
	qLS.50	N/3	41,234,338	Gm03_41234338_T_C	(−Log10(p) = 4.32)	10.80	Greenhouse	
	qLS.51	N/3	41,236,752	Gm03_41236752_T_C	(−Log10(p) = 4.25)	10.10	Greenhouse	
	qLS.52	N/3	41,245,330	Gm03_41245330_C_T	(−Log10(p) = 4.02)	9.50	Greenhouse	
	qLS.53	N/3	41,258,466	Gm03_41258466_T_C	(−Log10(p) = 6.27)	17.80	Greenhouse	
	qLS.54	N/3	41,301,876	Gm03_41301876_A_C	(−Log10(p) = 5.69)	17.10	Greenhouse	
	qLS.55	N/3	44,583,239	Gm03_44583239_A_C	(−Log10(p) = 4.04)	10.30	Greenhouse	
	qLS.56	A1/5	9,081,936	Gm05_9081936_A_G	(−Log10(p) = 3.51)	7.90	Greenhouse	
	qLS.57	C2/6	8,039,368	Gm06_8039368_C_T	(−Log10(p) = 3.64)	8.90	Greenhouse	
	qLS.58	A2/8	18,707,057	Gm08_18707057_T_C	(−Log10(p) = 3.45)	7.60	Greenhouse	
	qLS.59	A2/8	18,772,502	Gm08_18772502_C_T	(−Log10(p) = 3.45)	7.60	Greenhouse	
	qLS.60	G/18	56,568,068	Gm18_56568068_C_T	(−Log10(p) = 3.08)	7.00	Greenhouse	
	qLS.61 + A:I	G/18	56,571,714	Gm18_56571714_C_T	(−Log10(p) = 3.1)	7.10	Greenhouse	
	qLS.62	G/18	58,234,790	Gm18_58234790_A_G	(−Log10(p) = 3.25)	7.40	Greenhouse	
LCC	*qLC.1*	N/3	18–18.4	Barcsoyssr_3_1310-QS08064	*	*	China	Guan et al. (2014a)
LCC	*GmSALT3*	N/3	17.5 kb	QS100001-QS1119	*	*	China	Guan et al. (2014b)
LCC	*GmSALT3*	N/3	388 kb	Satt237-Satt312	*	*		Qi et al. (2014)

LR-STI	*qLR-STR_1*	A1/5	76.4–78.4	Sat_267	6.51	1.81	China	Zhang et al. (2014)
	qLR-STR_2	A1/5	91.9–93.9	Satt2000	4.21	0.49	China	
	qLR-STR_3	A2/8	8.2–10.2	Satt390	11.09	0.07	China	
	qLR-STR_4	A2/8	50.5–52.5	Satt632	4.64	0.76	China	
	qLR-STR_5	B1/11	31.5–33.5	Satt509	12.36	0.28	China	
	qLR-STR_6	B1/11	122.9–124.9	Satt453	3.09	2.12	China	
	qLR-STR_7	C1/4	0–2	Satt565	3.23	2.94	China	
	qLR-STR_8	C2/6	111.3–113.3	Satt289	4.19	0.70	China	
	qLR-STR_9	D1b/2	45.9–47.9	Sat_254	6.73	2.09	China	
	qLR-STR_10	D2/17	84.2–86.2	Satt226	4.84	2.79	China	
	qLR-STR_11	D2/17	91.2–93.7	Satt615-Sat_354	7.30	5.90	China	
	qLR-STR_12	E/15	45.7–47.7	Satt045	4.39	1.74	China	
	qLR-STR_13	E/15	63.2–65.2	Sat_381	3.05	0.48	China	
	qLR-STR_14	F/13	20.1–22.1	AW186493	3.41	0.42	China	
	qLR-STR_15	F/13	32.2–34.2	Satt160	4.90	2.55	China	
	qLR-STR_16	G/18	3.5–5.5	Satt309	4.92	2.21	China	
	qLR-STR_17	G/18	86.9–88.9	AF162283	5.41	0.47	China	
	qLR-STR_18	H/12	67.1–69.1	Satt222	3.85	0.46	China	
	qLR-STR_19	J/16	22.9–23.9	Sat_228	3.24	0.48	China	
LR-ATI	*qLR-ATI_1*	N/3	42.5–44.5	Sat_280	5.78	9.71	China	
	qLR-ATI_2	N/3	73.9–75.9	Satt237	4.48	1.28	China	
	qLR-ATI_3	N/3	101–103	Satt022	10.29	13.89	China	

(continued)

Table 5.1 (Continued)

Trait	*QTL*	LG/Chr	Interval (cM)	Marker/interval	LOD, *P, or F*	R^2 (%)	Environment	References
FWR-STI	*qFWR-STI_1*	C2/6	111.3–113.3	Satt289	7.28	1.13	China	
	qFWR-STI_2	D1b/2	51.6–53.6	Satt296	7.08	0.99	China	
	qFWR-STI_3	D2/17	92.7–94.7	Sat_354	5.99	6.65	China	
	qFWR-STI_4	D2/17	113.9–115.9	Satt672	3.20	0.36	China	
	qFWR-STI_5	F/13	4.4–6.4	Satt649	4.57	3.71	China	
	qFWR-STI_6	F/13	25.7–27.7	Satt659	3.28	3.11	China	
	qFWR-STI_7	G/18	49.5–51.5	Satt352	3.20	4.04	China	
	qFWR-STI_8	I/20	111.7–113.7	Satt440	3.49	4.06	China	
	qFWR-STI_9	M/7	49.1–51.1	Satt463	4.46	1.41	China	
	qFWR-STI_10	M/7	73.5–75.5	Sat_256	11.55	6.12	China	
	qFWR-STI_11	N/3	101–103	Satt022	9.30	4.20	China	
	qFWR-STI_12	O/10	56.1–58.1	Satt262	3.92	0.44	China	
	qFWR-STI_13	M/7	73.5–75.5	Satt656-Sat_256	3.40	15.80	China	
FWR-ATI	*qFWR-STI_1*	A1/5	18.4–20.4	Sat_344	3.66	15.00	China	
	qFWR-STI_2	A2/8	50.5–52.5	Satt632	3.38	5.11	China	
	qFWR-STI_3	B2/14	5.1–7.1	Satt577	3.09	4.53	China	
	qFWR-STI_4	C1/4	73.8–75.8	AW277661	4.93	5.61	China	
	qFWR-STI_5	D2/17	92.7–94.7	Sat_354	7.23	15.26	China	
	qFWR-STI_6	E/15	44.1–46.1	Satt452	3.09	2.77	China	
	qFWR-STI_7	F/13	33.2–35.2	Satt160	6.28	21.18	China	
	qFWR-STI_8	G/18	49.5–51.5	Satt352	4.70	5.43	China	

DWR-ATI	*qDWR-ATI_1*	C2/6	120.3–122.3	Satt307	4.28	0.47	China	
	qDWR-ATI_2	D1b/2	75.1–77.1	Satt282	3.81	6.55	China	
	qDWR-ATI	D2/17	92.7–94.7	Sat_354	6.19	13.17	China	
	qDWR-ATI_4	E/15	44.4–46.4	Satt263	4.10	4.93	China	
	qDWR-ATI_5	F/13	17.1–19.1	Satt149	3.81	1.26	China	
	qDWR-ATI_6	H/12	85.5–87.5	Satt142	7.14	1.88	China	
	qDWR-ATI_7	J/16	38.2–40.2	Satt132	3.08	6.92	China	
	qDWR-ATI_8	K/9	45.1–46.7	Satt417	4.11	4.51	China	
	qDWR-ATI_9	K/9	77.4–80.1	Sct_190-Satt260	4.10	2.40	China	
	qDWR-ATI_10	M/7	49.1–51.1	Satt463	4.23	15.36	China	
	qDWR-ATI_11	M/7	56.3–58.6	Satt220-Satt626	3.50	5.20	China	
	qDWR-ATI_12	M/7	73.5–75.5	Sat_256	3.97	10.68	China	
	qDWR-ATI_13	O/10	99.4–101.4	Satt592	3.31	9.23	China	
LH-STI	*qLH-STI_1*	A1/5	91.9–93.9	Satt200	5.52	0.53	China	
	qLH-STI_2	A2/8	50.5–52.5	Satt632	4.88	4.06	China	
	qLH-STI_3	A2/8	66.9–68.9	aw132402	12.53	3.20	China	
	qLH-STI_4	B2/14	65.2–67.2	Sat_355	4.58	1.83	China	
	qLH-STI_5	C2/6	118.8–120.8	Staga001	8.20	2.64	China	
	qLH-STI_6	D1a/1	103.3–105.3	Sat_160	5.15	3.68	China	
	qLH-STI_7	E/15	11.9–13.9	Satt411	8.64	1.33	China	
	qLH-STI_8	E/15	45.7–47.7	Satt045	13.30	5.73	China	
	qLH-STI_9	F/13	32.2–34.2	Satt160	10.25	4.50	China	
	qLH-STI_10	K/9	79.1–81.1	Satt260	4.25	1.60	China	
LH-ATI	*qLH-ATI_1*	N/3	75.5–77.5	Satt255	3.60	0.24	China	

(continued)

Table 5.1 (Continued)

Trait	*QTL*	LG/Chr	Interval (cM)	Marker/interval	LOD, *P, or F*	R^2 (%)	Environment	References
BS-STI	*qBS-STI_1*	A1/5	25.4–27.4	Satt382	3.50	2.79	China	
	qBS-STI_2	A2/8	66.9–68.9	aw132402	3.56	1.06	China	
	qBS-STI_3	B1/11	122.9–124.9	satt453	4.13	1.03	China	
	qBS-STI_4	B2/14	71.8–73.8	satt070	5.28	2.11	China	
	qBS-STI_5	B2/14	86.6–88.6	satt534	4.21	3.08	China	
	qBS-STI_6	C1/4	73.8–75.8	AW277661	4.05	0.81	China	
	qBS-STI_7	C2/6	29.5–31.5	satt640	3.20	0.37	China	
	qBS-STI_8	C2/6	43.7–45.7	satt422	3.73	0.73	China	
	qBS-STI_9	C2/6	111.3–113.3	satt289	3.93	0.62	China	
	qBS-STI_10	D1b/2	51.6–53.6	satt296	4.02	0.40	China	
	qBS-STI_11	D2/17	92.7–94.7	sat_354	5.39	3.82	China	
	qBS-STI_12	E/15	45.1–46.7	Satt452-Satt045	5.63	1.36	China	
	qBS-STI_13	F/13	25.7–27.7	satt659	5.81	1.56	China	
	qBS-STI_14	F/13	32.2–34.2	satt160	8.59	4.09	China	
	qBS-STI_15	G/18	3.5–5.5	satt309	6.91	2.24	China	
	qBS-STI_16	I/20	49.1–51.1	satt270	6.90	0.88	China	
	qBS-STI_17	J/16	22.9–23.9	sat_228	4.02	0.23	China	
	qBS-STI_18	M/7	73.5–75.5	sat_256	5.28	3.80	China	
	qBS-STI_19	M/7	106.7–108.7	satt250	3.51	0.44	China	

ST-IR	*qST-IR_1*	A2/8	7,716,379	BARC-028361-05839	(−Log(p) = 2.24)	*	China	Kan et al. (2015)
	qST-IR_2	A2/8	9,622,940	BARC-041663-08059	(−Log(p) = 2.13)	*	China	
	qST-IR_3	C2/6	16,660,196	BARC-025705-05001	(−Log(p) = 2.14)	*	China	
ST-GI	*qST-GI_1*	G/18	57,209,838	BARC-017681-03105	(−Log(p) = 2.07)	*	China	
ST-GI/ ST-GR	*qST-GI/ ST-GR_1*	B2/14	3,997,614	BARC-024525-04961	(−Log(p) = 2.37)	*	China	
	qST-GI/ ST-GR_2	D1b/2	47,313,970	BARC-026065-05238	(−Log(p) = 3.37)	*	China	
	qST-GI/ ST-GR_3	D1b/2	47,313,970	BARC-026065-05239	(−Log(p) = 3.37)	*	China	
	qST-GI/ ST-GR_4	D1b/2	47,547,417	BARC-017895-02427	(−Log(p) = 3.36)	*	China	
	qST-GI/ ST-GR_5	F/13	28,836,642	BARC-042991-08492	(−Log(p) = 3.18)	*	China	
	qST-GI/ ST-GR_6	H/12	38,675,904	BARC-027816-06683	(−Log(p) = 3.6)	*	China	
	qST-GI/ ST-GR_7	K/9	210,160	BARC-021347-04042	(−Log(p) = 3.59)	*	China	
	qST-GI/ ST-GR_8	N/3	2,993,385	BARC-028645-05979	(−Log(p) = 3.47)	*	China	
ST-GR	*qST-GR_1*	D2/17	1,785,893	BARC-031827-07220	(−Log(p) = 2.22)	*	China	
Using 'Kefeng 1' X 'Nannong 1138-2' RIL population (n = 184)								Kan et al. (2016)
ST-IR	*qST-IR-8*	A2/8	21.4–33.0	Sat_162	7.19	25.94	China	
	qST-IR-18	G/18	66.7–95.0	Satt217	2.61	5.28	China	
ST-GI	*qST-GI-2-1*	D1b/2	125.1–160.1	Satt428	2.42	17.80	China	
	qST-GI-2-2	D1b/2	287.5–291.8	Sat_351	3.37	7.60	China	

(continued)

Table 5.1 (Continued)

Trait	*QTL*	LG/Chr	Interval (cM)	Marker/interval	LOD, *P, or F*	R^2 (%)	Environment	References
ST-GR	*qST-GR-8*	A2/8	19.8–32.4	Sat_162	5.17	15.62	China	
	qST-GR-7	M/7	36.2–59.4	Satt245	2.85	6.10	China	
ST-IR	*qST-IR-8*	A2/8	9.0–31.8	Sat_162	5.11	15.87	China	
	qST-IR-18	G/18	72.1–99.0	Satt130	2.78	7.80	China	
ST-GI	*qST-GI-17*	D2/17	61.7–108.51	Sat_292	2.25	5.41	China	
	qST-GI-10-1	O/10	4.0–47.9	Sat_231	2.35	9.98	China	
ST-GR	*qST-GR-8*	A2/8	6.0–31.8	Sat_162	5.15	8.45	China	
	qST-GR-7	M/7	37.2–57.4	Satt245	3.21	7.85	China	
ST-IR	*qST-IR-8*	A2/8	17.4–32.0	Sat_162	5.9	20.94	China	
	qST-IR-2	D1b/2	150.4–191.6	Satt266	2.3	4.53	China	
ST-GI	*qST-GI-8*	A2/8	158.7–188.6	Sat_294	2.12	4.49	China	
	qST-GI-10-2	O/10	220.8–233.6	BE801128	3.19	9.25	China	
ST-GR	*qST-GR-8*	A2/8	9.6–36.2	Sat_162	3.6	9.57	China	
Using the natural population of landraces (n = 196)								
ST-IR	*qST-IR_1*	C2/6	106.6–108.6	Satt277	(–Log(p) = 2.19)	*	China	
	qST-IR_2	D1b/2	29.7–31.7	BE475343	(–Log(p) = 6.20)	*	China	
	qST-IR_3	J/16	14.7–16.7	Satt287	(–Log(p) = 3.77)	*	China	
	qST-IR_4	L/19	86.4–88.4	Sat_286	(–Log(p) = 5.90)	*	China	
	qST-IR_5	L/19	55.1–57.1	Satt156	(–Log(p) = 5.41)	*	China	
	qST-IR_6	M/7	12.6–14.6	Satt201	(–Log(p) = 2.25)	*	China	
	qST-IR_7	M/7	7.0–9.0	CSSR306	(–Log(p) = 2.55)	*	China	
	qST-IR_8	M/7	4.0–6.0	Satt636	(–Log(p) = 3.36)	*	China	

ST-GI	*qST-GI_1*	L/19	91.7–93.7	Satt664	(−Log(p) = 3.43)	*	China	
	qST-GI_2	L/19	55.1–57.1	Satt156	(−Log(p) = 3.33)	*	China	
	qST-GI_3	M/7	12.6–14.6	Satt20	(−Log(p) = 2.28)	*	China	
	qST-GI_4	M/7	32.5–34.5	Satt567	(−Log(p) = 2.51)	*	China	
	qST-GI_5	M/7	4.0–6.0	Satt636	(−Log(p) = 2.70)	*	China	
ST-GR	*qST_GR_1*	C2/6	97.1–99.1	Satt363	(−Log(p) = 2.62)	*	China	
	qST_GR_2	D1a/1	47.1–49.1	Satt342	(−Log(p) = 4.07)	*	China	
	qST_GR_3	D1b/2	29.7–31.7	BE475343	(−Log(p) = 2.46)	*	China	
	qST_GR_4	L/19	91.7–93.7	Satt664	(−Log(p) = 3.43)	*	China	
	qST_GR_5	L/19	55.1–57.1	Satt156	(−Log(p) = 3.17)	*	China	
	qST_GR_6	M/7	7.0–9.0	CSSR306	(−Log(p) = 3.34)	*	China	
	qST_GR_7	M/7	32.5–34.5	Satt567	(−Log(p) = 2.51)	*	China	
	qST_GR_8	M/7	4.0–6.0	Satt636	(−Log(p) = 3.26)	*	China	
	qST_GR_9	N/3	78.5–80.5	Sat_091	(−Log(p) = 3.00)	*	China	
LSS	*qLSS_1*	N/3	40,632,077	SNP (G/C)	(−Log(p) = 22.62)	63.00	Missouri	Patil et al. (2016)
	qLSS_2	N/3	40,528,845	SNP (C/T)	(−Log(p) = 21.50)	61.00	Missouri	
	qLSS_3	N/3	40,528,865	SNP (C/T)	(−Log(p) = 21.50)	61.00	Missouri	
	qLSS_4	N/3	40,493,449	SNP (C/T)	(−Log(p) = 21.22)	60.00	Missouri	
	qLSS_5	N/3	40,500,559	SNP (T/G)	(−Log(p) = 21.22)	60.00	Missouri	
SPAD	*qSPAD_1*	N/3	40,528,845	SNP (C/T)	(−Log(p) = 17.32)	53.00	Missouri	
	qSPAD_2	N/3	40,528,865	SNP (C/T)	(−Log(p) = 17.32)	53.00	Missouri	
	qSPAD_3	N/3	40,483,159	SNP (C/T)	(−Log(p) = 17.07)	52.00	Missouri	
	qSPAD_4	N/3	40,519,933	SNP (T/C)	(−Log(p) = 17.07)	52.00	Missouri	
	qSPAD_5	N/3	40,534,592	SNP (G/A)	(−Log(p) = 17.06)	52.00	Missouri	

(continued)

Table 5.1 (Continued)

Trait	*QTL*	LG/Chr	Interval (cM)	Marker/interval	LOD, *P, or F*	R^2 (%)	Environment	References
Leaf Cl-	*qLCl_1*	N/3	71.7–73.7	Gm03_38469714	$p < 0.001$	34.00	Arkansas	Zeng (2016)
	qLCl_2	N/3	72.6–74.6	Gm03_38761991	$p < 0.001$	35.00	Arkansas	
	qLCl_3	N/3	70.9–72.9	Gm03_38862467	$p < 0.001$	33.00	Arkansas	
	qLCl_4	N/3	71.6–73.6	Gm03_38931849	$p < 0.001$	35.00	Arkansas	
	qLCl_5	N/3	77.4–79.4	Gm03_39491355	$p < 0.001$	36.00	Arkansas	
	qLCl_6	N/3	79.7–81.7	Gm03_39574966	$p < 0.001$	42.00	Arkansas	
	qLCl_7	N/3	79.2–81.2	Gm03_39796778	$p < 0.001$	42.00	Arkansas	
	qLCl_8	N/3	78.8–80.8	Gm03_39843152	$p < 0.001$	43.00	Arkansas	
	qLCl_9	N/3	79.1–81.1	Gm03_39945298	$p < 0.001$	42.00	Arkansas	
	qLCl_10	N/3	80.2–82.2	Gm03_39998708	$p < 0.001$	38.00	Arkansas	
	qLCl_11	N/3	78.6–80.6	Gm03_40052612	$p < 0.001$	42.00	Arkansas	
	qLCl_12	N/3	80.2–82.2	Gm03_40197155	$p < 0.001$	39.00	Arkansas	
	qLCl_13	N/3	80.9–82.9	Gm03_40270199	$p < 0.001$	47.00	Arkansas	
	qLCl_14	N/3	82.2–84.2	Gm03_40417269	$p < 0.001$	48.00	Arkansas	
	qLCl_15	N/3	82.4–84.4	Gm03_40600088	$p < 0.001$	54.00	Arkansas	
	qLCl_16	N/3	83.7–85.7	Gm03_40613405	$p < 0.001$	56.00	Arkansas	
	qLCl_17	N/3	84.4–86.4	Gm03_40663609	$p < 0.001$	54.00	Arkansas	
	qLCl_18	N/3	91.7–93.7	Gm03_41605831	$p < 0.001$	32.00	Arkansas	
	qLCl_19	N/3	93.7–95.7	Gm03_41984976	$p < 0.001$	31.00	Arkansas	
	qLCl_20	N/3	94.6–96.6	Gm03_42148379	$p < 0.001$	28.00	Arkansas	
	qLCl_21	F/13	0–2	Gm13_27665585	0.0041	11.00	Arkansas	
	qLCl_22	F/13	0–4	Gm13_27846120	0.006	10.00	Arkansas	
	qLCl_23	F/13	0.5–2.5	Gm13_28206014	$p < 0.001$	19.00	Arkansas	
	qLCl_24	E/15	27.6–29.6	Gm15_6646246	0.04	6.00	Arkansas	
	qLCl_25	E/15	28–30	Gm15_6674572	0.0327	8.00	Arkansas	
	qLCl_26	E/15	28–30	Gm15_6823009	0.029	8.00	Arkansas	
	qLCl_27	E/15	29–31	Gm15_7023081	0.0574	8.00	Arkansas	

Leaf Cl-/ SPAD	*qLeaf-Cl-/ SPAD_1*	D1b/2	13,098,947	ss715581136	(–Log(p) = 7.8)	6.70	Arkansas	Zeng (2016) and
	qLeaf-Cl-/ SPAD_2	N/3	37,902,931	ss715585865	(–Log(p) = 10.9)	10.90	Arkansas	Zeng et al. (2017)
	qLeaf-Cl-/ SPAD_3	N/3	37,984,044	ss715585870	(–Log(p) = 11.8)	15.20	Arkansas	
	qLeaf-Cl-/ SPAD_4	N/3	38,037,948	ss715585874	(–Log(p) = 15.0)	19.20	Arkansas	
	qLeaf-Cl-/ SPAD_5	N/3	38,045,898	ss715585876	(–Log(p) = 14.5)	18.50	Arkansas	
	qLeaf-Cl-/ SPAD_6	N/3	38,047,240	ss715585877	(–Log(p) = 13.4)	17.80	Arkansas	
	qLeaf-Cl-/ SPAD_7	N/3	38,048,300	ss715585878	(–Log(p) = 15.0)	19.20	Arkansas	
	qLeaf-Cl-/ SPAD_8	N/3	38,103,441	ss715585883	(–Log(p) = 20.1)	25.10	Arkansas	
	qLeaf-Cl-/ SPAD_9	N/3	38,139,117	ss715585889	(–Log(p) = 14.8)	19.20	Arkansas	
	qLeaf-Cl-/ SPAD_10	N/3	38,140,216	ss715585890	(–Log(p) = 14.8)	19.00	Arkansas	
	qLeaf-Cl-/ SPAD_11	N/3	38,183,067	ss715585896	(–Log(p) = 15.7)	20.10	Arkansas	
	qLeaf-Cl-/ SPAD_12	N/3	38,197,459	ss715585899	(–Log(p) = 8.1)	10.40	Arkansas	
	qLeaf-Cl-/ SPAD_13	N/3	38,480,306	ss715585930	(–Log(p) = 49.1)	50.90	Arkansas	
	qLeaf-Cl-/ SPAD_14	N/3	38,532,213	ss715585934	(–Log(p) = 38.0)	42.60	Arkansas	

(continued)

Table 5.1 (Continued)

Trait	QTL	LG/Chr	Interval (cM)	Marker/interval	LOD, *P, or F*	R^2 (%)	Environment	References
	qLeaf-Cl-/ SPAD_15	N/3	38,543,691	ss715585936	(−Log(p) = 37.4)	42.40	Arkansas	
	qLeaf-Cl-/ SPAD_16	N/3	38,551,155	ss715585937	(−Log(p) = 55.3)	55.50	Arkansas	
	qLeaf-Cl-/ SPAD_17	N/3	38,571,016	ss715585942	(−Log(p) = 36.8)	41.90	Arkansas	
	qLeaf-Cl-/ SPAD_18	N/3	38,579,634	ss715585943	(−Log(p) = 55.9)	55.30	Arkansas	
	qLeaf-Cl-/ SPAD_19	N/3	38,583,144	ss715585944	(−Log(p) = 13.6)	17.60	Arkansas	
	qLeaf-Cl-/ SPAD_20	N/3	38,585,840	ss715585945	(−Log(p) = 13.9)	17.90	Arkansas	
	qLeaf-Cl-/ SPAD_21	N/3	38,591,888	ss715585948	(−Log(p) = 55.4)	54.90	Arkansas	
	qLeaf-Cl-/ SPAD_22	N/3	38,649,377	ss715585953	(−Log(p) = 19.4)	24.30	Arkansas	
	qLeaf-Cl-/ SPAD_23	N/3	39,303,211	ss715586057	(−Log(p) = 8.5)	6.00	Arkansas	
	qLeaf-Cl-/ SPAD_24	N/3	39,357,229	ss715586063	(−Log(p) = 10.7)	13.90	Arkansas	
	qLeaf-Cl-/ SPAD_25	N/3	39,403,852	ss715586070	(−Log(p) = 9.5)	12.50	Arkansas	
	qLeaf-Cl-/ SPAD_26	N/3	39,510,751	ss715586082	(−Log(p) = 10.0)	13.00	Arkansas	
	qLeaf-Cl-/ SPAD_27	N/3	39,600,402	ss715586086	(−Log(p) = 7.8)	10.10	Arkansas	
	qLeaf-Cl-/ SPAD_28	N/3	39,601,253	ss715586087	(−Log(p) = 7.7)	10.00	Arkansas	

	qLeaf-Cl-/ SPAD_29	N/3	39,693,677	ss715586094	(−Log(p) = 7.7)	10.00	Arkansas	
	qLeaf-Cl-/ SPAD_30	N/3	39,708,387	ss715586095	(−Log(p) = 7.4)	9.60	Arkansas	
	qLeaf-Cl-/ SPAD_31	N/3	39,759,678	ss715586102	(−Log(p) = 8.5)	11.10	Arkansas	
	qLeaf-Cl-/ SPAD_32	N/3	40,440,832	ss715586154	(−Log(p) = 7.6)	10.10	Arkansas	
	qLeaf-Cl-/ SPAD_33	M/7	38,606,673	ss715597790	(−Log(p) = 5.1)	6.50	Arkansas	
	qLeaf-Cl-/ SPAD_34	M/7	38,623,703	ss715597794	(−Log(p) = 5.3)	6.70	Arkansas	
	qLeaf-Cl-/ SPAD_35	M/7	38,800,856	ss715597821	(−Log(p) = 7.1)	9.20	Arkansas	
	qLeaf-Cl-/ SPAD_36	A2/8	37,018,844	ss715601563	(−Log(p) = 6.3)	8.00	Arkansas	
	qLeaf-Cl-/ SPAD_37	O/10	44,402,011	ss715607372	(−Log(p) = 6.8)	8.80	Arkansas	
	qLeaf-Cl-/ SPAD_38	O/10	44,420,445	ss715607376	(−Log(p) = 6.8)	8.90	Arkansas	
	qLeaf-Cl-/ SPAD_39	F/13	39,607,602	ss715616115	(−Log(p) = 8.4)	10.90	Arkansas	
	qLeaf-Cl-/ SPAD_40	B2/14	2,684,853	ss715618138	(−Log(p) = 8.5)	11.20	Arkansas	
	qLeaf-Cl-/ SPAD_41	B2/14	2,714,940	ss715618149	(−Log(p) = 7.6)	10.00	Arkansas	
	qLeaf-Cl-/ SPAD_42	B2/14	4,123,566	ss715618712	(−Log(p) = 6.7)	8.70	Arkansas	

(continued)

Table 5.1 (Continued)

Trait	*QTL*	LG/Chr	Interval (cM)	Marker/interval	LOD, *P, or F*	R^2 (%)	Environment	References
	qLeaf-Cl-/SPAD_43	B2/14	4,161,326	ss715618731	(−Log(p) = 8.1)	10.50	Arkansas	
	qLeaf-Cl-/SPAD_44	J/16	33,415,484	ss715624611	(−Log(p) = 6.5)	8.40	Arkansas	
	qLeaf-Cl-/SPAD_45	I/20	34,422,775	ss715637438	(−Log(p) = 7.0)	9.20	Arkansas	
LSS	*qLSS_1*	N/3	85.1–88.7	Salt20-Gm03_41135466	19.08	48.20	Greenhouse	Do et al. (2018)
CCR	*qCCR_1*	N/3	85.1–88.7	Salt20-Gm03_41135466	11.01	31.30	Greenhouse	
LSC	*qLSC_1*	N/3	85.1–88.7	Salt20-Gm03_41135466	7.69	20.60	Greenhouse	
LCC	*qLCC_1*	N/3	85.1–88.7	Salt20-Gm03_41135466	25.59	58.90	Greenhouse	
LSC	*qLSC_2*	F/13	138.5–148.9	Gm13_37204738	4.56	11.50	Greenhouse	
LSS	*QTL-LSS_1*	N/3	83.5–89.2	snp680-snp675	98.17	48.80	Greenhouse	Lopez et al. (2018)
PDP	*QTL-PDP_1*	N/3	83.5–95.7	snp680-snp686	86.10	37.00	Greenhouse	
Chlorophyll	*QTL-Chlorophyll_1*	N/3	83.5–89.5	snp680-snp676	72.10	46.50	Greenhouse	
Chloride	*QTL-Chloride_1*	N/3	81–87.7	snp674-snp673	57.20	43.00	Greenhouse	
Chlorophyll	*QTL-Chlorophyll_2*	L/19	20.3–31.3	snp4994-snp5024	8.60	4.80	Greenhouse	
STR	*qSTR_1-(SMR)*	N/3	61.2–64.9	Sat_091	19.80	44.30	Greenhouse	Shi et al. (2018a, b)
	qSTR_1-(CIM)	N/3	61.2–64.9	Sat_091	19.80	44.30	Greenhouse	
	qSTR_1-(MIM)	N/3	61.2–64.9	Sat_091	20.10	44.70	Greenhouse	
SPAD	*qSPAD_1-(SMR)*	N/3	61.2–64.9	Sat_091	10.70	27.00	Greenhouse	
	qSPAD_1-(CIM)	N/3	61.2–64.9	Sat_091	10.70	27.00	Greenhouse	
	qSPAD_1-(MIM)	N/3	61.2–64.9	Satt237-Satt255	11.00	27.60	Greenhouse	

RIL population ('Kefeng No. 1' by 'Nannong 1138-2', n = 184)								Zhang et al. (2019)
ST-IR	*qSTIR-2*	D1b/2	167.9–169.9	Satt428-Rsc_7	3.23	7.03	China	
	qSTIR-8	M/7	52.4–54.4	Satt590-Sat_148	3.01	10.69	China	
	qSTIR-8	A2/8	23.8–25.8	BE820148-AW132402	17.06	46.82	China	
	qSTIR-8	A2/8	17.8–19.8	BE820148-AW132402	3.39	6.94	China	
	qSTIR-8	A2/8	17.8–19.8	BE820148-AW132402	3.14	6.73	China	
ST-GI	*qSTGI-8*	A2/8	17.8–19.8	BE820148-AW132402	3.72	8.46	China	
	qSTGI-8	A2/8	17.8–19.8	BE820148-AW132402	3.34	7.11	China	
	qSTGI-15	E/15	41.7–43.7	Satt606-Sat_331	2.64	6.02	China	
	qSTGI-17	D2/17	76.7–78.7	Satt669-Sat_292	2.56	10.94	China	
ST-GP	*qSTGP-1-2*	D1a/1	48.3–50.3	Satt468-Sat_160	4.12	22.95	China	
	qSTGP-2	D1b/2	113.2–115.2	Satt611-Satt428	3.21	9.23	China	
	qSTGP-7	M/7	56.4–58.4	satt590-sat_148	3.04	9.08	China	
	qSTGP-8-1	A2/8	18.8–20.8	BE820148-AW132402	3.94	8.66	China	
	qSTGP-8-1	A2/8	18.8–20.8	BE820148-AW132402	7.63	15.61	China	
	qSTGP-8-2	A2/8	111–113	Sat_310-Sat_232	2.84	6.25	China	
	qSTGP-10	O/10	54.9–56.9	Satt331-Sat_196	4.44	33.40	China	
	qSTGP-17-1	D2/17	78.7–80.7	Satt669-Sat_292	3.04	16.98	China	
	qSTGP-17-2	D2/17	98.5–100.5	Satt669-Sat_292	2.85	6.08	China	
ST-GR	*qSTGR-1*	D1a/1	61.3–63.3	Satt468-Sat_160	3.70	13.12	China	
	qSTGR-7	M/7	35.2–37.2	Satt590-Sat_148	3.15	16.55	China	
	qSTGR-8	A2/8	17–18	BE820148-AW132402	8.85	22.41	China	
	qSTGR-8	A2/8	18.8–20.8	BE820148-AW132402	8.41	16.66	China	
	qSTGR-10	O/10	29.6–31.6	Sat_231-Satt331	4.80	10.29	China	
	qSTGR-17	D2/17	79–81	Satt669-Sat_292	3.93	13.73	China	
	qSTGR-18	G/18	46.6–48.6	Sat_358-Sat_290	2.50	6.74	China	

(continued)

Table 5.1 (Continued)

Trait	*QTL*	LG/Chr	Interval (cM)	Marker/interval	LOD, *P, or F*	R^2 (%)	Environment	References
Natural population (n = 211, GWAS)								
ST-IR	*qST-IR-1*	A2/8	8,032,589	AX-94048021	(–Log(p) = 4.93)	13.92	China	
	qST-IR-2	A2/8	8,055,772	AX-93751752	(–Log(p) = 4.85)	13.71	China	
	qST-IR-3	A2/8	8,062,435	AX-94048035	(–Log(p) = 4.77)	13.51	China	
	qST-IR-4	A2/8	8,068,777	AX-93751758	(–Log(p) = 4.85)	13.71	China	
	qST-IR-5	A2/8	8,075,710	AX-93751760	(–Log(p) = 4.94)	13.95	China	
	qST-IR-6	A2/8	8,081,879	AX-93751762	(–Log(p) = 4.94)	13.96	China	
	qST-IR-7	A2/8	8,087,610	AX-93751765	(–Log(p) = 5.16)	14.53	China	
	qST-IR-8	A2/8	8,089,799	AX-94048045	(–Log(p) = 5.05)	14.21	China	
	qST-IR-9	A2/8	8,101,015	AX-94048049	(–Log(p) = 5.05)	14.21	China	
	qST-IR-10	A2/8	8,112,374	AX-94048056	(–Log(p) = 4.94)	13.96	China	
	qST-IR-11	A2/8	8,131,400	AX-93751782	(–Log(p) = 4.80)	13.60	China	
	qST-IR-12	A2/8	8,148,738	AX-94048071	(–Log(p) = 4.77)	13.50	China	
	qST-IR-13	A2/8	8,157,061	AX-94048072	(–Log(p) = 5.18)	14.59	China	
	qST-IR-14	A2/8	8,162,127	AX-93751793	(–Log(p) = 4.88)	13.80	China	
	qST-IR-15	A2/8	8,195,444	AX-93751799	(–Log(p) = 5.08)	14.32	China	
	qST-IR-16	A2/8	8,216,666	AX-94048089	(–Log(p) = 4.98)	14.06	China	
	qST-IR-17	A2/8	8,228,424	AX-94048095	(–Log(p) = 5.13)	14.45	China	
	qST-IR-18	A2/8	8,229,952	AX-93751817	(–Log(p) = 5.25)	14.78	China	
	qST-IR-19	A2/8	8,231,340	AX-93930665	(–Log(p) = 5.08)	14.33	China	
	qST-IR-20	B1/11	11,288,981	AX-94086348	(–Log(p) = 4.51)	8.89	China	
	qST-IR-21	B1/11	11,290,853	AX-93789646	(–Log(p) = 4.71)	9.35	China	
	qST-IR-22	J/16	33,171,582	AX-94152460	(–Log(p) = 5.21)	10.52	China	
	qST-IR-23	G/18	1,919,656	AX-93868447	(–Log(p) = 4.76)	13.48	China	
	qST-IR-24	G/18	1,920,903	AX-93868448	(–Log(p) = 4.60)	13.06	China	
	qST-IR-25	G/18	1,963,410	AX-94165811	(–Log(p) = 4.87)	13.77	China	

ST-GI	*qST-GI-1*	A2/8	8,020,851	AX-93751738	(−Log(p) = 4.70)	17.02	China	
	qST-GI-2	A2/8	8,171,247	AX-94048075	(−Log(p) = 4.90)	17.50	China	
	qST-GI-3	A2/8	8,188,411	AX-94048076	(−Log(p) = 5.16)	18.14	China	
	qST-GI-4	A2/8	8,195,444	AX-93751799	(−Log(p) = 4.99)	17.72	China	
	qST-GI-5	A2/8	8,198,079	AX-94048081	(−Log(p) = 5.16)	18.14	China	
	qST-GI-6	A2/8	8,208,999	AX-94048086	(−Log(p) = 5.14)	18.10	China	
	qST-GI-7	A2/8	8,216,666	AX-94048089	(−Log(p) = 4.54)	16.62	China	
	qST-GI-8	A2/8	8,228,424	AX-94048095	(−Log(p) = 5.11)	18.02	China	
	qST-GI-9	A2/8	8,229,952	AX-93751817	(−Log(p) = 5.27)	18.41	China	
	qST-GI-10	A2/8	8,231,340	AX-93930665	(−Log(p) = 4.94)	17.59	China	
	qST-GI-11	A2/8	8,241,052	AX-94048102	(−Log(p) = 5.16)	18.14	China	
	qST-GI-12	A2/8	8,244,331	AX-93634502	(−Log(p) = 5.16)	18.14	China	
	qST-GI-13	A2/8	8,251,703	AX-94048104	(−Log(p) = 4.90)	17.50	China	
	qST-GI-14	A2/8	8,255,343	AX-94048105	(−Log(p) = 4.90)	17.50	China	
	qST-GI-15	A2/8	8,259,822	AX-93751829	(−Log(p) = 5.16)	18.14	China	
	qST-GI-16	A2/8	8,262,788	AX-93751830	(−Log(p) = 4.66)	16.91	China	
	qST-GI-17	A2/8	8,264,914	AX-93751831	(−Log(p) = 5.16)	18.14	China	
	qST-GI-18	B2/14	2,065,545	AX-94269323	(−Log(p) = 4.58)	16.73	China	

(continued)

Table 5.1 (Continued)

Trait	*QTL*	LG/Chr	Interval (cM)	Marker/interval	LOD, *P, or F*	R^2 (%)	Environment	References
ST-GP	*qST-GP-1*	D1a/1	9,714,061	AX-94279570	(−Log(p) = 4.61)	11.68	China	
	qST-GP-2	D1a/1	20,005,543	AX-93965565	(−Log(p) = 4.95)	12.43	China	
	qST-GP-3	A2/8	8,095,794	AX-93751768	(−Log(p) = 4.99)	15.95	China	
	qST-GP-4	A2/8	8,099,353	AX-93751769	(−Log(p) = 4.81)	15.49	China	
	qST-GP-5	A2/8	8,229,952	AX-93751817	(−Log(p) = 4.51)	19.26	China	
	qST-GP-6	A2/8	8,229,952	AX-93751817	(−Log(p) = 4.56)	14.85	China	
	qST-GP-7	A2/8	8,245,529	AX-93751824	(−Log(p) = 4.69)	19.71	China	
	qST-GP-8	A2/8	8,267,537	AX-94048110	(−Log(p) = 4.70)	15.20	China	
	qST-GP-9	A2/8	8,276,381	AX-93930669	(−Log(p) = 4.70)	15.21	China	
	qST-GP-10	A2/8	8,277,581	AX-93751836	(−Log(p) = 4.64)	15.06	China	
	qST-GP-11	A2/8	8,279,616	AX-94048117	(−Log(p) = 4.70)	15.21	China	
	qST-GP-12	A2/8	8,285,888	AX-93634506	(−Log(p) = 4.70)	15.21	China	
	qST-GP-13	A2/8	43,594,998	AX-94055981	(−Log(p) = 5.21)	16.48	China	
	qST-GP-14	K/9	5,210,383	AX-93762878	(−Log(p) = 4.57)	11.60	China	
	qST-GP-15	E/15	4,404,931	AX-93945068	(−Log(p) = 4.52)	14.77	China	
	qST-GP-16	J/16	6,155,245	AX-93848002	(−Log(p) = 4.90)	12.47	China	
	qST-GP-17	G/18	20,372,435	AX-93874537	(−Log(p) = 4.67)	15.13	China	
	qST-GP-18	G/18	20,826,679	AX-93874705	(−Log(p) = 4.59)	14.92	China	
	qST-GP-19	L/19	242,004	AX-94183050	(−Log(p) = 4.81)	12.13	China	

ST-GR	*qST-GR-1*	A2/8	7,741,884	AX-93751616	(−Log(p) = 5.25)	23.81	China	
	qST-GR-2	A2/8	8,421,388	AX-93751888	(−Log(p) = 4.81)	22.70	China	
	qST-GR-3	A2/8	9,458,964	AX-93752242	(−Log(p) = 4.51)	21.98	China	
	qST-GR-4	F/13	2,467,421	AX-93807682	(−Log(p) = 4.57)	14.65	China	
	qST-GR-5	F/13	40,020,985	AX-93940737	(−Log(p) = 4.51)	10.10	China	
	qST-GR-6	F/13	40,031,889	AX-93818619	(−Log(p) = 4.86)	11.13	China	
	qST-GR-7	F/13	40,086,888	AX-94115465	(−Log(p) = 4.78)	10.89	China	
	qST-GR-8	F/13	40,099,809	AX-94115470	(−Log(p) = 4.56)	10.25	China	
	qST-GR-9	F/13	40,114,029	AX-93940742	(−Log(p) = 4.71)	10.68	China	
	qST-GR-10	E/15	12,667,304	AX-93837619	(−Log(p) = 4.94)	23.03	China	
	qST-GR-11	G/18	6,252,000	AX-93870124	(−Log(p) = 4.83)	22.76	China	
	qST-GR-12	G/18	6,279,712	AX-94167497	(−Log(p) = 4.59)	22.18	China	

total variation of STR and SPAD, respectively, and derived the beneficial alleles from the wild-type line, JWS156-1 (Fig. 2.1; Table 5.1) (Hamwieh and Xu 2008).

Lee et al. (2009) investigated ST in PI483463, a wild soybean accession. They tested F_1 and F_2 plants derived from a cross between 'PI483463' and 'Hutcheson' lines ($n = 216$) in 100 mM salt solutions in the greenhouse and found that the segregation ratios were 3 (tolerant, T) to 1 (sensitive, S) in F_1 plants, 1 (all T) to 2 (3T and 1S) to 1 (all S) in F_2 plants, and 3T:1S in F_3 plants. The authors concluded that ST in PI483463 is controlled by a single dominant gene (allele) and named it *Ncl2* (Lee et al. 2009).

Do et al. (2010) grew a RIL population (F_6, $n = 112$) and an F_2 population ($n = 149$) derived from a cross between 'Jackson' and 'JWS156-1' lines in the greenhouse. For QTL data analysis, they used CIM of MapQTL software (Van Ooijen 2004). Two major QTL that control both STR and SPAD (leaf chlorophyll content) values have been identified and mapped on chromosome 17. The QTL explained 50% and 43% of the total variation of STR and SPAD, respectively (Fig. 2.1; Table 5.1) (Do et al. 2010).

Hamwieh et al. (2011) grew two RIL populations. The first RIL population was derived from a cross between 'FT-Abyara' and 'C01' lines ($n = 96$) and was from Brazil, and the second RIL population was derived from a cross between 'Jin dou No. 6' and '0197' lines ($n = 81$) and was from China. FT-Abyara by C01 RIL population was genotyped with 172 SSR markers, and Jin dou No. 6 by 0197 RIL population was genotyped by 85 SSR markers. One major QTL that controls both STR and SPAD was identified and mapped on chromosome 3 in both RIL populations. The QTL explained about 44% and 38% and 47.1% and 40.9% of variation in STR and SPAD in FT-Abyara X C01 and Jin dou No. 6 X 0197, respectively. Moreover, the QTL has been validated using a set of near-isogenic lines (NILs) and spanned about 72.1 cM interval (Satt080–Sat_306) (Fig. 2.1; Table 5.1) (Hamwieh et al. 2011).

Xu and Do (2012) summarized about 7 years (2004–2011) of studies that reported QTL for both saline and sodic (alkaline) salt stress in soybean (Lee et al. 2004; Chen et al. 2008; Hamwieh and Xu 2008; Hamwieh et al. 2011); however, the authors did not mention other QTL for salt stress reported in the same period and summarized here in this chapter (Lee et al. 2009; Do et al. 2010).

Ha et al. (2013) grew a RIL population derived from a cross between 'PI483463' and 'Hutcheson' lines ($n = 106$) in the greenhouse and a field in Portageville, MO, in 2007–2008; genotyped it with 1536 SNP markers; and used IM and multiple QTL mapping (MQM) of MapQTL (Van Ooijen 2004) for ST QTL data analysis. The study found a major QTL for leaf scorch on chromosome 3. This QTL alone explained 56.5% of the total variation of leaf scorch. The major QTL was further confirmed and validated using a set of 66 isogenic lines and explained 94.8% of the total variation of leaf scorch. This major QTL was narrowed from 72.1 cM interval (Satt080–Sat_306) (Hamwieh et al. 2011) to only 1.6–7.3 cM in this (Satt312–Satt 255) (Fig. 2.1; Table 5.1) (Ha et al. 2013).

Huang (2013) screened a population of soybean accessions ($n = 192$) using GWAS analysis to identify markers associated with ST estimated as leaf scorch after 3 weeks of treatment with 120-mM salt solution. The population consisted of 94 tolerant lines and 87 sensitive lines and was genotyped with 37,281 SNP markers. The study identified 62 SNP markers associated with six QTL that control ST (leaf scorch) and mapped on chromosomes 2, 3, 5, 6, 8, and 18 (Table 5.1). Moreover, ten candidate genes have been identified within these QTL regions. These genes are *Glyma03g32410* encoding for a pre-mRNA processing factor 39-like, *Glyma03g32800* encoding for a 26S proteasome regulatory subunit, *Glyma03g32850* encoding for a heat shock protein 70 kDa, *Glyma03g32890* encoding for a K^+/H^+ antiporter, *Glyma03g32900* encoding for a K^+/H^+ antiporter, *Glyma03g32940* encoding for a UNC93-like protein, *Glyma03g33240* encoding for a Ca^{2+}-transporting ATPase, *Glyma05g09310* encoding for a pyruvate kinase, *Glyma08g24580* encoding for a $Ca2^+$-transporting ATPase, and *Glyma18g46940* gene encoding for a ubiquitin-conjugating enzyme E2 5-like (Table 5.2) (Huang 2013). These same QTL reported in an MS thesis (Huang 2013) have been reported in a peer-reviewed publication thereafter (Table 5.1) (Huang et al. 2018). The new study confirmed two major QTL previously reported on chromosome 3 (Lee et al. 2004; Hamwieh and Xu 2008; Hamwieh et al. 2011) and chromosome 18 (Chen et al. 2008) and identified four new minor QTL regions on chromosomes 2, 5, 6, and 8 that explained 7–8.4% of the total variation of leaf scorch (Huang et al. 2018).

Do et al. (2013) used the same RIL population derived from a cross between ‘Jackson’ (salt-sensitive) and the wild accession ‘JWS156-1’ (salt-tolerant) lines and developed NILs to validate the major QTL for alkaline ST (STR and SPAD) on chromosome 17 between the markers GM17-11.6 and Satt447. They self-pollinated the heterozygote line RHL46 and developed a segregating population ($n = 1109$). The QTL region was validated and narrowed to 3.33 cM (GM17-11.6–Satt447) which is equivalent to about 771 kb. Moreover, the QTL had a high LOD score of 166.4, explained about 53.8% of the total variation of STR, and had a dominance effect of 0.83 (Do et al. 2013).

Guan et al. (2014a) grew an F_2 population ($n = 239$) and a validation RIL population ($n = 230$) derived from crosses between ‘Tiefeng 8’ (salt-tolerant) and ‘85–140’ (salt-sensitive) lines in China in 2009. They genotyped them with 42 SSR markers, a sequence-characterized amplified region (SCAR) marker, and four insertion/deletion polymorphism (InDel) markers around the major QTL for ST on chromosome 3. The QTL was narrowed to a genomic region of about 0.4 cM which is equivalent to about 209 kb (QS08064–Barsoyssr_3_3011) (Fig. 5.1) (Guan et al. 2014a). In a subsequent study, Guan et al. (2014b) identified the gene *GmSALT3* that limits Na^+ accumulation in shoots in the same region on chromosome 3. The gene *GmSALT3* was approximately 17.5 kb, encodes a cation/H^+ exchange protein located in the ER (endoplasmic reticulum), and is expressed in the xylem and phloem root cells of the salt-tolerant line Tiefeng 8. Moreover, by analyzing 172 landraces and 57 wild soybeans from China, they identified a significantly associated haplotype with ST in Tiefeng 8, named it haplotype 1 (H1), and concluded that natural and/or artificial selection favored it (Guan et al. 2014b).

Using whole genome de novo sequencing and QTL mapping by resequencing methods on a Chinese wild soybean (W05), Qi et al. (2014) identified the gene

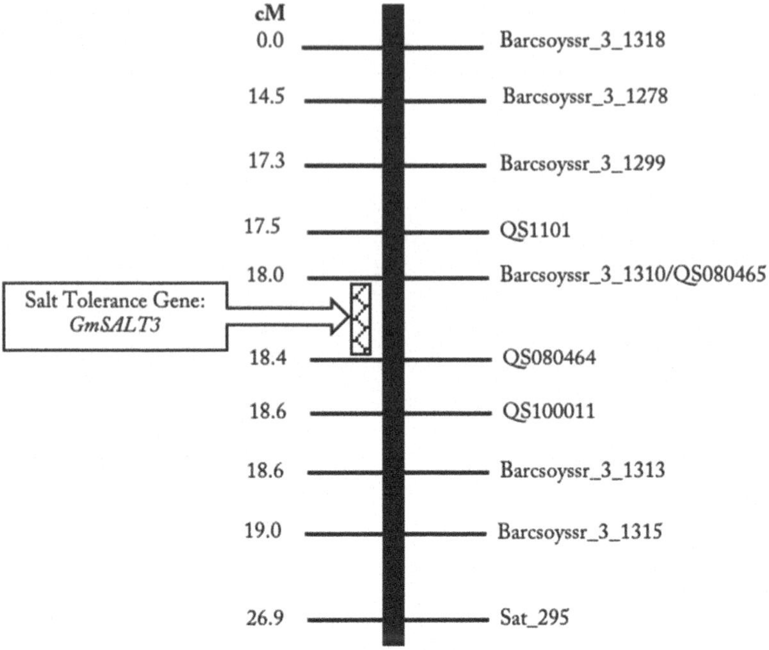

Fig. 5.1 Chromosome 3 and the location of the narrowed ST QTL region. (Adopted form Guan et al. (2014a))

GmCHX1 involved in ST and encoding for an iron protein. It spans about 978 kb and overlaps with *Ncl* locus (Lee et al. 2004; Hamwieh and Xu 2008; Hamwieh et al. 2011; Ha et al. 2013; Do et al. 2016; Valliyodan et al. 2016). A gain-of-function test was performed for *GmCHX1* gene in hairy roots of another soybean accession C08 and tobacco BY-2 cells under salt stress and confirmed that it alleviates significantly salt stress symptoms via ion homeostasis process specifically reducing the ratio of Na^{+}/K^{+} ions (Table 5.1) (Qi et al. 2014).

Zhang et al. (2014) screened a population of Chinese soybean accessions ($n = 257$) using GWAS analysis to identify markers associated with alkaline and ST estimated as the fresh weight of the roots (FWR), length of the main root (LR), dry weight of the roots (DWR), and calculated alkaline salt tolerance index (ATI) and salt tolerance index (STI) after treatment with 100 mM of NaCl solution. Seventeen (17) QTL that control LR-STI (*qLR-STI_1–qLR-STI_19*) have been identified and mapped on chromosomes 2, 4, 6, 8, 11, 12, 13, 15, 16, 17, and 18; three QTL that control LR-ATI (*qLR-ATI_1–qLR-ATI_3*) have been identified and mapped on chromosome 3; 13 QTL that control FWR-STI (*qFWR-STI_1–qFWR-STI_13*) have been identified and mapped on chromosomes 2, 3, 6, 7, 10, 13, 17, 18, and 20; eight QTL that control FWR-ATI (*qFWR-ATI_1–qFWR-ATI_8*) have been identified and mapped on chromosomes 4, 5, 8, 13, 14, 15, 17, and 18; 13 QTL that control DWR-ATI (*qDWR-ATI_1–qDWR-ATI_13*) have been identified and mapped on chromosomes 2, 6, 7, 9, 10, 12, 13, 15, 17, and 17; ten QTL that control LH-STI (*qLH-STI_1–qLH-STI_10*) have been identified and mapped on chromosomes 1, 5, 6, 8, 9, 13, 14, and 15; one QTL that controls LH-ATI (*qLH-ATI_1*) has been identified and mapped on chromosome 3; and 19 QTL that control BS-STI (*qBS-STI_1–qBS-STI_19*) have been identified and mapped on chromosomes 2, 4, 5, 6, 7, 8, 11, 13, 14, 15, 16, 17, 18, and 20 (Fig. 2.1; Table 5.1) (Zhang et al. 2014).

Kan et al. (2015) grew a natural population of landraces ($n = 191$) in four environments in China, genotyped it with 1142 SNP markers, and used GWAS analysis to identify markers associated with ST at the seed germination stage. ST was estimated as IR (inhibition rate) = (W2 – W1)/W1 [W1 = seed dry weight before imbibition; W2 = seed weight after imbibition for 24 h], GI (germination index) = $\sum$Gt/Dt [Gt = number of germinated seeds at N days; Dt = N days of germinated seeds], GR (germination rate) % = (number of germinated seeds/total seed number) × 100, and GR = $N_t/N_0 \times 100\%$ at days 5 and 10, where N_t is the number of germinated seeds at day t and N_0 is the total number of seeds. ST was estimated as the ratio of IR, GI, and GR under control and salt stress conditions. Three SNPs associated with QTL that control ST-IR (*qST-IT_1–qST-IR_3*) have been identified and mapped on chromosomes 6 and 8; one SNP associated with a QTL that controls ST-GI (*qST-GI_1*) has been identified and mapped on chromosome 18; eight SNPs associated with QTL that control ST-GI/ST-GR (*qST-GI/ST-GR_1–qST-GI/ST-GR_8*) have been identified and mapped on chromosomes 2, 3, 9, 12, 13, and 14; and one SNP associated with a QTL that controls ST-GR (*qST-GR_1*) has been identified and mapped on chromosome 17 (Fig. 2.1; Table 5.1). Moreover, the study identified nine candidate genes within these QTL regions: *Glyma08g09730.1* (SOS1) gene encodes for a cell membrane Na^+/H^+ antiporter, *Glyma08g12400.1* (SOS6) gene encodes for osmotic stress tolerance by regulation of reactive oxygen species (ROS) (Zhu et al. 2010; Sun et al. 2016; Chen et al. 2019), *Glyma06g20170.1* (CPK11) gene encodes for a calcium-dependent protein kinase (Zhu et al. 2007), *Glyma18g47140.1* (MPK4) gene encodes for a mitogen-activated protein kinase involved in plant immunity (Colcombet et al. 2013), *Glyma02g41850.1* (P5CS) gene encodes for a pyrroline 5-carboxylate synthase (Signorelli and Monza 2017), *Glyma09g00460.1* (ABA3) gene encodes for molybdenum cofactor (MoCo) required for the activation of aldehyde oxidase (AO) involved in abscisic acid (ABA) biosynthesis (Watanabe et al. 2018), *Glyma09g00490.3* (ETR1) gene encodes for an ethyl receptor (Gamble et al. 2002), *Glyma12g34880.1* (LEA) gene encodes for late embryogenesis abundant protein (Shao et al. 2005), and *Glyma17g02560.1* (NHX1) gene encodes for the Na^+/H^+ antiporter (Bassil et al. 2011); however, only five (*Glyma09g00490.3* [ETR1], *Glyma09g00460.1* [ABA3], *Glyma08g12400.1* [SOS6], *Glyma18g47140.1* [MPK4], and *Glyma08g09730.1* [SOS1]) of these nine genes are associated with salt stress at the germination stage (Table 5.2) (Kan et al. 2015).

Kan et al. (2016) grew a RIL population derived from a cross between 'Kefeng 1' and 'Nannong 1138-2' lines ($n = 184$) and a natural population of landraces ($n = 196$) in four environments (E1-E4) in China in 2012–2013 under no salt or salt stress (150 mM salt solution). ST was estimated IR, GI, and GR, and CIM of

WinQTL cartographer was used for QTL data analysis. The RIL population was genotyped with 221 SSR markers, three EST-SSRs, and *R* gene, and the derived genetic linkage map covered 2, 625.9 cM. For the natural population, they used GWAS analysis, and this population was genotyped with 205 SSR markers. In environment E1 and using the RIL ($n = 184$), two QTL that control ST-IR (*qST-IR-8* and *qST-IR-18*) have been identified and mapped on chromosomes 8 and 18; two QTL that control ST-GI (*qST-GI-2-1* and *qST-GI-2-2*) have been identified and mapped on chromosome 2; and two QTL that control ST-GR (*qST-GR-7* and *qST-GR-8*) have been identified and mapped on chromosomes 7 and 8. In environment E2 and using the RIL ($n = 184$), two QTL that control ST-IR (*qST-IR-8* and *qST-IR-18*) have been identified and mapped on chromosomes 8 and 18; two QTL that control ST-GI (*qST-GI-10* and *qST-GI-17*) have been identified and mapped on chromosomes 10 and 17; and two QTL that control ST-GR (*qST-GR-7* and *qST-GR-8*) have been identified and mapped on chromosomes 7 and 8. In combined environments (E1 and E2) and using the RIL ($n = 184$), two QTL that control ST-IR (*qST-IR-2* and *qST-IR-8*) have been identified and mapped on chromosomes 2 and 8; two QTL that control ST-GI (*qST-GI-8* and *qST-GI-10*) have been identified and mapped on chromosomes 8 and 10; and one QTL that controls ST-GR (*qST-GR-8*) has been identified and mapped on chromosome 8 (Fig. 2.1; Table 5.1). Using the natural population ($n = 196$) and GWAS analysis, eight SSR markers associated with QTL that control ST-IR have been identified and mapped on chromosomes 2, 6, 7, 16, and 19; five SSR markers associated with QTL that control ST-GI have been identified and mapped on chromosomes 7 and 19; and nine SSR markers associated with QTL that control ST-GR have been identified and mapped on chromosomes 1, 2, 3, 6, 7, and 19 (Fig. 2.1; Table 5.1) (Kan et al. 2016).

Liu et al. (2016) generated three sets of near-isogenic lines (NILs) from a cross between the salt-sensitive (S) variety '85–140' and salt-tolerant (T) variety 'Tiefeng 8' to investigate the effect of *GmSALT3* gene on soybean growth and development. Using eight markers, they identified a genomic region of 1098 kb, containing *GmSALT3* gene, common to these NILs. The salt-tolerant NILs with Tiefeng 8 allele have been named (NIL-T; 782-T, 820-T, 820-T) and the salt-sensitive NILs with 85–140 allele have been named (NIL-S; 782-S, 820-S, 820-S). The study concluded that early vigor or seedling emergence rate has not been improved by *GmSALT3*; however, a significantly less accumulation of Na^+ ions in leaves was observed in NIL-T lines compared to NIL-S lines under salt stress (200 mM of NaCl solution). Also, more Cl^- ions were accumulated in NIL-T lines' roots and less in leaves compared to NIL-S lines. It also seemed that GmSALT3 gene conferred yield advantage to NIL-T lines since their 100-seed weight (100-SW) and total seed weight (YSW) were greater than those of NIL-S lines under salt stress (Liu et al. 2016).

Patil et al. (2016) performed whole genome resequencing (WGRS) and GWAS analysis on a soybean population ($n = 106$) and a RIL population derived from a cross between 'PI483463' and 'Hutcheson' lines ($n = 104$) to investigate ST estimated as chlorophyll content (SPAD value), level of leaf scorch score (LSS), and Na+ ion accumulation in plants under salt stress. Three major structural variants within the region of *GmCHX1* gene (*Glyma03g32900*) encoding for an iron protein (Qi et al. 2014) have been identified by the study. Five SNP markers associated with QTL that control each of LSS (*qLSS_1–qLSS_5*) and SPAD values (*qSPAD_1–qSPAD_5*) have been identified and mapped on chromosome 3 (Table 5.1) (Patil et al. 2016).

Zeng (2016) grew a RIL population derived from a cross between 'RA-452' (salt-sensitive) and 'Osage' (salt-tolerant) cultivars ($n = 124$) in a field in Fayetteville, AR, in 2013; genotyped it with 5403 SNP markers; evaluated for ST under salt stress (120 mM of NaCl solution) estimated as leaf chloride concentration; and used CIM of WinQTL cartographer for QTL data analysis. Twenty-seven SNP markers associated with QTL that control leaf Cl^- ion content have been identified and mapped on chromosomes 3, 13, and 15 (Zeng 2016). Twenty of these QTL are clustered on chromosome 3 that contain major ST genes such as *GmSALT3* and *GmCHX1* previously reported (Lee et al. 2004; Liu et al. 2016; Patil et al. 2016; Valliyodan et al. 2016). The author also performed a GWAS on a population of soybean plant introductions (PIs) ($n = 283$) and used the cultivars 'RA-452' and 'Osage' as checks in salt stress study. They identified 45 SNP markers associated with QTL that control both leaf chlorophyll (SPAD) and leaf chloride contents and mapped them on chromosomes 2, 3, 7, 8, 10, 16, and 20. In addition to confirming *GmSALT3* and *GmCHX1* genes on chromosome 3 with 18 SNP marker associations, two additional major QTL that control ST (leaf chloride concentration and SPAD values) have been identified on chromosomes 2 (one SNP association) and 7 (two SNP associations) (Table 5.1). Moreover, six candidate genes *Glyma.02G228100* encoding for a glutamine-dependent asparagine synthase (Lam et al. 2003), *Glyma.03G031400* encoding for a cytochrome P450 (Barlier et al. 2000), *Glyma.04G180300* encoding for a leucine-rich repeat (LRR) protein kinase family protein (Jones and Jones 1997), *Glyma.04G180400* encoding for an RUBP domain-containing protein (Harshavardhan et al. 2014), *Glyma.05G204600* encoding for the osmotin 34 (Capelli et al. 1997), and *Glyma.17G173200* encoding for a dihydroflavonol 4-reductase (White and Rawlinson 2001) have been identified by RNA-seq analysis of RA-452 x Osage RILs (Zeng 2016). These same QTL reported in a PhD dissertation (Zeng 2016) have been reported in a peer-reviewed publication shortly thereafter (Table 5.1) (Zeng et al. 2017).

To further investigate the ST *Ncl* locus previously identified on chromosome 3 (Lee et al. 2004; Hamwieh and Xu 2008; Chen et al. 2008; Hamwieh et al. 2011; Ha et al. 2013), Do et al. (2016) used map-based cloning approach on a population derived from a cross between two heterozygous lines at the QTL region, 'FT-Abraya' and 'C01' lines ($n = 1053$). ST was estimated as leaf chlorophyll content (SPAD), STR, and ion (Na^+, K^+, and Cl^-) content under 100 mM NaCl or KCl solutions. They performed a series of self-pollinations, selected RILs, analyzed them, and narrowed the QTL region to genomic region of approximately 16.6 kb containing the *Ncl* locus (BARCSOYSSR_03_1342–BARCSOYSSR_03_1338). After analysis of its function, they found that the *Ncl* gene is involved in ion (Na^+, K^+, and Cl^-) regulation in leaves (Do et al. 2016). Cao et al. (2019) investigated the expression of the

Ncl gene in 51 soybean accessions from Indonesia and found that its expression is necessary for these Indonesian soybeans (Cao et al. 2019).

Do et al. (2018) grew a F_2 population derived from a cross between 'Fiskeby' (salt-tolerant) and 'Williams 82' (salt-sensitive) lines ($n = 132$) in the greenhouse in 2014, genotyped it with 2158 SNPs, and used CIM of WinQTL cartographer to identify QTL for ST estimated as chlorophyll content ratio (CCR), leaf sodium (LSC), leaf chloride content (LCC), and leaf scorch score (LSS) of plants under salt stress (120 mM of salt solution). One QTL that controls LSS (qLSS_1), CCR (qCCR_1), LSC (qLSC_1), and LCC (qLCC_1) has been identified and mapped on chromosome 3, and one QTL that controls LSC (qLSC_2) has been identified and mapped on chromosome 13 (Fig. 2.1; Table 5.1) (Do et al. 2018).

Genetic studies, genome-wide association studies (GWAS), and advances in breeding for ST in soybean from 2002 to 2018 have been well summarized (Chen et al. 2018a, b; Cao et al. 2018).

Lee et al. (2018) screened a population of soybean accessions ($n = 216$) from Korea, China, and Japan in order to investigate the possibility of identifying novel haplotypes of the ST gene *GmSALT3* previously identified (Guan et al. 2014a, b; Qi et al. 2014). Forty haplotypes have been identified including three previously reported (H1, H2, and H5) but not the other six haplotypes (H3, H4, H6, H7, H8, and H9) (Guan et al. 2014a, b; Qi et al. 2014) and 37 newly discovered. Of these 40 haplotypes, eight were salt-tolerant and 32 were salt-sensitive. Furthermore, expression analyses of *GmSALT3* in the new haplotypes showed very low expression levels in salt-sensitive haplotypes compared to salt-sensitive haplotypes (Lee et al. 2018).

Lopez et al. (2018) grew an F_2 population derived from a cross between 'Ozark' (salt susceptible) and 'Jake' (salt-tolerant) lines ($n = 269$) in the greenhouse in 2014, genotyped it with 5402 SNPs, and used single marker analysis (SMA) and CIM of WinQTL cartographer to identify QTL for ST estimated as percentage of dead plants (PDP), leaf scorch score (LSS), leaf chloride content, and leaf chlorophyll content of plants grown under salt stress (120 mM of salt solution). The linkage map used for QTL analysis was composed of 1156 SNPs and covered approximately 2356.5 cM. One QTL for each of LSS (*QTL-LSS_1*), PDP (*QTL-PDP_1*), and leaf chlorophyll content (*QTL-Chlorophyll_1*) has been identified and mapped on chromosome 3, and one QTL that controls leaf chlorophyll content (*QTL-Chlorophyll_2*) has been identified and mapped on chromosome 19 (Fig. 2.1; Table 5.1) (Lopez et al. 2018).

Shi et al. (2018a, b) grew a RIL population derived from a cross between 'Jidou 12' (salt-tolerant) and 'Ji NF 58' (salt-sensitive) lines ($n = 156$) in the greenhouse, genotyped it with 118 SSR markers, and used single marker regression (SMR), CIM, and MIM of QGene 4.3.10 (Nelson 1997) to identify QTL for ST estimated as leaf chlorophyll content (SPAD values) and salt stress tolerance (STR) of plants grown under salt stress (150 mM of salt solution). One major QTL that controls both STR (*qSTR_1*) and SPAD (*qSPAD_1*) has been identified by SMR, CIM, and MIM methods and mapped on chromosome 3. The QTL explained about 44.7% and

27.6% of the total variation in STR and SPAD values, respectively (Fig. 2.1; Table 5.1) (Shi et al. 2018a, b).

Do et al. (2019) grew a natural population of soybean accessions ($n = 234$) in the greenhouse in 2016 under salt stress (120 mM of NaCl solution), genotyped it with 37,573 SNPs and several other markers (Salt14056, Salt-20, Salt11655, and GBMs), and performed a genome-wide association study (GWAS) on the population to identify SNP markers associated with ST estimated as leaf scorch score (LSS), leaf sodium content (LSC), chlorophyll content ratio (CRR), and leaf chloride content (LCC). Two hundred seventeen SNPs associated with QTL that control LSS, 136 SNPs associated with QTL that control LSC, 190 SNPs associated with QTL that control CCR, and 278 SNPs associated with QTL that control LCC have been identified; however, eight markers (five SNPs and three SSRs) significantly associated with QTL that control ST traits (LSS, LSC, CRR, and LCC) have been identified and mapped on chromosomes 1, 2, 3, 5, 6, 8, 14, 15, 16, 18, 19, and 20 (Table 5.1). Moreover, authors identified many candidate genes within these QTL regions including *Glyma.02g204300* that encodes for an ion transport protein, *Glyma.02g247100* that encodes for a MYB-like DNA-binding domain, *Glyma.08g145600* that encodes for a cellulose synthase, *Glyma.08g145900* that encodes for an alcohol dehydrogenase GroES-like domain, *Glyma.14g211300* that encodes for a universal stress protein family, *Glyma.15g143800* that encodes for a cAMP-regulated phosphoprotein fendosulfine conserved region, *Glyma.16g182900* that encodes for a leucine-rich repeat N-terminal domain, *Glyma.18g043100* that encodes for a cytochrome c oxidase assembly protein CtaG FCox11, *Glyma.19g165500* that encodes for a C5HC2 zinc finger, *Glyma.20g132300* that encodes for a 4F5 protein family, and many other genes (Table 5.2) (Do et al. 2019).

Zhang et al. (2019) grew a RIL population derived from a cross between ‘Kefeng No. 1’ and ‘Nannong 1138-2’ lines ($n = 184$) and a natural population of soybean accessions ($n = 211$) in four Chinese environments in 2012–2013 to identify QTL that control ST at the seed germination stage estimated as GI, GR, imbibition rate (IR), germination potential (GP), and the ratios of these indices under no salt stress and salt stress conditions (150 mM of NaCl solution). The RIL population was genotyped with 221 SSR markers, three SSRs, and the R gene, and the derived genetic linkage map coverage was 2625.9 cM. Five QTL that control ST-IR have been identified and mapped on chromosomes 2, 7, and 8; four QTL that control ST-GI have been identified and mapped on chromosomes 8, 15, and 17; nine QTL that control ST-GP have been identified and mapped on chromosomes 1, 2, 7, 8, 10, and 17; and seven QTL that control ST-GR have been identified and mapped on chromosomes 1, 7, 8, 10, 17, and 18 (Fig. 2.1; Table 5.1). The natural population was genotyped with 207,608 SNP markers, and GWAS analysis identified 74 SNPs associated with QTL that control ST indices. Twenty-five (25) SNPs associated with ST-IR have been identified and mapped on chromosomes 8, 11, 16, and 18; 18 SNPs associated with ST-GI have been identified and mapped on chromosomes 8 and 14; 19 SNPs associated with ST-GP have been identified and mapped on chromosomes 1, 8, 9, 15, 16, 18, and 19; and 12 SNPs associated with ST-GR have been identified and mapped on chromosomes 8, 13, 15, and 18. Moreover, they identified

Table 5.2 Candidate genes identified within the genomic regions harboring QTL that control ST from 2003 to 2019

Gene	LG/ Chr	Position	Annotation	References
Glyma17g11790	D2/17	*	Purple acid phosphatase, enhancing ROS scavenging	Liao et al. (2003)
Glyma07g18150	M/7	*	Cation/proton exchanger	Luo et al. (2005a, b)
Glyma14g09320	B2/14	*	Transcription factors (TFs)	Gao et al. (2005)
Glyma20g37370	I/20	*	Vacuolar Na+/H+ antiporter	Sun et al. (2006)
Glyma06g04490	C2/6	*	Transcription factors (TFs)	Chen et al. (2007)
Glyma02g01300	D1b/2	*	Transcription factors (TFs)	Liao et al. (2008a, b)
Glyma04g03200	C1/4	*	Transcription factors (TFs)	
Glyma10g01450	O/10	*	Transcription factors (TFs)	Zhou et al. (2008)
Glyma15g13030	E/15	*	Plasma membrane and vacuolar Na+/H+ antiporter	Zhou et al. (2009)
Glyma03g42450	N/3	*	Transcription factors (TFs)	Zhang et al. (2009)
Glyma13g30990	F/13	*	Transcription factors (TFs)	Zhang et al. (2010)
Glyma05g01270	A1/5	*	Ubiquitin-conjugating enzyme	Zhou et al. (2010)
Glyma01g00530	D1a/1	*	Na + and K+ transporters	Chen et al. (2011)
Glyma19g28476	L/19	*	Transcription factors (TFs)	Hao et al. (2011)
Glyma05g05580	A1/5	*	Ca2+ sensors; calcineurin B-like proteins	Li et al. (2012)
Glyma07g04950	M/7	*	Transcription factors (TFs)	Zhai et al. (2013a, b)
Glyma05g14760	A1/5	*	Vacuolar Cl – transporter protein	Sun et al. (2013)

Glyma03g32410	N/3	40,154,304	Pre-mRNA processing factor 39-like	Huang (2013)
Glyma03g32800	N/3	40,557,941	26S proteasome regulatory subunit	
Glyma03g32850	N/3	40,585,266	Heat shock protein 70 kDa	
Glyma03g32890	N/3	40,613,405	K+/H+ antiporter	
Glyma03g32900	N/3	40,613,405	K+/H+ antiporter	
Glyma03g32940	N/3	40,661,459	UNC93-like protein	
Glyma03g33240	N/3	40,895,159	Ca^{2+}-transporting ATPase	
Glyma05g09310	A1/5	9,081,936	Pyruvate kinase	
Glyma08g24580	A2/8	18,772,502	Ca2 +-transporting ATPase	
Glyma18g46940	G/18	56,571,714	Ubiquitin-conjugating enzyme E2 5-like	
Glyma07g15590	M/7	*	Na^+ and K^+ transporters	Chen et al. (2014a, b)
Glyma10g32190	O/10	*	Ca^{2+} sensors, calmodulins	Rao et al. (2014)
Glyma08g01860	A2/8	*	Osmolytes; plasma membrane intrinsic proteins	Zhou et al. (2014)
Glyma12g07190	H/12	*	Antioxidants to protect cells from ROS	Yan et al. (2014)
Glyma14g05470	B2/14	*	Transcription factors (TFs)	Dong et al. (2015)
Glyma15g00570	E/15	*	Transcription factors (TFs)	Wang et al. (2015a, b)
Glyma08g09730	A2/8	*	Plasma membrane Na^+/H^+ antiporter	Nie et al. (2015)
Glyma02g41850.1	D1b/2	47,313,970	Pyrroline 5-carboxylate synthase	Kan et al. (2015)
Glyma06g20170.1	C2/6	16,660,196	Calcium-dependent protein kinase	
Glyma08g09730.1	A2/8	7,716,379	Cell membrane Na^+/H^+ antiporter	
Glyma08g12400.1	A2/8	9,622,940	Osmotic stress tolerance ROS	
Glyma09g00460.1	K/9	210,160	Molybdenum cofactor (MoCo)/activates aldehyde oxidase (AO)	
Glyma09g00490.3	K/9	210,160	Ethyl receptor	
Glyma12g34880.1	H/12	38,675,904	Late embryogenesis abundant protein	
Glyma17g02560.1	D2/17	1,785,893	Na^+/H^+ antiporter	
Glyma18g47140.1	G/18	57,209,838	Mitogen-activated protein kinase	

(continued)

Table 5.2 (Continued)

Gene	LG/ Chr	Position	Annotation	References
Upregulated				Zeng (2016)
Glyma.02G228100	D1b/2	41,497,138	Glutamine-dependent asparagine synthase 1	
Glyma.05G204600	A1/5	38,801,527	Osmotin 34	
Glyma.08G189600	A2/8	15,235,196	Lipoxygenase 1	
Glyma.16G174300	J/16	33,491,254	Disease resistance family protein/LRR family protein	
Glyma.19G055000	L/19	9,202,974	Disease resistance protein (TIR-NBS-LRR class) family	
Downregulated				
Glyma.03G031000	N/3	3,483,333	Cytochrome P450, family 83, subfamily B, polypeptide 1	
Glyma.03G219100	N/3	42,265,890	Signal transduction histidine kinase	
Glyma.03G226000	N/3	42,818,552	Glycosyl hydrolase superfamily	
Glyma.13G033400	F/13	10,549,358	Receptor serine/threonine kinase, putative	
Glyma.13G042200	F/13	13,542,394	2-Oxoglutarate (2OG) and Fe(II)-dependent oxygenase superfamily protein	
Consistent				
Glyma.03G031400	N/3	3,522,164	Cytochrome P450, family 83, subfamily B, polypeptide 1	
Glyma.04G180300	C1/4	44,611,967	Leucine-rich repeat protein kinase family protein	
Glyma.04G180400	C1/4	44,623,574	BURP domain-containing protein	
Glyma.08G261500	A2/8	25,674,084	Integrase-type D-binding superfamily protein	
Glyma.15G140700	E/15	11,520,846	HCP-like superfamily protein	
Glyma.16G119900	J/16	26,773,780	Cytidine/deoxycytidylate deaminase family protein	
Glyma.17G173200	D2/17	17,882,584	Dihydroflavonol 4-reductase	
Glyma.18G299000	G/18	57,670,461	PEBP (phosphatidyl ethanolamine-binding protein) family protein	
0 h vs. *12 h*				
Glyma.03G005800	N/3	545,101	MATE efflux family protein	
Glyma.03G113200	N/3	32,000,184	D(P)-binding Rossmann fold superfamily protein	

Glyma.03G157700	N/3	37,314,519	ATPase E1-E2 type family protein/haloacid dehalogenase-like hydrolase family protein	
Glyma.04G054400	C1/4	4,415,292	17.6 kDa class II heat shock protein	
Glyma.04G208300	C1/4	48,062,885	-C (no apical meristem, NAM) domain transcriptional regulator superfamily protein	
Glyma.05G036800	A1/5	3,241,817	Basic helix-loop-helix (bHLH) D-binding superfamily protein	
Glyma.06G157400	C2/6	12,937,295	-C (NAM) domain transcriptional regulator superfamily protein	
Glyma.06G202200	C2/6	18,712,865	Heat shock protein 101	
Glyma.08G201100	A2/8	16,266,130	HPT phosphotransmitter 4	
Glyma.12G116800	H/12	11,703,332	Protein phosphatase 2C family protein	
Glyma.13G068800	F/13	16,864,170	Cytochrome P450, family 82, subfamily C, polypeptide 4	
Glyma.13G279900	F/13	38,124,291	-C domain-containing protein 3	
Glyma.16G103900	J/16	21,014,275	Pinoresinol reductase 1	
Glyma.17G044300	D2/17	3,285,533	Lysine-ketoglutarate reductase/saccharopine dehydrogenase bifunctional enzyme	
Glyma.17G092800	D2/17	7,253,510	Gibberellin-regulated family protein	
Glyma.20G015900	I/20	1,444,191	17.6 kDa class II heat shock protein	
0 h vs. *24 h*				
Glyma.15G062700	E/15	4,807,822	Basic pathogenesis-related protein 1	
Glyma.19G168500	L/19	42,928,529	Auxin-induced protein 13	
Glyma03g32900	N/3		Cation/proton exchanger	Do et al. (2016)
Glyma08g12170	A2/8		Transcription factors (TFs)	Xu et al. (2016)
Glyma.02g143000	D1b/2	14,735,028	Guanylate-binding protein C N-terminal domain	Do et al. (2019)

(continued)

Table 5.2 (Continued)

Gene	LG/ Chr	Position	Annotation	References
Glyma.02g204300	D1b/2	38,958,949	Ion transport protein, cyclic nucleotide-binding domain	
Glyma.02g247400	D1b/2	43,512,032	AAR2 protein	
Glyma.02g247100	D1b/2	43,492,461	Myb-like DNA-binding domain	
Glyma.02g247300	D1b/2	43,504,539	CLASP N terminal	
Glyma.08g145600	A2/8	11,066,152	Cellulose synthase	
Glyma.08g145700	A2/8	11,083,230	WD domain C G-beta repeat	
Glyma.08g145800	A2/8	11,088,887	Prolamin-like (original Pfam: PF06915)	
Glyma.08g145900	A2/8	11,092,669	Alcohol dehydrogenase GroES-like domain, zinc-binding dehydrogenase	
Glyma.08g146000	A2/8	11,103,557	TFIIS helical bundle-like domain	
Glyma.08g146100	A2/8	11,119,305	EamA-like transporter family	
Glyma.08g146300	A2/8	11,136,129	PAP_fibrillin	
Glyma.08g146400	A2/8	11,139,657	BolA-like protein	
Glyma.08g146500	A2/8	11,145,877	Actin	
Glyma.08g146600	A2/8	11,157,526	Zinc finger C C3HC4 type (RING finger)	
Glyma.08g146700	A2/8	11,166,509	Autophagy-related protein 13	
Glyma.08g146800	A2/8	11,173,407	RNA polymerase Rpb4	
Glyma.08g146900	A2/8	11,183,844	GRAS domain family	
Glyma.08g156500	A2/8	12,100,644	NAM protein	
Glyma.08g157000	A2/8	12,140,034	Ras family	
Glyma.08g157100	A2/8	12,144,886	Unknown	
Glyma.08g157200	A2/8	12,146,949	SF0 – ring finger protein 5	
Glyma.08g157400	A2/8	12,162,921	SF9 – callose synthase-related	
Glyma.08g157500	A2/8	12,199,277	SF29 – P-loop-containing nucleoside triphosphate hydrolase family proteins	

Glyma.08g157700	A2/8	12,213,783	Citrate synthase	
Glyma.08g157800	A2/8	12,226,940	Cyclic pyranopterin phosphate synthase/molybdenum cofactor biosynthesis protein 1	
Glyma.08g157900	A2/8	12,235,183	VQ motif	
Glyma.08g158300	A2/8	12,260,220	Zinc finger C C3HC4 type (RING finger)	
Glyma.08g158400	A2/8	12,264,351	ATPase family associated with various cellular activities (AAA)	
Glyma.08g158600	A2/8	12,290,252	SF34 – major facilitator superfamily domain-containing protein	
Glyma.08g222200	A2/8	18,038,277	haloacid dehalogenase-like hydrolase, cation transporter FATPase C N-terminus	
Glyma.08g222300	A2/8	18,059,686	GDP-fucose protein O-fucosyltransferase	
Glyma.08g222400	A2/8	18,068,218	Adaptin N terminal region, coatomer beta C-terminal region	
Glyma.08g222500	A2/8	18,075,163	Exo70 exocyst complex subunit	
Glyma.08g222800	A2/8	18,085,127	Glycosyl transferase family 2	
Glyma.08g222900	A2/8	18,090,096	Mitochondrial ATP synthase epsilon chain	
Glyma.08g223000	A2/8	18,093,196	Vps51 FVps67	
Glyma.08g223200	A2/8	18,103,948	RNA recognition motif	
Glyma.08g223300	A2/8	18,107,199	ACT domain, D-isomer-specific 2-hydroxyacid dehydrogenase C NAD-binding domain	
Glyma.08g223400	A2/8	18,127,350	Protein kinase domain	
Glyma.08g223600	A2/8	18,138,052	Aldose 1-epimerase	
Glyma.08g223700	A2/8	18,147,188	Protein kinase domain	
Glyma.08g223800	A2/8	18,162,779	XPC-binding domain, UBA FTS-N domain, ubiquitin family	
Glyma.08g223900	A2/8	18,169,513	YGGT family	
Glyma.08g224200	A2/8	18,187,905	Ergosterol biosynthesis ERG4 FERG24 family	
Glyma.08g224300	A2/8	18,193,065	Eukaryotic family of unknown function (DUF1754)	
Glyma.08g224400	A2/8	18,196,846	V-type H+-transporting ATPase subunit A (ATPeV1A, ATP6A)	

(continued)

Table 5.2 (Continued)

Gene	LG/ Chr	Position	Annotation	References
Glyma.08g224500	A2/8	18,204,060	ATPase family associated with various cellular activities (AAA)	
Glyma.08g224600	A2/8	18,208,948	PAP_fibrillin	
Glyma.08g224800	A2/8	18,220,717	Protein of unknown function (DUF1666)	
Glyma.08g225000	A2/8	18,232,440	Mitochondrial carrier protein	
Glyma.08g225100	A2/8	18,245,614	Auxin-responsive protein	
Glyma.08g225200	A2/8	18,246,538	Auxin-responsive protein	
Glyma.08g225500	A2/8	18,260,357	SF11 – pyrophosphatase-energized membrane proton pump	
Glyma.08g225800	A2/8	18,290,918	Prenyltransferase and squalene oxidase repeat	
Glyma.08g225900	A2/8	18,303,275	Protein tyrosine kinase, leucine-rich repeat, leucine-rich repeat N-terminal domain	
Glyma.08g226000	A2/8	18,321,377	START domain	
Glyma.08g226100	A2/8	18,339,384	YT521-B-like domain	
Glyma.08g226400	A2/8	18,370,226	Sugar (and other) transporter	
Glyma.08g226500	A2/8	18,384,328	Pyruvate kinase C barrel domain	
Glyma.08g226600	A2/8	18,391,180	Complex I intermediate-associated protein 30 (CIA30), NmrA-like family	
Glyma.08g226700	A2/8	18,405,334	PHD finger, zinc finger C ZZ type, histone acetylation protein	
Glyma.08g226800	A2/8	18,444,007	Cupin	
Glyma.08g227000	A2/8	18,464,236	bZIP transcription factor	
Glyma.08g227200	A2/8	18,494,370	Protein kinase domain, EF hand	
Glyma.08g227300	A2/8	18,507,623	Domain of unknown function (DUF3546), arsenite-resistance protein 2	
Glyma.08g227400	A2/8	18,516,376	Core histone H2A FH2B FH3 FH4	
Glyma.08g227500	A2/8	18,518,682	FAR1 DNA-binding domain, SWIM zinc finger	
Glyma.08g227600	A2/8	18,535,293	Isoprenylcysteine carboxyl methyltransferase (ICMT) family	
Glyma.08g227700	A2/8	18,539,712	AP2 domain	
Glyma.08g228000	A2/8	18,584,819	Vacuolar protein sorting protein 36 Vps36	
Glyma.08g228100	A2/8	18,598,216	WD domain C G-beta repeat	
Glyma.08g228200	A2/8	18,601,501	Cupin	

Glyma.08g228300	A2/8	18,607,555	Oxidoreductase FAD-binding domain, oxidoreductase NAD-binding domain	
Glyma.08g228400	A2/8	18,621,259	S locus-related glycoprotein 1 binding pollen coat protein (SLR1-BP)	
Glyma.08g228600	A2/8	18,648,914	S locus-related glycoprotein 1 binding pollen coat protein (SLR1-BP)	
Glyma.08g228700	A2/8	18,660,718	Serine carboxypeptidase	
Glyma.08g228800	A2/8	18,684,920	Haloacid dehalogenase-like hydrolase, E1-E2 ATPase	
Glyma.08g228900	A2/8	18,695,544	Dof domain C zinc finger	
Glyma.08g229100	A2/8	18,702,845	Protein kinase domain, leucine-rich repeat	
Glyma.08g229400	A2/8	18,720,040	Double-stranded RNA binding motif, helicase-associated domain (HA2)	
Glyma.08g229600	A2/8	18,766,704	SART-1 family	
Glyma.08g229700	A2/8	18,795,227	F-box domain	
Glyma.08g229800	A2/8	18,800,851	WD domain C G-beta repeat	
Glyma.08g230000	A2/8	18,819,571	Hsp20/alpha crystallin family	
Glyma.08g230100	A2/8	18,824,513	Pathogenesis-related protein Bet v I family	
Glyma.08g230400	A2/8	18,847,360	Pathogenesis-related protein Bet v I family	
Glyma.08g230500	A2/8	18,862,871	Pathogenesis-related protein Bet v I family	
Glyma.08g230600	A2/8	18,879,029	BURP domain	
Glyma.08g230700	A2/8	18,897,439	Albumin I	
Glyma.08g230800	A2/8	18,911,649	Protein of unknown function (DUF3531)	
Glyma.08g230900	A2/8	18,924,711	WD domain C G-beta repeat	
Glyma.08g231000	A2/8	18,933,180	HAD superfamily C subfamily IIIB (acid phosphatase)	
Glyma.08g231100	A2/8	18,949,122	Protein kinase domain	
Glyma.08g231300	A2/8	18,969,636	Helicase conserved C-terminal domain	
Glyma.08g231400	A2/8	18,981,579	C3HC zinc finger-like	
Glyma.08g231500	A2/8	18,999,077	DnaJ domain	

(continued)

Table 5.2 (Continued)

Gene	LG/ Chr	Position	Annotation	References
Glyma.08g231600	A2/8	19,016,678	Metallo-beta-lactamase superfamily	
Glyma.08g231800	A2/8	19,053,865	Remorin C C-terminal region	
Glyma.08g231900	A2/8	19,064,682	Protein of unknown function (DUF1138)	
Glyma.08g232000	A2/8	19,071,849	SPFH domain F band 7 family	
Glyma.08g232100	A2/8	19,083,679	Cyclic nucleotide-binding domain	
Glyma.14g211300	B2/14	47,625,904	Universal stress protein family	
Glyma.15g143800	E/15	11,824,186	cAMP-regulated phosphoprotein fendosulfine conserved region	
Glyma.15g144000	E/15	11,843,111	Myb-like DNA-binding domain	
Glyma.16g185500	J/16	34,731,233	Leucine-rich repeat, leucine-rich repeat N-terminal domain	
Glyma.16g182700	J/16	34,393,343	Leucine-rich repeat	
Glyma.16g182900	J/16	34,436,482	Leucine-rich repeat, leucine-rich repeat N-terminal domain	
Glyma.16g183000	J/16	34,451,223	Leucine-rich repeat, leucine-rich repeat N-terminal domain	
Glyma.16g183200	J/16	34,497,329	Glycosyl transferase family 90	
Glyma.16g183300	J/16	34,500,477	Leucine-rich repeat	
Glyma.16g183400	J/16	34,513,882	Leucine-rich repeat, leucine-rich repeat N-terminal domain	
Glyma.16g183500	J/16	34,530,582	Leucine-rich repeat	
Glyma.16g183600	J/16	34,564,893	Leucine-rich repeat	
Glyma.16g183900	J/16	34,579,110	Glycosyl transferase family 90	
Glyma.16g184000	J/16	34,585,888	Leucine-rich repeat	
Glyma.16g184100	J/16	34,595,483	Protease inhibitor Fseed storage FLTP family	
Glyma.16g184200	J/16	34,604,592	Leucine-rich repeat, leucine-rich repeat N-terminal domain	
Glyma.16g184300	J/16	34,615,252	Leucine-rich repeat N-terminal domain, leucine-rich repeat	
Glyma.16g184400	J/16	34,621,915	Protease inhibitor Fseed storage FLTP family	
Glyma.16g184500	J/16	34,647,692	Leucine-rich repeat	
Glyma.16g184700	J/16	34,665,898	Leucine-rich repeat	
Glyma.16g184800	J/16	34,673,796	Leucine-rich repeat, leucine-rich repeat N-terminal domain	

Glyma.16g185000	J/16	34,681,294	Glycosyl transferase family 90	
Glyma.16g185100	J/16	34,686,428	Leucine-rich repeat	
Glyma.16g185200	J/16	34,692,049	Leucine-rich repeat, leucine-rich repeat N-terminal domain	
Glyma.16g185300	J/16	34,701,489	Leucine-rich repeat, leucine-rich repeat N-terminal domain	
Glyma.16g185400	J/16	34,712,976	Leucine-rich repeat, leucine-rich repeat N-terminal domain	
Glyma.16g185700	J/16	34,742,361	Protease inhibitor Fseed storage FLTP family	
Glyma.16g185800	J/16	34,747,498	Leucine-rich repeat, leucine-rich repeat N-terminal domain	
Glyma.16g185900	J/16	34,757,274	CAP160 repeat	
Glyma.16g186000	J/16	34,761,064	Glycosyl transferase family 90	
Glyma.16g186100	J/16	34,768,992	Leucine-rich repeat, leucine-rich repeat N-terminal domain	
Glyma.16g186200	J/16	34,780,811	Leucine-rich repeat	
Glyma.16g186400	J/16	34,799,381	Leucine-rich repeat N-terminal domain, leucine-rich repeat	
Glyma.16g186500	J/16	34,820,205	Leucine-rich repeat, leucine-rich repeat N-terminal domain	
Glyma.16g186600	J/16	34,913,608	Leucine-rich repeat	
Glyma.16g186700	J/16	34,921,963	Leucine-rich repeat, leucine-rich repeat N-terminal domain	
Glyma.16g186900	J/16	34,940,125	Leucine-rich repeat	
Glyma.18g043100	G/18	3,666,225	Cytochrome c oxidase assembly protein CtaG FCox11	
Glyma.18g202200	G/18	48,297,961	Serine carboxypeptidase	
Glyma.18g202300	G/18	48,298,637	Wound-induced protein	
Glyma.18g279800	G/18	56,074,203	GDSL FSGNH-like acyl-esterase family	
Glyma.18g279900	G/18	56,084,932	GDSL FSGNH-like acyl-esterase family	
Glyma.18g293800	G/18	57,166,496	PAP2 superfamily	
Glyma.18g293900	G/18	57,173,334	Protein of unknown function (DUF674)	
Glyma.18g294000	G/18	57,176,110	Protein of unknown function (DUF674)	
Glyma.18g294100	G/18	57,178,773	Weak chloroplast movement under blue light	
Glyma.18g294200	G/18	57,197,200	Protease inhibitor Fseed storage FLTP family	

(continued)

Table 5.2 (Continued)

Gene	LG/ Chr	Position	Annotation	References
Glyma.18g294300	G/18	57,205,035	Peptidase C26	
Glyma.18g294400	G/18	57,209,293	Protein kinase domain	
Glyma.19g165500	L/19	42,636,072	C5HC2 zinc finger, jmjN domain, JmjC domain C hydroxylase	
Glyma.19g175600	L/19	43,561,808	LNS2 (lipin FNed1 FSmp2), lipin C N-terminal conserved region	
Glyma.20g132100	I/20	37,250,648	SWIM zinc finger, FAR1 DNA-binding domain	
Glyma.20g132300	I/20	37,263,193	4F5 protein family	
Glyma.20g132400	I/20	37,266,529	Leucine-rich repeat, leucine-rich repeat N-terminal domain	
Glyma.20g132500	I/20	37,270,462	Mpv17 F PMP22 family	
Glyma.02G228100	D1b/2	*	Glutamine-dependent asparagine synthase 1	Zeng et al. (2019)
Glyma.03G031000	N/3	*	Cytochrome P450	
Glyma.03G226000	N/3	*	Glycosyl hydrolase superfamily protein	
Glyma.03G031400	N/3	*	Cytochrome P450	
Glyma.04G180300	C1/4	*	Leucine-rich repeat protein kinase family protein	
Glyma.04G180400	C1/4	*	BURP domain-containing protein	
Glyma.05g204600	A1/5	*	Osmotin 34	
Glyma.08G189600	A2/8	*	Lipoxygenase 1	
Glyma.13G042200	F/13	*	2-Oxoglutarate and Fe(II)-dependent oxygenase (OFE) superfamily protein	
Glyma.17G173200	D2/17	*	Dihydroflavonol 4-reductase	
Glyma.08g101200	A2/8	7,780,496	RCAR3 interacting F-box protein 1, RIFP1	Zhang et al. (2019)
Glyma.08g101300	A2/8	7,783,001	Myo-inositol oxygenase 4	
Glyma.08g101500	A2/8	7,801,075	ATP-binding cassette C3; multidrug resistance-associated protein 3	
Glyma.08g102000	A2/8	7,829,324	Cation efflux family protein	
Glyma.08g102100	A2/8	7,835,493	3-Oxoacyl-[acyl-carrier-protein] reductase 4	
Glyma.08g102200	A2/8	7,841,728	Protein of unknown function	
Glyma.08g102800	A2/8	7,882,269	Protein of unknown function	

Glyma.08g103000	A2/8	7,894,577	Urease	
Glyma.08g103200	A2/8	7,918,246	Glucosidase 1	
Glyma.08g103700	A2/8	7,958,283	Ribosomal protein L22p/L17e family protein	
Glyma.08g103800	A2/8	7,965,387	Exostosin family protein	
Glyma.08g103900	A2/8	7,973,259	Caffeoyl-CoA 3-O-methyltransferase	
Glyma.08g104100	A2/8	7,995,013	Cytochrome P450, family 78, subfamily A, polypeptide 5	
Glyma.08g104400	A2/8	8,017,355	HCP-like superfamily protein with MYND-type zinc finger	
Glyma.08g104500	A2/8	8,041,451	Protein of unknown function	
Glyma.08g104800	A2/8	8,062,435	Galactose oxidase/kelch repeat superfamily protein	
Glyma.08g105200	A2/8	8,096,014	Calmodulin-binding transcription activator protein with CG-1 and ankyrin domains	
Glyma.08g105300	A2/8	8,109,382	Protein of unknown function (DUF 3339)	
Glyma.08g105400	A2/8	8,113,537	K-box region and MADS-box transcription factor family protein	
Glyma.08g105800	A2/8	8,171,247	MAC/perforin domain-containing protein	
Glyma.08g106000	A2/8	8,191,127	Amidase family protein	
Glyma.08g106100	A2/8	8,194,373	Amidase family protein	
Glyma.08g106200	A2/8	8,195,807	Amidase family protein	
Glyma.08g106300	A2/8	8,197,293	Ubiquitin carboxyl-terminal hydrolase family protein	
Glyma.08g106400	A2/8	8,200,378	Phragmoplast-associated kinesin-related protein, putative	
Glyma.08g106500	A2/8	8,208,301	Tetratricopeptide repeat (TPR)-like superfamily protein	
Glyma.08g106800	A2/8	8,221,589	Ribosomal protein S19	
Glyma.08g107100	A2/8	8,233,645	Pentatricopeptide repeat (PPR) superfamily protein	
Glyma.08g107500	A2/8	8,267,782	Indole-3-acetate beta-D-glucosyltransferase	

several candidate genes within the QTL region on chromosome 8 including *Glyma.08g102100* that encodes for a 3-oxoacyl-[acyl-carrier-protein] reductase 4, *Glyma.08g105200* that encodes for a calmodulin-binding transcription activator protein with CG-1 and ankyrin domains, *Glyma.08g103000* that encodes for a urease, *Glyma.08g103200* that encodes for a glucosidase 1, *Glyma.08g103700* that encodes for a ribosomal protein L22p/L17e family protein, *Glyma.08g10380* that encodes for an exostosin family protein, *Glyma.08g103900* that encodes for a caffeoyl-CoA 3-O-methyltransferase, Glyma.08g104100 that encodes for a cytochrome P450, *Glyma.08g105800* that encodes for a MAC/perforin domain-containing protein, *Glyma.08g106000* that encodes for an amidase family protein, *Glyma.08g107500* that encodes for an indole-3-acetate beta-D-glucosyltransferase, and many other genes (Table 5.2) (Zhang et al. 2019).

5.2 Conclusion

Since the first QTL for salt stress tolerances have been identified and mapped on the soybean genome (Lee et al. 2004; Zhang et al. 2014, 2019; Kan et al. 2015, 2016), only few studies identified candidate genes within these QTL regions (Liao et al. 2003; Luo et al. 2005a, b; Aoki et al. 2005; Sun et al. 2006, 2013; Chen et al. 2007, 2011; Liao et al. 2008a, b; Zhou et al. 2008, 2009, 2010; Zhang et al. 2009; Hao et al. 2011; Li et al. 2012; Zhai et al. 2013a, b; Huang 2013; Qi et al. 2014; Kan et al. 2015; Zeng 2016; Cao et al. 2018, Cai et al. 2019; Zeng et al. 2019; Zhang et al. 2019).

5.2.1 Candidate Genes Involved in ST

Aoki et al. (2005) discovered the novel gene *GmTDF-5* that encodes for a cytosolic leucine-zipper-like protein highly expressed in stems and lower-expanded leaves of soybean plants grown under salt stress (100 mM of NaCl solution) suggesting that the gene may be involved in coping with salt stress (Aoki et al. 2005).

Huang (2013) screened a population of soybean accessions (n = 192) using GWAS analysis and identified QTL associated with ST. Within these QTL regions, ten candidate genes have been identified including *Glyma03g32410* encoding for a pre-mRNA processing factor 39-like, *Glyma03g32800* encoding for a 26S proteasome regulatory subunit, *Glyma03g32850* encoding for a heat shock protein 70 kDa, *Glyma03g32890* encoding for a K^+/H^+ antiporter, *Glyma03g32900* encoding for a K^+/H^+ antiporter, *Glyma03g32940* encoding for a UNC93-like protein, *Glyma03g33240* encoding for a Ca^{2+}-transporting ATPase, *Glyma05g09310* encoding for a pyruvate kinase, *Glyma08g24580* encoding for a Ca2+-transporting ATPase, and *Glyma18g46940* gene encoding for a ubiquitin-conjugating enzyme E2 5-like (Table 5.2) (Huang 2013).

Using whole genome de novo sequencing and QTL mapping by resequencing methods on a Chinese wild soybean (W05), Qi et al. (2014) identified the gene *GmCHX1* involved in salt tolerance and encoding for an iron protein. It spans about 978 kb and overlaps with *Ncl* locus (Lee et al. 2004; Hamwieh and Xu 2008; Hamwieh et al. 2011; Ha et al. 2013). A gain-of-function test was performed for *GmCHX1* gene in hairy roots of another soybean accession C08 and tobacco BY-2 cells under salt stress and confirmed that it alleviates significantly salt stress symptoms via ion homeostasis process specifically reducing the ratio of Na^+/K^+ ions (Qi et al. 2014).

Kan et al. (2015) screened a natural population of landraces ($n = 191$) in four environments in China and identified QTL that control salt stress and nine candidate genes within these QTL regions including *Glyma08g09730.1* (SOS1) gene that encodes for a cell membrane Na^+/H^+ antiporter, *Glyma08g12400.1* (SOS6) gene that encodes for osmotic stress tolerance by regulation of reactive oxygen species (ROS) (Zhu et al. 2010; Sun et al. 2016; Chen et al. 2019), *Glyma06g20170.1* (CPK11) gene that encodes for a calcium-dependent protein kinase (Zhu et al. 2007), *Glyma18g47140.1* (MPK4) gene that encodes for a mitogen-activated protein kinase involved in plant immunity (Colcombet et al. 2013), *Glyma02g41850.1* (P5CS) gene that encodes for a pyrroline 5-carboxylate synthase (Signorelli and Monza 2017), *Glyma09g00460.1* (ABA3) gene that encodes for molybdenum cofactor (MoCo) required for the activation of aldehyde oxidase (AO) involved in abscisic acid (ABA) biosynthesis (Watanabe et al. 2018), *Glyma09g00490.3* (ETR1) gene that encodes for an ethyl receptor (Gamble et al. 2002), *Glyma12g34880.1* (LEA) gene that encodes for late embryogenesis abundant protein (Shao et al. 2005), and *Glyma17g02560.1* (NHX1) gene that encodes for the Na^+/H^+ antiporter (Bassil et al. 2011); however, only five (*Glyma09g00490.3* [ETR1], *Glyma09g00460.1* [ABA3], *Glyma08g12400.1* [SOS6], *Glyma18g47140.1* [MPK4], and *Glyma08g09730.1* [SOS1]) of these nine genes are associated with salt stress at the germination stage (Table 5.2) (Kan et al. 2015).

Zeng (2016) screened a RIL population derived from a cross between 'RA-452' (salt-sensitive) and 'Osage' (salt-tolerant) cultivars ($n = 124$) in Fayetteville, AR, in 2013 and identified 45 SNP markers associated with QTL that control salt stress. Within these QTL regions, they confirmed *GmSALT3* and *GmCHX1* genes on chromosome 3 and identified six candidate genes *Glyma.02G228100* encoding for a glutamine-dependent asparagine synthase (Lam et al. 2003), *Glyma.03G031400* encoding for a cytochrome P450 (Barlier et al. 2000), *Glyma.04G180300* encoding for a leucine-rich repeat (LRR) protein kinase family protein (Jones and Jones 1997), *Glyma.04G180400* encoding for an RUBP domain-containing protein (Harshavardhan et al. 2014), *Glyma.05G204600* encoding for the osmotin 34 (Capelli et al. 1997), and *Glyma.17G173200* encoding for a dihydroflavonol 4-reductase (White and Rawlinson 2001) within these QTL regions (Table 5.2) (Zeng 2016).

A good number of genes involved in salt stress tolerance in soybean and other plants have been identified including *Glyma17g11790* (*GmPAP*) that encodes for a purple acid phosphatase that enhances ROS scavenging (Liao et al. 2003);

Glyma07g18150 (*GmCAX1*) that encodes for a cation/proton exchanger (Luo et al. 2005a, b); *Glyma14g09320* (*GmDREB*), *Glyma06g04490* (*GmDREB*), *Glyma02g01300* (*GmMYB*), *Glyma10g01450* (*GmWRKY*), *Glyma04g03200* (*GmbZIP*), *Glyma13g30990* (*GmERF*), *Glyma19g28476* (*GmNAC*), *Glyma03g42450* (*GmERF*), and *Glyma07g04950* (*GmERF*) that encode for several transcription factors (TFs) (Chen et al. 2007; Liao et al. 2008a, b; Zhou et al. 2008; Zhang et al. 2009, 2010; Hao et al. 2011; Zhai et al. 2013a, b); *Glyma20g37370* (*GmNHX*) that encodes for a vacuolar Na^+/H^+ antiporter (Sun et al. 2006); *Glyma15g13030* (*GmNHX*) that encodes for a plasma membrane and vacuolar Na^+/H^+ antiporter (Zhou et al. 2009); *Glyma05g01270* (*GmUBC*) that encodes for a ubiquitin-conjugating enzyme (Zhou et al. 2010); *Glyma01g00530* (*GmHKT*) that encodes for a Na^+ and K^+ transporters (Chen et al. 2011); *Glyma05g05580* (*GmCBL*) that encodes for Ca^{2+} sensors and calcineurin B-like proteins (Li et al. 2012); and Glyma05g14760 (*GmCLC*) that encodes for a vacuolar Cl^- transporter protein (Sun et al. 2013). For more information, summaries of these ST genes identified in soybean from 2003 to 2016 can be found in Cao et al. (2018).

Cai et al. (2019) investigated VQ gene family in 57 angiosperm genomes including representatives of dicots and monocots and found a variation of the number of VQ genes from 7 for *Selaginella moellendorffii* to 74 for soybean (*Glycine max*). The VQ motif (FxxhVQxhTG) has five conserved amino acids in its sequence. The "h" is a hydrophobic residue and "x" can be any amino acid (Wang et al. 2010a). VQ genes are involved in different plant developmental processes and are expressed during flower and leaf development as well as in abiotic stress responses. In soybean, two VQ genes (*GmVQ19* and *GmVQ27*) are expressed in developing leaves. Several VQ genes showed altered expression patterns under drought, cold, and salt stresses (Cai et al. 2019).

Zeng et al. (2019) selected the most tolerant and most sensitive RILs from the 'RA-452' by 'Osage' mapping population to identify differentially expressed genes (DEGs) between the two lines using RNA sequencing (RNA-Seq) after treating the plants with 250-mM NaCl solution. At 0, 6, 12, and 24 h after treatment, they found 2374, 998, 1746, and 630 DEGs, respectively, between the two lines. They further looked at the expression of 154 DEGs common to all these treatments (0, 6, 12, and 24 h) and discovered six unregulated DEGs and seven downregulated DEGs. Among these DEGs are *Glyma.02G228100* that encodes for a glutamine-dependent asparagine synthase 1, *Glyma.03G226000* that encodes for a glycosyl hydrolase superfamily protein, *Glyma.03G031000* and *Glyma.03G031400* that encode for a cytochrome P450, *Glyma.04G180300* that encodes for a leucine-rich repeat protein kinase family protein, *Glyma.04G180400* that encodes for a BURP domain-containing protein, *Glyma.05g204600* that encodes for osmotin 34, *Glyma.08G189600* that encodes for a lipoxygenase 1, *Glyma.13G042200* that encodes for a 2-oxoglutarate and Fe(II)-dependent oxygenase (OFE) superfamily protein, and *Glyma.17G173200* that encodes for a dihydroflavonol 4-reductase (Table 5.2) (Zeng et al. 2019).

Zhang et al. (2019) identified QTL that control ST-IR, ST-GI, ST-GP, ST-GR, and ST-GP as well as SNP markers associated with QTL for these ST indices on chromosomes 8, 9, 11, 13, 14, 16, and 18, and a major QTL was found on chromosome 8. Within this QTL region, they identified several candidate genes including *Glyma.08g102100* that encodes for a 3-oxoacyl-[acyl-carrier-protein] reductase 4, *Glyma.08g105200* that encodes for a calmodulin-binding transcription activator protein with CG-1 and ankyrin domains, *Glyma.08g103000* that encodes for a urease, *Glyma.08g103200* that encodes for a glucosidase 1, *Glyma.08g103700* that encodes for a ribosomal protein L22p/L17e family protein, *Glyma.08g10380* that encodes for an exostosin family protein, *Glyma.08g103900* that encodes for a caffeoyl-CoA 3-O-methyltransferase, Glyma.08g104100 that encodes for a cytochrome P450, *Glyma.08g105800* that encodes for a MA/perforin domain-containing protein, *Glyma.08g106000* that encodes for an amidase family protein, *Glyma.08g107500* that encodes for an indole-3-acetate beta-D-glucosyltransferase, and many other genes (Table 5.2) (Zhang et al. 2019).

The *J*-allele of soybean is an ortholog of the *AtELF3* (early flowering) gene of *Arabidopsis thaliana*. Cheng et al. (2020) investigated the role of J-allele in salt stress tolerance. The study found that salt stress induced highly the expression of the nuclear J gene. They further developed NILs around the J locus and found that soybean plants carrying the J-allele (NIL-J) are highly tolerant to salt stress than soybean plants carrying the j-allele (NIL-j) and transgenic hairy roots expressing *J* were also highly salt tolerant. Also, expression levels of several genes such as *GmWRKY12* (Shi et al. 2018a, b), *GmWRKY54* (Zhou et al. 2008), *GmWRKY27* (Wang et al. 2015a), *GmNAC* (Hussain et al. 2017; Melo et al. 2018), and *GmSIN1* (Li et al. 2019) involved in salt and drought stress responses have been positively induced by the gene *J* (Cheng et al. 2020).

References

An, P., S. Inanaga, Y. Cohen, U. Kafkafi, and Y. Sugimoto. 2001. Different effect of humidity on growth and salt tolerance of two soybean cultivars. *Biologa Plantarum* 44: 405–410.

———. 2002. Salt tolerance in two soybean cultivars. *Journal of Plant Nutrition* 25: 407–423.

Aoki, A., A. Kanegami, M. Mihara, T. Kojima, M. Shiraiwa, and H. Takahara. 2005. Molecular characterization of a novel soybean gene encoding a leucine-zipper-like protein induced to salt stress. *Gene* 356: 135–145.

Barlier, I., M. Kowalczyk, A. Marchant, K. Ljung, R. Bhalerao, M. Bennett, G. Sandberg, and C. Bellini. 2000. The SUR2 gene of Arabidopsis thaliana encodes the cytochrome P450 CYP83B1, a modulator of auxin homeostasis. *PNAS* 97 (26): 14819–14824. https://doi.org/10.1073/pnas.260502697.

Bassil, E., H. Tajima, Y.C. Liang, M.A. Ohto, K. Ushijima, R. Nakano, T. Esumi, A. Coku, M. Belmonte, and E. Blumwald. 2011. The Arabidopsis Na^+/H^+ Antiporters NHX1 and NHX2 control vacuolar pH and K+ homeostasis to regulate growth, flower development, and reproduction. *Plant Cell* 23 (9): 3482–3497. https://doi.org/10.1105/tpc.111.089581.

Cai, H., M. Zhang, Y. Liu, Q. He, M. Chai, L. Liu, F. Chen, Y. Huang, M. Yan, H. Zhao, J. Hu, and Y. Qin. 2019. Genome-wide classification and evolutionary and functional analyses of the VQ family. *Tropical Plant Biology*. https://doi.org/10.1007/s12042-019-09224-4.

Cao, D., Y. Li, B. Liu, F. Kong, and L.S.P. Tran. 2018. Adaptive mechanisms of soybean growth on salt-affected soils. *Land Degradation and Development* 29 (4): 1054–1064. https://doi.org/10.1002/ldr.2754.

Cao, C., Y.L. Yan, and D.H. Xu. 2019. Assessment of salt tolerance and analysis of the salt tolerance gene *Ncl* in Indonesian soybean germplasm. *Plant Genetic Resources: Characterization and Utilization* 17 (3): 265–271. https://doi.org/10.1017/S1479262118000618.

Capelli, N., T. Diogon, H. Greppin, and P. Simon. 1997. Isolation and characterization of a cDNA clone encoding an osmotin-like protein from *Arabidopsis thaliana*. *Gene* 191 (1): 51–56.

Chen, M., Q. Wang, X. Cheng, Z. Xu, L. Li, X. Ye, L. Xia, and Y. Ma. 2007. GmDREB2, a soybean DRE-binding transcription factor, conferred drought and high-salt tolerance in transgenic plants. *Biochemical and Biophysical Research Communications* 353: 299–305. https://doi.org/10.1016/j.bbrc.2006.12.027.

Chen, H., S. Cui, S. Fu, J. Gai, and D. Yu. 2008. Identification of quantitative trait loci associated with salt tolerance during seedling growth in soybean (*Glycine max* L.). *Australian Journal of Agricultural Research* 59: 1086–1091. https://doi.org/10.1071/AR08104.

Chen, H., H. He, and D. Yu. 2011. Overexpression of a novel soybean gene modulating Na plus and K plus transport enhances salt tolerance in transgenic tobacco plants. *Physiologia Plantarum* 141: 11–18. https://doi.org/10.1111/j.1399-3054.2010.01412.x.

Chen, G., W. Yan, L. Yang, J. Gai, and Y. Zhu. 2014a. Overexpression of StNHX1, a novel vacuolar Na^+/H^+ antiporter gene from *Solanum torvum*, enhances salt tolerance in transgenic vegetable soybean. *Horticulture Environment and Biotechnology* 55: 213–221. https://doi.org/10.1007/s13580-014-0003-z.

Chen, H., X. Chen, H. Gu, B. Wu, H. Zhang, X. Yuan, and X. Cui. 2014b. GmHKT1;4, a novel soybean gene regulating Na^+/K^+ ratio in roots enhances salt tolerance in transgenic plants. *Plant Growth Regulation* 73: 299–308. https://doi.org/10.1007/s10725-014-9890-3.

Chen, H., H. Ye, T.D. Do, J. Zhou, B. Valliyodan, G.J. Shannon, P. Chen, X. Chen, and H.T. Nguyen. 2018a. Advances in genetics and breeding of salt tolerance in soybean. In *Salinity responses and tolerance in plants*, ed. V. Kumar, S. Wani, P. Suprasanna, and L.S. Tran, vol. Volume 2. Cham: Springer. https://doi.org/10.1007/978-3-319-90318-7_9.

Chen, H., X.Q. Liu, H.M. Zhang, X.X. Yuan, H.P. Gu, X. Cui, and X. Chen. 2018b. Advances in salinity tolerance of soybean: Genetic diversity, heredity, and gene identification contribute to improving salinity tolerance. *Journal of Integrative Agriculture* 17 (10): 2215–2221. https://doi.org/10.1016/S2095-3119(17)61864-1.

Chen K., M. Song, Y. Guo, L. Liu, H. Xue, H. Dai, and Z. Zhang. 2019. MdMYB46 could enhance salt and osmotic stress tolerance in apple by directly activating stress-responsive signals. *Plant Biotechnology Journal* 17 (12): 2341–2355. https://doi.org/10.1111/pbi.13151.

Cheng, Q., Z. Gan, Y. Wang, S. Lu, Z. Hou, H. Li, H. Xiang, B. Liu, F. Kong, and L. Dong. 2020. The soybean gene *J* contributes to salt stress tolerance by up-regulating salt-responsive genes. *Frontiers in Plant Science* 11: 272. https://doi.org/10.3389/fpls.2020.00272.

Colcombet, J., S. Berriri, and H. Hirt. 2013. Constitutively active MPK4 helps to clarify its role in plant immunity. *Plant Signaling and Behavior* 8: e22991. http://dx.doi.org10.4161/psb.22991.

Deshmukh R., H. Sonah, G. Patil, W. Chen, S. Prince, R. Mutava, T. Vuong, B. Valliyodan, and H.T. Nguyen. 2014. Integrating omic approaches for abiotic stress tolerance in soybean. *Frontiers in Plant Science*. https://doi.org/10.3389/fpls.2014.00244.

Do, T.D., S.K. Lal, and D.H. Xu. 2010. Identification of a major QTL allele from wild soybean (*Glycine soja* Sieb. & Zucc.) for increasing alkaline salt tolerance in soybean. *Theoretical and Applied Genetics* 121: 229–236. https://doi.org/10.1007/s00122-010-1304-y.

Do, T.D., H.M. Zhang, and D.H. Xu. 2013. Validation and high-resolution mapping of a major quantitative trait locus for alkaline salt tolerance in soybean using residual heterozygous line. *Molecular Breeding* 31: 79–86. https://doi.org/10.1007/s11032-012-9771-2.

Do, T.D., H. Chen, H.T.T. Vu, A. Hamwieh, T. Yamada, T. Sato, Y. Yan, H. Cong, M. Shono, K. Suenaga, and D. Xu. 2016. *Ncl* synchronously regulates Na^+, K^+, and Cl^- in soybean and greatly increases the grain yield in saline field conditions. *Scientific Reports* 6: 19147. https://doi.org/10.1038/srep19147.

Do, T.D., T.D. Vuong, D. Dunn, S. Smothers, G. Patil, D.C. Yungbluth, P. Chen, A. Scaboo, D. Xu, T.E. Carter, H.T. Nguyen, and J.G. Shannon. 2018. Mapping and confirmation of loci for salt tolerance in a novel soybean germplasm, Fiskeby III. *Theoretical and Applied Genetics* 131: 513–524. https://doi.org/10.1007/s00122-017-3015-0.

Do, T.D., T.D. Vuong, D. Dunn, M. Clubb, B. Valliyodan, G. Patil, P. Chen, D. Xu, H.T. Nguyen, and J.G. Shannon. 2019. Identification of new loci for salt tolerance in soybean by high-resolution genome-wide association mapping. *BMC Genomics* 20: 318. https://doi.org/10.1186/s12864-019-5662-9.

Dong L., Y. Cheng, J. Wu, Q. Cheng, W. Li, S. Fan, L. Jiang, Z. Xu, F. Kong, D. Zhang, P. Xu, and S. Zhang. 2015. Overexpression of GmERF5, a new member of the soybean EAR motif-containing ERF transcription factor, enhances resistance to Phytophthora sojae in soybean. *Journal of Experimental Botany* 66: 2635–2647. https://doi.org/10.1093/jxb/erv078.

Essa, T.A. 2002. Effect of salinity stress on growth and nutrient composition of three soybean (*Glycine max* L. Merrill) cultivars. *Journal of Agronomy and Crop Science* 188: 86–93.

Flowers, T.J., and A.R. Yeo. 1995. Breeding for salinity resistance in crop plants. Where next? *Australian Journal of Plant Physiology* 22: 875–884.

Gamble, R.L., X. Qu, and G. Schaller. 2002. Mutational analysis of the ethylene receptor ETR1. Role of the histidine kinase domain in dominant ethylene insensitivity. *Plant Physiology* 128 (4): 1428–1438. https://doi.org/10.1104/pp.010777.

Gao, S., H. Xu, X. Cheng, M. Chen, Z. Xu, L. Li, X. Ye, L. Du, X. Hao, and Y. Ma. 2005. Improvement of wheat drought and salt tolerance by expression of a stress-inducible transcription factor GmDREB of soybean (*Glycine max*). *Chinese Science Bulletin* 50: 2714–2723. https://doi.org/10.1007/BF02899641.

Grant D., R.T. Nelson, S.B. Cannon, and R.C. Shoemaker. 2010. SoyBase, the USDA-ARS soybean genetics and genomics database. *Nucleic Acids Research* 38 (suppl 1): D843–D846. https://doi.org/10.1093/nar/gkp798.

Guan, R., J. Chen, J. Jiang, G. Liu, Y. Liu, L. Tian, L. Yu, R. Chang, and L.J. Qiu. 2014a. Mapping and validation of a dominant salt tolerance gene in the cultivated soybean (*Glycine max*) variety Tiefeng 8. *The Crop Journal* 2: 358–365. https://doi.org/10.1016/j.cj.2014.09.001.

Guan, R., Y. Qu, Y. Guo, L. Yu, Y. Liu, J. Jiang, J. Chen, Y. Ren, G. Liu, L. Tian, L. Jin, Z. Liu, H. Hong, R. Chang, M. Gilliham, and L. Qiu. 2014b. Salinity tolerance in soybean is modulated by natural variation in *GmSALT3*. *The Plant Journal* 80: 937–950. https://doi.org/10.1111/tpj.12695.

Ha, B.K., T.D. Vuong, V. Velusamy, H.T. Nguyen, J.G. Shannon, and J.D. Lee. 2013. Genetic mapping of quantitative trait loci conditioning salt tolerance in wild soybean (Glycine soja) PI 483463. *Euphytica* 193: 79–88. https://doi.org/10.1007/s10681-013-0944-9.

Hamwieh, A., and D.H. Xu. 2008. Conserved salt tolerance quantitative trait locus (QTL) in wild and cultivated soybeans. *Breeding Science* 58: 355–359.

Hamwieh, A., T.D. Do, H. Cong, E.R. Benitez, R. Takahashi, and D.H. Xu. 2011. Identification and validation of a major QTL for salt tolerance in soybean. *Euphytica* 179: 451–459. https://doi.org/10.1007/s10681-011-0347-8.

Hao, Y., W. Wei, Q. Song, H. Chen, Y. Zhang, F. Wang, H. Zou, G. Lei, A. Tian, W. Zhang, B. Ma, J. Zhang, and S. Chen. 2011. Soybean NAC transcription factors promote abiotic stress tolerance and lateral root formation in transgenic plants. *Plant Journal* 68: 302–313. https://doi.org/10.1111/j.1365-313X.2011.04687.x.

Harshavardhan, V.T., L.V. Son, C. Seiler, A. Junker, K. Weigelt-Fischer, C. Klukas, T. Altmann, N. Sreenivasulu, H. Bäumlein, and M. Kuhlmann. 2014. AtRD22 and AtUSPL1, members of the plant-specific BURP domain family involved in *Arabidopsis thaliana* drought tolerance. *PLoS One* 9 (10): e110065. https://doi.org/10.1371/journal.pone.0110065.

Huang, L. 2013. *Genome-wide association mapping identifies QTLs and candidate genes for salt tolerance in soybean*. MS thesis. University of Arkansas, Fayetteville, AR.

Huang, L., A. Zeng, P. Chen, C. Wu, D. Wang, and Z. Wen. 2018. Genome-wide association analysis of salt tolerance in soybean [*Glycine max* (L.) Merr.]. *Plant Breeding* 137 (5): 714–720. https://doi.org/10.1111/pbr.12623.

Hussain, R.M., M. Ali, X. Feng, and X. Li. 2017. The essence of NAC gene family to the cultivation of drought-resistant soybean (*Glycine max* L. Merr.) cultivars. *BMC Plant Biology* 17 (1): 55. https://doi.org/10.1186/s12870-017-1001-y.

Jones, D.A., and J.D.G. Jones. 1997. The role of leucine-rich repeat proteins in plants. *Advances in Botanical Research* 24: 89–167.

Kan, G., W. Zhang, W. Yang, D. Ma, D. Zhang, D. Hao, Z. Hu, and D. Yu. 2015. Association mapping of soybean seed germination under salt stress. *Molecular Genetics and Genomics* 290: 2147–2162. https://doi.org/10.1007/s00438-015-1066-y.

Kan, G., L. Ning, Y. Li, Z. Hu, W. Zhang, X. He, and D. Yu. 2016. Identification of novel loci for salt stress at the seed germination stage in soybean. *Breeding Science*. http://doi.or/10.1270/jsbbs.15147.

Kao, W.Y., T.T. Tsai, H.C. Tsai, and C.N. Shih. 2006. Response of three Glycine species to salt stress. *Environmental and Experimental Botany* 56: 120–125.

Lam, H.M., P. Wong, H.K. Chan, K.M. Yam, L. Chen, C.M. Chow, and G.M. Coruzzi. 2003. Overexpression of the ASN1 gene enhances nitrogen status in seeds of Arabidopsis. *Plant Physiology* 132: 926–935. http://doi.org/.1104/pp.103.020123.

Lee, G.J., H.R. Boerma, M.R. Villagarcia, X. Zhou, T.E. Carter, Z. Li, and M.O. Gibbs. 2004. A major QTL conditioning salt tolerance in S-100 soybean and descendent cultivars. *Theoretical and Applied Genetics* 109: 1610–1619.

Lee, J.D., J.G. Shannon, T.D. Vuong, and H.T. Nguyen. 2009. Inheritance of salt tolerance in wild soybean (Glycine soda Sieb. And Zucc.) accession PI483463. *Journal of Heredity* 100 (6): 798–801. https://doi.org/10.1093/jhered/esp027.

Lee, S., J.H. Kim, J. Sundaramoorthy, G.T. Park, J.D. Lee, J.H. Kim, G. Chung, H.S. Seo, and J.T. Song. 2018. Identification of *GmSALT3* haplotypes and development of molecular markers based on their diversity associated with salt tolerance in soybean. *Molecular Breeding* 38: 86. https://doi.org/10.1007/s11032-018-0845-7.

Li, H., G. Ye, and J. Wang. 2007. A modified algorithm for the improvement of composite interval mapping. *Genetics* 175: 361–374.

Li, Z., Z. Xu, G. He, G. Yang, M. Chen, L. Li, and Y. Ma. 2012. Overexpression of soybean GmCBL1 enhances abiotic stress tolerance and promotes hypocotyl elongation in Arabidopsis. *Biochemical and Biophysical Research Communications* 427: 731–736. https://doi.org/10.1016/j.bbrc.2012.09.128.

Li, S., N. Wang, D. Ji, W. Zhang, Y. Wang, Y. Yu, S. Zhao, M. Lyu, J. You, Y. Zhang, L. Wang, X. Wang, Z. Liu, J. Tong, L. Xiao, M.Y. Bai, and F. Xiang. 2019. A GmSIN1/GmNCED3s/GmRbohBs feed-forward loop actsas a signal amplifier that regulates root growth in soybean exposed to salt stress [OPEN]. *The Plant Cell* 31: 2107–2130.

Liao, H., F. Wong, T. Phang, M. Cheung, W.F. Li, S. Shao, X. Yan, and H. Lam. 2003. GmPAP3, a novel purple acid phosphatase-like gene in soybean induced by NaCl stress but not phosphorus deficiency. *Gene* 318: 103–111. https://doi.org/10.1016/S0378-1119(03)00764-9.

Liao, Y., H. Zou, H. Wang, W. Zhang, B. Ma, J. Zhang, and S. Chen. 2008a. Soybean GmMYB76, GmMYB92, and GmMYB177 genes confer stress tolerance in transgenic Arabidopsis plants. *Cell Research* 18: 1047–1060. https://doi.org/10.1038/cr.2008.280.

Liao, Y., H. Zou, W. Wei, Y. Hao, A. Tian, J. Huang, Y. Liu, J. Zhang, and S. Chen. 2008b. Soybean GmbZIP44, GmbZIP62 and GmbZIP78 genes function as negative regulator of ABA signaling and confer salt and freezing tolerance in transgenic Arabidopsis. *Planta* 228: 225–240. https://doi.org/10.1007/s00425-008-0731-3.

Liu, Y., L. Yu, Y. Qu, J. Chen, X. Liu, H. Hong, Z. Liu, R. Chang, M. Gilliham, L. Qiu, and R. Guan. 2016. GmSALT3, which confers improved soybean salt tolerance in the field, increases leaf Cl^- exclusion prior to Na^+ exclusion but does not improve early vigor under salinity. *Frontiers in Plant Science* 7: 1485. https://doi.org/10.3389/fpls.2016.01485.

Lopez, C., M. Orazalya, L. Mozzonia, K.L. Korthb, and P. Chen. 2018. Quantitative trait loci for salt tolerance in soybean. *Journal of Crop Improvement* 32 (6): 766–780. https://doi.org/10.1080/15427528.2018.1516261.

Luo, Q., B. Yu, and Y. Liu. 2005a. Differential sensitivity to chloride and sodium ions in seedlings of *Glycine max* and *G. soja* under NaCl stress. *Journal of Plant Physiology* 162: 1003–1012. https://doi.org/10.1016/j.jplph.2004.11.008.

Luo, G., H. Wang, J. Huang, A. Tian, Y. Wang, J. Zhang, and S. Chen. 2005b. A putative plasma membrane cation/proton antiporter from soybean confers salt tolerance in Arabidopsis. *Plant Molecular Biology* 59: 809–820. https://doi.org/10.1007/s11103-005-1386-0.

Melo, B.P., O.T. Fraga, J.C.F. Silva, D.O. Ferreira, O.J.B. Brustolini, P.A. Carpinetti, J.P.B. Machado, P.A.B. Reis, and E.P.B. Fontes. 2018. Revisiting the soybean GmNAC superfamily. *Frontiers in Plant Science* 9: 1864. https://doi.org/10.3389/fpls.2018.01864.

Nelson, J.C. 1997. QGENE software for marker-based genomic analysis and breeding. *Molecular Breeding* 3: 239–245.

Nie, W., L. Xu, and B. Yu. 2015. A putative soybean GmsSOS1 confers enhanced salt tolerance to transgenic Arabidopsis sos1-1 mutant. *Protoplasma* 252: 127–134. https://doi.org/10.1007/s00709-014-0663-7.

Pathan, M.S., J.D. Lee, J.G. Shannon, and H.T. Nguyen. 2007. Recent advances in breeding for drought and salt stress tolerance in soybean. In *Advances in molecular breeding towards drought and salt tolerance in crops*, ed. M.A. Jenks, P.M. Hasegawa, and S.M. Jain, 739–774. Dordrecht: Springer.

Patil, G., T.D. Do, T.D. Vuong, B. Valliyodan, J.D. Lee, J. Chaudhary, J.G. Shannon, and H.T. Nguyen. 2016. Genomic-assisted haplotype analysis and the development of high-throughput SNP markers for salinity tolerance in soybean. *Scientific Reports* 6: 19199. https://doi.org/10.1038/srep19199.

Phang, T.H., G. Shao, and H.M. Lam. 2008. Salt tolerance in soybean. *Journal of Integrative Plant Biology* 50 (10): 1196–1212. https://doi.org/10.1111/j.1744-7909.2008.00760.x.

Qi, X., M.W. Li, M. Xie, X. Liu, M. Ni, G. Shao, C. Song, A.K.Y. Yim, Y. Tao, F.L. Wong, S. Isobe, C.F. Wong, K.S. Wong, C. Xu, C. Li, Y. Wang, R. Guan, F. Sun, G. Fan, Z. Xiao, F. Zhou, T.H. Phang, X. Liu, S.W. Tong, T.F. Chan, S.M. Yiu, S. Tabata, J. Wang, X. Xu, and H.M. Lam. 2014. Identification of a novel salt tolerance gene in wild soybean by whole-genome sequencing. *Nature Communications* 5: 4340. https://doi.org/10.1038/ncomms5340.

Rao, S.S., M.H. El-Habbak, W.M. Havens, A. Singh, D.M. Zheng, L. Vaughn, J.S. Haudenshield, G.L. Hartman, S.S. Korban, and S.A. Ghabrial. 2014. Overexpression of GmCaM4 in soybean enhances resistance to pathogens and tolerance to salt stress. *Molecular Plant Pathology* 15: 145–160. https://doi.org/10.1111/mpp.12075.

Shao, H.B., Z.S. Liang, and M.A. Shao. 2005. LEA proteins in higher plants: Structure, function, gene expression and regulation. *Colloids and Surfaces. B, Biointerfaces* 45 (3–4): 131–135. https://doi.org/10.1016/j.colsurfb.2005.07.017.

Shi, W.Y., Y.T. Du, J. Ma, D.H. Min, L.G. Jin, J. Chen, M. Chen, Y.B. Zhou, Y.Z. Ma, Z.S. Xu, and X.H. Zhang. 2018a. The WRKY transcription factor GmWRKY12 confers drought and salt tolerance in soybean. *International Journal of Molecular Sciences* 19 (12): 4087. https://doi.org/10.3390/ijms19124087.

Shi, X., L. Yan, C.Y. Yang, W.W. Yan, D.O. Moseley, T. Wang, B.Q. Liu, R. Di, P.Y. Chen, and M.C. Zhang. 2018b. Identification of a major quantitative trait locus underlying salt tolerance in 'Jidou 12' soybean cultivar. *BMC Research Notes* 11: 95. https://doi.org/10.1186/s13104-018-3202-3.

Signorelli, S., and J. Monza. 2017. Identification of Δ1-pyrroline 5-carboxylate synthase (P5CS) genes involved in the synthesis of proline in *Lotus japonicus*. *Plant Signaling and Behavior* 12 (11): e1367464. https://doi.org/10.1080/15592324.2017.1367464.

Sun, Y., D. Wang, Y.L. Bai, N.N. Wang, and Y. Wang. 2006. Studies on the overexpression of the soybean GmNHX1 in Lotus corniculatus: The reduced Na+ level is the basis of the increased salt tolerance. *Chinese Science Bulletin* 51: 1306–1315. https://doi.org/10.1007/s11434-006-1306-y.

Sun, X., Q. Yu, L. Tang, W. Ji, X. Bai, H. Cai, X. Liu, X. Ding, and Y. Zhu. 2013. GsSRK, a G-type lectin S-receptor-like serine/threonine protein kinase, is a positive regulator of plant tolerance to salt stress. *Journal of Plant Physiology* 170: 505–515. https://doi.org/10.1016/j.jplph.2012.11.017.

Sun, X., M. Sun, B. Jia, Z. Qin, K. Yang, C. Chen, Q. Yu, and Y. Zhu. 2016. A *Glycine soja* methionine sulfoxide reductase B5a interacts with the Ca^{2+}/CAM-binding kinase GsCBRLK and activates ROS signaling under carbonate alkaline stress. *Plant Journal* 86: 514–529. https://doi.org/10.1111/tpj.13187.

Van Ooijen, J.M. 2004. Software for the mapping of quantitative trait loci in experimental populations. MapQTL5 Kyazma BV, Wageningen. The Netherlands.

Valliyodan, B., D. Qiu, G. Patil, P. Zeng, J. Huang, Dai Lu, C. Chen, Y. Li, T. Joshi, L. Song, T.D. Vuong, T.A. Musket, D. Xu, J.G. Shannon, C. Shifeng, X. Liu, and H.T. Nguyen. 2016. Landscape of genomic diversity and trait discovery in soybean. *Scientific Reports* 6: 23598. https://doi.org/10.1038/srep23598.

Wang, A., D. Garcia, H. Zhang, K. Feng, A. Chaudhury, F. Berger, W.J. Peacock, E.S. Dennis, and M. Luo. 2010a. The VQ motif protein IKU1 regulates endosperm growth and seed size in Arabidopsis. *The Plant Journal* 63 (4): 670–679. https://doi.org/10.1111/j.1365-313X.2010.04271.x.

Wang, Z., M. Libault, T. Joshi, B. Valliyodan, H.T. Nguyen, D. Xu, G. Stacey, and J. Cheng. 2010b. SoyDB: A knowledge database of soybean transcription factors. *BMC Plant Biology* 10. https://doi.org/10.1186/1471-2229-10-14.

Wang, F., H. Chen, Q. Li, W. Wei, W. Li, W. Zhang, B. Ma, Y. Bi, Y. Lai, X. Liu, W. Man, J. Zhang, and S. Chen. 2015a. GmWRKY27 interacts with GmMYB174 to reduce expression of GmNAC29 for stress tolerance in soybean plants. *Plant Journal* 83: 224–236. https://doi.org/10.1111/tpj.12879.

Wang, F., H.W. Chen, Q.T. Li, W. Wei, W. Li, W.K. Zhang, B. Ma, Y.D. Bi, Y.C. Lai, X.L. Liu, W.Q. Man, J.S. Zhang, and S.Y. Chen. 2015b. GmWRKY27 interacts with GmMYB174 to reduce expression of GmNAC29 for stress tolerance in soybean plants. *The Plant Journal* 83 (2): 224–236.

Watanabe, S., M. Sato, Y. Sawada, M. Tanaka, A. Matsui, Y. Kanno, M.Y. Hirai, M. Seki, A. Sakamoto, and M. Seo. 2018. Arabidopsis molybdenum cofactor sulfurase ABA3 contributes to anthocyanin accumulation and oxidative stress tolerance in ABA-dependent and independent ways. *Scientific Reports* 8: 16592. https://doi.org/10.1038/s41598-018-34862-1.

White, P.A., and W.D. Rawlinson. 2001. Current status of the aadA and dfr gene cassette families. *Journal of Antimicrobial Chemotherapy* 47 (4): 495–496. https://doi.org/10.1093/jac/47.4.495.

Xu, D., and T.D. Do. 2012. Genetic studies on saline and sodic tolerances in soybean. *Breeding Science* 61: 559–656. https://doi.org/10.1270/jsbbs.61.559.

Xu, Z., Z. Ali, L. Xu, X.L. He, Y. Huang, J.X. Yi, H. Shao, H. Ma, and D. Zhang. 2016. The nuclear protein GmbZIP110 has transcription activation activity and plays important roles in the response to salinity stress in soybean. *Scientific Reports* 6: 20366. https://doi.org/10.1038/srep20366.

Yan, J., B. Wang, Y. Jiang, L. Cheng, and T. Wu. 2014. GmFNSII-controlled soybean flavone metabolism responds to abiotic stresses and regulates plant salt tolerance. *Plant and Cell Physiology* 55: 74–86. https://doi.org/10.1093/pcp/pct159.

Zeng, A. 2016. *Quantitative trait loci mapping, genome-wide association analysis, and gene expression of salt tolerance in soybean*. Theses and Dissertations, University of Arkansas. http://scholarworks.uark.edu/etd/1778.

Zeng, A., P. Chen, K. Korth, F. Hancock, A. Pereira, K. Brye, C. Wu, and A. Shi. 2017. Genome-wide association study (GWAS) of salt stress tolerance in worldwide soybean germplasm lines. *Molecular Breeding*. https://doi.org/10.1007/s11032-017-0634-8.

Zeng, A., P. Chen, K.L. Korth, J. Ping, J. Thomas, C. Wu, S. Srivastava, A. Pereira, F. Hancock, K. Brye, and J. Ma. 2019. RNA sequencing analysis of salt tolerance in soybean (*Glycine max*). *Genomics* 114 (4): 629–635. https://doi.org/10.1016/j.ygeno.2018.03.020.

Zhai, Y., J. Li, X. Li, T. Lei, F. Yan, Y. Zhao, Y. Li, L. Su, Y. Wang, and Q. Wang. 2013a. Isolation and characterization of a novel transcriptional repressor GmERF6 from soybean. *Biologia Plantarum* 57: 26–32. https://doi.org/10.1007/s10535-012-0146-7.

Zhai, Y., Y. Wang, Y. Li, T. Lei, F. Yan, L. Su, X. Li, Y. Zhao, X. Sun, J. Li, and Q. Wang. 2013b. Isolation and molecular characterization of GmERF7, a soybean ethylene-response factor that increases salt stress tolerance in tobacco. *Gene* 513: 174–183. https://doi.org/10.1016/j.gene.2012.10.018.

Zhang, G., M. Chen, L. Li, Z. Xu, X. Chen, J. Guo, and Y. Ma. 2009. Overexpression of the soybean GmERF3 gene, an AP2/ERF type transcription factor for increased tolerances to salt, drought, and diseases in transgenic tobacco. *Journal of Experimental Botany* 60: 3781–3796. https://doi.org/10.1093/jxb/erp214.

Zhang, W.J., Y. Niu, S.H. Bu, M. Li, J.Y. Feng, J. Zhang, S.X. Yang, M.M. Odinga, S.P. Wei, X.F. Liu, and Y.M. Zhang. 2014. Epistatic association mapping for alkaline and salinity tolerance traits in the soybean germination stage. *PLoS One* 9 (1): e84750. https://doi.org/10.1371/journal.pone.0084750.

Zhang, W., X. Liao, Y. Cui, W. Ma, X. Zhang, H. Du, Y. Ma, L. Ning, H. Wang, F. Huang, H. Yang, G. Kan, and D. Yu. 2019. A cation diffusion facilitator, *GmCDF1*, negatively regulates salt tolerance in soybean. *PLoS Genetics* 15 (1): e1007798. https://doi.org/10.1371/journal.pgen.1007798.

Zhou, Q.Y., A.G. Tian, H.F. Zou, Z.M. Xie, G. Lei, J. Huang, C.M. Wang, H.W. Wang, J.S. Zhang, and S.Y. Chen. 2008. Soybean WRKY-type transcription factor genes, GmWRKY13, GmWRKY21, and GmWRKY54, confer differential tolerance to abiotic stresses in transgenic Arabidopsis plants. *Plant Biotechnology Journal* 6 (5): 486–503. https://doi.org/10.1111/j.1467-7652.2008.00336.x.

Zhou, G., R. Guan, Y. Li, R. Chang, and L. Qiu. 2009. Molecular characterization of GmNHX2, a Na^+/H^+ antiporter gene homolog from soybean, and its heterologous expression to improve salt tolerance in Arabidopsis. *Chinese Science Bulletin* 54: 3536–3545. https://doi.org/10.1007/s11434-009-0564-x.

Zhou, G., R. Chang, and L. Qiu. 2010. Overexpression of soybean ubiquitin-conjugating enzyme gene GmUBC2 confers enhanced drought and salt tolerance through modulating abiotic stress-responsive gene expression in Arabidopsis. *Plant Molecular Biology* 72: 357–367. https://doi.org/10.1007/s11103-009-9575-x.

Zhou, L., C. Wang, R. Liu, Q. Han, R.K. Vandeleur, J. Du, S. Tyerman, and H. Shou. 2014. Constitutive overexpression of soybean plasma membrane intrinsic protein GmPIP1;6 confers salt tolerance. *BMC Plant Biology* 14: 181. https://doi.org/10.1186/1471-2229-14-181.

Zhu, S.Y., X.C. Yu, X.J. Wang, R. Zhao, Y. Li, R.C. Fan, Y. Shang, S.Y. Du, X.F. Wang, F.Q. Wu, Y.H. Xu, X.Y. Zhang, and D.P. Zhang. 2007. Two calcium-dependent protein kinases, CPK4 and CPK11, regulate abscisic acid signal transduction in Arabidopsis. *Plant Cell* 19 (10): 3019–3036. https://doi.org/10.1105/tpc.107.050666.

Zhu, J., B.H. Lee, M. Dellinger, X. Cui, C. Zhang, S. Wu, E.A. Nothnagel, and J.K. Zhu. 2010. A cellulose synthase-like protein is required for osmotic stress tolerance in Arabidopsis. *Plant Journal* 63 (1): 128–140. https://doi.org/10.1111/j.1365-313X.2010.04227.x.

Chapter 6
Isoflavone Biosynthetic Pathways and Methods of Quantification

Moulay Abdelmajid Kassem

6.1 Isoflavone Biosynthesis

Phytoestrogens or isoflavones are secondary metabolites abundant in soybean (*Glycine max* (L.) Merr.) and other legumes' seeds but also in several nonlegume plants. Among the most studies, isoflavones in recent decades are genistein, daidzein, and glycitein because of their pharmacological properties and benefits to human health (Hoeck et al. 2000; Dixon and Summer 2003; Jiang et al. 2010; Wahajuddin et al. 2013; Wang et al. 2009; Chen et al. 2014; Dhaubhadel et al. 2003; Lozovaya et al. 2004a, b, 2005; Mebrahtu et al. 2004). It is common knowledge that isoflavones including genistein, daidzein, and glycitein are synthesized through the phenylpropanoid biosynthetic pathway (Fig. 6.1) (Graham 1991; Dixon and Paiva 1995; Cassidy et al. 2000; Winkel-Shirley 2001; Yu et al. 2003; Graham et al. 2008; Wang 2010; Du et al. 2010).

Isoflavones have several roles in plant-microbe interactions including defense against pathogens, attracting *Rhizobia* bacteria and inducing the expression of *nod* genes involved in nodule formation or root nodulation (Van Rhijn and Vanderleyden 1995; Dixon 2001; Yu et al. 2003; Pueppke 1996; Zhang and Smith 1996; Pueppke et al. 1998; Subramanian et al. 2006; Du et al. 2010). They also serve as phytoalexin precursors (Du et al. 2010).

The phenylpropanoid biosynthetic pathway gives rise to nine major subgroups of compounds: the chalcones, phlobaphene pigments, flavones, flavandiols, condensed tannins, flavonols, aurones, anthocyanins, and isoflavonoids (Winkel-Shirley 2001; Dixon 2001; Wang 2010; Du et al. 2010). The biosynthesis of the specific

M. A. Kassem (✉)
Plant Genomics and Biotechnology Lab, Department of Biological and Forensic Sciences, Fayetteville State University, Fayetteville, NC, USA
e-mail: mkassem@uncfsu.edu

M. A. Kassem (ed.), *Soybean Seed Composition*,
https://doi.org/10.1007/978-3-030-82906-3_6

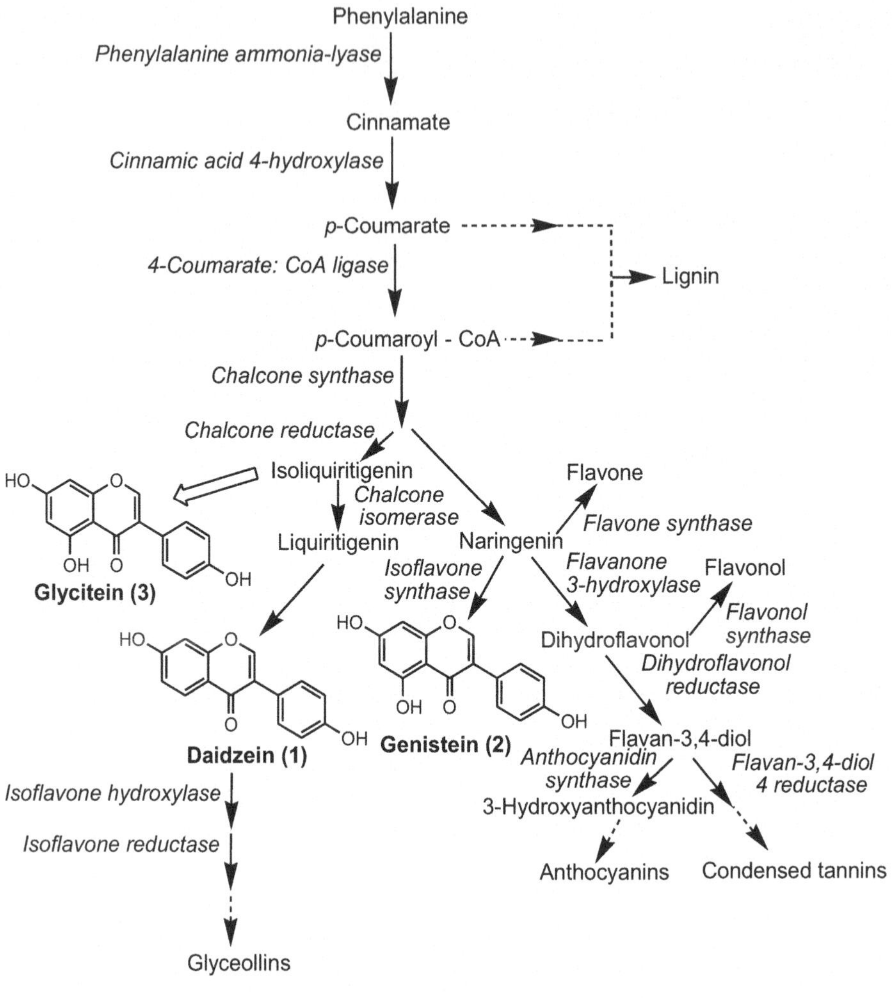

Fig. 6.1 The phenylpropanoid biosynthetic pathway showing the major isoflavones daidzein, genistein, and glycitein. (Adopted from Yu et al. (2000))

isoflavones genistein, daidzein, and glycitein is illustrated in Fig. 6.1 (Yu et al. 2003; Wang 2010; Barnes 2010).

The key enzyme *phenylalanine ammonia lyase* (PAL) converts the amino acid phenylalanine (Phe) into cinnamic acid or cinnamate which is converted into p-coumaric acid or p-coumarate by the enzyme *cinnamate-4-hydroxylase* (C4H) which is converted in turn to p-coumaroyl-CoA and malonyl-CoA by the enzyme *4-coumaroyl:CoA-ligase* (4CL). The p-coumaroyl-CoA is converted into 4,2′,4′-trihydroxychalcone by the enzymes *chalcone synthase* and *chalcone reductase* or into

4,2′,4′,6′-tetrahydroxychalcone by the enzyme *chalcone synthase*. The 4,2′,4′-trihydroxychalcone and the 4,2′,4′,6′-tetrahydroxychalcone are converted by the enzyme *chalcone isomerase* into 7,4′-dihydroxyflavanone (liquiritigenin) and 5,7,4′-trihydroxyflavanone (naringenin), respectively. The 7,4′-dihydroxyflavanone (liquiritigenin) and 5,7,4′-trihydroxyflavanone (naringenin) are then converted by *isoflavone synthase* (IFS) into 7,4′-dihydroxyisoflavone (daidzein (1)) and 5,7,4′-trihydroxyisoflavone (genistein (2)), respectively. The p-coumaroyl-CoA is converted into 4,2′,4′-trihydroxychalcone by the enzymes chalcone synthase and chalcone reductase or into an intermediate compound called isoliquiritigenin which is converted into glycitein (3) (Fig. 6.1) (Yu et al. 2000, 2003). A series of four chemical reactions are required to convert the 3-hydroxy-4-methoxybenzaldehyde into glycitein in the microwave-mediated method which gave a 26% yield (Fig. 6.2). Daidzein and genistein have been first isolated and characterized by Walter (1941), while glycitein has been isolated and characterized first by Naim et al. (1973). The chemical structures of the three isoflavone aglycones differ only in their radicals R_1 and R_2. Details of all enzymes involved in the phenylpropanoid biosynthetic pathway and their functions are described in Du et al. (2010). A conjugate-hydrolyzing β-glucosidase catalyzes the release of free isoflavones from their conjugates (Suzuki et al. 2006).

Several studies described the metabolic engineering processes to increase the contents of these isoflavones in yeast (*Saccharomyces cerevisiae*), *Arabidopsis thaliana*, tobacco (*Nicotiana tabacum*), corn (*Zea mays*), alfalfa (*Medicago sativa*), lettuce (*Lactuca sativa*), petunia, soybean (*Glycine max*), and the bacterium *E. coli* (Dixon and Steele 1999; Jung et al. 2000; Yu et al. 2000, 2003; Ralston et al. 2005; Liu et al. 2007; Du et al. 2010).

For example, Jung et al. (2000) identified the gene encoding the enzyme of isoflavone synthase (IFS) and expressed it in several legume species including white clover, red clover, alfalfa, lentil, lupine, and mung bean as well as in sugar beet and *Arabidopsis thaliana* (Jung et al. 2000).

Forkmann and Martens (2001) summarized several aspects of metabolic engineering and applications of flavonoids in general. They reported a summary of enzymes involved in flavonoid biosynthesis and metabolic engineering of genes involved in flower color in 14 plant species (Forkmann and Martens 2001). The gene encoding for the first enzyme in the isoflavone biosynthesis, IFS, was isolated from *Glycyrrhiza echinata* and was confirmed to be a member of the cytochrome P450 oxygenase family (Akashi et al. 1999). Later on IFS gene was also isolated from soybean (Steele et al. 1999; Winkel-Shirley 2001) and yeast (Jung et al. 2000).

Another study reported metabolic engineering of the genes involved in the phenylpropanoid biosynthetic pathway (Yu et al. 2003) using the chimeric gene, CRC (Bruce et al. 2000). Results showed an overall increase in daidzein and a decrease in genistein contents in soybean seeds expressing CRC genes (Yu et al. 2003). Another study reported the possibility of increasing genistein content by expression

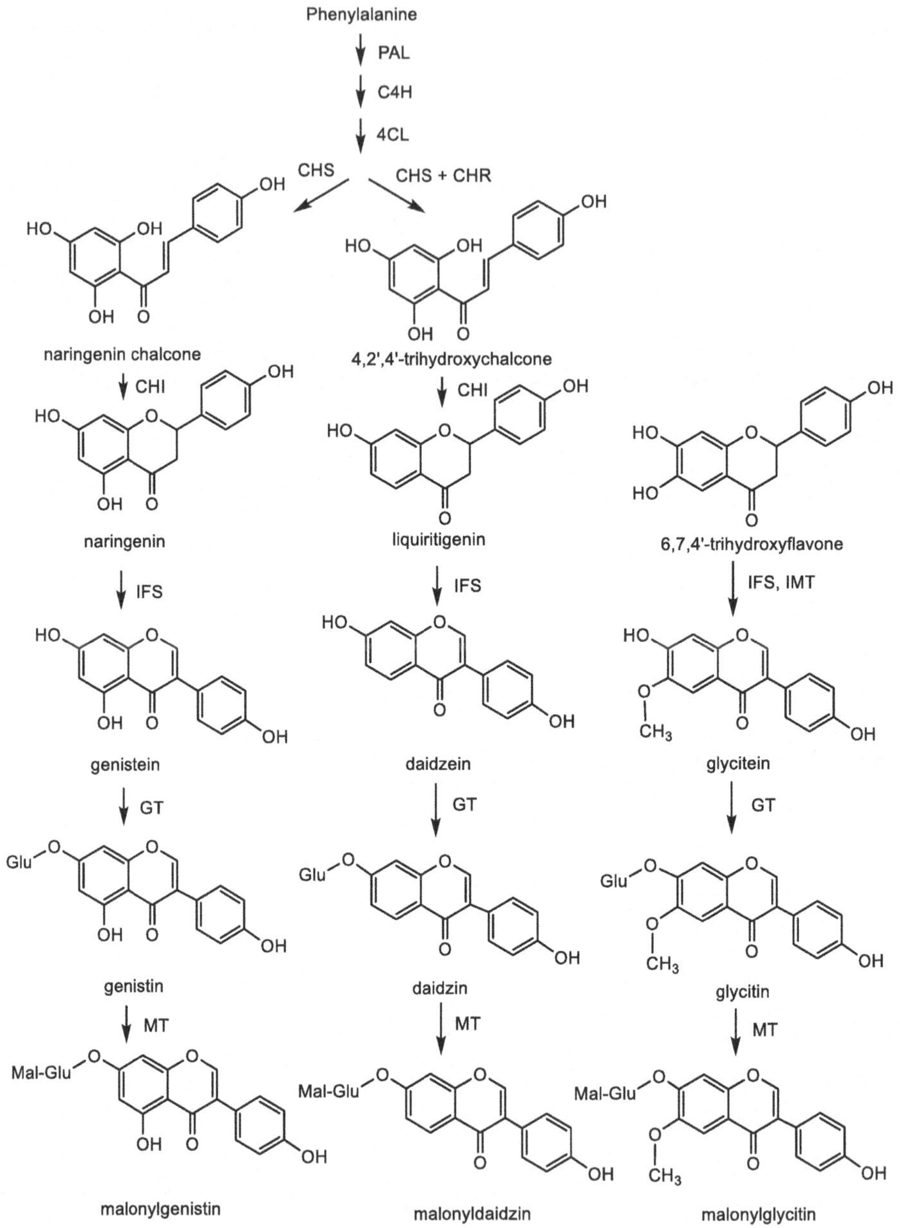

Fig. 6.2 Soybean isoflavone biosynthetic pathways. Involved enzymes: *4CL* 4-coumarate-CoA-ligase, *CHI* chalcone isomerase, *C4H* cinnamate-4-hydroxylase, *CHR* chalcone reductase, *F6H* flavonone-6-hydroxylase, *CHS* chalcone synthase, *GT* glucosyl-transferase, *IMT isoflavone methyl-transferase*, *IFS* 2-hydroxyisoflavanone synthase, *MT* malonyl-transferase, *PAL* phenylalanine ammonia lyase. (Adopted from Dhaubhadel et al. (2003))

of IFS in *Arabidopsis F3H* mutants (Liu et al. 2002). In another study, five chalcone isomerase genes (CHIs) have been cloned from soybean and expressed in yeast (Ralston et al. 2005).

Deavours and Dixon (2005) reported that they were able to increase genistein contents of up to 50 nmol g^{-1} FW in genetically engineered leaves of *Medicago sativa* by expressing the IFS gene from *Medicago truncatula* in those GM *Medicago sativa* leaves (Deavours and Dixon 2005). However, overexpression of IFS in GM tobacco plants resulted in an increase in genistein content in flowers rather than leaves (Yu et al. 2000). In this study, genistein was produced maize, *Arabidopsis*, and tobacco (Yu et al. 2000).

In another study, virus-induced gene silencing (VIGS) method was used to silence genes involved in flavonoid biosynthetic pathway in soybean (Nagamatsu et al. 2007). Several soybean cultivars have yellow seed coat phenotypes caused by a lack of anthocyanin pigmentation due to the degradation of chalcone synthase (CHS). The color of soybean seeds infested with VIGS vectors bearing CHS gene sequences turn from brown to yellow which demonstrate the effectiveness of VIGS RNA silencing method (Nagamatsu et al. 2007).

6.2 Methods of Isoflavone Extraction and Quantification

Among the most efficient methods of isoflavone extraction are the (1) ultrasonification extraction, (2) supercritical fluid extraction, (3) soxhlet extraction, (4) pressurized-liquid extraction (PLE) or pressurized solvent extraction (PSE), (5) ultrasound-assisted extraction, (6) supercritical carbon dioxide fluid extraction (SFE), and (7) eco-friendly methods of extraction (Vinatoru 2001; Romdhane and Gourdon 2002; Rostagno et al. 2002, 2003, 2004; Albu et al. 2004; Lee and Lin 2007; Araujo et al. 2007; Lante et al. 2018). To quantify the isoflavones genistein, daidzein, and glycitein and total isoflavones in seeds and other parts of soybean plants, several methods have been used including high-performance liquid chromatography (HPLC) (Wang and Murphy 1994; Vyn et al. 2002; Lin and Giusti 2005; Klejdus et al. 2005; Gutierrez-Gonzalez et al. 2009; Yuan et al. 2009; Sun et al. 2011), gas chromatography (GC) (Liggins et al. 2002; Grace et al. 2003), and near-infrared reflectance spectroscopy (NIRS) (Morrison et al. 2008; Sato et al. 2008).

6.2.1 Extraction Methods

In this section, we will briefly describe these methods of isoflavone extraction in Boxes 6.1, 6.2, 6.3, 6.4, 6.5 and 6.6.

Box 6.1: Ultrasonification Extraction (Rostagno et al. 2002)

- Mix 1 g of flour with water and methanol (20:80, v/v^{-1}) at room temperature and keep in an ultrasonic bath for 1 h.
- Repeat for three times (60 ml total volume).
- Concentrate the crude extract by reduced evaporation (30 °C).
- Centrifuge for 20 min at 800 g.
- Filter 2 ml of supernatant in a 20-μm PTFE filter.
- Add the standards and the mixture is ready for HPLC analysis.

Box 6.2: Supercritical Fluid Extraction – SFE (Rostagno et al. 2002)

- Hp model 7680A SFE module was used in extraction experiments.
- Mix 1 g of flour with water and methanol (30:70, v/v^{-1}) in 7 ml thimbles with CO_2 concentrations of 0, 5, and 10 molar % passing through the system.
- Apply pressures of 200–360 bars to the system at 40–70 °C temperatures.
- Carry out extractions using static followed by dynamic periods of 10 and 20 min, respectively, for 30 min total with a 1-ml min^{-1} CO_2 flow rate.
- Add the standards and the mixture is ready for HPLC analysis.

Box 6.3: Soxhlet Extraction (Rostagno et al. 2002)

- Mix 20 g of flour with water and methanol (20:80, v/v^{-1}) in 500-ml soxhlet thimble.
- Keep the crude extract for 9 h at the solvent (10 mg g^{-1}) boiling point (70–80 °C).
- Concentrate the crude extract by reduced pressure evaporation (30 °C).
- Centrifuge for 20 min at 800 g.
- Filter 2 ml of supernatant in a 20-μm PTFE filter.
- Add the standards and the mixture is ready for HPLC analysis.

Box 6.4: Pressurized-Liquid Extraction (Rostagno et al. 2004)

- Dionex ASE 200 Extractor was used in extraction experiments.
- Mix freeze-dried samples with sea sand in 11-ml stainless steel extraction cells with a cellulose filter at the bottom.
- Add a mixture of water and ethanol or methanol (30–80%, v/v^{-1}).
- Apply pressures of 100–200 atm to the system and heat at 60–200 °C temperatures.
- Carry out extractions using static followed by dynamic periods of 5, 7, or 10 min.
- Rinse cells with fresh solvent and a flow of N_2 for 300 s.
- Collect extracts in 60-ml vials.
- Filter 25 ml of extracts in a 0.45-μm nylon syringe filter.
- Add the standards and the mixture is ready for HPLC analysis.

Box 6.5: Ultrasound-Assisted Extraction (Lee and Lin 2007)

- Mix 3–6 g of flour with water and methanol (5:95, v/v^{-1}) and stir in 250-ml beaker (30 ml of total volume).
- Apply an ultrasound probe (Sonics, Ultrasonic Processor VCX 750, CT, USA) to the mixture at 20 kHz and 750 W with a 19-mm probe.
- Carry out extractions at temperatures <37 °C by applying 20, 40, 60, and 80 MJ corresponding to 27, 53, 80, and 102 min, respectively.
- Filter extracts in 0.45-μm filters.
- Add the standards and the mixture is ready for HPLC analysis.

Box 6.6: Eco-Friendly Method (Lante et al. 2018)

- Grinding of seeds and obtaining non-defatted flour.
- Mix with 7 ml of 80% methanol.
- Place in 1-cm-thick layer and irradiate at 253.7 nm with 30-W power at room temperature of 25 °C for increasing times (e.g., 30, 60, 90, and 120 min).
- Make several dilutions (e.g., 1:5, 1:10, 1:15, and 1:20 flour weight to solvent solution (g ml^{-1})).
- Heat samples for 6 min at time intervals like 1, 5, 10, 15, and 20 h at 25 °C or 40 °C while keeping them in constant steering mode at 50–70 rpm.
- Centrifuge at 6000 rpm for 12 min to separate the pellets (insoluble) from the extracts.
- The samples are ready for HPLC analysis.

6.2.2 *Quantification Methods*

Likewise, to quantify the isoflavones, we will briefly describe these quantification methods of isoflavones in Boxes 6.7, 6.8 and 6.9.

Box 6.7: High-Performance Liquid Chromatography (HPLC, Walsh et al. 2003)

- In a 50-ml centrifuge tube, a mixture of 0.5 g of seed sample (fine paste), 2 ml of HCl, 10 ml of acetonitrile, and 3 ml of water is prepared, shacked for 2 h, and centrifuged for 30 min at 430 g.
- From the supernatant, transfer a 1-ml aliquot to a glass vial, dry at room temperature under nitrogen, and resolubilize in 1 ml of 100% methanol.
- Vortex and filter the mixture and it is ready for HPLC analysis.
- To quantify isoflavone contents, a Waters 2996 Photodiode Array Detector (PDA) and a Waters 2695 Separations Module (Milford, MA) are used.
- A Waters Nova-Pak C18 reversed-phase column with a Nova-Pak C18 guard column was used in isoflavonoid separation.
- The mobile phase consists of 100% acetonitrile (solvent B) and 1.0% acetic acid in water (v/v) (solvent A) at a flow rate of 0.6 ml/min, and the injection volume is 10 μL.
- These solvent gradients were 15% B, 0–5 min; 15–29% B, 5–36 min; 35% B, 36–44 min 29; 15% B, 44–45 min; and re-equilibration for 5 min at 15% B before the next injection.

Box 6.8: Gas Chromatography – GC (Grace et al. 2003)

- Equipment:
- Gas Chromatograph GC 2000 series (Thermo Finnigan, San Jose, CA) and
- MD800 Mass Spectrometer (Thermo Finnigan).
- Use a GC-1, 100% dimethylpolysiloxane column (15 m × 0.25 mm i.d., 0.25-μm film) and the following operating conditions:
- Injector port temperature 330 °C,
- Injection volume 1 μl,
- Carrier gas He at a constant pressure of 35 kPa,
- Oven program temperature 160 °C (increase to 300 °C by 10 °C/min and hold temperature at 300 °C for 1 min, and run samples for 15 min),
- GC-MS interface temperature 300 °C,
- Source temperature 200 °C,
- Ionization by electron ionization (70 eV), and
- Emission current of 150 μA.
- The isoflavones were quantified by standard curves using internal standards (667 ng/ml).

Box 6.9: Liquid Chromatography Mass Spectrometry – LCMS (Doerge et al. 1999)

- Equipment:
- Quattro LC triple-quadrupole mass spectrometer (Micromass, Altrincham, UK)
- Micromass Platform single-quadrupole spectrometer with an ion source temperature of 150 °C.
- Positive ions in full scan (*m/z* 100–400 in 1-s cycle time) were used for MS measurements.
- A collision cell gas pressure (Ar) of 2–4 3 1023 mbar was used for MS/MS measurements.
- To confirm the presence of isoflavone conjugates, a constant neutral loss (*m/z* 100–600) and precursor ion scans (*m/z* 300–600) were used.
- To determine collision energies, multiple reaction monitoring transitions were optimized by directly infusing isoflavone standards.
- Two time functions were used for aglycone analysis using LCMS: The (M 1 H)1 ions for daidzein (*m/z* 255) and daidzein-d3 at a sampling cone-skimmer potential of 30 V were monitored by the first time function (0–7 min), and the (M 1 H)1 ions for genistein (*m/z* 271) and genistein-d4 (*m/z* 275) at 30 V were monitored by the second time function (7–12 min) monitored.

Table 6.1 Isoflavone contents in μg/g in whole soybean flour (WSF) and autoclaved whole soybean flour (AWSF)

Isoflavones	WSF	AWSF
Daidzin	143.65	153.21
Glycitin	23.95	ND
Genistin	279.32	318.07
Total glycosides	446.91	471.28
Daidzein	ND	ND
Glycitein	ND	16.16
Geistein	12.27	18.97
Total aglycones	12.27	35.13
Total isoflavones	459.19	506.41

Adopted from Ghafoor et al. (2013)

A recent study showed that soybean seed isoflavone contents ranged from 459.19 μg/g in WSF to 506.41 μg/g in AWSF in whole soybean flour (WSF) and autoclaved whole soybean flour (AWSF) (Table 6.1; Ghafoor et al. 2013).

For more details on isoflavone content quantification using HPLC, refer to Coward et al. (1993), Murphy et al. (1999), Franke et al. (1994, 1999), Nguyenle et al. (1995), Araujo et al. (2007), Griffith and Collison (2001), Walsh et al. (2003), Rostagno et al. (2003), Lin and Giusti (2005), Yuan et al. (2009), Lante et al. (2018).

References

Akashi, T., T. Aoki, and S. Ayabe. 1999. Cloning and functional expression of a cytochrome P450 cDNA encoding 2-hydroxyisoflavone synthase involved in biosynthesis of the isoflavonoid skeleton in licorice. *Plant Physiology* 121: 821–828.

Albu, S., E. Joyce, L. Paniwnyk, J.P. Lorimer, and T.J. Masson. 2004. Potential for the use of ultrasound in the extraction of antioxidants from *Rosmarinus officinalis* for the food and pharmaceutical industry. *Ultrasonics Sonochemistry* 11: 261–265.

Araujo, J.M.A., M.V. Silva, and J.B.P. Chaves. 2007. Supercritical fluid extraction of daidzein and genistein isoflavones from soybean hypocotyl after hydrolysis with endogenous β–glucosidases. *Food Chemistry* 105: 266–272.

Barnes, S. 2010. The biochemistry, chemistry and physiology of the isoflavones in soybeans and their food products. *Lymphatic Research and Biology* 8: 89–97.

Bruce, W., O. Folkerts, C. Garnaat, O. Crasta, B. Roth, and B. Bowen. 2000. Expression profiling of the maize flavonoid pathway genes controlled by estradiol-inducible transcription factors CRC and P. *The Plant Cell* 12: 65–79.

Cassidy, A., B. Hanley, and R.M. Lamuela-Raventos. 2000. Isoflavones, lignans and stilbenes – Origins, metabolism, and potential importance to human health. *Journal of Science and Food Agriculture* 80: 1044–1062.

Chen, M., Y. Ra, Y. Zheng, S. Wei, Y. Li, T. Guo, and P. Yin. 2014. Association between soy isoflavone intake and breast cancer risk for pre- and post-menopausal women: A meta-analysis of epidemiological studies. *PLoS One* 9 (2): e89288. https://doi.org/10.1371/journal.pone.0089288.

Coward, L., N.C. Barnes, K.D.R. Setchell, and S. Barnes. 1993. Genistein, daidzein, and their β-glycoside conjugates: Anti-tumor isoflavones in soybean foods from American and Asian diets. *Journal of Agricultural and Food Chemistry* 41: 1961–1967.

Deavours, B.E., and R.A. Dixon. 2005. Metabolic engineering of isoflavonoid biosynthesis in alfalfa. *Plant Physiology* 138: 2245–2259.

Dhaubhadel, S., B.D. McGarvey, R. Williams, and M. Gijzen. 2003. Isoflavonoid biosynthesis and accumulation in developing soybean seeds. *Plant Molecular Biology* 53: 733–743. https://doi.org/10.1023/b:plan.0000023666.30358.ae.

Dixon, R.A. 2001. Natural products and plant disease resistance. *Nature* 411: 843–847.

Dixon, R.A., and N.L. Paiva. 1995. Stress-induced phenylpropanoid metabolism. *The Plant Cell* 7 (7): 1085–1097.

Dixon, R.A., and C.L. Steele. 1999. Flavonoids and isoflavonoids – A gold mine for metabolic engineering. *Trends in Plant Science* 4 (10): 394–400.

Dixon, R.A., and L.W. Sumner. 2003. Legume natural products: Understanding and manipulating complex pathways for human and animal health. *Plant Physiology* 131: 878–885. https://doi.org/10.1104/pp.102.017319.

Doerge, D.R., H.C. Chang, M.I. Churchwell, and C.L. Holder. 1999. Analysis of soy isoflavone conjugation in vitro and in human blood using liquid-chromatography-mass spectrometry. *Drug Metabolism and Disposition* 28 (3): 298–307.

Du, Y., H. Chu, I.K. Chu, and C. Lo. 2010. CYP93G2 is a flavanone 2-hydroxylase required for C-glycosylflavone biosynthesis in rice. *Plant Physiology* 154: 324–333. https://doi.org/10.1104/pp.110.161042.

Forkmann, G., and S. Martens. 2001. Metabolic engineering and applications of flavonoids. *Current Opinion in Biotechnology* 12: 155–160.

Franke, A.A., L.J. Custer, C.M. Cerna, and K.K. Narala. 1994. Quantification of phytoestrogens in legumes by HPLC. *Journal of Agricultural and Food Chemistry* 42: 1905–1913.

Franke, A.A., J.H. Hankin, M.C. Yu, G. Masskarinec, S.H. Low, and L.J. Custer. 1999. Isoflavone levels in soy foods consumed by multiethnic populations in Singapore and Hawaii. *Journal of Agricultural and Food Chemistry* 47: 977–986.

Ghafoor, K., F.Y. Al-Juhaimi, and J. Park. 2013. The chemistry/biochemistry of the bioconversion of isoflavones in food preparation. In *Isoflavones: Chemistry, analysis, function, and effects*, ed. V.R. Preedy. Cambridge: The Royal Society of Chemistry (RSC) Press.

Grace, P.B., J.I. Taylor, N.P. Botting, T. Fryatt, M.F. Oldfield, and S.A. Bingham. 2003. Quantification of isoflavones and lignans in urine using gas chromatography/mass spectrometry. *Analytical Biochemistry* 315 (1): 114–121.

Graham, T.L. 1991. Flavonoid and isoflavonoid distribution in developing soybean seedling tissues and in seed and root exudates. *Plant Physiology* 95: 594–603.

Graham, T., M. Graham, and O. Yu. 2008. Genomics of secondary metabolites in soybean. In *Genetics and genomics of soybean*, ed. G. Stacy, 211–241. New York: Springer.

Griffith, A.P., and M.W. Collison. 2001. Improved methods for the extraction and analysis of isoflavones from soy-containing foods and nutritional supplements by reversed-phase high-performance liquid chromatography and liquid chromatography-mass spectrometry. *Journal of Chromatography. A* 913: 397–413.

Gutierrez-Gonzalez, J.J., X. Wu, J. Zhang, J.D. Lee, M. Ellersieck, J.G. Shannon, O. Yu, H.T. Nguyen, and D.A. Sleper. 2009. Genetic control of soybean seed isoflavone content: Importance of statistical model and epistasis in complex traits. *Theoretical and Applied Genetics* 119 (6): 1069–1083. https://doi.org/10.1007/s00122-009-1109-z.

Hoeck, J.A., W.R. Fehr, P.A. Murphy, and G.A. Welke. 2000. Influence of genotype and environment on isoflavone contents of soybean. *Crop Science* 40: 48–51.

Jiang, Q., F. Payton-Stewart, S. Elliott, J. Driver, L.V. Rhodes, Q. Zhang, S. Zheng, D. Bhatnagar, S.M. Boue, B.M. Collins-Burow, J. Sridhar, C. Stevens, J.A. McLachlan, T.E. Wiese, M.E. Burow, and G. Wang. 2010. Effects of 7-O substitutions on estrogenic and antiestrogenic activities of daidzein analogues in MCF-breast cancer cells. *Journal of Medicinal Chemistry* 53: 6153–6163.

Jung, W., O. Yu, S.M.C. Lau, D.P. O'Keefe, J. Odell, G. Fader, and B. McGonigle. 2000. Identification and expression of isoflavone synthase, the key enzyme for biosynthesis of isoflavones in legumes. *Nature Biotechnology* 18: 208–212.

Klejdus, B., R. Mikelova, J. Petrlova, D. Potesil, V. Adam, M. Stiborova, P. Hodek, J. Vacek, R. Kizek, and V. Kuban. 2005. Evaluation of isoflavone aglycon and glycoside distribution in soy plants and soybeans by fast column high-performance liquid chromatography coupled with a diode-array detector. *Journal of Agricultural and Food Chemistry* 53: 5848–5852. https://doi.org/10.1021/jf0502754.

Lante, A., G. Barion, S. Zannoni, M.R. Pastore, F. Tinello, C.D. Cortivo, T. Vamerali, and G. Mosca. 2018. An ecofriendly procedure to extract isoflavones from soybean seeds. *Journal of Cleaner Production* 170: 1102–1110.

Lee, M.H., and C.C. Lin. 2007. Comparison of techniques for extraction of isoflavones from the root of Radix Puerrariae: Ultrasonic and pressurized solvent extractions. *Food Chemistry* 105: 223–228.

Liggins, J., L.J.C. Bluck, W.A. Coward, and S.A. Bingham. 2002. Extraction and quantification of daidzein and genistein in food. *Analytical Biochemistry* 264 (1): 1–7.

Lin, F., and M.M. Giusti. 2005. Effects of solvent polarity and acidity on the extraction efficiency of isoflavones from soybeans (*Glycine max*). *Journal of Agricultural and Food Chemistry* 53: 3795–3800.

Liu, C.J., J.W. Blount, C.L. Steele, and R.A. Dixon. 2002. Bottlenecks for metabolic engineering of isoflavone glycoconjugates in Arabidopsis. *PNAS* 99 (22): 14578–14583.

Liu, R., Y. Hu, J. Li, and Z. Lin. 2007. Production of soybean isoflavone genistein in non-legume plants via genetically modified secondary metabolism pathway. *Metabolic Engineering* 9: 1–7.

Lozovaya, V.V., A.V. Lygin, P.V. Zernova, S.X. Li, G.L. Hartman, and J.M. Widholm. 2004a. Isoflavonoid accumulation in soybean hairy roots upon treatment with Fusarium solani. *Plant Physiology and Biochemistry* 42 (7–8): 671–679. https://doi.org/10.1016/j.plaphy.2004.06.007.

Lozovaya, V.V., A.V. Lygin, S. Li, G.L. Hartman, and J.M. Widholm. 2004b. Biochemical response of soybean roots to *Fusarium solani* f. sp. glycines infection. *Crop Science* 44 (3): 819–826. https://doi.org/10.2135/cropsci2004.8190.

Lozovaya, V.V., A.V. Lygin, A.V. Ulanov, R.L. Nelson, J. Dayde, and J.M. Widhohm. 2005. Effect of temperature and soil moisture status during seed development on soybean seed isoflavone concentration and composition. *Crop Science* 45: 1934–1940. https://doi.org/10.2135/cropsci2004.0567.

Mebrahtu, T., A. Mohamed, C.Y. Wang, and T. Andebrhan. 2004. Analysis of isoflavone contents in vegetable soybeans. *Plant Foods for Human Nutrition* 59: 55–61. https://doi.org/10.1007/s11130-004-0023-4.

Morrison, J., E.R. Cober, M.F. Saleem, N.B. McLaughlin, J. Fregeau-Reid, B.L. Ma, W. Yan, and L. Woodrow. 2008. Changes in isoflavone concentration with 58 years of genetic improvement of short-season soybean cultivars in Canada. *Crop Science* 48 (6): 2201–2208.

Murphy, P.A., T. Song, G. Buseman, K. Barua, G.R. Beecher, D. Trainer, and J. Holden. 1999. Isoflavones in retail and institutional soy foods. *Journal of Agricultural and Food Chemistry* 47: 2697–2740.

Nagamatsu, A., C. Masuta, M. Senda, H. Matsuura, A. Kasai, J.S. Hong, K. Kitamura, J. Abe, and A. Kanazawa. 2007. Functional analysis of soybean genes involved in flavonoid biosynthesis by virus-induced gene silencing. *Plant Biotechnology Journal* 5: 778–790.

Naim, M., B. Gestetner, I. Kirson, Y. Birk, and A. Bondi. 1973. A new isoflavone from soya beans. *Phytochemistry* 12 (1): 169–170.

Nguyenle, T., E. Wang, and A.P. Cheung. 1995. An investigation on the extraction and concentration of isoflavones in soy-based products. *Journal of Pharmaceutical and Biomedical Analysis* 14: 221–232.

Pueppke, S.G. 1996. The genetic and biochemical basis for nodulation of legumes by rhizobia. *Critical Reviews in Biotechnology* 16: 1–51.

Pueppke, S.G., M.C. Bolanos-Vasquez, D. Werner, M.P. Bec-Ferte, J.C. Prome, and H.B. Krishnan. 1998. Release of flavonoids by the soybean cultivars McCall and Peking and their perception as signals by the nitrogen-fixing symbiont Sinorhizobium fredii. *Physiology* 117: 599–608.

Ralston, L., S. Subramanian, M. Matsuno, and O. Yu. 2005. Partial reconstitution of flavonoid and isoflavonoid biosynthesis in yeast using soybean Type I and Type II chalcone isomerases. *Plant Physiology* 137: 1375–1388.

Romdhane, M., and C. Gourdon. 2002. Investigation in solid–liquid extraction: Influence of ultrasound. *Chemical Engineering Journal* 87 (1): 11–19.

Rostagno, M.A., J.M.A. Araujo, and D. Sandi. 2002. Supercritical fluid extraction of isoflavones from soybean flour. *Food Chemistry* 78: 111–117.

Rostagno, M.A., M. Palma, and C.G. Barosso. 2003. Ultrasound-assisted extraction of soy isoflavones. *Journal of Chromatography A* 1012 (2): 119–128.

———. 2004. Pressurized liquid extraction of isoflavones from soybeans. *Analytica Chimica Acta* 522: 169–177.

Sato, T., K. Eguchi, T. Hatano, and Y. Nishiba. 2008. Use of near-infrared reflectance spectroscopy for the estimation of isoflavone contents of soybean seeds. *Plant Production Science* 11 (4): 481–486.

Steele, C.L., M. Gijzen, D. Qutob, and R.A. Dixon. 1999. Molecular characterization of the enzyme catalyzing the aryl migration reaction of isoflavonoid biosynthesis in soybean. *Archives of Biochemistry and Biophysics* 367: 146–150.

Subramanian, S., G. Stacey, and O. Yu. 2006. Endogenous isoflavones are essential for the establishment of symbiosis between soybean and *Bradyrhizobium japonicum*. *The Plant Journal* 48: 261–273. https://doi.org/10.1111/j.1365-313x.2006.02874.x.

Sun, J., B. Sun, F. Han, S. Yan, H. Yang, and K. Akio. 2011. Rapid HPLC method for determination of 12 isoflavone components in soybean seeds. *Agricultural Sciences in China* 10 (1): 70–77.
Suzuki, H., S. Takahashi, R. Watanabe, Y. Fukushima, N. Fujita, A. Noguchi, R. Yokoyama, K. Nishitani, T. Nishino, and T. Nakayama. 2006. An Isoflavone Conjugate-hydrolyzing β-Glucosidase from the Roots of Soybean (Glycine max) Seedlings. *Journal of Biological Chemistry* 281 (40): 30251–30259. https://doi.org/10.1074/jbc.M605726200.
Van Rhijn, P., and J. Vanderleyden. 1995. The rhizobium-plant symbiosis. *Microbiological Review* 59 (1): 124–142.
Vinatoru, M. 2001. An overview of the ultrasonically assisted extraction of bioactive principles from herbs. *Ultrasonics Sonochemistry* 8: 303–313.
Vyn, T.J., X. Yin, T.W. Bruulsema, C.J.C. Jackson, I. Rajcan, and S.M. Brouder. 2002. Potassium fertilization effects on isoflavone concentrations in soybean [*Glycine max* (L.) Merr.]. *Journal of Agricultural and Food Chemistry* 50 (12): 3501–3506.
Wahajuddin, I. Taneja, S. Arora, K.S. Raju, and N. Siddiqui. 2013. Disposition of pharmacologically active dietary isoflavones in biological systems. *Current Drug Metabolism* 14 (4): 369–380.
Walsh, K.R., Y.C. Zhang, Y. Vodovotz, S.J. Schwartz, and M.L. Faila. 2003. Stability and bioaccessibility of isoflavones from soy bread during in vitro digestion. *Journal of Agricultural and Food Chemistry* 51: 4603–4609.
Walter, E.D. 1941. Genistin (an isoflavone glucoside) and its aglucone, genistein from soybean. *Journal of the American Oil Chemists' Society* 63: 3273–3276.
Wang, X. 2010. Structure, function, and engineering of enzymes in isoflavonoid biosynthesis. *Functional and Integrative Genomics* 11 (1): 13–22.
Wang, H., and P. Murphy. 1994. Isoflavone composition of American and Japanese soybeans in Iowa: Efferents of variety, crop year, and location. *Journal of Agricultural and Food Chemistry* 42: 1674–1677. https://doi.org/10.1021/jf00044a017.
Wang, B.F., J.S. Wang, J.F. Lu, T.H. Kao, and B.H. Chen. 2009. Antiproliferation effect and mechanism of prostate cancer cell lines as affected by isoflavones from soybean cake. *Journal of Agricultural and Food Chemistry* 57: 2221–2232.
Winkel-Shirley, B. 2001. Update on flavonoid biosynthesis. Flavonoid biosynthesis. A colorful model for genetics, biochemistry, cell biology, and biotechnology. *Plant Physiology* 126: 485–493.
Yu, O., W. Jung, J. Shi, R.A. Croes, G.M. Fader, B. McGonigle, and J.T. Odell. 2000. Production of the isoflavones genistein and daidzein in non-legume dicot and monocot tissues. *Plant Physiology* 124: 781–794.
Yu, O., J. Shi, A.O. Hession, C.A. Maxwell, B. McGonigle, and J.T. Odell. 2003. Metabolic engineering to increase isoflavone biosynthesis in soybean seed. *Phytochemistry* 63: 753–763.
Yuan, J.P., Y.B. Liu, J. Peng, J.H. Wang, and X. Liu. 2009. Changes of isoflavone profile in the hypocotyls and cotyledons of soybeans during dry heating and germination. *Journal of Agricultural and Food Chemistry* 57: 9002–9010.
Zhang, F., and D. Smith. 1996. Genistein accumulation in soybean (*Glycine max* [L.] Merr.) root systems under suboptimal root zone temperatures. *Journal of Experimental Botany* 47 (299): 785–792.

Chapter 7
Isoflavone Locations and Variations in Seeds, Roots, Leaves, and Other Plant Parts

Moulay Abdelmajid Kassem

7.1 Variation of Isoflavone Contents in Seeds

Many studies showed that isoflavone contents vary in soybean seeds depending on the genotype, environmental factors, growth conditions, and seed developmental stages (Tsukamoto et al. 1995; Kaufman et al. 1997; Mazur et al. 1998; Njiti et al. 1999; Hoeck et al. 2000; Wang et al. 2000; Romani et al. 2003; Primomo et al. 2005a, b; Lee et al. 2008; Morrison et al. 2008; Sakthivelu et al. 2008; Graham et al. 2008; Murphy et al. 2009; Mujic et al. 2011; Kim and Chung 2007; Devi et al. 2009).

For example, Tsukamoto et al. (1995) grew seven soybean cultivars in Japan in 1991 and analyzed their seed isoflavones. They found that daidzein, genistein, malonyldaidzein, malonylgenistein, and total seed isoflavone contents along with the ratio of linoleic plus linolenic acids to the total fatty acids decreased significantly for all the seven varieties tested. Averages of cotyledon and hypocotyl isoflavone contents were 80–90% and 10–20%, respectively (Tsukamoto et al. 1995). Kaufman et al. (1997) quantified daidzein and genistein contents in over 80 taxa of legumes and found that the leaves of *Psoralea corylifolia* contained high contents of genistein (2 g kg^{-1} dry weight) and the roots of *Vicia faba* contained high contents of daidzein (1 g kg^{-1} dry weight). Soybean (*Glycine max*) seeds contained 13 and 10 mg kg^{-1} dry weight of daidzein and genistein, respectively (Kaufman et al. 1997).

Mazur et al. (1998) developed an isotope dilution GC-MS (gas chromatography-mass spectrometry) method and quantified the isoflavones formononetin, biochanin A, daidzein, and genistein, and lignans in seeds of 52 legumes (Mazur et al. 1998). They found that these legume seeds contained high concentrations of isoflavonoids (0–1853.35 mg.kg^{-1} dry weight) and low concentrations of lignans (0–15.85 mg.

M. A. Kassem (✉)
Plant Genomics and Biotechnology Lab, Department of Biological and Forensic Sciences, Fayetteville State University, Fayetteville, NC, USA
e-mail: mkassem@uncfsu.edu

M. A. Kassem (ed.), *Soybean Seed Composition*,
https://doi.org/10.1007/978-3-030-82906-3_7

kg^{-1} dry weight) (Mazur et al. 1998). They also found that individuals who lived in regions or countries with lower risks of cancer and cardiovascular diseases had the highest concentrations of these phytoestrogen compounds in their plasma which made them strong candidates for cancer prevention (Mazur et al. 1998).

Njiti et al. (1999) demonstrated that seed isoflavone (daidzein, genistein, glycitein, and total isoflavone) contents varied among 100 recombinant inbred lines (RILs) derived from a cross between 'Essex' and 'Forrest' cultivars (Njiti et al. 1999). Subsequent studies on ExF RIL population showed that the distribution of daidzein and genistein contents was skewed toward the cultivar Forrest, while glycitein content was not skewed, and heritability was 22%, 79%, and 80% for genistein, daidzein, and glycitein, respectively (Table 7.1) (Meksem et al. 2001; Kassem et al. 2004, 2006).

Wang et al. (2000) showed that the total isoflavone contents varied from 1161 to 2743 μg g^{-1} in 210 soybean cultivars from maturity groups 0 to II grown in South Dakota (Wang et al. 2000). They also studied the relationship between isoflavone contents and hilum color, seed weight, yield, maturity (in days), disease resistance, and plant height and found that maturity group I varieties have higher amounts of total isoflavones when compared to varieties of maturity group 0 (Wang et al. 2000).

Vyn et al. (2002) evaluated potassium (K) effect on isoflavone contents in Canadian soils with low to high K contents. The seed isoflavone contents varied from 702 to 573 μg g^{-1} for daidzein, 641 to 1117 μg g^{-1} for genistein, 104 to 174 μg g^{-1} for glycitein, and 1248 to 2650 μg g^{-1} for total isoflavones. Their results showed that genistein, daidzein, and total isoflavone contents increased significantly in low K soils but not much in medium to high K soils. They have demonstrated that K application could be an efficient way of increasing seed isoflavones in soybean varieties grown under low K soils (Vyn et al. 2002). Romani et al. (2003) quantified coumarins, flavonols, phenolics, and isoflavones in seeds, roots, pods, stems, and

Table 7.1 Comparison of total isoflavone and phenolic contents (μmol g^{-1}) in small, medium, and large soybean varieties in Korea

		Yeoncheon	Yeoju	Chuncheon	Yesan
Seed size		Isoflavones	Isoflavones	Isoflavones	Isoflavones
Small	No. of varieties	15	33	5	31
	Max (μmol g^{-1})	4.026	2.401	2.015	2.883
	Min (μmol g^{-1})	0.956	1.104	1.226	1.333
	Mean (μmolg^{-1})	2.491	1.753	1.621	2.108
Medium	No. of varieties	19	19	3	25
	Max (μmol g^{-1})	3.171	2.195	1.804	2.797
	Min (μmol g^{-1})	1.288	1.179	1.183	0.886
	Mean (μmol g^{-1})	2.300	1.687	1.494	1.842
Large	No. of varieties	42	–	19	29
	Max (μmol g^{-1})	2.482	–	2.258	2.611
	Min (μmol g^{-1})	0.945	–	1.540	0.982
	Mean (μmol g^{-1})	1.714	–	1.899	1.797

Adopted from Lee et al. (2008)

leaves of three soybean cultivars and found that seeds contained the highest amounts (12.61 g kg^{-1} in cv. Emiliana, 8.97 g kg^{-1} in cv. Elvir, and 4.49 g kg^{-1} in cv. Kure). Coumarins existed only in roots and flavonols in leaves, stems, and pods (Romani et al. 2003). A study on biosynthesis and accumulation of isoflavones in soybean seeds showed that these compounds are present in all plant organs but their concentrations depend on the developmental stages and tissues with the highest amounts in leaves and mature seeds (Dhaubhadel et al. 2003). The study also found that the 2-hydroxyisoflavone synthase gene *IFS1* was highly expressed in the roots and seed coats, while *IFS2* was highly expressed in pods and embryos (Dhaubhadel et al. 2003).

Bennett et al. (2004) showed that seed isoflavone contents varied significantly in four soybean cultivars from maturity groups IV and V grown in early soybean planting system (ESPS) and conventional soybean planting system (CSPS). A 1.3-fold increase was observed for total isoflavone contents in plants grown in ESPS vs. CSPS. An increase of up to 2.5-fold was observed for genistein, daidzein, and total isoflavone contents in irrigated plants grown in both ESPS and CSPS (Bennett et al. 2004).

Another study investigated the relationships between isoflavones, protein, and oil contents and seed yield in 13 trials (Yin and Vyn 2005). The study found that daidzein content increased by 249 Mg ha^{-1} of increased yield, genistein content increased by 164 Mg ha^{-1} of increased yield, glycitein content increased by 11 Mg ha^{-1} of increased yield, and total isoflavone content increased by 427 Mg ha^{-1} of increased yield (Yin and Vyn 2005).

Zhu et al. (2005) studied the influence of seed germination on isoflavone contents in Hutcheson and Caviness cultivars. The highest contents of daidzein, genistein, and total isoflavones were obtained from 0.5 mm to 2.5 mm hypocotyl length germinated seeds of Hutcheson and Caviness, respectively (Zhu et al. 2005). Immediately after seed imbibition, daidzein and genistein contents were the highest, while glycitein and its β-glucoside conjugate levels remained steady throughout the seed germination (Zhu et al. 2005).

Primomo et al. (2005b) showed that seed isoflavone contents were significantly different among soybean lines derived from three recombinant inbred line populations (Primomo et al. 2005a). They determined significantly different "high" and "low" maturity phenotypic classes in these populations, and their seed weight had minimal effects on protein, oil, and isoflavone contents (Primomo et al. 2005b).

Kim et al. (2006) identified seed composition of isoflavones, free sugars, lipids, and proteins in two Korean soybean cultivars. They found that malonylglucoside isoflavone contents were the highest followed by the acetyl glucosides, glucosides, and aglycone isoflavone forms. Moreover, lipid and free sugar contents were high at early stages of seed development and remained constant. Total and conjugated isoflavone contents showed positive correlations with free sugars, lipid, and protein contents (Kim et al. 2006).

In another study, tissues from embryos, seed coats, and cotyledons of nine soybean varieties were analyzed for their isoflavone contents (Kim et al. 2007). Findings showed that embryos, whole seeds, cotyledons, and seed coats contain approximately 2887, 575, 325, and 33 μg g^{-1} of isoflavones, respectively (Kim et al. 2007).

Kim and Chung (2007) showed that seed isoflavone contents increased rapidly with seed growth stages from 145 μg.g^{-1} in R5 (beginning) to 611 μg.g^{-1} in R6 (green full-sized) to 669 μg.g^{-1} in R7 (full maturity) growth stages (Kim and Chung 2007).

Lee et al. (2008) found that seed isoflavone and phenolic compound contents of 322 Korean cultivars from six locations were significantly different (Lee et al. 2008). They divided these cultivars into three groups according to their 100-seed weight: small, medium, and large. They found the high amounts of isoflavones in small and medium seeds of cultivars grown in Yesan, Yeoncheon, and Milyang and the lowest amounts of isoflavones in varieties grown in Chuncheon. Meanwhile, large seeds of varieties grown at Chuncheon exhibited the highest amounts of isoflavones (Table 7.1) (Lee et al. 2008).

Morrison et al. (2008) found that daidzein, genistein, and total isoflavone contents and yields of several short-season soybean cultivars grown in Canada increased over a period of 58 years (Morrison et al. 2008). Sakthivelu et al. (2008) quantified seed isoflavone contents of 11 Indian and four Bulgarian soybean cultivars and found the ranges to be 558.2–1048.6 μg g^{-1} and 627.9–1716.9 μg g^{-1} for Indian and Bulgarian cultivars, respectively (Sakthivelu et al. 2008). Another study showed that soybean lines can be classified as high, intermediate, and low isoflavone content lines in two recombinant inbred line (RIL) populations grown in Ontario and Quebec, Canada (Murphy et al. 2009). Devi et al. (2009) quantified isoflavone contents including daidzein and genistein contents in four soybean cultivar seeds and 26 soybean products. They reported that the ranges of total isoflavones were 525–986 mg kg^{-1} and 32.9–795 mg kg^{-1} in cultivar seeds and soybean products, respectively. The lowest isoflavone contents were found in soy sauce, followed by soy meals, soy milk, soy flour, soy seeds, and soy sprouts, respectively (Devi et al. 2009). They also observed significant differences in daidzein and genistein contents with the soybean cultivars as well as soy products tested (Devi et al. 2009). Mujic et al. (2011) found significant differences in total seed isoflavone contents of five Croatian soybean cultivars ranging from 807 to 2136 μg.g^{-1} of seed weight with genistein being the most abundant isoflavone in these cultivars (Mujic et al. 2011). In a recent study, we used the ‘PI 438489B’ by ‘Hamilton’ RIL population and found high variation in daidzein, glycitein, genistein, and total isoflavone contents among the parents (Table 7.1) and RIL seeds, and we mapped QTL for these traits (see Chap. 5, Akond et al. 2013). Similar results have been found among ‘MD96–5722’ and ‘Spencer’ parents (Table 7.2), and RIL seeds and QTL for these traits have been mapped (see Chap. 10, Akond et al. 2014). We also found variations of these isoflavone contents in seeds of PI 438489B by Hamilton RILs planted in two different row spaces 25 and 50 cm, respectively (Table 7.3) (Ragin et al. 2014).

In a recent study, we grew the Hamilton by Spencer RIL population in a field in Fayetteville, North Carolina, and found differences in individual and total isoflavone contents among the parents Hamilton and Spencer and their RILs (Tables 7.4 and 7.5) and mapped QTL for these traits (see Chap. 5; Akond et al. 2015). RIL

Table 7.2 Individual (daidzein, glycitein, and genistein) and total isoflavone contents (μg/g) in seeds of several soybean cultivars planted in Illinois and North Carolina

Cultivar	Year, environment	Daidzein	Glycitein	Genistein	Tot. isof.
Essex	2001, Carbondale, IL	350	255	225	830
Forrest	2001, Carbondale, IL	329	289	143	761
PI 438489B	2011, Fayetteville, NC	650	350	320	1320
Hamilton	2011, Fayetteville, NC	426	685	645	1756
MD96-5722	2012, Fayetteville, NC	284	436	427	1147
Spencer	2012, Fayetteville, NC	364	674	574	1612

coefficient variations (CV) were 7.85%, 11.71%, 8.59%, and 8.48% for daidzein, glycitein, genistein, and total isoflavones, respectively. Hamilton had higher mean values of all isoflavone contents than Spencer.

Another study quantified daidzein (DZ), genistein (GT), glycitein (GC), and total isoflavone (TI) contents in seeds of 135 soybean varieties from China and found a mean of 587.15 μg/g and range of 219.67–1346.96 μg/g for DZ (VC = 45.11%), a mean of 801.72 μg/g and range of 224.11–1620.08 μg/g for GT (CV = 41.84%), a mean of 190.27 μg/g and range of 3.87 to 122.45 μg/g for GC (15.9%), and a mean of 1579.14 μg/g and range of 479.32–2721.44 μg/g for TI (38.66%) (Wang et al. 2016).

Other recent studies quantified DZ, GT, and TI contents in seeds of three RIL populations derived from the cross of the two soybean cultivars 'Zhongdou 27' (DZ = 1865 μg/g, GT = 1614 μg/g, GC = 311 μg/g, and TI = 3791 μg/g) and 'Jiunong 20' (DZ = 844 μg/g, GT = 1046 μg/g, GC = 193 μg/g, and TI = 2061 μg/g) and grown in seven environments in China (Zeng et al. 2009). They found highly significant differences of all isoflavone contents among the parents and RIL seeds across all seven environments (Zeng et al. 2009; Han et al. 2015; Wang et al. 2015).

Another recent study quantified DZ, GT, GC, and TI contents in seeds of 11 cultivars plus 109 RILs derived from the cross of 'DH4202' and 'RCAT1004' grown in four environments in 2015 and 2016 in Canada. The author found that DZ content range was 480–828 μg/g and 450–921 μg/g in 2015 and 2016, respectively. GT content range was 795–1129 μg/g and 738–1503 μg/g in 2015 and 2016, respectively. TI contents range was 1304–2099 μg/g and 1284–2521 μg/g in 2015 and 2016, respectively (Carter 2017).

Cai et al. (2018) grew 196 RILs derived from the cross of 'Huachun 2' and 'Wayao' cultivars in four environments in China and quantified DZ, GT, GC, and TI in their seeds and found that TI range was 817–6976 μg/g with a continuous distribution (Cai et al. 2018).

Azam et al. (2020) quantified isoflavones in seeds of a natural population of different accessions from China (n = 1168) and found that seed isoflavones varied among these accessions, depending on their geographical origin, up to sevenfold and ranged between 745 μg g^{-1} and 5253.98 μg g^{-1}. They also found positive correlations between individual and total isoflavones in several cases (Azam et al. 2020).

Table 7.3 Individual (daidzein, glycitein, and genistein) and total isoflavone contents (μg g^{-1}) in seeds of PI 438489B, Hamilton, and several of their RILs planted in two different row spaces 25 and 50 cm in Fayetteville, North Carolina

Parent/ RIL	Daidzein (25 cm)	Daidzein (50 cm)	Glycitein (25 cm)	Glycitein (50 cm)	Genistein (25 cm)	Genistein (50 cm)
PI 438489B	6.0	16.8	2.7	1.2	1.3	2.6
Hamilton	1.4	9.5	3.2	5.4	5.0	7.0
RIL 1	1.3	6.0	1.0	3.3	2.0	2.0
RIL 2	5.0	10.3	1.0	4.0	2.0	4.0
RIL 3	1.0	8.3	1.0	26.5	1.0	4.7
RIL 4	5.7	23.0	3.0	19.0	3.7	11.0
RIL 5	2.1	28.3	16.7	20.7	26.0	10.3
RIL 6	38.5	12.3	6.5	3.3	29.0	16.7
RIL 7	105.0	58.0	1.0	15.7	10.0	4.0
RIL 8	16.0	20.3	2.7	19.0	4.3	10.0
RIL 9	14.0	17.7	4.0	46.3	6.7	13.0
RIL 10	166.0	29.0	36.0	24.3	15.7	22.0
RIL 11	9.0	11.3	1.0	11.0	2.3	18.7
RIL 12	57.3	13.0	11.00	6.0	1.0	6.7
RIL 13	23.5	8.0	1.0	7.0	13.0	3.0
RIL 14	4.0	36.3	1.0	2.0	2.0	3.0
RIL 15	2.0	15.0	3.0	19.7	12.3	7.0
RIL 16	4.0	22.0	1.0	1.7	2.0	5.7
RIL 17	2.0	45.7	6.0	28.0	2.0	2.0
RIL 18	0.7	7.3	10.0	24.0	10.0	5.0
RIL 19	18.0	22.7	6.0	11.3	8.0	2.3
RIL 20	14.0	79.7	2.0	6.7	2.0	9.3
RIL 21	31.0	75.3	5.3	23.0	16.0	23.3
RIL 22	94.3	24.3	1.3	21.0	10.0	1.0
RIL 23	38.0	87.0	6.0	15.0	10.0	4.0
RIL 24	9.0	26.0	1.0	17.0	2.0	83.6
RIL 25	68.0	56.3	3.7	17.3	13.3	68.5
RIL 26	38.3	59.7	1.0	11.3	4.0	5.0
RIL 27	27.0	56.0	2.7	68.0	12.0	16.0
RIL 28	32.0	109.3	3.7	61.3	10.3	49.0
RIL 29	1.3	6.0	1.0	3.3	2.0	2.0
RIL 30	5.0	10.3	1.0	4.0	2.0	4.0

7.2 Variation of Isoflavone Contents in Roots

It was reported that isoflavones including genistein and daidzein play a primordial role in nitrogen-fixation symbiosis between *Bradyrhizobium japonicum* and soybean and induce *nod* genes in *B. japonicum* and other legumes (Peters et al. 1986; Banfalvi et al. 1988; Van De Sande and Bisseling 1997; Geurts and Bisseling 2002;

Table 7.4 Summary statistics of isoflavone concentrations (values in μg g^{-1}) of 'Hamilton' by 'Spencer' RIL populations

	Daidzein	Glycitein	Genistein	Total isoflavones
RIL mean values (n = 93)				
Range	284–423	445–785	427–683	1167–1808
Mean	357.08	608.10	561.41	1526.59
Std Dev	28.04	71.23	48.21	129.40
Skewness	−0.12	0.01	0.05	−0.12
Kurtosis	−0.30	−0.13	0.31	−0.30
Variance	786.44	5073.11	2324.14	16743.53
CV	7.85	11.71	8.59	8.48
Parents mean values (n = 4)				
Hamilton	427.75	711.50	625.50	1764.75
Std Dev	11.87	35.58	18.70	29.80
Spencer	388.75	676.75	585	1650.5
Std Dev	42.93	51.08	26.65	44.86

Adopted from Akond et al. (2015)

Wasson et al. 2006; Subramanian et al. 2007; Ha et al. 2019). It was also reported that *nod*C and *nod*X genes are induced by coumestrol, genistein, and daidzein in strain USDA257 of *Sinorhizobium fredii*, while glycitein has no effect (Pueppke et al. 1998). The large excretion of isoflavones from the roots of legumes including soybean was well established (Graham 1991; Buchanan et al. 2002; Suzuki et al. 2006).

Zhang and Smith (1996) found that genistein contents in roots of soybean plants maintained at suboptimal root zone temperatures (RZT) of 13–17 °C were lower than those in roots of plants maintained at RZTs higher than 17 °C. Moreover, before the establishment of nitrogen fixation, root genistein contents increased; however, it declined afterward (Zhang et al. 1995; Zhang and Smith 1994, 1995, 1996).

Graham (1991) identified isoflavonoid distribution in soybean seedlings' seeds and root exudates and found that daidzein and genistein are abundant in seedling roots, dry seeds, cotyledons, and hypocotyls; however, 3 days after germination, isoflavones glycosides, quercetin, kaempferol, and isorhamnetin become predominant, but cotyledons maintain high levels of these compounds. Root tips of seedlings contained the highest levels of daidzein and its conjugates. Light plays a major role in isoflavone contents. In the dark, the seedling root tips had the lowest isoflavone contents (Graham 1991).

Cho and Harper (1991) investigated the possible relationship between the root production of isoflavones and hypernodulation in several soybean hypernodulating mutants (NOD1-3, NOD2-4, and NOD3-7) and non-nodulating (NN5) mutant. Twelve (12) days after transplanting inoculated and noninoculated plants, all lines showed highly increased amounts of root daidzein genistein and coumestrol compared to the cultivar Williams. The roots of the three noninoculated hypernodulating

Table 7.5 Individual (daidzein, glycitein, and genistein) and total isoflavone contents (μmol g^{-1}) in seeds of Hamilton, Spencer, and several of their RILs planted in Fayetteville, North Carolina

Parents/RIL	Daidzein	Glycitein	Genistein	Total isoflavones
Hamilton	427.75	711.50	625.50	1764.75
Spencer	388.75	676.75	585	1650.5
1	317	445	503	1265
2	338	635	554	1527
3	354	530	554	1435
4	337	498	537	1372
5	352	527	553	1432
6	353	533	587	1473
7	356	613	571	1535
8	389	575	625	1589
9	391	576	627	1591
10	311	577	625	1409
11	355	611	567	1533
12	342	515	562	1419
13	392	654	631	1677
14	341	518	537	1396
15	375	593	603	1571
16	323	525	526	1374
17	297	557	516	1370
18	358	568	573	1499
19	309	574	522	1405
20	359	579	605	1543
21	389	651	629	1675
22	314	513	515	1342
23	352	587	572	1511
24	388	689	615	1692
25	341	497	541	1385
26	311	581	511	1414
27	386	598	671	1655
28	317	587	519	1419
29	341	557	557	1455
30	379	582	611	1572

mutants did not show any significant differences in daidzein genistein and coumestrol concentrations when compared to Williams (Cho and Harper 1991).

Lozovaya et al. (2004a, b) quantified genistein, daidzein, and glycitein in hairy roots of the soybean cultivars 'Spencer' and 'PI567.374' which are resistant and susceptible to sudden death syndrome (SDS) caused by the fungal *Fusarium virguliforme*, respectively. They found that Spencer hairy roots contained the highest amounts of daidzein compared to PI567.374 hairy roots (Lozovaya et al. 2004a, b).

Ha et al. (2019) investigated coumestrol (CMS), a polyphenol abundant in soybean leaves (Yuk et al. 2011), in two genotypes of soybean: the high CMS-producing genotype (Daewonkong) and the low CMS-producing genotype (SS0903-21-1-2). CMS is biosynthesized from daidzein and accumulates in roots in response to drought stress (Xie et al. 1995; Tripathi et al. 2016). The study identified 14 candidate genes, among which are genes encoding for oxidoreductases, MLP-like proteins, and MYB domain proteins (Ha et al. 2019).

7.3 Variation of Isoflavone Contents in Leaves and Other Plant Parts

Soybean leaves normally contain kaempferol-3-glycosides but no isoflavones; however, few studies reported that leaves contain isoflavones (Cosio et al. 1985; Kaufman et al. 1997; Graham and Graham 1991; Ho et al. 2002; Dhaubhadel et al. 2003; Klejdus et al. 2005; Yuan et al. 2009).

Cosio et al. (1985) sprayed the whole leaves of soybean plants with the herbicide diphenyl ether and maintained them under light conditions and found that they contained pterocapans and both isoflavone glucoside and aglycone forms. They also noticed enzymatic activities of chalcone synthase (CHS) and phenylalanine ammonia lyase (PAL) (Cosio et al. 1985), the two major enzymes in the phenylpropanoid biosynthetic pathway from which isoflavones are synthesized (Yu et al. 2003; Wang et al. 2010; Wang 2011; Barnes 2010). Moreover, they found no CHS or PAL activity in mesophyll cells of control leaves, while they detected 25% of isoflavone glucosides and 24% and 12% of CHS and PAL enzymatic activity, respectively, in the leaves of treated plants (Cosio et al. 1985).

Graham and Graham (1991) investigated the effects of glyceollin elicitors on isoflavone metabolism in soybean cotyledons and found that wall glycans from *Phytophthora megasperma* boosted daidzein and genistein amounts to levels higher than 4000 nmoles g^{-1} and glyceollin levels to 1800 nmoles g^{-1}. In the lower cells of the cotyledons, the same wall glycans boosted daidzein and genistein levels to more than 9000 nmoles g^{-1} (Graham and Graham 1991).

Kaufman et al. (1997) quantified daidzein and genistein contents in over 80 taxa of legumes and found that the leaves of *Psoralea corylifolia* contained high contents of genistein (2 g kg^{-1} dry weight) and the roots of *Vicia faba* contained high contents of daidzein (1 g kg^{-1} dry weight). Soybeans (*Glycine max*) contained 13 and 10 mg kg^{-1} dry weight of daidzein and genistein, respectively (Kaufman et al. 1997).

Ho et al. (2002) investigated the isoflavonoid and isoflavone profiles of soybean and soy leaves and found only trace concentrations of malonylgenistin and genistin. They concluded that soybeans contain mainly isoflavone glycosides and their derivatives, while soy leaves contain mainly kaempferol glycosides (Ho et al. 2002).

Table 7.6 Mean values of individual and total isoflavones in soybean hypocotyls, cotyledons, and seed coats in mg/g

Isoflavone type	Hypocotyls (mg/g)	Cotyledons (mg/g)	Seed coat (mg/g)
Malonyldaidzin	8.5856	0.6652	–
Daidzin	2.9926	0.2622	0.5429
Daidzein	0.3135	0.0405	0.0751
Malonylgenistin	2.0306	1.1199	–
Genistin	0.5846	0.3916	0.9021
Genistein	0.0755	0.0428	0.0578
Total Isoflavones	12.3926	1.5943	1.6601
Daidzein (%)	59.6	38.1	38.6
Genistein (%)	12.8	61.9	61.4
Glycitein (%)	26.6	0	0
Malonyl (%)	69.1	69.4	0
Acetyl (%)	0	0	38.1
β-Glycoside (%)	27.1	25.4	53.9
Aglycone (%)	3.8	5.2	8.0

Adopted from Yuan et al. (2009)

Dhaubhadel et al. (2003) studied the biosynthesis and accumulation of isoflavones in soybean seeds and demonstrated that these compounds are present in all plant organs but their concentrations depend on the developmental stages and tissues with the highest amounts in leaves and mature seeds (Dhaubhadel et al. 2003).

Klejdus et al. (2005) used an HPLC/UV method and found genistein, daidzein, glycitein, genistin, daidzin, glycitin, ononin, biochanin, formononetin, and sissortin in the pods, stems, roots, and leaves of five varieties of soybean; however, the highest concentrations of isoflavones were found in the roots (12.5 μg g^{-1} dry weight) (Klejdus et al. 2005).

Yuan et al. (2009) identified the contents of genistein, daidzein, glycitein, and their conjugated and nonconjugated forms (malonyl-, acetyl-, and nonconjugated β-glycosides) in hypocotyls and cotyledons of soybean plants under dry heating conditions (Table 7.6). They found that isoflavone contents were 7.8-fold higher in hypocotyls than in cotyledons and that daidzein and its glycoside conjugate contents were more abundant and higher (59.6%) than glycitein (26.6%) and genistein (13.8%) (Yuan et al. 2009).

With this high genotype and environmental variation, it is possible to genetically map quantitative trait loci (QTL) that control seed isoflavone contents which was achieved using different mapping populations and molecular markers (Njiti et al. 1999; Meksem et al. 2001; Kassem et al. 2004, 2006; Primomo et al. 2005b; Zeng et al. 2009; Cheng et al. 2008; Gutierrez-Gonzalez et al. 2009, 2010, 2011; Yang et al. 2011; Smallwood 2012; Akond et al. 2013, 2014, 2015; Smallwood et al. 2014; Zhang et al. 2014; Wang et al. 2014, 2015; Zhao et al. 2015; Han et al. 2015, 2016). These QTL mapping studies will be discussed extensively in Chap. 5.

References

Akond, M., R. Bazzelle, B. Ragin, H. Herrera, U. Kaodi, C. Akbay, S.K. Kantartzi, V. Njiti, A. Barakat, K. Meksem, D.A. Lightfoot, and M.A. Kassem. 2013. Additional quantitative trait loci and candidate genes for seed isoflavone content in soybean. *Journal of Agricultural Science* 5 (11): 20–33. https://doi.org/10.5539/jas.v5n11p20.

Akond, M., S. Liu, S.K. Kantartzi, K. Meksem, N. Bellaloui, D.A. Lightfoot, J. Yuan, D. Wang, and M.A. Kassem. 2014. Quantitative trait loci for seed isoflavone contents in 'MD96-5722' by 'Spencer' recombinant inbred lines of soybean. *Journal of Agricultural and Food Chemistry* 62 (7): 1464–1468. dx.doi.org. https://doi.org/10.1021/jf4040173.

Akond, M., S. Liu, S.K. Kantartzi, K. Meksem, N. Bellaloui, D.A. Lightfoot, J. Yuan, D. Wang, J. Anderson, D.A. Lightfoot, and M.A. Kassem. 2015. A SNP genetic linkage map based on the 'Hamilton' by 'Spencer' recombinant inbred line (RIL) population of soybean [*Glycine max* (L.) Merr.] identified QTL for seed isoflavone contents. *Plant Breeding* 134 (5): 580–588. https://doi.org/10.1111/pbr.12298.

Azam, M., S. Zhanga, A.M. Abdelghany, A.S. Shaibu, Y. Feng, Y. Li, Y. Tian, H. Hong, B. Li, and J. Sun. 2020. Seed isoflavone profiling of 1168 soybean accessions from major growing ecoregions in China. *Food Research International* 130 (2020): 108957. https://doi.org/10.1016/j.foodres.2019.108957.

Banfalvi, Z., A. Nieuwkoop, M. Schell, L. Besl, and G. Stacey. 1988. Regulation of nod gene expression in *Bradyrhizobium japonicum. Molecular & General Genetics* 214: 420–424.

Barnes, S. 2010. The Biochemistry, Chemistry and Physiology of the Isoflavones in Soybeans and their Food Products. *Lymphatic Research and Biology* 8 (1): 89–98. https://doi.org/10.1089/lrb.2009.0030.

Bennett, J.O., O. Yu, L.G. Heatherly, and H.B. Krishnan. 2004. Accumulation of genistein and daidzein, soybean isoflavones implicated in promoting human health, is significantly elevated by irrigation. *Journal of Agricultural and Food Chemistry* 52 (25): 7574–7579. https://doi.org/10.1021/jf049133k.

Buchanan, B., W. Gruissen, and R. Jones, eds. 2002. *Biochemistry and molecular biology of plants*, 796–815. New York: Wiley.

Cai, Z., Y. Cheng, Z. Ma, X. Liu, Q. Ma, Q. Xia, G. Zhang, Y. Mu, and H. Nian. 2018. Fine mapping of QTLs for individual and total isoflavone content in soybean (*Glycine max* L.) using a high-density genetic map. *Theoretical and Applied Genetics* 131 (3): 555–568. https://doi.org/10.1007/s00122-017-3018-x.

Carter, A. 2017. *Identification of Quantitative Trait Loci (QTL) and genes associated with seed isoflavone concentration in soybean (Glycine max [L] Merrill.)*. PhD Dissertation. The University of Guelph, Canada.

Cheng, H., O. Yu, and D.Y. Yu. 2008. Polymorphisms of IFS1 and IFS2 gene are associated with isoflavone concentrations in soybean seeds. *Plant Science* 175: 505–512.

Cho, M.J., and J.E. Harper. 1991. Effect of inoculation and nitrogen on isoflavonoid concentration in wild-type and nodulation-mutant soybean roots. *Plant Physiology* 95: 435–442.

Cosio, E.G., G. Weissenbock, and J.W. McClure. 1985. Acifluorfen-induced isoflavonoids and enzymes of their biosynthesis in mature soybean leaves. *Plant Physiology* 78: 14–19.

Devi, M.K.A., M. Gondi, G. Sakthivelu, P. Giridhar, T. Rajasekaran, and G.A. Ravishankar. 2009. Functional attributes of soybean seeds and products, with reference to isoflavone content and antioxidant activity. *Food Chemistry* 114: 771–776.

Dhaubhadel, S., B.D. McGarvey, R. Williams, and M. Gijzen. 2003. Isoflavonoid biosynthesis and accumulation in developing soybean seeds. *Plant Molecular Biology* 53: 733–743. https://doi.org/10.1023/b:plan.0000023666.30358.ae.

Geurts, R., and T. Bisseling. 2002. Rhizobium nod factor perception and signalling. *Plant Cell* 14: S239–S249.

Graham, T.L. 1991. Flavonoid and isoflavonoid distribution in developing soybean seedling tissues and in seed and root exudates. *Plant Physiology* 95: 594–603.
Graham, M.Y., and T.L. Graham. 1991. Rapid accumulation of anionic peroxidases and phenolic polymers in soybean cotyledon tissues following treatment with *Phytophthora megasperma* f.sp. glycinea wall glucan. *Plant Physiology* 97: 1445–1455.
Graham, T., M. Graham, and O. Yu. 2008. Genomics of secondary metabolites in soybean. In *Genetics and genomics of soybean*, ed. G. Stacy, 211–241. New York: Springer.
Gutierrez-Gonzalez, J.J., X. Wu, J. Zhang, J.D. Lee, M. Ellersieck, J.G. Shannon, O. Yu, H.T. Nguyen, and D.A. Sleper. 2009. Genetic control of soybean seed isoflavone content: Importance of statistical model and epistasis in complex traits. *Theoretical and Applied Genetics* 119 (6): 1069–1083. http://doi.org/10.1007/s00122-009-1109-z.
Gutierrez-Gonzalez, J.J., X.L. Wu, J.D. Gillman, J.D. Lee, R. Zhong, O. Yu, J.G. Shannon, H.T. Nguyen, and D.A. Sleper. 2010. Intricate environment-modulated genetic networks control isoflavone accumulation in soybean seeds. *BMC Plant Biology* 10: 105–120. http://doi.org/10.1186/1471-2229-10-105.
Gutierrez-Gonzalez, J.J., T.D. Vuong, R. Zhong, O. Yu, J.D. Lee, G. Shannon, M. Ellersieck, H.T. Nguyen, and D.A. Sleper. 2011. Major locus and other novel additive and epistatic loci involved in modulation of isoflavone concentration in soybean seeds. *Theoretical and Applied Genetics* 123: 1375–1385. http://doi.org/10.1007/s00122-011-1673-x.
Ha, J., Y.G. Kang, T. Lee, M. Kim, M.Y. Yoon, E. Lee, X. Yang, D. Kim, Y.J. Kim, T.R. Lee, M.Y. Kim, and S.H. Lee. 2019. Comprehensive RNA sequencing and co-expression network analysis to complete the biosynthetic pathway of coumestrol, a phytoestrogen. *Scientific Reports* 9: 1934. https://doi.org/10.1038/s41598-018-38219-6.
Han, Y., W. Teng, Y. Wang, X. Zhao, L. Wu, D. Li, and W. Li. 2015. Unconditional and conditional QTL underlying the genetic interrelationships between soybean seed isoflavone, and protein or oil contents. *Plant Breeding* 134 (3): 300–309. https://doi.org/10.1111/pbr.12259.
Han, Y., D. Li, G. Zhao, X. Zhao, Z. Jiang, H. Hu, L. Wu, Y. Wang, Y. Gao, Y. Li, G. Zeng, F. Meng, W. Teng, and W. Li. 2016. Dynamic quantitative trait loci underlies isoflavone accumulation in soybean seed. *Plant Breeding*. https://doi.org/10.1111/pbr.12354.
Ho, H.M., R.Y. Chen, L.K. Leung, F.L. Chan, Y. Huang, and Z.Y. Chen. 2002. Difference in flavonoid and isoflavone profile between soybean and so leaf. *Biomedicine & Pharmacotherapy* 56: 289–295.
Hoeck, J.A., W.R. Fehr, P.A. Murphy, and G.A. Welke. 2000. Influence of genotype and environment on isoflavone contents of soybean. *Crop Science* 40: 48–51.
Kassem, M.A., K. Meksem, M.J. Iqbal, V.N. Njiti, W.J. Banz, T.A. Winters, A.J. Wood, and D.A. Lightfoot. 2004. Definition of soybean genomic regions that control seed phytoestrogen amounts. *Journal of Biomedicine & Biotechnology* 1: 52–60. https://doi.org/10.1155/S1110724304304018.
Kassem, M.A., J. Shultz, K. Meksem, Y. Cho, A. Wood, M.J. Iqbal, and D.A. Lightfoot. 2006. An updated 'Essex' by 'Forrest' linkage map and first composite interval map of QTL underlying six soybean traits. *Theoretical and Applied Genetics* 113: 1015–1026.
Kaufman, P.B., J.A. Duke, H. Brielmann, J. Boik, and J.E. Hoyt. 1997. A comparative survey of leguminous plants as sources of the isoflavones, genistein and daidzein: Implications for human nutrition and health. *The Journal of Alternative and Complementary Medicine* 3 (1): 7–12.
Kim, J.A., and I.M. Chung. 2007. Change in isoflavone concentration of soybean (*Glycine max* L.) seeds at different growth stages. *Journal of the Science of Food and Agriculture* 87: 496–503.
Kim, J.A., S.B. Hong, W.S. Jung, C.Y. Yu, K.H. Ma, J.G. Gwag, and I.M. Chung. 2007. Comparison of isoflavones composition in seed, embryo, cotyledon and seed coat of cooked-with-rice and vegetable soybean (*Glycine max* L.) varieties. *Food Chemistry* 102 (3): 738–744. https://doi.org/10.1016/j.foodchem.2006.06.061.
Kim, E.H., S.H. Kim, J.I. Chung, H.Y. Chi, J.A. Kim, and I.M. Chung. 2006. Analysis of phenolic compounds and isoflavones in soybean seeds (*Glycine max* (L.) Merill.) and sprouts grown under different conditions. *European Food Research and Technology* 222 (1): 201–208.

Klejdus, B., R. Mikelova, J. Petrolova, D. Potesil, V. Adam, M. Stiborova, P. Hodek, J. Vacek, R. Kizek, and V. Kuban. 2005. Evaluation of isoflavone aglycon and glycoside distribution in soy plants and soybeans by fast column high-performance liquid chromatography coupled with a diode-array detector. *Journal of Agricultural and Food Chemistry* 53 (15): 5848–5852.

Lee, S.J., J.J. Kim, H.I. Moon, J.K. Ahn, S.C. Chun, W.S. Jung, O.K. Le, and I.M. Chung. 2008. Analysis of isoflavones and phenolic compounds in Korean soybean [Glycine max (L.) Merr.] seeds of different seed weights. *Journal of Agricultural and Food Chemistry* 56 (8): 2751–2758. https://doi.org/10.1021/jf073153f.

Lozovaya, V.V., A.V. Lygin, P.V. Zernova, S.X. Li, G.L. Hartman, and J.M. Widholm. 2004a. Isoflavonoid accumulation in soybean hairy roots upon treatment with Fusarium solani. *Plant Physiology and Biochemistry* 42 (7–8): 671–679. https://doi.org/10.1016/j.plaphy.2004.06.007.

Lozovaya, V.V., A.V. Lygin, S. Li, G.L. Hartman, and J.M. Widholm. 2004b. Biochemical response of soybean roots to *Fusarium solani* f. sp. glycines infection. *Crop Science* 44 (3): 819–826. https://doi.org/10.2135/cropsci2004.8190.

Mazur, W.M., J.A. Duke, K. Wahala, S. Rasku, and H. Adlercreutz. 1998. Isoflavonoids and lignans in legumes: Nutritional and health aspects in humans. *Nutritional Biochemistry* 9: 193–200.

Meksem, K., V.N. Njiti, W.J. Banz, M.J. Iqbal, M.A. Kassem, D.L. Hyten, J. Yuan, T.A. Winters, and D.A. Lightfoot. 2001. Genomic regions that underlie soybean seed isoflavone content. *Journal of Biomedicine & Biotechnology* 1: 38–44.

Morrison, J., E.R. Cober, M.F. Saleem, N.B. McLaughlin, J. Fregeau-Reid, B.L. Ma, W. Yan, and L. Woodrow. 2008. Changes in isoflavone concentration with 58 years of genetic improvement of short-season soybean cultivars in Canada. *Crop Science* 48 (6): 2201–2208.

Mujic, I., E. Sertovic, S. Jokic, Z. Saric, V. Alibabic, S. Vidovic, and V. Alibabic. 2011. Isoflavone content and antioxidant properties of soybean seeds. *Croatian Journal of Food Science and Technology* 3 (1): 16–20.

Murphy, S.E., E.A. Lee, L. Woodrow, P. Seguin, J. Kumar, I. Rajcan, and G.R. Ablett. 2009. Genotype × Environment interaction and stability for isoflavone content in soybean. *Crop Science* 49 (4): 1313–1321. https://doi.org/10.2135/cropsci2008.09.0533.

Njiti, V.N., K. Meksem, J. Yuan, D.A. Lightfoot, W.J. Banz, and T.A. Winters. 1999. DNA markers associated with loci underlying seed phytoestrogen content in soybeans. *Journal of Medicinal Food* 2 (3–4): 185–187.

Peters, N.K., J.W. Frost, and S.R. Long. 1986. A plant flavone, luteolin, induces expression of rhizobium meliloti nodulation genes. *Science* 233: 977–980.

Primomo, V.S., V. Poysa, G.R. Ablett, C.J. Jackson, M. Gijzen, and I. Rajcan. 2005a. Mapping QT for individual and total isoflavone content in soybean seeds. *Crop Science* 45: 2454–2462. https://doi.org/10.2135/cropsci2004.0672.

Primomo, V.S., V. Poysa, G.R. Ablett, C.G. Jackson, and I. Rajcan. 2005b. Agronomic performance of recombinant inbred line populations segregating for isoflavone content in soybean seeds. *Crop Science* 45 (6): 2203–2211.

Pueppke, S.G., M.C. Bolanos-Vasquez, D. Werner, M.P. Bec-Ferte, J.C. Prome, and H.B. Krishnan. 1998. Release of flavonoids by the soybean cultivars McCall and Peking and their perception as signals by the nitrogen-fixing symbiont *Sinorhizobium fredii*. *Physiology* 117: 599–608.

Ragin, B., M. Akond, S.K. Kantartzi, K. Meksem, H. Herrera, C. Akbay, D.A. Lightfoot, and M.A. Kassem. 2014. Effect of row spacing on seed isoflavone contents in soybean [*Glycine max* (L.) Merr.]. *American Journal of Plant Science* 5: 4003–4010. https://doi.org/10.4236/ajps.2014.526418.

Romani, A., P. Vignolini, C. Galardi, C. Aroldi, C. Vazzana, and D. Heimler. 2003. Polyphenolic content in different plant parts of soy cultivars grown under natural conditions. *Journal of Agricultural and Food Chemistry* 51 (18): 5301–5306. https://doi.org/10.1021/jf0212136.

Sakthivelu, G., M.K.A. Devi, P. Giridhar, T. Rajasekaran, G.A. Ravishankar, M.T. Nikolova, G.B. Angelov, R.M. Todorova, and G.P. Kosturkova. 2008. Isoflavone composition, phenol content, and antioxidant activity of soybean seeds from India and Bulgaria. *Journal of Agricultural and Food Chemistry* 56 (6): 2090–2095. https://doi.org/10.1021/jf072939a.

Smallwood, C.J. 2012. Detection of quantitative trait loci for marker-assisted selection of soybean isoflavone genistein. Master's thesis, University of Tennessee.

Smallwood, C.J., C.N. Nyinyi, D.A. Kopsell, C.E. Sams, D.R. West, P. Chen, S.K. Kantartzi, P.B. Cregan, D.L. Hyten, and V.R. Pantalone. 2014. Detection and confirmation of quantitative trait loci for soybean seed isoflavone. *Crop Science* 54: 595–606.

Subramanian, S., G. Stacey, and O. Yu. 2007. Distinct, crucial roles of flavonoids during legume nodulation. *Trends in Plant Science* 12 (7): 282–285. https://doi.org/10.1016/j.tplants.2007.06.006.

Suzuki, H., S. Takahashi, R. Watanabe, Y. Fukushima, N. Fujita, A. Noguchi, R. Yokoyama, K. Nishitani, T. Nishino, and T. Nakayama. 2006. An isoflavone conjugate-hydrolyzing β-glucosidase from the roots of soybean (Glycine max) seedlings. *The Journal of Biological Chemistry* 281 (40): 30251–30259.

Tripathi, P., R.C. Rabara, R.N. Reese, M.A. Miller, J.S. Rohila, S. Subramanian, Q.J. Shen, D. Morandi, H. Bucking, V. Shulaev, and P.J. Rushton. 2016. A toolbox of genes, proteins, metabolites and promoters for improving drought tolerance in soybean includes the metabolite coumestrol and stomatal development genes. BMC Genomics 17, 102 (2016). https://doi.org/10.1186/s12864-016-2420-0.

Tsukamoto, C., S. Shimada, K. Igita, S. Kudou, M. Kokubun, K. Okubo, and K. Kitamura. 1995. Factors affecting isoflavone content in soybean seeds: Changes in isoflavones, saponins, and composition of fatty acids at different temperatures during seed development. *Journal of Agricultural and Food Chemistry* 43 (5): 1184–1192. https://doi.org/10.1021/jf00053a012.

Van De Sande, K., and T. Bisseling. 1997. Signalling in symbiotic root nodule formation. *Essays in Biochemistry* 32: 127–142.

Vyn, T.J., X. Yin, T.W. Bruulsema, C.J.C. Jackson, I. Rajcan, and S.M. Brouder. 2002. Potassium fertilization effects on isoflavone concentrations in soybean [*Glycine max* (L.) Merr.]. *Journal of Agricultural and Food Chemistry* 50 (12): 3501–3506.

Wang, X. 2011. Structure, function, and engineering of enzymes in isoflavonoid biosynthesis. *Functional and Integrative Genomics* 11: 13–22. https://doi.org/10.1007/s10142-010-0197-9.

Wang, Y., J. Li, and R. Xia. 2010. Expression of chalcone synthase and chalcone isomerase genes and accumulation of corresponding flavonoids during fruit maturation of Guoqing No. 4 satsuma mandarin (Citrus unshiu Marcow). *Scientia Horticulturae* 125 (2): 110–116. https://doi.org/10.1016/j.scienta.2010.02.001.

Wang, C., M. Sherrard, S. Pagadala, R. Wixon, and R.A. Scott. 2000. Isoflavone content among maturity group 0 to II soybeans. *Journal of the American Oil Chemists' Society* 77 (5): 483–487.

Wang, Y., Y. Han, W. Teng, X. Zhao, Y. Li, L. Wu, D. Li, and W. Li. 2014. Expression quantitative trait loci infer the regulation of isoflavone accumulation in soybean (Glycine max L. Merr.) seed. *BMC Genomics* 15: 680. https://doi.org/10.1186/1471-2164-15-680.

Wang, Y., Y. Han, X. Zhao, Y. Li, W. Teng, D. Li, Y. Zhan, and W. Li. 2015. Mapping isoflavone QTL with main, epistatic and QTL × environment effects in recombinant inbred lines of soybean. *PLoS One* 10 (3): e0118447. https://doi.org/10.1371/journal.pone.0118447.

Wang, C., Y. Guo, J. Hu, J. Zhang, L. Yang, B. Guo, J. Han, and H. Xie. 2016. Exploring elite alleles for seed isoflavones concentration in soybean by association analysis. *Emirates Journal of Food and Agriculture* 28 (2): 85–92. https://doi.org/10.9755/ejfa.2015-09-787.

Wasson, A.P., F.I. Pellerone, and U. Mathesius. 2006. Silencing the flavonoid pathway in *Medicago truncatula* inhibits root nodule formation and prevents auxin transport regulation by rhizobia. *The Plant Cell* 18: 1617–1162.

Xie, Z.P., C. Staehelin, H. Vierheilig, A. Wiemken, S. Jabbouri, W. J. Broughton, R. Vogeli-Lange, and T. Boller. 1995. Rhizobial Nodulation Factors Stimulate Mycorrhizal Colonization of Nodulating and Nonnodulating Soybeans. *Plant Physiology* 108: 1519–1525.

Yang. K., J.K. Moon, N. Jeong, H.K. Chun, S.T. Kang, K. Back, and S.C. Jeong. 2011. Novel major quantitative trait loci regulating the content of isoflavone in soybean seeds. *Genes & Genomics* 33 (6): 685–692.

Yin, X., and T.J. Vyn. 2005. Relationships of isoflavone, oil, and protein in seed with yield of soybean. *Agronomy Journal* 97 (5): 1314–1321.

Yu, O., J. Shi, A.O. Hession, C.A. Maxwell, B. McGonigle, and J.T. Odell. 2003. Metabolic engineering to increase isoflavone biosynthesis in soybean seed. *Phytochemistry* 63: 753–763. https://doi.org/10.1016/S0031-9422(03)00345-5.

Yuk, H.J., J.H. Lee, M.J. Curtis-Long, J.W. Lee, Y.S. Kim, H.W. Ryu, C.G. Park, T.S. Jeong, and K.H. Park. 2011. The most abundant polyphenol of soy leaves, coumestrol, displays potent α-glucosidase inhibitory activity. *Food Chemistry* 126 (3): 1057–1063. https://doi.org/10.1016/j.foodchem.2010.11.125.

Yuan, J.P., Y.B. Liu, J. Peng, J.H. Wang, and X. Liu. 2009. Changes of isoflavone profile in the hypocotyls and cotyledons of soybeans during dry heating and germination. *Journal of Agricultural and Food Chemistry* 57: 9002–9010.

Zeng, G., D. Li, Y. Han, W. Teng, J. Wang, L. Qiu, and W. Li. 2009. Identification of QTL underlying isoflavone contents in soybean seeds among multiple environments. *Theoretical and Applied Genetics* 118: 1455–1463.

Zhang, F., and D.L. Smith. 1994. Effects of low root zone temperature on the early stages of symbiosis establishment between soybean [*Glycine max* (L.) Merr.] and *Bradyrhizobium japonicum*. *Journal of Experimental Botany* 45: 1467–1473.

Zhang, F., and D.L. Smith. 1995. Preincubation of *Bradyrhizobium japonicum* with genistein accelerates nodule development of soybean at suboptimal root zone temperatures. *Plant Physiology* 108 (3): 961–968. https://doi.org/10.1104/pp.108.3.961.

Zhang, F., and D.L. Smith. 1996. Genistein accumulation in soybean (*Glycine max* [L.] Merr.) root systems under suboptimal root zone temperatures. *Journal of Experimental Botany* 47 (6): 785–792. https://doi.org/10.1093/jxb/47.6.785.

Zhang, F., D.H. Lynch, and D.L. Smith. 1995. Impact of low root zone temperatures in soybean (*Glycine max* L. Merr.) on nodulation and nitrogen fixation. *Environmental and Experimental Botany* 35: 279–285.

Zhang, H.J., J.W. Li, Y.J. Liu, W.Z. Jiang, X.L. Du, L. Li, X.W. Li, L.T. Su, Q.Y. Wang, and Y. Wang. 2014. Quantitative trait loci analysis of individual and total isoflavone contents in soybean seeds. *Journal of Genetics* 93 (2): 331–338.

Zhao, G., Z. Jiang, D. Li, Y. Han, H. Hu, L. Wu, and Y. Wang. 2015. Molecular loci associated with seed isoflavone content may underlie resistance to soybean pod borer (*Leguminivora glycinivorella*). *Plant Breeding* 178: 78–84.

Zhu, D., N.S. Hettiarachchy, R. Horax, and P. Chen. 2005. Isoflavone contents in germinated soybean seeds. *Plant Foods for Human Nutrition* 60: 147–151.

Chapter 8
Isoflavones' Positive and Negative Effects on Humans, Animals, and Plants

Moulay Abdelmajid Kassem

8.1 Positive and Negative Effects on Human Health

8.1.1 Introduction

During the past four to five decades, numerous studies reported effects of isoflavones on human and animal health. In humans, these effects include the prevention of certain types of cancers such as liver, prostate, and breast cancers, prevention of cardiovascular diseases, reduction of menopausal symptoms, positive effects on menstrual cycles in women, positive effects on overall thyroid function, obesity reduction, boosting immunity, and other beneficial effects (Pinchera et al. 1965; Setchell et al. 1980; Murphy 1982; Labib et al. 1989; Messina and Barnes 1991; Coward et al. 1993; Phipps et al. 1993; Adlercreutz 1995; Anderson et al. 1995; Anderson and Garner 1997; Potter 1995; Knight and Eden 1996; Chorazy et al. 1995; Reinli and Block 1996; Divi et al. 1997; Kaufman et al. 1997; Mazur et al. 1998; Ishida et al. 1998; Lichttenstein 1998; Tham et al. 1998; Adlercreutz 1998; Setchell and Cassidy 1999; Thigpen et al. 1999; Messina 1999; Anderson et al. 1999; Cassidy et al. 2000a, b; Guerra et al. 2000; Tikkanen and Adlercreutz 2000; Forkmann and Martens 2001; Ren et al. 2001; Brown and Setchell 2001; Doerge and Chang 2002; Doerge and Sheehan 2002; Lichtenstein et al. 2002; Degen et al. 2002; Messina and Hughes 2003; Fritz et al. 2003; Dixon and Sumner 2003; Mezei et al. 2003; Sirtori et al. 2004; Ali et al. 2004; Lee et al. 2005; Sacks et al. 2006; Cassidy et al. 2006; Song et al. 2007; Orgaard and Jensen 2008; Banerjee et al. 2008; Sarkaki et al. 2008; Xiao 2008; Chandrareddy et al. 2008; Pop et al. 2008; Cavallini et al. 2009; Cederroth and Nef 2009; Cederroth et al. 2010, 2012; Kwon

M. A. Kassem (✉)
Plant Genomics and Biotechnology Lab, Department of Biological and Forensic Sciences, Fayetteville State University, Fayetteville, NC, USA
e-mail: mkassem@uncfsu.edu

M. A. Kassem (ed.), *Soybean Seed Composition*,
https://doi.org/10.1007/978-3-030-82906-3_8

2014) and contribute to nitrogen fixation and defense against pathogens in many plant species (Banfalvi et al. 1988; Smit et al. 1992; Loh et al. 1994; Kaufman et al. 1997; Begum et al. 2001; Cseke et al. 2006; Subramanian et al. 2006, 2007).

Other studies reported that soy isoflavones can reduce the symptoms of menopause in women by prolonging the menstrual cycle length (Phipps et al. 1993; Setchell and Cassidy 1999), the risks of breast, ovarian, prostate, and neuronal cancers (Adlercreutz et al. 1991, 1993; Adlercreutz 1995, 1998; Barnes 1995; Lamartiniere et al. 1995; Lamartiniere 2000; Watanabe et al. 2002; Lo et al. 2007; Xiao 2008; Wang et al. 2009; Jiang et al. 2010; Fritz et al. 2013; Hamilton-Reeves et al. 2013; Chen et al. 2014), and the risk of cardiovascular diseases (Stampfer et al. 1991; O'Brian and Nguyen 1997; Setchell and Cassidy 1999; Tikkanen and Adlercreutz 2000; Watanabe et al. 2002); improve immunity (Sakai and Kogiso 2008; Wei et al. 2012); lower triglycerides and low-density lipoprotein (LDL) cholesterol and total cholesterol; and improve weight and fat loss (Cederroth and Nef 2009). It has been reported that isoflavones are generally produced by plants under environmental stresses such as the lack of nutrients, drought, or pathogen infection and they have a high affinity to bind estrogen receptors, thus affecting estrogen-regulated systems such as the central nervous system, cardiovascular, skeletal, reproductive, metabolic systems, etc., in humans and animals.

8.1.2 Cancer Prevention

Genistein possesses higher antioxidant properties than daidzein because it has three hydroxyl groups and both prevent gastric cancer cells from proliferating. Moreover, gastric cancer cells are arrested at the G2/M phase of the cell cycle by genistein (Park and Ko 2013; Preedy 2013). Genistein inhibits the enzyme protein tyrosine kinase (PTK) expressed in gastric cancer cells (Lin et al. 2000) which reduces carcinogenesis (Piontek et al. 1993; Preedy 2013). Kaufman et al. (1997) quantified daidzein and genistein contents in over 80 taxa of legumes and found that the leaves of *Psoralea corylifolia* contained high contents of the anticancer genistein (2 g kg^{-1} dry weight) and the roots of *Vicia faba* contained high contents of daidzein (1 g kg^{-1} dry weight). Soybeans (*Glycine max*) contained 13 and 10 mg kg^{-1} dry weight of daidzein and genistein, respectively (Kaufman et al. 1997). Mazur et al. (1998) developed an isotope dilution GC-MS (gas chromatography-mass spectrometry) method and quantified the isoflavones formononetin, biochanin A, daidzein, and genistein and lignans in seeds of 52 legumes (Mazur et al. 1998). They found that these legume seeds contained high concentrations of isoflavonoids (0–1853.35 mg.kg^{-1} dry weight) and low concentrations of lignans (0–15.85 mg.kg^{-1} dry weight) (Mazur et al. 1998). They also found that individuals who lived in regions or countries with lower risks of cancer and cardiovascular diseases had the highest concentrations of these phytoestrogen compounds in their plasma which made them strong candidates for cancer prevention (Mazur et al. 1998).

Many studies reported that soy products containing isoflavones prevent breast cancer in cell lines of rats (Barnes et al. 1990) and humans (Messina and Barnes 1991; Peterson and Barnes 1991; Coward et al. 1993; Adlercreutz 1995; Tham et al. 1998; Jiang et al. 2010) and prostate cancer cell lines in humans (Peterson and Barnes 1993; Rokhlin and Cohen 1995; Adlercreutz 1998; Tham et al. 1998). In an experiment in vitro, Rokhlin and Cohen (1995) studied the effect of genistein, a tyrosine protein kinase inhibitor, and other protein kinase and phosphatase inhibitors on the growth of several cell lines of human prostate cancer (PC3, DU145, ND1, ALVA31, JCA1, and LNCaP) and found that the proliferation of these prostate cancer cells lines is highly reduced but the reduction is dose-dependent (Rokhlin and Cohen 1995). There is ample evidence that tyrosine kinases especially play major roles in increasing many types of cancers including gastric cancers (Rokhlin and Cohen 1995; Dalu et al. 1998; Lin et al. 2000). Increased expression of genes such as erbB2/neu, k-sam, and c-met encoding for tyrosine kinases has been reported to increase the progression of gastric cancers in humans (Lin et al. 2000). Other studies reported a low incidence of breast cancer in Asian women which is due to the fact that they consume large amounts of soy isoflavones (Setchell et al. 1984; Barnes et al. 1990; Adlercreutz et al. 1991; Lee et al. 1991; Coward et al. 1993). In a commentary about the workshop on the role of soy products in cancer prevention sponsored by the National Cancer Institute, Messina and Barnes (1991) reported progress made on the involvement of soy products in reducing cancer risk (Messina and Barnes 1991). Animal models have been used to demonstrate the role of phytoestrogens in reducing breast cancer risk (Barnes et al. 1990). Another study reported that because they consume large amounts of soy products (containing isoflavones), oriental women have low risks of breast cancer than their American counterparts (Nagasawa 1980). Several scholars argued that the antiestrogenic activity of isoflavones might not be the only mechanism of their anticancer action but also their inhibition of tyrosine protein kinase activity (Akiyama et al. 1987), ribosomal S6 kinase (Linassier et al. 1990), and DNA topoisomerase II (Okura et al. 1988). Coward et al. (1993) described a method to separate and analyze isoflavones β-glycoside conjugates and aglycones in various soy foods and concluded that the concentrations of these compounds are similar in American and Asian foods. They found that β-glycoside conjugates are abundant in both American and Asian foods, while isoflavones aglycones are abundant in fermented Asian soy foods. Therefore, isoflavones are involved in lowering the mortality incidence and breast cancer in Asian women (Coward et al. 1993). Adlercreutz (1995) reported that bacteria can convert plant lignans and isoflavonoid glycosides into hormone-like substances with anti-oxidative and weak estrogenic activities, thus preventing certain types of cancers. These soy-derived isoflavones affect cell differentiation, sex hormone metabolism, growth factor action, malignant cell proliferation, protein synthesis, and angiogenesis making them natural substances preventing cancers (Adlercreutz 1995). In a recent study, authors treated the breast cancer cell lines MCF-7 cells with daidzein analogs and observed the inhibition of their proliferation and transcriptional activity of their genes especially the progesterone receptor (PgR) (Jiang et al. 2010). The study also found that the analog 2 (substitution of 7-hydroxyl hydrogen of daidzein by pyrano- group) was the most efficient in suppressing MCF-7 cell tumorigenesis in xenograft mouse (Jiang et al. 2010).

A review concluded that many studies showed the inhibitory effect of genistein on carcinogenesis in animals via modulation of genes involved in cell cycle control and apoptosis (Banerjee et al. 2008). In another study, authors studied the effect of daidzein and genistein on the cell cycle and cell proliferation of two prostate cancer cell lines (LNCaP and PC-3) (Wang et al. 2009). They observed that daidzein, genistein, and their combination as well as extracts from soybean cake were very efficient in inhibiting LNCaP and PC-3 cell proliferation and growth and increasing the G2/M ratio of LNCaP cell lines. A combination of genistein and daidzein and genistein alone caused the same effects in PC-3 cell lines. Also, both LNCaP and PC-3 cell lines showed a decrease in B1 cyclin expression for all isoflavone treatments (Wang et al. 2009). Knight and Eden (1996) summarized the effects of phytoestrogens on human health in a review article. They reviewed 861 articles on the subject from 1980 to 1995 and reported that phytoestrogens inhibit cancer cell line growth in humans and animals and reduce the risks of osteoporosis and cholesterol levels in humans (Knight and Eden 1996). Other studies reported that most cancer cells' growths are inhibited by genistein in vitro including breast cancer cells (Peterson and Barnes 1991) by blocking their cell cycle (Nguyen et al. 2006), prostate cancer dells (Wang et al. 2002), and colonic cancer cells (Messina and Bennink 1998). It is noteworthy to mention that colon cancer only killed over 500,000 people in 1996 and is the third and fourth leading cause of death for men and women, respectively (Messina and Bennink 1998). In a clinical and comprehensive review, Tham et al. (1998) discussed the role of isoflavones in reducing the menopausal symptoms, chronic diseases such as atherosclerosis, several types of cancers, hypercholesterolemia, osteoporosis, and coronary heart disease (Wilcox and Blumenthal 1995; Tham et al. 1998). There is substantial evidence that the lower rates of cardiovascular diseases and hormone-related cancers such as breast and prostate cancers in Asian populations are mainly due to the high consumption of isoflavones (Keys et al. 1984; Rose et al. 1986; Parkin 1989; Ziegler et al. 1993; Artaud-Wild et al. 1993; Tham et al. 1998). Messina (1999) summarized findings of more than 188 studies on the effects of soybean isoflavones on cancers including breast and prostate cancers and bones' health (Messina 1999). It has been also reported that isoflavone consumption at adequate levels exerts a positive effect on increasing bone density and mass in humans (Anderson and Garner 1997). Ren et al. (2001) summarized results from many clinical and epidemiological studies that show that soybean isoflavones lower incidences of osteoporosis, cardiovascular diseases, and certain cancers especially prostate and breast cancers (Ren et al. 2001).

8.1.3 Cardiovascular Disease Prevention

Several studies showed that soy proteins (containing isoflavones) lower cholesterol (Crouse et al. 1998; Lichttenstein 1998; Anthony and Clarkson 1998; Cederroth and Nef 2009) and reduce the symptoms of atherosclerosis (Wilcox and Blumenthal

1995; Honore et al. 1997; Setchell and Cassidy 1999; Cano et al. 2010) and osteoporosis (Cheng et al. 1994; Setchell and Cassidy 1999; Watanabe et al. 2002; Jiang et al. 2010; Wang et al. 2009). After ingestion, isoflavones undergo several changes especially deglycosylation (Heinonen et al. 2002). Isoflavone metabolism and effects on human health have been summarized previously (Cassidy et al. 2000a, b; Heinonen et al. 2002; Brouns 2002; Cederroth and Nef 2009).

Anderson et al. (1995) investigated the effect of soy protein intact on serum lipid concentrations and summarized the results of 38 publications of clinical trials in humans. They found significant decreases in concentrations of triglycerides, low-density lipoprotein (LDL) cholesterol, and total cholesterol in humans who consume soy proteins rather than animal proteins (Anderson et al. 1995). Other studies reported that isoflavones have similar effects of lowering triglycerides, low-density lipoprotein (LDL) cholesterol, and total cholesterol and improving weight and fat loss (Potter 1995; Cederroth and Nef 2009). Potter (1995) focused on the mechanisms of hypocholesterolemic effect of soy and stated that several studies suggested that (1) impairment of cholesterol absorption and/or bile acid reabsorption, (2) changes in endocrine status especially the alteration of insulin/glucagon ratio, (3) thyroid hormone concentration, and (4) saponins, fiber, minerals, phytic acid, and isoflavones are all potential causes of lowering triglycerides, low-density lipoprotein (LDL) cholesterol, and total cholesterol in humans and animals consuming these soy products (Potter 1995). Lichttenstein (1998) reviewed the effects of soy protein and isoflavones genistein, daidzein, and glycitein on reducing the risk of cardiovascular diseases from 1940 to 1998 (Lichttenstein 1998). The effect of soy proteins on blood cholesterol levels was reported in rabbits (Khosla et al. 1991; Kritchevsky 1995), rats (Potter 1995; Balmir et al. 1996), hamsters (Balmir et al. 1996), guinea pigs, pigs, monkeys, and baboons (Carroll and Kurowska 1995) as well as humans (Crouse et al. 1998, 1999). Balmir et al. (1996) investigated the effect of soy flour extracts on serum cholesterol and thyroid hormones in rats and hamsters. They found that the soy extracts reduced serum total cholesterol levels and increased thyroxine levels in rats; however, the soy extracts reduced serum total cholesterol levels but increased thyroxine levels in hamsters (Balmir et al. 1996). Tikkanen and Adlercreutz (2000) investigated the effects of soy isoflavones on coronary heart disease prevention and found that in subjects with high cholesterol levels, isoflavones tend to lower the plasma cholesterol levels and benefit arterial endothelial function. They and others (Lee et al. 2005; Lee 2006) demonstrated that isoflavones exert an antioxidant effect on low-density lipoproteins (LDLs). Moreover, isoflavone-containing LDLs slowed the proliferation of U937 cell lines in vitro (Tikkanen and Adlercreutz 2000). Ren et al. (2001) summarized results from many clinical and epidemiological studies that show that soybean isoflavones lower incidences of osteoporosis, cardiovascular diseases, and certain cancers especially prostate and breast cancers (Ren et al. 2001). Isoflavones contribute to reducing the risk of cardiovascular diseases by lowering LDL cholesterol levels in subjects consuming high amounts of these isoflavones (Carroll and Kurowska 1995; Setchell and Cassidy 1999). This was demonstrated in animals including rhesus monkeys and rats (Sirtori et al. 1993; Anthony et al. 1996, 1997; Setchell and Cassidy 1999).

Another study found that individuals who consumed >50 g/day of soy protein had their blood's total and LDL cholesterol and triglycerides significantly reduced; however, no significant effect was observed when these individuals consumed soybean isoflavones (Lichtenstein et al. 2002). Other studies concluded that isoflavones lower triglycerides, low-density lipoprotein (LDL) cholesterol, and total cholesterol and improve weight and fat loss (Cederroth and Nef 2009). There might be an overlap between the actions of soybean seed isoflavones and estrogen on cardiovascular disease (CVD) and coronary heart disease (CHD); however, it is well established that isoflavones reduce the risk of CVD, CHD, cerebral infraction (CI), and myocardial infarction (MI) (Kokubo et al. 2007).

8.1.4 Osteoporosis Prevention

Osteoporosis increases with age in both men and women. Several studies showed that soy proteins (containing isoflavones) reduce the symptoms osteoporosis (Ho et al. 1993; Cheng et al. 1994; Kao and Peng 1995; Arjmandi et al. 1996; Ishida et al. 1998; Setchell and Cassidy 1999; Watanabe et al. 2002). Ho et al. (1993) found that the rates of hip fractures are lower in men and women over 85 years old in age living in Hong Kong compared to their American counterpart and concluded that these facts may be related to the high consumption of isoflavones (Ho et al. 1993). Similarly, another study found that osteoporosis-related fractures are lower in Asian populations compared to Western populations (Kao and Peng 1995). Another study demonstrated that ipriflavone (IP) stimulates human bone osteoblasts (HBO) and thus bone formation. IP structure is similar to the structures of genistein and daidzein, and its ingestion at a dose of 600 mg/d stimulates bone formation and drastically reduces bone loss (Cheng et al. 1994). Another study reported the effects of soy protein on bone loss in an ovariectomized rat model (Arjmandi et al. 1996). They found that rats fed on soy had significantly high bone density than the ovariectomized (ovx) rats fed casein (Cheng et al. 1994). In menopausal women, it is very important to allow hormone replacement therapy in treating women with high cholesterol levels (hypercholesterolemia); however, the treatment must be on an individual basis to avoid negative side effects (O'Brian and Nguyen 1997). Ishida et al. (1998) investigated the effect of the isoflavones genistein and daidzein on bone loss in ovariectomized rats fed a calcium-deficient diet; the rats were fed genistein and daidzein orally for 4 weeks. They found that the rats' femurs have significantly lower calcium and phosphorus contents, strength, bone density, and ash weigh and atrophy of their uteruses. However, these effects were reversed by administering daidzein (Ishida et al. 1998). Messina (1999) summarized findings of more than 188 studies on the effects of soybean isoflavones on cancers including breast and prostate cancers and bones' health (Messina 1999). Ren et al. (2001) summarized results from many clinical and epidemiological studies that show that soybean isoflavones lower incidences of osteoporosis, cardiovascular diseases, and certain cancers especially prostate and breast cancers (Ren et al. 2001). In another study, authors fed young women isoflavone-rich tablets containing 20 and 40 mg of daidzein per day and observed longer menstrual cycles and increase in bone density (Watanabe et al. 2002).

8.1.5 Menopausal Symptom Prevention and Menstrual Cycles

Many studies reported that soybean isoflavone ingestion improved menopausal symptoms and menstrual cycles in women (Phipps et al. 1993; Setchell and Cassidy 1999; Anderson et al. 1999; Messina and Hughes 2003; Pop et al. 2008; Cederroth and Nef 2009). Phipps et al. (1993) investigated the effect of flaxseed powder ingestion on the menstrual cycle of several women with normal menstrual cycles. They found that longer luteal phase (LP) was associated with flax cycles. No significant differences in estradiol or estrone concentrations were observed. Higher LP progesterone/estradiol ratios during flax cycles were observed, and the study concluded that lignans play a significant role in the relationships between steroid action and sex and the risk of breast and other types of cancers (Phipps et al. 1993). Setchell and Cassidy (1999) reviewed literature on isoflavones and their effects on human health and summarized their effects on reducing symptoms of menopause, the risk of cancers, cardiovascular diseases, osteoporosis, and their effects on infants (Setchell and Cassidy 1999). Over a period of 12 weeks, 40–55% of women living on a diet rich in isoflavones had reduced hot flashes compared to women living on normal diet who had only 20–35% reduction in hot flashes (Wilcox et al. 1990; Baird et al. 1995; Setchell and Cassidy 1999). Anderson et al. (1999) investigated the health effects of soy isoflavones in menopausal women and found that isoflavones had beneficial effects on humans' arterial dimensions and blood lipids and bone parameters in rodents (Anderson et al. 1999). Messina and Hughes (2003) summarized data from 13 clinical trials involving more than 1700 menopausal women with hot flushes and found that there is a significant decrease in hot flushes in women consuming larger amounts of isoflavones in their diet; therefore, the authors recommended women to consume diets with isoflavones to reduce hot flushes (Messina and Hughes 2003). Another study demonstrated that even a high dose of up to 900 mg/day of soy isoflavone intake is safe and causes no significant effects on DNA damage nor apoptosis in postmenopausal women (Pop et al. 2008). Cederroth and Nef (2009) summarized findings of many studies showing the effects of isoflavones on human metabolism (Cederroth and Nef 2009).

8.1.6 Effects on Thyroid Function

To function properly, thyroid gland produces the hormones calcitonin, 3,5,3′-tri-iodothroin (T3), and 3,5,3′,5′-tetra-iodothroin (T4 or thyroxine), and iodine is essential for both T3 and T4 functions. Thyroid hormones play major roles in the development of the human skeleton, the cardiovascular, and central nervous systems (Biondi and Cooper 2008; Squadrito and Bitto 2013).

Several studies reported that soybean isoflavones may alter thyroid function (Ishizuki et al. 1991; Divi et al. 1997; Fritzpatrick 2000; Doerge and Chang 2002; Doerge and Sheehan 2002; Messina and Redmond 2006). Thirty-seven (37) healthy human adults with no iodine deficiency were given 30 g of soybean per day and showed an increase in the thyroid-stimulating hormone (TSH) and goiter (Ishizuki et al. 1991). Divi et al. (1997) investigated the role of soybean isoflavones on thyroid function in rodents. They found that the thyroid peroxidase (TPO) was inhibited by the isoflavones genistein and daidzein that act as alternate substrates in the presence of iodide ions. They also found that thyroxine synthesis can be inhibited by genistein using the substrate human goiter thyroglobulin (Divi et al. 1997). Several studies reported that porcine thyroid peroxidase (TPO) is deactivated by several naturally occurring flavonoids (Divi and Doerge 1996; Divi et al. 1997; Doerge and Chang 2002; Doerge and Sheehan 2002). Doerge and Chang (2002) administered 0, 5, 100, and 500 ppm of genistein to pregnant female rats and observed high levels of genistein in their blood and thyroid tissues (0.1–1.2 nmol/g) which translated into the loss of up to 80% of thyroid function in rats fed 500 ppm genistein (Doerge and Chang 2002). Therefore, they concluded that soybean isoflavones may cause thyroid problems or goiter by extrapolating their findings from animal essays to humans who may consume large amounts of isoflavones (Divi et al. 1997). However, this claim is refuted by many studies proving the health benefits of soy isoflavones and by the fact that many humans consuming large amounts of these isoflavones do not suffer from any symptoms of goiter or thyroid problems. Antithyroid effects of soybean isoflavones have been reported in infants consuming soy-based formulas; however, these effects can be reversed by feeding these infants with cow milk or diets supplemented with iodine (Pinchera et al. 1965; Labib et al. 1989; Chorazy et al. 1995). In a review paper, Messina and Redmond (2006) summarized data from 14 clinical trials containing data on the potential effects of soybean isoflavones on thyroid function and concluded that (1) there is no evidence that thyroid dysfunction is caused by soy isoflavones despite the fact that isoflavones have been reported to inhibit thyroid peroxidase (TPO) (Divi and Doerge 1996; Chang and Doerge 2000) and (2) thyroid abnormalities are not caused by soy isoflavones and (3) suggested a sufficient iodine intake among individuals consuming soy food regularly (Messina and Redmond 2006).

Infant formulas supplemented with iodine have been commercialized since the 1960s. In a study, Chorazy et al. (1995) warned of the excessive use of soy-based formulas for infant feeding and demanded close monitoring of those infants with thyroid function issues (Setchell et al. 1987; Chorazy et al. 1995; Fritzpatrick 2000). Fritzpatrick (2000) reported that the New Zealand Ministry of Health (MOH) issued a warning about the excessive use of soy-based formulas for infant feeding. The phytoestrogens can undergo some degrees of toxicity (Fritzpatrick 2000). Likewise, the director of the US National Centre for Toxicological Research, Food and Drug Administration (FDA), also warned about the risks of infant formulas based on soy isoflavones or phytoestrogens. In adults, excessive use of soy isoflavone-based or phytoestrogen-based diets may affect thyroid function negatively (Fritzpatrick 2000).

8.1.7 *Other Isoflavone Effects*

Many studies reported other human health benefits of soybean isoflavones such as antioxidant effect, effect on memory, improvement of overall cognitive functions, effects on impotency, improvement of erectile dysfunction, obesity prevention, anti-apoptotic action of neurons, lowering blood sugars, alleviating the symptoms of Alzheimer's disease, and many other benefits (Mitchell et al. 2001; Drewes et al. 2002, 2004; Fielden et al. 2003; Dixon and Sumner 2003; Duffy et al. 2003; Fritz et al. 2003; Dixon and Sumner 2003; File et al. 2005; Song et al. 2006; Assinder et al. 2007; Orgaard and Jensen 2008; Sarkaki et al. 2008; Chavarro et al. 2008; Bhatia and Greer 2008; Hooper et al. 2009; Kim et al. 2010; Bansal and Parle 2010; Santos-Galduroz et al. 2010; Cederroth et al. 2010, 2012; Akingbemi 2013; Wrenn 2013).

8.1.7.1 Effects on Memory Improvement

Isoflavones are believed to affect the brain, learning, and memory in three ways. They can impair learning and memory and interfere with brain development, prevent aging diseases such as Alzheimer's disease, or improve cognitive function in general (Drewes et al. 2004; Dixon and Sumner 2003; Wrenn 2013; Preedy 2013). In a review of legume natural products, Dixon and Sumner (2003) reported that isoflavones improved cognitive functions of students asked to perform complex mental tasks (File et al. 2005) and that Zulus tribe (South Africa) use plants containing phenylated isoflavones to treat impotency. These active compounds are known to improve erectile dysfunction (Drewes et al. 2004; Dixon and Sumner 2003). In experiments using male and female rats fed or not with dietary isoflavones, several studies observed an improvement of females' memory performance compared to males which indicate the beneficial effects of dietary isoflavones on enhancing memory performance at least in females (Lund et al. 2001; Ball et al. 2010). Sarkaki et al. (2008) studied the effect of soybean isoflavones on Alzheimer's disease using ovx rats and found that isoflavones may protect neurons and help in alleviating the symptoms of Alzheimer's disease (Sarkaki et al. 2008). Another study fed dietary isoflavones (0.23 mg g^{-1} of genistein and 0.25 mg g^{-1} of daidzein) to the first group of rats for 60 days before ovx and to the second group of rats for 30 days after ovx. The treatment (ovx) impaired the memory performance significantly; however, genistein and daidzein improved the memory performance in both groups (Monteiro et al. 2008). Li et al. (2010) observed that rat models of Alzheimer's disease injected with β-amyloid, a cause of Alzheimer's disease, have their memory impaired; however, when fed the isoflavone puerarin for 1 month, their memory improved to a level comparable to rats fed the Alzheimer's disease medication donepezil (Li et al. 2010; Wrenn 2013). Scopolamine, an acetylcholine antagonist, injected to rat models of Alzheimer's disease caused impairment of their memory; however, this memory deficit has been shown to be reversed by the oral administration of daidzein or daidzein at 5 mg g^{-1} (Kim et al. 2010; Wrenn 2013). Bansal and Parle (2010) fed dietary isoflavones to young and mature mice for 2 months among which a group fed scopolamine. Dietary isoflavones reversed significantly the scopolamine-induced memory deficit in these animals and protected them against memory impairment (Bansal and Parle 2010). A recent study showed that daidzein and genistein exert an antiapoptotic action in neurons of people infected with HIV-1 virus (Adams et al. 2012). Other studies reported that a diet with soy supplements for 6–12 weeks improved cognition and memory function in women aged 50–60 years (Duffy et al. 2003; File et al. 2005; Wrenn 2013). In another study, women were fed 80 mg g^{-1} of dietary isoflavones for 16 weeks, and authors observed an enhancement in these women's executive function (Santos-Galduroz et al. 2010; Wrenn 2013).

8.1.7.2 Effect on Obesity Reduction

Obesity epidemic is spreading in the USA and worldwide as it affects all ages, genders, and races or ethnic groups regardless of smoking status or educational levels (Kim et al. 2006; Kohno et al. 2006; Maesta et al. 2007; Aubertin-Leheudre et al. 2007; Orgaard and Jensen 2008; Mokdad et al. 1999; WHO 2017). The body mass index (BMI) of a person is the weight (in kg) divided by the height squared (in m^2) [BMI = W/H^2]. A person with BMI = 18.5–24.99, BMI = 25.0–29.99, and BMI > 30.0 is normal, overweight, and obese, respectively. Today, 1.9 billion people are overweight, and 650 million are obese. The epidemic is also affecting children among which 41 million are overweight (WHO 2017). The roles of isoflavones in reducing BMI and visceral body fat have been reported in several studies (Kim et al. 2006; Kohno et al. 2006; Maesta et al. 2007; Aubertin-Leheudre et al. 2007; Christie et al. 2010; Doonan et al. 2013). Fat pad and body weight as well as food intake have been lowered by genistein administered orally to *ovx* female mice (Kim et al. 2006). Another study demonstrated that β-conglycinin (a soybean isoflavone) consumed in candies by over 100 human volunteers over 12 weeks reduced significantly their visceral body fat and overall BMI (Kohno et al. 2006). In another study, 46 Brazilian overweight postmenopausal women took soy proteins and exercised for 16 weeks, and authors observed a significant reduction in their BMI (Maesta et al. 2007). In a similar study conducted in Canada, exercises combined with isoflavones reduced significantly BMI, body weight, and abdominal fat in obese postmenopausal women (Aubertin-Leheudre et al. 2007). Orgaard and Jensen (2008) reviewed literature on the effects of soy isoflavones on obesity which leads in many cases to type 2 diabetes, cardiovascular diseases, and other health problems. They concluded that many studies showed that isoflavones inhibit the enlargement of the adipose tissue in vivo and prevent obesity (Orgaard and Jensen 2008). Another recent study reported the significant reduction of BMI of 33 obese postmenopausal African-American and Caucasian women who consumed soy supplements containing isoflavones (Christie et al. 2010).

8.1.7.3 Effects on Reproduction

The male phenotype and fertility are maintained by sex hormones produced in testis that also produce sperm cells. Several studies reported that isoflavones interfere with the sperm quality, development, and production of sperm cells (Mitchell et al. 2001; Fielden et al. 2003; Song et al. 2006; Assinder et al. 2007; Chavarro et al. 2008; Bhatia and Greer 2008; Hooper et al. 2009; Cederroth et al. 2010, 2012; Akingbemi 2013). Fielden et al. (2003) fed sexual mature mice low levels of genistein (0, 0.1, 0.5, 2.5, and 10 mg kg^{-1} day^{-1}) and observed an increase in fertility but not sperm quantity or mobility at the highest dose of 10 mg kg^{-1} day^{-1} (Fielden et al. 2003). However, another study reported a reduction in the sperm number of rats fed high phytoestrogens for 24 days (Assinder et al. 2007). Likewise, a reduction in sperm number but not fertility was observed in male mice maintained on a soy-based diet of daidzein (150 ppm) and genistein (190 ppm) (Cederroth et al. 2010). Cederroth et al. (2012) reviewed literature on the effects of soy isoflavones on the human's reproductive system. Soy-based formulas have been used to feed infants without any apparent effect on their reproductive system for the past 40+ years although the American Academy of Pediatrics (AAP) does not recommend using such formulas exclusively (Bhatia and Greer 2008). Women who consume soy isoflavones may have their menstrual cycle's length increased (Hooper et al. 2009; Cederroth et al. 2012). Other studies reported conflicting results concerning males' fertility (Mitchell et al. 2001; Song et al. 2006; Chavarro et al. 2008). Mitchell et al. (2001) showed no influence of soy isoflavone consumption (40 mg/day) on sexual hormones or semen quality. Song et al. (2006) found that isoflavone consumption improved sperm count and motility. Chavarro et al. (2008) found that isoflavone consumption was inversely proportional to the concentration of sperm.

8.1.7.4 Effects on Bone Density

Glycitein, the third major seed isoflavone after daidzein and genistein, is also abundant in soybean seeds and has tremendous benefits on human and animal health (Lichttenstein 1998; Kinjo et al. 2004; Lee et al. 2010; Stephens and Bomser 2013). Several studies reported that glycitein may play a role in enhancing bone density and health in postmenopausal women (Yoshida et al. 2001; Gilad et al. 2006; Taku et al. 2010; Stephens and Bomser 2013); however, more studies are needed to confirm these findings.

8.1.7.5 Effects on Diabetes

Insulin resistance and abnormal beta-cells not able to secrete enough insulin are the main causes of type 2 diabetes (T2D). It has been reported that consumption of soybean isoflavones lowers the levels of sugars in the blood in both humans (Jayagopal et al. 2002; Cheng et al. 2004) and animals (Lavigne et al. 2000; Mezei et al. 2003; Ali et al. 2004). It is also well known that genistein stimulates high secretion of insulin in the presence of glucose in pancreatic cells of mice (Liu et al. 2006), increases the production of cyclic adenosine monophosphate (cAMP) in intercellular spaces of pancreatic tissues, and stimulates the activation of protein kinase A (Liu 2013).

8.1.7.6 Effects on Immunity and Other Effects

Several studies reported that soybean isoflavones or phytoestrogens can boost immunity in humans (Rapaport et al. 1997; Record et al. 1997; Yellayi et al. 2002; Sakai et al. 2006; Sakai and Kogiso 2008; Wei et al. 2012). Genistein effect on immunity was extensively studied especially in ovariectomized (OVX) mice; however, studies using non-OVX mice report different outcomes. Among the notable effects of a diet high in genistein in mice are a decrease in thymus apoptosis, a decrease in thymus weight and the number of peripheral lymphocytes (Sakai and Kogiso 2008), a reduction of certain types of T cells, suppression of antigen (Ag)-induced antibody (Ab) synthesis (Rapaport et al. 1997; Yellayi et al. 2002; Sakai and Kogiso 2008), inhibition of tumor growth by enhancing the function of immune cells (Record et al. 1997; Sakai and Kogiso 2008), reduction in arthritis (Verdrengh et al. 2003), and suppression of dermatitis (Sakai et al. 2006).

Fritz et al. (2003) investigated the in vivo antioxidant effect of isoflavones in ten healthy women (18–35 years of age) and found significantly lower levels of six urinary nonpolar compounds (NPC) (Fritz et al. 2003).

8.2 Positive and Negative Effects on Animals

The effects of isoflavones on animal health have also been well documented since 1946 (Bennetts et al. 1946). These effects also include the prevention of certain types of cancers such as prostate and breast cancers, prevention of cardiovascular diseases, reproduction, positive effects on thyroid gland function, bone density, obesity reduction, and other effects (Nagasawa 1980; Barnes et al. 1990; Messina and Barnes 1991; Okura et al. 1988; Lamartiniere et al. 1995; Adams 1995; Balmir et al. 1996; Divi et al. 1997; Pollard and Luckert 1997; Ishida et al. 1998; Lichttenstein 1998; Khosla et al. 1991; Kritchevsky 1995; Potter 1995; Balmir et al. 1996; Carroll and Kurowska 1995; Sharma et al. 1992; Tham et al. 1998; Thigpen et al. 1999; Fitzsimons et al. 1989; Ashby et al. 2000; Brown and Setchell 2001; Doerge and

Chang 2002; Degen et al. 2002; Mezei et al. 2003; Sirtori et al. 2004; Ali et al. 2004; Cederroth and Nef 2009; Cavallini et al. 2009). Many of these studies have been conducted on small lab animals such as rats, mice, baboons, and hamsters as well as monkeys which are fed different diets containing soybean isoflavones and/or soy proteins. For example, Brown and Setchell (2001) demonstrated that the levels of serum isoflavones were 30,000- to 60,000-fold higher in adult rats (2613 ± 873 ng g^{-1}) and mice (2338 ± 531 ng g^{-1}) fed soy-containing diet compared to the controls which pose a concern for animal experiments because these isoflavones may have biochemical and morphological influences on these lab animals (Brown and Setchell 2001). Several studies warned about studies that use casein-based and phytoestrogen-based diets in animals (Thigpen et al. 1999; Ashby et al. 2000). Thigpen et al. (1999) reviewed many studies using casein-based and isoflavone-based or phytoestrogen-based diets on small lab animals especially rodents and warned that these diets available in the market might affect the outcomes or results of these studies especially studies dealing with endocrine system's disruption and reproductive toxicity (Thigpen et al. 1999). Shortly thereafter, Ashby et al. (2000) warned also that the phytoestrogen/soy content ratio of the diet cannot solely determine its estrogenic activity (Ashby et al. 2000).

8.2.1 Effects on Cancer Prevention

In a commentary about the workshop on the role of soy products in cancer prevention sponsored by the National Cancer Institute, Messina and Barnes (1991) reported progress made on the involvement of soy products in reducing cancer risk (Messina and Barnes 1991). Animal models have been used to demonstrate the role of phytoestrogens in reducing breast cancer risk (Barnes et al. 1990). Another study reported that because they consume large amounts of soy products (containing isoflavones), oriental women have low risks of breast cancer than their American counterparts (Nagasawa 1980). Several scholars argued that the antiestrogenic activity of isoflavones might not be the only mechanism of their anticancer action but also their inhibition of tyrosine protein kinase activity (Akiyama et al. 1987), ribosomal S6 kinase (Linassier et al. 1990), and DNA topoisomerase II (Okura et al. 1988). In another study, inflammation of lateral prostate was observed in rats feeding on soy-free diet for 11 weeks compared to rats feeding on a soy diet that have no prostate inflammation (Sharma et al. 1992; Tham et al. 1998). A study reported that the soy isoflavone genistein inhibits erbB2/neu expression in rats' prostate (Dalu et al. 1998). In another study, genistein and dimethylsulfoxide (control) have been administered to rats exposed to dimethylbenz(a)anthracene (carcinogen) where mammary caner was induced. The study found a high reduction in mammary tumor growth and multiplication in rats fed genistein compared to those fed dimethylsulfoxide which demonstrates a protection of genistein against mammary cancer (Lamartiniere et al. 1995). Thigpen et al. (1999) assayed 12 diets, genistein, daidzein, and other compounds on rodents and found high daidzein and genistein contents in the majority of these diets. Liver cancer is reduced significantly in mice fed 30% soy-based diet (Fitzsimons et al. 1989), and rats fed soy-based diet showed a 40% reduction in mammary tumors compared to the control group (Barnes et al. 1990; Lamartiniere et al. 1995; Pollard and Luckert 1997). In another study, Pollard and Luckert (1997) fed the genetically susceptible Lobund-Wistar (L-W) rats with high amounts of daidzein and genistein (soy protein) and found that the prostate cancer is highly reduced in these rats compared to the control rats fed low amounts of daidzein and genistein. Moreover, the period where these rats were free of prostate cancer was prolonged by 27% compared to the control group (Pollard and Luckert 1997).

8.2.2 Effects on Cardiovascular Disease Prevention

Lichttenstein (1998) reviewed the effects of soy protein and isoflavones genistein, daidzein, and glycitein on reducing the risk of cardiovascular diseases from 1940 to 1998 (Lichttenstein 1998). The effect of soy proteins on blood cholesterol levels was reported in rabbits (Khosla et al. 1991; Kritchevsky 1995), rats (Potter 1995; Balmir et al. 1996), hamsters (Balmir et al. 1996), guinea pigs, pigs, monkeys, and baboons (Carroll and Kurowska 1995). Balmir et al. (1996) investigated the effect of soy flour extracts on serum cholesterol and thyroid hormones in rats and hamsters. They found that the soy extracts reduced serum total cholesterol levels and increased thyroxine levels in rats; however, the soy extracts reduced serum total cholesterol levels but increased thyroxine levels in hamsters (Balmir et al. 1996). Another study examined the effect of soy isoflavones on atherosclerotic female and male monkeys. These monkeys have been put on high-isoflavone and low-isoflavone diets for a period of 6 months, and their coronary arteries were examined. They found that acetylcholine causes a constriction of male arteries, while arteries of females on high-isoflavone diet dilated more compared to those of females on low-isoflavone diet (6.4% ± 1.2%, mean ± SEM) (Honore et al. 1997). Another study demonstrated that although white lupin is low in isoflavones, it lowered cholesterol levels in rats when they are fed lupin protein and that lower cholesterol stimulated LDL receptors (Sirtori et al. 2004). Another study found that soybean isoflavones reduced total LDL and steroid hormone levels in lean and obese rats; however, no effect was observed for thyroid hormones (Ali et al. 2004).

8.2.3 Effects on Thyroid Gland Function

Divi et al. (1997) investigated the role of soybean isoflavones on thyroid function in rodents. They found that the thyroid peroxidase (TPO) was inhibited by the isoflavones genistein and daidzein that act as alternate substrates in the presence of iodide ions. They also found that thyroxine synthesis can be inhibited by genistein using the substrate human goiter thyroglobulin. Therefore, they concluded that soybean isoflavones may cause thyroid problems or goiter by extrapolating their findings from animal essays to humans who may consume large amounts of isoflavones (Divi et al. 1997). Cederroth and Nef (2009) summarized findings of many studies showing various metabolic effects on rodents injected or fed isoflavones or soy proteins (Cederroth and Nef 2009). Doerge and Chang (2002) reviewed literature on the antithyroid effects of isoflavones in both humans and animals. Thyroid peroxidase (PTO) catalyzes thyroid hormone synthesis, and lactoperoxidase (LPO) is the synthetic commercial model structurally similar to PTO. Thyroxine (T4) and triiodothyronine (T3) are released in thyroglobulin after tyrosyl residue iodination and oxidative coupling. Genistein and daidzein inhibit TPO-catalyzed iodination and coupling (Doerge and Chang 2002). The reduction of PTO activity can reach up to 80% in rats, and females are affected more than males (Doerge and Chang 2002). Other studies showed that rats fed 20% soy-containing diet had a severe thyroid demonstrated by an increase in thyroid weight, a sharp decrease in T4, and increase in thyroid-stimulating hormone (TSH) levels (Kimura et al. 1976; Ikeda et al. 2000).

8.2.4 Effect on Obesity

Mezei et al. (2003) investigated the effects of isoflavones on lipid and glucose metabolism in obese Zucker (OZR) rats which are used as type 2 diabetes. Like in humans treated with antidiabetic PPAR (peroxisome-proliferator activated receptor) agonists, improved metabolism was observed in these OZR rats fed on high-isoflavone soy proteins. The expression of lipid and glucose metabolism genes is regulated by PPARs (Mezei et al. 2003).

8.2.5 Effects on Reproduction

Adams (1995) found impaired function of the ovaries and the loss of embryos in ewes and cows feeding on forage with phytoestrogens. Infertility is temporary in cows but may be permanent in ewes with prolonged periods of feeding on forage with phytoestrogens (Adams 1995). Other studies investigated the role of isoflavones in uterine weight (uterotrophic response) in rodents and found that ovx rats who consumed a diet free of isoflavones had higher uterine weights compared to the controls on a standard diet containing genistein (Ashby et al. 2000; Degen et al. 2002).

8.2.6 Effects on Bone Density, Diabetes, Immunity, and Other Effects

Ishida et al. (1998) investigated the effect of the isoflavones genistein and daidzein on bone loss in ovariectomized (ovx) rats fed a calcium-deficient diet; the rats were fed genistein and daidzein orally for 4 weeks. They found that the rats' femurs have significantly lower calcium and phosphorus contents, strength, bone density, and ash weight and atrophy of their uteruses. However, these effects were reversed by administering daidzein (Ishida et al. 1998).

Another study showed that streptozotocin (STZ)-induced diabetic rats fed on genistein and isolated soy protein supplement (ISP) had their glucokinase levels increased but their glucose-6-phosphatase activity decreased and concluded that genistein and ISP had correction properties for hyperglycemia and prevent diabetes (Lee 2006). Another study demonstrated that soy isoflavones lowered total cholesterol in hypercholesterolemic rabbits (Cavallini et al. 2009).

Wei et al. (2012) showed that the immune sensitization of mice was suppressed by the administration of daidzein and genistein. This response is most likely due to a suppression of dendritic cell (DC cell) function. In mice treated with isoflavones, dendritic cell number, T cell proliferation, and cytokine synthesis from T cells are reduced. This shows that isoflavones especially daidzein and genistein play a role in immunity regulation (Wei et al. 2012).

8.3 Positive Effects on Plants Especially Legumes

Many studies showed that root exposure to isoflavones especially daidzein and genistein leads to more nodulation, nodule formation, gene expression, and efficient nitrogen fixation in many plant species (Banfalvi et al. 1988; Smit et al. 1992; Loh et al. 1994; Kaufman et al. 1997; Begum et al. 2001; Cseke et al. 2006; Subramanian et al. 2006, 2007). Isoflavones also cause an increase in resistance to pathogens including fungi attacks (Kaufman et al. 1997). Banfalvi et al. (1988) demonstrated that the construct nod:: lacZ was induced by both genistein and daidzein. The activity of lacZ fused with nod gene was induced by soybean seed extract which is rich in isoflavones (Banfalvi et al. 1988). Few studies reported that some flavonoids can inhibit nodulation in certain types of rhizobia bacteria (Kosslak et al. 1990). Morris and Ward (1992) found that daidzein and genistein produced by the soybean roots act as chemoattractants to *Phytophthora sojae* zoospores (Morris and Ward 1992). Smit et al. (1992) demonstrated that nodYABCSUIJ operon of *Bradyrhizobium japonicum* is highly induced by both daidzein and genistein. Other soybean seed extracts induce nodYABCSUIJ operon in synergy with genistein but not alone (Smit et al. 1992). Begum et al. (2001) investigated the effect of numerous compounds including daidzein, genistein, hesperetin, rutin, apigenin, naringenin, luteolin, etc., on nodulation (nod) gene expression in lentil and pea in a growth chamber under controlled environments. They found that hesperetin combined with naringenin caused high activities of nod genes. Moreover, treatment with hesperetin and pre-induced rhizobium bacteria caused an increase of 60% in nodule numbers in both lentil and pea plants (Begum et al. 2001). Subramanian et al. (2006) demonstrated that isoflavones induce nodulation (nod) genes in *Bradyrhizobium japonicum* species. The key enzyme in the isoflavone biosynthetic pathways, isoflavone synthase (IFS), is highly induced by *B. japonicum*, and its gene silencing caused a high reduction in nodulation; however, when chalcone synthase is silenced, no effect on nodulation was observed but led to increased genistein and low daidzein contents. Moreover, nodulation can occur only with genistein (Subramanian et al. 2006, 2007).

8.4 Conclusion

The overwhelming majority of studies discussed above prove that soybean isoflavones and proteins have tremendous benefits on human and animal health despite the fact that several recent studies tried to deny, be cautious about, or minimize their beneficial effects (Sacks et al. 2006; Cassidy et al. 2006; Song et al. 2007; Xiao 2008; Chandrareddy et al. 2008). Kwon (2014) stated that the data on the effects of isoflavones on breast cancer is mixed, not conclusive, and mechanisms of inhibition or promotion of cancer cell line growth by soybean isoflavones need to be identified and fully understood (Kwon 2014). Sacks et al. 2006 summarized data from 22 trials and argued that the low reduction in LDL levels observed (3%) is mostly due to the high soy proteins used in these studies (50 g) which is half of daily intake. They also stated that they observed no significant evidence of soy proteins, rich in isoflavones, in lowering LDL, blood pressure, lipoprotein A, or triglycerides. Furthermore, they claimed that soy isoflavones had no effect on lowering symptoms of menopause, postmenopausal bone loss, breast, prostate, or endometrium cancers (Sacks et al. 2006). These claims contradict the overwhelming majority of studies discussed at length in this chapter that demonstrated numerous health benefits of consuming soybean isoflavones. For example, Cassidy et al. (2006) did a comprehensive review of literature on possible isoflavone effects on postmenopausal women's health. They concluded that soybean isoflavones had (1) a positive role in bone health, (2) no effect on blood lipid levels and blood pressure, and (3) a positive effect on reducing hot flushes. They also stated that they could not reach a conclusion on soybean isoflavone effects on diabetes, cognitive function, and breast, prostate, and colon cancers because of the low number of randomized controlled trials (RCT) (Cassidy et al. 2006). Song et al. (2007) argued whether the positive effects of soy isoflavones on lowering cholesterol, triglyceride levels, LDL cholesterol, and breast cancer are due to isoflavones or not. They further stated that the only adverse effects of isoflavones have been reported in animals and that caution must be taken in using isoflavones as alternative to estrogen in postmenopausal women (Song et al. 2007). In a review, Xiao (2008) stated that the American Heart Association through its Nutrition Committee concluded that soy isoflavone consumption had no effect on HDL cholesterol, blood pressure, or triglycerides but did lower slightly LDL cholesterol. Furthermore, several studies suggested to be extra cautious about consuming extra soy products because of possible adverse health effects (Xiao 2008). Another study reported possible adverse effects of isoflavones in three women. All three women had uterine bleeding. The first had uterine polyp and an increase in endometrium, the second had severe dysmenorrhea, and the third had uterine leiomyomata and secondary infertility. These symptoms are reversed with soy removal from diet (Chandrareddy et al. 2008).

References

Adams, N.R. 1995. Detection of the effects of phytoestrogens on sheep and cattle. *Journal of Animal Science* 73 (5): 1509–1515.

Adams, S.M., M.V. Aksenova, M.Y. Aksenov, C.F. Mactutus, and R.M. Booze. 2012. Soy isoflavones genistein and daidzein exert anti-apoptotic actions via a selective ER-mediated mechanism in neurons following HIV-1 Tat1–86 exposure. *PLoS One* 7 (5): e37540. https://doi.org/10.1371/journal.pone.0037540.

Adlercreutz, H. 1995. Phytoestrogens: Epidemiology and a possible role in cancer protection. *Environmental Health Perspectives* 103: 103–111.

Adlercreutz, M. 1998. Epidemiology of phytoestrogens. *Baillière's Clinical Endocrinology and Metabolism* 12: 605–623.

Adlercreutz, H., H. Honjo, A. Higashi, T. Fotsis, E. Hamalainen, T. Hasegawa, et al. 1991. Urinary excretion of lignans and isoflavonoid phytoestrogens in Japanese men and women consuming a traditional Japanese diet. *The American Journal of Clinical Nutrition* 54: 1093–1100.

Adlercreutz, H., T. Fotsis, J. Lampe, K. Wahala, T. Makela, G. Brunow, et al. 1993. Quantitative determination of lignans and isoflavones in plasma of omnivorous and vegetarian women by isotope-dilution gas chromatography mass spectrometry. *Scandinavian Journal of Clinical and Laboratory Investigation* 53: 5–18.

Akingbemi, B.T. 2013. Soy isoflavones and testicular function. In *Isoflavones: Chemistry, analysis, function, and effects*, ed. V.R. Preedy. Cambridge: The Royal Society of Chemistry (RSC) Press.

Akiyama, T., J. Ishida, S. Nakagawa, H. Ogawara, S. Watanabe, N. Itoh, M. Shibuya, and Y. Fukami. 1987. Genistein, a specific inhibitor of tyrosine-specific protein kinases. *The Journal of Biological Chemistry* 262: 5592–5595.

Ali, A.A., M.T. Velasquez, C.T. Hansen, A.I. Mohamed, and S.J. Bhathena. 2004. Effects of soybean isoflavones, probiotics, and their interactions on lipid metabolism and endocrine system in an animal model of obesity and diabetes and endocrine system in an animal model of obesity and diabetes. *The Journal of Nutritional Biochemistry* 15 (10): 583–590.

Anderson, J.J.B., and S.C. Garner. 1997. The effects of phytoestrogens on bone. *Nutrition Research* 17 (10): 1617–1632.

Anderson, J.M., B.M. Johnstone, and M.E. Cook-Newell. 1995. Mata-analysis of the effects of soy protein intake on serum lipids. *The New England Journal of Medicine* 333: 276–282.

Anderson, J.J.B., M.S. Anthony, J.M. Cline, S.A. Washburn, and S.C. Garner. 1999. Health potential of soy isoflavones for menopausal women. *Public Health Nutrition* 2 (4): 489–504.

Anthony, M.S., and T.B. Clarkson. 1998. Comparison of soy phytoestrogens and conjugated equine estrogens on atherosclerosis progression in postmenopausal monkeys. *Circulation* 97: 829.

Anthony, M.S., T.B. Clarkson, C.L. Hughes, T.M. Morgan, and G.L. Burke. 1996. Soybean isoflavones improve cardiovascular risk factors without affecting the reproductive system of peripubertal rhesus monkeys. *The Journal of Nutrition* 126: 43–50.

Anthony, M.S., T.B. Clarkson, B.C. Bullock, and J. Wagner. 1997. Soy protein versus soy phytoestrogens in the prevention of diet-induced coronary artery atherosclerosis of male Cynomolgous monkeys. *Arteriosclerosis, Thrombosis, and Vascular Biology* 17: 2524–2531.

Arjmandi, B.H., L. Alekel, B.W. Hollis, D. Amin, M. Stacewicz-Sapuntzakis, P. Guo, and S.C. Kukreja. 1996. Dietary soybean protein prevents bone loss in an ovariectomized rat model of osteoporosis. *The Journal of Nutrition* 126: 161–167.

Artaud-Wild, S.M., S.L. Connor, G. Sexton, and W.E. Connor. 1993. Differences in coronary mortality can be explained by differences in cholesterol and saturated fat intakes in 40 countries but not in France and Finland. A paradox. *Circulation* 88: 2771–2779.

Ashby, J., H. Tinwell, and J. Odum. 2000. Uterotrophic activity of a 'phyto-estrogen-free' rat diet. *Environmental Health Perspectives* 108: A12–A13.

Assinder, S., R. Davis, M. Fenwick, and A. Glover. 2007. Adult-only exposure of male rats to a diet of high phytoestrogen content increases apoptosis of meiotic and post-meiotic germ cells. *Reproduction* 133: 11–19.

Aubertin-Leheudre, M., C. Lord, A. Khalil, and I. Dionne. 2007. Six months of isoflavone supplement increases fat-free mass in obese-sarcopenic post-menopausal women: A randomized double-blind controlled trail. *European Journal of Clinical Nutrition* 61 (12): 1442–1444.

Baird, D., D. Umbach, and L. Lansdell. 1995. Dietary intervention study to assess estrogenicity of dietary soy, among postmenopausal women. *Clinical Endocrinology 80 (5): 1685–1690.*

Ball, E.R., M.K. Caniglia, J.L. Wilcox, K.A. Overton, M.J. Burr, B.D. Wolfe, B.J. Sanders, A.B. Wisniewski, and C.C. Erenn. 2010. Effects of genistein in the maternal diet on reproductive development and spatial learning in male rats. *Hormones and Behavior* 57: 313–322.

Balmir, F., R. Staack, E. Jeffrey, M.D.B. Jimenez, L. Wang, and S.M. Potter. 1996. An extract of soy flour influences serum cholesterol and thyroid hormones in rats and hamsters. *The Journal of Nutrition* 126: 3046–3053.

Banerjee, S., Y. Li, Z. Wang, and F.H. Sarkar. 2008. Multi-targeted therapy of cancer by genistein. *Cancer Letters* 269 (2): 226–242.

Banfalvi, Z., A. Nieuwkoop, M. Schell, L. Besl, and G. Stacey. 1988. Regulation of nod gene expression in *Bradyrhizobium japonicum. Molecular & General Genetics* 214: 420–424.

Bansal, N., and M. Parle. 2010. Soybean supplementation helps reverse age- and scopolamine-induced memory deficits in mice. *Journal of Medicinal Food* 13: 1293–1300.

Barnes, S. 1995. Effect of genistein on in vitro and in vivo models of cancer. *Journal of Nutrition* 125: 777S–783S.

Barnes, S., C. Grubbs, K.D.R. Setchell, and J. Carlson. 1990. Soybeans inhibit mammary tumors in models of breast cancer. *Progress in Clinical and Biological Research* 347: 239–253.

Begum, A.A., S. Leibovitch, P. Migner, and F. Zhang. 2001. Specific flavonoids induced nod gene expression and pre-activated nod genes of *Rhizobium leguminosarum* increased pea (*Pisum sativum* L.) and lentil (Lens culinaris L.) nodulation in controlled growth chamber environments. *Journal of Experimental Botany* 52 (360): 1537–1543.

Bennetts, H.W., E.J. Underwood, and F.L. Shier. 1946. A specific breeding problem of sheep on subterranean clover pastures in Western Australia. *Australian Veterinary Journal* 22: 2–12.

Bhatia, J., and F. Greer. 2008. Use of soy protein-based formulas in infant feeding. *Pediatrics* 121: 1062–1068.

Biondi, B., and D.S. Cooper. 2008. The clinical significance of subclinical thyroid dysfunction. *Endocrinology Review* 29: 76–131.

Brouns, F. 2002. Soya isoflavones a new and promising ingredient for the health food sector. *Food Research International* 35 (2–3): 187–193. https://doi.org/10.1016/s0963-9969(01)00182-x.

Brown, N.M., and K.D.R. Setchell. 2001. Animal models impacted by phytoestrogens in commercial chow: Implications for pathways influenced by hormones. *Laboratory Investigation* 81 (5): 735–747.

Cano, A., M.A. Garcia-Perez, and J.J. Tarin. 2010. Isoflavones and cardiovascular disease. *Maturitas* 67: 219–226.

Carroll, K.K., and E.M. Kurowska. 1995. Soy consumption and cholesterol reduction: Review of animal and human studies. *The Journal of Nutrition* 125: 594S–597S.

Cassidy, A., B. Hanley, and R.M. Lamuela-Raventos. 2000a. Isoflavones, lignans and stilbenes – Origins, metabolism and potential importance to human health. *Journal of Science and Food Agriculture* 80: 1044–1062.

Cassidy, A., H. Bryn, and M. Rosa. 2000b. Isoflavones, lignans and stilbenes origins, metabolism and potential importance to human health. *Journal of the Science of Food and Agriculture* 80: 1044–1062.

Cassidy, A., P. Albertazzi, I.L. Nielsen, W. Hall, G. Williamson, I. Tetens, S. Atkins, H. Cross, Y. Manios, A. Wolk, C. Steiner, and F. Branca. 2006. Critical review of health effects of soybean phytoestrogens in post-menopausal women. *Proceedings of the Nutrition Society* 65: 76–92.

Cavallini, D.C.U., R. Bedani, L.Q. Bomdespacho, R.C. Vendramini, and E.A. Rossi. 2009. Effects of probiotic bacteria, isoflavones and simvastatin on lipid profile and atherosclerosis in cholesterol-fed rabbits: A randomized double-blind study. *Lipids in Health and Disease* 8: 1. https://doi.org/10.1186/1476-511X-8-1.

Cederroth, C.R., and S. Nef. 2009. Soy, phytoestrogens and metabolism: A review. *Molecular and Cellular Endocrinology* 304 (1–2): 30–42. https://doi.org/10.1016/j.mce.2009.02.021.
Cederroth, C.R., C. Zimmermann, J.L. Beny, O. Schaad, C. Combepine, P. Descombes, D.R. Doerge, F.P. Pralong, J.D. Vassalli, and S. Nef. 2010. Potential detrimental effects of a phytoestrogen-rich diet on male fertility in mice. *Molecular and Cellular Endocrinology* 321: 152–160.
Cederroth, C.R., C. Zimmermann, and S. Nef. 2012. Soy, phytoestrogens and their impact on reproductive health. *Molecular and Cellular Endocrinology* 355: 192–200.
Chandrareddy, A., O. Muneyyirci-Delale, S.I. McFarlane, and O.M. Murad. 2008. Adverse effects of phytoestrogens on reproductive health: A report of three cases. *Complementary Therapies in Clinical Practice* 14: 132–135.
Chang, C. and D.R. Doerge. 2000. Dietary Genistein Inactivates Rat Thyroid Peroxidase in Vivo without an Apparent Hypothyroid Effect. *Toxicology and Applied Pharmacology 168 (3): 244–252.*
Chavarro, J.E., T.L. Toth, S.M. Sadio, and R. Hauser. 2008. Soy food and isoflavone intake in relation to semen quality parameters among men from an infertility clinic. *Human Reproduction* 23: 2584–2590.
Chen, M., Y. Ra, Y. Zheng, S. Wei, Y. Li, T. Guo, and P. Yin. 2014. Association between soy isoflavone intake and breast cancer risk for pre- and post-menopausal women: A meta-analysis of epidemiological studies. *PLoS One* 9 (2): e89288. https://doi.org/10.1371/journal.pone.0089288.
Cheng, S.L., S.F. Zhang, T.L. Nelson, P.M. Warlow, and R. Civitelli. 1994. Stimulation of human osteoblast differentiation and function by Ipriflavone and its metabolites. *Calcified Tissue International* 55: 356–362.
Cheng, S.Y., N.S. Shaw, K.S. Tsai, and C.Y. Chen. 2004. The hypoglycemic effects of soy isoflavones on postmenopausal women. *Journal of Women's Health* 13: 1080–1086.
Chorazy, P.A., S. Himelhoch, N.J. Hopwood, N.G. Greger, and D.C. Postellon. 1995. Persistent hypothyroidism in an infant receiving a soy formula: Case report and review of the literature. *Pediatrics* 96 (1): 148.
Christie, D.R., J. Grant, B.E. Darnell, V.R. Chapman, A. Gastaldelli, and C.K. Sites. 2010. Metabolic effects of soy supplementation in post-menopausal Caucasian and Africa-American women: A randomized, placebo-controlled trail. *American Journal of Obstetrics and Gynecology* 203 (2): 153e1–153e9.
Coward, L., N.C. Barnes, K.D.R. Setchell, and S. Barnes. 1993. Genistein, daidzein, and their β-glycoside conjugates: Anti-tumor Isoflavones in soybean foods from American and Asian diets. *Journal of Agricultural and Food Chemistry* 41: 1961–1967.
Crouse, J.R., J.G. Terry, T.M. Morgan, B.L. McGill, D.H. Davis, T. King, J.E. Ellis, and G.L. Burke. 1998. Soy protein containing isoflavones reduces plasma concentrations of lipids and lipoproteins. *Circulation* 97: 816.
Crouse, J.R., T. Morgan, J.G. Terry, and J. Ellis. 1999. A randomized trail comparing the effects of casein with that of soy protein containing varying amounts of isoflavones on plasma concentrations of lipids and lipoproteins. *Archives of Internal Medicine* 159: 2070–2076.
Cseke, L.J., A. Kirakosyan, P.B. Kaufman, S.L. Warber, A.J. Duke, and H.L. Brielmann. 2006. *Natural products from plants*. 2nd ed. Boca Raton: CRC Press-Taylor & Francis Group.
Dalu, A., J.F. Haskell, L. Coward, and C.A. Lamartiniere. 1998. Genistein, a component of soy, inhibits the expression of the EGF and ErbB2/Neu receptors in the rat dorsolateral prostate. *The Prostate* 37 (1): 36–43.
Degen, G.H., P. Janning, P. Diel, and H.M. Bolt. 2002. Estrogenic isoflavones in rodent diets. *Toxicology Letters* 128: 145–157.
Divi, R.L. and D.R. Doerge. 1996. Inhibition of thyroid peroxidase by dietary flavonoids. *Chemical Research in Toxicology 9: 16–23.*
Divi, R.L., H.C. Chang, and D.R. Doerge. 1997. Anti-thyroid isoflavones from soybean: Isolation, characterization, and mechanisms of action. *Biochemical Pharmacology* 54: 1087–1096.
Dixon, R.A., and L.W. Sumner. 2003. Legume natural products: Understanding and manipulating complex pathways for human and animal health. *Plant Physiology* 131: 878–885.

Doerge, D.R., and H.C. Chang. 2002. Inactivation of thyroid peroxidase by soy isoflavones, in vitro and in vivo. *Journal of Chromatography. B, Analytical Technologies in the Biomedical and Life Sciences* 777: 269–279.

Doerge, D.R., and D.M. Sheehan. 2002. Goitrogenic and estrogenic activity of soy isoflavones. *Environmental Health Perspectives* 110 (3): 349–353.

Doonan, B.B., E. Wu, and J.M. Wu. 2013. Prevention and management of obesity by isoflavones. In *Isoflavones: Chemistry, analysis, function, and effects*, ed. V.R. Preedy. Cambridge: The Royal Society of Chemistry (RSC) Press.

Drewes, S.E., M.H. Horn, O.Q. Munro, J.T.B. Dhlamini, J.J.M. Meyer, and N.C. Rakuambo. 2002. Pyrano-isoflavones with erectile-dysfunction activity from Eriosema kraussianum. *Phytochemistry 59 (7): 739–747.* https://doi.org/10.1016/S0031-9422(02)00035-3.

Drewes, S.E., M.M. Horn, F. Khan, O.Q. Munro, J.T.B. Dhlamini, C. Rakuambo, and J.J.M. Meyer. 2004. Minor pyrano-isoflavones from Eriosema kraussianum: Activity-, structure-, and chemical reaction studies. *Phytochemistry* 65 (13): 1955–1961.

Duffy, R., H. Wiseman, and S.E. File. 2003. Improved cognitive function in postmenopausal women after 12 weeks of consumption of a soya extract containing isoflavones. *Pharmacology, Biochemistry, and Behavior* 75: 721–729.

Fielden, M.R., S.M. Samy, K.C. Chou, and T.R. Zacharrewski. 2003. Effect of human dietary exposure levels of genistein during gestation and lactation on long-term reproductive development and sperm quality in mice. *Food and Chemical Toxicology* 41: 447–454.

File, S.E., D.E. Hartley, S. Elsabagh, R. Duffy, and H. Wiseman. 2005. Cognitive improvement after 6 weeks of soy supplements in postmenopausal women is limited to frontal lobe function. *Menopause* 12: 193–201.

Fitzsimons, J.T., N.V. Orson, and A.A. El-Aaser. 1989. Effects of soyabean and ascorbic acid on experimental carcinogenesis. *Comparative Biochemistry and phys$iology. A, Comparative Physiology* 93 (1): 285–290. https://doi.org/10.1016/0300-9629(89)90218-1.

Forkmann, G., and S. Martens. 2001. Metabolic engineering and applications of flavonoids. *Current Opinion in Biotechnology* 12: 155–160.

Fritz, K.L., C.M. Seppanen, M.S. Kurzer, and A.S. Csallany. 2003. The in vivo antioxidant activity of soybean isoflavones in human subjects. *Nutrition Research* 23 (2003): 479–487.

Fritz, H., D. Seely, G. Flower, B. Skidmore, and R. Fernandes. 2013. Soy, red clover, and isoflavones and breast cancer: A systematic review. *PLoS One* 8 (11): e81968. https://doi.org/10.1371/journal.pone.0081968.

Fritzpatrick, M. 2000. Soy formulas and the effects of isoflavones on the thyroid. *New Zealand Medical Journal* 113 (1103): 24–26.

Gilad, L.A., O. Tirosh, and B. Schwartz. 2006. Phytoestrogens regulate transcription and translation of vitamin D receptor in colon cancer cells. *Journal of Endocrinology* 191 (2): 387–398.

Guerra, M.C., E. Speroni, M. Broccoli, M. Cangini, P. Pasini, A. Minghetti, N. Crespi-Perellino, M. Mirasoli, G. Gantelli-Forti, and M. Paolini. 2000. Comparison between Chinese medical herb Pueraria lobata crude extract and its main isoflavone puerarin antioxidant properties and effects on rat liver CYP-catalyzed drug metabolism. *Life Sciences* 67: 2997–3006.

Hamilton-Reeves, J.M., S. Banerjee, S.K. Banerjee, J.M. Holzbeierlein, J.B. Thrasher, S. Kambhampati, J. Keighley, and P.V. Veldhuizen. 2013. Short-term soy isoflavone intervention in patients with localized prostate cancer: A randomized, double-blind, placebo-controlled trial. *PLoS One* 8 (7): e68331. https://doi.org/10.1371/journal.pone.0068331.

Heinonen, S.M., K. Wahala, and H. Adlercreutz. 2002. Metabolism of isoflavones in human subjects. *Phytochemistry Reviews* 1: 175–182.

Ho, S.C., W.E. Bacon, T. Harris, A. Looker, and S. Maggi. 1993. Hip fracture rates in Hong Kong and the United States, 1988 through 1989. *American Journal of Public Health* 83: 694–697.

Honore, E.K., J.K. Williams, M.S. Anthony, and T.B. Clarkson. 1997. Soy isoflavones enhance coronary vascular reactivity in atherosclerotic female macaques. *Fertility and Sterility* 67: 148–154.

Hooper, L., J.J. Ryder, M.S. Kurzer, J.W. Lampe, M.J. Messina, W.R. Phipps, and A. Cassidy. 2009. Effects of soy protein and isoflavones on circulating hormone concentrations in pre- and post-menopausal women: A systematic review and meta-analysis. *Human Reproduction Update* 15: 423–440.

Ikeda, T., A. Nishikawa, T. Imazawa, S. Kimura, and M. Hirose. 2000. Dramatic synergism between excess soybean intake and iodine deficiency on the development of rat thyroid hyperplasia. *Carcinogenesis* 21 (4): 707–713.

Ishida, H., T. Uesugi, K. Hirai, T. Toda, H. Nukaya, K. Yokotsuka, and K. Tsuji. 1998. Preventive effects of the plant isoflavones, daidzein and genistein, on bone loss in ovariectomized rats fed a calcium-deficient diet. *Biological and Pharmaceutical Bulletin* 21 (1): 62–66.

Ishizuki, Y., Y. Hirooka, Y. Murata, and K. Togashi. 1991. The effects on the thyroid gland of soybeans administered experimentally in healthy subjects. *Nippon Naibunpi Gakkai Zasshi* 67: 622.

Jayagopal, V., P. Albertazzi, E.S. Kilpatrick, E.M. Howarth, P.E. Jennings, D.A. Hepburn, and S.L. Atkin. 2002. Beneficial effects of soy phytoestrogen intake in postmenopausal women with type 2 diabetes. *Diabetes Care* 25: 1709–1714.

Jiang, Q., F. Payton-Stewart, S. Elliott, J. Driver, L.V. Rhodes, Q. Zhang, S. Zheng, D. Bhatnagar, S.M. Boue, B.M. Collins-Burow, J. Sridhar, C. Stevens, J.A. McLachlan, T.E. Wiese, M.E. Burow, and G. Wang. 2010. Effects of 7-O substitutions on estrogenic and antiestrogenic activities of daidzein analogues in MCF-breast cancer cells. *Journal of Medicinal Chemistry* 53: 6153–6163.

Kao, P.C., and F.K. Peng. 1995. How to reduce the risk factors of osteoporosis in Asia. *Chinese Medical Journal* 55: 209–213.

Kaufman, P.B., J.A. Duke, H. Brielmann, J. Boik, and J.E. Hoyt. 1997. A comparative survey of leguminous plants as sources of the isoflavones, genistein and daidzein: Implications for human nutrition and health. *The Journal of Alternative and Complementary Medicine* 3 (1): 7–12.

Keys, A., A. Menotti, C. Aravanis, H. Blackburn, B.S. Djordevič, R. Buzina, A.S. Dontas, F. Fidanza, M.J. Karvonen, N. Kimura, I. Mohaček, S. Nedeljkovič, V. Puddu, S. Punsar, H.L. Taylor, S. Conti, D. Kromhout, and H. Toshima. 1984. The seven countries study: 2,289 deaths in 15 years. *Preventive Medicine* 13: 141–154.

Khosla, P., S. Samman, and K. Carroll. 1991. Decreased receptor-mediated LDL catabolism in casein-fed rabbits precedes the increase in plasma cholesterol levels. *The Journal of Nutritional Biochemistry* 2: 203–209.

Kim, H.K., C. Nelson-Dooley, M.A. Della-Fera, J.Y. Yang, W. Zhang, J. Duan, D.L. Hartzell, M.W. Hamrick, and C.A. Baile. 2006. Genistein decreases food intake, body weight, and fat pad weight and causes adipose tissue apoptosis in ovariectomized female mice. *Journal of Nutrition* 136 (2): 409–414.

Kim, D., H. Jung, S. Park, J. Kim, S. Lee, J. Choi, J. Cheong, K. Ko, and J. Ryu. 2010. The effects of daidzin and its aglycon, daidzein, on the scopolamine induced memory impairment in male mice. *Archives of Pharmacy Research, Pharmaceutical Society of Korea* 33: 1685–1690.

Kimura, S., J. Suwa, M. Ito, and H. Sato. 1976. Development of malignant goiter by defatted soybean with iodine-free diet in rats. *Gann* 67: 763.

Kinjo, J., R. Tsuchihashi, K. Moriko, T. Hirose, T. Aomori, T. Nagao, H. Okabe, T. Nohara, and Y. Masamune. 2004. Interactions of phytoestrogens with estrogen receptors α and β (III). Estrogenic activities of soy isoflavone aglycones and their metabolites isolated from human urine. *Biological and Pharmaceutical Bulletin* 27 (2): 185–188.

Knight, D.C., and J.A. Eden. 1996. A review of clinical effects of phytoestrogen. *Obstetrics and Gynecology* 87: 897–904.

Kohno, M., M. Hirotsuka, M. Kito, and Y. Matsuzawa. 2006. Decreases in serum triacylglycerol and visceral fat mediated by dietary soybean beta-conglycinin. *Journal of Atherosclerosis and Thrombosis* 13 (5): 247–255.

Kokubo, Y., H. Iso, J. Ishihara, K. Okada, and M. Inoue. 2007. Association of dietary intake of soy, beans, and isoflavones with risk of cerebral and myocardial infarctions in Japanese populations. *The Japan Public Health Center-Based (JPHC) Study Cohort I*. https://doi.org/10.1161/CIRCULATIONAHA.106.683755.

Kosslak, R.M., R.S. Joshi, B.A. Bowen, H.E. Paaren, and E.R. Appelbaum. 1990. Strain-specific inhibition of nod gene induction in *Bradyrhizobium japonicum* by flavonoid compounds. *Applied and Environmental Microbiology* 56: 1333–1341.

Kritchevsky, D. 1995. Dietary protein, cholesterol, and atherosclerosis: A review of the early history. *The Journal of Nutrition* 125: 589S–593S. https://doi.org/10.1093/jn/125.suppl_3.589S.

Kwon, Y. 2014. Effect of soy isoflavones on the growth of human breast tumors: Findings from preclinical studies. *Food Science & Nutrition* 2 (6): 613–622. https://doi.org/10.1002/fsn3.142.

Labib, M., R. Gama, J. Wright, V. Marks, and D. Robins. 1989. Dietary maladvice as a cause of hypothyroidism and short stature. *British Medical Journal* 293 (6668): 232–233.

Lamartiniere, C.A. 2000. Protection against breast cancer with genistein: A component of soy. *The American Journal of Clinical Nutrition* 71: 1705S–1707S.

Lamartiniere, C.A., J. Moore, M. Holland, and S. Barnes. 1995. Neonatal genistein chemoprevents mammary cancer. *Proceedings of the Society for Experimental Biology and Medicine* 208: 120–123.

Lavigne, C., A. Marette, and H. Jacques. 2000. Cod and soy protein compared with casein improve glucose tolerance and insulin sensitivity in rats. *American Journal of Physics* 278: E491–E500.

Lee, J.S. 2006. Effects of soy protein and genistein on blood glucose, antioxidant enzyme activities, and lipid profile in streptozotocin-induced diabetic rats. *Life Sciences* 79: 1578–1584.

Lee, H.P., J. Lee, L. Gourley, S.W. Duffy, N.E. Day, and J. Estève. 1991. Dietary effects on breast-cancer risk in Singapore. *The Lancet* 337 (8751): 1197–1200.

Lee, C.H., L. Yang, J.Z. Xu, S.Y.V. Yeung, Y. Huang, and Z.Y. Chen. 2005. Relative antioxidant activity of soybean isoflavones and their glycosides. *Food Chemistry* 90: 735–741.

Lee, E.J., S.Y. Kim, J.W. Hyun, S.W. Min, D.H. Kim, and H.S. Kim. 2010. Glycitein inhibits glioma cell invasion through down-regulation of MMP-3 and MMP-9 gene expression. *Chemico-Biological Interactions* 185 (1): 18–24.

Li, J., G. Wang, J. Liu, L. Zhou, M. Dong, R. Wang, X. Li, X. Li, C. Lin, and Y. Niu. 2010. Puerarin attenuates amyloid-c-induced cognitive impairment through suppression of apoptosis in rat hippocampus in vivo. *European Journal of Pharmacology* 649: 195–201.

Lichtenstein, A.H., S.M. Jalbert, H. Adlercreutz, B.R. Goldin, H. Rasmussen, E.J. Schaefer, and L.M. Ausman. 2002. Lipoprotein response to diets high in soy or animal moderately hypercholesterolemic subjects. *Arteriosclerosis, Thrombosis, and Vascular Biology* 22: 1852–1858.

Lichttenstein, A.H. 1998. Soy protein, isoflavones and cardiovascular disease risk. *The Journal of Nutrition* 128 (10): 1589–1592. https://doi.org/10.1093/jn/128.10.1589.

Lin, W., H.W. Kao, D. Robinson, H.J. Kung, C.W. Wu, and C.H. Chen. 2000. Tyrosine kinases and gastric cancer. *Oncogene* 19: 5680–5689.

Linassier, C., M. Pier, J.B. Le Pecq, and J. Pierre. 1990. Mechanisms of action in NIH 3T3 cells of genistein, an inhibitor of EGF receptor tyrosine kinase activity. *Biochemical Pharmacology* 39: 187–193.

Liu, D. 2013. Genistin and insulin secretory function. In *Isoflavones: Chemistry, analysis, function, and effects*, ed. V.R. Preedy. Cambridge: The Royal Society of Chemistry (RSC) Press.

Liu, D., W. Zhen, Z. Yang, J.D. Carter, H. Si, and K.A. Reynolds. 2006. Genistein acutely stimulates insulin secretion in pancreatic β-cells through a cAMP-dependent protein kinase pathway. *Diabetes* 55 (4): 1043–1050. https://doi.org/10.2337/diabetes.55.04.06.db05-1089.

Lo, F.H., N.L. Mak, and K.N. Leung. 2007. Studies on the anti-tumor activities of the soy isoflavone daidzein on murine neuroblastoma cells. *Biomedicine & Pharmacotherapy* 61: 591–595.

Loh, J.T., S.C. Ho, J.L. Wang, and M. Schindler. 1994. Carbohydrate binding activities of *Bradyrhizobium japonicum*: IV. Effect of lactose and flavones on the expression of the lectin, BJ38. *Glycoconjugate Journal* 11: 363–370.

Lund, T.D., T.W. West, L.Y. Tian, L.H. Bu, D.L. Simmons, K.D.R. Setchell, H. Adlercreutz, and E.D. Lephart. 2001. Visual spatial memory is enhanced in female rats (but inhibited in males) by dietary soy phytoestrogens. *BMC Neuroscience* 2: 20.

Maesta, N., E.A. Nahas, J. Nahas-Neto, F.L. Orsatti, C.E. Fernandes, P. Traiman, and R.C. Burini. 2007. Effects of soy protein and resistance exercise on body composition and blood lipids in postmenopausal women. *Maturitas* 56 (4): 350–358.

Mazur, W.M., J.A. Duke, K. Wahala, S. Rasku, and H. Adlercreutz. 1998. Isoflavonoids and lignans in legumes: Nutritional and health aspects in humans. *The Journal of Nutritional Biochemistry* 9: 193–200.

Messina, M. 1999. Legumes and soybeans: Overview of their nutritional profiles and health effects. *The American Journal of Clinical Nutrition* 70 (suppl): 439S–450S.

Messina, M., and S. Barnes. 1991. The role of soy products in reducing risk of cancer. *Workshop Report Commentary* 83 (8): 541–547.

Messina, M., and M. Bennink. 1998. Ten soyfoods, isoflavones and risk of colonic cancer: A review of the in vitro and in vivo data. *Baillikre's Clinical Endocrinology and Metabolism* 12 (4): 707–728.

Messina, M., and C. Hughes. 2003. Efficacy of soy foods and soybean isoflavone supplements for alleviating menopausal symptoms is positively related to initial hot flush frequency. *Journal of Medicinal Food* 6 (1): 1–11.

Messina, M., and G. Redmond. 2006. Effects of soy protein and soybean isoflavones on thyroid function in healthy adults and hypothyroid patients: A review of the relevant literature. *Thyroid* 16 (3): 249–258.

Mezei, O., W.J. Banz, R.W. Steger, M.R. Peluso, T.A. Winters, and N. Shay. 2003. Soy isoflavones exert antidiabetic and hypolipidemic effects through the PPAR pathways in obese Zucker rats and murine RAW 264.7 cells. *Journal of Nutrition* 133: 1238–1243.

Mitchell, J.H., E. Cawood, D. Kinniburgh, A. Provan, A.R. Collins, and D.S. Irvine. 2001. Effect of a phytoestrogen food supplement on reproductive health in normal males. *Clinical Science (London, England)* 100: 613–618.

Mokdad, A.H., M.K. Serdula, W.H. Dietz, B.A. Bowman, J.S. Marks, and J.P. Koplan. 1999. The spread of the obesity epidemic in the United States, 1991–1998. *Journal of the American Medical Association–JAMA* 282 (16): 1519–1522.

Monteiro, S., C. de Mattos, J. Ben, C. Netto, and A. Wyse. 2008. Ovariectomy impairs special memory: Prevention and reversal by a soy isoflavone diet. *Metabolic Brain Disease* 23: 243–253.

Morris, P.F., and E.W.B. Ward. 1992. Chemoattraction of zoospores of the soybean pathogen, *Phytophthora sojae*, by isoflavones. *Physiological and Molecular Plant Pathology* 40 (1): 17–22.

Murphy, P.A. 1982. Phytoestrogen content of processed soybean products. *Food Technology* 34: 60–64.

Nagasawa, H. 1980. Nutrition and breast cancer: A survey of experimental and epidemiological evidence. *IRCS Journal of Medical Science* 8: 317–325.

Nguyen, D.T., E. Hernandez-Montes, D. Vauzour, A.H. Schonthal, C. Rice-Evans, E. Cadenas, and J.P. Spencer. 2006. The intracellular genistein metabolite 5,7,3′, 4′-tetrahydroxyisoflavone mediates G2–M cell cycle arrest in cancer cells via modulation of the p38 signaling pathway. *Free Radical Biology and Medicine* 41 (8): 1225–1239.

O'Brian, T., and T.T. Nguyen. 1997. Lipids and lipoproteins in women. *Mayo Clinic Proceedings* 72 (3): 235–244. https://doi.org/10.4065/72.3.235.

Okura, A., H. Arakawa, H. Oka, T. Yoshinari, and Y. Monden. 1988. Effects of genistein on topoisomerase activity and the growth of [val 12] Ha-ras-transformed NIH 3T3 cells. *Biochemical and Biophysical Research Communications* 157: 183–189.

Orgaard, A., and L. Jensen. 2008. The effects of soy isoflavones on obesity. *Experimental Biology and Medicine* 233: 1066–1080.

Park, S.K., and K.P. Ko. 2013. Isoflavones against gastric cancer: Function and effects. In *Isoflavones: Chemistry, analysis, function, and effects*, ed. V.R. Preedy. Cambridge: The Royal Society of Chemistry (RSC) Press.

Parkin, D.M. 1989. Cancers of the breast, endometrium, and ovary: Geographic correlations. *European Journal of Cancer & Clinical Oncology* 25: 1917–1925.

Peterson, G., and S. Barnes. 1991. Genistein inhibition of the growth of human breast cancer cells: Independence from estrogen receptors and the multi-drug resistance gene. *Biochemical and Biophysical Research Communications* 179 (1): 661–667.

———. 1993. Genistein and biochanin a inhibit the growth of human prostate cancer cells but not epidermal growth factor receptor tyrosine autophosphorylation. *The Prostate* 22: 335–345.

Phipps, W.R., M.C. Martini, J.W. Lampe, J.L. Slavin, and M.S. Kurzer. 1993. Effect of flaxseed ingestion on the menstrual cycle. *The Journal of Clinical Endocrinology* 77: 1215–1219.

Pinchera, A., M.H. MacGillivray, J.D. Crawford, and A.G. Freeman. 1965. Thyroid refractoriness in an athyreotic cretin fed soybean formula. *The New England Journal of Medicine* 265 (2): 83–87.

Piontek, M., K.J. Hengels, R. Porschen, and G. Strohmeyer. 1993. Antiproliferation effect of tyrosine kinase inhibitors in epidermal growth factor-stimulated growth of human gastric cancer cells. *Anticancer Research* 13: 2119–2123.

Pollard, M., and P. Luckert. 1997. Influence of isoflavones in soy protein isolates on development of induced prostate-related cancers in L-W rats. *Nutrition and Cancer* 28: 41–45.

Pop, E.A., L.M. Fischer, A.D. Coan, M. Gitzinger, J. Nakamura, and S.H. Zeisel. 2008. Effects of a high daily dose of soy isoflavones on DNA damage, apoptosis and estrogenic outcomes in healthy, postmenopausal women – A Phase I clinical trial. *Menopause* 15 (4): 684–692. https://doi.org/10.1097/gme.0b013e318167b8f2.

Potter, S.M. 1995. Overview of proposed mechanisms for the hypocholesterolemic effect of soy. *The Journal of Nutrition* 125: 606S–611S.

Preedy, V.R. 2013. *Isoflavones: Chemistry, analysis, function, and effects*. Cambridge: The Royal Society of Chemistry (RSC) Press.

Rapaport, F.T., T. Terashima, K. Tsukuda, K. Kono, T. Takayama, and K. Malinowski. 1997. Suppression of lectin, alloantigen, and xenoantigen-induced T-cell proliferation by genistein. *Transplantation Proceedings* 29: 1261–1264.

Record, I.R., J.L. Boroadbent, R.A. King, I.E. Dreosti, R. Head, and A.L. Tonkin. 1997. Genistein inhibits growth of B16 melanoma cells in vitro and in vivo and promotes differentiation in vitro. *International Journal of Cancer* 72: 860–864.

Reinli, K., and G. Block. 1996. Phytoestrogen content of foods-compendium of literature values. *Nutrition and Cancer* 26: 123–148.

Ren, M.Q., G. Kuhn, J. Wegner, and J. Chen. 2001. Isoflavones, substances with multi-biological and clinical properties. *European Journal of Nutrition* 40: 135–146.

Rokhlin, O.W., and M.B. Cohen. 1995. Differential sensitivity of human prostatic cancer cell lines to the effects of protein kinase and phosphatase inhibitors. *Cancer Letters* 98: 103–110.

Rose, D.P., A.P. Boyar, and E.L. Wynder. 1986. International comparisons of mortality rates for cancer of the breast, ovary, prostate and colon, and per capita food consumption. *Cancer* 58: 2363–2371.

Sacks, F.M., A. Lichtenstein, L. Van Horn, W. Harris, P. Kris-Etherton, and M. Winston. 2006. Soy protein, isoflavones, and cardiovascular health: An American Heart Association Science Advisory for professionals from the Nutrition Committee. *Circulation* 113: 1034–1044.

Sakai, T., and M. Kogiso. 2008. Soy isoflavones and immunity. *The Journal of Medical Investigation* 55: 167–173.

Sakai, T., M. Kogiso, K. Mitsuya, T. Komatsu, and S. Yamamotoi. 2006. Genistein suppresses development of spontaneous atopic-like dermatitis in NC/Nga mice. *Journal of Nutritional Science and Vitaminology* 52: 293–296.

Santos-Galduroz, R.F., J.C.F. Galduroz, R.L. Facco, H. Hachul, and S. Tufik. 2010. Effects of isoflavone on the learning and memory of women in menopause: A double-blind placebo-controlled study. *Brazilian Journal of Medical and Biological Research* 43: 1123–1126.

Sarkaki, A., R. Amani, M. Badavi, A.Z. Moghaddam, H. Aligholi, M. Safahani, and M. Haghighizadeh. 2008. Pre-treatment effect of different doses of soy isoflavones on spatial learning and memory in an ovariectomized animal model of Alzheimer's disease. *Pakistan Journal of Biological Sciences* 11 (8): 1114–1119.

Setchell, K.D.R., and A. Cassidy. 1999. Dietary isoflavones: Biological effects and relevance to human health. *The Journal of Nutrition* 129: 758S–767S.

Setchell, K.D.R., A.M. Lawson, F.L. Mitchell, H. Adlercreutz, D.N. Kirk, and M. Axelson. 1980. Lignans in man and animal species. *Nature (London)* 287: 740–742.

Setchell, K.D., S.P. Borriello, P. Hulme, D.N. Kirk, and M. Axelson. 1984. Nonsteroidal estrogens of dietary origin: Possible roles in hormone-dependent disease. *The American Journal of Clinical Nutrition* 40 (3): 569–578. https://doi.org/10.1093/ajcn/40.3.569.

Setchell, K.D.R., M.B. Welsh, and C.K. Lim. 1987. HPLC analysis of phytoestrogens in soy protein preparations with ultraviolet, electrochemical and thermospray mass spectrometric detection. *Journal of Chromatography A* 368: 315–323.

Sharma, O.P., H. Adlercrueutz, J.D. Strandberg, B.R. Zirkin, D.S. Coffey, and L.L. Ewing. 1992. Soy of dietary source plays a preventive role against the pathogenesis of prostatitis in rats. *The Journal of Steroid Biochemistry and Molecular Biology* 43: 557–564.

Sirtori, C.R., R. Even, and M.R. Lovati. 1993. Soybean protein diet and plasma cholesterol: From therapy to molecular mechanisms. *Annals of the New York Academy of Sciences* 676: 188–201.

Sirtori, C.R., M.R. Lovati, C. Manzoni, S. Castiglioni, M. Duranti, C. Magni, S. Morandi, A. D'Agostina, and Anna Arnoldi. 2004. Proteins of white lupin seed, a naturally isoflavone-poor legume, reduce cholesterolemia in rats and increase LDL receptor activity in HepG2 cells. *The Journal of Nutrition* 134: 18–23.

Smit, G., V. Puvanesarajah, R.W. Carlson, W.M. Barbour, and G. Stacey. 1992. Bradyrhizobium japonicum nodD1 can be specifically induced by soybean flavonoids that do not induce the nodYABCSUIJ operon. *The Journal of Biological Chemistry* 267: 310–318.

Song, G., L. Kochman, E. Andolina, R.C. Herko, K.J. Brewer, and V. Lewis. 2006. Beneficial effects of dietary intake of plant phytoestrogens on semen parameters and sperm DNA integrity in infertile men. *Fertility and Sterility* 86: S49.

Song, W.O., O.K. Chun, I. Hwang, H.S. Shin, B.G. Kim, K.S. Kim, S.Y. Lee, D. Shin, and S.G. Lee. 2007. Soy isoflavones as safe functional ingredients. *Journal of Medicinal Food* 10 (4): 571–580.

Squadrito, F., and A. Bitto. 2013. Isoflavones and thyroid function: An overview. In *Isoflavones: Chemistry, analysis, function, and effects*, ed. V.R. Preedy. Cambridge: The Royal Society of Chemistry (RSC) Press.

Stampfer, M.J., G.A. Colditz, W.C. Willett, J.E. Manson, B. Rosner, F.E. Speizer, and C.H. Hennekens. 1991. Postmenopausal estrogen therapy and cardiovascular disease. *The New England Journal of Medicine* 325: 756–762.

Stephens, B.R., and J.A. Bomser. 2013. Glycitein in health. In *Isoflavones: Chemistry, analysis, function, and effects*, ed. V.R. Preedy. Cambridge: The Royal Society of Chemistry (RSC) Press.

Subramanian, S., G. Stacey, and O. Yu. 2006. Endogenous isoflavones are essential for the establishment of symbiosis between soybean and *Bradyrhizobium japonicum*. *The Plant Journal* 48: 261–273. https://doi.org/10.1111/j.1365-313x.2006.02874.x.

———. 2007. Distinct, crucial roles of flavonoids during legume nodulation. *Trends in Plant Science* 12 (7): 282–285. https://doi.org/10.1016/j.tplants.2007.06.006.

Taku, K., M.K. Melby, M.S. Kurzer, S. Mizuno, S. Watanabe, and Y. Ishimi. 2010. Effects of soy isoflavone supplements on bone turnover markers in menopausal women: Synthetic review and meta-analysis of randomized controlled trials. *Bone* 47 (2): 413–423.

Tham, D.M., C.D. Gardner, and W.L. Haskell. 1998. Potential health benefits of dietary phytoestrogens: A review of the clinical, epidemiological, and mechanistic evidence. *The Journal of Clinical Endocrinology and Metabolism* 83: 2223–2235.

Thigpen, J.E., K.D.R. Setchell, K.B. Ahlmark, J. Locklear, T. Spahr, G.F. Cavines, M.F. Goelz, J.K. Haseman, R.R. Newbold, and D.B. Forsyth. 1999. Phytoestrogen content of purified, open- and closed-formula laboratory animal diets. *Laboratory Animal Science* 49 (5): 530–536.

Tikkanen, M.J., and H. Adlercreutz. 2000. Dietary soy-derived isoflavone phytoestrogens. *Biochemical Pharmacology* 60: 1–5.

Verdrengh, M., I.M. Johnson, R. Holdahl, and A. Tarkowski. 2003. Genistein as an anti-inflammatory agent. *Inflammation Research* 52: 341–346.

Wang, J., I.E. Eltoum, and C.A. Lamartiniere. 2002. Dietary genistein suppresses chemically induced prostate cancer in Lobund-Wistar rats. *Cancer Letters* 186 (1): 11–18.

Wang, B.F., J.S. Wang, J.F. Lu, T.H. Kao, and B.H. Chen. 2009. Antiproliferation effect and mechanism of prostate cancer cell lines as affected by isoflavones from soybean cake. *Journal of Agricultural and Food Chemistry* 57: 2221–2232.

Watanabe, S., S. Uesugi, and Y. Kikuchi. 2002. Isoflavones for prevention of cancer, cardiovascular diseases, gynecological problems and possible immune potentiation. *Biomedicine & Pharmacotherapy* 56 (6): 302–312.

Wei, J., S. Bhatt, L.M. Chang, H.A. Sampson, and M. Masilamani. 2012. Isoflavones, genistein and daidzein, regulate mucosal immune response by suppressing dendritic cell function. *PLoS One* 7 (10): e47979. https://doi.org/10.1371/journal.pone.0047979.

WHO (World Health Organization). 2017. http://www.who.int/mediacentre/factsheets/fs311/en/. Accessed on 22 Dec 2017.

Wilcox, G., M.L. Wahlqvist, and H.G. Burger. 1990. Oestrogenic effects of plant foods of postmenopausal women. *British Medical Journal 301:905–906.*

Wilcox, J.N., and B.F. Blumenthal. 1995. Thrombotic mechanisms in atherosclerosis: Potential impact of soy proteins. *The Journal of Nutrition* 125 (Suppl): 631S–638S.

Wrenn, C.C. 2013. Dietary isoflavones and learning and memory. In *Isoflavones: Chemistry, analysis, function, and effects*, ed. V.R. Preedy. Cambridge: The Royal Society of Chemistry (RSC) Press.

Xiao, C.W. 2008. Health effects of soy protein and isoflavones in humans. *The Journal of Nutrition* 138: 1244S–1249S.

Yellayi, S., A. Naaz, M.A. Szewczkowski, T. Sato, J.A. Woods, J. Chang, M. Serge, C.D. Allred, W.G. Helferich, and P.S. Cooke. 2002. The phytoestrogen genistein induces thymic and immune changes: A human health concern. *PNAS* 99: 7616–7621.

Yoshida, H., T. Teramoto, K. Ikeda, and Y. Yamori. 2001. Glycitein effect on suppressing the proliferation and stimulating the differentiation of osteoblastic MC3T3-E1 cells. *Bioscience, Biotechnology, and Biochemistry* 65 (5): 1211–1213.

Ziegler, R.G., R.N. Hoover, M.C. Pike, A. Hildesheim, A.M.Y. Nomura, D.W. West, A.H. Wu-Williams, L.N. Kolonel, P.L. Horn-Ross, J.F. Rosenthal, and M.B. Hyer. 1993. Migration patterns and breast cancer risk in Asian-American women. *Journal of the National Cancer Institute* 85: 1819–1827.

Chapter 9
Environmental Factors Affecting Isoflavone Contents

Moulay Abdelmajid Kassem

9.1 Introduction

In addition to being genotype-dependent, many studies showed that soybean seed isoflavone contents depend also on environmental and growth conditions including temperature, light, rainfall, seed storage, seed size, water availability, growing season, row spacing, and other environmental factors (Tsukamoto et al. 1995; Lee et al. 2003a, b, 2007, 2008; Vyn et al. 2002; Duke et al. 2003; Bennett et al. 2004; Kim et al. 2005; Caldwell et al. 2005; Al-Tawaha et al. 2005; Lozovaya et al. 2005; Subramanian et al. 2007; Morrison et al. 2008; Phommalth et al. 2008; Rasolohery et al. 2008; Gutierrez-Gonzalez et al. 2010a, b; Shi et al. 2010; Ragin et al. 2014). Several studies also showed that seed isoflavones accumulate in soybeans due to biotic stresses from various pathogen infections such as bacteria, fungi, and viruses (Stafford 1997; Treutte 2006; Dwivedi et al. 2016).

9.2 Abiotic Factors

9.2.1 Influence of Genotype, Environment, Growth Season, Temperature, and Light

Many studies demonstrated that soybean seed isoflavones and other compound contents vary depending on the genotype, growth season, and environmental conditions such as the availability of adequate water supply, light, temperature, soil conditions,

M. A. Kassem (✉)
Plant Genomics and Biotechnology Lab, Department of Biological and Forensic Sciences, Fayetteville State University, Fayetteville, NC, USA
e-mail: mkassem@uncfsu.edu

M. A. Kassem (ed.), *Soybean Seed Composition*,
https://doi.org/10.1007/978-3-030-82906-3_9

etc. (Tsukamoto et al. 1995; Wang et al. 2000; Lee et al. 2003a, b, 2007; Bennett et al. 2004; Kim et al. 2005; Caldwell et al. 2005; Lozovaya et al. 2005; Yin and Vyn 2005; Kirakosyan et al. 2006; Morrison et al. 2008; Phommalth et al. 2008; Rasolohery et al. 2008; Jeong et al. 2008; Yuan et al. 2009; Goes-Favoni et al. 2010; Shi et al. 2010). Tsukamoto et al. (1995) grew seven soybean cultivars (four had low and three controls which had normal isoflavone contents) in temperature-controlled growth chambers that are then transferred into two fields in Japan (Kyushu and Tsukuba); the plants were sown at different dates both in early soybean planting system (ESPS) and conventional soybean planting system (CSPS) (April to July). For all varieties tested, a sharp decrease in isoflavone contents and in the ratio of linoleic and linolenic acid to the total fatty acids was observed in plants grown in higher temperatures. Cotyledons contained 80–90% of total seed isoflavones and hypocotyls contained 10–20% and cotyledon isoflavones showed the highest changes in response to exposure to higher temperatures during seed development (Tsukamoto et al. 1995). Wang et al. (2000) grew 210 soybean cultivars from maturity groups 0 (41 cultivars), I (96 cultivars), and II (73 cultivars) in South Dakota and compared their isoflavone contents among other agronomic traits such as plant height, seed yield, seed weight, days of maturity, hilum color, and disease resistance. Isoflavone content range for the tested varieties was 1161–2743 μg/g. Their results indicated that seeds of varieties of maturity group 0 showed significantly lower total isoflavone contents compared to varieties of maturity group I. Moreover, plant height, days to maturity, and yield were negatively correlated with genistein content, while disease resistance profile had no effect on isoflavone contents (Wang et al. 2000). Lee et al. (2003a) observed significant differences in seed isoflavone contents and a strong negative environmental association between genistein and daidzein in 15 soybean cultivars grown in three sites in Korea. They showed that total seed isoflavone contents of these 15 cultivars were significantly different depending on the year and increased with seed storage over a period of 3 years (Lee et al. 2003a). In another study, they showed that genotype by environment effects was responsible for the significant variation of soybean seed isoflavone contents observed in those 15 cultivars grown in three Korean locations (Lee et al. 2003b). Bennett et al. (2004) showed that seed isoflavone contents varied significantly in four soybean cultivars from maturity groups IV and V grown in early soybean planting system (ESPS) and conventional soybean planting system (CSPS). A 1.3-fold increase was observed for total isoflavone contents in plants grown in ESPS vs. CSPS. An increase of up to 2.5-fold was observed for genistein, daidzein, and total isoflavone contents in irrigated plants grown in both ESPS and CSPS (Bennett et al. 2004). In another study, researchers quantified isoflavones in seeds stored at low and room temperatures of eight soybean cultivars over a period of 3 years. They found that cultivar 'Dawson' showed the highest isoflavone contents (2581.6 $\mu g\ g^{-1}$) and the highest variation in between seeds stored at room temperature and those stored at low temperature over a period of 3 years. The cultivar 'Geomjeong 1' had the lowest isoflavone contents (699.7 $\mu g\ g^{-1}$). Their results supported the possibility of improving the preservation method of seed isoflavones over a long period of time (Kim et al. 2005). Caldwell et al. (2005) found that the total seed isoflavone contents of growth chamber-grown plants can decrease by 65% and 90% when the temperatures are increased from 18 °C to 23 °C and 28 °C, respectively. Moreover, drought-stressed plants grown at 23 °C with elevated CO_2 (700 ppm) can have their seed total isoflavone contents reversed to the levels of plants grown in normal conditions (18 °C and 400 ppm CO_2) which is interesting in predicting the effect of climate change on seed isoflavone contents (Caldwell et al. 2005). Lozovaya et al. (2005) grew two French and three US soybean cultivars under controlled environment under low, intermediate, and high temperatures along with different soil moistures and found that plants grown in low temperatures and high soil moisture produced the highest amounts of seed isoflavones and temperature changes had a higher effect. All cultivars grown at low temperatures had high seed isoflavone contents compared to plants grown at low temperatures (Lozovaya et al. 2005). Investigating the relationships of isoflavone, oil, and protein contents with various seed yield levels, a study found that daidzein, genistein, glycitein, and total isoflavone contents increased when yield increases. Daidzein content increased by 249 mg kg^{-1} for each 1 Mg ha^{-1} increase in seed yield, genistein content increased by 164 mg kg^{-1} for each 1 Mg ha^{-1} increase in seed yield, and glycitein content increased by 11 mg kg^{-1} for each 1 Mg ha^{-1} increase in seed yield. Oil contents decreased by 4.2 g kg^{-1} for each 1 Mg ha^{-1} increase in seed yield. Also, seed yield correlated positively with daidzein, genistein, glycitein, total isoflavone, protein, and oil contents (Yin and Vyn 2005). Kirakosyan et al. (2006) investigated the effect of phytochrome-mediated light treatments on shoot and root isoflavone contents of five soybean cultivars and found that isoflavone contents are higher in seedlings' roots and shoots compared to seeds. Moreover, isoflavones were higher in dark-grown plants compared to light-grown plants after 10 days of light treatment for three cultivars out of five, and in general far-red light treatment enhanced root and shoot isoflavone contents compared to red light treatment. Therefore, the phytochrome plays a role in isoflavone biosynthesis regulation during seedling development, and this regulation is genotype-dependent (Kirakosyan et al. 2006). Lee et al. (2007) investigated the effect of dark and light conditions on seed isoflavones in roots, cotyledons, and hypocotyls of 17 soybean sprout cultivars and found that isoflavone contents vary among both the organs and cultivars with the highest amounts in roots and then cotyledons and hypocotyls. They also found no significant differences in isoflavone contents between green and yellow sprouts despite the fact that other studies (Hwang 2004; Chi et al. 2005) showed the effect of light quantity and quality on seed isoflavone contents (Lee et al. 2007). A study investigated the influence of nearly six decades (1934–1992) of breeding programs for yield on seed isoflavone contents of 14 short-season cultivars in Canada, and the results showed that daidzein content increased by 1.04% per year, genistein content increased by 1.47% per year, total isoflavones increased by 0.98% per year, yield increased by 0.43% per year, and oil content increased by 0.24% per year, but protein content decreased by 0.15% per year. Moreover, recently developed cultivars had higher seed isoflavone contents and better adaptation to environmental fluctuations (Morrison et al. 2008). Phommalth et al. (2008) investigated the possibility of increasing seed isoflavone contents in four soybean cultivars (Aga4, Cheongjakong,

Pungsannamulkong, and Aga3) by growing them under ultraviolet light, fluorescent light, dark, and greenhouse and showed that the cultivar Aga3 had the highest isoflavone contents among all four cultivars. Moreover, seeds of growing Aga3 plants under a combination of greenhouse light (12 h/day) and ultraviolet light (40 min/day) had the highest isoflavone contents. Also, Aga3 seeds of plants grown under dark contained 1.53-fold isoflavones than seeds of other plants (Phommalth et al. 2008). Rasolohery et al. (2008) found a three- to sixfold increase in total isoflavones, genistein, and daidzein, in cotyledons of plants grown in high temperatures compared to cotyledons of plants grown in low temperatures; however, no effect on isoflavone contents and composition was observed for water supply or availability (Rasolohery et al. 2008). Yuan et al. (2009) developed an efficient HPLC method and quantified daidzein, genistein, daidzein, glycitein, and their malonyl-, acetyl-, and nonconjugated β-glycosides in hypocotyls and cotyledons of soybean plants under dry heating conditions and germination (Yuan et al. 2009). Their results showed the following contents of daidzein and its glycoside conjugates (59.6%), glycitein (26.6%), and genistein (13.8%), respectively. Isoflavone contents in hypocotyls are approximately 7.8-fold higher than the cotyledons' contents (Yuan et al. 2009).

Conflicting results have been reported on the effects of soaking germinating seeds in water on isoflavone contents (Lee et al. 2007; Jeong et al. 2008; Goes-Favoni et al. 2010; Shi et al. 2010). Three soybean cultivars ('Taekwang', 'Daehwang', and 'Ilmi') have been used to study the effect of germination and osmopriming on isoflavone contents, and results showed high isoflavone contents in cultivars Ilmi and Taekwang after 12 h of germination at 20 °C and that their total isoflavone contents were not significantly different (Jeong et al. 2008). Shi et al. (2010) quantified isoflavones, proteins, oil, and other compounds during a 7-day period of soybean seed germination with and without light exposure and found that light exposure had no significant effect on isoflavone contents during early seed development stages despite a slight increase in daidzein and genistein contents after 5 days of seed germination under light conditions (Shi et al. 2010).

9.2.2 Influence of Soil Conditions, Elicitors, Nitrogen Application, Irrigation, and Row Spacing

Other studies demonstrated the effects of soil conditions, presence or absence of elicitors, nitrogen application, irrigation, and row spacing (plant density) (Vyn et al. 2002; Duke et al. 2003; Al-Tawaha et al. 2005; Kim et al. 2005; Caldwell et al. 2005; Gutierrez-Gonzalez et al. 2010a; Al-Tawaha and Seguin 2006; Anaele and Bishnoi 1992; Rahman et al. 2011; Ouertani et al. 2011; Ragin et al. 2014). Vyn et al. (2002) studied the effect of soil potassium (K) on seed isoflavone contents and found that the application of deep-banded or residual surface-applied K to low-K soils significantly increased genistein, daidzein, glycitein, and total isoflavone contents. Moreover, all these isoflavones correlated positively with seed K, seed yield, and leaf K; therefore, isoflavone contents could be increased by K application to low-K soils (Vyn et al. 2002). Duke et al. (2003) quantified the levels of aminomethylphosphonic acid, glyphosate, and isoflavones in seeds of glyphosate-resistant soybean plants treated with different concentrations of glyphosate in two locations (Stoneville, MS and Columbia, MO). At Columbia, they found no significant effect of glyphosate treatments on isoflavone contents; however, in Stoneville, they observed an increase in daidzein content after glyphosate application (Duke et al. 2003). Another study investigated the effect of elicitors (lipo-chitooligosaccharides [LCO], chitosan, actinomycetes spores [*Streptomyces melanosporofaciens* strain EF-76], and yeast extract) on seed isoflavone contents of plants grown in the greenhouse. The results showed a significant increase ranging from 21% to 84% in daidzein, genistein, glycitein, and total isoflavone contents of mature seeds of plants foliar-treated with LCO, actinomycetes, and chitosan. Foliar application of 2 mg L^{-1} of yeast extract was sufficient to increase total isoflavone contents by 56% (Al-Tawaha et al. 2005). Another study investigated the influence of temperature, nitrogen (N) application, and high CO_2 concentrations on seed isoflavone contents and demonstrated that treatment of plants with high concentrations of CO_2 and no N application caused the highest total isoflavone contents (1383.0 $\mu g.g^{-1}$) in seeds of these plants, while these contents dropped (414.4 $\mu g.g^{-1}$) in seeds of plants treated with N application and high temperatures. Temperature and CO_2 concentration treatment of plants had the highest effects on seed isoflavone content increase, while N application had no effect. Soybean adaptation to growth at high temperatures and concentrations of CO_2 might help in increasing seed isoflavone contents and yield (Kim et al. 2005). Many studies showed that isoflavone contents are negatively affected by drought or water deficit stress during seed developmental stages; however, very little is known about the molecular mechanisms underlying these effects (Caldwell et al. 2005; Gutierrez-Gonzalez et al. 2010a). A study showed that total seed isoflavone contents of plants grown in a growth chamber can decrease by 65% and 90% when the temperatures are increased from 18 °C to 23 °C and 28 °C, respectively; however, drought-stressed plants grown at 23 °C with elevated CO_2 (700 ppm) can have their seed total isoflavone contents reversed to the levels of plants grown in normal conditions (18 °C and 400 ppm CO_2) which is interesting in predicting the effect of climate change on seed isoflavone contents (Caldwell et al. 2005). Different levels of water stress conditions have been applied in field trials and controlled environments on soybean plants during seed development stages, and the results showed that only prolonged periods of drought and its intensity negatively affect seed isoflavone contents, and the response is genotype-dependent. Moreover, there was a positive correlation in the expression of *CHS, CHS8, and IFS2* (key enzymes in isoflavone biosynthetic pathway) with isoflavone accumulation in seeds, and *IFS2* contributes to isoflavone decrease during water stress (Gutierrez-Gonzalez et al. 2010a). Al-Tawaha and Seguin (2006) investigated the effects of row spacing (20, 40, and 60 cm), seeding date (late May and mid-June), and presence or absence of weed on seed isoflavone contents and found that isoflavone contents increased by 38% in seeds of plants grown in mid-June compared to the ones grown in late May and by 9% in the presence of weed. Seed yield was higher in plants grown in 20- and 40-cm row spacing compared to 60-cm row spacing, but it is negatively correlated with isoflavone contents (Al-Tawaha and Seguin 2006), and, in general, seed yield is higher in plants grown in low row spaces compared to high row spaces (Anaele and Bishnoi 1992; Rahman et al. 2011; Ouertani et al. 2011). In a recent study, we grew the PI 438489B by 'Hamilton' recombinant inbred line (RIL) population (PIxH, n = 50) in a field in Fayetteville, NC, in two row spaces (25 cm and 50 cm) and studied the effect of row spacing on seed isoflavone contents. The two row spaces generated 250,000 plants/ha (25 cm) and 90,000 plants/ha (50 cm). We showed significant differences in daidzein, genistein, glycitein, and total isoflavone contents between seeds of plants grown with 25-cm vs. 50-cm row spaces, and plants grown with 50-cm row spaces had higher isoflavone contents than those grown with 25-cm row spaces (Table 9.1, Figs. 9.1 and 9.2) (Ragin et al. 2014).

In a recent study, we also grew the Hamilton by Spencer RIL population in a field in Fayetteville, North Carolina, and found differences in individual and total isoflavone contents among the parents Hamilton and Spencer and their RILs and mapped QTL for these traits (Akond et al. 2015). RIL coefficient variations (CV) were 7.85%, 11.71%, 8.59%, and 8.48% for daidzein, glycitein, genistein, and total isoflavones, respectively. Hamilton had higher mean values of all isoflavone contents than Spencer (Table 9.2, Fig. 9.3).

9.2.3 Influence of Climate Change and Other Factors

Studies of the effects of climate change on soybean seed isoflavone contents are rare; however, several studies reported the effects of increased levels of CO_2, high temperatures, short rainfall, and other factors on soybean yield and seed composition in general (Polley 2002; Thomas et al. 2003; Prasad et al. 2005; Hay and Porter 2006; Long et al. 2004, 2006; Taub et al. 2008; Challinor and Wheeler 2008a, b; Challinor 2009; Lee et al. 2008; Kim et al. 2011; Yendrek et al. 2015; Bellaloui et al. 2016). Exposure to high levels of CO_2 usually leads to a decrease in activities of photosynthetic enzymes, decrease in photosynthetic rate, and sugar accumulation in leaves in many crops; however, other studies reported no effect of elevated CO_2 levels on field-grown crops such as rice (Baker et al. 1997; Polley 2002), wheat (Kimball et al. 1995; Polley 2002), and soybean (Campbell et al. 1990; Polley 2002). In a study, authors exposed soybean plants to increased temperatures (28 °C, 32 °C, 36 °C, 40 °C, 44 °C) and CO_2 levels (350, 700 μmol mol^{-1}) to investigate their effects on seed composition and found that seed oil, linolenic acid, and carbohydrate contents decreased when temperatures increased; however, seed oleic acid, N, and P contents increased when temperatures increased, while CO_2 increase had a minimal effect on the alteration of seed composition (Thomas et al. 2003). Other studies however predict alteration of seed yield and composition due to global climate changes including high CO_2 levels that will affect stomatal conductance and

Table 9.1 Individual (daidzein, glycitein, and genistein) and total isoflavone contents (μmol g^{-1}) in seeds of PI 438489B, Hamilton, and several of their RILs planted in three different row spaces 25, 50, and 75 cm in Fayetteville, North Carolina

Parent/ RIL	Daidzein (25 cm)	Daidzein (50 cm)	Glycitein (25 cm)	Glycitein (50 cm)	Genistein (25 cm)	Genistein (50 cm)
PI 438489B	6.0	16.8	2.7	1.2	1.3	2.6
Hamilton	1.4	9.5	3.2	5.4	5.0	7.0
RIL 1	903	6.0	1.0	3.3	2.0	2.0
RIL 2	507	10.3	1.0	4.0	2.0	4.0
RIL 3	728	8.3	1.0	26.5	1.0	4.7
RIL 4	944	23.0	3.0	19.0	3.7	11.0
RIL 5	819	28.3	16.7	20.7	26.0	10.3
RIL 6	911	12.3	6.5	3.3	29.0	16.7
RIL 7	240	58.0	1.0	15.7	10.0	4.0
RIL 8	786	20.3	2.7	19.0	4.3	10.0
RIL 9	164	17.7	4.0	46.3	6.7	13.0
RIL 10	877	29.0	36.0	24.3	15.7	22.0
RIL 11	372	11.3	1.0	11.0	2.3	18.7
RIL 12	360	13.0	11.00	6.0	1.0	6.7
RIL 13	136	8.0	1.0	7.0	13.0	3.0
RIL 14	914	36.3	1.0	2.0	2.0	3.0
RIL 15	137	15.0	3.0	19.7	12.3	7.0
RIL 16	751	22.0	1.0	1.7	2.0	5.7
RIL 17	785	45.7	6.0	28.0	2.0	2.0
RIL 18	142	7.3	10.0	24.0	10.0	5.0
RIL 19	787	22.7	6.0	11.3	8.0	2.3
RIL 20	478	79.7	2.0	6.7	2.0	9.3
RIL 21	843	75.3	5.3	23.0	16.0	23.3
RIL 22	919	24.3	1.3	21.0	10.0	1.0
RIL 23	571	87.0	6.0	15.0	10.0	4.0
RIL 24	662	26.0	1.0	17.0	2.0	83.6
RIL 25	399	56.3	3.7	17.3	13.3	68.5

photosynthetic rates (Long et al. 2004; Prasad et al. 2005; Hay and Porter 2006; Taub et al. 2008). In another recent study, authors exposed soybean plants to increased temperatures (26 °C and 45 °C) and CO_2 levels (360 and 700 μmol mol^{-1}) to investigate their effects on seed composition and found that seed protein, linolenic acid, and mineral nutrient (P, K, N, Fe, Zn, and B) contents decreased when temperature and CO_2 levels increased; however, seed oil and oleic acid contents increased when temperatures and CO_2 levels increased. Moreover, they found an increase in seed fructose, glucose, and sucrose contents with increased temperature and CO_2 levels, while seed raffinose and stachyose contents remain unchanged (Bellaloui et al. 2016). Future exposure to high levels of CO_2 can change proteins, oil, and fatty acids, and several mineral nutrients such as iron, iodine, and zinc can

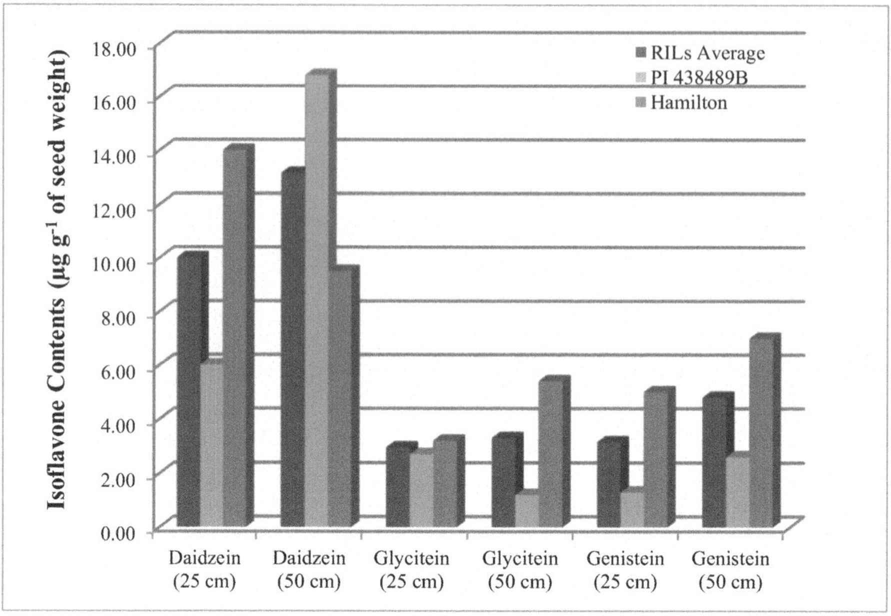

Fig. 9.1 Seed daidzein, glycitein, and genistein contents in PI 438489B, Hamilton, and their RILs grown in 25- and 50-cm row spaces in a field in Fayetteville, NC, in 2012

change in plant tissues due to changes in photosynthetic rates and stomatal movements (Long et al. 2006; Taub et al. 2008). Therefore, it is important to understand the effects of these climate changes on seed yield and composition in order to adjust farmers' practices to these new climatic conditions and to optimize seed yield and composition including isoflavone production (Challinor and Wheeler 2008a, b). A recent study investigated the physiological effects of atmospheric pollution with ozone (O_3) in several species of legumes including pea, common bean, and soybean. They found that garden pea plants resisted the most exposure to elevated levels of O_3 when compared to common bean and soybean plants; however, all species showed reduced leaf weight and carbohydrate amounts (Yendrek et al. 2015). Another study found that there are significant differences of isoflavones and phenolic contents in relationship to seed size and environmental conditions (Lee et al. 2008). The study found that both contents are significantly higher in small and medium seeds compared to large seeds (Lee et al. 2008). Another recent study showed that seed isoflavone contents dropped dramatically due to water stress and high solar radiations and temperatures (Carrera and Dardanelli 2015). Likewise, exposure of plants to high UV-B radiations reduced not only isoflavone contents but also phenolic compounds (Kim et al. 2011).

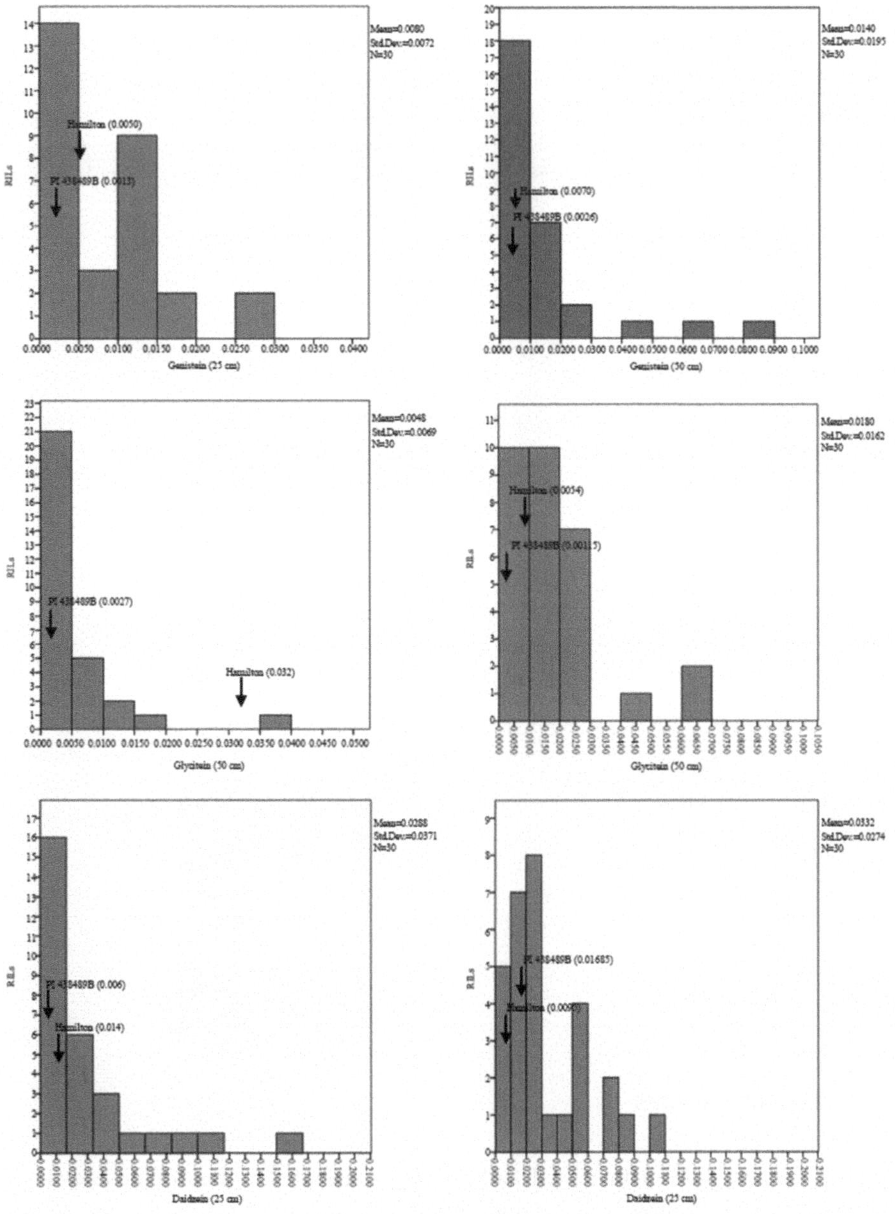

Fig. 9.2 Frequency distributions of daidzein, glycitein, genistein, and total seed isoflavone contents in the PI 438489B by Hamilton RIL population grown in two row spaces (25 cm and 50 cm) in a field in Fayetteville, NC, in 2012. Seed isoflavone concentrations are expressed in μg.g^{-1} of seed weight. (Adopted from Ragin et al. (2014))

Table 9.2 Individual (daidzein, glycitein, and genistein) and total isoflavone contents (μmol g^{-1}) in seeds of Hamilton, Spencer, and several of their RILs planted in Fayetteville, North Carolina

Parent/RIL	Daidzein	Glycitein	Genistein	Total isoflavones
Hamilton	428	712	626	1765
Spencer	389	377	585	1651
31	328	589	524	1441
32	326	594	534	1454
33	363	697	575	1635
34	316	584	515	1415
35	342	495	544	1381
36	309	485	512	1306
37	335	543	533	1411
38	384	668	620	1672
39	349	674	574	1597
40	341	672	537	1550
41	390	748	592	1730
42	323	546	492	1361
43	343	614	540	1497
44	358	636	582	1576
45	364	655	580	1599
46	331	557	520	1408
47	383	638	559	1580
48	386	715	633	1734
49	363	643	564	1570
50	312	577	459	1348
51	356	583	528	1467
52	354	625	558	1537
53	384	694	570	1648
54	371	687	596	1654
55	355	586	524	1465
56	348	575	514	1437
57	357	648	539	1544
58	355	623	557	1535

9.3 Biotic Factors

Several studies showed that seed isoflavones contribute to plant defense against pathogen infections; however, the underlying mechanisms of action remain largely unknown (Stafford 1997; Ndakidemi and Dakora 2003; Treutte 2006; Dwivedi et al. 2016). Ndakidemi and Dakora (2003) demonstrated that the levels of terpenoids, alkaloids, and isoflavones in the seeds largely determine the efficiency and control of these pathogen infections as well as insect damages in legumes (Ndakidemi and Dakora 2003). Others found that high amounts of phenolics are toxic and provide resistance to soybean and other legumes (Mapope and Dakora 2013; Sharma and

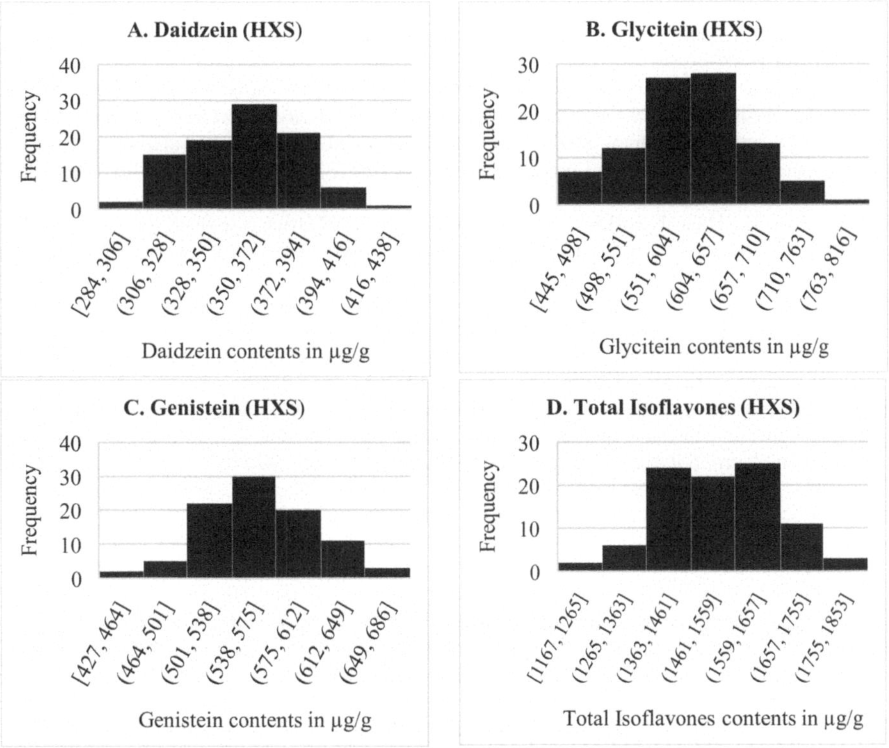

Fig. 9.3 Frequency distributions of (**a**) daidzein, (**b**) glycitein, (**c**) genistein, and (**d**) total seed isoflavone contents in the Hamilton by Spencer RIL population grown in a field in Fayetteville, NC, in 2012

Thakur 2014; Rubiales et al. 2015). Sharma and Thakur (2014) investigated the effect of bioactive compounds from 13 soybean, 10 cowpea, and 8 chickpea cultivars on bruchid insects (*Callosobruchus* species: *C. maculatus* and *C. chinensis*) and found that all soybean cultivars were relatively resistant to both species of bruchids when compared to cowpea and chickpea (Sharma and Thakur 2014); however, the results are not conclusive on the role of isoflavones in this resistance.

References

Akond, M., S. Liu, S.K. Kantartzi, K. Meksem, N. Bellaloui, D.A. Lightfoot, J. Yuan, D. Wang, J. Anderson, D.A. Lightfoot, and M.A. Kassem. 2015. A SNP genetic linkage map based on the 'Hamilton' by 'Spencer' recombinant inbred line (RIL) population of soybean [*Glycine max* (L.) Merr.] identified QTL for seed isoflavone contents. *Plant Breeding* 134 (5): 580–588. https://doi.org/10.1111/pbr.12298.

Al-Tawaha, A.M., and P. Seguin. 2006. Seeding date, row spacing, and weed effects on soybean isoflavone concentrations and other seed characteristics. *Canadian Journal of Plant Science* 86: 1079–1087.

Al-Tawaha, A.M., P. Seguin, D.L. Smith, and C. Beaulieu. 2005. Biotic elicitors as a means of increasing isoflavone concentration of soybean seeds. *Annals of Applied Biology* 146: 303–310.

Anaele, A.O., and U.R. Bishnoi. 1992. Effects of tillage, weed control method and row spacing on soybean yield and certain soil properties. *Soil and Tillage Research* 23: 333–340. https://doi.org/10.1016/0167-1987(92)90079-Q.

Baker, J.T., L.H. Allen, K.J. Boote, and N.B. Pickering. 1997. Rice responses to drought under carbon dioxide enrichment. 2. Photosynthesis and evapotranspiration. *Global Change Biology* 3: 129–138.

Bellaloui, N., Y. Hu, A. Mengistu, H.K. Abbas, M.A. Kassem, and M. Tigabu. 2016. Elevated atmospheric carbon dioxide and temperature affect seed composition, mineral nutrition, and 15N and 13C dynamics in soybean genotypes under controlled environments. *Atlas Journal of Plant Biology* 2016: 56–65. https://doi.org/10.5147/ajpb.2016.0157.

Bennett, J.O., O. Yu, L.G. Heatherly, and B. Krishnan. 2004. Accumulation of genistein and daidzein, soybean isoflavones implicated in promoting human health, is significantly elevated by irrigation. *Journal of Agricultural and Food Chemistry* 52: 7574–7579.

Caldwell, C.R., S.J. Britz, and R.M. Mirecki. 2005. Effect of temperature, elevated carbon dioxide, and drought during seed development on the isoflavone content of dwarf soybean (Glycine max (L.) Merrill) grown in controlled environments. *Journal of Agricultural and Food Chemistry* 53: 1125–1129.

Campbell, W.J., L.H. Allen, and G. Bowes. 1990. Response of soybean canopy photosynthesis to CO_2 concentration, light, and temperature. *Journal of Experimental Botany* 41: 427–433.

Carrera, C.S., and J.L. Dardanelli. 2015. Changes in the relationship between temperature during the seed-filling period and soybean seed isoflavones under water-deficit conditions. *Journal of Agronomy and Crop Science*. https://doi.org/10.1111/jac.12147.

Challinor, A. 2009. Towards the development of adaptation options using climate and crop yield forecasting at seasonal to multi-decadal timescales. *Environmental Science & Policy* 12: 453–465.

Challinor, A.J., and T.R. Wheeler. 2008a. Use of a crop model ensemble to quantify CO2 stimulation of water-stressed and well-watered crops. *Agricultural and Forest Meteorology* 148: 1062–1077.

Challinor, A.J., and T.R. Wheeler. 2008b. Crop yield reduction in the tropics under climate change: Processes and uncertainties. *Agricultural and Forest Meteorology* 148: 3433–4356. https://doi.org/10.1016/j.agrformet.2007.09.015.

Chi, H.Y., J.S. Roh, J.T. Kim, S.J. Lee, M.J. Kim, S.J. Hahn, and I.M. Chung. 2005. Light quality on nutritional composition and isoflavones content in soybean sprouts. *Korean Journal of Crop Science* 50: 415–418.

Duke, S.O., A.M. Rimando, P.F. Pace, K.N. Reddy, and R.J. Smeda. 2003. Isoflavone, glyphosate, and aminomethylphosphonic acid levels in seeds of glyphosate-treated, glyphosate-resistant soybean. *Journal of Agricultural and Food Chemistry* 51 (1): 340–344. https://doi.org/10.1021/jf025908i.

Dwivedi, S.L., H.D. Upadhyaya, I.M. Chung, P. De Vita, S. Garcia-Lara, D. Guajardo-Flores, J.A. Gutiérrez-Uribe, S.O. Serna-Saldívar, G. Rajakumar, K.L. Sahrawat, J. Kumar, and R. Ortiz. 2016. Exploiting phenylpropanoid derivatives to enhance the nutraceutical values of cereals and legumes. *Frontiers in Plant Science* 7: 763. https://doi.org/10.3389/fpls.2016.00763.

Goes-Favoni, S.P., M.C. Carrao-Panizzi, and A. Beleia. 2010. Changes of isoflavone in soybean cotyledons soaked in different volumes of water. *Food Chemistry* 19: 1605–1612.

Gutierrez-Gonzalez, J.J., S.K. Guttikonda, L.S.P. Tran, D.L. Aldrich, R. Zhong, O. Yu, H.T. Nguyen, and D.A. Sleper. 2010a. Differential expression of isoflavone biosynthetic genes in soybean during water deficits. *Plant Cell Physiology* 51 (6): 936–948. https://doi.org/10.1093/pcp/pcq065.

Gutierrez-Gonzalez, J.J., X.L. Wu, J.D. Gillman, J.D. Lee, R. Zhong, O. Yu, J.G. Shannon, H.T. Nguyen, and D.A. Sleper. 2010b. Intricate environment-modulated genetic networks control isoflavone accumulation in soybean seeds. *BMC Plant Biology* 10: 105–120. https://doi.org/10.1186/1471-2229-10-105.

Hay, R., and J. Porter. 2006. *The physiology of crop yield*. 2nd ed. Oxford: Blackwell.

Hwang, Y.H. 2004. Present status and future developmental direction of soy-related industries in Korea. *Korea Soybean Digest* 21: 28–44.

Jeong, P.H., D.H. Shin, and Y.S. Kim. 2008. Effects of germination and osmopriming treatment on enhancement of isoflavone contents in various soybean cultivars and cheonggukjang (fermented unsalted soybean paste). *Journal of Food Science* 73: H187–H194.

Kim, J.J., S.H. Kim, S.J. Hahn, and I.M. Chung. 2005. Changing soybean isoflavone composition and concentrations under two different storage conditions over three years. *Food Research International* 38 (2005): 435–444.

Kim, E.H., P. Seguin, J.E. Lee, C.G. Yoon, H.K. Song, J.K. Ahn, and I.M. Chung. 2011. Elevated ultraviolet-B radiation reduces concentrations of isoflavones and phenolic compounds in soybean seeds. *Journal of Agronomy and Crop Science* 197: 75–80. https://doi.org/10.1111/j.1439-037X.2010.00444.x.

Kimball, B.A., P.J. Pinter, R.L. Garcia, R.L. LaMorte, G.W. Wall, D.J. Hunsaker, G. Wechsung, F. Wechsung, and T. Kartschall. 1995. Productivity and water use of wheat under free-air CO2 enrichment. *Global Change Biology* 1: 429–442.

Kirakosyan, A., P. Kaufman, R.L. Nelson, M.J. Kasperbauer, J.A. Duke, E. Seymour, S.C. Chang, S. Warber, and S. Bolling. 2006. Isoflavone levels in five soybean (*Glycine max*) genotype are alter by phytochrome-mediated light treatment. *Journal of Agricultural and Food Chemistry* 54: 54–58.

Lee, S.J., J.K. Ahn, S.H. Kim, J.T. Kim, S.J. Hahn, M.Y. Jung, and I.M. Chung. 2003a. Variation in isoflavone of soybean cultivars with location and storage duration. *Journal of Agricultural and Food Chemistry* 51: 3382–3389.

Lee, S.J., W. Yan, J.K. Ahn, and I.M. Chung. 2003b. Effects of year, site, genotype and their interactions on various soybean isoflavones. *Field Crops Research* 81: 181–192.

Lee, S.J., J.K. Ahn, T.D. Khanh, S.C. Chun, S.L. Kim, H.M. Ro, H.K. Song, and I.M. Chung. 2007. Comparison of isoflavone concentrations in soybean (Glycine max (L.) Merrill) sprouts grown under two different light conditions. *Journal of Agricultural and Food Chemistry* 55: 9415–9421.

Lee, S.J., J.J. Kim, H.I. Moon, J.K. Ahn, S.C. Chun, W.S. Jung, O.K. Le, and I.M. Chung. 2008. Analysis of isoflavones and phenolic compounds in Korean soybean [*Glycine max* (L.) Merr.] seeds of different seed weights. *Journal of Agricultural and Food Chemistry* 56 (8): 2751–2758. https://doi.org/10.1021/jf073153f.

Long, S.P., E.A. Ainsworth, A. Rogers, and D.R. Ort. 2004. Rising atmospheric carbon dioxide: Plants FACE the future. *Annual Review of Plant Biology* 55: 591–628.

Long, S.P., E.A. Ainsworth, A.D.B. Leakey, J. Nosberger, and D.R. Ort. 2006. Food for thought: Lower-than-expected crop yield stimulation with rising CO2 concentrations. *Science* 312 (5782): 1918–1921. https://doi.org/10.1126/science.1114722.

Lozovaya, V.V., A.V. Lygin, A.V. Ulanov, R.L. Nelson, J. Dayde, and J.M. Widhohm. 2005. Effect of temperature and soil moisture status during seed development on soybean seed isoflavone concentration and composition. *Crop Science* 45: 1934–1940. https://doi.org/10.2135/cropsci2004.0567.

Mapope, N., and F.D. Dakora. 2013. Role of flavonoid and isoflavonoid molecules in symbiotic functioning and host-plant defense in the Leguminosae. In *The chemistry for sustainable development in Africa*, ed. A. Gurib-Fakim and J.N. Eloff, 33–48. Berlin/Heidelberg: Springer.

Morrison, J., E.R. Cober, M.F. Saleem, N.B. McLaughlin, J. Fregeau-Reid, B.L. Ma, W. Yan, and L. Woodrow. 2008. Changes in isoflavone concentration with 58 years of genetic improvement of short-season soybean cultivars in Canada. *Crop Science* 48 (6): 2201–2208.

Ndakidemi, P.A., and F.D. Dakora. 2003. Legume seed flavonoids and nitrogenous metabolites as signals and protectants in early seedling development. *Functional Plant Biology* 30: 729–745. https://doi.org/10.1071/FP03042.

Ouertani, K., E. Washington, P. Lage, S.K. Kantartzi, D.A. Lightfoot, and M.A. Kassem. 2011. Comparison of early and conventional soybean production systems for yield and other agronomic traits. *Atlas Journal of Plant Biology* 1 (1): 1–5. https://doi.org/10.5147/ajpb.2011.0008.

Phommalth, S., Y.S. Jeong, Y.H. Kim, K.H. Dhakal, and Y.H. Hwang. 2008. Effects of light treatment on isoflavone content of germinated soybean seeds. *Journal of Agricultural and Food Chemistry* 56: 10123–10128.

Polley, H.W. 2002. Implications of atmospheric and climatic change for crop yield and water use efficiency. *Crop Science* 42: 131–140.

Prasad, P., V. Vara, L.H. Allen, and K.J. Boote. 2005. Crop responses to elevated carbon dioxide and interaction with temperature: Grain legumes. *Journal of Crop Improvement* 13: 113–155.

Ragin, B., M. Akond, S.K. Kantartzi, K. Meksem, H. Herrera, C. Akbay, D.A. Lightfoot, and M.A. Kassem. 2014. Effect of row spacing on seed isoflavone contents in soybean [*Glycine max* (L.) Merr.]. *American Journal of Plant Science* 5: 4003–4010. https://doi.org/10.4236/ajps.2014.526418.

Rahman, M., M. Hossain, M. Anwar, and A.S. Juraimi. 2011. Plant density influence on yield and nutritional quality of soybean seed. *Asian Journal of Plant Sciences* 10: 125–132. https://doi.org/10.3923/ajps.2011.125.132.

Rasolohery, C.A., M. Berger, A.V. Lygin, V.V. Lozovaya, R.L. Nelson, and J. Daydé. 2008. Effect of temperature and water availability during late maturation of the soybean seed on germ and cotyledon isoflavone content and composition. *Journal of Science and Food Agriculture* 88 (2): 218–228.

Rubiales, D., S. Fondevilla, W. Chen, L. Gentzbittel, T.J.V. Higgins, M.A. Castillejo, K.B. Singh, and N. Rispail. 2015. Achievements and challenges in legume breeding for pest and disease resistance. *Critical Reviews in Plant Science* 34 (1–3): 195–236. https://doi.org/10.1080/07352689.2014.898445.

Sharma, S., and D.R. Thakur. 2014. Biochemical basis for bruchid resistance in cowpea, chickpea and soybean genotypes. *American Journal of Food Technology* 9: 318–324. https://doi.org/10.3923/ajft.2014.318.324.

Shi, H., P.K. Nam, and M.A. Yinfa. 2010. Comprehensive profiling of isoflavones, phytosterols, tocopherols, minerals, crude protein, lipid, and sugar during soybean (*Glycine max*) germination. *Journal of Agricultural and Food Chemistry* 58: 4970–4976. https://doi.org/10.1021/jf100335j.

Stafford, H.A. 1997. Roles of flavonoids in symbiotic and defense functions in legume roots. *The Botanical Review* 63 (1): 27–39.

Subramanian, S., G. Stacey, and O. Yu. 2007. Distinct, crucial roles of flavonoids during legume nodulation. *Trends in Plant Science* 12 (7): 282–285. https://doi.org/10.1016/j.tplants.2007.06.006.

Taub, D., B. Miller, and H. Allen. 2008. Effects of elevated CO_2 on the protein concentration of food crops: A meta-analysis. *Global Change Biology* 14: 565–575.

Thomas, J.M.G., K.J. Boote, L.H. Allen, M. Gallo-Meagher, and J.M. Davis. 2003. Elevated temperature and carbon dioxide effects on soybean seed composition and transcript abundance. *Crop Science* 43: 1548–1557.

Treutte, D. 2006. Significance of flavonoids in plant resistance: A review. *Environmental Chemistry Letters* 4: 147–157. https://doi.org/10.1007/s10311-006-0068-8.

Tsukamoto, C., S. Shimada, K. Igita, S. Kudou, M. Kokubun, K. Okubo, and K. Kitamura. 1995. Factors affecting isoflavone content in soybean seeds: Changes in isoflavones, saponins, and composition of fatty acids at different temperatures during seed development. *Journal of Agricultural and Food Chemistry* 43 (5): 1184–1192. https://doi.org/10.1021/jf00053a012.

Vyn, T.J., X. Yin, T.W. Bruulsema, C.J.C. Jackson, I. Rajcan, and S.M. Brouder. 2002. Potassium fertilization effects on isoflavone concentrations in soybean [*Glycine max* (L.) Merr.]. *Journal of Agricultural and Food Chemistry* 50 (12): 3501–3506.

Wang, C., M. Sherrard, S. Pagadala, R. Wixon, and R.A. Scott. 2000. Isoflavone content among maturity group 0 to II soybeans. *Journal of the American Oil Chemists' Society* 77 (5): 483–487.

Yendrek, C.R., R.P. Koester, and E.A. Ainsworth. 2015. A comparative analysis of transcriptomic, biochemical, and physiological responses to elevated ozone identifies species-specific mechanisms of resilience in legume crops. *Journal of Experimental Botany* 66 (22): 7101–7112.

Yin, X., and T.J. Vyn. 2005. Relationships of isoflavone, oil, and protein in seed with yield of soybean. *Agronomy Journal* 97 (5): 1314–1321.

Yuan, J.P., Y.B. Liu, J. Peng, J.H. Wang, and X. Liu. 2009. Changes of isoflavone profile in the hypocotyls and cotyledons of soybeans during dry heating and germination. *Journal of Agricultural and Food Chemistry* 57: 9002–9010.

Chapter 10
Two Decades of QTL Mapping of Isoflavone in Soybean Seed

Moulay Abdelmajid Kassem

10.1 Seed Isoflavone Content QTL Mapping

For the past three decades, many different molecular markers such as amplified fragment length polymorphisms (AFLPs), restriction fragment length polymorphisms (RFLPs), simple sequence repeats (SSRs or microsatellites), and single nucleotide polymorphisms (SNPs) have been used to map quantitative trait loci (QTL) for many important agronomic traits including isoflavone contents in many different soybean populations. The main approaches are the use of biparental recombinant inbred lines (RILs) and genome-wide association studies (GWAS). For QTL analysis, single marker analysis (ANOVA), interval mapping (IM), composite interval mapping (CIM), multiple interval mapping (MIM), and several other methods have been routinely used to identify QTL positions (Njiti et al. 1999; Meksem et al. 2001; Kassem et al. 2004, 2006; Primomo et al. 2005; Matsumura et al. 2005; Cheng et al. 2008; Gutierrez-Gonzalez et al. 2009, 2010, 2011; Zeng et al. 2009; Grant et al. 2010; Liang et al. 2010; Yoshikawa et al. 2010; Meng et al. 2011; Yang et al. 2011; Smallwood 2012; Smallwood et al. 2014; Akond et al. 2013, 2014, 2015; Zhang et al. 2014; Li et al. 2014; Wang et al. 2014, 2015; Bolon et al. 2014; Zhao et al. 2015; Han et al. 2015, 2016; Wang et al. 2016; Chu et al. 2017; Carter 2017; Carter et al. 2018; Cai et al. 2018; Pei et al. 2018; Li et al. 2018; Watanabe et al. 2019; Wu et al. 2020). In this chapter, we will summarize QTL that control soybean seed isoflavone contents (daidzein, genistein, glycitein, and total isoflavone) that have been identified for over two decades, from 1999 to 2020.

In the first study of mapping QTL for seed isoflavone contents, the authors used a population of 40 RILs derived from the cross of 'Essex' and 'Forrest' cultivars

M. A. Kassem (✉)
Plant Genomics and Biotechnology Lab, Department of Biological and Forensic Sciences, Fayetteville State University, Fayetteville, NC, USA
e-mail: mkassem@uncfsu.edu

M. A. Kassem (ed.), *Soybean Seed Composition*,
https://doi.org/10.1007/978-3-030-82906-3_10

(ExF, $n = 40$) and DNA-based markers such as random amplified polymorphic DNA (RAPD), RFLP, and DNA amplification fingerprinting (DAF) (Njiti et al. 1999). They found one QTL that had a major effect on glycitein contents but minor effect on genistein contents (Njiti et al. 1999). The QTL region contained six markers linked to either genistein or glycitein contents. The study showed also that 11 unlinked markers may be potentially linked to phytoestrogen contents, among which seven markers are significantly associated with glycitein contents; however, the study did not specify what these markers (both linked and unlinked) are (Njiti et al. 1999).

In Meksem et al. (2001), we have used the same 'Essex' by 'Forrest' RIL population but increased the number of RILs to 100 (ExF, $n = 100$; Lightfoot et al. 2005) and 150 mainly SSR markers. The frequency distributions of daidzein, genistein, and glycitein departed significantly from normality; however, all distributions were unimodal and continuous. We subdivided the RILs into two subgroups of 40 and 60 RILs in which we quantified seed genistein, glycitein, daidzein, and total isoflavone contents and found six QTL that control seed isoflavone contents in this population. The SSR marker Satt276 (QTL-1) on linkage group (LG) A1 (Chr_5) was significantly associated with daidzein contents (Table 10.1). Satt197 and Satt251 on LG B1 (Chr_11) were associated with glycitein contents (QTL-2); Satt302 on LG H (Chr_12) was associated with glycitein contents (QTL-3); Satt337 and [Satt116 and Satt337] on LG K (Chr_9) were associated with daidzein and (glycitein and genistein) contents, respectively (QTL-4); and Satt080 and Satt237 on LG N (Chr_3) were associated with daidzein (QTL-5) and glycitein contents (QTL-6), respectively (Fig. 2.1; Table 10.1) (Meksem et al. 2001).

In Kassem et al. (2004), we have used the same ExF RIL population ($n = 100$) but increased the number of markers to 240 SSR markers and identified two additional QTL for seed isoflavone contents. Satt063 on LG B2 (Chr_14) was associated with genistein and daidzein contents (QTL-7), and Satt129 on D1a (Chr_1) was associated with glycitein contents (QTL-8) (Kassem et al. 2004). In Kassem et al. (2006), we identified three new QTL for seed isoflavone contents using the same population. Satt097 and Sat_129 on LG A2 (Chr_8) were associated with glycitein contents (QTL-9); Satt323 and Satt463 on LG M (Chr_7) were associated with daidzein contents (QTL-10); and Satt356 and Satt130 on LG G (Chr_18) were associated with glycitein contents (QTL-11) which brought the number of QTL that control seed isoflavone contents in ExF RIL population to 11 QTL (Fig. 2.1; Table 10.1) (Kassem et al. 2006).

In another study, Primomo et al. (2005) used 207 RILs derived from the cross of 'AC756' and 'RCAT Angora' grown in two Canadian locations and only 99 SSR markers. They found seven QTL associated with seed isoflavone contents and significant genotype by environment (GxE) interactions (Primomo et al. 2005). Satt129 on LG D1a (Chr_1) was associated with glycitein contents, Satt200 on LG A1 (Chr_5) was associated with daidzein and total isoflavone contents, and Satt289 on LG C2 (Chr_6) was associated with genistein and total isoflavone contents. On LG M (Chr_7), they found a cluster of 11 QTL for daidzein, genistein, glycitein, and total isoflavone contents spanning the marker interval Satt201-Satt245 which is approximately 50.5 cM (Primomo et al. 2005). Satt196 on LG K (Chr_9) was associated with glycitein contents, Satt181 on LG H (Chr_12) was associated with daidzein and total isoflavone contents, Satt516 on LG F (Chr_13) was associated with glycitein contents, and Satt547 on LG J (Chr_16) was associated with glycitein contents (Fig. 2.1; Table 10.1) (Primomo et al. 2005).

Matsumura et al. (2005) used 150 RILs derived from the cross of the Japanese cultivar 'Misuzudaizu' and the Chinese cultivar 'Moshidou Gong 503' and 360 markers (150 SSRs, 177 RFLP, 28 AFLP, and 5 morphological markers) (Matsumura et al. 2005). The study aimed at mapping the locations of chalcone synthase (CHS) multigene family (CHS1–8) (Sangeeta et al. 2007) using mainly cleaved amplified polymorphic sequence (CAPS), derived cleaved amplified polymorphic sequence (dCAPS), and amplicon length polymorphism (ALP). They mapped CHS2 41.8–48.8 cM on LG A1 (Chr_5); dCHS1, CHS1-3, CHS4, and CHS5 as a cluster 32.8–55.3 cM on LG A2 (Chr_8); CHS8 0.00–13.2 cM on LG B1 (Chr_11); CHS7 109.2–136.4 cM on LG D1a (Chr_1); and CHS6 53.3–55.4 cM on LG K (Chr_10) (Matsumura et al. 2005).

In another study, Gutierrez-Gonzalez et al. (2009) used 196 RILs derived from the cross of 'Essex' and PI 437654 grown in two locations at the Columbia, MO, and 276 SSR and RFLP markers. They identified and mapped 26 QTL for various seed isoflavone contents on 13 soybean LGs or chromosomes. The marker interval (Satt174-Sat_271) on LG A1 (Chr_5) was associated with daidzein, genistein, and total isoflavone contents; the marker interval (Satt187-EAGGMCTT205) on LG A2 (Chr_8) was associated with daidzein and total isoflavone contents; the marker interval (Satt687-Sat_424) on LG B2 (Chr_14) was associated with genistein contents; the marker interval (Sat_279-EACAMCTT123) on LG D1b (Chr_2) was associated with daidzein, genistein, and total isoflavone contents; the marker interval (Sat281-Satt291) on LG C2 (Chr_6) was associated with glycitein contents; the marker interval (Sctt009-Satt541) on LG H (Chr_12) was associated with daidzein and total isoflavone contents; the marker interval (Sat_112-Satt691) on LG E (Chr_15) was associated with glycitein, genistein, and total isoflavone contents; the marker interval (Satt552-Satt499) on LG K (Chr_9) was associated with glycitein contents; the marker intervals (Satt458-Sat_284) and (Satt186-Sat_086) on LG D2 (Chr_17) were associated with glycitein contents; the marker interval (Satt612-Sat_117) on LG G (Chr_18) was associated with glycitein contents; the marker interval (Satt490-Sat_197) on LG F (Chr_13) was associated with genistein contents; the marker interval (Satt697-Satt551) on LG M (Chr_7) was associated with daidzein, genistein, and total isoflavone contents; and the marker interval (Sat_113-Sat_286) on LG L (Chr_19) was associated with glycitein contents. Clusters of QTL controlling at least three types of isoflavone contents have been observed on chromosomes 2, 5, 7, 15, and 19 (Fig. 2.1; Table 10.1) (Gutierrez-Gonzalez et al. 2009).

In another study, Zeng et al. (2009) used 130 RILs derived from the cross of 'Zhongdou 27' and 'Jiunong 20', grown in seven Chinese locations, and 99 SSR markers. The study found three QTL [*qDZF_1* identified by Satt144 on LG F (Chr_13); *qDZI_1* identified by Satt587 on LG I (Chr_20); and *qDZK_1* identified

Table 10.1 Details of QTL that control seed isoflavone contents (daidzein, genistein, glycitein, and total isoflavone) reported and not reported in SoyBase (www.soybase.org; Grant et al. 2010) from 1999 to 2020

Trait	QTL	LG/Chr	Interval (cM)	Markers Interval	LOD/*P*	R^2 (%)	Environment	References
Glycitein	*	Unknown	Not reported	Six markers not reported	Unknown	Unknown	Carbondale, IL	Njiti et al. (1999)
	*	Unknown	Not reported	Seven markers not reported	Unknown	Unknown	Carbondale, IL	
Genistein	*	Unknown	Not reported	Six unknown markers	Unknown	Unknown	Carbondale, IL	
	*	Unknown	Not reported	One unknown marker	Unknown	Unknown	Carbondale, IL	
Glycitein	*qGLY1*	B1/11	38.80–49.07	Satt251-Satt197	5.50	50.00	Carbondale, IL	Meksem et al. (2001)
	qGLY2	K/9	*	*	*	*	Carbondale, IL	
	qGLY3	H/12	78.95–82.99	Satt302-Satt293	2.00	19.50	Carbondale, IL	
	qGLY4	N/3	43.69–61.24	Satt387-Satt237	2.30	11.10	Carbondale, IL	
Daidzein	*qDAID1*	K/9	47.99–60.88	Satt240-Satt260	2.10	–	Carbondale, IL	
	qDAID2	N/3	37.64–43.69	Satt080-Satt387	3.20	10.30	Carbondale, IL	
	qDAID3	A1/5	18.91–22.68	Satt276-Satt165	2.70	9.60	Carbondale, IL	
Daidzein	*Seed daidzein 4-1*	B2/14	86.81–100.15	Satt122-Satt063	2.00	17.00	Carbondale, IL	Kassem et al. (2004)
	Seed daidzein 4-2	A1/5	12.16–22.16	Satt276	2.70	9.60	Carbondale, IL	
	Seed daidzein 4-3	K/9	73.87–86.37	Satt260	2.10	21.30	Carbondale, IL	
	Seed daidzein 4-4	N/3	39.30–50.96	Satt080-Satt387	3.50	10.30	Carbondale, IL	
Genistein	*qGEN*	B2/14	63.44–80.78	Satt122-Satt063	2.90	29.50	Carbondale, IL	
Glycitein	*qGLY*	D1a/1	90.07–93.50	Satt408-Satt129	3.60	32.00	Carbondale, IL	

Daidzein	*Seed daidzein 2-1*	A1/5	92.90–96.10	Satt200	2.10	3.80	Harrow, Ontario	Primomo et al. (2005)
	Seed daidzein 2-2	H/12	84.00–91.10	Satt181	2.60	8.10	Harrow, Ontario	
	Seed daidzein 2-3	M/7	13.60–25.00	Satt201	3.40	6.70	Harrow, Ontario	
	Seed daidzein 2-4	M/7	43.00–56.40	Satt245	3.10	6.90	Harrow, Ontario	
	Seed daidzein 2-5	A2/8	58.60–160.00	Satt538	2.30	3.60	Harrow, Ontario	
	Seed daidzein 2-6	F/13	97.00–99.00	Satt490	p = 0.0085	2.70	Harrow, Ontario	
	Seed daidzein 2-7	J/16	10.70–12.70	Satt249	p = 0.0057	4.00	Harrow, Ontario	
Genistein	*Seed genistein 2-1*	C2/6	112.40–113.40	Satt289	2.70	5.50	Harrow, Ontario	
	Seed genistein 2-2	M/7	13.60–28.00	Satt201	7.60	10.50	Harrow, Ontario	
	Seed genistein 2-3	M/7	37.00–50.00	Satt540	4.00	7.90	Harrow, Ontario	
	Seed genistein 2-4	A1/5	31.70–33.70	Satt155	p = 0.0046	4.00	Harrow, Ontario	
	Seed genistein 2-5	A1/5	91.90–93.90	Satt200	p = 0.0080	3.60	Harrow, Ontario	
	Seed genistein 2-6	H/12	90.10–92.10	Satt181	p = 0.0031	4.50	Harrow, Ontario	
	Seed genistein 2-7	H/12	46.00–48.00	Satt442	p = 0.0097	3.40	Harrow, Ontario	
Glycitein	*Seed glycitein 2-1*	J/16	69.00–77.00	Satt547	4.80	8.60	Harrow, Ontario	
	Seed glycitein 2-2	D1a/1	109.00–109.60	Satt129	2.30	3.40	Harrow, Ontario	
	Seed glycitein 2-3	F/13	44.00–53.00	Satt516	2.90	4.00	Harrow, Ontario	
	Seed glycitein 2-4	K/9	88.00–102.00	Satt196	p = 0.0037	4.40	Harrow, Ontario	
	Seed glycitein 2-5	G/18	45.00–58.00	Sat_088	2.00	3.50	Harrow, Ontario	
	Seed glycitein 2-6	M/7	14.00–23.00	Satt201	3.10	5.30	Harrow, Ontario	
	Seed glycitein 2-7	M/7	25.00–43.00	Satt540	3.30	7.70	Harrow, Ontario	
	Seed glycitein 2-8	H/12	90.10–92.10	Satt181	p = 0.0057	3.90	Harrow, Ontario	
	Seed glycitein 2-9	N/3	64.50–66.50	Satt521	p = 0.0064	3.70	Harrow, Ontario	

(continued)

Table 10.1 (Continued)

Trait	QTL	LG/Chr	Interval (cM)	Markers Interval	LOD/*P*	R^2 (%)	Environment	References
Total isoflavone	*Seed TI 1*	A1/5	91.90–93.90	Satt200	p = 0.0044	–	Harrow, Ontario	
	Seed TI 2	C2/6	112.40–113.40	Satt289	p = 0.0052	–	Harrow, Ontario	
	Seed TI 3	H/12	84.00–91.10	Satt181	p = 0.0002	6.50	Harrow, Ontario	
	Seed TI 4	J/16	10.70–12.70	Satt249	p = 0.0032	–	Harrow, Ontario	
	Seed TI 5	M/7	13.60–28.00	Satt201	p < 0.0001	21.90	Harrow, Ontario	
	Seed TI 6	M/7	43.00–56.40	Satt245	p = 0.0003	7.70	Harrow, Ontario	
	Seed TI 7	M/7	37.00–50.00	Satt540	p = 0.0022	–	Harrow, Ontario	
Glycitein	*qGLY2*	A2/8	71.20–73.20	Satt097-Sat_129	3.36	10.23	Carbondale, IL	Kassem et al. (2006)
	qGLY3	B1/11	45.4–47.4	Satt251-Satt415	15.02	48.71	Carbondale, IL	
	qGLY4	D1a/1	42.84–44.84	Satt368-Satt605	3.47	8.92	Carbondale, IL	
Daidzein60	*qDAID60-1*	N/3	37.1–38.1	Satt485-Satt080	2.63	55.91	Carbondale, IL	
	qDAID60-2	A2/8	–	CAA13-CAA16	4.69	16.26	Carbondale, IL	
	qDAID60-3	M/7	59.04–61.04	Satt323-Satt463	4.02	24.01	Carbondale, IL	
Genistein60	*qGEN60-1*	A2/8	–	CAA13	2.86	9.93	Carbondale, IL	
	qGEN60-2	M/7	59.04–61.04	Satt323-Satt463	3.75	22.74	Carbondale, IL	
Glycitein60	*qGLY60-1*	A2/8	–	CAA16-CAA13	3.66	12.71	Carbondale, IL	
	qGLY60-2	G/18	11.18–13.18	Satt356-Satt130	2.89	10.71	Carbondale, IL	
	qGLY60-3	M/7	59.04–61.04	Satt323-Satt463	2.51	14.91	Carbondale, IL	
Tot. genist.40	*qTGEN40-1*	A2/8	–	CAA13	3.74	13.50	Carbondale, IL	
	qTGEN40-2	M/7	59.04–61.04	Satt323-Satt463	2.62	17.42	Carbondale, IL	
Tot. glycitein40	*qTGLY40-1*	A2/8	–	CAA13	2.85	10.94	Carbondale, IL	
Tot. phytoest.	*qTPHY1*	G/18	11.18–13.18	Satt356	2.78	10.24	Carbondale, IL	

DZ, GN, GL, TIF	*IFS1* (gene)	*	*	SNP(A/G)	*	*	Nanjing, China	Cheng et al. (2008)
		*	*	SNP(A/G)	*	*	Nanjing, China	
		*	*	SNP(T/C)	*	*	Nanjing, China	
	IFS2 (gene)	*	*	SNP(T/C)	*	*	Nanjing, China	
		*	*	SNP(G/A)	*	*	Nanjing, China	
Genistein	*gen-A1*	A1/5	29.4–36.4	Satt236-Sat_271	3.50	9.40	Columbia, MO	Gutierrez-Gonzalez et al. (2009)
	gen-B2	B2/14	0.0–3.0	Satt687-Sat_424	3.10	6.00	Columbia, MO	
	gen-D1b	D1b/2	0.0–10.0	Sat_279-EACAMCTT123	3.80	8.40	Columbia, MO	
	gen-E	E/15	8.0–11.0	Sat_112-Satt691	7.60	7.70	Columbia, MO	
	gen-F	F/13	3.4–10.0	Satt490-Sat_197	4.50	5.40	Columbia, MO	
	gen-M	M/7	87.4–93.4	Satt697-Satt551	*	*	Columbia, MO	
	gen-L	L/19	141.4–160.9	Sat_113-Sat_286	3.00	6.90	Columbia, MO	
Daidzein	*dai-A1*	A1/5	22.5–36.4	Satt236-Sat_271	5.70	13.90	Columbia, MO	
	dai-A2	A2/8	23.0–27.1	Satt187-EAGGMCTT205	3.20	5.40	Columbia, MO	
	dai-D1b	D1b/2	0.0–8.0	Sat_279-EACAMCTT123	3.00	5.70	Columbia, MO	
	dai-H	H/12	0.0–7.0	Sct009-Satt541	3.20	5.80	Columbia, MO	
	dai-M	M/7	87.4–94.4	Satt697-Satt551	4.30	4.90	Columbia, MO	
	dai-L	L/19	142.8–161.9	Sat_113-Sat_286	5.00	6.50	Columbia, MO	
Glycitein	*gly-C2*	C2/6	41.0–48.6	Satt281-Satt291	4.10	7.60	Columbia, MO	
	gly-E	E/15	1.0–11.0	Sat_112-Satt691	4.30	7.90	Columbia, MO	
	gly-K	K/9	68.6–78.3	Satt552-Satt499	3.10	3.60	Columbia, MO	
	gly-G	G/18	81.0–92.5	Satt612-Sat_117	4.70	7.30	Columbia, MO	
	gly-D2_1	D2/17	13.6–26.8	Satt458-Sat_284	4.10	6.80	Columbia, MO	
	gly-D2_2	D2/17	0.0–3.2	Satt186-Sat_086	3.20	4.40	Columbia, MO	

(continued)

Table 10.1 (Continued)

Trait	QTL	LG/Chr	Interval (cM)	Markers Interval	LOD/*P*	R^2 (%)	Environment	References
Total isoflavone	*tot-A1*	A1/5	29.4–36.4	Satt236-Sat_271	3.50	9.50	Columbia, MO	
	tot-D1b	D1b/2	0.0–7.0	Sat_279-EACAMCTT123	5.00	9.50	Columbia, MO	
	tot-E	E/15	8.0–11.3	Sat_112-Satt691	3.30	6.40	Columbia, MO	
	tot-H	H/12	0.0–10.0	Sct_009-Satt541	3.20	6.30	Columbia, MO	
	tot-L	L/19	143.8–159.9	Sat_286-Satt229	7.40	7.40	Columbia, MO	
	tot-M	M/7	87.4–93.4	Satt697-Satt551	4.10	8.60	Columbia, MO	
	tot-A2	A2/8	23.0–27.1	Satt187-EAGGMCTT205	3.10	3.40	Columbia, MO	
Daidzein	*QDZF_1 (3-1)*	F/13	101.08–103.08	Satt144	p(0.0001)	8.54	Harbin, China	Zeng et al. (2009)
	QDZI_1 (3-2)	I/20	30.49–32.49	Satt587	p(0.0003)	11.29	Hulan, China	
	QDZK_1 (3-3)	K/9	45.18–47.18	Satt124	p(0.001)	8.49	Harbin, China	
Genistein	*QGTA2_1 (3-1)*	A2/8	115.63–117.63	Sat_040	p(0.0004)	8.59	Harbin, China	
	QGTC2_1 (3-2)	C2/6	116.76–118.76	Satt460	p(0.00032)	3.80	Harbin, China	
	QGTF_1 (3-3)	F/13	101.08–103.08	Satt144	p(0.0003)	6.79	Harbin, China	
	QGTM_1 (3-4)	M/7	34.84–36.84	Satt540	p(0.0024)	4.39	Harbin, China	
Glycitein	*QGCI_1 (3-1)*	I/20	76.83–78.83	Satt330	p(0.0001)	7.34	Harbin, China	
	QGCM_1 (3-2)	M/7	34.84–36.84	Satt540	p(0.0003)	8.45	Harbin, China	
	QGCO_1 (3-3)	O/10	99.37–101.37	Satt592	p(0.0001)	5.24	Harbin, China	

Total isoflavone	*QTID2_1*	D2/17	56.10–58.10	Satt208	p(<0.0001)	11.24	Harbin, China	
	QTIF_1	F/13	101.08–103.08	Satt144	p(0.0006)	8.48	Harbin, China	
	QTIG_1	G/18	70.60–72.60	Satt288	p(<0.0001)	12.39	Hulan, China	
	QTIO_1	O/10	*	Satt241	p(0.002)	7.00	Harbin, China	
	QTIM_1	M/7	34.84–36.84	Satt540	p(<0.0001)	5.38	Harbin, China	
Isoflavone	*Seed isoflavone 1*	J/16	0–0.5	Satt249-Satt285	4.14	25.20	China	Liang et al. (2010)
	Seed isoflavone 2	N/3	25.02–27.02	Satt009-Satt152	3.14	33.08	China	
	Seed isoflavone 3	N/3	33.61–35.61	Satt152-Satt521	2.97	32.29	China	
	Seed isoflavone 4	D2/17	176.53–178.53	Satt386-Satt256	5.34	25.19	China	
	Seed isoflavone 5	G/18	99.15–101.15	Satt131-Satt501	4.60	12.14	China	
	Seed isoflavone 6	J/16	0–0.5	Satt249-Satt285	7.70	55.20	China	
Genistein	*qGEN12*	H/12	0.0–6.0	Satt009-Satt541	p(0.00000)	1.20	Columbia, MO	Gutierrez-Gonzalez et al. (2010)
	qGEN20	I/20	72.5–74.5	GMLPSI2-Sct_189	p(0.00000)	0.40	Columbia, MO	
	qGEN4	C1/4	6.2–17.6	Sat_337-Sat_140	p(0.00000)	3.30	Columbia, MO	
	qGEN2_2	D1b/2	107.6–126.2	Sat_139-Sat_289	p(0.00001)	5.30	Columbia, MO	
Daidzein	*qDAI4*	C1/4	0.0–6.0	Satt396-Sat_337	p(0.00000)	1.20	Columbia, MO	
	qDAI1	D1a/1	34.7–42.6	Sat_106-AW781285	p(0.00000)	1.20	Columbia, MO	
	qDAI16	J/16	0.0–15.0	Sat_339-Satt280	p(0.00000)	2.30	Columbia, MO	
	qDAI7_2	M/7	145.7–151.7	Sat_147-Sat_330	p(0.00000)	1.00	Columbia, MO	

(continued)

Table 10.1 (Continued)

Trait	QTL	LG/Chr	Interval (cM)	Markers Interval	LOD/*P*	R^2 (%)	Environment	References
Glycitein	*qGLY5*	A1/5	26.4–35.4	Satt236-Sat_271	p(0.00000)	1.40	Columbia, MO	
	qGLY6_2	C2/6	41.2–48.8	Satt319-EAACMCAC113	p(0.00000)	1.00	Columbia, MO	
	qGLY2	D1b/2	0.0–5.0	Sat_279-EACAMCTT123	p(0.00000)	0.90	Columbia, MO	
	qGLY9	K/9	25.1–36.3	Satt242-EAACMCAC227	p(0.00000)	0.90	Columbia, MO	
Total isoflavone	*qTOT4*	C1/4	3.0–9.2	Sat_337-Sat_140	p(0.00000)	1.10	Columbia, MO	
	qTOT2_2	D1b/2	107.6–118.2	Sat_139-Sat_289	p(0.00000)	1.90	Columbia, MO	
	qTOT16	J/16	3.0–19.0	sat_339-Satt280	p(0.00000)	1.30	Columbia, MO	
Daidzein	*Seed daidzein 8-1*	A2/8	66.86–68.86	AW132402	9.22	20.40	Kyoto-2003	Yoshikawa et al. (2010)
	Seed daidzein 8-2	L/19	106.23–108.23	Satt373	3.42	6.20	Kyoto-2003	
	Seed daidzein 8-3	A1/5	96.76–98.76	Sat_271	4.01	13.60	Osaka-2003	
	Seed daidzein 8-4	A2/8	52.75–54.75	Sat_215	5.10	9.80	Osaka-2003	
	Seed daidzein 8-5	C1/4	81.51–83.51	Sat_042	3.48	10.00	Osaka-2003	
	Seed daidzein 8-6	D2/17	66.70–68.70	Satt669	3.18	12.00	Nagano-2003	
	Seed daidzein 8-7	E/15	18.29–20.29	Satt284	3.91	0.20	Nagano-2003	
	Seed daidzein 8-8	F/13	62.68–64.68	Satt114	2.00	13.10	Osaka-2004	

Genistein	*Seed genistein 8-1*	A2/8	47.84–49.84	I	5.84	19.00	Kyoto-2003	
	Seed genistein 8-2	C1/4	75.00–77.00	Sat_077	4.90	15.10	Kyoto-2003	
	Seed genistein 8-3	D2/17	66.70–68.70	Satt669	4.96	8.10	Kyoto-2003	
	Seed genistein 8-4	C1/4	72.32–74.32	Satt190	4.89	15.20	Osaka-2003	
	Seed genistein 8-5	G/18	104.75–106.75	Sat_372	4.92	15.30	Osaka-2003	
	Seed genistein 8-6	A1/5	91.88–93.88	Satt200	4.50	7.20	Nagano-2003	
	Seed genistein 8-7	A2/8	66.86–68.86	AW132402	5.31	20.00	Nagano-2003	
	Seed genistein 8-8	K/9	16.11–18.11	Sat_119	3.06	5.20	Osaka-2004	
Glycitein	*Seed glycitein 8-1*	A2/8	47.84–49.84	*I*	7.72	23.70	Kyoto-2003	
	Seed glycitein 8-2	B1/11	31.50–33.50	Satt509	4.45	13.40	Kyoto-2003	
	Seed glycitein 8-3	B1/11	101.55–103.55	Satt359	3.35	9.50	Kyoto-2003	
	Seed glycitein 8-4	D2/17	78.23–80.23	Satt389	3.65	8.90	Kyoto-2003	
	Seed glycitein 8-5	G/18	72.41–74.41	Sat_143	3.07	7.50	Kyoto-2003	
	Seed glycitein 8-6	H/12	7.48–9.48	Satt353	3.32	2.90	Kyoto-2003	
	Seed glycitein 8-7	B1/11	45.38–47.38	Satt197	8.75	38.40	Osaka-2003	
	Seed glycitein 8-8	D2/17	56.50–58.50	Sat_092	3.45	12.80	Osaka-2003	
	Seed glycitein 8-9	A2/8	52.75–54.75	Sat_215	6.99	18.60	Osaka-2003	
	Seed glycitein 8-10	C1/4	72.32–74.32	Satt190	3.03	2.70	Nagano-2003	
	Seed glycitein 8-11	C2/6	116.45–118.45	Sat_238	7.21	11.30	Nagano-2003	
	Seed glycitein 8-12	G/18	106.75–108.75	Sat_372	3.55	9.50	Nagano-2003	
	Seed glycitein 8-13	A1/5	16.16–18.16	Satt276	3.86	10.10	Nagano-2003	

(continued)

Table 10.1 (Continued)

Trait	QTL	LG/Chr	Interval (cM)	Markers Interval	LOD/*P*	R^2 (%)	Environment	References
Total isoflavone	*KT_1*	A2/8	51.9–53.9	Sat_215	8.28	26.00	Kyoto-2003	
	KT_2	C1/4	66.9–68.9	Satt190	4.32	12.60	Kyoto-2003	
		D1b/2	79.5–81.5	Satt282	3.01	6.10	Kyoto-2003	
	KT_3	D2/17	58.0–60.0	Satt669	4.16	9.30	Kyoto-2003	
	OT_1	A1/5	103.6–105.6	Sat_271	4.60	7.90	Osaka-2003	
	OT_2	A2/8	51.9–53.9	Sat_215	5.43	12.80	Osaka-2003	
		B2/14	72.2–74.2	Satt066	3.23	3.00	Osaka-2003	
	NT_1	A2/8	50.9–52.9	I	14.21	41.00	Nagano-2003	
		E/15	0.0–0.5	Satt213	3.05	2.70	Nagano-2003	
	OT_3	A1/5	103.64–105.64	Sat_271	4.11	11.50	Osaka-2004	
	OT_4	A2/8	50.86–52.86	I	7.32	21.40	Osaka-2004	
		C1/4	73.26–75.26	Sat_042	3.00	5.90	Osaka-2004	
		D1b/2	100.38–102.38	Satt546	3.01	0.90	Osaka-2004	
	NT_2	A2/8	62.74–64.74	AW132402	7.75	26.00	Nagano-2004	
		D2/17	58.02–60.02	Satt669	3.67	9.70	Nagano-2004	
Genistein	*qGm05 (qGEN5)*	A1/5	80.0–83.7	BARC-041257-07953	*F*(93.5)	36.00	Columbia, MO	Gutierrez-Gonzalez et al. (2011)
	qGm06 (qGEN6)	C2/6	79.2–84.0	BARC-063661-18416	*F*(12.6)	4.00	Columbia, MO	
	qgm08 (qGEN8)	A2/8	51.7–58.9	BARC-031627-07124	*F*(12.1)	2.00	Columbia, MO	
	qGm11 (qGEN11)	B1/11	45.6–61.2	BARC-042999-08498	*F*(10.7)	7.00	Columbia, MO	

Daidzein	*qGm05 (qDAI5)*	A1/5	80.0–83.7	BARC-041257-07953	*F*(112.2)	35.00	Columbia, MO	
	qGm06 (qDAI6)	C2/6	84.5–86.2	BARC-040475-07751	*F*(23.9)	8.00	Columbia, MO	
	qGm11 (qDAI11)	B1/11	44.6–58.2	BARC-042999-08498	*F*(9.7)	7.00	Columbia, MO	
Glycitein	*qGm05 (qGLY5)*	A1/5	81.0–84.7	BARC-060051-16321	*F*(57.0)	12.00	Columbia, MO	
	qGm05_2 (qGLY5_2)	C2/6	6.2–26.8	BARC-048025-10453	*F*(9.5)	25.80	Columbia, MO	
	qGm11 (qGLY11)	B1/11	59.2–72.9	BARC-054421-12081	*F*(8.7)	3.00	Columbia, MO	
Total isoflavone	*qGm05 (qTOT5)*	A1/5	80.0–83.7	BARC-060051-16321	*F*(97.5)	37.00	Columbia, MO	
	qGm06 (qTOT6)	C2/6	79.2–84.5	BARC-063661-18416	*F*(17.2)	5.00	Columbia, MO	
	qGm08 (qTOT8)	A2/8	51.7–59.9	BARC-042491-08277	*F*(9.8)	2.00	Columbia, MO	
	qGm11 (qTOT11)	B1/11	45.6–60.2	BARC-042999-08498	*F*(10.9)	7.00	Columbia, MO	
Daidzein	*QDZGT1*	A1/5	101.1	Sat_217	9.41	28.30	Suwon, Korea	Yang et al. (2011)
Genistein	*QDZGT1*	A1/5	101.1	Sat_217	6.03	15.20	Suwon, Korea	
	QDZGT2	A2/8	55.9	SM306	3.82	16.30	Suwon, Korea	
Glycitein	*QGC1*	N/3	29.5	Sat_033	4.60	18.70	Suwon, Korea	
Total isoflavone	*QDZGT2*	A1/5	101.1	Sat_217	6.83	23.80	Suwon, Korea	
Daidzein	*QDZGT1*	A1/5	101.1	Sat_217	9.55	27.10	Daejeon, Korea	
	QDZGT2	A2/8	56.9	SM306	3.49	9.30	Daejeon, Korea	
Genistein	*QDZGT1*	A1/5	101.1	Sat_217	5.52	15.80	Daejeon, Korea	
	QDZGT2	A2/8	57.9	SM306	6.08	18.80	Daejeon, Korea	
Glycitein	*QGC2*	A2/8	56.9	SM306	3.32	11.70	Daejeon, Korea	
	QGC3	B1/11	43.8	SM003	3.03	15.60	Daejeon, Korea	
	QGC4	C1/4	29.1	Sat_140	2.52	12.20	Daejeon, Korea	
	QGC5	J/16	37.8	Satt693	2.47	12.70	Daejeon, Korea	
Total isoflavone	*QDZGT1*	A1/5	101.1	Sat_217	8.94	23.40	Daejeon, Korea	
	QDZGT2	A2/8	58.9	SM306	4.00	13.60	Daejeon, Korea	

(continued)

Table 10.1 (Continued)

Trait	QTL	LG/Chr	Interval (cM)	Markers Interval	LOD/*P*	R^2 (%)	Environment	References
Daidzein	*QDZGT1*	A1/5	101.1	Sat_217	5.86	17.70	Ochang, Korea	
Genistein	*QDZGT1*	A1/5	101.1	Sat_217	4.92	11.70	Ochang, Korea	
	QDZGT2	A2/8	57.9	SM306	7.00	18.60	Ochang, Korea	
	QGT1	M/7	73.0	Satt697	3.85	11.40	Ochang, Korea	
Glycitein	*QGC6*	F/13	56.4	Sat_375	3.93	11.80	Ochang, Korea	
	QGC7	H/12	49.1	Satt192	3.65	11.70	Ochang, Korea	
	QGC8	K/9	43.3	A636.p1	3.90	9.90	Ochang, Korea	
	QGC9	N/3	48.3	Satt521	4.88	12.80	Ochang, Korea	
Total isoflavone	*QDZGT1*	A1/5	101.1	Sat_217	5.14	15.50	Ochang, Korea	
	QDZGT3	M/7	75.8	Satt697	3.64	10.40	Ochang, Korea	
	QDZGT4	N/3	50.3	Satt521	5.23	16.30	Ochang, Korea	

Genistein	*GEN1*	C2/6	209.4	BARC-023517-05442	7.90	16.00	TN, IL, AR	Smallwood (2012)
	GEN2	K/9	95.1	BARC-038909-07393	10.50	23.00	TN, IL, AR	
	GEN3	F/13	32.4	BARC-030359-06859	6.00	14.00	TN, IL, AR	
	GEN4	A1/5	186.3	BARC-0428853-08438	4.00	13.00	TN, IL, AR	
	GEN5	D2/17	21.6	BARC-030909-06973	3.20	10.00	TN, IL, AR	
	GEN6	L/19	137.2	BARC-035235-07156	7.60	22.00	TN, IL, AR	
	GEN7	K/9	48.3	BARC-048623-10678	3.30	9.00	TN, IL, AR	
Daidzein	*DAI1*	C2/6	209.4	BARC-023517-05442	5.80	13.00	TN, IL, AR	
	DAI2	K/9	93.4	BARC-038909-07393	7.10	16.00	TN, IL, AR	
	DAI3	F/13	28.7	BARC-030359-06859	4.30	11.00	TN, IL, AR	
	DAI4	L/19	141.1	BARC-024345-04854	4.80	10.00	TN, IL, AR	
	DAI5	A1/5	184.2	BARC-042853-08438	4.80	16.00	TN, IL, AR	
Glycitein	*GLY1*	C2/6	200.0	BARC-031337-07051	4.20	15.00	TN, IL, AR	
	GLY2	K/9	77.6	BARC-014813-01678	4.60	20.00	TN, IL, AR	
	GLY3	I/20	105.0	BARC-053725-11957	3.50	17.00	TN, IL, AR	
Tot. isoflavone	*ISO1*	C2/6	209.4	BARC-023517-05442	6.40	13.00	TN, IL, AR	
	ISO2	K/9	94.7	BARC-038909-07393	9.30	21.00	TN, IL, AR	
	ISO3	F/13	31.4	BARC-030359-06859	4.80	11.00	TN, IL, AR	
	ISO4	L/19	141.1	BARC-024345-04854	4.50	9.00	TN, IL, AR	
	ISO5	A1/5	187.2	BARC-059081-15595	4.30	9.00	TN, IL, AR	
	ISO6	A1/5	26.6	BARC-019415-03923	3.30	9.00	TN, IL, AR	

(continued)

Table 10.1 (Continued)

Trait	QTL	LG/Chr	Interval (cM)	Markers Interval	LOD/*P*	R^2 (%)	Environment	References
Genistein	*GEN1*	C2/6	209.4	BARC-023517-05442	7.90	16.00	TN, IL, AR	Smallwood et al. (2014)
	GEN2	K/9	95.1	BARC-038909-07393	10.50	23.00	TN, IL, AR	
	GEN3	F/13	32.4	BARC-030359-06859	6.00	14.00	TN, IL, AR	
	GEN4	A1/5	186.3	BARC-0428853-08438	4.00	13.00	TN, IL, AR	
	GEN5	D2/17	21.6	BARC-030909-06973	3.20	10.00	TN, IL, AR	
	GEN6	L/19	137.2	BARC-035235-07156	7.60	22.00	TN, IL, AR	
	GEN1	*C2/6*	*205.9*	*BARC-066175-19800*	*3.10*	*9.00*	*TN, IL, AR*	
	GEN7	K/9	48.3	BARC-048623-10678	3.30	9.00	TN, IL, AR	
	GEN6	*L/19*	*138.8*	*BARC-016145-02292*	*3.30*	*9.00*	*TN, IL, AR*	
Daidzein	*DAI1*	C2/6	209.4	BARC-023517-05442	5.80	13.00	TN, IL, AR	
	DAI2	K/9	93.4	BARC-038909-07393	7.10	16.00	TN, IL, AR	
	DAI3	F/13	28.7	BARC-030359-06859	4.30	11.00	TN, IL, AR	
	DAI4	L/19	141.1	BARC-024345-04854	4.80	10.00	TN, IL, AR	
	DAI5	A1/5	184.2	BARC-042853-08438	4.80	16.00	TN, IL, AR	
	DAI4	*L/19*	*144.5*	*BARC-026069-05243*	*8.00*	*30.00*	*TN, IL, AR*	
	DAI2	*L/19*	*98*	*BARC-038909-07393*	*4.10*	*13.00*	*TN, IL, AR*	
	DAI4	*L/19*	*138.8*	*BARC-016145-02292*	*6.30*	*18.00*	*TN, IL, AR*	
Glycitein	*GLY1*	C2/6	200.0	BARC-031337-07051	4.20	15.00	TN, IL, AR	
	GLY2	K/9	77.6	BARC-014813-01678	4.60	20.00	TN, IL, AR	
	GLY3	I/20	105.0	BARC-053725-11957	3.50	17.00	TN, IL, AR	
Tot. isoflavone	*ISO1*	C2/6	209.4	BARC-023517-05442	6.40	13.00	TN, IL, AR	
	ISO2	K/9	94.7	BARC-038909-07393	9.30	21.00	TN, IL, AR	
	ISO3	F/13	31.4	BARC-030359-06859	4.80	11.00	TN, IL, AR	
	ISO4	L/19	141.1	BARC-024345-04854	4.50	9.00	TN, IL, AR	
	ISO5	A1/5	187.2	BARC-059081-15595	4.30	9.00	TN, IL, AR	
	ISO6	A1/5	26.6	BARC-019415-03923	3.30	9.00	TN, IL, AR	
	ISO4	*L/19*	*138.8*	*BARC-016145-02292*	*3.90*	*11.00*	*TN, IL, AR*	

Daidzein	*qDAID001-2010*	M/7	66.6–70.6	ss107915334	2.60	11.00	Fayetteville, NC	Akond et al. (2013)
	qDAID002-2010	D2/17a	19.9–22.8	ss107930758	2.80	14.00	Fayetteville, NC	
Glycitein	*qGLY001-2010*	D1b/2	0–1.4	ss107930023	5.30	24.00	Fayetteville, NC	
	qGLY002-2010	A2/8	31.7–40.9	ss107912848	4.40	22.00	Fayetteville, NC	
Genistein	*qGEN001-2010*	A2/8	26.1–32.4	ss107914369	4.40	23.00	Fayetteville, NC	
	qGEN002-2010	H/12	0–1	ss107924882	2.60	12.00	Fayetteville, NC	
	qGEN003-2010		7.9–26.6	ss107914072	2.60	12.00	Fayetteville, NC	
Daidzein	*qDAID001-2011*	C2/6	18.7–33.3	ss107921568	4.79	3.00	Saint Pauls, NC	
	qDAID002-2011	F/13b	13.9–37.9	ss107912991	6.62	37.00	Saint Pauls, NC	
Glycitein	*qGLY001-2011*	D1a/1	34.3–40.1	ss107914319	2.51	7.00	Saint Pauls, NC	
	qGLY002-2011	D2/17c	11.5–17.0	ss107927109	2.72	25.00	Saint Pauls, NC	
Genistein	*qGLY001-2011*	N/3	30.8–32.50	ss107912938	7.87	44.00	Saint Pauls, NC	
	qGEN002-2011	A2/8	9.3–13.4	ss107930668	2.88	7.00	Saint Pauls, NC	
	qGEN003-2011	K/9	60.5–65.1	ss107918120	6.16	34.00	Saint Pauls, NC	
	qGEN004-2011	D2/17a	11.9–13.2	ss107921332	2.58	10.00	Saint Pauls, NC	
	qGEN005-2011	G/18	33.6–33.9	ss107931011	3.37	8.00	Saint Pauls, NC	
Daidzein	*qDAID001*	A1/5	11.7	ss245748832	4.55	2.30	Fayetteville, NC	Akond et al. (2014)
Glycitein	*qGLY001*	K/9	0.6	ss24687662	2.66	0.20	Fayetteville, NC	
	qGLY002	B2/14	20.2	ss248275008	2.97	1.00	Fayetteville, NC	
Daidzein	*qDAID*	C2/6	7–9.5	ss246100375	28.23	77.00	Fayetteville, NC	
Glycitein	*qGLY*	G/18	0–3.8	ss249909538	3.10	13.00	Fayetteville, NC	
Tot. isoflavone	*qISO*	C2/6	7–9.5	ss246100375	25.10	76.00	Fayetteville, NC	

(continued)

Table 10.1 (Continued)

Trait	QTL	LG/Chr	Interval (cM)	Markers Interval	LOD/*P*	R^2 (%)	Environment	References
Daidzein (DC)	*QDTIC9_1*	K/9	0–21.6	Sat_319-Sat_344	2.85	5.81	China	Zhang et al. (2014)
	QDGlTIC3_1	N/3	0–4.2	Satt009-Sat_239	2.63	4.87	China	
	QDGeGlTIC4_1	C1/4	79.1–97.5	Satt338-Satt524	2.78	6.70	China	
	QDTIC4_2	C1/4	97.5–118.8	Satt524-Sat_140	4.17	7.37	China	
	QDGlTIC12_1	H/12	30–31.4	Satt353-Satt434	2.57	5.11	China	
Genistein (GeC)	*QGeC13_1*	F/13	107.88–109.88	Satt395	2.77	5.26	China	
	QDGeGlTIC4_1	C1/4	79.1–97.5	Satt338-Satt524	3.48	7.87	China	
	QGeC7_1	M/7	54.87–56.87	Satt323	2.92	5.78	China	
	QGeC10_1	O/10	–	Satt479	4.01	8.83	China	
Glycitein (GlC)	*QGlC4_1*	C1/4	0–13.1	Satt150-Satt164	3.98	5.53	China	
	QGlC17_1	D2/17	13.7–18	Satt311-Satt488	2.92	5.90	China	
	QDGlTIC3_1	N/3	0–4.2	Satt009-Sat_239	3.27	7.29	China	
	QDGeGlTIC4_1	C1/4	79.1–97.5	Satt338-Satt524	2.97	4.87	China	
	QDGlTIC12_1	H/12	30–31.4	Satt353-Satt434	2.97	4.92	China	
	QGlC17_1	D2/17	13.7–18	Satt311-Satt488	2.77	5.90	China	
Tot. isoflavone (TIC)	*QDTIC9_1*	K/9	0–21.6	Sat_319-Sat_344	2.69	5.32	China	
	QDGlTIC12_1	H/12	30–31.4	Satt353-Satt434	2.69	4.87	China	
	QDGlTIC3_1	N/3	0–4.2	Satt009-Sat_239	2.68	5.75	China	
	QDGeGlTIC4_1	C1/4	79.1–97.5	Satt338-Satt524	3.40	6.63	China	
	QDTIC4_2	C1/4	97.5–118.8	Satt524-Sat_140	2.76	5.09	China	
	QDGlTIC12_1	H/12	30–31.4	Satt353-Satt434	2.58	4.48	China	

Daidzin	*qD4*	C1/4	14.67–18.75	M634469-M625652	6.98	16.00	China	Li et al. (2014)
Isoflavone	*qIF7*	M/7	33.21–38.13	M519691-M491368	8.17	14.30	China	
	qIF8	A2/8	101.44–101.76	M1045556-M1053803	5.73	16.90	China	
	qIF13	F/13	45.77–51.16	M153314-M163112	4.91	12.50	China	
	qIF16-1	J/16	20.85–21.19	M889900-M897989	3.96	8.56	China	
	qIF16-2	J/16	59.53–63.53	M895322-M899870	5.12	15.10	China	
	qIF20-2	I/20	3.83–8.43	M943408-M941848	9.11	19.62	China	
Malonylglycitin	*qMGL7*	M/7	3.62–4.24	M523495-M518225	4.59	13.30	China	
	qMGL14	B2/14	6.83–10.60	M542764-M545951	4.00	8.61	China	
	qMGL14	B2/14	0.00–2.57	M575406-M531913	3.50	9.54	China	
Glycitin	*qGL8-2*	A2/8	130.48–132.34	M1060220-M1053451	12.40	28.00	China	
pQTL								Wang et al. (2014)
DZ	*qDZF_1*	F/13	188.34	Sat_103-Sat_262	2.00	10.57	Harbin06, China	
GC	*qGCA2_1*	A2/8	38.46	Sat_040-Satt233	2.65	6.01	Hulan06, China	
	qGCD1b_1	D1b/2	215.67	Satt546-Sat_459	2.38	3.12	Harbin06, China	
					2.21	4.17	Harbin07, China	
GT	*qGTD2_1*	D2/17	50.81	Satt186-Satt226	2.00	3.41	Harbin05, China	
					2.36	5.23	Harbin06, China	
					5.76	10.98	Hulan06, China	
					3.09	8.23	Harbin07, China	
	qGTF_2	F/13	41.23	Satt149-Sat_234	2.00	1.56	Harbin05, China	
					2.49	4.17	Hulan06, China	
					4.03	5.47	Suihua07, China	

(continued)

Table 10.1 (Continued)

Trait	QTL	LG/Chr	Interval (cM)	Markers Interval	LOD/*P*	R^2 (%)	Environment	References
TI	*qTIF_1*	F/13	6.01	Satt423-Satt569	4.59	3.21	Hulan07, China	
					2.15	4.20	Suihua07, China	
eQTL								
PAL	*qPALB2_1*	B2/14	0.01	Satt560-Satt556	3.39	8.11	Harbin11, China	
	qPALD2_1	D2/17	15.9	Sat_209-Sat_022	4.24	6.67	Harbin11, China	
CHS	*qCHSA1_1*	A1/5	0.01	Satt236-D26A	5.48	4.21	Harbin11, China	
	qCHSD1a_1	D1a/1	0.01	Satt436-Sat_345	8.64	2.71	Harbin11, China	
	qCHSD1b_1	D1b/2	214.8	Satt546-Satt459	2.55	2.13	Harbin11, China	
	qCHSD1b_2	D1b/2	211.22	Satt546-Satt459	2.18	3.90	Harbin11, China	
	qCHSD2_1	D2/17	10.74	Satt528-Satt256	2.73	2.07	Harbin11, China	
	qCHSF_1	F/13	46.17	Sat_234-Satt149	2.72	3.57	Harbin11, China	
	qCHSL_1	L/19	14.00	Satt278-Sat_134	2.40	15.65	Harbin11, China	
	qCHSL_2	L/19	24.51	Sat_134-Sct_010	2.09	9.98	Harbin11, China	
IFS	*qIFSA2_1*	A2/8	55.45	Sat_129-Sat_181	7.46	17.67	Harbin11, China	
	qIFSC1_1	C1/4	6.67	Sat042-Satt524	5.63	22.80	Harbin11, China	
	qIFSD2_1	D2/17	54.88	Satt186-Satt226	8.87	16.67	Harbin11, China	
	qIFSF_1	F/13	6.97	Satt569-Satt423	3.09	15.84	Harbin11, China	
	qIFSF_2	F/13	56.01	Sat_234-Satt149	10.92	17.89	Harbin11, China	
	qISFH_1	H/12	0.01	Satt302-Satt279	3.23	7.27	Harbin11, China	
	qIFSL_1	L/19	20.99	Sat_134-Satt278	7.26	16.12	Harbin11, China	
	qIFSL_2	L/19	43.95	Sct_010-Sat_134	9.75	17.97	Harbin11, China	
	qIFSN_1	N/3	6.67	Satt152-Satt530	2.50	27.42	Harbin11, China	
	qIFSN_2	N/3	29.53	Satt530-Satt152	2.50	12.80	Harbin11, China	
	qIFSO_1	O/10	6.00	Satt345-Satt592	9.43	19.43	Harbin11, China	
	qIFSO_2	O/10	88.39	Sat_341-Satt585	9.78	15.69	Harbin11, China	

F3H	*qF3HC2_1*	C2/6	57.64	Satt322-Satt658	2.27	2.37	Harbin11, China	
	qF3HD1b_1	D1b/2	25.71	Satt157-Satt271	3.62	10.01	Harbin11, China	
	qF3HD1b_2	D1b/2	30.28	Sat_135-Sat_096	7.53	14.32	Harbin11, China	
	qF3HD1b_3	D1b/2	168.62	Sat_069-Sat_279	2.41	8.49	Harbin11, China	
	qF3HD1b_4	D1b/2	185.58	Satt459-Sat_069	2.18	5.54	Harbin11, China	
	qF3HD2_1	D2/17	0.01	Satt031-Sat_326	2.67	1.05	Harbin11, China	
	qF3HE_1	E/15	22.09	Sat_112-Sat_380	2.10	4.85	Harbin11, China	
	qF3HF_1	F/13	101.22	Sat_262-Sat_103	2.63	2.24	Harbin11, China	
	qF3HK_1	K/9	141.56	Satt349-Satt518	2.08	1.57	Harbin11, China	
	qF3HN_1	N/3	41.45	Sat_084-Sat_304	4.70	6.10	Harbin11, China	
	qF3HO_1	O/10	27.54	Satt592-Satt633	2.62	2.32	Harbin11, China	
Daidzein (DZ)	*qDZC2_1 (11-1)*	C2/6	30.79–35.79	Sat_062	2.09	5.60	Harbin, China	Wang et al. (2015)
	qDZF_1 (11-2)	F/13	9.69–15.69	Sat_262	2.00	10.57	Harbin, China	
	qDZI-2 (11-3)	I/20	36.90–60.82	Satt239	3.49	7.82	Harbin, China	
	qDZF_2 (11-4)	F/13	3.35–5.35	Satt569	4.15	8.01	Harbin, China	
	qDZI_1 (11-5)	I/20	31.49–40.51	Satt623	4.90	9.23	Hulan, China	
	qDZK_1 (11-6)	K/9	50.69–56.84	Satt725	2.46	1.64	Harbin, China	
Glycitein (GC)	*qGCA2_1 (11-1)*	A2/8	100.08–118.63	Satt233	2.65	6.01	Hulan, China	
	qGCD1b_1 (11-2)	D1b/2	87.19–118.62	Sat_459	2.38	3.12	Harbin, China	
	qGCE_1 (11-3)	E/15	15.86–43.29	Sat_380	2.34	2.33	Harbin, China	
	qGCN_1 (11-4)	N/3	32.84–40.84	Sat_304	2.41	5.21	Suihua, China	
	qGCM_2 (11-5)	M/7	60.04–73.83	Satt323	3.20	7.01	Harbin, China	
	qGCI_1 (11-6)	I/20	36.93–56.84	Satt239	3.89	5.01	Harbin, China	
	qGCM_1 (11-7)	M/7	35.84–48.86	Sat_244	2.77	2.02	Harbin, China	

(continued)

Table 10.1 (Continued)

Trait	QTL	LG/Chr	Interval (cM)	Markers Interval	LOD/*P*	R^2 (%)	Environment	References
Genistein (GT)	*qGTC2_1 (11-1)*	C2/6	51.84–60.26	Sat_336	2.48	15.16	Hulan, China	
	qGTD2_1 (11-2)	D2/17	85.15–105.44	Satt226	2.00	3.41	Harbin, China	
	qGTE_1 (11-3)	E/15	15.86–43.29	Sat_380	2.58	5.37	Suihua, China	
	qGTF_2 (11-4)	F/13	15.86–43.29	Sat_234	2.00	1.56	Harbin, China	
	qGTH_1 (11-5)	H/12	66.55–67.16	Satt253	2.97	17.05	Suihua, China	
	qGTL_1 (11-6)	L/19	32.36–34.40	Sat_320	3.00	12.01	Harbin, China	
	qGTO_1 (11-7)	O/10	51.00–63.95	Satt241	3.21	5.01	Harbin, China	
	qGTF_1 (11-8)	F/13	3.35–20.55	Satt569	2.12	2.01	Harbin, China	
	qGTM_1 (11-9)	M/7	35.84–65.04	Satt244	2.36	2.67	Harbin, China	
	qGTA2_1 (11-10)	A2/8	100.08–118.63	Satt233	3.45	7.01	Harbin, China	
Tot. isoflavone (TI)	*qTIA2_1 (9-1)*	A2/8	14.99–20.73	Satt470	2.16	1.39	Suihua, China	
	qTIE_1 (9-2)	E/15	15.86–43.29	Sat_380	2.35	1.20	Harbin, China	
	qTIF_1 (9-3)	F/13	3.35–20.55	Satt569	4.59	3.21	Hulan, China	
	qTIH_1 (9-4)	H/12	67.16–72.18	Satt629	2.33	4.74	Hulan, China	
	qTIH_2 (9-5)	H/12	67.16–78.05	Satt334	3.26	7.03	Harbin, China	
	qTIK_1 (9-6)	K/9	46.20–48.20	Sat_044	2.15	11.62	Suihua, China	
	qTIL_1 (9-7)	L/19	32.36–34.40	Sat_320	2.00	1.50	Harbin, China	
	qTID2_1 (9-8)	D2/17	32.36–34.40	Satt208	3.12	1.46	Harbin, China	
	qTIM_1 (9-9)	M/7	35.48–48.86	Sat_244	4.32	7.34	Suihua, China	
	qTIO_1 (9-10)	O/10	51.00–65.18	Sat_221	2.24	5.62	Harbin, China	
	qTIF_2 (9-11)	F/13	3.35–5.35	Satt569	3.17	2.03	Harbin, China	

Daidzein	*QDZD2_1*	D2/17	56.10–58.10	Satt208	2.33	5.57	Harbin, China	Zhao et al. (2015)
	QDZD2_2	D2/17	73.50–75.50	Satt528	2.11	4.87	Harbin, China	
	QDZF_1	F/13	77.90–79.90	Satt144	3.80	8.32	Harbin, China	
Glycitein	*QGTF_1*	F/13	110.64–112.64	Sat_074	5.27	6.63	Harbin, China	
	QGTM_1	M/7	34.27–46.19	Satt540-Sat_244	2.56	11.76	Harbin, China	
Genistein	*QGCO_1*	O/10	51.10–53.10	Satt345	3.18	4.40	Harbin, China	
	QGCM_1	M/7	34.27–46.19	Satt540-Sat_244	6.20	5.59	Harbin, China	
Total isoflavone	*QDZD2_1*	D2/17	56.10–58.10	Satt208	5.71	3.31	Harbin, China	
	QTIF_1	F/13	77.90–79.90	Satt144	4.44	10.88	Harbin, China	
	QTIO_1	O/10	51.10–53.10	Satt345	2	7.91	Harbin, China	
	QTIM_1	M/7	34.27–46.19	Satt540-Sat_244	2.06	5.33	Harbin, China	
Daidzein	*qCDZ|proN_1*	N/3	58.38–87.34	Sat_033-Sat_239	*	*	Harbin05, China	Han et al. (2015)
	qDZ|proC2_1	C2/6	30.79–35.79	Sat_252-Sat_062	*	*	Harbin05, China	
	qDZ|proK_1	K/9	46.18–56.84	Satt124-Satt725	*	*	Harbin06, China	
	qCDZ|proO_1	O/10	55.58–56.93	Satt633-Satt188	*	*	Hulan06, China	
	qDZ|proF_2	F/13	3.35–20.55	Satt144-Satt569	*	*	Harbin06, China	
	qCDZ|proD2_1	D2/17	112.84–124.30	Sat_326-Satt256	*	*	Hulan07, China	
	qCDZ|proG_1	G/18	76.76–96.57	Satt191-Satt288	*	*	Harbin05, China	
	qDZ|proI_1	I/20	31.49–40.51	Satt587-Satt623	*	*	Harbin05, China	
	qDZ|proI_2	I/20	36.93—77.83	Satt330-Satt239	*	*	Hulan06, China	
	qCDZ|oilD1b_1	D1b/2	0.00–20.61	Sat_096-Sat_351	*	*	Hulan06, China	
	qCDZ|oilC1_1	C1/4	85.37–100/12	Satt524-Satt670	*	*	Hulan06, China	
	qCDZ|oilA1_1	A1/5	30.93–38.58	Satt174-Satt300	*	*	Hulan06, China	
	qCDZ|oilH_1	H/12	68.50–81.04	Satt302-Satt279	*	*	Hulan06, China	

(continued)

Table 10.1 (Continued)

Trait	QTL	LG/Chr	Interval (cM)	Markers Interval	LOD/*P*	R^2 (%)	Environment	References
Glycitein	*qGC|proM_1*	M/7	35.84–48.86	Satt540-Sat_244	*	*	Suihua06, China	
	qCG|proK_1	K/9	14.35–51.00	Satt710-Satt242	*	*	Suihua06, China	
	qCGC|proO_1	O/10	56.93–66.37	Satt592-Satt633	*	*	Harbin06, China	
	qGC|proE_1	E/15	15.86–43.29	Sat_124-Sat_380	*	*	Harbin06, China	
	qCGC|proE_1	E/15	39.16–70.23	Sat_136-Satt231	*	*	Harbin06, China	
	qGC|proI_1	I/20	36.93–56.84	Satt330-Satt239	*	*	Hulan06, China	
	qGC|oilN_1	N/3	58.38–67.34	Sat_033-Sat_239	*	*	Hulan06, China	
	qGC|oilC2_1	C2/6	82.23–98.62	Satt658-Satt322	*	*	Hulan06, China	
	qGC|oilH_1	H/12	67.16–72.18	Satt253-Satt629	*	*	Hulan06, China	
Genistein	*qCGT|proN_1*	N/3	87.34–103.33	Sat_239-Sat_125	*	*	Suihua06, China	
	qGT|proM_1	M/7	35.84–65.04	Satt540-Satt244	*	*	Harbin06, China	
	qGT|proA2_1	A2/8	100.08–118.63	Sat_040-Satt233	*	*	Harbin06, China	
	qCGT|proA2_1	A2/8	33.95–40.95	Satt437-Satt589	*	*	Suihua06, China	
	qCGT|proO_1	O/10	55.58–56.93	Satt188-Satt153	*	*	Harbin06, China	
	qGT|proF_1	F/13	3.35–20.55	Satt144-Satt569	*	*	Harbin06, China	
	qGT|proF_2	F/13	18.12–26.55	Satt149-Sat_234	*	*	Harbin07, China	
	qGT|proD2_1	D2/17	85.15–105.44	Satt186-Satt226	*	*	Harbin07, China	
	qGT|proL_1	L/19	32.36–34.40	Sat_113-Sat_320	*	*	Suihua06, China	
	qCGT|oilA1_1	A1/5	30.93–38.58	Satt174-Satt300	*	*	Suihua06, China	
	qCGT|oilE_1	E/15	39.16–45.63	Satt716-Sat_136	*	*	Suihua06, China	
	qCGT|oilG_1	G/18	76.76–96.57	Satt191-Satt288	*	*	Suihua06, China	

Total isoflavone	*qCTI|proD1b_1*	D1b/2	54.57–56.57	Satt634-Satt271	*	*	Harbin06, China	
	qCTI|proC1_1	C1/4	62.13–64.13	Satt524-Satt670	*	*	Harbin05, China	
	qCTI|proC2_1	C2/6	51.51–53.51	Satt336-Satt457	*	*	Harbin05, China	
	qTI|proM_1	M/7	35.84–65.04	Satt540-Sat_244	*	*	Suihua06, China	
	qTI|proA2_1	A2/8	12.84–14.84	Sct_067	*	*	Suihua06, China	
	qTI|proO_1	O/10	45.21–47.21	Sat_221	*	*	Suihua06, China	
	qCTI|proO_2	O/10	56.93–66.37	Satt592-Satt633	*	*	Suihua06, China	
	qCTI|proH_1	H/12	67.16–72.18	Satt253-Satt334	*	*	Harbin06, China	
	qTI|proF_2	F/13	3.35–20.55	Satt144-Satt569	*	*	Harbin06, China	
	qTI|proE_1	E/15	49.77–51.77	Sat_124-Sat_380	*	*	Hulan06, China	
	qTI|proD2_1	D2/17	95.12–97.12	Sat_022-Satt208	*	*	Harbin05, China	
	qCTI|proD2_2	D2/17	109.05–111.05	Satt256-Satt528	*	*	Harbin06, China	
	qTI|proL_1	L/19	29.20–31.20	Sat_113-Sat_320	*	*	Harbin05, China	
Daidzein	*qDZ1*	C2/6	7.00–9.50	ss246100375	28.23	77.00	Fayetteville, NC	Akond et al. (2015)
Glycitein	*qGT1*	G/18	0.00–3.80	ss249909538	3.10	13.00	Fayetteville, NC	
Tot. isoflavone	*qTI1*	C2/6	7.00–9.50	ss246100375	25.10	76.00	Fayetteville, NC	
Unconditional QTL								Han et al. (2016)
Daidzein (DZ)	*QDZA_1*	A1/5	*	Satt236-D26A	3.01	3.12	China	
	QDZF_1	F/13	*	Satt144-Satt569	5.65	5.78	China	
	QDZK_1	K/9	*	Satt124-Satt725	2.98	1.87	China	
Genistein (GT)	*QGTA2_1*	A2/8	*	Sat_040-Satt233	2.57	17.92	China	
	QGTC2_1	C2/6	*	Satt460-Satt289	3.99	3.88	China	
	QGTH_1	H/12	*	Sat_334-Satt253	2.05	0.78	China	
	QGTE_1	E/15	*	Sat_124-Sat_380	2.59	11.54	China	

(continued)

Table 10.1 (Continued)

Trait	QTL	LG/Chr	Interval (cM)	Markers Interval	LOD/*P*	R^2 (%)	Environment	References
Glycitein (GC)	*QGCI_1*	I/20	*	Satt330-Satt239	3.34	24.64	China	
	QGCM_1	M/7	*	Satt540-Sat_244	6.7	9	China	
	QGCO_1	O/10	*	Satt592-Satt345	3.87	8.77	China	
	QGCO_2	O/10	*	Sat_341-Satt585	2.18	5.79	China	
	QGCK_1	K/9	*	Satt417-Sat_044	2	1	China	
	QGCH_1	H/12	*	Satt253-Satt629	5.09	1.08	China	
Tot. isoflavone (TI)	*QTID2_1*	D2/17	*	Satt208-Sat_022	6.57	7.68	China	
	QTIF_1	F/13	*	Satt144-Satt569	2	5.66	China	
	QTIG_1	G/18	*	Satt288	4.21	3.03	China	
	QTIO_1	O/10	*	Satt241-Sat_221	5.59	6	China	
	QTIM_1	M/7	*	Satt540-Sat_244	4.33	4.7	China	
	QTIB2_1	B2/14	*	Satt560-Satt556	2.24	1.9	China	
Conditional QTL								
Daidzein (DZ)	*QDZA1_1*	A1/5	*	Satt236-D26A	2.87	1.89	China	
Genistein (GT)	*QDZF_1*	F/13	*	Satt144-Satt569	4.65	5.34	China	
	QDZI_1	I/20	*	Satt587-Satt623	2.57	4.3	China	
	QDZK_1	K/9	*	Satt124-Satt725	2.07	5.88	China	
	QDZL_1	L/19	*	Sat_113-Sat_329	2.97	5	China	
	QDZC2_1	C2/6	*	Satt658-Satt322	2.51	4.01	China	
	QGTA2_1	A2/8	*	Sat_040-Satt233	3.44	5	China	
	QGTC2_1	C2/6	*	Satt460-Satt289	2.65	4.9	China	
	QGTF_1	F/13	*	Satt144-Satt569	3.4	3.2	China	
	QGTF_2	F/13	*	Satt149-Sat_234	5.51	3.69	China	
	QGTE_1	E/15	*	Sat_124-Sat_380	2.99	8.85	China	

Glycitein (GC)	*QGCI_1*	I/20	*	Satt330-Satt239	2.78	7.8	China	
	QGCM_1	M/7	*	Satt540-Sat_244	4.6	7.44	China	
	QGCF_1	F/13	*	Satt144-Satt569	4.04	8.89	China	
	QGCK_1	K/9	*	Satt417-Sat_044	2.7	1.92	China	
	QGCH_1	H/12	*	Satt253-Satt629	4.04	0.24	China	
Tot. isoflavone (TI)	*QTID2_1*	D2/17	*	Satt208-Sat_022	6.4	8.79	China	
	QTIF_1	F/13	*	Satt144-Satt569	2.76	12.33	China	
	QTIG_1	G/18	*	Satt288-Satt330	5.5	2.11	China	
	QTIO_1	O/10	*	Satt241-Sat_221	3.06	1.29	China	
	QTIO_2	O/10	*	Sat_341-Satt585	5.34	2.22	China	
	QTIC2_1	C2/6	*	Satt658-Satt322	3.9	3.12	China	
	QTID1b_1	D1b/2	*	Sat_279-Sat_069	3.9	3.12	China	
Daidzein (DZ)	*Seed daidzein*	I/20	*	Satt440	0.00026	5.00	China	Wang et al. (2016)
	Seed daidzein	O/10	*	Satt592	0.00034	10.30	China	
	Seed daidzein	A1/5	*	Satt155	0.00030	13.40	China	
	Seed daidzein	C2/6	*	Satt376	0.00025	9.80	China	
	Seed daidzein	D2/17	*	Satt186	0.00027	4.90	China	
	Seed daidzein	E/15	*	Satt384	0.00036	5.80	China	
	Seed daidzein	G/18	*	Satt038	0.00048	5.40	China	
	Seed daidzein	J/16	*	Satt621	0.00032	6.00	China	
	Seed daidzein	K/9	*	Satt375	0.00020	6.70	China	
	Seed daidzein	M/7	*	Satt540	0.00013	8.60	China	
	Seed daidzein	O/10	*	Satt241	0.00015	7.10	China	
	Seed daidzein	F/13	*	Sat_120	0.00045	10.70	China	

(continued)

Table 10.1 (Continued)

Trait	QTL	LG/Chr	Interval (cM)	Markers Interval	LOD/*P*	R^2 (%)	Environment	References
Glycitein (GC)	*Seed glycitein*	B2/14	*	Satt687	0.00001	18.50	China	
	Seed glycitein	C1/4	*	Sat_337	0.00019	6.70	China	
	Seed glycitein	H/12	*	Sctt009	0.00023	7.60	China	
	Seed glycitein	H/12	*	Satt629	0.00010	33.60	China	
	Seed glycitein	J/16	*	Satt431	0.00010	10.50	China	
	Seed glycitein	A2/8	*	Satt187	0.00025	6.30	China	
	Seed glycitein	G/18	*	Satt038	0.00023	6.50	China	
	Seed glycitein	G/18	*	Satt288	0.00035	10.10	China	
	Seed glycitein	D1b/2	*	Satt546	0.00027	7.40	China	
	Seed glycitein	J/16	*	Satt621	0.00310	6.00	China	
	Seed glycitein	K/9	*	Satt264	0.00010	17.30	China	
	Seed glycitein	D1b/2	*	Sat_096	0.00015	7.10	China	
	Seed glycitein	D2/17	*	Satt002	0.00010	18.30	China	
	Seed glycitein	F/13	*	Satt114	0.00024	10.80	China	
Genistein (GT)	*Seed genistein*	D1a/1	*	Satt129	0.00036	12.90	China	
	Seed genistein	O/10	*	Satt592	0.00027	9.90	China	
	Seed genistein	C2/6	*	Satt376	0.00049	8.00	China	
	Seed genistein	M/7	*	Satt540	0.00032	9.60	China	
	Seed genistein	D1b/2	*	Satt546	0.00026	6.90	China	
	Seed genistein	I/20	*	Satt587	0.00016	8.70	China	
	Seed genistein	K/9	*	Satt375	0.00011	7.00	China	
	Seed genistein	M/7	*	Satt245	0.00025	7.00	China	
	Seed genistein	M/7	*	Satt323	0.00049	6.00	China	
	Seed genistein	N/3	*	Satt080	0.00020	16.20	China	
	Seed genistein	D1a/1	*	Satt516	0.00045	7.10	China	
	Seed genistein	D2/17	*	Satt543	0.00036	4.20	China	

Total isoflavone	*Seed tot. isoflavone*	O/10	*	Satt592	0.00023	10.60	China	
	Seed tot. isoflavone	C2/6	*	Satt376	0.00015	10.30	China	
	Seed tot. isoflavone	D1b/2	*	Satt546	0.00037	9.80	China	
	Seed tot. isoflavone	H/12	*	Sat_218	0.00026	7.20	China	
	Seed tot. isoflavone	I/20	*	Satt587	0.00020	8.80	China	
	Seed tot. isoflavone	K/9	*	Satt375	0.00040	8.60	China	
	Seed tot. isoflavone	M/7	*	Satt540	0.00030	9.40	China	
	Seed tot. isoflavone	D2/17	*	Satt543	0.00045	4.10	China	
Total isoflavone	*qTIC2-1*	C2/6	0.0–4.0	s6_105140	3.33	12.9	Chatham-15	Carter (2017) and Carter et al. (2018)
	qTIO-1	O/10	20–23.7	s10_48369966	*	*	Houston-15	
	qTIJ-1	J/16	78.0–79.0	s16_3262274	2.3	9.1	Rodney-15	
	qTIJ-2	J/16	4.8–6.4	s16_37301586	2.22	8.8	Houston-15	
Daidzein	*qDaidC2-1*	C2/6	0.0–4.5	s6_105140	3.53	13.6	Chatham-15	
	qDaidJ-1	J/16	4.8–6.4	s16_37301586	2.2	8.7	Houston-15	
	qDaidC2-2	C2/6	1.0–3.0	s6_105140	2.65	10.4	Non-SCN-15	
	qDaidJ-2	J/16	75.3–79.1	s16_3267638	2.53	10	Woodstock-16	
Genistein	*qGenC2-1*	C2/6	1.0–4.0	s6_105140	2.98	11.6	Chatham-15	
	qGenJ-1	J/16	4.8–6.4	s16_37301586	2.24	8.9	Houston-15	

(continued)

Table 10.1 (Continued)

Trait	QTL	LG/Chr	Interval (cM)	Markers Interval	LOD/*P*	R^2 (%)	Environment	References
Total isoflavone	*qTIC2-1*	C2/6	1.0–4.0	s6_105140 - s6_567370	3.13	12.1	Chatham-15	
	qTIO-1	O/10	20–23.7	s10_48369966	2.38	8.3	Houston-15	
	qTIJ-1	J/16	78.0–79.0	s16_3262274	*	*	Rodney-15	
	qTIJ-2	J/16	4.8–6.4	s16_37301586	*	*	Houston-15	
Daidzein	*qDaidC2-1*	C2/6	1.0–3.0	s6_105140	3.31	12.7	Chatham-15	
	qDaidJ-1	J/16	4.8–6.4	s16_37301586	*	*	Houston-15	
	qDaidC2-2	C2/6	1.0–3.0	s6_105140	*	*	Non-SCN-15	
	qDaidJ-2	J/16	75.3–77.2	s16_3267638	2.54	9.8	Woodstock-16	
Genistein	*qGenC2-1*	C2/6	2.0–3.0	s6_105140	2.67	10.3	Chatham-15	
	qGenJ-1	J/16	4.8–6.4	s16_37301586	2.34	9.2	Houston-15	
GLC	*qGLC-1*	A1/5	36,904,003	AX-93722354	2.55E-06	14.98	China	Chu et al. (2017)
TIC	*qTIC-1*	C2/6	41,283,321	AX-93734476	3.66E-06	10.98	China	
DAC	*qDAC-1*	C2/6	41,283,321	AX-93734476	1.44E-06	11.64	China	
GLC	*qGLC-2*	B1/11	8,247,884	AX-93788682	3.27E-06	12.16	China	
GLC	*qGLC-3*	B1/11	8,248,245	AX-93788683	3.27E-06	12.16	China	
GLC	*qGLC-4*	B1/11	8,252,466	AX-94286338	2.34E-06	12.72	China	
GLC	*qGLC-5*	B1/11	8,254,189	AX-93936943	1.35E-06	13.17	China	
GLC	*qGLC-6*	B1/11	8,262,066	AX-93788688	2.70E-06	11.60	China	
GLC	*qGLC-7*	B1/11	8,262,730	AX-93788689	8.03E-07	12.90	China	
GLC	*qGLC-8*	B1/11	8,268,689	AX-93788692	9.14E-07	12.76	China	
GLC	*qGLC-9*	B1/11	8,269`,763	AX-94085385	1.31E-06	12.37	China	
GLC	*qGLC-10*	B1/11	8,271,460	AX-93788694	9.19E-07	12.76	China	
GLC	*qGLC-11*	B1/11	8,284,464	AX-94085391	1.35E-06	12.34	China	
TIC	*qTIC-2*	I/20	43,281,039	AX-94292181	3.10E-06	10.55	China	

TIC	*qTIC-3*	I/20	43,283,579	AX-94208339	2.09E-06	10.91	China	
DAC	*qDAC-2*	I/20	43,283,579	AX-94208339	4.01E-06	10.72	China	
TIC	*qTIC-4*	I/20	43,285,450	AX-93910368	2.19E-06	10.85	China	
TIC	*qTIC-5*	I/20	43,285,529	AX-93661412	1.88E-06	11.10	China	
TIC	*qTIC-6*	I/20	43,294,213	AX-94208344	1.50E-06	11.22	China	
DAC	*qDAC-3*	I/20	43,294,213	AX-94208344	3.56E-06	10.84	China	
TIC	*qTIC-7*	I/20	43,428,004	AX-94208372	1.44E-06	11.46	China	
DAC	*qDAC-4*	I/20	43,428,004	AX-94208372	4.35E-06	10.64	China	
TIC	*qTIC-8*	I/20	43,452,456	AX-93661429	2.18E-06	11.11	China	
TIC	*qTIC-9*	I/20	43,453,176	AX-93910415	1.61E-06	11.19	China	
TIC	*qTIC-10*	I/20	43,458,721	AX-93910416	1.14E-06	12.22	China	
DAC	*qDAC-5*	I/20	43,458,721	AX-93910416	4.57E-06	10.95	China	
TIC	*qTIC-11*	I/20	43,460,939	AX-93910417	6.10E-07	12.12	China	
DAC	*qDAC-6*	I/20	43,460,939	AX-93910417	2.74E-06	11.11	China	
TIC	*qTIC-12*	I/20	43,465,595	AX-93661433	2.19E-06	10.85	China	
TIC	*qTIC-13*	I/20	43,485,102	AX-94292195	4.30E-07	12.47	China	
DAC	*qDAC-7*	I/20	43,485,102	AX-94292195	2.42E-06	11.38	China	
TIC	*qTIC-14*	I/20	43,490,475	AX-93910432	4.30E-07	12.47	China	
DAC	*qDAC-8*	I/20	43,490,475	AX-93910432	2.11E-06	11.38	China	
TIC	*qTIC-15*	I/20	43,497,182	AX-93910435	6.39E-07	12.07	China	
DAC	*qDAC-9*	I/20	43,497,182	AX-93910435	2.11E-06	10.92	China	
TIC	*qTIC-16*	I/20	43,502,141	AX-94208409	3.47E-07	12.69	China	
DAC	*qDAC-10*	I/20	43,502,141	AX-94208409	1.80E-06	11.54	China	
TIC	*qTIC-17*	I/20	43,517,749	AX-93957635	4.25E-06	10.28	China	

(continued)

Table 10.1 (Continued)

Trait	QTL	LG/Chr	Interval (cM)	Markers Interval	LOD/*P*	R^2 (%)	Environment	References
Acetyldaidzin, AD	*qIF2-1*	D1b/2	71.6–72.9	Bin214-215	2.53–6.74	2.12–11.24	Zengcheng-13	Cai et al. (2018)
		D1b/2	71.6–72.9		2.53–6.74	2.12–11.24	Guangdong-14	
		D1b/2	71.6–72.9		2.53–6.74	2.12–11.24	MEAN	
Acetylglycine, AGL		D1b/2	71.6–72.9		2.53–6.74	2.12–11.24	Guangdong-14	
Daidzin, D		D1b/2	71.6–72.9		2.53–6.74	2.12–11.24	Guangdong-14	
		D1b/2	71.6–72.9		2.53–6.74	2.12–11.24	MEAN	
Genistin, G		D1b/2	71.6–72.9		2.53–6.74	2.12–11.24	Guangdong-14	
Malonyldaidzin, MD		D1b/2	71.6–72.9		2.53–6.74	2.12–11.24	Guangdong-14	
Malonylgenistin, MG		D1b/2	71.6–72.9		2.53–6.74	2.12–11.24	Zengcheng-13	
		D1b/2	71.6–72.9		2.53–6.74	2.12–11.24	Guangdong-14	
		D1b/2	71.6–72.9		2.53–6.74	2.12–11.24	MEAN	
Total isoflavone		D1b/2	71.6–72.9		2.53–6.74	2.12–11.24	Zengcheng-13	
		D1b/2	71.6–72.9		2.53–6.74	2.12–11.24	Guangdong-14	

		D1b/2	71.6–72.9		2.53–6.74	2.12–11.24	Guangdong-15	
		D1b/2	71.6–72.9		2.53–6.74	2.12–11.24	MEAN	
Acetyldaidzin, AD	*qIF5-1*	A1/5	236.4–237.1	Bin799-800	2.96–45.63	6.37–59.95	Zengcheng-13	
		A1/5	236.4–237.1		2.96–45.63	6.37–59.95	Guangdong-15	
		A1/5	236.4–237.1		2.96–45.63	6.37–59.95	MEAN	
Daidzin, D		A1/5	236.4–237.1		2.96–45.63	6.37–59.95	Zengcheng-13	
		A1/5	236.4–237.1		2.96–45.63	6.37–59.95	Zengcheng-14	
		A1/5	236.4–237.1		2.96–45.63	6.37–59.95	Guangdong-14	
		A1/5	236.4–237.1		2.96–45.63	6.37–59.95	Guangdong-15	
		A1/5	236.4–237.1		2.96–45.63	6.37–59.95	MEAN	
Daidzein, DE		A1/5	236.4–237.1		2.96–45.63	6.37–59.95	Guangdong-14	

(continued)

Table 10.1 (Continued)

Trait	QTL	LG/Chr	Interval (cM)	Markers Interval	LOD/*P*	R^2 (%)	Environment	References
Genistin, G		A1/5	236.4–237.1		2.96–45.63	6.37–59.95	Zengcheng-13	
		A1/5	236.4–237.1		2.96–45.63	6.37–59.95	Zengcheng-14	
		A1/5	236.4–237.1		2.96–45.63	6.37–59.95	Guangdong-14	
		A1/5	236.4–237.1		2.96–45.63	6.37–59.95	Guangdong-15	
		A1/5	236.4–237.1		2.96–45.63	6.37–59.95	MEAN	
Genistein, GE		A1/5	236.4–237.1		2.96–45.63	6.37–59.95	Guangdong-14	
		A1/5	236.4–237.1		2.96–45.63	6.37–59.95	Guangdong-15	
Glycitin, GL		A1/5	236.4–237.1		2.96–45.63	6.37–59.95	Zengcheng-13	
		A1/5	236.4–237.1		2.96–45.63	6.37–59.95	Guangdong-14	
		A1/5	236.4–237.1		2.96–45.63	6.37–59.95	Guangdong-15	
		A1/5	236.4–237.1		2.96–45.63	6.37–59.95	MEAN	

Malonyldaidzin, MD		A1/5	236.4–237.1		2.96–45.63	6.37–59.95	Zengcheng-13	
		A1/5	236.4–237.1		2.96–45.63	6.37–59.95	Guangdong-14	
		A1/5	236.4–237.1		2.96–45.63	6.37–59.95	Guangdong-15	
		A1/5	236.4–237.1		2.96–45.63	6.37–59.95	MEAN	
Malonylgenistin, MG		A1/5	236.4–237.1		2.96–45.63	6.37–59.95	Zengcheng-13	
		A1/5	236.4–237.1		2.96–45.63	6.37–59.95	Zengcheng-14	
		A1/5	236.4–237.1		2.96–45.63	6.37–59.95	Guangdong-14	
		A1/5	236.4–237.1		2.96–45.63	6.37–59.95	Guangdong-15	
		A1/5	236.4–237.1		2.96–45.63	6.37–59.95	MEAN	
Malonylglycitin, MGL		A1/5	236.4–237.1		2.96–45.63	6.37–59.95	Zengcheng-13	
		A1/5	236.4–237.1		2.96–45.63	6.37–59.95	Guangdong-14	
		A1/5	236.4–237.1		2.96–45.63	6.37–59.95	Guangdong-15	

(continued)

Table 10.1 (Continued)

Trait	QTL	LG/Chr	Interval (cM)	Markers Interval	LOD/*P*	R^2 (%)	Environment	References
Total isoflavone		A1/5	236.4–237.1		2.96–45.63	6.37–59.95	Zengcheng-13	
		A1/5	236.4–237.1		2.96–45.63	6.37–59.95	Zengcheng-14	
		A1/5	236.4–237.1		2.96–45.63	6.37–59.95	Guangdong-14	
		A1/5	236.4–237.1		2.96–45.63	6.37–59.95	Guangdong-15	
		A1/5	236.4–237.1		2.96–45.63	6.37–59.95	MEAN	
Acetylgenistin, AG	*qIF5-2*	A1/5	234.9–235.6	Bin801-802	3.98–32.36	7.17–48.07	MEAN	
Acetylglycitin, AGL		A1/5	234.9–235.6		3.98–32.36	7.17–48.07	Guangdong-15	
DE		A1/5	234.9–235.6		3.98–32.36	7.17–48.07	Zengcheng-13	
		A1/5	234.9–235.6		3.98–32.36	7.17–48.07	MEAN	
Genistein, GE		A1/5	234.9–235.6		3.98–32.36	7.17–48.07	MEAN	
Malonyldaidzin, MD		A1/5	234.9–235.6		3.98–32.36	7.17–48.07	Zengcheng-14	

Acetyldaidzin, AD	*qIF5-3*	A1/5	232.5	Bin805	2.84–22.08	4.61–31.06	Zengcheng-14	
		A1/5	232.5		2.84–22.08	4.61–31.06	Guangdong-14	
Acetylglycitin, AGL		A1/5	232.5		2.84–22.08	4.61–31.06	Zengcheng-14	
		A1/5	232.5		2.84–22.08	4.61–31.06	Guangdong-14	
Malonylglycitin, MGL		A1/5	232.5		2.84–22.08	4.61–31.06	MEAN	
Daidzin, D	*qIF7-1*	M/7	7.2	Bin1047	3.02–4.68	2.59–4.23	Guangdong-15	
Genistin, G		M/7	7.2		3.02–4.68	2.59–4.23	Guangdong-15	
Malonyldaidzin, MD		M/7	7.2		3.02–4.68	2.59–4.23	Guangdong-15	
TOT		M/7	7.2		3.02–4.68	2.59–4.23	Guangdong-15	
Malonyldaidzin, MD	*qIF9-1*	K/9	11.4	Bin1400	2.53–4.37	1.88–4.41	Guangdong-15	
Malonylgenistin, MG		K/9	11.4		2.53–4.37	1.88–4.41	Guangdong-015	
TOT		K/9	11.4		2.53–4.37	1.88–4.41	Guangdong-14	
		K/9	11.4		2.53–4.37	1.88–4.41	Guangdong-15	

(continued)

Table 10.1 (Continued)

Trait	QTL	LG/Chr	Interval (cM)	Markers Interval	LOD/*P*	R^2 (%)	Environment	References
Acetylgenistin, AG	*qIF10-1*	O/10	29.3–34.6	Bin1637-1639	2.87–6.28	4.00–11.50	Guangdong-15	
		O/10	29.3–34.6		2.87–6.28	4.00–11.50	MEAN	
Genistin, G		O/10	29.3–34.6		2.87–6.28	4.00–11.50	Guangdong-14	
		O/10	29.3–34.6		2.87–6.28	4.00–11.50	Guangdong-15	
		O/10	29.3–34.6		2.87–6.28	4.00–11.50	MEAN	
Glycitein, GLE		O/10	29.3–34.6		2.87–6.28	4.00–11.50	Guangdong-14	
		O/10	29.3–34.6		2.87–6.28	4.00–11.50	MEAN	
Acetyldaidzin, AD	*qIF10-2*	O/10	170.4	Bin1727	2.57–9.73	2.80–10.39	Guangdong-14	
Daidzin, D		O/10	170.4		2.57–9.73	2.80–10.39	Zengcheng-14	
		O/10	170.4		2.57–9.73	2.80–10.39	Guangdong-15	
		O/10	170.4		2.57–9.73	2.80–10.39	MEAN	
Genistin, G		O/10	170.4		2.57–9.73	2.80–10.39	Zengcheng-14	
		O/10	170.4		2.57–9.73	2.80–10.39	Guangdong-15	
		O/10	170.4		2.57–9.73	2.80–10.39	MEAN	

Genistin, G		O/10	170.4		2.57–9.73	2.80–10.39	Zengcheng-14	
		O/10	170.4		2.57–9.73	2.80–10.39	Guangdong-14	
Malonyldaidzin, MD		O/10	170.4		2.57–9.73	2.80–10.39	Guangdong-14	
Malonylglycitin, MGL		O/10	170.4		2.57–9.73	2.80–10.39	Zengcheng-13	
Malonylgenistin, MG		O/10	170.4		2.57–9.73	2.80–10.39	Guangdong-14	
		O/10	170.4		2.57–9.73	2.80–10.39	Guangdong-14	
Total isoflavone		O/10	170.4		2.57–9.73	2.80–10.39	Guangdong-14	
		O/10	170.4		2.57–9.73	2.80–10.39	MEAN	
Acetyldaidzin, AD	*qIF10-3*	O/10	178–185.5	Bin1729-1735	2.75–6.46	3.09–7.08	Zengcheng-13	
		O/10	178–185.5		2.75–6.46	3.09–7.08	Zengcheng-14	
		O/10	178–185.5		2.75–6.46	3.09–7.08	Guangdong-15	
		O/10	178–185.5		2.75–6.46	3.09–7.08	MEAN	
Daidzein, DE		O/10	178–185.5		2.75–6.46	3.09–7.08	Zengcheng-14	
Genistin, G		O/10	178–185.5		2.75–6.46	3.09–7.08	Guangdong-15	
		O/10	178–185.5		2.75–6.46	3.09–7.08	MEAN	
Genistein, GE		O/10	178–185.5		2.75–6.46	3.09–7.08	Zengcheng-14	
		O/10	178–185.5		2.75–6.46	3.09–7.08	Guangdong-14	
		O/10	178–185.5		2.75–6.46	3.09–7.08	Guangdong-15	

(continued)

Table 10.1 (Continued)

Trait	QTL	LG/Chr	Interval (cM)	Markers Interval	LOD/*P*	R^2 (%)	Environment	References
Malonyldaidzin, MD		O/10	178–185.5		2.75–6.46	3.09–7.08	Zengcheng-13	
		O/10	178–185.5		2.75–6.46	3.09–7.08	Guangdong-15	
Malonylgenistin, MG		O/10	178–185.5		2.75–6.46	3.09–7.08	Zengcheng-13	
		O/10	178–185.5		2.75–6.46	3.09–7.08	Guangdong-15	
		O/10	178–185.5		2.75–6.46	3.09–7.08	MEAN	
Total isoflavone		O/10	178–185.5		2.75–6.46	3.09–7.08	Zengcheng-13	
		O/10	178–185.5		2.75–6.46	3.09–7.08	Zengcheng-14	
		O/10	178–185.5		2.75–6.46	3.09–7.08	Guangdong-15	
Genistein, GE	*qIF11-1*	B1/11	99.9–101.6	Bin1806-1809	2.60–5.06	3.59–8.14	Guangdong-14	
Glycitin, GL		B1/11	99.9–101.6		2.60–5.06	3.59–8.14	Guangdong-15	
Malonylglycitin, MGL		B1/11	99.9–101.6		2.60–5.06	3.59–8.14	Zengcheng-13	
		B1/11	99.9–101.6		2.60–5.06	3.59–8.14	Guangdong-15	
		B1/11	99.9–101.6		2.60–5.06	3.59–8.14	MEAN	
Acetyldaidzin, AD	*qIF13-1*	F/13	91.8–100.3	Bin2136-2144	2.74–5.21	2.08–6.73	Zengcheng-14	
Daidzin, D		F/13	91.8–100.3		2.74–5.21	2.08–6.73	Zengcheng-13	
		F/13	91.8–100.3		2.74–5.21	2.08–6.73	MEAN	
Malonylgenistin, MG		F/13	91.8–100.3		2.74–5.21	2.08–6.73	Zengcheng-14	
		F/13	91.8–100.3		2.74–5.21	2.08–6.73	Guangdong-15	
		F/13	91.8–100.3		2.74–5.21	2.08–6.73	MEAN	
Total isoflavone		F/13	91.8–100.3		2.74–5.21	2.08–6.73	Zengcheng-14	
		F/13	91.8–100.3		2.74–5.21	2.08–6.73	MEAN	

Acetyldaidzin, AD	*qIF13-3*	F/13	204.2–206.3	Bin2245-2247	2.52–4.86	2.18–5.25	Guangdong-15	
		F/13	204.2–206.3		2.52–4.86	2.18–5.25	MEAN	
Acetylgenistin, AG		F/13	204.2–206.3		2.52–4.86	2.18–5.25	Guangdong-14	
Daidzin, D		F/13	204.2–206.3		2.52–4.86	2.18–5.25	Zengcheng-13	
		F/13	204.2–206.3		2.52–4.86	2.18–5.25	Guangdong-15	
		F/13	204.2–206.3		2.52–4.86	2.18–5.25	MEAN	
Genistin, G		F/13	204.2–206.3		2.52–4.86	2.18–5.25	Zengcheng-13	
Total isoflavone		F/13	204.2–206.3		2.52–4.86	2.18–5.25	Guangdong-14	
		F/13	204.2–206.3		2.52–4.86	2.18–5.25	Guangdong-15	
Acetyldaidzin, AD	*qIF16*	J/16	0–0.8	Bin2544-2545	2.56–4.80	2.08–4.55	Guangdong-14	
Malonyldaidzin, MD		J/16	0–0.8		2.56–4.80	2.08–4.55	Zengcheng-14	
Malonylgenistin, MG		J/16	0–0.8		2.56–4.80	2.08–4.55	Guangdong-14	
Total isoflavone		J/16	0–0.8		2.56–4.80	2.08–4.55	Zengcheng-14	
Daidzin, D	*qIF17-1*	D2/17	3.0–7.0	Bin2749-2754	2.50–5.45	2.02–3.77	Guangdong-14	
		D2/17	3.0–7.0		2.50–5.45	2.02–3.77	MEAN	
Malonyldaidzin, MD		D2/17	3.0–7.0		2.50–5.45	2.02–3.77	Guangdong-15	
		D2/17	3.0–7.0		2.50–5.45	2.02–3.77	MEAN	
Acetyldaidzin, AD	*qIF19-2*	L/19	97.6–102.4	Bin3219-3220	2.64–5.98	3.94–8.51	Zengcheng-13	
Malonylgenistin, MG		L/19	97.6–102.4		2.64–5.98	3.94–8.51	Zengcheng-13	
Malonylglycitin, MGL		L/19	97.6–102.4		2.64–5.98	3.94–8.51	Zengcheng-13	
		L/19	97.6–102.4		2.64–5.98	3.94–8.51	Guangdong-14	
Acetyldaidzin, AD	qIF5-1	A1/5	237.1	Bin799	26.16	33.55	Zengcheng-13	
		A1/5	237.1		24.46	34.18	Guangdong-15	

(continued)

Table 10.1 (Continued)

Trait	QTL	LG/Chr	Interval (cM)	Markers Interval	LOD/*P*	R^2 (%)	Environment	References
Acetylgenistin, AG		A1/5	237.1		20.95	34.55	Guangdong-14	
		A1/5	237.1		5.07	8.79	Guangdong-15	
Acetylglycitin, AGL		A1/5	237.1		2.96	6.37	MEAN	
Daidzin, D		A1/5	237.1		28.26	41.99	Zengcheng-13	
		A1/5	237.1		36.75	49.48	Guangdong-14	
Glycitin, GL		A1/5	237.1		6.86	12.25	Zengcheng-13	
		A1/5	237.1		5.36	9.1	Guangdong-14	
		A1/5	237.1		6.59	11.04	Guangdong-15	
		A1/5	237.1		4.19	6.77	MEAN	
Malonyldaidzin, MD		A1/5	237.1		37.03	51.25	Zengcheng-13	
Malonylgenistin, MG		A1/5	237.1		31.41	42.32	Zengcheng-13	
		A1/5	237.1		20.96	32.63	Zengcheng-14	
		A1/5	237.1		26.63	38.46	Guangdong-15	
Malonylglycitin, MGL		A1/5	237.1		11.77	17.95	Zengcheng-13	
		A1/5	237.1		8.32	13.37	Guangdong-14	
		A1/5	237.1		7.63	13.42	Guangdong-15	
Total isoflavone		A1/5	237.1		39.33	49.38	Zengcheng-13	
		A1/5	237.1		28.14	43.27	Zengcheng-14	
		A1/5	237.1		36.07	46.59	Guangdong-14	
		A1/5	237.1		32.85	45.15	Guangdong-15	
		A1/5	237.1		41.26	52.5	MEAN	

Daidzin, D		A1/5	236.4	Bin800	36.45	50.58	MEAN	
Daidzein, DE		A1/5	236.4		15.94	26.17	Guangdong-14	
Genistein, GE		A1/5	236.4		10.45	17.49	Guangdong-14	
Daidzin, D		A1/5	236.4		27.63	40.92	Zengcheng-14	
		A1/5	236.4		33.27	46.5	Guangdong-15	
Genistin, G		A1/5	236.4		23.93	38.28	Zengcheng-13	
		A1/5	236.4		13.87	23.37	Zengcheng-14	
		A1/5	236.4		21.49	30.62	Guangdong-14	
		A1/5	236.4		23.47	33.35	Guangdong-15	
		A1/5	236.4		25.52	36.74	MEAN	
Genistein, GE		A1/5	236.4		13.05	20.59	Guangdong-15	
Malonyldaidzin, MD		A1/5	236.4		37.81	48.61	Guangdong-14	
		A1/5	236.4		35.67	49.24	Guangdong-15	
		A1/5	236.4		45.63	59.95	MEAN	
Malonylgenistin, MG		A1/5	236.4		31.63	40.21	Guangdong-14	
		A1/5	236.4		36.44	48.33	MEAN	
Daidzin, D	*qDl6*	J/16	19–21	Gm16_7314	7.7	*	2009SY/2010SY	Pei et al. (2018)
	qD20	I/20	0–1	Gm20_49565	7.4	*	2009SY/2010SY	
Genistin, G	*qG8*	A2/8	0–1	Gm08_424	8.7	*	2009SY/2011SY	
	qG9	K/9	133–135	Gm09_47520	4.9	*	2009SY/2011SY	
	qG20	I/20	2–4	Gm20_19811	13.3	*	2009SY/2010SY	
Glycitin, GL	*qGL5*	A1/5	56–58	Gm05_8127	10.6	*	2009CP/2010SY	

(continued)

Table 10.1 (Continued)

Trait	QTL	LG/Chr	Interval (cM)	Markers Interval	LOD/*P*	R^2 (%)	Environment	References
Malonyldaidzin, MD	*qMD2*	D1b/2	101–103	Gm02_446	11.7	*	2010SY/2011SY	
	qMD3	N/3	11–13	Gm03_36302	6.2	*	2009CP/2009SY	
	qMD7	M/7	51–53	Gm07_39984	8.1	*	2009CP/2011SY	
	qMD13	F/13	64–66	Gm13_34575	10.1	*	2010SY/2011SY	
	qMD15-1	E/15	34–36	Gm15_36421	8.5	*	2010SY/2011SY	
	qMD15-2	E/15	91–93	Gm15_69377	9.5	*	2010SY/2011SY	
	qMD19	L/19	13–15	Gm19_18007	9.2	*	09SY/10SY/11SY	
	qMD20	I/20	0–1	Gm20_49565	11.9	*	09CP/09SY/10SY	
Malonylgenistin, MG	*qMG14*	B2/14	22–25	Gm14_1880	12.2	*	2009CP/2011SY	
	qMG16	J/16	19021	Gm16_7314	12.2	*	2010SY/2011SY	
	qMG18	G/18	157–159	Gm18_1940	5.8	*	2009SY/2010SY	
	qMG20	I/20	2–4	Gm20_19811	10.7	*	09CP/09SY/10SY	
Total isoflavone	*qTIF2*	D1b/2	114–116	Gm02_76087	17.9	*	2009SY/2010SY	
	qTIF7	M/7	54–56	Gm07_18835	8.2	*	2010SY/2011SY	
	qTIF16	J/16	19021	Gm16_7314	8	*	2010SY/2011SY	
	qTIF18	G/18	169–171	Gm18_49681	8.9	*	2009SY/2011SY	
	qTIF19	L/19	13–15	Gm19_18007	9.2	*	2010SY/2011SY	
	qTIF20	I/20	0–1	Gm20_49565	17.6	*	09CP/09SY/10SY	
Isoflavone content QTL in ZL RIL population								Li et al. (2018)

Daidzein, D	*qDA1_1*	A1/5	22.45–28.25	Satt211–ss715591655	*	19.43	AUH Breeding Ctre	
		A1/5	22.15–27.95		*	22.66	Cangzhou City	
Genistin, G	*qGA1_1*	A1/5	21.28–28.35		*	16.11	AUH Breeding Ctre	
		A1/5	22.85–28.35		*	18.45	Cangzhou City	
Total isoflavone	*qTIA1_1*	A1/5	20.85–27.15		*	19.21	AUH Breeding Ctre	
		A1/5	22.15–28.335		*	23.74	Cangzhou City	
Daidzein, D	*qDA1_2*	A1/5	29.55–33.95		*	19.81	Qing-yuan	
Total isoflavone	*qTIA1_2*	A1/5	28.85–33.95	ss715591655	*	28.59	Baoding City	
Daidzein, D	*qDA1_3*	A1/5	30.05–3.85		*	23.37	Baoding City	
Genistin, G	*qGA1_2*	A1/5	34.55–39.25	ss715591780	*	14.46	Qing-yuan	
		A1/5	35.15–39.85		*	25.99	Baoding City	
Total isoflavone	*qTIA1_3*	A1/5	34.35–38.65		*	14.09	Qing-yuan	
Glycitin, GL	*qGLB1_1*	B1/11	151.35–153.45	ss715611230	*	29.83	Cangzhou City	
	qGLB1_2	B1/11	151.65–153.55	ss715611213	*	20.29	Qing-yuan	
	qGLB1_3	B1/11	152.85–154.45	ss715611204	*	16.52	Baoding City	
	qGLB1_4	B1/11	153.95–154.95	ss715611194	*	18.86	AUH Breeding Ctre	

(continued)

Table 10.1 (Continued)

Trait	QTL	LG/Chr	Interval (cM)	Markers Interval	LOD/*P*	R^2 (%)	Environment	References
Genistin, G	*qGC2*	C2/6	100.35–101.75	ss715593753	*	4.68	AUH Breeding Ctre	
		C2/6	100.35–101.75		*	3.78	Cangzhou City	
Total isoflavone	*qTIC2*	C2/6	100.35–101.75		*	3.31	AUH Breeding Ctre	
		C2/6	100.35–101.75		*	4.00	Cangzhou City	
Total isoflavone	*qTID1b*	D1b/2	176.95–177.45	ss71558123–Satt537	*	3.27	Qing-yuan	
		D1b/2	177.05–177.45		*	3.16	Cangzhou City	
Daidzein, D	*qDA1_4*	A1/5	47.75–50.65	ss715592231–Satt545	*	3.95	Cangzhou City	
Glycitin, GL	*qGLA1_2*	A1/5	47.85–50.15		*	6.32	Baoding City	
Total isoflavone	*qTIA1_4*	A1/5	47.85–50.15		*	4.50	Cangzhou City	
Daidzein, D	*qDF_1*	F/13	243.56–244.56	ss715616114	*	3.45	Qing-yuan	
	qDF_2	F/13	243.56–244.56		*	3.60	Qing-yuan	
	qDF_3	F/13	229.05–239.76	ss715616040	*	6.12	Baoding City	
Total isoflavone	*qTIF_2*	F/13	227.85–239.76		*	4.83	Baoding City	
Daidzein, D	*qDF_4*	F/13	185.75–195	ss715615453	*	3.60	AUH Breeding Ctre	
Genistin, G	*qGF_1*	F/13	187.95–201.45		*	5.74	Cangzhou City	

Daidzein, D	*qDO*	O/10	37.75–39.85	ss715607450	*	3.27	Qing-yuan	
	qGLO	O/10	37.75–39.85		*	3.29	AUH Breeding Ctre	
Total isoflavone	*qTID2_1*	D2/17	86.05–86.45	ss715626244	*	4.02	Cangzhou City	
Glycitin, GL	*qGD2*	D2/17	86.5–87.15	ss715626239	*	4.17	Cangzhou City	
Glycitin, GL	*qGLM_1*	M/7	178.45–184.35	ss715597502–Satt680	*	4.51	AUH Breeding Ctre	
	qGLM_2	M/7	185.85–199.35	Satt680–ss715597917	*	7.30	Baoding City	
Daidzein, D	*qDG_1*	G/18	200.75–204.65	ss715632208	*	7.86	Baoding City	
	qDG_2	G/18	204.65–207.45	ss715632241	*	3.80	Qing-yuan	
	qDG_3	G/18	204.65–207.45	ss715632285	*	4.83	Cangzhou City	
Total isoflavone	*qTIF_1*	F/13	195.55–207.45	ss715615594	*	3.66	Cangzhou City	
Glycitin, GL	*qGLA1_1*	A1/5	14.25–27.85	Satt211–ss715591655	*	3.87	AUH Breeding Ctre	
Isoflavone content QTL in ZQ RIL population								
Daidzin, D	*qDA1'*	A1/5	98.25–98.70	ss715591790	*	20.18	E1 = Qingxian15	
Genistin, G	*qGA1'*	A1/5	98.25–98.70		*	15.67	E1 = Qingxian15	
Total isoflavone	*aTIA1'*	A1/5	98.25–98.70		*	17.22	E1 = Qingxian15	
Glycitin, GL	*qGLB1*	B1/11	95.05–99.25	ss715611194	*	13.08	E1 = Qingxian15	
Daidzin, D	*qDF'*	F/13	120.75–127.05	ss715616158	*	4.33	E1 = Qingxian15	
Total isoflavone	*qTIF'*	F/13	120.75–127.15		*	4.79	E1 = Qingxian15	

(continued)

Table 10.1 (Continued)

Trait	QTL	LG/Chr	Interval (cM)	Markers Interval	LOD/*P*	R^2 (%)	Environment	References
Genistin, G	*qGF'*	F/13	120.75–127.65	ss715616173	*	4.64	E1 = Qingxian15	
Genistin, G	*qGD1b'*	D1b/2	136.45–140.25	ss715581320	*	4.17	E1 = Qingxian15	
Total isoflavone	*qTID1b'*	D1b/2	136.45–140.15		*	4.18	E1 = Qingxian15	
Daidzin, D	*qDD1b'*	D1b/2	123.95–131.85	ss715581329	*	3.25	E1 = Qingxian15	
Genistin, G	*qGN'*	N/3	173.35–183.75	ss715584758	*	3.42	E1 = Qingxian15	
Malonylglycitin, MGL	qMGly_11	B1/11	20–28.2	M11 – M14	7.6	27.9	Saga, Japan	Watanabe et al. (2019)
	qMGly_15	E/15	33.63–37.76	M57 – M68	4.4	15.1	Saga, Japan	
Malonylgenistin, MGN	qMGe_12	H/12	*	*	5.3	19.7	Saga, Japan	
Daidzein	*rs42737801*	A2/8	42,737,801	*	8.73	0.39	China	Wu et al. (2020)
	rs42737801	A2/8	42,737,801	*	7.88	0.88	China	
	rs42737801	A2/8	42,737,801	*	8.85	0.63	China	
Total isoflavone	*rs42737801*	A2/8	42,737,801	*	4.90	0.37	China	
	rs42737801	A2/8	42,737,801	*	3.80	1.16	China	
	rs42737801	A2/8	42,737,801	*	5.10	0.71	China	
Daidzein	*qDZ8_2*	A2/8	42,868,659	*	2.59	12.15	China	
	qDZ8_2	A2/8	42,868,659	*	4.98	22.18	China	
	qDZ8_2	A2/8	42,868,659	*	3.09	12.98	China	
	qDZ8_2	A2/8	42,868,659	*	4.08	14.26	China	
Total isoflavone	*qTI8_1*	A2/8	42,868,659	*	2.61	9.06	China	
	qTI8_1	A2/8	42,868,659	*	3.75	13.64	China	
	qTI8_1	A2/8	42,868,659	*	8.23	27.27	China	
	qTI8_1	A2/8	42,868,659	*	4.67	16.46	China	

by Satt124 on LG K (Chr_9)] associated with daidzein contents; four QTL [*qGTA2_1* identified by Sat_040 on LG A2 (Chr_8); *qGTC2_1* identified by Satt460 on LG C2 (Chr_6); *qGTF_1* identified by Satt144 on LG F (Chr_13); and *qGTM_1* identified by the marker Satt540 on LG M (Chr_7)] associated with genistein; three QTL [*qGCI_1* identified by Satt330 on LG I (Chr_20); *qGCM_1* identified by Satt540 on LG M (Chr_7); and *qGCO_1* identified by Satt592 on LG O (Chr_10)] associated with glycitein contents; and five QTL [*qTID2_1* identified by Satt208 on LG D2 (Chr_17); *qTIF_1* identified by Satt144 on LG F (Chr_13); *qTIG_1* identified by Satt288 on LG G (Chr_18); *qTIO_1* identified by Satt241 on LG O (Chr_10); and *qTIM_1* identified by Satt540 on LG M (Chr_7)] associated with total isoflavone contents (Table 10.1, Fig. 2.1) (Zeng et al. 2009). Interestingly, four of these QTL (*QGTA2_1, DZF_1, QGTF_1,* and *QTIF_1*) were found to be highly associated with two QTL for aphid resistance (*qRa_1* and *qRa_2*) which demonstrates that soybean cultivars with resistance to aphids and high isoflavone contents can be developed (Meng et al. 2011).

Liang et al. (2010) grew a RIL population derived from a cross between 'Jindou 23' and 'Huibuzhi' line ($n = 474$) in China in 2008, genotyped it with SSR markers, and used CIM of WinQTL cartographer to identify QTL that control seed protein, oil, and isoflavone contents. Three QTL that control seed protein content have been identified and mapped on chromosomes 6, 12, and 18; five QTL that control seed oil content have been identified and mapped on chromosomes 5, 6, 8, 14, and 17; and three QTL that control seed isoflavone contents have been identified and mapped on chromosomes 3 and 16 (Fig. 2.1; Table 10.1) (Liang et al. 2010).

Gutierrez-Gonzalez et al. (2010) used the same RIL population derived from the cross of 'Essex' and PI 437654 but increased the number of RILs from 196 RILs (Gutierrez-Gonzalez et al. 2009) to 205 RILs (Gutierrez-Gonzalez et al. 2010). They grew the population in two locations over 2 years and used the same genetic map comprised of 276 SSR and RFLP markers (Gutierrez-Gonzalez et al. 2009). This new study identified 35 QTL controlling daidzein, glycitein, genistein, and total isoflavone contents; many epistatic interactions among these QTL; and candidate genes in several of these QTL regions. Similarly, clusters of QTL controlling at least three types of isoflavone have been observed on chromosomes 2, 4, 5, 7, 12, 15, and 16 (Gutierrez-Gonzalez et al. 2010). In a subsequent study, the same research group used 188 RILs derived from the cross of 'Magellan' and PI 437654 and a genetic linkage map comprised of 451 single nucleotide (SNP) and SSR markers to map QTL for seed isoflavone contents. They identified five QTL controlling seed isoflavone contents: 1 QTL that controls both seed genistein, daidzein, glycitein, and total isoflavone contents (*qGEN5, qDAI5, qGLY5,* and *qTOT5*) and second QTL that controls seed glycitein content (*qGLY5_2*) on Chromosome 5; 1 QTL that controls seed genistein (*qGEN6*), daidzein (*qDAI6*), and total isoflavone (*qTOT6*) contents on chromosome 6 ; 1 QTL that controls seed genistein (*qGEN8*) and total isoflavone (*qTOT8*) contents on chromosome 8; and 1 QTL that controls seed genistein (*qGEN11*), daidzein (*qDAI11*), glycitein (*qGLY11*), and total isoflavone (*qTOT11*) contents on chromosome 11 (Fig. 2.1; Table 10.1) (Gutierrez-Gonzalez et al. 2011).

Yoshikawa et al. (2010) used a 96 RIL population derived from a cross between 'Peking' and 'Tamahomare' (PxT, $n = 96$) grown in three locations (Kyoto, Osaka, and Nagano) over 2 years (2003 and 2004) in Japan. The genetic linkage map based on this population was composed of 344 SSR markers and covered 2652.5 cM (Sayama et al. 2009) and was used to map QTL for daidzein, genistein, glycitein, and total isoflavone contents. Their study identified about 36 QTL controlling daidzein, glycitein, genistein, and total isoflavone contents in 2003 (Fig. 2.1; Table 10.1) (Yoshikawa et al. 2010). In Kyoto (2003), they identified four QTL controlling total isoflavone content on chromosomes 2, 4, 8, and 17; two QTL controlling daidzein content on chromosomes 8 and 19; three QTL controlling genistein content on chromosomes 4, 8, and 17; and six QTL controlling glycitein content on chromosomes 8, 11, 12, 17, and 18. In Osaka (2003), they identified three QTL controlling total isoflavone content on chromosomes 5, 8, and 14; three QTL controlling daidzein content on chromosomes 4, 5, and 8; two QTL controlling genistein content on chromosomes 4 and 8; and three QTL controlling glycitein content on chromosomes 8, 11, and 17. In Nagano location (2003), they identified two QTL controlling total isoflavone content on chromosomes 8 and 15; two QTL controlling daidzein content on chromosomes 8 and 17; two QTL controlling genistein content on chromosomes 8 and 18; and four QTL controlling glycitein content on chromosomes 4, 6, 8, and 11 (Fig. 2.1; Table 10.1) (Yoshikawa et al. 2010).

In another study, Yang et al. (2011) used 113 RILs derived from the cross of 'Hwangkeum' and *Glycine soja* grown in three Korean locations (Suwon 2004, Daejeon, and Ochang) and 447 markers (Yang et al. 2008). They identified ten QTL on chromosomes 3, 4, 5, 7, 8, 9, 11, 12, 13, and 16. *QGC1* identified by the marker Sat_033 on Chr_3 controls glycitein contents; *QGC4* identified by the marker Sat_140 on Chr_4 controls glycitein contents; *QDZGT1* identified by the marker Sat_217 on Chr_5 controls daidzein, genistein, and total isoflavone contents; *QGT1* identified by the marker Sat697 on Chr_7 controls genistein contents; *QDZGT2* identified by the marker SM306 on Chr_8 controls daidzein, genistein, and total isoflavone contents; *QGC8* identified by the marker A636.p1 on Chr_8 controls glycitein contents; *QGC3* identified by the marker SM003 on Chr_11 controls glycitein contents; *QGC7* identified by the marker Satt196 on Chr_12 controls glycitein contents; *QGC6* identified by the marker Sat_375 on Chr_13 controls glycitein contents; and *QGC5* identified by the marker Sat693 on Chr_16 controls glycitein contents. *QDZGT1* and *QDZGT2* are two major QTL explaining 11.7–28.3% of the total variation of seed isoflavone contents in the population (Fig. 2.1; Table 10.1) (Yang et al. 2011).

In his MS degree thesis, Smallwood (2012) used 274 RILs derived from the cross between 'Essex' and 'Williams 82' and 480 SNPs. He identified 21 QTL. Seven QTL controlling seed genistein contents have been identified and mapped on chromosomes 5, 6, 9, 13, 17, and 19; five QTL controlling seed daidzein contents have been identified and mapped on chromosomes 5, 6, 9, 13, and 19; three QTL controlling seed glycitein content have been identified and mapped on chromosomes 6, 9, and 20; and six QTL for total isoflavone content have been identified and mapped on chromosomes 5, 6, 9, 13, and 19 (Smallwood 2012). The same data was published in a subsequent study using the same RILs and a genetic linkage map comprised of 480 SNPs (Smallwood et al. 2014).

In a recent study, we used the 'PI 438489B' by 'Hamilton' RIL population and found high variation in daidzein, glycitein, genistein, and total isoflavone contents among the parents (Table 10.1) and RIL seeds, and we mapped QTL for these traits (Akond et al. 2013). Similar results have been found among 'MD96-5722' and 'Spencer' parents and RIL seeds, and QTL for these traits have been mapped (Akond et al. 2014). We also found variations of these isoflavone contents in seeds of PI 438489B by Hamilton RILs planted in three different row spaces 25, 50, and 75 cm, respectively (Figs. 2.1 and 10.1; Table 5.1) (Ragin et al. 2014). In this population, daidzein content showed a bimodal distribution, while glycitein and genistein showed unimodal distributions but skewed toward low concentrations (Fig. 10.2).

Using the 'PI 438489B' by 'Hamilton' RIL population (PIxH, $n = 50$) and its SNP-based genetic linkage map of 648 SNPs and 1524.7 cM (Kassem et al. 2012), we identified 15 QTL for seed isoflavone contents in 2010 and 2011. Two QTL for daidzein have been mapped on chromosomes 2 and 17; two QTL for glycitein have been mapped on chromosomes 2 and 8; and two QTL for genistein have been mapped on chromosomes 8 and 12 in the RIL population grown in 2010. Two QTL for daidzein have been mapped on chromosomes 6 and 13; two QTL for glycitein have been mapped on chromosomes 1 and 17; and five QTL for genistein have been mapped on chromosomes 3, 8, 9, and 18 in the RIL population grown in 2011 (Akond et al. 2013).

Using the 'MD96-5722' by 'Spencer' RIL population (MDxS, $n = 92$) and its SNP-based genetic linkage map of 550 SNPs and 201.57 cM (Akond et al. 2013), we identified three major QTL for seed isoflavone contents. One QTL for daidzein has been mapped on chromosome 5, and two QTL for glycitein have been mapped on chromosomes 9 and 14 (Akond et al. 2014).

Zhang et al. (2014) used 184 RILs derived from the cross of 'Xiaoheidou' and 'GR8836' grown in China and 87 SSR markers. They identified 21 QTL that control seed genistein, glycitein, daidzein, and total isoflavone contents on chromosomes 3, 4, 7, 9, 10, 12, 13, and 17 (Zhang et al. 2014). *QDTI9_1, QDGlTIC3_1, QDGeGlTIC4_1, QDTIC4_1,* and *QDGlTIC12_1* control daidzein contents; *QGeC13_1, QDGeGlTIC4_1, QGeC7_1,* and *QGeC10_1* control genistein contents; *QGlC4_1, QGlC17_1, QDGlTIC3_1, QDGeGlTIC4_1, QDGlTIC12_1,* and *QGlC17* control glycitein contents; and *QDTIC9_1, QDGlTIC12_1, QDGlTIC3_1, QDGeGlTIC4_1, QDTIC4_2,* and *QDGlTIC12_1* control total isoflavone contents (Fig. 2.1; Table 10.1) (Zhang et al. 2014).

Li et al. (2014) used 110 RILs derived from a cross between 'Luheidu2' and 'Nanhuizao' and 5785 specific length amplified fragment sequencing (SLAF) markers. The relatively dense genetic map covered 2255.18 cM with and 0.43 cM average distance between adjacent markers. They identified 41 QTL that underlie soybean seed isoflavone contents including 11 stable QTL identified in multiple environments, and six among them were novel QTL (Li et al. 2014). *qIF20-2* in

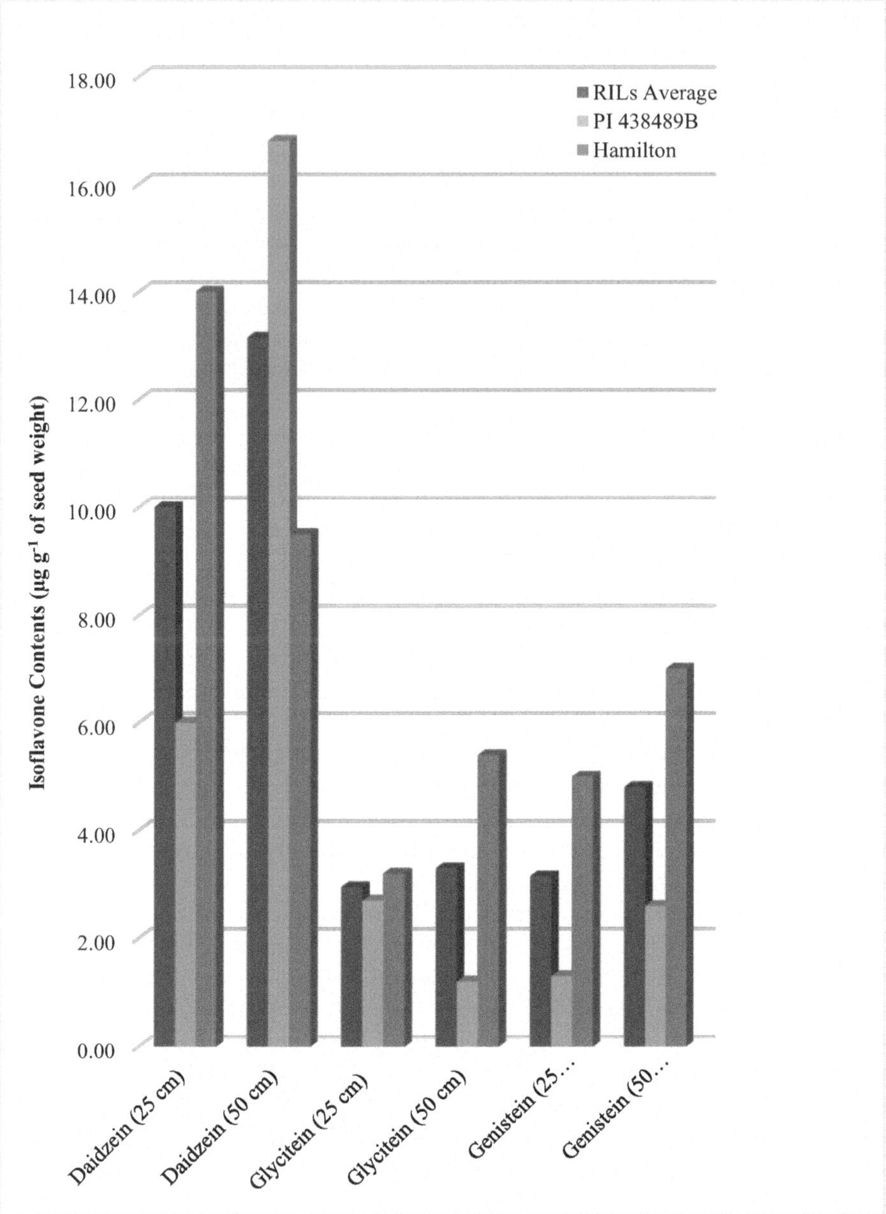

Fig. 10.1 Seed daidzein, glycitein, genistein, and total isoflavone contents in PI 438489B, Hamilton, and their RILs grown in a field in Fayetteville, NC, in 25-cm and 75-cm row spaces

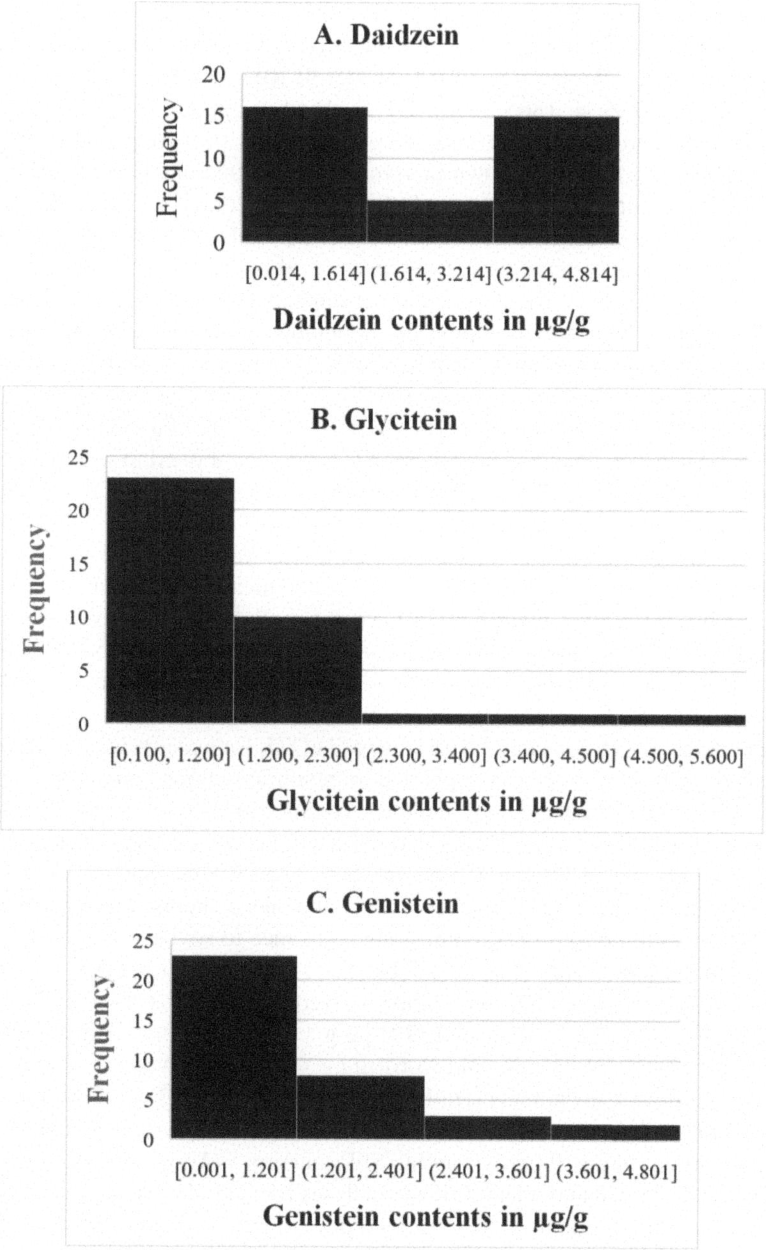

Fig. 10.2 Frequency distributions of daidzein (**a**), glycitein (**b**), and genistein (**c**) contents (µg/g) in the PI 438489B by Hamilton RIL population (PIxH; $n = 50$) grown in 2010 in Fayetteville, NC

particular is a major isoflavone QTL identified in multiple environments and explained about 8.7–35.3% of isoflavone phenotypic variation (Li et al. 2014).

In another study, Wang et al. (2014) used 130 RILs derived from the cross of 'Zhongdou 27' (DZ = 1865 μg/g, GT = 1614 μg/g, GC = 311 μg/g, and TI = 3791 μg/g) and 'Jiunong 20' (DZ = 844 μg/g, GT = 1046 μg/g, GC = 193 μg/g, and TI = 2061 μg/g) that are grown in seven environments in China in 2011. They found highly significant differences of all isoflavone contents among the parents and RIL seeds across all seven environments (Han et al. 2015; Wang et al. 2015). They also found highly significant differences of all isoflavone contents among the parents and RIL seeds across all seven environments (Zeng et al. 2009; Han et al. 2015; Wang et al. 2015). Genotyping the population with 194 SSR markers, they identified expression quantitative trait loci (eQTL) of four key enzymes involved in isoflavone pathway: chalcone synthase (CHS), 2-hydroxyisoflavone synthase (IFS) (Jung et al. 2000), phenylalanine ammonia-lyase (PAL), and flavonone 3-hydroxylase (F3H). They also analyzed how these eQTL interact or overlap with phenotypic QTL (pQTL) (Wang et al. 2014). They identified and mapped 33 eQTL that underlie CHS, PAL, IFS, and F3H transcript abundances on 15 soybean chromosomes and found an overlap between five eQTL and the pQTL (Wang et al. 2014). In a subsequent study using the same RILs but grown in three locations in China over 3 years (2005, 2006, 2007), they identified 34 QTL for isoflavone contents including 23 new ones: six QTL that control daidzein contents, seven QTL that control glycitein contents, ten QTL that control genistein contents, and 11 QTL that control total isoflavone contents (Fig. 2.1; Table 10.1) (Wang et al. 2015).

In another study, Bolon et al. (2014) mapped eQTL for genes related to flavonoid biosynthetic pathway which corresponded to genomic regions containing repeated genes of chalcone synthase (CHS) and the seed color locus (Bolon et al. 2014).

Zhao et al. (2015) used 130 RILs derived from the cross of 'Zhongdou 27' and 'Jiunong 20', grown in Harbin, China, in 2009, and 99 SSR markers (Zeng et al. 2009) and identified 11 QTL that control isoflavone contents, among which ten QTL overlap with the four QTL controlling resistance to soybean pod borer (SPB) identified in the same study (Fig. 2.1; Table 10.1) (Zhao et al. (2015).

Han et al. (2015) grew a RIL population derived from a cross between 'Zhongdou 27' (42.84% protein, 18.73% oil, and 3791 μg.g^{-1} total isoflavone) and 'Jiunong 20' (34.05% protein, 21.42% oil, and 2061 μg.g^{-1} total isoflavone contents) lines (n = 130) in three environments (Harbin, Hulan, and Suihua, 2005–2007) in China in 2005–2007 and genotyped it with 606 SSR markers; however, the genetic linkage map used was based on 194 polymorphic SSR markers and covered 2020 cM. They performed conditional and unconditional analyses for QTL data using the MINQUE method (Zhu 1992) to identify QTL for seed protein, oil, daidzein, genistein, glycitein, and total isoflavone contents. Five QTL for seed daidzein (DZ) content have been identified and mapped on chromosomes 3, 6, 9, 10, 13, 17, 18, and 20; three QTL for seed genistein (GC) contents have been identified and mapped on chromosomes 7, 9, 10, 15, and 20; six QTL for seed glycitein (GT) contents have been identified and mapped on chromosomes 3, 7, 8, 10, 13, 17, and 19; and seven QTL for seed total isoflavone (TI) contents have been identified and mapped on chromosomes 2, 4, 6, 7, 8, 10, 12, 13, 15, and 19. However, when the QTL for isoflavone contents are conditioned with QTL for protein and oil contents, they identified 37 QTL: nine QTL controlling daidzein content, six QTL controlling glycitein content, nine QTL controlling genistein content, and 13 QTL controlling total isoflavone content. Seven among these QTL control also see protein and oil contents (Han et al. 2015). These QTL were identified and mapped on all the 20 chromosomes of soybean along with QTL for protein and oil contents, and clusters of these QTL have been observed on most chromosomes (Han et al. 2015).

In Akond et al. (2015), we used 93 RILs derived from the cross of ‘Hamilton’ and ‘Spencer’, grown in two environments in North Carolina and 1502 SNP markers (Akond et al. 2015). The frequency distributions of daidzein, genistein, glycitein, and total isoflavone were normal (Fig. 10.3). The RIL coefficient variations (CV) were 7.85%, 11.71%, 8.59%, and 8.48% for daidzein, glycitein, genistein, and total isoflavone, respectively. Hamilton had higher mean values of all isoflavone contents than Spencer (Akond et al. 2015). We identified two QTL for seed isoflavone contents: one major QTL that underlies both daidzein and total isoflavone

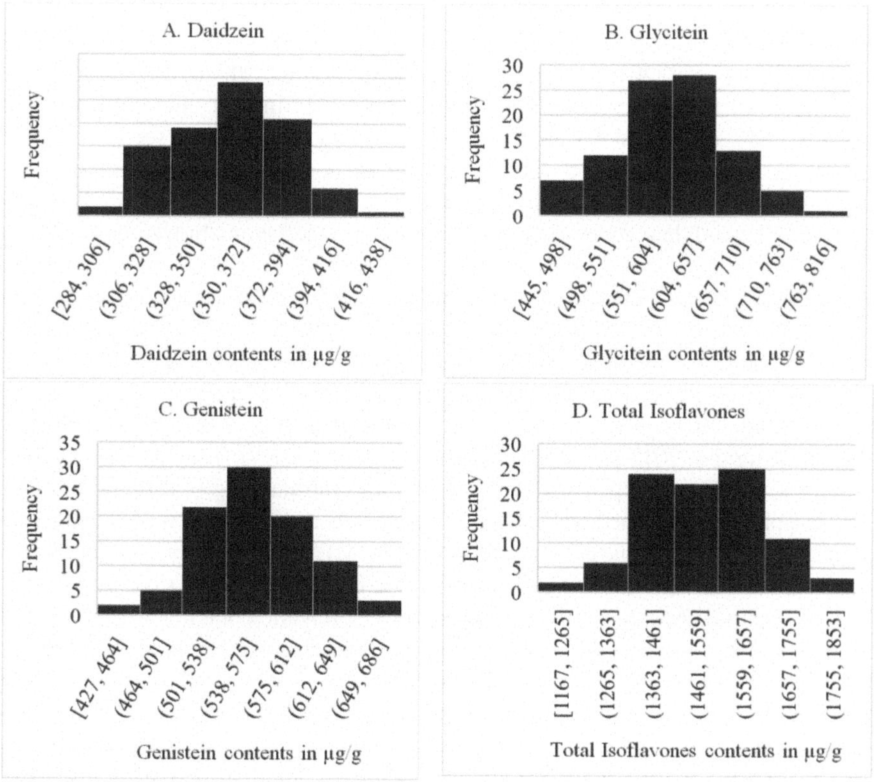

Fig. 10.3 Frequency distributions of (**a**) daidzein, (**b**) glycitein, (**c**) genistein, and (**d**) total isoflavone contents (µg/g) in the Hamilton by Spencer RIL population (HXS; $n = 93$) grown in two environments in Fayetteville, North Carolina. (From Akond et al. (2015))

contents on chromosome 6 (LG C2) identified by the SNP marker interval (ss246100375-ss245879277) and the second QTL that underlies glycitein contents on chromosome 18 (LG G) identified by the SNP marker interval (ss249909538-ss249919445) (Figs. 2.1 and 5.4; Table 10.1) (Akond et al. 2015).

Han et al. (2016) used 112 RILs derived from the cross of 'Zhongdou 27' and 'Jiunong 20' to map conditional QTL for genistein (GT), daidzein (DZ), glycitein (GT), and total isoflavone (TI) contents using the CIM method and mixed genetic model. They identified three unconditional QTL and six conditional QTL for DZ, six unconditional and five conditional QTL for GC, four unconditional and five conditional QTL for GT, and six unconditional and seven conditional QTL for TI among R3 and R8 developmental stages which suggests that isoflavone gene expression may be time-dependent (Fig. 2.1; Table 10.1) (Han et al. 2016).

Wang et al. (2016) used association mapping method, genotyped 135 Chinese cultivars with 100 microsatellite markers, and found 31 markers associated with seed daidzein (DZ), glycitein (GC), genistein (GT), and total isoflavone (TI). They also found a mean of 587.15 μg/g and range of 219.67–1346.96 μg/g for DZ (VC = 45.11%), a mean of 801.72 μg/g and range of 224.11–1620.08 μg/g for GT (CV = 41.84%), a mean of 190.27 μg/g and range of 3.87 to 122.45 μg/g for GC (15.9%), and a mean of 1579.14 μg/g and range of 479.32–2721.44 μg/g for TI (38.66%; Wang et al. 2016) (Fig. 2.1; Table 10.1) (Wang et al. 2016).

Meng et al. (2016) screened 366 Chinese landraces in four environments with 116,769 SNPs using GWAS analysis and identified 44 QTL for seed isoflavone content. These QTL explained 72% of the total phenotypic variation of these isoflavone, and significant variation was observed in eco-regions. Furthermore, the study identified 55 candidate genes closely related to seed isoflavone biosynthesis (Meng et al. 2016).

Carter (2017) and Carter et al. (2018) screened 109 F_4-derived RILs from a cross between the cultivars 'RCAT 1004' and 'DH 4202' grown in Ontario, Canada, in 2015 and 2016. The author found that DZ content range was 480–828 μg/g and 450–921 μg/g in 2015 and 2016, respectively. GT content range was 795–1129 μg/g and 738–1503 μg/g in 2015 and 2016, respectively. TI content range was 1304–2099 μg/g and 1284–2521 μg/g in 2015 and 2016, respectively (Carter 2017). They used 1,344 SNP markers with 648 unmapped SNPs, composite interval mapping (CIM) and multiple QTL mapping (MQM) methods, and and single marker analysis (ANOVA) and found ten QTL for seed isoflavone contents (CIM and MQM) and ten markers associated with seed isoflavone content (ANOVA) (Fig. 2.1; Table 10.1) (Carter 2017).

In a recent GWAS, 28 SNPs associated with isoflavone contents were identified using 196 Chinese accessions, and one of the SNPs was within the 5′-UTR of the R2R3-type MYB transcription factor (GmMYB29). Further analyses showed that GmMYB29 is localized in the nucleus and activates the promoters of *CHS8* (chalcone synthase 8) and *IFS2* (isoflavone synthase 2). The study did not identify any known genes in the isoflavone biosynthetic pathways but identified many transcription factors (TFs) such as MYB, bZIP, WRKY, and NAC (Chu et al. 2017).

Chu et al. (2017) grew a natural population of Chinese soybean accessions (n = 196) in two environments in 2014 in China, genotyped it with 207,608 SNPs, and used GWAS to identify QTL that control seed contents of 12 isoflavones (daidzin (D), genistin (G), glycitin (GL), glycitein (GLE), daidzein (DE), malonyldaidzin (MD), genistein (GE), malonylglycitin (MGL), acetyldaidzin (AD), malonylgenistin (MG), acetylgenistin (AG), and acetylglycitin (AGL)) in addition to total isoflavone content. Daidzein content (DAC), glycitein content (GLC), genistein content (GEC), and total isoflavone content (TIC) have been estimated for all individual plants in the population (TIC = 5445.74 μg g^{-1}, DAC = 3596.17 μg g^{-1}, GEC = 946.73 μg g^{-1}, and GLC = 423.85 μg g^{-1}). Twenty-eight (28) SNPs associated with QTL that control seed isoflavone contents have been identified and mapped on chromosomes 5, 6, 11, and 20 (Table 10.1). Within one of these QTL regions, they identified the candidate gene GmMYB29. GmMYB29 expression analysis showed that it activates CHS8 (chalcone synthase 8) and IFS2 (isoflavone synthase 2) promoters (Chu et al. 2017).

In another recent study, a 196 $F_{7:8–10}$ RIL population derived from the cultivars 'Huachun 2' and 'Wayao' was evaluated for isoflavone contents across four environments in China. The study quantified DZ, GT, GC, and TI in their seeds and found that TI range was 817–6976 μg/g with a continuous distribution. The 196 RIL population was screened by 3469 restriction site-associated DNA tag sequencing (RAD-seq) markers and identified 15 QTL for seed isoflavone contents that were mapped on chromosomes 2, 5, 7, 9, 10, 11, 13, 16, 17, and 19 (Fig. 2.1; Table 10.1). One major QTL (qIF5-1) on chromosome 5 explained 6.37–59.95% of all isoflavone components across all four environments (Cai et al. 2018).

In a recent study, Li et al. (2018) screened 345 $F_{6:7}$ RILs derived from the cross of Zheng 92116 and Liaodou14 (ZL) with 1739 SNP and 127 SSR markers and identified and mapped 48 QTL on chromosomes 2 (LG D1b), 5 (LG A1), 6 (LG C2), 7 (LG M), 11 (LG B1), 13 (LG F), 17 (LG D2), and 18 (LG G) of the soybean genetic linkage map over 2 years and in four environments. Authors also validated eight QTL for genistin, daidzin, and total isoflavone on chr 5 in another RIL population derived from the cross between Zheng 92116 and Qihuang 30 (ZQ). In total, ten QTL for individual and total isoflavone contents have been validated in the Zheng 92116 by Qihuang 30 (ZQ) RIL population (Fig. 2.1; Table 10.1). Within the QTL from both RIL populations, they identified several candidate genes such as MYB transcription factor (*MYB84*) identified earlier (Chu et al. 2017), cytochrome P450 CYP2 genes, and O-methyltransferase gene (Li et al. 2018). *GmMYB39* was reported to act as an inhibitor in isoflavone biosynthetic pathways (Liu et al. 2013).

Likewise, Pei et al. (2018) screened 200 $F_{5:7}$ RILs derived from a cross of 'Luheidou 2' (LHD2) and 'Nanhuizao' (NHZ) cultivars with 3541 specific-locus amplified fragment sequencing (SLAF-seq) markers and identified 24 QTL for seed isoflavone contents across four environments, among which are four new and stable QTL for total isoflavone, malonyldaidzin, genistein, and malonylgenistin (*qTIF19, qMD19*, *qG8*, and *qMG18*). They also discovered three genes (*GmCHR, Gm4CL*, and *GmIFR*) and 13 *MYB-like* genes, involved in the biosynthetic pathways of these seed isoflavones, within these QTL intervals (Pei et al. 2018).

In a recent study, Watanabe et al. (2019) used two RIL populations. The first population was derived from the cross of 'Aokimame' (Chinese landrace) and the

Japanese cultivar 'Fukuyutaka' (FUK), and the second population was derived from the cross of 'Kumaji-1' (KUM, Japanese landrace) and FUK. They also used a third mini core population of 158 lines (Kaga et al. 2012). These populations were grown in a field in Japan, and they identified three QTL for seed isoflavone contents: two QTL for malonylglycitin (M_Gly) on chromosomes 11 (*qMGly_11*) and 15 (*qMGly_15*) and one QTL for malonylgenistin (M_Ge) on chromosome 12 (*qMGe_12*). *qMGly_11* was a major QTL that explained approximately of the phenotypic variation of malonylglycitin (Watanabe et al. 2019).

Wu et al. (2020) grew a natural population of soybean accessions ($n = 200$), of which 179 were from China and 21 from other parts of the world. From these accessions, they selected two cultivars and developed a RIL population. The RIL population was derived from a cross between 'Zhongdou 27' (high isoflavone; DZ = 1865 μg g^{-1}; GT = 1614 μg g^{-1}; GC = 311 μg g^{-1}; and TI = 3791 μg g^{-1}) and 'Jiunong 20' (low isoflavone; DZ = 844 μg g^{-1}; GT = 1046 μg g^{-1}; GC = 193 μg g^{-1}; and TI = 2061 μg g^{-1}) lines ($n = 150$). They also developed a BC2F4 population to fine-map identified QTL for seed isoflavone content. They grew these populations in China in 2015–2018. The natural population ($n = 200$) was genotyped with 28,926 SNP markers, and the RIL population ($n = 150$) was genotyped with 343,907 SNPs. The authors used GWAS CIM of ICIMapping (Meng et al. 2015) to identify QTL that control seed isoflavone contents. Using GWAS analysis, 26 SNPs associated with QTL that control seed daidzein content, 28 SNPs associated with QTL that control seed genistein content, 15 SNPs associated with QTL that control seed glycitein content, and 18 SNPs associated with QTL that control seed total isoflavone content have been identified and mapped on different chromosomes. Using the RIL population, 37 QTL that control seed daidzein, genistein, glycitein, and total isoflavone have been also identified and mapped on different soybean chromosomes (Table 10.1) (Wu et al. 2020). Among these QTL, a major QTL on chromosome 8 (*qISO8-1*) identified with both GWAS and CIM methods was further analyzed using the BC_2F_4 backcrossing population. Two NIL populations were screened with 89 SSRs close to *qISO-8-1* QTL which was narrowed to a 99.5 kb region (SSR_08_1651-SSR_08_1656). Seven candidate genes have been identified in the region including *Glyma.08G309500* that encodes for a mitogen-activated protein kinase (MPK) renamed *GmMPK1* which is involved in secondary metabolite biosynthesis and signal transduction pathways (Wu et al. 2020).

10.2 Conclusion

In conclusion, QTL for seed isoflavone contents have been identified in all soybean chromosomes (Njiti et al. 1999; Meksem et al. 2001; Kassem et al. 2004, 2006; Primomo et al. 2005; Matsumura et al. 2005; Gutierrez-Gonzalez et al. 2009, 2010; Zeng et al. 2009; Grant et al. 2010; Liang et al. 2010; Meng et al. 2011; Yang et al.

2011; Smallwood 2012; Zhang et al. 2014; Smallwood et al. 2014; Li et al. 2014; Wang et al. 2014, 2015; Bolon et al. 2014; Zhao et al. 2015; Akond et al. 2015; Han et al. 2015, 2016; Wang et al. 2016; Chu et al. 2017; Cai et al. 2018; Li et al. 2018) using single marker analysis, SIM, CIM, and MQM. In general, these QTL span genomic distances of several centimorgans (cM) (Fig. 2.1; Table 10.1); the genes underlying these QTL are mostly unknown, and further research efforts are needed to identify the candidate genes (Meng et al. 2016; Chu et al. 2017; Li et al. 2018) responsible for their biosynthetic pathways (Pei et al. 2018; Ha et al. 2019; Li et al. 2018; Wu et al. 2020).

Pei et al. (2018) identified three candidate genes (*GmCHR, Gm4CL*, and *GmIFR*) and 13 *MYB-like* genes (Pei et al. 2018). Daidzein is a precursor of coumestrol (CMS) which plays a major role in nodulation during soybean-rhizobia symbiosis (Lee et al. 2012; Ha et al. 2019). Fourteen (14) candidate genes involved in CMS biosynthesis have been identified, and seven of them encode for enzymes oxidoreductases known to transfer electrons from daidzein which may indicate the role of CMS and daidzein in stress. Two genes encoding for MLP-like protein and a MYB domain protein have also been identified (Ha et al. 2019).

Li et al. (2018) identified two major QTL for seed isoflavone contents, the first on chromosome 5 (LG A1) and the second on chromosome 11 (LG B1) and several caudate genes (Li et al. 2018). The region of 1–2.1 cM on chromosome 11, identified earlier (Kassem et al. 2004; Yoshikawa et al. 2010; Yang et al. 2011), contains several candidate genes including bZIP transcription factors, MYB transcription factors, and PLATZ transcription factors. MYBZ2 close to the region is responsible for the regulation of the expression of CHI, CHS, C4H, PAL, and 4CL which are key enzymes in the isoflavone biosynthetic pathways (Li et al. 2018). MYB transcription factor (*MYB84*) identified earlier (Chu et al. 2017), cytochrome P450 CYP2 genes, and O-methyltransferase gene have been also identified in these QTL regions (Li et al. 2018). *GmMYB39* was reported to act as a general inhibitor in isoflavone biosynthetic pathways (Liu et al. 2013).

Mapping the locations of genes involved in seed isoflavone biosynthesis will help soybean breeders to develop new cultivars with improved seed isoflavone through marker-assisted selection. Recently, Sarkar et al. (2020) developed a mutant line F333ES017D9 from a 3331-mutant-line library of 'Fukuyutaka' using ethyl methanesulfonate (EMS) (Xia et al. 2012). Within this mutant's (F333ES017D9) seeds, they identified a gene (*GmCHR5*) that boosts genistein-to-daidzein ratio (Sarkar et al. 2020). One single-base mutation in this gene (*GmCHR5*) was sufficient enough to shift the production of more genistein in the isoflavone biosynthetic pathways. Moreover, they overexpressed *GmCHR5* in both the mutant and the wild-type plants and observed a high increase in daidzein and its derivatives (Sarkar et al. 2020). Therefore, using this approach can be beneficial to develop soybean lines with highly desirable isoflavone and other nutritionally beneficial compounds. Another recent study identified a major QTL (*qISO8-1*) on chromosome 8, and within this region, they identified a candidate gene (*GmMPK1*) (Wu et al. 2020).

References

Akond, M., R. Bazzelle, B. Ragin, H. Herrera, U. Kaodi, C. Akbay, S.K. Kantartzi, V. Njiti, A. Barakat, K. Meksem, D.A. Lightfoot, and M.A. Kassem. 2013. Additional quantitative trait loci and candidate genes for seed isoflavone content in soybean. *Journal of Agricultural Science* 5 (11): 20–33. https://doi.org/10.5539/jas.v5n11p20.

Akond, M., S. Liu, S.K. Kantartzi, K. Meksem, N. Bellaloui, D.A. Lightfoot, J. Yuan, D. Wang, and M.A. Kassem. 2014. Quantitative trait loci for seed isoflavone contents in 'MD96-5722' by 'Spencer' recombinant inbred lines of soybean. *Journal of Agricultural and Food Chemistry* 62 (7): 1464–1468. dx.doi.org. https://doi.org/10.1021/jf4040173.

Akond, M., S. Liu, S.K. Kantartzi, K. Meksem, N. Bellaloui, D.A. Lightfoot, J. Yuan, D. Wang, J. Anderson, D.A. Lightfoot, and M.A. Kassem. 2015. A SNP genetic linkage map based on the 'Hamilton' by 'Spencer' recombinant inbred line (RIL) population of soybean [*Glycine max* (L.) Merr.] identified QTL for seed isoflavone contents. *Plant Breeding* 134 (5): 580–588. https://doi.org/10.1111/pbr.12298.

Bolon, Y.T., D.L. Hyten, J.H. Orf, C.P. Vance, and G.J. Muehlbauer. 2014. eQTL networks reveal complex genetic architecture in the immature soybean seed. *The Plant Genome* 7: 1–14. https://doi.org/10.3835/plantgenome2013.08.0027.

Cai, Z., Y. Cheng, Z. Ma, X. Liu, Q. Ma, Q. Xia, G. Zhang, Y. Mu, and H. Nian. 2018. Fine mapping of QTLs for individual and total isoflavone content in soybean (*Glycine max* L.) using a high-density genetic map. *Theoretical and Applied Genetics* 131 (3): 555–568. https://doi.org/10.1007/s00122-017-3018-x.

Carter, A. 2017. *Identification of Quantitative Trait Loci (QTL) and genes associated with seed isoflavone concentration in soybean (Glycine max [L] Merril.)*. PhD Dissertation. The University of Guelph, Canada.

Carter, A., A. Tenuta, I. Rajcan, T. Welacky, L. Woodrow, and M. Eskandari. 2018. Identification of quantitative trait loci for seed isoflavone concentration in soybean (*Glycine max*) against soybean cyst nematode stress. *Plant Breeding*. https://doi.org/10.1111/pbr.12627.

Cheng, H., O. Yu, and D.Y. Yu. 2008. Polymorphisms of IFS1 and IFS2 gene are associated with isoflavone concentrations in soybean seeds. *Plant Science* 175: 505–512.

Chu, S., J. Wang, Y. Zhu, S. Liu, X. Zhou, H. Zhang, C. Wang, W. Yang, Z. Tian, H. Cheng, and D. Yu. 2017. An R2R3-type MYB transcription factor, GmMYB29, regulates isoflavone biosynthesis in soybean. *PLoS Genetics* 13 (5): e1006770. https://doi.org/10.1371/journal.pgen.1006770.

Grant, D., R.T. Nelson, S.B. Cannon, and R.C. Shoemaker. 2010. SoyBase, the USDA-ARS soybean genetics and genomics database. *Nucleic Acids Research* 38 (suppl 1): D843–D846. https://doi.org/10.1093/nar/gkp798.

Gutierrez-Gonzalez, J.J., X. Wu, J. Zhang, J.D. Lee, M. Ellersieck, J.G. Shannon, O. Yu, H.T. Nguyen, and D.A. Sleper. 2009. Genetic control of soybean seed isoflavone content: Importance of statistical model and epistasis in complex traits. *Theoretical and Applied Genetics* 119 (6): 1069–1083. https://doi.org/10.1007/s00122-009-1109-z.

Gutierrez-Gonzalez, J.J., X.L. Wu, J.D. Gillman, J.D. Lee, R. Zhong, O. Yu, J.G. Shannon, H.T. Nguyen, and D.A. Sleper. 2010. Intricate environment-modulated genetic networks control isoflavone accumulation in soybean seeds. *BMC Plant Biology* 10: 105–120. https://doi.org/10.1186/1471-2229-10-105.

Gutierrez-Gonzalez, J.J., T.D. Vuong, R. Zhong, O. Yu, J.D. Lee, G. Shannon, M. Ellersieck, H.T. Nguyen, and D.A. Sleper. 2011. Major locus and other novel additive and epistatic loci involved in modulation of isoflavone concentration in soybean seeds. *Theoretical and Applied Genetics* 123: 1375–1385. https://doi.org/10.1007/s00122-011-1673-x.

Ha J., Y.G. Kang, T. Lee, M. Kim, M.Y. Yoon, E. Lee, X. Yang, D. Kim, Y.J. Kim, T.R. Lee, M.Y. Kim, and S.H. Lee. 2019. Comprehensive RNA sequencing and co-expression network analysis to complete the biosynthetic pathway of coumestrol, a phytoestrogen. *Scientific Reports* 9: 1934. https://doi.org/10.1038/s41598-018-38219-6.

Han, Y., W. Teng, Y. Wang, X. Zhao, L. Wu, D. Li, and W. Li. 2015. Unconditional and conditional QTL underlying the genetic interrelationships between soybean seed isoflavone, and protein or oil contents. *Plant Breeding* 134 (3): 300–309. https://doi.org/10.1111/pbr.12259.

Han, Y., D. Li, G. Zhao, X. Zhao, Z. Jiang, H. Hu, L. Wu, Y. Wang, Y. Gao, Y. Li, G. Zeng, F. Meng, W. Teng, and W. Li. 2016. Dynamic quantitative trait loci underlies isoflavone accumulation in soybean seed. *Plant Breeding*. https://doi.org/10.1111/pbr.12354.

Jung W., O. Yu, S.M.C. Lau, D.P. O'Keefe, J. Odell, G. Fader, and B. McGonigle. 2000. Identification and expression of isoflavone synthase, the key enzyme for biosynthesis of isoflavones in legumes. *Nature Biotechnology* 18: 208–212.

Kaga, A., T. Shimizu, S. Watanabe, Y. Tsubokura, Y. Katayose, K. Harada, D.A. Vaughan, and N. Tomooka. 2012. Evaluation of soybean germplasm conserved in NIAS genebank and development of mini core collections. *Breeding Science* 61: 566–592.

Kassem, M.A., K. Meksem, M.J. Iqbal, V.N. Njiti, W.J. Banz, T.A. Winters, A.J. Wood, and D.A. Lightfoot. 2004. Definition of soybean genomic regions that control seed phytoestrogen amounts. *Journal of Biomedicine & Biotechnology* 1: 52–60. https://doi.org/10.1155/S1110724304304018.

Kassem, M.A., J. Shultz, K. Meksem, Y. Cho, A. Wood, M.J. Iqbal, and D.A. Lightfoot. 2006. An updated 'Essex' by 'Forrest' linkage map and first composite interval map of QTL underlying six soybean traits. *Theoretical and Applied Genetics* 113: 1015–1026.

Kassem M.A., L. Ramos, L. Leandro, G. Mbofung, D.L. Hyten, S.K. Kantartzi, R.L. Grier, V.N. Njiti, S. Cianzio, and K. Meksem. 2012. The 'PI 438489B' by 'Hamilton' SNP-based genetic linkage map of soybean [Glycine max (L.) Merr.] Identified quantitative trait loci that underlie seedling SDS resistance. *Journal of Plant Genome Sciences* 1 (1): 18–30. https://doi.org/10.5147/jpgs.2012.0053.

Lee H.I., J.H. Lee, K.H. Park, D. Sangurdekar, and W.S. Chang. 2012. Effect of soybean coumestrol on bradyrhizobium japonicum nodulation ability, biofilm formation, and transcriptional profile. *Applied and Environmental Microbiology* 78 (8): 2,896–2,903. https://doi.org/10.1128/AEM.07336-11.

Li, B., L. Tian, J. Zhang, L. Huang, F. Han, S. Yan, L. Wang, H. Zheng, and J. Sun. 2014. Construction of a high-density genetic map based on large-scale markers developed by specific length amplified fragment sequencing (SLAF-seq) and its application to QTL analysis for isoflavone content in Glycine max. *BMC Genomics* 15: 1086. https://doi.org/10.1186/1471-2164-15-1086.

Li, X.H., S. Kamala, R. Tian, H. Du, W. Li, Y. Kong, and C. Zhang. 2018. Identification and validation of quantitative trait loci controlling seed isoflavone content across multiple environments and backgrounds in soybean. *Molecular Breeding* 38: 8. https://doi.org/10.1007/s11032-017-0768-8.

Liang, H.Z., Y.L. Yu, S.F. Wang, Y. Lian, T.G. Wang, T.L. Wei, P.T. Gong, X.Y. Liu, X.J. Fang, and M.C. Zhang. 2010. QTL mapping of isoflavone, oil and protein contents in soybean (Glycine max L. Merr.). *Agricultural Sciences in China* 9 (8): 1108–1116.

Lightfoot, D.A., V.N. Njiti, P.T. Gibson, M.A. Kassem, J.M. Iqbal, and K. Meksem. 2005. Registration of the essex by forrest Recombinant Inbred Line (RIL) Mapping Population. *Crop Science* 45 (4): 1678–1781.

Liu X., L. Yuan, L. Xu, Z. Xu, Y. Huang, X. He, H. Ma, J. Yi, and D. Zhang. 2013. Over-expression of GmMYB39 leads to an inhibition of the isoflavonoid biosynthesis in soybean (Glycine max. L). *Plant Biotechnology Reports* 7: 445–455. https://doi.org/10.1007/s11816-013-0283-2.

Matsumura, H., S. Watanabe, K. Harada, M. Senda, S. Akada, S. Kawasaki, E.G. Dubouzet, N. Minaka, and R. Takahashi. 2005. Molecular linkage mapping and phylogeny of the chalcone synthase multigene family in soybean. *Theoretical and Applied Genetics* 110: 1203–1209.

Meksem, K., V.N. Njiti, W.J. Banz, M.J. Iqbal, M.A. Kassem, D.L. Hyten, J. Yuan, T.A. Winters, and D.A. Lightfoot. 2001. Genomic regions that underlie soybean seed isoflavone content. *Journal of Biomedicine & Biotechnology* 1: 38–44.

Meng, F.L., Y.P. Han, W.L. Teng, Y.G. Li, and W.B. Li. 2011. QTL underlying the resistance to soybean aphid (Aphis glycines Matsumura) through isoflavone-mediated antibiosis in soybean cultivar 'Zhongdou 27'. *Theoretical and Applied Genetics* 123: 1459–1465. https://doi.org/10.1007/s00122-011-1680-y.

Meng L., H. Li, L. Zhang, and J. Wang. 2015. QTL IciMapping: Integrated software for genetic linkage map construction and quantitative trait locus mapping in biparental populations. *Crop Journal* 3: 269–283. https://doi.org/10.1016/j.cj.2015.01.001.

Meng S., J. He, T. Zhao, G. Xing, Y. Li, S. Yang, J. Lu, Y. Wang, and J. Gai. 2016. Detecting the QTL-allele system of seed isoflavone content in Chinese soybean landrace population for optimal cross design and gene system exploration. *Theoretical and Applied Genetics* 129: 1557–1576. https://doi.org/10.1007/s00122-016-2724-0.

Njiti, V.N., K. Meksem, J. Yuan, D.A. Lightfoot, W.J. Banz, and T.A. Winters. 1999. DNA markers associated with loci underlying seed phytoestrogen content in soybeans. *Journal of Medicinal Food* 2 (3–4): 185–187.

Pei, R., J. Zhang, L. Tian, S. Zhang, F. Han, S. Yan, L. Wang, B. Li, and J. Sun. 2018. Identification of novel QTL associated with soybean isoflavone content. *The Crop Journal* 6 (3): 244–252.

Primomo, V.S., V. Poysa, G.R. Ablett, C.J. Jackson, M. Gijzen, and I. Rajcan. 2005. Mapping QTL for individual and total isoflavone content in soybean seeds. *Crop Science* 45 (6): 2454–2464. https://doi.org/10.2135/cropsci2004.0672.

Ragin B., M. Akond, S.K. Kantartzi, K. Meksem, H. Herrera, C. Akbay, D.A. Lightfoot, and M.A. Kassem. 2014. Effect of row spacing on seed isoflavone contents in soybean [Glycine max (L.) Merr.]. *American Journal of Plant Sciences* 5 (26): 4003–4010. https://doi.org/10.4236/ajps.2014.526418.

Sangeeta D., G. Mark, M. Pat, and F. Mana. 2007. Transcriptome analysis reveals a critical role of CHS7and CHS8 genes for isoflavonoid synthesis in soybean seeds. *Plant Physiology* 143: 326–338.

Sarkar, M.A.R., W. Otsu, A. Suzuki, F. Hashimoto, T. Anai, and S. Watanabe. 2020. Single-base deletion in *GmCHR5* increases the genistein-to-daidzein ratio in soybean seed. *Breeding Science*: 1–12. https://doi.org/10.1270/jsbbs.19134.

Sayama, T., T. Nakazaki, G. Ishikawa, K. Yagasaki, N. Yamada, N. Hirota, K. Hirata, T. Yoshikawa, M. Teraishi, Y. Okumoto, T. Tsukiyama and T. Tanisaka. 2009. QTL analysis of seed-flooding tolerance in soybean (Glycine max (L.) Merr.). *Plant Science* 176: 514–521.

Smallwood, C.J. 2012. *Detection of quantitative trait loci for marker-assisted selection of soybean isoflavone genistein*. Master's thesis, University of Tennessee.

Smallwood, C.J., C.N. Nyinyi, D.A. Kopsell, C.E. Sams, D.R. West, P. Chen, S.K. Kantartzi, P.B. Cregan, D.L. Hyten, and V.R. Pantalone. 2014. Detection and confirmation of quantitative trait loci for soybean seed isoflavone. *Crop Science* 54: 595–606.

Wang, Y., Y. Han, W. Teng, X. Zhao, Y. Li, L. Wu, D. Li, and W. Li. 2014. Expression quantitative trait loci infer the regulation of isoflavone accumulation in soybean (*Glycine max* L. Merr.) seed. *BMC Genomics* 15: 680. https://doi.org/10.1186/1471-2164-15-680.

Wang, Y., Y. Han, X. Zhao, Y. Li, W. Teng, D. Li, Y. Zhan, and W. Li. 2015. Mapping isoflavone QTL with main, epistatic and QTL × environment effects in recombinant inbred lines of soybean. *PLoS One* 10 (3): e0118447. https://doi.org/10.1371/journal.pone.0118447.

Wang, C., Y. Guo, J. Hu, J. Zhang, L. Yang, B. Guo, J. Han, and H. Xie. 2016. Exploring elite alleles for seed isoflavone concentration in soybean by association analysis. *Emirates Journal of Food and Agriculture* 28 (2): 85–92. https://doi.org/10.9755/ejfa.2015-09-787.

Watanabe, S., R. Yamada, H. Kanetake, A. Kaga, and T. Anai. 2019. Identification and characterization of a major QTL underlying soybean isoflavone malonylglycitin content. *Breeding Science*: 1–9. https://doi.org/10.1270/jsbbs.19027.

Wu, D., D. Li, X. Zhao, Y. Zhan, W. Teng, L. Qiu, H. Zheng, W. Li, and Y. Han. 2020. Identification of a candidate gene associated with isoflavone content in soybean seeds using genome-wide association and linkage mapping. *The Plant Journal* 104 (4): 950–963. https://doi.org/10.1111/tpj.14972.

Xia, X., S. Watanabe, T. Yamada, Y. Tsubokura, H. Nakashima, H. Zhai, T. Anai, S. Sato, T. Yamazaki, S. Lü, H. Wu, S. Tabata, and K. Harada. 2012. Positional cloning and characterization reveal the molecular basis for soybean maturity locus *E1* that regulates photoperiodic flowering. *PNAS* 109 (32): E2155–E2164. https://doi.org/10.1073/pnas.1117982109.

Yang K., J.K. Moon, N. Jeong, K. Back, H.M. Kim and S.C. Jeong. 2008. Genome structure in soybean revealed by a genomewide genetic map constructed from a single population. *Genomics* 92: 52–59.

Yang, K., J.K. Moon, N. Jeong, H.K. Chun, S.T. Kang, K. Back, and S.C. Jeong. 2011. Novel major quantitative trait loci regulating the content of isoflavone in soybean seeds. *Genes & Genomics* 33 (6): 685–692.

Yoshikawa, T., Y. Okumoto, D. Ogata, T. Sayama, M. Teraishi, M. Terai, T. Toda, K. Yamada, K. Yagasaki, N. Yamada, T. Tsukiyama, T. Yamada, and T. Tanisaka. 2010. Transgressive segregation of isoflavone contents under the control of four QTLs in a cross between distantly related soybean varieties. *Breeding Science* 60 (3): 243–254.

Zeng, G., D. Li, Y. Han, W. Teng, J. Wang, L. Qiu, and W. Li. 2009. Identification of QTL underlying isoflavone contents in soybean seeds among multiple environments. *Theoretical and Applied Genetics* 118: 1455–1463.

Zhang, H.J., J.W. Li, Y.J. Liu, W.Z. Jiang, X.L. Du, L. Li, X.W. Li, L.T. Su, Q.Y. Wang, and Y. Wang. 2014. Quantitative trait loci analysis of individual and total isoflavone contents in soybean seeds. *Journal of Genetics* 93 (2): 331–338.

Zhao, G., Z. Jiang, D. Li, Y. Han, H. Hu, L. Wu, and Y. Wang. 2015. Molecular loci associated with seed isoflavone content may underlie resistance to soybean pod borer (*Leguminivora glycinivorella*). *Plant Breeding* 178: 78–84.

Zhu, J. 1992. Mixed model approaches for estimating genetic variance and covariance. *Journal of Biomathematics* 7: 1–11.

Chapter 11
Bioactive Anticancer Peptides in Soybean Seeds

Jiazheng Yuan, Meriam Bousselham, and Moulay Abdelmajid Kassem

11.1 Introduction

Soybean [*Glycine max* (L.) Merrill] is a valuable crop in the USA in terms of planting acreage and cash value. A total of 33.9 million hectares of land were planted with soybean and more than $40 billion of soybean in 2014 had been produced nationwide (USDA–NASS 2015a, b). The data from different studies show that soybean consumption has been associated with beneficial effects on human health, such as reducing obesity and preventing cardiovascular disease, immune disorders, and certain type of cancers (Velasquez and Bhathena 2007; He and Chen 2013; de Lumen 2005). The diets rich in soy products are associated with reduced colon, breast, and prostate cancer mortalities, suggesting that soybean components play a significant role in carcinogenesis prevention (Greenwald et al. 2001; Messina et al. 1994). Studies indicated that anticancer activity of soybean seeds is mainly attributed to the antioxidant activity and estrogenic effects of isoflavones (McCue and Shetty 2004). In addition, the therapeutic function of soybean has also been recognized due to the bioactive peptides present in seeds or derived from gastrointestinal digestion (Chatterjee et al. 2018). These similar structural amino acids are easily identified in many bioactive peptides including arginine, lysine, proline, and hydrophobic amino acids (Kim et al. 2000). As anticancer bioactive molecules, BBI (Bowman-Birk protease inhibitor) and lunasin exhibit unique anticancer activities (Kennedy 1998; Galvez et al. 2001; Liu et al. 2014; Singh et al. 2014; Chatterjee et al. 2018) and they have demonstrated anticancer effects without showing toxicity

J. Yuan · M. Bousselham · M. A. Kassem (✉)
Plant Genomics and Biotechnology Lab, Department of Biological and Forensic Sciences, Fayetteville State University, Fayetteville, NC, USA
e-mail: mkassem@uncfsu.edu

M. A. Kassem (ed.), *Soybean Seed Composition*,
https://doi.org/10.1007/978-3-030-82906-3_11

(Kennedy 1998). BBI naturally occurs in soybean plant and plays key regulatory roles in the biological pathways including plant defense and hormone processing (Qi et al. 2005). Lunasin has shown remarkable cancer-preventive properties and antioxidant and hypocholesterolemic effects in animal and in vitro trials (Galvez and de Lumen 1999; Galvez et al. 2001; Lam et al. 2003; Jeong et al. 2003; Liu et al. 2014; Wan et al. 2017; Ďúranová et al. 2019). The anticancer ability of the bioactive peptides originates from the tryptic hydrolyzates of soybean proteins, which act as agents to stimulate superoxide anions thereby triggering nonspecific immune defense mechanisms (Kitts and Weiler 2003). Lunasin is considered a promising peptide with anticancer properties since its isolation from soybean seeds in 1981 (Liu et al. 2014). It also plays an important role in regulating the biosynthesis of body cholesterol with its inherent antioxidative and anti-inflammatory effects (Lule et al. 2015). Soybean peptides display both antioxidant and antiobesity effects (Pena-Romas and Xiong 2002; Nakamori 2002). Therefore, soy products can be used in functional foods designs. Overall, tremendous progress has been made in understanding the anticancer bioactive function for lunasin and the role it will play in preventing and treating cancer diseases.

11.2 Anticancer Peptides in Soybean Seeds

Soybean contains a variety of phytochemicals that have potential health benefits. As bioactive molecules in soybean seeds, BBI and lunasin exhibit special anticancer activities (Kennedy 1998; Galvez et al. 2001; Liu et al. 2014; Singh et al. 2014; Chatterjee et al. 2018). BBI is a low molecular weight protein of 8 kDa with 71 amino acids (Losso 2008). It was first isolated by Bowman (1946) in soybean and it was characterized nearly 20 years later (Birk et al. 1963) and has demonstrated anticancer effects without showing toxicity (Kennedy 1998). As a protease inhibitor, BBI naturally occurs in soybean plant and plays key regulatory roles in different biological pathways including plant defense and hormone processing (Qi et al. 2005). BBI inhibits the function of trypsin and chymotrypsin both in in vivo and in vitro experiments (Kennedy 1998). Lunasin is composed of 43 amino acids and weighs 5.5 kDa (Galvez et al. 1997). The name lunasin is derived from the word "Lunas" (Tagalog), which means "cure," and the short peptide (43 aa) is encoded by the gene *Gm2S-1* (Galvez et al. 1997). Lunasin contains a chromatin-binding proteins' conserved region, a cell adhesion motif (Arg-Gly-Asp), and 9 units of Asp (Galvez and de Lumen 1999; Galvez et al. 2001). Lunasin has antioxidant and anticancer properties and has exhibited hypocholesterolemic effects in animal and in vitro trials (Galvez and de Lumen 1999; Galvez et al. 2001; Lam et al. 2003; Jeong et al. 2003; Liu et al. 2014; Wan et al. 2017; Ďúranová et al. 2019). It is also associated with antihypertention, antiobesity, and anticancer properties that have been assessed both in in vitro and in vivo studies (Singh et al. 2014; Liu et al. 2014; Fernández-Tomé and Hernández-Ledesma 2019). Moreover, fermented soy peptides possess the inhibitory activity of angiotensin-converting enzymes and bind firmly to bile acid (Shin et al. 2001). BBI is required to prevent lunasin from digestion in the gastrointestinal tract and therefore lunasin can be digested by trypsin and chymotrypsin without BBI's protection (Park et al. 2007; Cruz-Huerta et al. 2015). The protective role of BBI offers a proper environment for lunasin to perform the anticancer function with a currently unknown mechanism. The molecular mechanisms of lunasin for anticancer effects have been increasingly investigated in various studies (Wan et al. 2017). Lunasin displays the abilities to regulate the cell cycle through induction of cell apoptosis, decrease of gene expression of cyclin, inhibition of oncogene expression, and reduction of mutation rate caused by carcinogen (Jeong et al. 2007; Hsieh et al. 2010; Hernandez-Ledesma et al. 2011). There was a significant reduction of pulmonary colonization after injection of highly metastatic melanoma cells in mice treated with lunasin compared to that of mice in the control group (Shidal et al. 2016). In vitro, lunasin inhibits the growth of cells of murine B16-F0 melanoma, murine LLC, and wild-type C57BL/6 mice (Devapatla et al. 2019). Lunasin treatment also prevents oncogene activation and chromatin acetylation in cells with a mutated tumor suppressor gene which blocks the transformation of normal embryo fibroblast cells into cancerous cells (de Mejia et al. 2003). Lunasin can also prevent metastasis and relapse of melanoma by the reduction of invasive potential of cancer initiation cells (CICs) both in in vitro and in vivo mice experimental metastasis model; therefore, lunasin has been considered an exceptional anticancer agent for patients who have developed resistance to current chemotherapies (Shidal et al. 2016). BBI and lunasin can be found naturally in foods, released after gastrointestinal digestion, or freed from source proteins (Cruz-Huerta et al. 2015; Fernández-Tomé and Hernández-Ledesma 2019).

11.3 Genetics of Anticancer Bioactive Peptides in Soybean

Studies have revealed that lunasin production varies depending on the genotypes of crops and is impacted by environmental conditions such as temperature and soil water potential (Wang et al. 2008; Nakurte et al. 2013). de Mejia et al. (2004) quantified lunasin in a population of USDA soybean accessions ($n = 144$) using enzyme-linked immunosorbent assay (ELISA) and found that lunasin concentration ranged from 0.10 to 1.33 g/100 g flour and its variation exceeded 100% among these accessions suggesting that differences in lunasin concentration were largely due to genetic variation. Soybean seeds have an array of anticancer phytochemicals such as lunasin, phytate, and isoflavone; however, hydrolyzed soybean proteins contain the highest level of lunasin and isoflavone-enriched products contain very little amounts of the peptide lunasin or none of it at all. Within 144 genetically diverse samples assessed, the highest values of lunasin were identified in the ancestral and exotic accessions in the USDA germplasm stock (de Mejia et al. 2004).

It is generally believed that the cultivated soybean *Glycine max* (L.) Merr. was derived from its wild progenitor *G. soja* Sieb. et Zucc. approximately 0.8 million years ago in Eastern Asia (Li et al. 2014). The landraces of soybean were subsequently selected from the adaptation of localized *Glycine max* after its domestication (Xu et al. 2002). The genetic sources of current elite cultivars in the USA were developed from small number accessions of landraces (Gizlice et al. 1994). In soybean, many studies used DNA-based markers and mapped quantitative trait loci (QTL) for seed protein content (Diers et al. 1992; Sebolt et al. 2000; Csanadi et al. 2001). Traditional breeding methods for soybean selection have greatly improved soybean production for economically important traits such as yield, protein, and oil contents, but they have not done in large scale to develop cultivars with therapeutic components or anticancer properties. Currently, the tarit of BBI has been described in soybean plant (SO:0002188, Soybase.org). However, the genetic bases of BBI and lunasin inheritance have not been fully understood. The quantitative trait loci (QTL) and molecular pathways for BBI and lunasin productions in soybean have not been defined to the best of our knowledge. More studies are required to elucidate the genetic mechanisms of production for BBI and lunasin in soybean seeds in order to seal the gap between advanced clinical therapeutic success and lack of genetic bases for BBI and lunasin inheritance.

The lack of key genetic information of BBI and lunasin productions in the US soybean germplasm collection will prevent the development of elite soybean cultivars with high contents of anticancer compounds. Mapped QTL and associated SNPs will be the primary step to understand the inheritance of traits for lunasin contents in soybean seeds. QTL or QTNs (quantitative trait nucleotides) for lunasin contents will help us to define candidate genes for these soybean peptides on soybean genome at the sequence level. Genome-wide association study (GWAS) is one of the most powerful methods alternative to the traditional biparental mapping approach to map the targeted QTL for complex traits with a high resolution because it accounts for population structure and historical recombination events in mapping panels in its mathematic model (Kump et al. 2011; Wen et al. 2014; Wen et al. 2018; Yuan et al. 2019). GWAS has been widely used to define quantitative trait loci because it will directly detect genetic association between single nucleotide polymorphism (SNP) markers and traits in the mapping panels of landraces and advanced breeding populations based on linkage disequilibrium (LD) information (Nordborg and Tavare 2002; Wen et al. 2014). In the GWAS approach, marker-estimated kinship (K) matrix, linkage disequilibrium, and population structure (Q) of the mapping panels will be used as genetic co-variances to determine the genomic regions that have affected the traits such as the contents of BBI and lunasin in soybean seeds. GWAS approach can complement the linkage mapping with high recombination in mapping panels of natural populations and identify QTL associated with target traits at genomic sequence level (Nordborg and Weigel 2008). The GWAS mapping panel of soybean accessions will represent large variations in the geographical origins, genetic diversity, and phenotypic characters including wild soybean accessions (*Glycine soja* Siebold & Zucc.). Due to natural selection, genetic drift, and population stratification, GWAS approach may produce higher LDs between non-linked loci; therefore, it should be combined with the biparental linkage-based mapping strategy to prevent the identification of false QTL that are derived from the factors due to high LDs.

11.4 Molecular Analysis of Lunasin

The lunasin gene has been cloned from mid-maturation soybean seed [named 2S albumin (*Gm2S-1*)] using a homolog search-based cDNA approach (Revilleza et al. 1996). The gene *Gm2S-1* (Fig. 11.1) encodes for large subunit methionine-rich protein, the small subunit peptide lunasin, a linker peptide, and a signal peptide (Jeong et al. 2003, Lule et al. 2015). The bioactivities of lunasin have been assessed both in synthetic and in natural lunasin (Jeong et al. 2003). Jeong et al. (2003) characterized lunasin using mass peptide mapping and measured the activities of various soybean seed proteins during their development. With soybean seed development, the lunasin peptide has been detected 5 weeks after flowering and it is continuously accumulated until seed maturity. Moreover, the amount of lunasin varies in different soybean varieties. Lunasin contains a chromatin-binding proteins' conserved region,

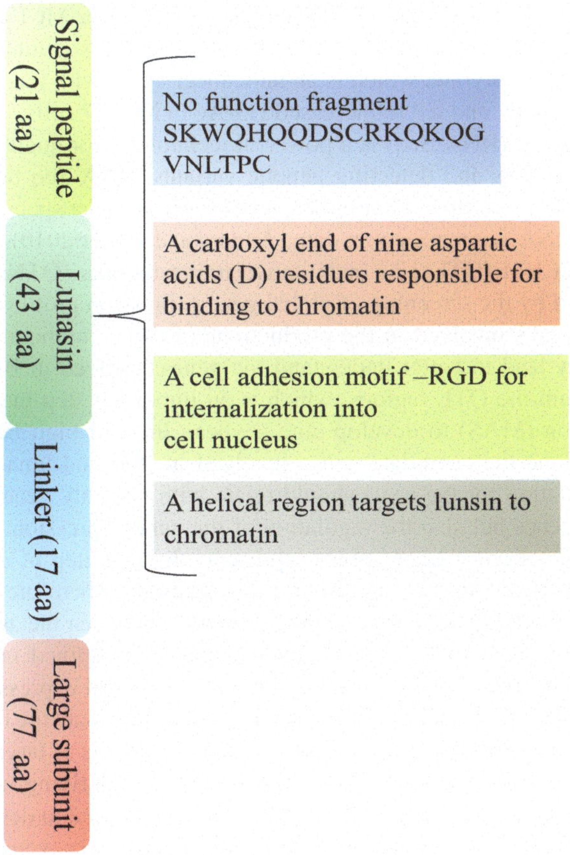

Fig. 11.1 Structure of 2S albumin (*Gm2s-1* gene encoded) protein in soybean seeds. Among the 43 residues of lunasin, the function of 1–21 aa is unknown. (Modified based on Lule et al. 2015)

a cell adhesion motif (Arg-Gly-Asp), and 9 units of Asp (Galvez and de Lumen 1999; Galvez et al. 2001). Lunasin has 43 amino acids (SKWQHQQDSCRKQKQGVNLTPCEKHI MEKIQGRGDDDDDDDDD), which are encoded within the soybean *Gm2S-1* gene (Chatterjee et al. 2018; Jeong et al. 2003). Natural lunasin extracted from soybean seeds can inhibit histone acetylation, which plays an important role in histone modification, phosphorylation, and DNA methylation. Lunasin also shows an antimitotic effect and results in cell death in transfected cell with lunasin gene (Galvez and de Lumen 1999). The peptide of poly-aspartyl carbonxyl (D) end of lunasin binds to hypoacetylated chromatin and leads to chromatin modification and thereby the control cell cycle and suppression of carcinogenesis. Experiments on synthetic lunasin have shown that RGD motif of lunasin acts as a chemopreventive agent due to its binding ability to chromatin (Galvez et al. 2001). The prevention of cell division and promotion of lipids, carbohydrates, and storage proteins biosynthesis may be attributed to lunasin antimitotic effect role. The same study showed that lunasin reduced H3 histone acetylation 50X in mice fibroblast C3H cells and 3X in human breast cancer MCF-7 cells (Galvez and de Lumen 1999). Jeong et al. (2003) also showed that soy lunasin completely neutralized sodium butyrate, which is an anticancer agent with inconsistent results in in vivo experiment, and led to increased H3-Ac in MCF-7 cells.

RNA sequencing (RNA-Seq) is a powerful technology for quantifying and profiling gene transcripts and detecting genetic variants. RNA-Seq has been widely used to assess the differentially expressed genes (DEGs) under specific treatment by estimating the RNA sequence count data (Anders and Huber 2010). By identifying quantitative trait loci (QTLs) or quantitative trait networks (QTNs) for BBI and lunasin followed by the screening of global gene expression profiles for the DEGs, molecular pathways involved in the productions of these compounds can be analyzed. This may lead to better understand the genes involved in BBI and lunasin production within the QTL regions, which is an important step in using markers-assisted selection (MAS) to develop elite soybean lines of high BBI and lunasin contents and clone the candidate genes involved in BBI and lunasin production. Molecular regulation for the traits should relate with not only genetic structure of the functional genes but also the regulation of the gene expression. Elucidation of the biological function of target genes represents the first step of analysis for the molecular mechanisms underlying the trait associations. Therefore, in addition to GWAS and biparental QTL mapping, the correlation between the associated SNPs and differentially expressed genes (DEGs) in the QTL-defined intervals can be investigated using RNA-Seq approach to analyze the global expression variants in coding genes. RNA-Seq can provide an efficient way to validate the loci that control the trait and thereby identify the candidate genes. The effects of target genes can be deciphered by their interactions among DEGs within underlying molecular pathways (Ideker and Nussinov 2017). RNA-Seq will be also used to detect gene-based markers that provide a useful resource for trait genetic improvement in soybean (Guo et al. 2018). In addition, the expression of putative genes associated with traits identified by RNA-Seq can be further confirmed using RT-qPCR.

The seed samples for RNA-Seq approach are harvested from pods of R6-7 stages and the soybean seeds are placed into a 1.5-mL microcentrifuge tube filled with RNALater solution (Thermal Fisher Scientific, Waltham, MA). Ten seeds from three pots of each line are pooled into one tube with three biological replicates from three different panels. In order to investigate the differential expression of genes (DEGs) at midmaturation of soybean seeds, seeds from three extremely high and low contents of BBI and lunasin containing lines can be selected. The seed samples in RNALater solution are first stored in −20 °C freezer prior total RNA extraction. Total RNA are extracted using QIAGEN RNeasy Plant Mini Kit (Qiagen, Valencia, CA) combining DNase treatment (Qiagen Rnase-Free Dnase Set, Cat #79254) from the collected tissues following the protocol provided by the manufacturer. To quantify total RNA for reverse transcription, a BioTek microplate reader (Seattle, WA) and a Picogreen RNA quantitation kit (Invitrogen, Eugene, OR) are used. The quantified RNA samples are sent by dry ice shipment to Genomic Sequencing facility for RNA sequencing. In order to identify the DEGs that are associated with BBI and lunasin contents at the seed stage, fragments per kilobase per million (FPKM) value is used to normalize the gene expression levels. All of the uniquely mapped reads can be used to estimate the genes' FPKM values. Based on DEGs, the international standard system Gene Ontology (GO) (http://www.geneontology.org/) is used to categorize the functions of the predicted genes: (1) biological processes, (2) cellular components, and (3) molecular functions. Other methods such as the sequence de novo assembly and Kyoto Encyclopedia of Genes and Genomes (KEGG) are carried out to categorize the functions of the predicted genes based on the DEGs. The function annotation for all DEGs can also be visualized using MapMan platform (Thimm et al. 2004), and Log_2 fold changes obtained from DESeq output will be loaded into the software. Custom-specific mapping file for MapMan based on the soybean sequence output is created using the Mercator (Lohse et al. 2014) pipeline (http://mapman.gabipd.org/web /guest/mercator). The dynamics of DEG expression is validated using RT-qPCR. For RT-qPCR reaction, the first strand cDNA can be synthesized using cDNA synthesis kit (cDNA Synthesis Kits - Ambion™, Cat#AM1722, ThermoFisher Scientific) according to the manufacturer's description. RT-qPCR can be performed in any real-time PCR platforms using Power SYBR Green Real-Time PCR Master Mixes (Thermo Fisher Sceintific # 4368577). Primers for RT-qPCR can be designed using Primer 3 (version 4.1, Koressaar et al. 2018) based on the DEG sequences of identified RNA-Seq and QTL intervals of soybean genome that are associated with BBI and lunasin production. The relative quantification strategy of $2^{-\Delta\Delta CT}$ method is used for quantitative (qPCR) data analysis. The metabolic pathways for BBI and lunasin productions are analyzed using KEGG and MapMan platforms. The co-evolution among the mapping panels for the BBI and lunasin contents can be depicted.

11.5 Quantification of Lunasin in Soybean Seeds

Based on Jeong et al. (2003), soybean proteins can be purchased from SPI Group, Inc. (San Leandro, CA). The standard lunasin can also be custom synthesized by American Peptide Co (Sunnyvale, CA). Polyclonal lunasin antibody is available from Zymed, Inc. (South San Francisco, CA). The following steps are used to extract lunasin from soybean seeds and quantify the concentration using ELISA method (Jeong et al. 2003):

(a) Grind soybean seeds into a fine powder by a grinder.
(b) Mix 30 g of soybean flour with fresh protease inhibitor cocktail (Sigma) added to 100 mL of phosphate-buffered saline (PBS, pH 7.4, 0.1 M).
(c) Centrifuge samples at 20,000 g to avoid the precipitates.
(d) Add 150 μL of tricine buffer (Bio-Rad Laboratories, Hercules, CA) to 50 μL of supernatant and boil for 5 min.
(e) Add 150 μL of sample buffer to 50 μL of lunasin solution (1 mg/mL; Alpha Diagnostics, Inc., Houston, TX) and boil for 5 min.
(f) Prepare a 16.5% Tris-Tricine polypeptide-ready electrophoresis gels (Bio-Rad).
(g) Load 10 μ of the protein standard solution (broad-range from Bio-Rad) into the first well and 10 μ of the lunasin standard solution (1 mg/mL) into the neighboring well.
(h) Run the electrophoresis gel in Mini Protean-3 Cells (Bio-Rad) with Tris-Tricine-SDS buffer.
(i) Stain the electrophoresis gel in Coomassie Brilliant G-250 stain and fix it in a solution of acetic acid/methanol/water (40:10:50%) for 30 min.
(j) Destain the electrophoresis gel 40% acetic acid for one hour and wash it with distilled water.
(k) Use a Kodak Imaging System 440 CF (Kodak, Rochester, NY) or similar to visualize the protein bands on the electrophoresis gel.
(l) Band intensities and molecular weights are estimated and calculated based on the standards.
(m) The antibody for enzyme linked immunosorbent assay (ELISA) is purchased from Abcam (Cambridge, MA) and the protein samples will be quantified.
(n) Recovered lunasin is quantified using ELISA and absorbance at 405 nm wavelength is used to define the relative lunasin concentration (Dia et al. 2009).

11.6 Conclusion

To reveal the genetic bases of the inheritance of bioactive peptides in soybean at molecular level and improve human and animal health through enhanced food quality will always be our grand goals for soybean research. The valuable information derived from the analysis of soybean anticancer bioactive peptides will be used to develop elite soybean cultivars that aid in curing or limiting certain types of cancer thereby improving human health. The genetic and molecular analyses of BBI and lunasin productions in the US soybean germplasm collection will promote to develop elite soybean cultivars with high anticancer compound contents. Mapped QTL and associated SNPs will be the primary step to understand the inheritance of traits for BBI and lunasin contents in soybean seeds. QTL for BBI and lunasin contents will help us to define candidate genes for these soybean peptides on soybean genome.

References

Anders, S., and W. Huber. 2010. Differential expression analysis for sequence count data. *Genome Biology* 11: R106.

Birk, Y., A. Gertler, and S. Khalef. 1963. A pure trypsin inhibitor from soya bean. *The Biochemical Journal* 87: 281–284.

Bowman, D.E. 1946. Differentiation of soybean antitryptic factor. *Proceedings of the Society for Experimental Biology and Medicine* 63: 547–550.

Chatterjee, C., S. Gleddie, and C.W. Xiao. 2018. Soybean bioactive peptides and their functional properties. *Nutrients* 10 (9): 1211. https://doi.org/10.3390/nu10091211.

Cruz-Huerta, E., S. Fernández-Tomé, M.C. Arques, L. Amigo, I. Recio, A. Clemente, and B. Hernández-Ledesma. 2015. The protective role of the Bowman-Birkprotease inhibitor in soybean lunasin digestion: The effect of released peptides on colon cancer growth. *Food & Function* 6: 2626–2635. https://doi.org/10.1039/c5fo00454c.

Csanádi, G.J., J. Vollmann, G. Stift, and T. Lelley. 2001. Seed quality QTLs identified in a molecular map of early maturing soybean. *Theoretical and Applied Genetics* 103: 912–919.

de Lumen, B.O. 2005. Lunasin: A cancer-preventive soy peptide. *Nutrition Reviews* 63: 16–21.

de Mejia, E.G., T. Bradford, and C. Hasler. 2003. The Anticarcinogenic Potential of Soybean Lectin and Lunasin. *Nutrition Reviews 61 (7): 239–246.* https://doi.org/10.1301/nr.2003.jul.239-246.

de Mejia, G.E., M. Vásconez, B.O. de Lumen, and R. Nelson. 2004. Lunasin concentration in different soybean genotypes, commercial soy protein, and isoflavone products. *Journal of Agricultural and Food Chemistry* 52: 5882–5887.

Devapatla, B., C. Shidal, K. Yaddanapudi, and K. Davis. 2019. *F1000Research.* https://doi.org/10.12688/f1000research.9661.2.

Dia, V.P., D. Torres, B.O. de Lumen, J.W. Erdman, and E.G. de Mejia. 2009. Presence of lunasin in plasma of men after soy protein consumption. *Journal of Agricultural and Food Chemistry* 57: 1260–1266.

Diers, B.W., P. Keim, W.R. Fehr, and R.C. Shoemaker. 1992. RFLP analysis of soybean seed protein and oil content. *Theoretical and Applied Genetics* 83: 608–612.

Ďúranová, H., V. Fialková, J. Bilčíková, N. Lukáč, and Z Kňažická. 2019 Lunasin and its versatile health-promoting actions. *Journal of Microbiology, Biotechnology and Food Sciences*. https://doi.org/10.15414/jmbfs.2019.8.4.1106-1110.

Fernández-Tomé, S., and B. Hernández-Ledesma. 2019. Current state of art after twenty years of the discovery of bioactive peptide lunasin. *Food Research International* 116: 71–78.

Galvez, A.F., and B.O. de Lumen. 1999. A soybean cDNA encoding a chromatin-binding peptide inhibits mitosis of mammalian cells. *Nature Biotechnology* 17: 495–500.

Galvez, A.F., M. Revilleza, and B.O. de Lumen. 1997. A novel methionine-rich protein from soybean cotyledon: Cloning and characterization of cDNA (accession No. AF005030). Plant gene register #PGR97-103. *Plant Physiology* 114: 1567–1569.

Galvez, A.F., N. Chen, J. Macasieb, and B.O. de Lumen. 2001. Chemopreventive property of a soybean peptide (lunasin) that binds to deacetylated histones and inhibits acetylation. *Cancer Research* 61: 7473–7478.

Gizlice, Z., T.E. Carter, and J.W. Burton. 1994. Genetic base for North American public soybean cultivars released between 1947 and 1988. *Crop Science* 34: 1143–1151.

Greenwald, P., C.K. Clifford, and J.A. Milner. 2001. Diet and cancer prevention. *European Journal of Cancer* 37 (8): 948–965.

Guo, Y., B. Su, J. Tang, F. Zhou, and L.J. Qiu. 2018. Gene-based SNP identification and validation in soybean using next-generation transcriptome sequencing. *Molecular Genetics and Genomics* 293 (3): 623–633.

He, F.J., and J.Q. Chen. 2013. Consumption of soybean, soy foods, soy isoflavones and breast cancer incidence: Differences between Chinese women and women in Western countries and possible mechanisms. *Food Science and Human Wellness* 2: 146–161.

Hernande-Ledesma, B., C.C. Hsieh, and B.O. de Lumen. 2011. Relationship between lunasin's sequence and its inhibitory activity of histones H3 and H4 acetylation. *Molecular Nutrition & Food Research* 55: 989–998.

Hsieh, C.C., B. Hernandez-Ledesma, and B.O. de Lumen. 2010. Soybean peptide lunasin suppresses *in vitro* and *in vivo* 7,12-dimethylbenz anthracene-induced tumorigenesis. *Journal of Food Science* 75: H311–H316.

Ideker, T., and R. Nussinov. 2017. Network approaches and applications in biology. *PLoS Computational Biology* 13: e1005771.

Jeong, H.J., J.H. Park, Y. Lam, and B.O. de Lumen. 2003. Characterization of lunasin isolated from soybean. *Journal of Agricultural and Food Chemistry* 51: 7901–7906.

Jeong, J.B., H.J. Jeong, J.H. Park, S.H. Lee, J.R. Lee, H.K. Lee, G.Y. Chung, J.D. Choi, and B.O. de Lumen. 2007. Cancer-preventive peptide lunasin from *Solanum nigrum* L. inhibits acetylation of core histones H3 and H4 and phosphorylation of retinoblastoma protein (Rb). *Journal of Agricultural and Food Chemistry* 55: 10707–10713.

Kennedy, A.R. 1998. The Bowman-Birk inhibitor from soybeans as an anticarcinogenic agent. *The American Journal of Clinical Nutrition* 68: 1406–1412.

Kim, S.E., H.H. Kim, J.Y. Kim, Y.I. Kang, H.J. Woo, and H.J. Lee. 2000. Anticancer activity of hydrophobic peptides from soy proteins. *BioFactors* 12: 151–155.

Kitts, D., and K. Weiler. 2003. Bioactive proteins and peptides from food sources: Applications of bioprocesses used in isolation and recovery. *Current Pharmaceutical Design* 9: 1309–1323.

Koressaar, T., M. Lepamets, L. Kaplinski, K. Raime, R. Andreson, and M. Remm. 2018. Primer3_masker: Integrating masking of template sequence with primer design software. *Bioinformatics* 34 (11): 1937–1938.

Kump, K.L., P.J. Bradbury, R.J. Wisser, E.S. Buckler, A.R. Belcher, M.A. Oropeza-Rosas, J.C. Zwonitzer, S. Kresovich, M.D. McMullen, D. Ware, P.J. Balint-Kurti, and J.B. Holland. 2011. Genome-wide association study of quantitative resistance to southern leaf blight in the maize nested association mapping population. *Nature Genetics* 43 (2): 163–168.

Lam, Y., A. Galvez, and B.O. de Lumen. 2003. Lunasin suppresses E1Amediated transformation of mammalian cells but does not inhibit growth of immortalized and established cancer cell lines. *Nutrition and Cancer* 47: 88–94.

Li, Y.H., G. Zhou, J. Ma, W. Jiang, L.G. Jin, Z. Zhang, Y. Guo, J. Zhang, Y. Sui, L. Zheng, S.S. Zhang, Q. Zuo, X.H. Shi, Y.F. Li, W.H. Zhang, Y. Hu, G. Kong, H.L. Hong, B. Tan, J. Song, Z.X. Liu, Y. Wang, H. Ruan, C.K.L. Yeung, J. Liu, H. Wang, L.J. Zhang, R.X. Guan, K.J. Wang, W.B. Li, S.Y. Chen, R.Z. Chang, Z. Jiang, S.A. Jackson, R. Li, and L.J. Qiu. 2014. De novo assembly of soybean wild relatives for pan-genome analysis of diversity and agronomic traits. *Nature Biotechnology* 32 (10): 1045–1052.

Liu, J., S.H. Jia, M. Kirberger, and N. Chen. 2014. Lunasin as a promising health-beneficial peptide. *European Review for Medical and Pharmacological Sciences* 18: 2070–2075.

Lohse, M., A. Nagel, T. Herter, P. May, M. Schroda, R. Zrenner, T. Tohge, A.R. Fernie, M. Stitt, and B. Usadel. 2014. Mercator: A fast and simple web server for genome scale functional annotation of plant sequence data. *Plant, Cell & Environment* 37 (5): 1250–1258.

Losso, J.N. 2008. The biochemical and functional food properties of the bowman-birk inhibitor. *Critical Reviews in Food Science and Nutrition* 48: 94–118. https://doi.org/10.1080/10408390601177589.

Lule, V.K., S. Garg, S.D. Pophaly, Hitesh, and S.D. Tomar. 2015. Potential health benefits of Lunasin: A multifaceted Soy—derived bioactive peptide. *Journal of Food Science* 80 (3): R485–R494. https://doi.org/10.1111/1750-3841.12786.

McCue, P., and K. Shetty. 2004. Health benefits of soy isoflavonoids and strategies for enhancement: A review. *Critical Reviews in Food Science and Nutrition* 44: 361–367.

Messina, M.J., V. Persky, K.D. Setchell, and S. Barnes. 1994. Soy intake and cancer risk: A review of the in vitro and in vivo data. *Nutrition and Cancer* 21 (2): 113–131.

Nakamori, T. 2002. Antiobesity effects of soy proteins and soy peptides. *Food Style* 21: 86–88.

Nakurte, I., I. Kirhnere, J. Namniece, K. Saleniece, L. Krigere, P. Mekss, Z. Vicupe, M. Bleidere, L. Legzdina, and R. Muceniece. 2013. Detection of the lunasin peptide in oats (*Avena sativa* L.). *Journal of Cereal Science* 57: e319–e324.

Nordborg, M., and S. Tavare. 2002. Linkage disequilibrium: What history has to tell us. *Trends in Genetics* 18: 83–90.

Nordborg, M., and D. Weigel. 2008. Next-generation genetics in plants. *Nature* 456: 720–723.

Park, J.H., H.J. Jeong, and B.O. de Lumen. 2007. In vitro digestibility of the cancer-preventive soy peptides lunasin and BBI. *Journal of Agricultural and Food Chemistry* 55: 10703–10706.

Pena-Ramos, E.A., and Y.L. Xiong. 2002. Antioxidant activity of soy protein hydrolysates in a liposomal system. *Journal of Food Science* 67: 2952–2956.

Qi, R.F., Z.W. Song, and C.W. Chi. 2005. Structural features and molecular evolution of Bowman–Birk protease inhibitors and their potential application. *Acta Biochimica et Biophysica Sinica* 37: 283–292.

Revilleza, M.J., A.F. Galvez, D.C. Krenz, and B.O. de Lumen. 1996. An 8 kDa methionine-rich protein from soybean (*Glycine max*) cotyledon: Identification, purification, and N-terminal sequence. *Journal of Agricultural and Food Chemistry* 44: 2930–2935.

Sebolt, A.M., R.C. Shoemaker, and B.W. Diers. 2000. Analysis of a Quantitative Trait Locus Allele from Wild Soybean That Increases Seed Protein Concentration in Soybean. *Crop Science 40 (5): 1438–1444.* https://doi.org/10.2135/cropsci2000.4051438x.

Shidal, C., N. Al-Rayyan, K. Yaddanapudi, and K.R. Davis. 2016. Lunasin is a novel therapeutic agent for targeting melanoma cancer stem cells. *Oncotarget* 7 (51): 84128–84141.

Shin, Z., R. Yu, S.A. Park, D.K. Chung, C.W. Ahn, H.S. Nam, K.S. Kim, and H.J. Lee. 2001. His-His-Leu, an angiotensin I converting enzyme inhibitory peptide derived from Korean soybean paste, exerts antihypertensive activity in vivo. *Journal of Agricultural and Food Chemistry* 49: 3004–3009.

Singh, B.P., S. Vij, and S. Hati. 2014. Functional significance of bioactive peptides derived from soybean. *Peptides* 54: 171–179.

Thimm, O., O. Bläsing, Y. Gibon, A. Nagel, S. Meyer, P. Krüger, J. Selbig, L.A. Müller, S.Y. Rhee, and M. Stitt. 2004. MAPMAN: A user-driven tool to display genomics data sets onto diagrams of metabolic pathways and other biological processes. *The Plant Journal* 37: 914–939. https://doi.org/10.1111/j.1365-313X.2004.02016.x.

USDA–NASS. 2015a. *Crop production: 2014 summary*. Washington, DC: USDA National Agricultural Statistics Service. http://www.nass.usda.gov.

———. 2015b. *Crop values: 2014 summary*. Washington, DC: USDA National Agricultural Statistics Service. http://www.nass.usda.gov.

Velasquez, M.T., and S.J. Bhathena. 2007. Role of dietary soy protein in obesity. *International Journal of Medical Sciences* 4: 72–82.

Wan, X., H. Liu, Y. Sun, J. Zhang, X. Chen, and N. Chen. 2017. Lunasin: A promising polypeptide for the prevention and treatment of cancer (Review). *Oncology Letters 13 (6): 3997–4001.* https://doi.org/10.3892/ol.2017.6017.

Wang, W., V.P. Dia, M. Vasconez, E.D. de Mejia, and R. Nelson. 2008. Analysis of soybean protein-derived peptides and the effect of cultivar, environmental conditions, and processing on lunasin concentration in soybean and soy products. *Journal of AOAC International* 91: 936–946.

Wen, Z., R. Tan, J. Yuan, C. Bales, W. Du, S. Zhang, M.I. Chilvers, C. Schmidt, Q. Song, P.B. Cregan, and D. Wang. 2014. Genome-wide association mapping of quantitative resistance to sudden death syndrome in soybean. *BMC Genomics* 15: 809. http://www.biomedcentral.com/1471-2164/15/809.

Wen, Z., R. Tan, S. Zhang, P. Collins, J. Yuan, W. Du Jiazheng, C. Gu, S. Ou, Q. Song, Y.Q.C. An, J.F. Boyse, M.I. Chilvers, and D. Wang. 2018. Integrating GWAS and gene expression data for functional characterization of resistance to white mold in soya bean. *Plant Biotechnology Journal* 16 (11): 1825–1835.

Xu, H., J. Abe, Y. Gai, and Y. Shimamoto. 2002. Diversity of chloroplast DNA SSRs in wild and cultivated soybeans: Evidence for multiple origins of cultivated soybean. *Theoretical and Applied Genetics* 105 (5): 645–653.

Yuan, J., U. Rosyara, Z. Wen, M. Lague, and B. Bizimungu. 2019. GWAS analysis of potato (*Solanum tuberosum* L.) resistance to common scab in Canadian potato germplasm. *Potato Research*. https://doi.org/10.1007/s11540-019-09437-w.

Chapter 12
Soybean Tocopherols: Biosynthesis, Factors Affecting Seed Content, QTL Mapping, and Candidate Genes

Moulay Abdelmajid Kassem

12.1 Introduction

The four naturally occurring antioxidants tocopherols and tocotrienols also known as vitamin E are biosynthesized in plastids of many plants including soybean (Ujiie et al. 2005; Carrao-Panizzi and Erhan 2007; Dwiyanti et al. 2007), canola (Marwede et al. 2004), rapeseed (*Brassica napus*) (Endrigkeit et al. 2009; Fritsche et al. 2012; Wang et al. 2012a, b; Gajardo et al. 2015), sunflower (Velasco et al. 2002; Del Moral et al. 2012; Haddadi et al. 2012), maize (Chander et al. 2008), lettuce (Hayashi et al. 2012), and many other plant species such as cashews, Brazil nuts, pecans, peanuts, pistachios, pines, walnuts, etc. (Kornsteiner et al. 2006). Tocopherols are synthesized via the shikimate biosynthetic pathways in seven steps starting with the combination of phosphoenolpyruvate (PEP) and erythrose4-phosphate (E4P) and ending with chorismate (see below for details). They exist in four forms α-, β-, γ-, and δ-tocopherols which differ in their chemical structures, bioactivities, and content depending on the plant species and with the same plant species, and it is well known that α-tocopherol (vitamin E) is the most bioactive among these forms (Kamal-Eldin and Appelqvist 1996; Marwede et al. 2004; Ujiie et al. 2005; Kornsteiner et al. 2006; Dwiyanti et al. 2007; Li et al. 2010, 2016; Carrera and Seguin 2016).

Plants including soybean respond to a wide range of abiotic stresses such as drought, cold, salinity, shilling, flooding, light, and heat stresses and biotic stresses from bacteria, fungi, virions, and viruses by producing a wide range of secondary metabolites including the antioxidants tocopherols that allow them to cope with

M. A. Kassem (✉)
Plant Genomics and Biotechnology Lab, Department of Biological and Forensic Sciences, Fayetteville State University, Fayetteville, NC, USA
e-mail: mkassem@uncfsu.edu

M. A. Kassem (ed.), *Soybean Seed Composition*,
https://doi.org/10.1007/978-3-030-82906-3_12

these stresses (Munne-Bosch 2005; Maeda and DellaPenna 2007; Li et al. 2008). It is well established that an increase in the content of α-tocopherol, located in the thylakoid membranes and chloroplast envelopes, deactivates reactive oxygen species (ROS) and prevents lipid peroxidation leading to stress tolerance (Munne-Bosch 2005; Maeda and DellaPenna 2007). Other studies showed that α-tocopherol is mainly involved in cyclic electron flow during photosynthesis (photooxidation of photosystem II during photorespiration), intracellular signaling, and membrane stability (Munne-Bosch and Alegre 2002; Havaux et al. 2003; Munne-Bosch and Falk 2004; Li et al. 2008). Other tocopherol forms such as γ-tocopherol were shown to be involved in seed desiccation tolerance by protecting polyunsaturated fatty acids (Sattler et al. 2004). Moreover, α- and γ-tocopherols are involved in reducing salt stress and sorbitol stress that causes osmotic stress (Abbasi et al. 2007; Li et al. 2008). Other studies showed that tocopherol biosynthesis is affected by plant hormones such as abscisic acid (ABA), jasmonic acid (JA), and salicylic acid (SA) during stress by activating chlorophyll degradation and increasing anthocyanin and antioxidant production including tocopherols (Munne-Bosch and Alegre 2002; Li et al. 2008). Sucrose export defective 1 (SXD1) gene, a homolog of tocopherol cyclase (VTE1), encodes for a protein which if mutated leads to an absence of tocopherol and is involved in chloroplast-to-nucleus signal transduction and carbohydrate metabolism regulation (Porfirova et al. 2002; Li et al. 2008).

Several studies reported that vitamin E (α-tocopherol) is involved in reducing the risk of cancer, vascular diseases, and atherosclerosis through the oxidation of low-density lipoprotein (LDL) (Upston et al. 2003). Because of this high oxidation activity and potential benefits, vitamin E is widely used as a supplement in human diets but also in animal diets because it improves meat quality in cattle and poultry (Morrissey et al. 1994). However, other studies reported that vitamin E may be harmful for humans (Robinson et al. 2006; Clarke et al. 2008).

Seed tocopherol contents vary considerably depending on the plant species, within the cultivars of the same species, and the environment the plant is grown (Ujiie et al. 2005; Carrao-Panizzi and Erhan 2007; Dwiyanti et al. 2007; Li et al. 2010, 2016; Carrera and Seguin 2016). Ujiie et al. (2005) quantified α-tocopherol contents in seeds of 1109 soybean cultivars and accessions and identified three varieties with high α-tocopherol content. For example, soybean seeds with α-tocopherol, β-tocopherol, γ-tocopherol, and δ-tocopherol contents are 4–10%, 1–3%, 60–65%, and 24–29%, respectively (Evans et al. 2002; MacDuff 2011). Another study analyzed α-, β-, γ-, and δ-tocopherol contents in 89 Brazilian soybean cultivars grown in Ponta Grossa and Parana states in 2001 and found high variabilities and ranges to be 11–191 ppm for α-tocopherol, 6–64 ppm for β-tocopherol, 304–1333 ppm for γ-tocopherol, and 174–580 ppm for δ-tocopherol (Carrao-Panizzi and Erhan 2007). Therefore, several studies identified quantitative trait loci (QTL) that control seed tocopherol contents in soybean (Dwiyanti et al. 2007, 2011; Li et al. 2010, 2016; Shaw and Rajcan 2017; Liu et al. 2017a, b; Park et al. 2019; Sui et al. 2020; Zhan et al. 2020), maize (Chander et al. 2008), rapeseed (*Brassica napus*) (Fritsche et al. 2012; Wang et al. 2012a, b; Gajardo et al. 2015), sunflower (Del Moral et al. 2012; Haddadi et al. 2012), and other plant species (Wang et al. 2015).

Here, we will summarize tocopherol biosynthetic pathways, factors affecting their content, QTL mapping, and candidate genes involved in their biosynthesis in soybean.

12.2 Tocopherols Biosynthesis

Tocopherols and tocotrienols are biosynthesized through the shikimate biosynthetic pathway that exists in both several microorganisms and higher plants (Dwiyanti et al. 2011; Fritsche et al. 2017). The pathway consists of seven biochemical reactions that start with the first reaction where the enzyme DAHP synthase combines phosphoenolpyruvate (PEP) with erythrose4-phosphate (E4P) and produce the 3-deoxy-D-arabino-heptulosonate 7-phosphate (DAHP) which is also knows as 1-carboxy-2-deoxy-α-D-glucose-6-phosphate. In the second biochemical reaction, DHQ synthase converts DAHP into 3-dehydroquinate (DHQ). In the third biochemical reaction, DHQ dehydratase converts DHQ into 3-dehydroshikimate (DHS). In the fourth biochemical reaction, NADP-dependent shikimate dehydrogenase converts DHS into shikimate. In the fifth biochemical reaction, shikimate kinase converts shikimate into shikimate 3-phosphate (S3P). In the sixth reversible biochemical reaction, EPSP synthase converts S3P and a second PEP into 5-enolpyruvylshikimate 3-phosphate (EPSP). In the seventh and last biochemical reaction, chorismate synthase converts EPSP into chorismate (Fig. 12.1). Other metabolic pathways are

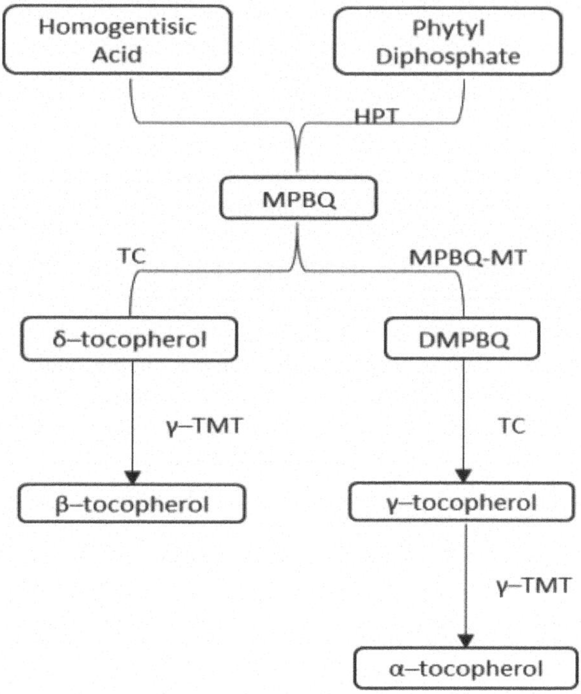

Fig. 12.1 Tocopherol biosynthetic pathway in plants. The homogentisic acid is synthesized via the shikimate biosynthetic pathway, and phytyl diphosphate is synthesized by the 2-C-methyl-d-erythritol-4-phosphate (MEP) pathway. *HPT* phytyltransferase, *MPBQ* methyl-6-phytyl-1,4-benzoquinone, *MPBQ-MT* MPBQ-methyltransferase, *DMPBQ* 2,3-dimethyl-6-phytyl-plastoquinol, *TC* tocopherol cyclase, *γ-TMT* methyltransferase. (Adapted from Dwiyanti et al. (2011))

branched from these steps of the shikimate biosynthetic pathway (Hermann and Weaver 1999; DellaPenna and Pogson 2006; Li et al. 2008).

12.3 Factors Affecting Seed Tocopherols Contents

The demand for α-tocopherol or vitamin E is increasing worldwide; therefore, developing crops with high seed α-Toc contents is the aim of scientists, plant breeders, and farmers. However, many studies showed that α-, β-, γ-, and δ-tocopherols vary considerably in soybean seeds depending on the genotype and environmental conditions (Table 12.1), and genotype-by-environment interactions are common (Marwede et al. 2004; Ujiie et al. 2005; Carrera and Seguin 2016; Dwiyanti et al. 2016; Zhang et al. 2017). Seed tocopherol content is a complex polygenic trait that varies depending on the plant genotype and abiotic and biotic stresses (environmental conditions) and is controlled by multiple QTL, and genotype-by-environment interactions are very commonly observed (Li et al. 2010, 2016; Dwiyanti et al. 2011, 2016; Carrera and Seguin 2016; Shaw and Rajcan 2017; Liu et al. 2017a, b; Park et al. 2019; Sui et al. 2020; Zhan et al. 2020). Among the factors affecting seed tocopherol content are soil fertility and moisture, planting date, solar radiation, row spacing, fertilization, irrigation methods, temperature during seed filling period, and crop management (Almonor et al. 1998; Dolde et al. 1999; Britz and Kremer 2002; Britz et al. 2008; Seguin et al. 2009, 2010; Whent et al. 2009; Chennupati et al. 2011; Carrera et al. 2014; Shaw et al. 2016; Carrera and Seguin 2016). Dwiyanti et al. (2016) grew a natural population of soybean accessions (n = 528) from different geographical regions of Japan and South Korea in Japan in 2010–2011; quantified α-, β-, γ-, and δ-tocopherol contents in their seeds; and found that the tocopherol contents are highly variable among these accessions and 16 accessions with stable and high seed α-tocopherol ratio (α-TR > 20%) (α-TR = α-tocopherol/α-tocopherol + γ-tocopherol + δ-tocopherol). Among other studies, this study showed that seed α-tocopherol ratios are affected by temperature during the seed maturation process (Britz and Kremer 2002; Britz et al. 2008; Seguin et al. 2010; Dwiyanti et al. 2016).

A study demonstrated that environmental factors, genotype-by-environment interactions, and genotype contributed 3%, 12%, and 81% of the total variation in seed tocopherol content, respectively (Dolde et al. 1999). Whent et al. (2009) found that the plant genotype alone contributed approximately 43%, 57%, 69%, and 70% of seed δ-Toc, α-Toc, T-Toc, and γ-Toc content variations, respectively (Whent et al. 2009). However, other studies found that the environment and genotype by environment contributed to the highest variations among seed α-Toc, γ-Toc, δ-Toc, and T-Toc contents (Seguin et al. 2009; Carrera et al. 2014; Shaw et al. 2016; Carrera and Seguin 2016). Seguin et al. (2009) grew a population of soybean genotypes (n = 20) in six locations across Canada; quantified seed α-Toc, γ-Toc, δ-Toc, and T-Toc contents among other agronomic traits (100-seed weight and seed protein content); observed genotype-by-environment interactions for α-Toc; and found that

Table 12.1 Variation of soybean seed α-Toc, γ-Toc, β-Toc, δ-Toc, and T-Toc contents depending on environmental conditions (country of planting)

Country	Tocopherol	Range (μg g^{-1} oil)	Reference
USA	α-Tocopherol (α-Toc)	44–158	Dolde et al. (1999)
	β-Tocopherol (β-Toc)	2–29	
	γ-Tocopherol (γ-Toc)	850–1559	
	δ-Tocopherol (δ-Toc)	254–477	
	Total tocopherol (T-Toc)	1363–2195	
Japan	α-Tocopherol (α-Toc)	5–43	Ujiie et al. (2005)
	β-Tocopherol (β-Toc)	*	
	γ-Tocopherol (γ-Toc)	27–69	
	δ-Tocopherol (δ-Toc)	11–54	
	Total tocopherol (T-Toc)	80–128	
Brazil	α-Tocopherol (α-Toc)	12–191	Carrao-Panizzi and Erhan (2007)
	β-Tocopherol (β-Toc)	6–64	
	γ-Tocopherol (γ-Toc)	174–580	
	δ-Tocopherol (δ-Toc)	304–1334	
	Total tocopherol (T-Toc)	561–1983	
Argentina	α-Tocopherol (α-Toc)	67–85	Izquierdo et al. (2011)
	β-Tocopherol (β-Toc)	15–27	
	γ-Tocopherol (γ-Toc)	765–1115	
	δ-Tocopherol (δ-Toc)	478–501	
	Total tocopherol (T-Toc)	1345–1705	
Canada	α-Tocopherol (α-Toc)	13.15–312.62	Shaw (2012)
	β-Tocopherol (β-Toc)	*	
	γ-Tocopherol (γ-Toc)	81.48–1024.13	
	δ-Tocopherol (δ-Toc)	21.12–394.54	
	Total tocopherol (T-Toc)	123.35–1296.81	

*Information/data is not provided in the articles

the range of variation among cultivars was 264–314%. α-Toc was the highest in all cultivars and locations, and its content range was 8.7–33.2 μg g^{-1} of seed (Seguin et al. 2009).

Many studies reported that high air temperatures, soil moisture, or drought during seed filling period affects seed α-, β-, γ-, δ-, and total tocopherols (Almonor et al. 1998; Dolde et al. 1999; Britz and Kremer 2002; Britz et al. 2008; Chennupati et al. 2011). Almonor et al. (1998) tested the effect of 18 °C/14 °C, 22 °C/18 °C, 26 °C/22 °C, and 30 °C/26 °C day/night temperatures on seed α-Toc, γ-Toc, δ-Toc, T-Toc, and oil contents of soybean cultivar 'Dare' (control) and lines N78–2245, PI-123440, and N85-2176. They found that both tocopherols and oil are deposited in parallel during seed development. Moreover, T-Toc contents are decreased by low temperatures; however, at maturity, seed α-Toc, γ-Toc, δ-Toc, and T-Toc contents were positively correlated with growth temperatures (Almonor et al. 1998). Britz and Kremer (2002) grew 'Essex', 'Forrest', and 'Williams' cultivars over 2 years (1999–2000) in a greenhouse at different temperatures and water regimes and quantified their seed α-Toc, γ-Toc, δ-Toc, and T-Toc contents. High temperatures caused an increase of 8–10% in seed T-Toc contents of Essex and Forrest but not Williams. However, under drought and high temperature (28 °C) conditions, seed α-Toc contents increased significantly by 128%, 142%, and 198% in Forrest, Essex, and Williams, respectively, compared to the controls grown at normal irrigation and temperature (23 °C). They also found that drought stress decreased seed γ-Toc and δ-Toc contents by 5% and 20%, respectively, while seed α-Toc content increased by 42% (Britz and Kremer 2002). Likewise, another study demonstrated that seed α-Toc content was increased by 245%, while both seed γ-Toc and δ-Toc contents decreased under drought stress conditions (Carrao-Panizzi and Erhan 2007).

Other studies investigated the effects of row spacing, cultivar, seeding rate, planting date, solar radiation, and P and K fertilization on seed α-Toc, γ-Toc, δ-Toc, and T-Toc contents (Britz et al. 2008; Seguin et al. 2010; Izquierdo et al. 2011; Carrera et al. 2014; Carrera and Seguin 2016). Britz et al. (2008) grew several soybean lines ($n = 18$) in various locations in Maryland in 1999–2001 and quantified seed α-Toc/T-Toc, δ-Toc/T-Toc, and T-Toc contents. They observed significant effects of planting date and location on seed T-Toc content but not temperature. However, seed relative α-Toc (α-Toc/T-Toc) content increased with warmer temperatures in 1999–2001. In 2002, they observed higher seed relative α-Toc (α-Toc/T-Toc) content for all lines in all locations, and significant effects of location, maturity group, and planting date were observed (Britz et al. 2008). Seguin et al. (2010) investigated the effects of row spacing, cultivar, seeding rate, planting date, and P and K fertilization on seed α-Toc, γ-Toc, δ-Toc, and T-Toc contents in five locations in Canada. They observed an increase of 4% in seed α-Toc content of plants grown at 40 seeds m^{-2} seeding rate compared to seeds of plants grown at 50 and 60 seeds m^{-2} seeding rates. Likewise, there is an increase of 6% in seed α-Toc content of plants grown at 36 cm or wider row spacing compared to seeds of plants grown at 18 cm row spacing. Seed γ-Toc content was not affected neither by seeding rate nor by row spacing; however, seed δ-Toc content was 4% higher in 18 cm row spacing compared to wider row spacings. Seed α-Toc content of plants grown at early seeding date (mid-May) was 45% higher than seed α-Toc content of plants grown at a later date (mid-June) and one location (Normandin, QC); they found a 183% difference in seed α-Toc content between plants grown in mid-June and mid-May. However, seed δ-Toc content was 24% higher in plants grown in mid-June (late seeding date) compared to plants grown in mid-May (early seeding date). The study also found that P and K fertilization had limited effect on all seed tocopherols in all locations (Seguin et al. 2010). Izquierdo et al. (2011) investigated the effect of intercepted solar radiation (ISR) on seed oil tocopherol contents in soybean, sunflower, rape, and maize during seed filling period. They found that an increase in solar radiation caused an increase in seed oil tocopherol contents in soybean, sunflower, and rape; however, reduced oil tocopherol contents were observed in soybean and sunflower when ISR is increased. Variation ranges of seed oil tocopherol due to variations in IRS during seed filling period were 98.3–189.5 mg, 38.5–63.7 mg, and 188.4–267.9 mg for soybean, sunflower, and maize, respectively (Izquierdo et al. 2011). Carrera et al.

(2014) grew a population of soybean germplasm (n = 23) in six locations across Argentina and quantified seed protein, oil, isoflavones, tocopherols, and other agronomic traits and followed the effects of air temperature, solar radiation, and precipitation on these traits. They found significant variabilities in all studied traits and genotypes including tocopherols. The ranges of seed α-Toc, β-Toc, γ-Toc, and δ-Toc contents were 114.1–139.1 mg kg^{-1} oil, 37.9–51.7 mg kg^{-1} oil, 939.8–1080.4 mg kg^{-1} oil, and 307.8–397.3 mg kg^{-1} oil, respectively (Carrera et al. 2014). Carrera and Seguin (2016) showed that seed α-Toc, β-Toc, γ-Toc, δ-Toc, and T-Toc contents ranged from 7 to 436 μg g^{-1} oil, 2 to 64 μg g^{-1} oil, 40 to 1559 μg g^{-1} oil, 18 to 1334 μg g^{-1} oil, and 78 to 2195 μg g^{-1} oil, respectively (Table 12.1); however, when expressed in μg g^{-1} of seed, the ranges were 2–127 μg g^{-1} seed (α-Toc), 2–10 μg g^{-1} seed (β-Toc), 14–362 μg g^{-1} seed (γ-Toc), 9–149 μg g^{-1} seed (δ-Toc), and 29–363 μg g^{-1} seed (T-Toc) (Carrera and Seguin 2016).

12.4 Tocopherol QTL Mapping

As stated earlier, several studies tried to map QTL that control seed α-, β-, γ-, δ-, and total tocopherol contents in soybean (Dwiyanti et al. 2007, 2011; Li et al. 2010, 2016; Shaw and Rajcan 2017; Liu et al. 2017a, b; Park et al. 2019; Sui et al. 2020; Zhan et al. 2020), maize (Rocheford et al. 2002; Wong et al. 2003; Chander et al. 2008), rapeseed (*Brassica napus*) (Fritsche et al. 2012; Wang et al. 2012a, b; Gajardo et al. 2015), sunflower (Del Moral et al. 2012; Haddadi et al. 2012), and other crops.

Dwiyanti et al. (2007) grew an F_2 population derived from a cross between 'Keszthelyi A.S.' (high α-tocopherol, 20–30%) and 'Ishihime' (low α-tocopherol, <10%) lines (n = 140) in the greenhouse in 2005, genotyped it with 67 SSR markers, and used chi-square goodness-of-fit test to identify SSR markers associated with QTL that control seed α-tocopherol content. Two SSR markers Sat_243 and Sat_167 associated with QTL that control seed α-tocopherol content have been identified and mapped on chromosome 9 (Table 12.2; Fig. 2.1) (Dwiyanti et al. 2007).

Li et al. (2010) grew a recombinant inbred line (RIL) population derived from a cross between 'OAC Bayfield' (high vitamin E content) and 'Hefeng 25' (low vitamin E content) lines (n = 144) in Harbin and Hulan, China, in 2007–2008, genotyped it with 107 SSR markers, and used single marker analysis (ANOVA) method to identify QTL that control seed vitamin E (α-, β-, γ-, δ-, and total vitamin E) content. Five QTL that control seed α-tocopherol (−Toc) content have been identified and mapped on chromosomes 2, 6, 8, 14, and 20; seven QTL that control seed α-tocopherol (γ-Toc) content have been identified and mapped on chromosomes 2, 4, 6, 10, 15, 16, and 18; four QTL that control seed α-tocopherol (δ-Toc) content have been identified and mapped on chromosomes 1, 8, 13, and 20; and five QTL that control seed total vitamin E (TVE) content have been identified and mapped on chromosomes 2, 3, 6, and 10 (Table 12.2; Fig. 2.1) (Li et al. 2010). In a subsequent study and using the same 'Hefeng 25' by 'OAC Bayfield' RIL population, Li et al. (2016) reported the same 18 QTL that control seed α-tocopherol, γ-tocopherol,

Table 12.2 Details of QTL that control seed α-, γ-, δ-, and total tocopherol contents in soybean

Trait	QTL	Chr.	Position (cM)	Marker/interval	LOD/P/chi-value	R^2 (%)	Environment	References
QTL identified in soybean								
α-Toc	*qα-Toc-1*	9	58.4–60.4	Sat_167	Chi-v = 18.7	*	Greenhouse	Dwiyanti et al. (2007)
	qα-Toc-2	9	64.1–66.1	Sat_243	Chi-v = 17.3	*	Greenhouse	
α-Toc	*QαB2_1*	14	4.9–6.9	Sat_177	<0.0001	6.70	Harbin-07	Li et al. (2010) And
					<0.0001	6.50	Harbin-08	Li et al. (2016)
	QαC2_1	6	89.8–91.8	Satt376	0.0012	7.40	Harbin-07	
					0.0002	8.60	Harbin-08	
					<0.0001	16.70	Suihua-08	
					<0.0001	5.80	Hulan-08	
	QαD1b_1	2	65.4–67.4	Satt266	0.0002	5.60	Harbin-07	

δ-tocopherol, and total vitamin E (TVE) contents. Moreover, they observed additive effect by additive effect, QTL by environment, additive effect by environment, and epistasis interactions among the QTL identified (Li et al. 2016).

Dwiyanti et al. (2011) grew the same F_2 population derived from a cross between 'Keszthelyi A.S.' (KAS) and 'Ishihime' lines ($n = 140$) (Dwiyanti et al. 2007) and heterogeneous inbred family lines (HIF lines) at the locus *γ-TMT3* and used multiple QTL mapping (MQM) method of MapQTL5 to identify QTL that control seed tocopherol contents. The genetic linkage map constructed was composed of 152 SSR markers and covered 3401 cM. In the F_2 seed population, one QTL that controls seed α-, γ-, and δ-tocopherols has been identified and mapped on chromosome 9, and another QTL that controls seed δ-tocopherols has been identified and mapped on chromosome 12. In the F_2 plant population, the same QTL on chromosome 9 was also identified to control seed α-, γ-, and δ-tocopherol contents. Another QTL that controls seed δ-tocopherol content has been identified and mapped on chromosome 7 but only in the 'F_2 plant' population (Table 12.2; Fig. 2.1). Moreover, within the QTL region on chromosome 9, ten candidate genes that control seed α-tocopherol, γ-tocopherol, and δ-tocopherol contents have been identified including Glyma09g35680.1 that encodes for a γ-tocopherol methyltransferase (γ-TMT) (Table 12.3) (Dwiyanti et al. 2011).

Shaw and Rajcan (2017) grew a RIL population derived from a cross between 'OAC Bayfield' (high vitamin E content) and 'OAC Shire' (low vitamin E content) lines ($n = 79$) over 2 years (2009–2010) across multi-environments in Canada, genotyped it with 151 SSR markers, and used both single marker analysis (one-way ANOVA) and interval mapping (IM) of MapQTL (Van Ooijen 2006) methods to identify QTL that control seed α-tocopherol, γ-tocopherol, δ-tocopherol, and total tocopherol contents. Forty-two (42) SSR markers associated with seed tocopherol contents have been identified by one-way ANOVA, and 26 SSR markers associated with seed tocopherol contents have been identified by IM. These QTL that control seed tocopherol contents have been mapped on 17 chromosomes of the 20 chromosomes of the soybean genome. Twelve QTL that control seed α-tocopherol (α-Toc) have been identified and mapped on chromosomes 2, 10, 13, 15, and 18; 21 QTL that control seed γ-tocopherol (γ-Toc) have been identified and mapped on chromosomes 3, 4, 6, 7, 8, 10, 11, 12, 15, 16, 17, and 18; 16 QTL that control seed δ-tocopherol (δ-Toc) have been identified and mapped on chromosomes 2, 5, 6, 7, 9, 12, 13, 15, 17, and 20; and 19 QTL that control seed total tocopherol (T-Toc) have been identified and mapped on chromosomes 2, 3, 4, 5, 6, 8, 9, 10, 12, 13, 15, 16, 17, and 18 (Table 12.2; Fig. 2.1) (Shaw and Rajcan 2017).

Liu et al. (2017a) grew a RIL population derived from a cross between 'Beifeng 9' (high vitamin E content) and 'Freeborn' (low vitamin E content) lines ($n = 238$) over 2 years (2013–2014) across three environments in China, genotyped it with 218 SSR markers, and used composite interval mapping (CIM) of WinQTL Cartographer (Wang et al. 2012a, b) to identify QTL that control seed α-tocopherol, γ-tocopherol, δ-tocopherol, and total tocopherol contents. Twenty-one QTL that control seed α-tocopherol content have been identified and mapped on

Table 12.3 Candidate genes identified within the genomic regions containing QTL that control seed α-, γ-, δ-, and total tocopherol contents in soybean

Trait	QTL	Chr.	Marker	Candidate gene	Annotation/pathway	References
α-, γ-, and δ-tocopherols	*qα-Toc%, qγ-Toc%, qδ-Toc%, qα-content, qγ-content*	9	Sat_243	Glyma09g35620.1	Auxin-responsive protein	Dwiyanti et al. (2011)
				Glyma09g35630.1	Auxin-responsive protein	
				Glyma09g35640.1	Diphtheria toxin resistance	
				Glyma09g35650.1	No function annotation	
				Glyma09g35660.1	Amidophosphoribosylpyrophosphate transferase domain	
				Glyma09g35670.1	Amidophosphoribosylpyrophosphate transferase domain	
				Glyma09g35680.1	γ-tocopherol methyltransferase (γ-TMT)	
				Glyma09g35690.1	No function annotation	
				Glyma09g35700.1	No function annotation	
				Glyma09g35710.1	DNA topoisomerase type I	
α-tocopherol	*qα-TC-9*	9	S09_43927286	Glyma09G22280	γ-tocopherol methyltransferase (γ-TMT)	Park et al. (2019)
α-, γ-, and δ-tocopherols	*qα-, qγ-, and qδ-Toc*	2	rs9337368	Glyma.02G098200	Transmembrane amino acid transporter family protein	Sui et al. (2020)
			rs9337368	Glyma.02G098300	Phytochrome and flowering time regulatory protein (PFT1)	
			rs9337368	Glyma.02G098400	Protein with RNI-like/FBD-like domains	
			rs9337368	Glyma.02G098500	Phytochrome and flowering time regulatory protein (PFT1)	
			rs9337368	Glyma.02G098600	Endosomal targeting BRO1-like domain-containing protein	
			rs9337368	Glyma.02G098700	Metallo-hydrolase/oxidoreductase superfamily protein	
			rs9337368	Glyma.02G098800	MYB-like transcription factor family protein	

Trait	QTL	Chr.	Marker	Candidate gene	Annotation/pathway	References
			rs9337368	Glyma.02G099000	ROTUNDIFOLIA-like 21	
			rs9337368	Glyma.02G099100	C2H2-like zinc finger protein	
			rs9337368	Glyma.02G099200	Allene oxide cyclase 4	
			rs9337368	Glyma.02G099300	Methionine aminopeptidase 1B	
			rs9337368	Glyma.02G099400	Glutathione S-transferase tau 7	
			rs9337368	Glyma.02G099500	AP2/B3 transcription factor family protein	
			rs9337368	Glyma.02G099800	Di-glucose binding protein with leucine-rich repeat domain	
			rs9337368	Glyma.02G099900	Galacturonosyltransferase-like 3	
			rs9337368	Glyma.02G100000	STT7 homolog STN7	
			rs9337368	Glyma.02G100100	ROTUNDIFOLIA-like 15	
			rs9337368	Glyma.02G100200	NAC (no apical meristem) domain transcriptional regulator superfamily protein	
			rs9337368	Glyma.02G100300	Protein kinase superfamily protein	
			rs9337368	Glyma.02G100400	Protein kinase superfamily protein	
			rs9337368	Glyma.02G100500	Leucine-rich repeat protein kinase family protein	
			rs9337368	Glyma.02G100800	Eukaryotic aspartyl protease family protein	
			rs9337368	Glyma.02G101300	Cytochrome b561/ferric reductase transmembrane protein family	
			rs9337368	Glyma.02G101400	2-Oxoglutarate (2OG) and Fe(II)-dependent oxygenase superfamily protein	
			rs9337368	Glyma.02G101500	Decapping 5	
			rs9337368	Glyma.02G101700	Decapping 5	
			rs9337368	Glyma.02G101800	Nuclear matrix constituent protein-related	
			rs9337368	Glyma.02G101900	Leucine-rich repeat (LRR) family protein	

(continued)

Table 12.3 (Continued)

Trait	QTL	Chr.	Marker	Candidate gene	Annotation/pathway	References
			rs9337368	Glyma.02G102000	Tetratricopeptide repeat (TPR)-like superfamily protein	
			rs9337368	Glyma.02G102400	Thioredoxin superfamily protein	
			rs9337368	Glyma.02G102500	Putative endonuclease or glycosyl hydrolase	
			rs9337368	Glyma.02G102600	Protein kinase superfamily protein	
			rs9337368	Glyma.02G102700	PLC-like phosphodiesterase superfamily protein	
			rs9337368	Glyma.02G102800	PLC-like phosphodiesterase superfamily protein	
			rs9337368	Glyma.02G102900	Plant U-box 26	
			rs9337368	Glyma.02G103200	Basic helix-loop-helix (bHLH) DNA-binding superfamily protein	
			rs9337368	Glyma.02G103300	RING/U-box superfamily protein	
			rs9337368	Glyma.02G103400	Zinc ion binding; nucleic acid binding	
			rs9337368	Glyma.02G103500	S-adenosyl-L-methionine-dependent methyltransferase superfamily protein	
			rs9337368	Glyma.02G103600	Ribosomal protein S5 domain 2-like superfamily protein	
			rs9337368	Glyma.02G103700	Plant protein of unknown function (DUF869)	
			rs9337368	Glyma.02G103800	RING/U-box superfamily protein	
α-, γ-, and δ-tocopherols	*qα-, qγ-, and qδ-Toc*	1	rs37505279	Glyma.01G110200	Golgi nucleotide-sugar transporter 3	Zhan et al. (2020)
			rs37505279	Glyma.01G110300	Nucleotide-sugar transporter family protein	
			rs37505279	Glyma.01G110400	Plant uncoupling mitochondrial protein 1	
			rs37505279	Glyma.01G110500	Reticulon family protein	
			rs37505279	Glyma.01G110600	Thioredoxin superfamily protein	

Trait	QTL	Chr.	Marker	Candidate gene	Annotation/pathway	References
			rs37505279	Glyma.01G110700	Transmembrane amino acid transporter family	
			rs55811336	Glyma.01G229300	Ribonuclease P family protein	
			rs55811336	Glyma.01G229400	P-loop containing nucleoside triphosphate hydrolase superfamily protein	
			rs55811336	Glyma.01G229500	Tudor/PWWP/MBT superfamily protein	
			rs55811336	Glyma.01G229600	Transducin family protein/WD-40 repeat family protein	
			rs55811336	Glyma.01G229700	Pre-mRNA-splicing factor 3	
			rs55811336	Glyma.01G229800	Rubredoxin family protein	
			rs55811336	Glyma.01G229900	Protein of unknown function (DUF604)	
			rs55811336	Glyma.01G230000	Eukaryotic aspartyl protease family protein	
			rs55811336	Glyma.01G230100	Eukaryotic aspartyl protease family protein	
			rs55811336	Glyma.01G230400	PLAC8 family protein	
			rs55811336	Glyma.01G230500	Tetratricopeptide repeat (TPR)-like superfamily protein	
			rs55811336	Glyma.01G230600	Pentatricopeptide repeat (PPR-like) superfamily protein	
			rs55811336	Glyma.01G230700	KH domain-containing protein	
			rs55811336	Glyma.01G230800	Esterase/lipase/thioesterase family protein	
			rs55811336	Glyma.01G230900	Cytochrome c-2	
			rs55811336	Glyma.01G231000	Integrase-type DNA-binding superfamily protein	
			rs55811336	Glyma.01G231200	Integrase-type DNA-binding superfamily protein	
			rs55811336	Glyma.01G231300	Nuclear transport factor 2 (NTF2) family protein	
			rs55811336	Glyma.01G231400	FK506 binding protein 53	

(continued)

Table 12.3 (Continued)

Trait	QTL	Chr.	Marker	Candidate gene	Annotation/pathway	References
			rs55811336	Glyma.01G231500	Tetratricopeptide repeat (TPR)-like superfamily protein	
			rs55811336	Glyma.01G231600	Alpha/beta-hydrolase superfamily protein	
			rs55811336	Glyma.01G231700	Light-mediated development protein 1/ deetiolated1 (DET1)	
			rs55811336	Glyma.01G231800	Protein of unknown function (DUF506)	
			rs55811336	Glyma.01G231900	Rieske (2Fe-2S) domain-containing protein	
			rs55811336	Glyma.01G232000	Integrase-type DNA-binding superfamily protein	
			rs55811336	Glyma.01G232100	Acyl activating enzyme 1	
			rs55811336	Glyma.01G232200	Acyl activating enzyme 1	
			rs55811336	Glyma.01G232300	RNA-binding (RRM/RBD/RNP motifs) family protein	
			rs55811336	Glyma.01G232400	4-Coumarate:CoA ligase 3	
			rs55811336	Glyma.01G232500	Cellulose synthase-like D3	
			rs55811336	Glyma.01G232600	Target of Myb protein 1	
		2	rs122109	Glyma.02G000300	NAD(P)H dehydrogenase 18	
			rs122109	Glyma.02G000400	Protein kinase family protein/ peptidoglycan-binding LysM domain-containing protein	
			rs122109	Glyma.02G000500	AUX/IAA transcriptional regulator family protein	
			rs122109	Glyma.02G000600	Pentatricopeptide repeat 596	
			rs122109	Glyma.02G000800	Basic helix-loop-helix (bHLH) DNA-binding superfamily protein	
			rs122109	Glyma.02G000900	BolA-like family protein	
			rs122109	Glyma.02G001000	P-loop containing nucleoside triphosphate hydrolases superfamily protein	

Trait	QTL	Chr.	Marker	Candidate gene	Annotation/pathway	References
			rs122109	Glyma.02G001100	HXXXD-type acyl-transferase family protein	
			rs122109	Glyma.02G001200	Neurochondrin family protein	
			rs122109	Glyma.02G001400	ARM repeat superfamily protein	
			rs122109	Glyma.02G001500	3-Ketoacyl-CoA synthase 11	
			rs122109	Glyma.02G001600	ARM repeat superfamily protein	
			rs122109	Glyma.02G001700	Protein of unknown function (DUF506)	
			rs7975863	Glyma.02G089800	Nucleic acid-binding, OB-fold-like protein	
			rs7975863	Glyma.02G089900	MATE efflux family protein	
			rs7975863	Glyma.02G090000	MATE efflux family protein	
			rs7975863	Glyma.02G090100	MATE efflux family protein	
			rs7975863	Glyma.02G090300	DNA ligase 6	
			rs7975863	Glyma.02G090400	DNA ligase 6	
			rs7975863	Glyma.02G090500	S-adenosyl-L-methionine-dependent methyltransferase superfamily protein	
			rs7975863	Glyma.02G090600	FKBP-like peptidyl-prolyl cis-trans isomerase family protein	
			rs7975863	Glyma.02G090800	Translation initiation factor IF2/IF5	
			rs7975863	Glyma.02G090900	RING/FYVE/PHD zinc finger superfamily protein	
			rs7975863	Glyma.02G091200	CLP protease proteolytic subunit 3	
			rs7975863	Glyma.02G091300	Phenylalanyl-tRNA synthetase class IIc family protein	
		6	rs37686178	Glyma.06G234200	Homogentisate 1,2-dioxygenase	
			rs37686178	Glyma.06G234300	RING/U-box superfamily protein	
			rs37686178	Glyma.06G234500	RING/U-box superfamily protein	
			rs37686178	Glyma.06G234700	Homeobox protein 52	

(continued)

Table 12.3 (Continued)

Trait	QTL	Chr.	Marker	Candidate gene	Annotation/pathway	References
		8	rs36359429	Glyma.08G274400	Serine transhydroxymethyltransferase 1	
			rs36359429	Glyma.08G274500	Cytochrome P450, family 718	
			rs36359429	Glyma.08G274600	Concanavalin A-like lectin protein kinase family protein	
			rs36359429	Glyma.08G274700	Concanavalin A-like lectin protein kinase family protein	
			rs36359429	Glyma.08G274800	Homogentisate phytyltransferase 1	
			rs36359429	Glyma.08G274900	Concanavalin A-like lectin protein kinase family protein	
			rs36359429	Glyma.08G275000	Concanavalin A-like lectin protein kinase family protein	
			rs36359429	Glyma.08G275100	Concanavalin A-like lectin protein kinase family protein	
			rs36359429	Glyma.08G275200	Nucleic acid-binding, OB-fold-like protein	
			rs36359429	Glyma.08G275300	PIF1 helicase	
			rs37094897	Glyma.08G276800	Leucine-rich repeat (LRR) family protein	
			rs37094897	Glyma.08G276900	tRNA/rRNA methyltransferase (SpoU) family protein	
			rs37094897	Glyma.08G277000	Deoxyxylulose-5-phosphate synthase	
			rs37094897	Glyma.08G277100	Deoxyxylulose-5-phosphate synthase	
			rs37094897	Glyma.08G277200	UDP-glycosyltransferase superfamily protein	
			rs37094897	Glyma.08G277300	Tetratricopeptide repeat (TPR)-containing protein	
			rs37094897	Glyma.08G277400	PIF1 helicase	

Trait	QTL	Chr.	Marker	Candidate gene	Annotation/pathway	References
		10	rs6898494	Glyma.10G069200	Ribosomal protein L31	
			rs6898494	Glyma.10G069400	Titan9	
			rs6898494	Glyma.10G069500	Leucine-rich repeat protein kinase family protein	
			rs6898494	Glyma.10G069600	RAB GTPase homolog A1F	
			rs6898494	Glyma.10G069700	WUSCHEL-related homeobox 8	
			rs6898494	Glyma.10G069800	Lysyl-tRNA synthetase 1	
			rs6898494	Glyma.10G069900	Small RNA degrading nuclease 3	
			rs6898494	Glyma.10G070000	Anaphase-promoting complex/cyclosome 2	
			rs6898494	Glyma.10G070100	Homogentisate phytyltransferase 1	
			rs6898494	Glyma.10G070200	Homogentisate phytyltransferase 1	
			rs7097743	Glyma.10G070300	Homogentisate phytyltransferase 1	
			rs7097743	Glyma.10G070400	Protein phosphatase 2A subunit A2	
			rs7097743	Glyma.10G070500	GDSL-like lipase/acylhydrolase superfamily protein	
			rs7097743	Glyma.10G070600	Peptidoglycan-binding LysM domain-containing protein	
			rs7097743	Glyma.10G070700	TBP-associated factor 12	
			rs7097743	Glyma.10G070900	A20/AN1-like zinc finger family protein	
			rs7097743	Glyma.10G071000	Purple acid phosphatase 22	
			rs7097743	Glyma.10G071100	Protein binding	
			rs7097743	Glyma.10G071200	Farnesylated protein-converting enzyme 2	

(continued)

Table 12.3 (Continued)

Trait	QTL	Chr.	Marker	Candidate gene	Annotation/pathway	References
		12	rs17125409	Glyma.12G139200	RNA-binding (RRM/RBD/RNP motifs) family protein	
			rs17125409	Glyma.12G139300	Beta-glucosidase 41	
			rs17125409	Glyma.12G139400	Ribosomal protein L30/L7 family protein	
			rs17125409	Glyma.12G139500	Beta-glucosidase 41	
			rs17125409	Glyma.12G139600	mRNA capping enzyme family protein	
			rs17125409	Glyma.12G139700	Uncharacterized conserved protein (UCP012943)	
			rs17125409	Glyma.12G139800	Uncharacterized conserved protein (UCP012943)	
			rs17125409	Glyma.12G139900	Isoprenoid F	
		13	rs330000	Glyma.13G000900	Plastid ribosomal protein l11	
			rs330000	Glyma.13G001000	Lung seven transmembrane receptor family protein	
			rs330000	Glyma.13G001200	Methionine gamma-lyase	
			rs4585748	Glyma.13G024400	PIF1 helicase	
			rs4585748	Glyma.13G024600	2-Isopropylmalate synthase 1	
			rs4585748	Glyma.13G024700	2-Isopropylmalate synthase 1	
			rs4585748	Glyma.13G024900	2-Isopropylmalate synthase 1	
			rs10200201	Glyma.13G030800	Alternative NAD(P)H dehydrogenase 2	
			rs10200201	Glyma.13G030900	NAC (no apical meristem) domain transcriptional regulator superfamily protein	
			rs10200201	Glyma.13G031100	RING/U-box superfamily protein	
			rs10200201	Glyma.13G031200	Proteasome beta subunit C1	
			rs10200201	Glyma.13G031300	SWITCH/sucrose nonfermenting 3C	
			rs10200201	Glyma.13G031400	Protein of unknown function (DUF784)	

Trait	QTL	Chr.	Marker	Candidate gene	Annotation/pathway	References
			rs10200201	Glyma.13G031500	Cyclin A1;1	
			rs10200201	Glyma.13G031600	Amino acid permease 2	
			rs10200201	Glyma.13G031700	Ribosomal protein S19 family protein	
			rs10200201	Glyma.13G031800	Protein kinase superfamily protein	
			rs21080726	Glyma.13G094700	Adenosylmethionine-8-amino-7-oxononanoate transaminases	
			rs21080726	Glyma.13G094800	Calmodulin-binding protein	
			rs21080726	Glyma.13G094900	Xyloglucan endotransglucosylase/hydrolase family protein	
			rs21080726	Glyma.13G095000	Xyloglucan endotransglucosylase/hydrolase family protein	
			rs21080726	Glyma.13G095100	Xyloglucan endotransglycosylase 6	
			rs21080726	Glyma.13G095200	Xyloglucan endotransglycosylase 6	
			rs21080726	Glyma.13G095300	Xyloglucan endotransglycosylase 6	
			rs21080726	Glyma.13G095400	Uncharacterized protein family (UPF0497)	
			rs21080726	Glyma.13G095500	Zinc finger protein 2	
			rs21080726	Glyma.13G095600	4-Coumarate:CoA ligase 2	
			rs21080726	Glyma.13G095800	Galactosyltransferase family protein	
			rs21080726	Glyma.13G095900	Arogenate dehydrogenase	
			rs21080726	Glyma.13G096000	Voltage-dependent anion channel 4	
			rs21080726	Glyma.13G096100	Integral membrane Yip1 family protein	
			rs21080726	Glyma.13G096200	P-loop containing nucleoside triphosphate hydrolase superfamily protein	
			rs21080726	Glyma.13G096300	SET domain-containing protein	
			rs21080726	Glyma.13G096400	Haloacid dehalogenase-like hydrolase (HAD) superfamily protein	

(continued)

Table 12.3 (Continued)

Trait	QTL	Chr.	Marker	Candidate gene	Annotation/pathway	References
			rs21080726	Glyma.13G096500	P450 reductase 2	
			rs21080726	Glyma.13G096600	Afadin/alpha-actinin-binding protein	
			rs21080726	Glyma.13G096700	OSBP (oxysterol binding protein)-related protein 4B	
			rs21080726	Glyma.13G096800	Replication protein A, subunit RPA32	
			rs43578642	Glyma.13G343500	S-adenosyl-L-methionine-dependent methyltransferase superfamily protein	
			rs43578642	Glyma.13G343600	Farnesyltransferase A	
			rs43578642	Glyma.13G343700	Endonuclease III 2	
			rs43578642	Glyma.13G343900	ARM repeat superfamily protein	
			rs43578642	Glyma.13G344000	Ribosomal L18p/L5e family protein	
			rs43578642	Glyma.13G344100	Far-red impaired responsive (FAR1) family protein	
			rs43578642	Glyma.13G344200	ARF-GAP domain 11	
			rs43578642	Glyma.13G344400	Crinkly4	
			rs43578642	Glyma.13G344500	pfkB-like carbohydrate kinase family protein	
			rs43578642	Glyma.13G344700	B-box zinc finger family protein	
			rs43578642	Glyma.13G344900	Protein of unknown function (DUF1336)	
			rs43578642	Glyma.13G345000	Tetratricopeptide repeat (TPR)-like superfamily protein	
			rs43578642	Glyma.13G345100	Calcium-dependent phosphotriesterase superfamily protein	
			rs43578642	Glyma.13G345200	Basic-leucine zipper (bZIP) transcription factor family protein	
			rs43578642	Glyma.13G345300	RNA binding	
			rs43578642	Glyma.13G345400	Glyoxalase 2–4	

Trait	QTL	Chr.	Marker	Candidate gene	Annotation/pathway	References
			rs43578642	Glyma.13G345500	Plant protein of unknown function (DUF936)	
			rs43578642	Glyma.13G345600	Ribosomal L38e protein family	
			rs43578642	Glyma.13G345700	GRAS family transcription factor	
			rs43578642	Glyma.13G345800	RNA-binding ASCH domain protein	
			rs43578642	Glyma.13G346000	Protein of unknown function (DUF3755)	
			rs43578642	Glyma.13G346100	Peroxidase superfamily protein	
			rs43578642	Glyma.13G346200	Ypt/Rab-GAP domain of gyp1p superfamily protein	
		17	rs4730199	Glyma.17G060800	Pumilio 6	
			rs4730199	Glyma.17G060900	S-adenosyl-L-methionine-dependent methyltransferase superfamily protein	
			rs4730199	Glyma.17G061000	Spermidine hydroxycinnamoyl transferase	
			rs4730199	Glyma.17G061100	Spermidine hydroxycinnamoyl transferase	
			rs4730199	Glyma.17G061200	Villin 4	
			rs4730199	Glyma.17G061300	Galactose mutarotase-like superfamily protein	
			rs4730199	Glyma.17G061400	Peroxidase superfamily protein	
			rs4730199	Glyma.17G061700	Sequence-specific DNA-binding transcription factors; transcription regulators	
			rs4730199	Glyma.17G061800	H(+)-ATPase 2	
			rs4730199	Glyma.17G061900	Homogentisate phytyltransferase 1	
			rs4730199	Glyma.17G062000	Galactose oxidase/kelch repeat superfamily protein	
			rs4730199	Glyma.17G062100	Hydroxyproline-rich glycoprotein family protein	

(continued)

Table 12.3 (Continued)

Trait	QTL	Chr.	Marker	Candidate gene	Annotation/pathway	References
			rs4730199	Glyma.17G062200	Transducin/WD40 repeat-like superfamily protein	
			rs4730199	Glyma.17G062300	Protein kinase superfamily protein	
			rs4730199	Glyma.17G062500	Vernalization5/VIN3-like	
			rs4730199	Glyma.17G062600	AINTEGUMENTA-like 6	
			rs4730199	Glyma.17G062700	Replication protein A, subunit RPA32	
		18	rs5680781	Glyma.18G061500	PHF5-like protein	
			rs5680781	Glyma.18G061600	Plant peptide containing sulfated tyrosine 1	
			rs5680781	Glyma.18G061700	Alternative NAD(P)H dehydrogenase 1	
			rs5680781	Glyma.18G061800	Mannose-binding lectin superfamily protein	
			rs5680781	Glyma.18G061900	Senescence-associated E3 ubiquitin ligase 1	
			rs5680781	Glyma.18G062000	Fatty acid desaturase 8	
			rs5680781	Glyma.18G062200	Zinc finger (CCCH-type) family protein	
			rs5680781	Glyma.18G062300	Double Clp-N motif-containing P-loop nucleoside triphosphate hydrolase superfamily protein	
			rs5680781	Glyma.18G062500	CRINKLY4-related 3	
			rs5680781	Glyma.18G062600	*Arabidopsis thaliana* protein of unknown function (DUF821)	
			rs5680781	Glyma.18G062700	RING/U-box superfamily protein	
			rs5680781	Glyma.18G062800	Zinc ion binding; nucleic acid binding	
			rs5680781	Glyma.18G062900	Cyclophilin 5	
			rs5680781	Glyma.18G063000	smr (small MutS related) domain-containing protein	
			rs5680781	Glyma.18G063100	HSP20-like chaperone superfamily protein	
			rs5680781	Glyma.18G063200	Cation exchanger 1	
		19	rs37037897	Glyma.19G114600	Protein kinases; ubiquitin-protein ligases	

Trait	QTL	Chr.	Marker	Candidate gene	Annotation/pathway	References
			rs37037897	Glyma.19G114800	Plant protein of unknown function (DUF639)	
			rs37037897	Glyma.19G114900	Transducin/WD40 repeat-like superfamily protein	
			rs37037897	Glyma.19G115000	BRI1-like 2	
			rs37037897	Glyma.19G115100	Polynucleotidyl transferase, ribonuclease H fold protein with HRDC domain	
			rs37037897	Glyma.19G115300	Protein of unknown function (DUF1637)	
			rs37037897	Glyma.19G115400	Peptide deformylase 1B	
			rs37037897	Glyma.19G115600	Protein of unknown function (DUF1637)	
			rs37037897	Glyma.19G115700	Endoribonucleases	
			rs37037897	Glyma.19G115800	Leucine-rich repeat protein kinase family protein	
			rs37037897	Glyma.19G115900	Formyl transferase	
			rs37037897	Glyma.19G116000	Peptidase M50 family protein	

chromosomes 10, 15, 18, and 19; 17 QTL that control seed γ-tocopherol content have been identified and mapped on chromosomes 9, 10, 12, 15, 18, and 19; 13 QTL that control seed δ-tocopherol content have been identified and mapped on chromosomes 10, 12, 15, and 19; and 15 QTL that control seed total vitamin E content have been identified and mapped on chromosomes 9, 10, 12, 15, 18, and 19. Moreover, the cultivar Beifeng 9 contributed 60 high-tocopherol beneficial alleles, while the cultivar Freeborn contributed only six high-δ-tocopherol-only alleles among the 66 QTL identified (Table 12.2; Fig. 2.1) (Liu et al. 2017a). In another study, Liu et al. (2017b) grew the same Beifeng 9 by Freeborn RIL population in six environments and found significant additive effects and/or additive by environment effects and epistatic interactions among the QTL identified (Liu et al. 2017b).

Park et al. (2019) grew a RIL population derived from a cross between breeding line 'TK780' (low α-Toc) and the wild accession 'B04009' (high α-Toc) ($n = 94$) in 2016 and 2018 in the greenhouse, genotyped it with 7710 single nucleotide polymorphism (SNP) markers, and used inclusive composite interval mapping (ICIM) of IciMapping (Meng et al. 2015) to identify QTL that control seed α-tocopherol, γ-tocopherol, δ-tocopherol, and total tocopherol contents. Seven QTL that control seed α-tocopherol content have been identified and mapped on chromosomes 5, 9, 11, and 12; four QTL that control seed γ-tocopherol content have been identified and mapped on chromosomes 6, 7, 13, and 14; three QTL that control seed δ-tocopherol content have been identified and mapped on chromosomes 4, 6, and 19; and one QTL that control seed total tocopherol content has been identified and mapped on chromosome 1. They also identified QTL for α-tocopherol/γ-tocopherol ratio on chromosomes 9, 11, and 17 (Table 12.2; Fig. 2.1). Moreover, they observed additive effects of the three QTL and identified the gene *γ-TMT3* (Glyma.09G22280) previously identified (Dwiyanti et al. 2011) within the QTL region (*qαTC-2*) on chromosome 9 (Park et al. 2019).

Sui et al. (2020) grew a RIL population derived from a cross between 'Hefeng 25' and 'OAC Bayfield' lines ($n = 144$) and a natural population of soybean accessions ($n = 180$) from outside and inside China from 2014 to 2017. They developed 22,611 SNPs through specific locus amplified fragment sequencing (SLAF-seq) method, genotyped the populations, and used GWAS analysis and composite interval mapping (ICIM) of IciMapping (Meng et al. 2015) to identify QTL that control seed α-tocopherol, γ-tocopherol, δ-tocopherol, and total tocopherol content. Using GWAS analysis, one SNP associated with seed α-tocopherol (α-Toc) content has been identified and mapped on chromosome 1; five SNPs associated with seed γ-tocopherol (γ-Toc) content have been identified and mapped on chromosomes 1, 6, 15, 17, and 20; 12 SNPs associated with seed δ-tocopherol (δ-Toc) content have been identified and mapped on chromosomes 1, 2, 7, 9, 11, 17, 18, and 20; and 15 SNPs associated with seed total tocopherol (T-Toc) content have been identified and mapped on chromosomes 1, 2, 6, 7, 9, 11, 15, 17, and 20 (Table 12.1). Using CIM and the RIL population, nine QTL that control seed α-tocopherol (α-Toc) content have been identified and mapped on chromosomes 2, 4, 10, 11, 13, 18, 19, and 20; three QTL that control seed γ-tocopherol (γ-Toc) content have been identified and mapped on chromosomes 2, 6, and 18; and six QTL that control seed δ-tocopherol (δ-Toc) content have been identified and mapped on chromosomes 2, 13, 15, and 18 (Table 12.2; Fig. 2.1). Furthermore, transcript abundance and expression analysis identified nine eQTL (expression QTL) for tocopherol contents on chromosomes 2, 4, 15, 18, and 20. The eQTL on chromosomes 2 (qMPBQD1b_1), 15 (qMPBQE_1), and 20 (qMPBQI_1) were associated with methyl-6-phytyl-1,4-benzoquinone methyltransferase (MPBQ-MT) transcript abundance. Within 200-kbp flanking genomic regions harboring QTL for tocopherols, they identified 170 candidate genes; however, 42 candidate genes all on chromosome 2 were related to tocopherol metabolic pathways (Table 12.2). Among these are Glyma.02G098200 that encodes for a transmembrane amino acid transporter family protein, *Glyma.02G099900* that encodes for a galacturonosyltransferase-like 3, *Glyma.02G100000* that encodes for an STT7 homolog STN7, *Glyma.02G100100* that encodes for a ROTUNDIFOLIA-like 15, *Glyma.02G100200* that encodes for a NAC (no apical meristem) domain transcriptional regulator superfamily protein, *Glyma.02G101800* that encodes for a nuclear matrix constituent protein-related, *Glyma.02G101900* that encodes for a leucine-rich repeat (LRR) family protein, *Glyma.02G102000* that encodes for a tetratricopeptide repeat (TPR)-like superfamily protein, *Glyma.02G102400* that encodes for a thioredoxin superfamily protein, *Glyma.02G102500* that encodes for a putative endonuclease or glycosyl hydrolase, *Glyma.02G102900* that encodes for a plant U-box 26, and *Glyma.02G103500* that encodes for an S-adenosyl-L-methionine-dependent methyltransferase superfamily protein, to name a few (Table 12.3) (Sui et al. 2020).

Zhan et al. (2020) grew a natural population of soybean accessions ($n = 180$) in six locations in China from 2014 to 2016, genotyped it with 22,611 SNPs, and used GWAS analysis to identify SNPs associated with quantitative trait nucleotides (QTNs) that control seed α-tocopherol (α-Toc), γ-tocopherol (γ-Toc), δ-tocopherol (δ-Toc), and total tocopherol (T-Toc) contents. They identified 18 QTNs significantly associated with seed α-Toc, γ-Toc, δ-Toc, and T-Toc contents. Fourteen QTL that control seed α-Toc content have been identified and mapped on chromosomes 2, 8, 12, 13, and 19; 12 QTL that control seed γ-Toc content have been identified and mapped on chromosomes 1, 8, 10, 12, 13, and 18; ten QTL that control seed δ-Toc content have been identified and mapped on chromosomes 1, 6, 13, and 17; and 12 QTL that control seed T-Toc content have been identified and mapped on chromosomes 1, 2, 6, 8, 10, 12, 13, and 18 (Table 12.2; Fig. 2.1). Within 200-kbp flanking genomic regions harboring QTL that control seed tocopherol content, they identified 216 candidate genes; however, only 37 candidate genes were with seed tocopherol contents (Table 12.2). Among these genes are *Glyma.01G110400* that encodes for a plant uncoupling mitochondrial protein 1, Glyma.01G230900 that encodes for a cytochrome c-2, *Glyma.02G090500* that encodes for an S-adenosyl-L-methionine-dependent methyltransferase superfamily protein, *Glyma.02G091300* that encodes for a phenylalanyl-tRNA synthetase class IIc family protein, *Glyma.06G234200* that encodes for a homogentisate 1,2-dioxygenase, *Glyma.06G234700* that encodes for a homeobox protein 52, *Glyma.08G274400* that encodes for a serine transhydroxymethyltransferase 1, *Glyma.08G274500* that encodes for a cytochrome P450,

Glyma.10G069500 that encodes for a leucine-rich repeat (LLR) protein kinase family protein, *Glyma.10G069600* that encodes for a RAB GTPase homolog A1F, *Glyma.12G139300* that encodes for a beta-glucosidase 41, *Glyma.12G139400* that encodes for a ribosomal protein L30/L7 family protein, *Glyma.13G000900* that encodes for a plastid ribosomal protein 111, *Glyma.13G024900* that encodes for a 2-isopropylmalate synthase 1, *Glyma.17G061000* that encodes for a spermidine hydroxycinnamoyl transferase, *Glyma.17G061900* that encodes for a homogentisate phytyltransferase 1, *Glyma.18G061600* that encodes for a plant peptide containing sulfated tyrosine 1, *Glyma.18G062000* that encodes for a fatty acid desaturase 8, *Glyma.19G114600* that encodes for a protein kinase, and Glyma.19G115100 that encodes for a polynucleotidyl transferase, to name a few (Table 12.3) (Zhan et al. 2020).

12.5 Candidate Genes Involved in Tocopherol Biosynthesis

Several studies identified candidate genes involved in the biosynthesis of α-Toc, γ-Toc, δ-Toc, and T-Toc within the genomic regions containing QTL for these traits (Dwiyanti et al. 2011; Park et al. 2019; Sui et al. 2020; Zhan et al. 2020).

For example, Dwiyanti et al. (2007) used the 'Keszthelyi A.S.' (KAS) by 'Ishihime' F_2 population and identified two QTL that control seed α-tocopherol content on chromosome 9 (Sat_167 and Sat_243). Within this QTL region, they identified the candidate gene *γ-TMT3* encoding for γ-tocopherol methyltransferase (Table 12.3) (Dwiyanti et al. 2011).

Park et al. (2019) used the 'TK780' by 'B04009' RIL population and identified seven QTL that control seed α-Toc, four QTL that control seed γ-Toc, three QTL that control seed δ-Toc contents, 8 QTL that control α-Toc/γ-Toc ratio, and the gene *γ-TMT3* (Glyma.09G22280) previously identified (Dwiyanti et al. 2011) within the QTL region (*qαTC-2*) on chromosome 9 (Table 12.3) (Park et al. 2019).

Sui et al. (2020) used the 'Hefeng 25' by 'OAC Bayfield' RIL population and a natural population, used both GWAS analysis and CIM, and identified SNPs associated with QTL that control seed α-Toc, γ-Toc, and δ-Toc contents. Within these genomic regions harboring QTL for tocopherols, they identified 170 candidate genes; however, 42 candidate genes all on chromosome 2 were related to tocopherol metabolic pathways (Table 12.3). Among these are Glyma.02G098200 that encodes for a transmembrane amino acid transporter family protein, *Glyma.02G099900* that encodes for a galacturonosyltransferase-like 3, *Glyma.02G100000* that encodes for an STT7 homolog STN7, *Glyma.02G100100* that encodes for a ROTUNDIFOLIA-like 15, *Glyma.02G100200* that encodes for a NAC (no apical meristem) domain transcriptional regulator superfamily protein, *Glyma.02G101800* that encodes for a nuclear matrix constituent protein-related, *Glyma.02G101900* that encodes for a leucine-rich repeat (LRR) family protein, *Glyma.02G102000* that encodes for a tetratricopeptide repeat (TPR)-like superfamily protein, *Glyma.02G102400* that encodes for a thioredoxin superfamily protein, *Glyma.02G102500* that encodes for a putative endonuclease or glycosyl hydrolase, *Glyma.02G102900* that encodes for a plant U-box 26, and *Glyma.02G103500* that encodes for an S-adenosyl-L-methionine-dependent methyltransferase superfamily protein, to name a few (Table 12.3) (Sui et al. 2020).

Zhan et al. (2020) used a natural population of soybean accessions and identified 18 QTNs associated with seed α-Toc, γ-Toc, δ-Toc, and T-Toc contents. Within 200-kbp flanking genomic regions harboring QTL that control seed tocopherol content, they identified 216 candidate genes; however, only 37 candidate genes were with seed tocopherol contents (Table 12.3). Among these genes are *Glyma.01G110400* that encodes for a plant uncoupling mitochondrial protein 1, Glyma.01G230900 that encodes for a cytochrome c-2, *Glyma.02G090500* that encodes for an S-adenosyl-L-methionine-dependent methyltransferase superfamily protein, *Glyma.02G091300* that encodes for a phenylalanyl-tRNA synthetase class IIc family protein, *Glyma.06G234200* that encodes for a homogentisate 1,2-dioxygenase, *Glyma.06G234700* that encodes for a homeobox protein 52, *Glyma.08G274400* that encodes for a serine transhydroxymethyltransferase 1, *Glyma.08G274500* that encodes for a cytochrome P450, *Glyma.10G069500* that encodes for a leucine-rich repeat (LLR) protein kinase family protein, *Glyma.10G069600* that encodes for a RAB GTPase homolog A1F, *Glyma.12G139300* that encodes for a beta-glucosidase 41, *Glyma.12G139400* that encodes for a ribosomal protein L30/L7 family protein, *Glyma.13G000900* that encodes for a plastid ribosomal protein 111, *Glyma.13G024900* that encodes for a 2-isopropylmalate synthase 1, *Glyma.17G061000* that encodes for a spermidine hydroxycinnamoyl transferase, *Glyma.17G061900* that encodes for a homogentisate phytyltransferase 1, *Glyma.18G061600* that encodes for a plant peptide containing sulfated tyrosine 1, *Glyma.18G062000* that encodes for a fatty acid desaturase 8, *Glyma.19G114600* that encodes for a protein kinase, and Glyma.19G115100 that encodes for a poly-nucleotidyl transferase, to name a few (Table 12.3) (Zhan et al. 2020).

From these few studies, it is obvious that further studies are needed to narrow these QTL regions and number of genes with them in order to identify specific genes involved in seed α-Toc, γ-Toc, δ-Toc, and T-Toc contents and characterize them not only in soybean but also in other crops and plant species.

References

Abbasi, A.R., M. Hajirezaei, D. Hofius, U. Sonnewald, and L.M. Voll. 2007. Specific roles of α– and γ–tocopherol in abiotic stress responses of transgenic tobacco. *Plant Physiology* 143: 1720–1173. http://doi.org/0.1104/pp.106.094771.

Almonor, G., G. Fenner, and R. Wilson. 1998. Temperature effects on tocopherol composition in soybeans with genetically improved oil quality. *Journal of the American Oil Chemists' Society* 75: 591–596.

Britz, S.J., and D.F. Kremer. 2002. Warm temperatures or drought during seed maturation increase free α-tocopherol in seeds of soybean (Glycine max [L.] Merr.). *Journal of Agricultural and Food Chemistry* 50: 6058–6063.

Britz, S.J., D.F. Kremer, and W.J. Kenworthy. 2008. Tocopherols in soybean seeds: genetic variation and environmental effects in field grown crops. *Journal of the American Oil Chemists' Society* 85: 931–936. https://doi.org/10.1007/s11746-008-1286-y.

Carrao-Panizzi, M.C., and S.Z. Erhan. 2007. Environmental and genetic variation of soybean tocopherol content under Brazilian growing conditions. *Journal of the American Oil Chemists' Society* 84: 921–928. https://doi.org/10.1007/s11746-007-1128-3.

Carrera, C.S., and P. Seguin. 2016. Factors affecting tocopherol concentrations in soybean seeds. *Journal of Agricultural and Food Chemistry* 2016 (64): 9465–9474. https://doi.org/10.1021/acs.jafc.6b03902.

Carrera, C.S., J.L. Dardanelli, and D.O. Soldini. 2014. Chemical compounds related to nutraceutical and industrial qualities of non-transgenic soybean genotypes. *Journal of the Science of Food and Agriculture* 94: 1463–1469.

Chander, S., Y.Q. Guo, X.H. Yang, J.B. Yan, Y.R. Zhang, T.M. Song, and J.S. Li. 2008. Genetic dissection of tocopherol content and composition in maize grain using quantitative trait loci analysis and the candidate gene approach. *Molecular Breeding* 22: 353–365. https://doi.org/10.1007/s11032-008-9180-8.

Chennupati, P., P. Seguin, and W. Liu. 2011. Effects of high temperature stress at different development stages on soybean isoflavone and tocopherol concentrations. *Journal of Agricultural and Food Chemistry* 59 (13): 081–13,088.

Clarke, M.W., J.R. Burnett, and K.D. Croft. 2008. Vitamin E in human health and disease. *Critical Reviews in Clinical Laboratory Sciences* 45 (5): 417–450. https://doi.org/10.1080/10408360802118625.

Del Moral, L., J.M. Fernández-Martínez, L. Velasco, and B. Pérez-Vich. 2012. Quantitative trait loci for seed tocopherol content in sunflower. *Crop Science* 52: 786–794. https://doi.org/10.2135/cropsci2011.08.0406.

DellaPenna, D., and B.J. Pogson. 2006. Vitamin synthesis in plants: Tocopherols and carotenoids. *Annual Reviews of Plant Biology* 57: 711–738. https://doi.org/10.1146/annurev.arplant.56.032604.144301.

Dolde, D., C. Vlahakis, and J. Hazebroek. 1999. Tocopherols in breeding lines and effects of planting location, fatty acid composition and temperature during development. *Journal of the American Oil Chemists' Society* 76: 349–355.

Dwiyanti, M.S., A. Ujiie, L.T.B. Thuy, T. Yamada, and K. Kitamura. 2007. Genetic analysis of high alpha-tocopherol content in soybean seed. *Breeding Science* 57: 23–28.

Dwiyanti, M.S., T. Yamada, M. Sato, J. Abe, and K. Kitamura. 2011. Genetic variation of γ-tocopherol methyltransferase gene contributes to elevated α-tocopherol content in soybean seeds. *BMC Plant Biology* 11: 152. http://www.biomedcentral.com/1471-2229/11/152.

Dwiyanti, M.S., S. Maruyama, M. Hirono, M. Sato, E. Park, S.H. Yoon, T. Yamada, and J. Abe. 2016. Natural diversity of seed α-tocopherol ratio in wild soybean (Glycine soja) germplasm collection. *Breeding Science Preview* 66 (4): 653–657. https://doi.org/10.1270/jsbbs.16028.

Endrigkeit, J., X. Wang, D. Cai, C. Zhang, Y. Long, J. Meng, and C. Jung. 2009. Genetic mapping, cloning, and functional characterization of the BnaX.VTE4 gene encoding a α-tocopherol methyltransferase from oilseed rape. *Theoretical and Applied Genetics* 119: 567–575. https://doi.org/10.1007/s00122-009-1066-6.

Evans, C.J., D.R. Kodali, and P.B. Addis. 2002. Optimal tocopherol concentrations to inhibit soybean oil oxidation. *Journal of the American Oil Chemists' Society* 79: 47–51.

Fritsche, S., X. Wang, J. Li, B. Stich, F.J. Kopisch-Obuch, J. Endrigkeit, G. Leckband, F. Dreyer, W. Friedt, J. Meng, and C. Jung. 2012. A candidate gene-based association study of tocopherol content and composition in rapeseed (Brassica napus). *Frontiers in Plant Science* 3: 129. https://doi.org/10.3389/fpls.2012.00129.

Fritsche, S., X. Wang, and C. Jung. 2017. Recent advances in our understanding of tocopherol biosynthesis in plants: An overview of key genes, functions, and breeding of vitamin E improved crops. *Antioxidants* 6: 99. https://doi.org/10.3390/antiox6040099.

Gajardo, H.A., B. Wittkop, B. Soto-Cerda, E.E. Higgins, I.A.P. Parkin, R.J. Snowdon, M.L. Federico, and F.L. Iniguez-Luy. 2015. Association mapping of seed quality traits in Brassica napus L. using GWAS and candidate QTL approaches. *Molecular Breeding* 35: 143. https://doi.org/10.1007/s11032-015-0340-3.

Haddadi, P., A. Ebrahimi, N.B. Langlade, B. Yazdi-samadi, M. Berger, A. Calmon, M.R. Naghavi, P. Vincourt, and A. Sarrafi. 2012. Genetic dissection of tocopherol and phytosterol in recombinant inbred lines of sunflower through quantitative trait locus analysis and the candidate gene approach. *Molecular Breeding* 29: 717–729. https://doi.org/10.1007/s11032-011-9585-7.

Havaux, M., C. Lutz, and B. Grimm. 2003. Chloroplast membrane stability in chlP transgenic tobacco plants deficient in tocopherols. *Plant Physiology* 132: 300–310.

Hayashi, E., Y. You, R. Lewis, M.C. Calderon, G. Wan, and D.W. Still. 2012. Mapping QTL, epistasis and genotype 3 environment interaction of antioxidant activity, chlorophyll content and head formation in domesticated lettuce (*Lactuca sativa*). *Theoretical and Applied Genetics* 124: 1487–1502. https://doi.org/10.1007/s00122-012-1803-0.

Hermann, K.M., and L.M. Weaver. 1999. The shikimate pathway. *Annual Reviews of Plant Physiology and Molecular Biology* 50: 473–503.

Izquierdo, N., S. Nolasco, C. Mateo, D. Santos, and L. Aguirrezaabal. 2011. Relationship between oil tocopherol concentration and oil weight per grain in several crop species. *Crop & Pasture Science* 62: 1088–1097.

Kamal-Eldin, A., and L.A. Appelqvist. 1996. The chemistry and anti-oxidant properties of tocopherols and tocotrienols. *Lipids* 31: 671–701.

Kornsteiner, M., K.H. Wagner, and I. Elmadfa. 2006. Tocopherols and total phenolics in 10 different nut types. *Food Chemistry* 98: 381–387.

Li, Y., Z. Wang, X. Sun, and K. Tang. 2008. Current opinions on the functions of tocopherol based on the genetic manipulation of tocopherol biosynthesis in plants. *Journal of Integrative Plant Biology* 50 (9): 1057–1069.

Li, H., H. Liu, Y. Han, X. Wu, W. Teng, G. Liu, and W. Li. 2010. Identification of QTL underlying vitamin E contents in soybean seed among multiple environments. *Theoretical and Applied Genetics* 120: 1405–1413. https://doi.org/10.1007/s00122-010-1264-2.

Li, H., Y. Wang, Y. Han, W. Teng, X. Zhao, Y. Li, and W. Li. 2016. Mapping quantitative trait loci (QTLs) underlying seed vitamin E content in soybean with main, epistatic and QTL 3 environment effects. *Plant Breeding* 135: 208–214. https://doi.org/10.1111/pbr.12346.

Liu, H., G. Cao, Y. Han, Z. Jiang, H. Zhao, and W. Li. 2017a. Identification of the QTL underlying the vitamin E content of soybean seeds. *Plant Breeding* 136 (2): 147–154. https://doi.org/10.1111/pbr.12454.

Liu, H., G. Gao, D. Wu, Z. Jiang, Y. Han, and W. Li. 2017b. Quantitative trait loci underlying soybean seed tocopherol content with main additive, epistatic and QTL 3 environment effects. *Plant Breeding* 2017: 1–15. https://doi.org/10.1111/pbr.12534.

MacDuff, M. 2011. *Validation Of QTL associated with tocopherol levels in three half-sib populations derived from Keszthelyi Aproszemu Sarga soybean*. MS thesis, The University of Guelph, Canada.

Maeda, H., and D. DellaPenna. 2007. Tocopherol functions in photosynthetic organisms. *Current Opinion in Plant Biology* 10 (3): 260–265. https://doi.org/10.1016/j.pbi.2007.04.006.

Marwede, V., A. Schierholt, C. Mollers, and H.C. Becker. 2004. Genotype environment interactions and heritability of tocopherol contents in canola. *Crop Science* 44: 728–731.

Meng, L., H. Li, L. Zhang, and J. Wang. 2015. QTL IciMapping: Integrated software for genetic linkage map construction and quantitative trait locus mapping in biparental populations. *The Crop Journal* 3 (3): 269–283. https://doi.org/10.1016/j.cj.2015.01.001.

Morrissey, P.A., D.J. Buckley, P.J.A. Sheehy, and F.J. Monahan. 1994. Vitamin E and meat quality. *Proceedings of the Nutrition Society (PNS)* 53: 289–295.

Munne-Bosch, S. 2005. The role of α–tocopherol in plant stress tolerance. *Journal of Plant Physiology* 162 (7): 743–748. https://doi.org/10.1016/j.jplph.2005.04.022.

Munne-Bosch, S., and L. Alegre. 2002. The function of tocopherols and tocotrienols in plants. *Critical Reviews in Plant Sciences* 21 (1): 31–57.

Munne-Bosch, S., and J. Falk. 2004. New insights into the function of tocopherols in plants. *Planta* 218: 323–326. https://doi.org/10.1007/s00425-003-1126-0.

Park, C., M.S. Dwiyanti, A.J. Nagano, B. Liu, T. Yamada, and J. Abe. 2019. Identification of quantitative trait loci for increased α-tocopherol biosynthesis in wild soybean using a high-density genetic map. *BMC Plant Biology* 19: 510. https://doi.org/10.1186/s12870-019-2117-z.

Porfirova, S., E. Bergmuller, S. Tropf, R. Lemke, and P. Dormann. 2002. Isolation of an Arabidopsis mutant lacking vitamin E and identification of a cyclase essential for tocopherol biosynthesis. *Proceedings of the National Academy of Sciences* 99: 12495–12500.

Robinson, I., D.G. de Serna, A. Gutierrez, and D.S. Schade. 2006. Vitamin E in humans: An explanation of clinical trial failure. *Endocrine Practice* 12 (5): 576–582. https://doi.org/10.4158/EP.12.5.576.

Rocheford, T.R., J.C. Wong, C.O. Egesel, and R.J. Lambert. 2002. Enhancement of vitamin E levels in corn. *Journal of the American College of Nutrition* 21: 191S–198S. https://doi.org/10.1080/07315724.2002.10719265.

Sattler, S.E., L.U. Gilliland, M. Magallanes-Lundback, M. Pollard, and D. DellaPenna. 2004. Vitamin E is essential for seed longevity and for preventing lipid peroxidation during germination. *Plant Cell* 16: 1419–1432. https://doi.org/10.1105/tpc.021360.

Seguin, P., P. Turcotte, G. Tremblay, D. Pageau, and W. Liu. 2009. Tocopherols concentration and stability in early maturing soybean genotypes. *Agronomy Journal* 101: 1153–1159.

Seguin, P., G. Tremblay, D. Pageau, and W. Liu. 2010. Soybean tocopherol concentrations are affected by crop management. *Journal of Agricultural and Food Chemistry* 58: 5495–5501.

Shaw, E.J. 2012. *Genetic basis of tocopherol accumulation in soybean (Glycine Max [L.] Merr.) seeds*. PhD Dissertation, The University of Guelph, Ontario, Canada.

Shaw, E.J., and I. Rajcan. 2017. Molecular mapping of soybean seed tocopherols in the cross 'OAC Bayfield' 3 'OAC Shire'. *Plant Breeding* 136 (1): 83–93. https://doi.org/10.1111/pbr.12437.

Shaw, E.J., Y. Kakuda, and I. Rajcan. 2016. Effect of genotype, environment, and genotype × environment interaction on tocopherol accumulation in soybean seed. *Crop Science* 56: 40–50.

Sui, M., Y. Jing, H. Li, Y. Zhan, J. Luo, W. Teng, L. Qiu, H. Zheng, W. Li, X. Zhao, and Y. Han. 2020. Identification of loci and candidate genes analyses for tocopherol concentration of soybean seed. *Frontiers in Plant Science* 11: 539460. https://doi.org/10.3389/fpls.2020.539460.

Ujiie, A., T. Yamada, K. Fujimoto, Y. Endo, and K. Kitamura. 2005. Identification of soybean varieties with high α–tocopherol content. *Breeding Science* 55: 123–125.

Upston, J.M., L. Kritharides, and R. Stocker. 2003. The role of vitamin E in atherosclerosis. *Progress in Lipid Research* 42: 405–422. https://doi.org/10.1016/S0163-7827(03)00024-9.

Van Ooijen, J.M. 2006. *Software for the mapping and calculations of quantitative trait loci and genetic linkage maps in experimental populations*. Wageningen: Kyazma, B.V.

Velasco, L., J.M. Fernandez-Martinez, R. Garcia-Ruiz, and J. Domínguez. 2002. Genetic and environmental variation for tocopherol content and composition in sunflower commercial hybrids. *The Journal of Agricultural Science* 139: 425–429. https://doi.org/10.1017/S0021859602002678.

Wang, S., C.J. Basten, and Z.B. Zeng. 2012a Windows QTL cartographer 2.5. Dept. of Statistics, North Carolina State University, Raleigh, NC. http://statgen.ncsu.edu/qtlcart/WQTLCart.htm.

Wang, X., C. Zhang, L. Li, S. Fritsche, J. Endrigkeit, W. Zhang, Y. Long, C. Jung, and J. Meng. 2012b. Unraveling the genetic basis of seed tocopherol content and composition in rapeseed (*Brassica napus* L.). *PLoS One* 7 (11): e50038. https://doi.org/10.1371/journal.pone.0050038.

Wang, X.Q., M.Y. Yoon, Q. He, T.S. Kim, W. Tong, B.W. Choi, Y.S. Lee, and Y.J. Park. 2015. Natural variations in OsγTMT contribute to diversity of the α-tocopherol content in rice. *Molecular Genetics and Genomics* 290: 2121–2135. https://doi.org/10.1007/s00438-015-1059-x.

Whent, M., J. Hao, M. Slavin, M. Zhou, J. Song, W. Kenworthy, and L. Yu. 2009. Effect of genotype, environment, and their interaction on chemical composition and antioxidant properties of low-linolenic soybeans grown in Maryland. *Journal of Agricultural and Food Chemistry* 57 (21): 10163–10174. https://doi.org/10.1021/jf902344h.

Wong, J.C., R.J. Lambert, Y. Tadmor, and T.R. Rocheford. 2003. QTL associated with accumulation of tocopherols in maize. *Crop Science* 43 (6): 2257–2266. https://doi.org/10.2135/cropsci2003.2257.

Zhan, Y., H. Li, M. Sui, X. Zhao, Y. Jing, J. Luo, W. Teng, L. Qiu, H. Zheng, W. Li, D. Yang, and Y. Han. 2020. Genome wide association mapping for tocopherol concentration in soybean seeds across multiple environments. *Industrial Crops and Products* 154 (2020): 112674. https://doi.org/10.1016/j.indcrop.2020.112674.

Zhang, G., P. Li, W. Zhang, and J. Zhao. 2017. Analysis of multiple soybean phytonutrients by near-infrared reflectance spectroscopy. *Analytical and Bioanalytical Chemistry* 409: 3515–3525. https://doi.org/10.1007/s00216-017-0288-8.

Index

M. A. Kassem (ed.), *Soybean Seed Composition*,
https://doi.org/10.1007/978-3-030-82906-3

Q

T

The manufacturer's authorised representative in the EU is Springer Nature Customer Service Centre GmbH, Europaplatz 3, 69115 Heidelberg, Germany. If you have any concerns regarding our products, please contact ProductSafety@springernature.com

Printed and bound by CPI Group (UK) Ltd, Croydon, CR0 4YY
30/09/2026
02225687-0015